Compound Semiconductor Device Physics

Compound Semiconductor Device Physics

Sandip Tiwari

IBM Thomas J. Watson Research Center
Yorktown Heights, New York

ACADEMIC PRESS
Harcourt Brace Jovanovich, Publishers

Boston   San Diego   New York
London   Sydney
Tokyo   Toronto

₀ 4566907

PHYSICS

This book is printed on acid-free paper. ∞

ACADEMIC PRESS, INC.
1250 Sixth Avenue, San Diego, CA 92101

United Kingdom Edition published by
ACADEMIC PRESS LIMITED
24–28 Oval Road, London NW1 7DX

Cover: Wave vector distribution of hot electrons in GaAs
shown in the first Brillouin zone. Carriers are from a section
of the base–collector depletion region of a bipolar transistor.

Library of Congress Cataloging-in-Publication Data

Tiwari, Sandip, 1955–
   Compound semiconductor device physics / Sandip Tiwari,
      p. cm.
   Includes bibliographical references and index.
   ISBN 0-12-691740-X (alk. paper)
   1. Compound semiconductors. 2. Semiconductors. I. Title.
QC611.8.C64T59   1992
537.6'22—dc20   91-17270
                                       CIP

Printed in the United States of America
91 92 93 94   9 8 7 6 5 4 3 2 1

To Kunal, Nachiketa, and Mari

From the unreal lead me to the real!
From darkness lead me to light!

—Brihadaranyaka Upanishad

Contents

Preface   **xv**

1 Introduction   **1**
  1.1 Outline of the Book   2
  1.2 Suggested Usage   4
  1.3 Comments on Nomenclature   5

2 Review of Semiconductor Physics, Properties, and Device Implications   **7**
  2.1 Introduction   7
  2.2 Electrons, Holes, and Phonons   9
  2.3 Occupation Statistics   26
    2.3.1 Occupation of Bands and Discrete Levels   26
    2.3.2 Band Occupation in Semiconductors   32
  2.4 Band Structure   36
  2.5 Phonon Dispersion in Semiconductors   43
  2.6 Scattering of Carriers   47
    2.6.1 Defect Scattering   49
    2.6.2 Carrier–Carrier Scattering 51
    2.6.3 Lattice Scattering   52
    2.6.4 Phonon Scattering Behavior   53
  2.7 Carrier Transport   54
    2.7.1 Majority Carrier Transport   55
    2.7.2 Two-Dimensional Effects on Transport   70
    2.7.3 Minority Carrier Transport   72
    2.7.4 Diffusion of Carriers   76
  2.8 Some Effects Related to Energy Bands   78
    2.8.1 Avalanche Breakdown   79
    2.8.2 Zener Breakdown   90
    2.8.3 Density of States and Related Considerations   92

2.8.4   Limiting and Operational Velocities of Transport . .   95
2.8.5   Tunneling Effects   99
2.9   Summary   100
General References   101
Problems   102

3   Mathematical Treatments   **105**
3.1   Introduction   105
3.2   Kinetic Approach   107
3.3   Boltzmann Transport Approach   113
3.3.1   Relaxation Time Approximation   116
3.3.2   Conservation Equations   122
3.3.3   Limitations   128
3.4   Monte Carlo Transport Approach   135
3.5   Drift-Diffusion Transport   149
3.5.1   Quasi-Static Analysis   153
3.5.2   Quasi-Neutrality   153
3.5.3   High Frequency Small-Signal Analysis   157
3.6   Boundary Conditions   163
3.6.1   Shockley Boundary Conditions   163
3.6.2   Fletcher Boundary Conditions   166
3.6.3   Misawa Boundary Conditions   168
3.6.4   Dirichlet Boundary Conditions   170
3.6.5   Neumann Boundary Conditions   172
3.7   Generation and Recombination   173
3.7.1   Radiative Recombination   174
3.7.2   Hall–Shockley–Read Recombination   175
3.7.3   Auger Recombination   180
3.7.4   Surface Recombination   182
3.7.5   Surface Recombination with Fermi Level Pinning . .   186
3.8   Summary   190
General References   191
Problems   193

4   Transport Across Junctions   **199**
4.1   Introduction   199
4.2   Metal–Semiconductor Junctions   200
4.2.1   Drift-Diffusion   208
4.2.2   Thermionic Emission   211
4.2.3   Field Emission and Thermionic Field Emission . . .   214
4.2.4   Thermionic Emission-Diffusion theory   219
4.3   Heterojunctions   222

	4.3.1	Thermionic Emission	226
	4.3.2	Tunneling	231
	4.3.3	Resonant Fowler–Nordheim Tunneling	239
4.4	Ohmic Contacts		241
4.5	p–n Junctions		248
	4.5.1	Depletion Approximation	248
	4.5.2	High-Level Injection	252
	4.5.3	Gummel–Poon Quasi-Static Model	257
	4.5.4	Abrupt and Graded Heterojunction p–n Diodes	280
4.6	Summary		292
	General References		292
	Problems		293

5 Metal–Semiconductor Field Effect Transistors   299

5.1	Introduction		299
5.2	Analytic Quasi-Static Models		301
	5.2.1	Gradual Channel Approximation	302
	5.2.2	Constant Mobility Model	302
	5.2.3	Constant Velocity Model	308
5.3	Constant Mobility with Saturated Velocity Model		311
	5.3.1	Transconductance	324
	5.3.2	Output Conductance	326
	5.3.3	Gate-to-Source Capacitance	327
	5.3.4	Gate-to-Drain Capacitance	327
	5.3.5	Drain-to-Source Capacitance	329
5.4	Accumulation–Depletion of Carriers		330
5.5	Sub-Threshold and Substrate Injection Effects		339
5.6	Sidegating Effects		340
	5.6.1	Injection and Conduction Effects	341
	5.6.2	Bulk-Dominated Behavior	348
	5.6.3	Surface-Dominated Behavior	354
5.7	Piezoelectric Effects		356
5.8	Signal Delay along the Gate		363
5.9	Small-Signal High Frequency Models		365
5.10	Limit Frequencies		385
5.11	Transient Analysis		390
5.12	Off-Equilibrium Effects		392
5.13	Summary		394
	General References		397
	Problems		398

6 Insulator and Heterostructure Field Effect Transistors   403
   6.1   Introduction 403
   6.2   Heterostructures 405
   6.3   Strained Heterostructures 408
   6.4   Band Discontinuities 410
   6.5   Band Bending and Subband Formation 412
   6.6   Channel Control in HFETs 426
   6.7   Quasi-Static Insulator FET Analysis 432
         6.7.1   Capacitance of the MIS Structure 447
         6.7.2   Flat-Band Voltage 449
         6.7.3   MISFET Models Based on Sheet Charge Approxima-
                 tion 451
   6.8   HFET Analysis 483
         6.8.1   Sub-Threshold Currents 501
         6.8.2   Intrinsic Capacitances 502
         6.8.3   Transconductance 506
   6.9   Quasi-Static Circuit Refinements 506
   6.10 Small-Signal Analysis 509
   6.11 Transient Analysis 520
   6.12 Hot Carrier Injection Effects 522
   6.13 Effects Due to DX Centers 528
   6.14 Off-Equilibrium Effects 535
   6.15 p-channel Field Effect Transistors 540
   6.16 Summary 541
   General References 542
   Problems 542

7 Heterostructure Bipolar Transistors   551
   7.1   Introduction 551
   7.2   Quasi-Static Analysis 554
         7.2.1   Extended Gummel–Poon Model 558
         7.2.2   Ebers–Moll Model 575
   7.3   Heterostructure Implications 580
         7.3.1   Charge Transport and Storage in the Base–Emitter
                 Junction 589
         7.3.2   Alloy Grading, Doping Design and Transport at the
                 Base–Emitter Junction 591
         7.3.3   Base–Emitter Capacitance 598
         7.3.4   Electron Quasi-Fields in Single Heterojunction Bipo-
                 lar Transistors 602
   7.4   High Current Considerations 604

7.4.1   Barriers and their Influence in Heterojunction Collectors 604

7.4.2   Collector Electron Quasi-Fields 610

7.4.3   Diffusion Capacitances 612

7.4.4   Current Gain Effects 615

7.5   Generation and Recombination Effects 616

    7.5.1   Bulk Effects 618

    7.5.2   Surface Effects 620

    7.5.3   Current Gain Behavior 626

7.6   Small-Signal Analysis 629

    7.6.1   Parameter Notation and Assumptions 631

    7.6.2   Static and Small-Signal Solutions 637

    7.6.3   Network Parameters and their Approximations . . . 645

    7.6.4   Frequency Figures of Merit 664

7.7   Small-Signal Effects of Alloy Grading 672

7.8   Transit Time Resonance Effects 681

7.9   Transient Analysis 683

7.10 Off-Equilibrium Effects 684

7.11 Summary 698

General References 699

Problems 700

8   Hot Carrier and Tunneling Structures   705

8.1   Introduction 705

8.2   Quantum-Mechanical Reflections 709

8.3   Hot Carrier Structures 714

8.4   Resonant and Sequential Tunneling 723

8.5   Transistors with Coupled Barrier Tunneling 743

8.6   Summary 746

General References 747

Problems 748

9   Scaling and Operational Limitations   755

9.1   Introduction 755

9.2   Operational Generalities 757

9.3   General Scaling Considerations 761

9.4   Limits from Operational Considerations 765

9.5   Scaling and Operational Considerations of FETs 767

    9.5.1   Limitations from Transport 776

9.6   Scaling and Operational Considerations of HBTs 779

9.7   Summary 780

General References 781
Problems 782

A Network Parameters and Relationships   783

B Properties of Compound Semiconductors   791

C Physical Constants, Units, and Acronyms   803

Glossary   807

Index   819

Preface

New books on the use of semiconductors in devices, of the last few years, have been either directed to the practitioner by emphasizing the state of the art or to the university student by using major simplifications in a treatment that achieves analytic and closed-form mathematical solutions. Presumably, a reason for this is the difficulty in describing numerical techniques and their validity in the restricted space of technical publications or the limited time in the classroom. Modern devices, with their small dimensions, however, require an understanding of the physics of reduced dimensions, the use of statistical methods, and the use of one-, two-, and three-dimensional analytic and numerical analysis techniques. These techniques also bring with them alternate approximations and simplifications. An understanding of these is a prerequisite to the appraisal of the results.

This book is an attempt at finding the common ground for the above, and the product of a desire to write it the way I advocate the teaching of this subject. The subject is not trivial and I have resisted oversimplifications. It is an intermediate-level text for graduate students interested in learning about compound semiconductor devices and analysis techniques for small dimension devices. It should be appealing to students who have already achieved an understanding of the principles of operation of field effect transistors and bipolar transistors. While the emphasis of the treatment is on compound semiconductor devices, and examples are mostly drawn from them, it should also be useful to those interested in silicon devices. The principles of devices are the same; compound semiconductor devices only bring with them more complications associated with negative differential mobility and stronger quantum-mechanical and off-equilibrium effects. As befitting a textbook, there is very little original material here, and references have been chosen for the inquisitive to find elaborations and complementary treatments. I have included original material only where I have felt a compelling reason to buttress and elaborate an argument whose results were not in doubt.

Both one-dimensional classical approximations and numerical procedures of better accuracy have been incorporated in order to clarify device behavior. I have also tried to use consistent device examples for both the analytic and the numerical approaches. This allows both a better appreciation of the approximations and a better feel for cause and effect. Students, I hope, will look at appearances of unusual features carefully and try to recognize their origin either in the underlying nature or the approximation of the model. Most problems at the end of the chapters have been designed to emphasize the concepts and to complement arguments or derivations in the chapters. The rest emphasize the nature of approximations by critically evaluating realistic conditions. These rely on the use of numerical techniques by the student. The reward is a deeper feel for the subject; a note of caution though, they require both will and an access to computers.

This effort owes its gratitude to many: family, teachers, authors of discerning reviews and books, and my colleagues, who have generously discussed, encouraged, and criticized. I thank J. East, T. N. Jackson, P. Mc-Cleer, P. Mooney, P. Solomon, W. Wang, and S. L. Wright for the conversations. M. Fischetti, D. Frank, S. Laux, and J. Tang, premier practitioners of the art and science of numerical techniques and their applications, have influenced this work in countless ways. This book was partly written, revised, and practiced during a sabbatical leave at the University of Michigan. For the delightful and refreshing time, I thank Professors P. Bhattacharya, G. I. Haddad, J. Singh, D. Pavlidis, F. L. Terry, and K. Tomizawa, and the students who attended my courses and provided invaluable feedback. S. Akhtar, J. East, S. Laux, P. Price, J. Singh, F. Stern, and J. Tsang have commented on parts of the manuscript in depth; I am very grateful to them for this and to F. L. Terry for many exchanges on the contents and teaching of semiconductor devices. This book would certainly not have been possible without the active encouragement and support of International Business Machines Corporation, the help of my colleagues, and the influence of the environment at Thomas J. Watson Research Center.

As the book has evolved, its many rough edges have worn away, but many remain. I welcome comments and suggestions on the latter.

Sandip Tiwari

Chapter 1

Introduction

Compound semiconductors have been a subject of semiconductor research for nearly as long as elemental semiconductors. Initial discoveries of the late 1940's and early 1950's, discoveries that began the use of semiconductors in our everyday life, were in germanium. With time, it has been supplanted by silicon—a more robust, reliable, and technologically well-behaved material with a stable oxide. Compound semiconductors, whose merit of superior transport was recognized as early as 1952 by Welker, have continued to be of interest since these early days although their success has been narrower in scope. The areas of significant applications include light sources (light emitting diodes and light amplification by stimulated emission of radiation), microwave sources (Gunn diodes, Impatt diodes, etc.), microwave detectors (metal–semiconductor diodes, etc.), and infrared detectors. All of these applications have been areas of the semiconductor endeavor to which compound semiconductors are uniquely suited. The success in optical, infrared, and microwave applications resulted from the direct bandgap of most compound semiconductors for the first, the small bandgap of some compound semiconductors for the second, and the hot carrier properties, superior electron transport characteristics, and unique band-structure effects for the last.

With the increasing demand placed on voice and data communications, transmitting, receiving, and processing information at high frequencies and high speeds using both microwave and optical means has become another area where compound semiconductor transistors (the field effect transistors and bipolar transistors) have also become increasingly important. This increased usage stems from their higher operating frequencies and speeds as well as from the functional appeal of integration of optical and electronic devices. This expansion in the use of compound semiconductors, in ap-

1

plications requiring the highest performance, is expected to continue even though they still remain difficult materials for electronic manufacturing.

1.1   Outline of the Book

The emphasis of the text is on the operating principles of compound semiconductor transistors: both field effect and bipolar devices. However, a rigorous treatment of the quasi-static, high frequency, and off-equilibrium behavior of small-sized devices requires a careful understanding of the underlying physics of transport and the mathematical methods used in the analysis of devices. While this is better handled as a separate course on semiconductor physics, quite often these courses are not aimed specifically towards understanding of device physics, and hence are difficult to unite with device physics. Additionally, students—the intended audience of this text,—arrive at an advanced course on devices with a very varied training and with one or more of the many diverse ways of analyzing semiconductor devices.

Compound semiconductors bring with them their own peculiarities. Effects related to high surface state density that manifest themselves in the form of Fermi level pinning and high surface recombination, negative differential mobility, etc., have been treated to varied levels of sophistication in the older texts. However, the use of compositional changes and heterostructures, and the emphasis on off-equilibrium behavior of small sized device structures, are relatively new developments and have lacked the attention to detail exemplified in the earlier texts. Thus, advanced semiconductor device courses, generally geared towards silicon, leave an insufficient grasp of the techniques and tools needed to understand small-dimension high frequency compound semiconductor devices. A book emphasizing the operation of compound semiconductor devices, however, does still have relevance to silicon devices. The operating principles are similar; the parameters and relative magnitude of effects is different. The examples here are largely chosen from compound semiconductors although the underlying discussion also stresses silicon and germanium.

To coherently develop the theory of device operation, especially because of the emphasis on small structures, the book breaks with the traditional treatment of device theory. It begins with a review and general development of semiconductor properties and their general relationship with devices. It then discusses the mathematical treatments that are either traditional or increasingly being employed. The stress here is on their underlying assumptions, their limitations, and how one would employ them in different parts of the device to derive the device characteristics of interest. A thorough discussion of two-terminal junctions has also been undertaken because they constitute building blocks for the transistors of interest, and because they are a very convenient means to develop and emphasize the methodology to be employed in the three-terminal device treatment. This is followed by a development of the operation of transistors based on the use of unipolar transport with metal–semiconductor junctions and heterojunctions and the use of bipolar transport. The second to last chapter of the book is devoted to a discussion of devices where hot carrier transport and tunneling is central to the operation. These are two- and three-terminal device structures using transit of carriers with a limited number of scattering events or externally tunable tunneling, two unique phenomena not conventionally employed in three-terminal device structures. They are compound semiconductor devices that need non-traditional device physics which ultimately may also be useful in description of the conventional devices as their sizes continue to shrink. The last chapter discusses this from a general perspective in order to show how present understanding of devices may be used to understand both the future development and the future limitations of devices.

The bulk of the book is devoted to the treatment of the mainstream compound semiconductor transistors: the metal–semiconductor field effect transistor (MESFET), the heterostructure field effect transistor (HFET), and the heterostructure bipolar transistor (HBT). Our discussion of these devices includes quasi-static behavior and development of the corresponding low frequency models, small-signal high frequency behavior and development of corresponding models useful to higher frequencies, transient behavior, the role of off-equilibrium effects in these devices, and various other phenomena important to the operation and usefulness of the device. Through-out the book, we will use the term "off-equilibrium" phenomenon to describe the local phenomenon resulting from loss of equilibrium between the carrier energy and the local electric field, a phenomenon that results in an overshoot effect in the velocity of carriers. This term should be distinguished from "non-equilibrium" or lack of thermal equilibrium, which always occurs when a bias is applied and results in the flow of current. Examples of the various other phenomena include effects such as sidegating— the effect of a remote terminal other than the three terminals of the device acting as an additional gate electrode with an effect on the channel transport; piezoelectric effect—the influence on device characteristics because of strain-induced piezo-effects; and surface recombination—recombination of electrons and holes at the surfaces of the bipolar transistor—and others.

Among the compound semiconductors, the discussion frequently takes GaAs as an example. This is largely because our understanding of GaAs and the technology of GaAs-based devices is more mature than that of the other compound semiconductors. However, many examples are cited from $Ga_{.47}In_{.53}As$ (the lattice matched composition with InP), InP itself, and InAs. Other compound semiconductors are cited when particular properties are specifically of interest. The discussion has largely been kept general even if referring to GaAs.

1.2 Suggested Usage

The text is best suited for a two semester course with the first covering through MESFETs. The development of a feel for a subject is best achieved by a simultaneous development of healthy skepticism. This requires a good understanding of the underlying assumptions and the methods of attack on the problem being analyzed, and an appreciation of the relevant general properties. Chapters 2 and 3 are an untraditional attempt at meeting these requirements, and the discussion of two-terminal diode structures and one example of a three-terminal transistor structure brings these concepts together coherently. The second course, constituting the rest of the book, then builds on this by analyzing HFETs and HBTs and then dwells on the new developments that are of considerable intellectual interest. The last chapter on scaling and operational limitations serves to continue a discussion begun in the previous chapter of the conventional development of devices with time and how one may look at the subject very broadly by going to the roots in electromagnetic and other classical equations.

The text could be used for a one semester course with the subject ranging from the treatment of MESFETs through the hot carrier and tunneling structures. I also recommend that small-signal treatment based on the small-signal drift-diffusion equation be excluded from such a course. This treatment is peripheral to the objectives that a one semester course would usually have.

The references occur in two places. General references, long articles and books that have extensive breadth and scope, have been placed at the end of chapters. The reader will find these advanced, complementary, and supplementary reading, and the vast reference lists in them useful for further perusal. References have also been placed in figures and footnotes, as part of the main text. These references are more specific; they supplement comments made in the text and are also in many instances sources for the material developed or reproduced in the text.

The problems need to be treated with caution. They range from questions that can be answered in one line to questions that require extensive use of computation facilities. The author believes in a balance of the two; success with both is indicative of a systematic and detailed understanding of the subject and of using the knowledge in uncharted regions. The reader should exercise appropriate restraint in what he or she attempts, keeping in mind the available facilities.

1.3 Comments on Nomenclature

We will generally use only MESFET, HFET, and HBT as the acronyms for metal–semiconductor field effect transistor, heterostructure field effect transistor, and heterostructure bipolar transistor. These are meant to define a general class of devices which have sufficiently different operational basis. These devices, however, can be implemented in many ways, largely because compound semiconductors can be grown in a variety of heterostructures with many variations in control of the bandgap and doping in the structures.

MESFETs, e.g., can be grown with a thin heterostructure underneath the gate or a thick heterostructure underneath the channel—one suppresses gate current while the other suppresses substrate injection current. They, however, all use a doped channel where quantization effects are unimportant and the operational basis is largely unchanged. HFETs similarly can be grown in a variety of ways. In such transistors a large bandgap material abuts a small bandgap material with an abrupt hetero-interface. HFETs exist with a gate made of metal, a gate made of a semiconductor[1], a doped large bandgap semiconductor, an undoped large bandgap semiconductor, and various heterostructures at the channel-substrate interface. These are all generically HFETs where the abruptness of the hetero-interface between the channel region and the control region is important to the operational basis of the device. HBTs, similarly, also have many variations—two examples being use of one heterostructure (in the emitter) and two heterostructures (in both emitter and collector) in the structure.

---

[1]A close compound semiconductor analog of the poly-silicon gate silicon MOSFET (metal–oxide–semiconductor field effect transistor) may, more generically, be called a semiconductor–insulator–semiconductor field effect transistor (SISFET). HFET is a superset of such devices.

Chapter 2

Review of Semiconductor Physics, Properties, and Device Implications

2.1 Introduction

An adequate description of different semiconductor devices requires an understanding of the physics of the semiconductor ranging all the way from the semi-classical description to the quantum-mechanical description. The nearly-free electron model of a carrier in a semiconductor is a semi-classical description. Along with other semi-classical phenomenology this description has been quite adequate in describing the functioning of several important devices—both the silicon bipolar transistor and the silicon MOSFET. However, this same carrier, when confined in a short space between two ideal abrupt heterostructure interfaces,[1] exhibits behavior in which the quantum-mechanical effects are of paramount importance and central to explaining many of the observations on the structure. In this, a particle in a box problem, the momentum of the carrier in the direction orthogonal to the heterostructure interface is no longer continuous. The particle exhibits a behavior in which quantum-mechanical effects are important. The applicability of these descriptions is not mutually exclusive. The quantum-mechanical description can be shown to reduce to the semi-classical in the proper limits. It just so happens that the sophistication and depth of the

---

[1] An abrupt interface between two dissimilar materials that exhibit long range lattice periodicity.

quantum-mechanical description is not necessary in what we view as the functioning of the devices cited. If we probe deeper, e.g., the silicon MOS-FET operating at 77 K or the compound semiconductor HFET operating at 300 K, we find that we need to incorporate the increased sophistication to explain many of the observations, and the rigorous quantum-mechanical treatment also has the beauty of reducing to the classical description in the classical limits.

The HFET provides quite an interesting example of the relationship between the accuracy of results sought and the necessary rigorousness in the description of the problem. This is a device in which a channel is formed for conduction in a small bandgap material by using a large bandgap material for the insulator and/or for providing the carriers. It may be viewed as one of the MOSFET-like implementations of compound semiconductor FETs. If the potential variation in the small bandgap material is rapid, and this rapid potential variation occurs at high carrier concentrations following Gauss's law, then one would expect quantum-mechanical effects resulting from confinement similar to that for a particle in a box. However, if carrier concentration is small, then the potential variation is slow and the sophisticated description is unnecessary. The device, in different bias regions, requires different levels of description, to explain observations. In fact, different regions of the device at the same bias may require different levels of sophistication. There are more carriers under normal operating conditions at the source end of the device where they are injected than at the drain end where they are collected. So, while quantum-mechanical effects may be necessary in the description of the device at the source end, they may not be at the drain end. Temperature is also important in the evaluation of the acceptable methodology of handling this problem. At room temperature, thermal energy is 25.9 meV; clearly if the confinement leads to energy effects much smaller than this energy, then the inclusion of confinement effects is not necessary. Silicon MOSFET, which also has a large discontinuity at the oxide-semiconductor interface, has a much smaller predicted effect for both electrons and holes at 300 K where the effect is ignored, but at 77 K, thermal energy and energy discretization is comparable, and at least for some problems, the discretization may not be ignored. So for conventional bipolar and field effect transistors, we need to make a judicious choice of the level of description needed. In devices, where the wave nature of the electron is central to their functioning (tunneling, e.g., in tunnel diodes) the quantum-mechanical description is unavoidable.

This chapter is a review that connects the basis of semiconductor theory in classical and quantum mechanics to its implications in devices, with an emphasis on examples from compound semiconductors. We discuss the particle-wave duality and the description of electrons in a lattice with the effect of lattice on the electron folded into the effective mass. We discuss lattice vibrations and the concept of phonons. This early description is made with respect to a one-dimensional lattice, which allows us to explore the formation of energy bands, the concept of holes, the specification the of energy-momentum relationship for both the electrons and the phonons, and the description of a quasi-continuum of allowed states via the density of states. Using these one-dimensional examples as a basis, we then discuss the behavior of the three-dimensional semiconductor crystals.

Since our primary interest is in devices, our discussion of the properties of semiconductor crystal is related to specifically those phenomena that influence electron transport and electron interactions. Transport, e.g., is intimately related to the electron's interaction with its environment, i.e. scattering. We consider the various mechanisms of scattering and their behavior in the semiconductor crystal. The carrier velocities are also related to the band structure. We discuss this and other band structure-related phenomena as the last part of this chapter.

This chapter is written as a general review with an emphasis on mathematics where it has been considered necessary. One example of this emphasis is the occupation of carriers at discrete energy levels, such as donors and acceptors, which brings out the importance of degeneracy; and, together with the description of band occupation, establishes the framework for semiconductor statistics to be developed further in the following chapters.

2.2 Electrons, Holes, and Phonons

Consider a free electron in the absence of any other interactions. The time-independent Schrödinger's equation characterizing the behavior of this single electron consists of only the kinetic energy term and is

$$\frac{\hbar^2}{2m}\nabla^2\psi(\boldsymbol{r}) = E\psi(\boldsymbol{r}). \tag{2.1}$$

Including the time-dependent component, the solution is of the form

$$\psi_k(\boldsymbol{r}, t) = \mathcal{A}\exp\left[j\left(\boldsymbol{k}.\boldsymbol{r} - \omega t\right)\right], \tag{2.2}$$

where $\mathcal{A}$ is a normalization pre-factor. This eigenfunction describes a plane wave. This plane wave, whose wave vector is $\boldsymbol{k}^2$, is associated with

---

[2]Throughout the book, we employ bold fonts to indicate vectors. The magnitude of these vectors is indicated using the normal font. Readers should refer to the Glossary for the definition of symbols.

an energy whose dispersion relation is a parabolic relationship,

$$E(\boldsymbol{k}) = \frac{\hbar^2 k^2}{2m}, \tag{2.3}$$

and whose momentum is given by

$$\boldsymbol{p} = \hbar \boldsymbol{k}. \tag{2.4}$$

The wavelength and angular frequency associated with this plane wave are

$$\lambda = \frac{h}{p}, \tag{2.5}$$

and

$$\omega = \frac{E}{\hbar}. \tag{2.6}$$

This wavelength is the de Broglie wavelength characterizing the wave particle duality.

For a single plane wave, the probability of finding the electron $\psi\psi^*$ is a constant in real space; the electron can be found anywhere with an identical likelihood. To describe a more localized electron, one may construct a wave packet by considering a group of plane waves with a finite amplitude in a narrow range of the wave vector. The function

$$\phi(r, t) = \int_{-\infty}^{\infty} \mathcal{A}(k) \exp\left[j(kr - \omega t)\right] dk, \tag{2.7}$$

which is a superposition of plane waves, or equivalently a Fourier expansion in the k-space, describes a wave packet if $\mathcal{A}(k)$ is non-zero for a range δ of wave vector k about k_0 such that $\delta \ll k_0$. This function now allows us to determine the probability of finding an electron; it is now higher in certain ranges of r and k.

Using a Taylor series expansion of the angular frequency ω about k_0, i.e.,

$$\omega = \omega_0 + (k - k_0) \left.\frac{d\omega}{dk}\right|_{k_0} + \cdots, \tag{2.8}$$

we can write

$$\phi(r, t) \approx \exp\left[j(k_0 r - \omega_0 t)\right] \int_{-\infty}^{\infty} \mathcal{A}(k) \exp\left[j(k - k_0)\left(r - \frac{d\omega}{dk}t\right)\right] dk$$

$$= \mathcal{B}\left(r - \frac{d\omega}{dk}t\right) \exp\left[j(k_0 r - \omega_0 t)\right], \tag{2.9}$$

where $\mathcal{B}(r - (d\omega/dk)t)$ is the envelope function of the resultant plane wave. The group velocity of this wave packet, the velocity of the electron, is given by

$$v_g = \frac{d\omega}{dk}, \tag{2.10}$$

and the phase velocity, the velocity of the plane wave, is

$$v_p = \frac{\omega_0}{k_0}. \tag{2.11}$$

In doing this we have shown that superposition of plane waves can be used to characterize localization of electrons and hence have related some of the consequences of the quantum-mechanical description with the classical description.

We have, until now, considered the free electron as possessing no potential energy, and solved the single body problem whose Hamiltonian contained only the kinetic energy term. In studying semiconductor devices, we are interested in electrons in semiconductors. This is a many-body problem. Even if the behavior of electrons is treated as a single electron problem unaffected by other electrons in the crystal (the one-electron approximation), we must consider several other factors. The Hamiltonian must now include several additional energy terms such as the kinetic energy of nuclei and core electrons (the ion core) that oscillate around a mean position, a movement whose effect is characterized by phonons; the potential energy of these ion cores; and the potential energy of the electrons due to influence of other electrons and the cores. In addition, the crystal is finite sized, albeit of a large size. There are explicit quantization effects associated with this confinement; there is a large but finite set of allowed values of the wave vector k which are characterized by a finite density of states for the electron. The crystal is also a periodic structure where the ion cores, on average, occupy a lattice with a specific lattice constant. Since the ions are considerably heavier than the electrons, the electron eigenfunction may be treated as being instantly responsive to the ion movement. This is the adiabatic approximation and it allows the decoupling of the Schrödinger equation into an ionic and an electronic equation. The one-electron approximation provides a reasonable description for this problem because both the core electrons and the valence electrons exist in close proximity to the cores and hence screen the interaction, since both the exclusion principle and repulsion cause separation of electrons and because electrons spend very little time near a core due to the large accelerations resulting from large forces.

Let $V(r)$ represent the potential energy term for the electron Hamiltonian. It has the periodicity of the lattice R—the direct lattice vector—and

$$V(r) = V(r + R). \tag{2.12}$$

Bloch's theorem states that for such a periodic potential, the eigenfunction solution of the Hamiltonian for the one-electron problem has a solution of the form

$$\psi_k(\boldsymbol{r}) = \exp(j\boldsymbol{k}.\boldsymbol{r})u_k(\boldsymbol{r}), \tag{2.13}$$

and

$$u_k(\boldsymbol{r}) = u_k(\boldsymbol{r} + \boldsymbol{R}). \tag{2.14}$$

The consequence of this is that

$$\psi_k(\boldsymbol{r} + \boldsymbol{R}) = \exp(j\boldsymbol{k}.\boldsymbol{R})\psi_k(\boldsymbol{r}), \tag{2.15}$$

and it brings us to the concept of reciprocal lattice. The wave vector $\boldsymbol{k}$ has the units of reciprocal length. For values of $\boldsymbol{k} = \boldsymbol{G}$, such that

$$\boldsymbol{G}.\boldsymbol{R} = 2n\pi, \tag{2.16}$$

where n is an integer, the eigenfunction ψ_G is periodic in $\boldsymbol{R}$. If the magnitude of the wave vector was equal to the reciprocal lattice vector $\boldsymbol{G}$, then the eigenfunction would be periodic in real space. That is the important consequence of Equation 2.15. The translating of the wave vector $\boldsymbol{k}$ by the reciprocal lattice vector, i.e.,

$$\boldsymbol{k} = \boldsymbol{G} + \boldsymbol{k}', \tag{2.17}$$

results in

$$\psi_k(\boldsymbol{r} + \boldsymbol{R}) = \exp(j\boldsymbol{k}'.\boldsymbol{R})\psi_k(\boldsymbol{r}). \tag{2.18}$$

This is a restatement of Bloch's theorem with a new wave vector $\boldsymbol{k}'$, i.e., all wave vectors differing from $\boldsymbol{k}$ by integer multiples of the reciprocal lattice vector satisfy the requirements of the eigenfunction. We make the wave vector unique by translating all the wave vectors to the first Brillouin zone, the primitive or Wigner–Seitz unit cell in reciprocal space.

This brings us to the definition of the reciprocal lattice. In general, the direct lattice vector may be expressed in terms of primitive unit cell vectors $\boldsymbol{a}_i$ which are not necessarily orthogonal to each other.[3] The primitive unit cell vectors of the reciprocal lattice $\boldsymbol{b}_i$, defining the Brillouin zone, are then given by

$$\boldsymbol{b}_1 = 2\pi \frac{\boldsymbol{a}_2 \times \boldsymbol{a}_3}{\boldsymbol{a}_1.(\boldsymbol{a}_2 \times \boldsymbol{a}_3)}, \tag{2.19}$$

---

[3] A common method of notation, to describe various planes in the unit cell, is to use the reciprocal of the intercepts in units of the primitive vectors. As an example, if a_1/h, a_2/k, and a_3/l are the three intercepts, then the Miller indices (hkl) identify the plane. The vector perpendicular to this plane $\langle hkl \rangle$ is the direction.

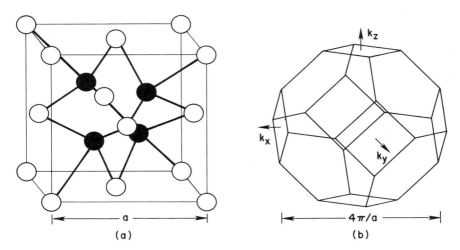

Figure 2.1: The cubic unit cell of the zinc blende structure is shown in the part (a). The open and closed circles denote the sites for the group III and group V atoms. Part (b) shows the corresponding first Brillouin zone—a truncated octahedron.

$$b_2 = 2\pi \frac{a_3 \times a_1}{a_1 \cdot (a_2 \times a_3)},$$ (2.20)

and

$$b_3 = 2\pi \frac{a_1 \times a_2}{a_1 \cdot (a_2 \times a_3)}.$$ (2.21)

Figure 2.1 shows the primitive cell of the zinc blende structure (the common form of occurrence for most compound semiconductors) and the first Brillouin zone of this face-centered cubic crystal. The first Brillouin zone is a truncated octahedron. The importance of this first Brillouin zone for the present discussion is that all wave vectors beyond the first Brillouin zone may be folded into the first Brillouin zone using a translation of the reciprocal lattice vector G, which is composed of integer multiples of the primitive vectors of the reciprocal cell. Certain positions in the Brillouin zone are of particular importance because of their symmetry, and the relation to other characteristics that this entails. We will return to a discussion of the significance of the Brillouin zone and of these symmetry points during our discussion of semiconductor properties.

We have now considered the form the eigenfunction solution may take for the one-electron approximation in periodic potential. Determination

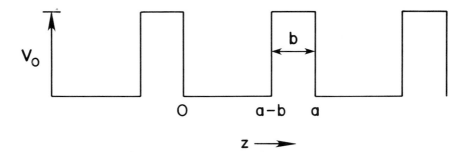

Figure 2.2: The periodic potential used in the Kronig–Penney model. It consists of periodic potential wells in a hypothetical one-dimensional crystal. The spatial periodicity is a, the barrier width is b, and potential barrier height is V_0.

of the actual eigenfunction and the eigenenergy can be considerably more complex. If the kinetic energy of the electrons is much larger than the periodic potential energy resulting from the lattice, then the behavior of the electron can be modelled approximately by a nearly free electron eigenfunction. The effect of the lattice potential is to make the electron respond to externally applied forces as if it has a differing effective mass (m^*) instead of the free electron mass. Note that the mass of the electron itself has not changed. To determine the response of the electron in the crystal to an externally applied stimulus such as an electric field, one need not consider the internal forces such as those due to lattice—their influence has already been folded into the effective mass which can be either smaller or larger than the free electron mass. Thus, the behavior of the nearly free electron is similar to that of the free electron discussed earlier; the difference is the change in effective mass.

The periodicity of the lattice potential has another significant effect. Standing waves formed due to this periodicity cause a shifting in energy because of the resultant charge movement. This leads to bands of energy that are allowed energies for an electron and an energy gap of disallowed energies.

A good example of this that can be solved explicitly is the Kronig–Penney model. Consider Figure 2.2, which shows the periodic potential in a hypothetical one-dimensional crystal. The spatial periodicity is a for a potential barrier whose height and width are V_0 and b respectively. We are interested in finding the energies and wave vectors allowed in this periodic structure. Schrödinger equation for the barrier region $(a - b < z < a$, etc.$)$

is

$$-\frac{\hbar^2}{2m}\frac{\partial^2\psi}{\partial z^2} + V_0 = E\psi, \qquad (2.22)$$

whose solution should be of the form

$$\psi_1(z) = A\exp(\alpha z) + B\exp(-\alpha z), \qquad (2.23)$$

where

$$\alpha = \left[\frac{2m(V_0 - E)}{\hbar^2}\right]^{1/2}. \qquad (2.24)$$

The Schrödinger equation for the well region ($0 < z < a - b$, etc.) is

$$-\frac{\hbar^2}{2m}\frac{\partial^2\psi}{\partial z^2} = E\psi, \qquad (2.25)$$

whose solution should be of the form

$$\psi_2(z) = C\exp(j\beta z) + D\exp(-j\beta z), \qquad (2.26)$$

where

$$\beta = \left(\frac{2mE}{\hbar^2}\right)^{1/2}. \qquad (2.27)$$

The constants can be evaluated based on continuity at $z = 0$ and consequences of Bloch's theorem. The consequence of the latter is that, for any k,

$$\psi_k(z + a) = \exp(-jka)\psi_k(z), \qquad (2.28)$$

and one could use this to establish continuity at $z = -b$. Thus the four boundary conditions that are periodic in nature are:

$$
\begin{aligned}
\psi_1(0) &= \psi_2(0), \\
\left.\frac{\partial\psi_1}{\partial z}\right|_{z=0} &= \left.\frac{\partial\psi_2}{\partial z}\right|_{z=0}, \\
\psi_1(-b) &= \exp(-jka)\psi_2(a - b), \\
\text{and} \qquad \left.\frac{\partial\psi_1}{\partial z}\right|_{z=-b} &= \exp(-jka)\left.\frac{\partial\psi_2}{\partial z}\right|_{z=0}.
\end{aligned} \qquad (2.29)
$$

This is a set of four equations, with four unknowns A, B, C, and D. A solution exists when the determinant of the coefficients of these unknowns vanishes, a condition that can be written as

$$\cos(ka) = \frac{\alpha^2 - \beta^2}{2\alpha\beta}\sinh(\alpha b)\sin\left(\beta(a - b)\right) + \cosh(\alpha b)\cos\left(\beta(a - b)\right). \qquad (2.30)$$

For real k, allowing for travelling waves, the left hand side will have a value between -1 and $+1$. The right hand side is an oscillating function with increasing energy, i.e., also β. Only those values of E that limit the magnitude of the right hand side between -1 and $+1$ allow for a wave solution or a pass-band in energy. Outside this range lie the values of E that are forbidden; these form the energy gap regions. We will consider the solution using a parameter P which is proportional to the area under the barrier,

$$P = \frac{mabV_0}{\hbar^2}. \tag{2.31}$$

We now consider, for constant P, barriers of infinitely small thickness, i.e., the condition where $b \to 0$ and the barriers are replaced by delta functions. Since $\sinh(\alpha b) \to \alpha b$ and $\cosh(\alpha b) \to 1$, the constraint of Equation 2.30 reduces to

$$\cos(ka) = \frac{P}{\beta a} \sin(\beta a) + \cos(\beta a). \tag{2.32}$$

The constant P representing the area under the potential barrier represents the strength of electron binding in the crystal. A large value of P is the tight binding limit while $P = 0$ is the free electron limit. Figure 2.3 shows the right hand side of Equation 2.32 plotted as a function of βa for a small and a large value of P. The transition between the allowed and forbidden bands occurs at $\beta = n\pi/a$ where n is an integer. For a one-dimensional crystal, these are the Bragg reflection conditions. At this transition, $k = n\pi/a$ and the electron wave function is a standing wave. Since β has a direct square-root dependence on the energy of the electron, Figure 2.3 also shows that for small P, the allowed energy bands are larger, together with smaller bandgaps, compared to large values of P. For any P, at higher energies, the allowed bands get broader and bandgaps smaller. So, for tight binding of an electron represented by a large P, the allowed energy bands are smaller, but for high energies in either of the limits, the passband approaches that of a free electron.

The energy versus wave vector relationship resulting from this analysis is shown in Figure 2.4. The figure shows, in both extended and reduced zone representation, the formation of allowed bands and their respective energies versus wave vector relationship. The reduced zone, also called folded zone representation, is the more popular form because it completely and succinctly describes the relationship of interest. It can be seen that near the band edge the E versus k relationship is close to parabolic (the second term of a Taylor series expansion). In the case of a free electron, it is exactly parabolic over the whole energy range.

The Kronig–Penney is a highly idealized model limited to a one-dimensional crystal; arbitrary potential forms can only be treated via sums of a

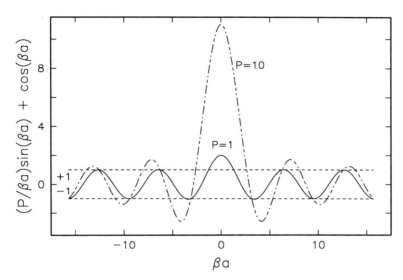

Figure 2.3: A plot representing Equation 2.32 as a function of the parameter βa for $P = 1$ (solid line) and $P = 10$ (dot-dashed line). Allowed bands occur where the function lies in between -1 and $+1$.

Fourier series. In a real crystal one would therefore resort to rather intensive numerical techniques using complicated wave functions to determine the energy-band diagram.

Some examples of such approaches[4] are the plane wave approach suitable for nearly-free electrons, the tightly bound electron approach, the orthogonalized plane wave approach and its specialized application using the pseudopotential approach. Nearly free electron models are quite approximate representations of the real crystals. One problem, e.g., is that near the ion core, the eigenfunction must differ substantially from a plane wave and hence several terms of the Fourier expansion of $u_k(\mathbf{r})$ with reciprocal lattice vectors are necessary to obtain any accuracy. The tightly bound electron model takes the opposite approach by assuming that the lattice potential energy is much larger than the kinetic energy. The tighter binding is particularly true for the electrons that are largely bound to the ion cores. The result is that one would expect the wave functions to be closer to the wave functions of the electrons when the atoms are separated from each other. The Bloch electron wave function can then be constructed using a linear

---

[4]W. A. Harrison, *Solid State Theory*, Dover, N.Y. (1979) has a rigorous treatment of these approaches.

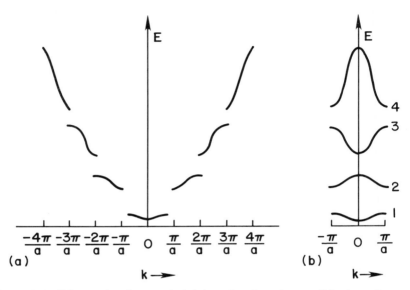

Figure 2.4: Schematic of extended (a) and reduced zone (b) plot of energy E versus wave vector k for the Kronig–Penney model.

combination of atomic orbitals to determine the energy-band structure. In the same vein, one may construct the electron wave functions using a linear combination of molecular orbitals. The necessity of using several terms of the Fourier expansion of $u_k(\boldsymbol{r})$ with reciprocal lattice vector in the plane wave approach is addressed by a suitable modification of the wave function in the orthogonalized plane wave method. This method allows development of a wave function which behaves approximately like a plane wave between ion cores and approximately like an atomic wave function near ion cores with the atomic wave function part orthogonal to the plane wave part. A development from the orthogonalized plane wave method is the use of pseudopotentials, which takes advantage of the reduction in effective potential for a valence electron due to the opposite effect of the real potential $V(\boldsymbol{r})$ and the atomic orbital effect.

The result of these complex calculations is the development of the energy-band diagram of the three-dimensional lattice—a description of the relationship between the energy and crystal momentum associated with the electron. While we will discuss these specifically for the semiconductors later in this chapter, it should come as no surprise that they are considerably different from that seen for the Kronig–Penney model. They are different in different directions in the k-space, and the energy minima or maxima of various bands do not have to occur at zone center, i.e., $k = 0$. The highest normally filled band at absolute zero temperature is the valence band and the lowest normally empty band at absolute zero temperature is the conduction band.

We return now for further discussion to the highly idealized Kronig–Penney model whose E–k relationship is shown in Figure 2.4. Let the second band be a valence band and the third band a conduction band. If we expand, for the conduction band, the E–k relationship via a Taylor series expansion, i.e.,

$$E = E_0 + ak^2 + bk^4 + \cdots, \tag{2.33}$$

at small values of k, the relationship is parabolic. So, at small values of k, this is like the free electron case, but has a differing mass m^* given by

$$m^* = \hbar^2 \left(\frac{\partial^2 E}{\partial k^2} \right)^{-1}, \tag{2.34}$$

with the energy relationship given by

$$E - E_0 = \frac{\hbar^2 k^2}{2m^*}. \tag{2.35}$$

The electron, in the periodic potential, behaves as if it is moving in a uniform potential whose magnitude is E_0/q and as if it has an effective mass m^*. The equation for effective mass has a much more general validity than implied above. Mass has the significance of being the proportionality factor between force and acceleration. Thus, for a force $\mathcal{F}$, and a group velocity $\boldsymbol{v}_g$,[5]

$$\mathcal{F} = m^* \frac{d\boldsymbol{v}_g}{dt}. \tag{2.36}$$

In addition, since

$$\mathcal{F} dt = d\boldsymbol{p} = \hbar d\boldsymbol{k}, \tag{2.37}$$

$$\mathcal{F} = \hbar \frac{d\boldsymbol{k}}{dt}. \tag{2.38}$$

Since the group velocity is given as

$$v_g = \frac{\partial \omega}{\partial k}, \tag{2.39}$$

---

[5]Effective mass and momentum will be discussed again when we consider real three-dimensional crystal structures. We have not yet stressed the distinction between crystal and electron momentum. Here, $p = \hbar k$ is, of course, the electron momentum.

from the E–k relationship, it follows in one dimension as

$$v_g = \frac{1}{\hbar} \frac{\partial E}{\partial k},$$ (2.40)

or more generally in the n-dimension case as

$$v_g = \frac{1}{\hbar} \boldsymbol{\nabla}_k E(\boldsymbol{k}).$$ (2.41)

For one dimension, the time rate of change of the group velocity because of the application of the force $\mathcal{F}$, is

$$\frac{dv_g}{dt} = \frac{1}{\hbar} \frac{d}{dt} \left(\frac{dE}{dk} \right) = \frac{1}{\hbar} \frac{d}{dk} \left(\frac{dk}{dt} \frac{dE}{dk} \right),$$ (2.42)

and hence, using Equation 2.38 and 2.42, the force is related as

$$\mathcal{F} = \hbar^2 \left(\frac{d^2 E}{dk^2} \right)^{-1} \frac{dv_g}{dt},$$ (2.43)

and therefore the effective mass is given by

$$\frac{1}{m^*} = \frac{1}{\hbar^2} \left(\frac{d^2 E}{dk^2} \right).$$ (2.44)

For a general three-dimensional situation where it may be different in different directions, it is related, by extension, as

$$\frac{1}{m_{ij}^*} = \frac{1}{\hbar^2} \frac{\partial^2 E}{\partial k_i \partial k_j}.$$ (2.45)

The second and the fourth band of our Kronig–Penney example have maximum in energy at zone center. The energy near these maxima is also expandable in a Taylor series form, and should have a parabolic relationship between E and k at the maximum. However, here, similar arguments as before yield a negative effective mass. An electron occupying one of the states near the band maximum would accelerate in the opposite direction. At normal temperatures the bands are partially occupied or filled, thus allowing movement of an electron from a filled state to an empty state and hence causing the flow of current. If we consider an electric field applied in the negative direction, an electron near the bottom of the conduction band, which has a positive m^* and a negative charge, feels a force in the positive direction and hence has an associated velocity, the group velocity, given by

$$v_g = \frac{\partial E}{\partial \boldsymbol{k}},$$ (2.46)

which is also positive. The electron at the bottom of the partially filled conduction band moves in the positive direction. Now, consider the valence band which is only partially empty. Since the current in a filled band is zero, the current in a band with a single unoccupied state is the negative of the current in a band with a single occupied state. Therefore, we may interpret the conduction in the partially empty band in terms of holes, "particles" representing absence of electrons and possessing a positive elementary charge as well as positive mass. Now, we may treat these particles, electrons in the conduction band and holes in the valence band, as having a positive mass but opposite charge.

The energy band diagrams describe the relationship between the energy and crystal momentum for electrons in the crystal. If an electron did not undergo scattering, then the effect of an applied field $\mathcal{E}$, assuming $-q$ as the electronic charge,[6] is to cause a change in momentum following

$$-q\mathcal{E} = \hbar \frac{d\mathbf{k}}{dt}, \qquad (2.47)$$

and the electron, in the reduced-zone approach, would be expected to transit the Brillouin zone, reach the Brillouin zone boundary, re-enter and transit again, an oscillatory phenomenon in k-space, and hence real space, a phenomenon termed Zener–Bloch oscillations. The angular frequency of such an oscillation would be

$$\omega = \frac{q\mathcal{E}a}{\hbar}, \qquad (2.48)$$

where a is the unit cell dimension. In reality, there are several phenomena that reduce the likelihood of this happening. We have considered the lattice to be perfectly periodic with no imperfections. Any disturbance from the idealized picture leads to a perturbation, which can cause the electron to change its state—a process that may or may not conserve energy and momentum. Impurities, defects in crystallinity, etc., can all cause changes in energy and momentum. So can the vibration of ion cores around their mean positions. All these processes, that we call scattering, dampen an unlimited change of momentum because any perturbation introduced by these processes can cause a change in the state of the electron and hence its momentum, and even in the most perfect of crystals, scattering due to lattice vibrations are always present.

Our discussion of electrons was based on idealized grounds in a one-dimensional framework. On the basis of this we made comments on how

---

[6]We use the notation q to represent the magnitude of elementary charge. The sign of the charge will be included explicitly.

one may go about developing accurate three-dimensional models that mimic
real crystals. The results of these will be the subject of discussion in the
latter part of this chapter from a device perspective. In a similar vein,
we will consider the lattice vibration in one dimension, to understand the
behavior of phonons—"particles" that characterize these vibrations. This
will establish the conceptual framework for us to discuss phonon dispersion
in three dimensions in the latter part of this chapter.

Consider Figure 2.5, which shows a one-dimensional lattice consisting
of two atoms of differing atomic mass m and M. Since the amplitude of the
vibrations of the atoms tends to be small, the force on the atoms can be
characterized by the first term of the functional expansion with position.
This is the restoring force similar to that of spring models. Denoting the
force constant as β, the position as z, we may write the force $\mathcal{F}$ for the two
species as:

$$\mathcal{F}_{2n} = m\frac{d^2 z_{2n}}{dt^2} = \beta\left(z_{2n+1} + z_{2n-1} - 2z_{2n}\right), \tag{2.49}$$

and

$$\mathcal{F}_{2n+1} = M\frac{d^2 z_{2n+1}}{dt^2} = \beta\left(z_{2n+2} + z_{2n} - 2z_{2n+1}\right). \tag{2.50}$$

The wave solution is of the form

$$z_{2n} = A\exp\left[j(\omega_1 t - 2nqa)\right], \tag{2.51}$$

and

$$z_{2n+1} = B\exp\left[j\left(\omega_2 t - (2n+1)qa\right)\right], \tag{2.52}$$

where q, like k for electrons, is the symbol for wave vector. Substituting,
and solving for the displacement relationship between z_{2n+1} and z_{2n}, we
get

$$z_{2n+1} = \frac{\beta\left[1 + \exp\left(-2jqa\right)\right]}{2\beta - M\omega_2^2}z_{2n}. \tag{2.53}$$

Since this can only be satisfied if the time dependence of z_{2n+1} is the same
as that of z_{2n}, the angular frequencies must be the same, i.e., $\omega_1 = \omega_2 = \omega_q$. Following a procedure similar to that for derivation of coefficients for
Kronig–Penney model, we have a unique solution for the constants only if
the determinant of coefficients in A and B is zero, i.e.,

$$(2\beta - M\omega_q^2)(2\beta - m\omega_q^2) - 4\beta^2\cos^2(qa) = 0. \tag{2.54}$$

The dispersion relation for phonons that characterize these vibrations,
then, is:

$$\left(\frac{E}{\hbar}\right)^2 = \omega_q^2 = \beta\left(\frac{1}{m} + \frac{1}{M}\right) \pm \beta\left[\left(\frac{1}{m} + \frac{1}{M}\right)^2 - \frac{4\sin^2(qa)}{mM}\right]^{1/2}. \tag{2.55}$$

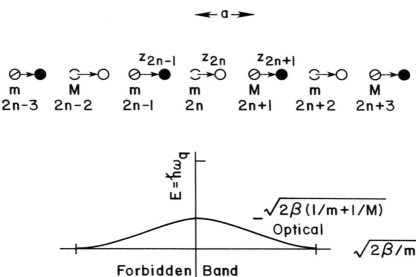

Figure 2.5: Mean and instantaneous position of atoms in a diatomic linear lattice is shown in the top part of the figure. The vibration shown corresponds to the acoustic branch. The lower part of the figure shows the dispersion relationship for this problem with $m < M$ as discussed in the text.

This dispersion relationship is shown in Figure 2.5. As with the energy-band behavior in Kronig–Penney model, there exists a forbidden zone of energy, or angular frequency, that the vibrations may not have. The vibrations in this periodic structure have been treated as for harmonic oscillators; they have more fundamental foundation in quantum mechanics akin to the treatment of photons. It is therefore quite useful to introduce the concept of phonons, as indistinguishable "particles" which characterize these vibrations. In many ways, this wave aspect of lattice vibrations in a periodic structure is also similar to that of the wave nature of electrons in the periodic structure. The modes of vibrations are discrete just as the states of the electrons are discrete. Phonons are particles that obey principles of conservation of energy and momentum; however, phonons themselves need not be conserved, since a change in temperature can increase or reduce the number of vibrations, unlike electrons in conduction or valence bands. These phonons, representing vibrations occurring at a frequency ω_q, have an energy $\hbar\omega_q$ where $\hbar$ is the reduced Planck's constant, and have a momentum $\hbar q$. The occupancy probability of the state (n_q) of energy $\hbar\omega_q$ at a temperature T is given by Bose–Einstein statistics:

$$n_q = \frac{1}{\exp\left(\hbar\omega_q/kT\right) - 1}. \tag{2.56}$$

Returning now to the dispersion relationship of phonon modes in the simple one-dimensional diatomic model, Figure 2.5 shows that there are two bands—the higher energy branch is called optical branch, and the lower energy branch is called acoustic branch. Acoustic branch results from in-phase vibration of neighboring atoms, while optical branch results from out-of-phase vibration of neighboring atoms. At the zone edge, however, they show a similar character. Figure 2.6 shows a schematic of the displacement of the atoms at the zone center and at the zone edge for the acoustic and optical branches for the diatomic lattice. Since the acoustic mode has in-phase vibration, it has a smaller frequency and hence smaller energy. The optical branch has a higher frequency and energy because of the out-of-phase vibration. The acoustic phonon is similar to a propagating acoustic wave; hence the use of the adjective. The term optical phonon originates from the excitation of these vibrations by photons in the infrared part of the spectrum. At the Brillouin zone boundary, the frequency of the optical branch reduces to $\sqrt{2\beta/m}$, i.e., that due to the lower mass ions, and, as Figure 2.6 shows the, heavier atoms remain stationary. Similarly, the frequency of the acoustic branch reduces to $\sqrt{2\beta/M}$, i.e., that due to the higher mass ion with the smaller mass ion stationary.

Interaction between phonons and electrons, alluded to in our comments

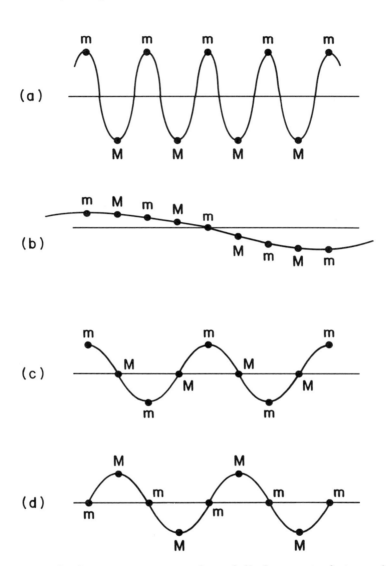

Figure 2.6: An instantaneous snap-shot of displacement of atoms during acoustic and optical mode vibrations for the diatomic one-dimensional lattice. Part (a) shows the zone center schematic for the optical branch, part (b) shows the same for the acoustic branch, part (c) shows the zone edge schematic for the optical branch, and part (d) shows the same for the acoustic branch.

regarding Zener–Bloch oscillations, is among the stronger scattering mechanisms and is crucial to understanding the transport of carriers in semiconductors. We will review this later in the chapter. Interactions can also occur between phonons and other phonons. Since phonons, with their basis in lattice vibrations, are one of the major means for heat transport in the lattice, phonon-phonon interactions are of a large interest in that subject. In semiconductors, this is of interest because heat dissipation has to be transported; however, it is not of primary importance. When two phonons interact with the sum of the momentum still within the first Brillouin zone, then the process, called a normal process, leaves the heat transport unaffected. However, if the resultant phonon wave vector is outside the first Brillouin zone, then the translation by the reciprocal lattice vector G results in a wave vector which is opposite in sign to the sum of the the wave vectors of the two incident phonons. This process is known as the umklapp[7] process, and since the equivalent wave vector is negative, it produces a thermal resistance.

2.3 Occupation Statistics

The occupation of carriers in the energy bands, as well as the behavior of impurities in the semiconductor, are all related to the statistics of occupation of energy levels by electrons. Electrons are fermions and hence the statistics are non-classical. Limitations placed by Pauli's exclusion principle should cause a behavior that is dependent on doping. Occupation of bands and discrete impurity energy levels is related to many effects of importance in semiconductor devices, most prominent among which is the behavior of degenerate material due to Fermi–Dirac statistics of occupation. The subject of occupation statistics arises not just in calculating this doping dependence, but also in generation-recombination effects, etc., This section is a review of the general principles of the approach as applied to both energy band states and impurity states.

2.3.1 Occupation of Bands and Discrete Levels

Electrons are independent and indistinguishable particles with a spin of $\pm 1/2$. Pauli's exclusion principle restricts the occupation of any state by only one electron. The statistics for this situation are commonly referred to as Fermi–Dirac statistics. Consider a closed system of energy E consisting of n particles that can occupy energy levels $\cdots E_j \cdots$ each of which has a degeneracy $\cdots g_j \cdots$. The number of distinguishable arrangements by

---

[7]To turn over; from the German word *umklappen*.

which n_j particles can occupy a jth state is given by the permutation term P_j:

$$P_j = \frac{g_j!}{n_j!\,(g_j - n_j)!}. \tag{2.57}$$

For the closed system, the total number of such arrangements is the product over all the states of the system. This is given by the permutation term P given by

$$P = \prod_j P_j = \prod_j \frac{g_j!}{n_j!\,(g_j - n_j)!}. \tag{2.58}$$

This permutation term is very large, since, in a usual problem such as electrons in the semiconductor, the number of states and carriers is large. We may use Sterling's approximation for the natural logarithm of a factorial, given by

$$\ln(x!) = x\ln(x) - x. \tag{2.59}$$

The system is in thermodynamic equilibrium, i.e., has time-independent macroscopic properties when the P is a maximum. This can be found by setting the derivative of P or equivalently of $\ln(P)$ to zero, with the condition that the number of particles n and the total energy E remain a constant. The total number of particles and the total energy of the system are therefore constrained by

$$n = \sum_j n_j \tag{2.60}$$

and

$$E = \sum_j n_j E_j. \tag{2.61}$$

Therefore, the maximum, using Lagrangian multipliers,[8] occurs when

$$\begin{aligned}
0 = d\ln(P) &= \sum_j \left[-1 - \ln(n_j) + \ln(g_j - n_j) + \left(\frac{g_j}{g_j - n_j} - \right. \right. \\
&\qquad \left. \left. \frac{n_j}{g_j - n_j} \right) + \mathcal{A} + \mathcal{B}E_j \right] dn_j \\
&= \sum_j \left[\ln\left(\frac{g_j - n_j}{n_j} \right) + \mathcal{A} + \mathcal{B}E_j \right] dn_j.
\end{aligned} \tag{2.62}$$

---

[8]The method of Lagrangian multipliers allows us to maximize and minimize a dependent variable when the maximization and minimization is performed under the constraint that another set of dependent variables, which are also a function of the same independent variables, remain constant. Let f be the former, and let ϕ and ψ be the latter; then, $df + \mathcal{A}d\phi + \mathcal{B}d\psi = 0$.

Therefore,

$$n_j = \frac{g_j}{\exp\left(-\mathcal{A} - \mathcal{B}E_j\right) + 1}. \tag{2.63}$$

The ratio of particles to states of this specific problem is referred to as the Fermi–Dirac distribution function, and is given by

$$f(E) = \frac{n_j}{g_j} = \frac{1}{\exp\left(E_j/kT\right) + 1} = \frac{1}{\exp\left[(E - \xi_f)/kT\right] + 1}, \tag{2.64}$$

where we have made the appropriate substitutions for constants $\mathcal{A}$ and $\mathcal{B}$ from the definition of macroscopic properties of the system. $E_j = E - \xi_f$, where E is the energy of the level of interest, and ξ_f is known as the Fermi energy. Fermi energy, based on the above relationship, has the following meaning: at absolute zero temperature, all the states above the Fermi energy are empty and all the states below it are filled. At any finite temperature the states at the Fermi energy are half filled.

Similar to the Fermi–Dirac statistics, the Maxwell–Boltzmann statistics apply to a system of independent particles, but unlike Fermi–Dirac statistics, these are distinguishable particles which are non-interacting. One may employ mathematics similar to the above with the distinction that now the number of arrangements of the jth particle state are $g_j^{n_j}/n_j!$, and hence the total number of permutations of the system are:

$$P = n! \prod_j \left(\frac{g_j^{n_j}}{n_j!}\right). \tag{2.65}$$

Since $g_1^{n_1} \times \cdots g_j^{n_j} \times \cdots$ is a constant, a similar procedure to the above (see Problem 1) leads the ratio of particles to states for the Maxwell–Boltzmann distribution function as

$$f(E) = \frac{n_j}{g_j} = \exp\left(-\frac{E_j}{kT}\right) = \exp\left(-\frac{E - \xi_f}{kT}\right). \tag{2.66}$$

This is the probability of the occupation of the state j; the exponential factor is the Maxwell–Boltzmann factor.

Both of these statistics are useful in describing the occupation of states in the energy bands. When the number of carriers is small, the number of available states is so much larger than the number of carriers that the Fermi–Dirac distribution reduces to the Maxwell–Boltzmann distribution. This conclusion also follows from the above relationships. If the energy level of interest is higher than the Fermi energy by $3kT$, then the exponential term is greater than 20, and use of the much simpler Maxwell–Boltzmann statistics leads to an error of less than 5%.

We now consider the statistics of donors, acceptors and traps. Donor and acceptor statistics become considerably important in the limits of degeneracy. Degeneracy implies a Fermi level above the donor level or below the acceptor level since the carrier densities are now close to or higher than the number of states available in the bands. This implies that only a fraction of the donors or acceptors contribute the carriers. The occupation of the donor energy levels is therefore central to the determination of the carrier density given a donor or acceptor impurity density in the high carrier density limit. This problem is actually quite complex. We are implicitly arguing that electrons at the donor levels, or lack of them at the acceptor levels, does not contribute to conduction and hence can not be included in the count of nearly free electrons or holes. In the limit of very high doping, impurity bands form because of energy level splitting caused by impurity–impurity interactions. Such impurity bands may coalesce with the conduction or valence bands. States are now available for change of momentum and hence conduction may take place through these levels. This is the reason that carrier freeze-out, the removal of carriers from conduction processes, does not occur as easily in highly doped material. Here, we will ignore such band formation. Thus, the derivation is strictly useful only in the limits of low degeneracy.

Let N_D be the total number of donor centers, N_D^+ the ionized donors and N_D^0 the neutral donor centers. N_D^+ electrons contributed by the ionized donors are distributed among energy levels $\cdots j \cdots$ of energy $\cdots E_j \cdots$ and degeneracy $\cdots g_j \cdots$; let these be occupied by $\cdots n_j \cdots$ electrons. In addition to occupying these energy levels in the conduction band, electrons also occupy N_D^0 sites at the donor energy level E_D. The number of ways that the jth level in the conduction band can be occupied is

$$Q_j = C_{n_j}^{g_j} = \frac{g_j!}{n_j!\,(g_j - n_j)!}, \tag{2.67}$$

and hence the total number of such arrangements is

$$R = \prod_j \frac{g_j!}{n_j!\,(g_j - n_j)!}. \tag{2.68}$$

Since a neutral donor center results from the addition of an electron to an ionized donor, we have N_D^0 electrons to be distributed among N_D donor centers at an energy E_D. In the lattice, the donor usually exists at a substitutional site, and ionizes by giving away an excess electron from its outer shell (see Figure 2.7). Since spin degeneracy is two, this excess

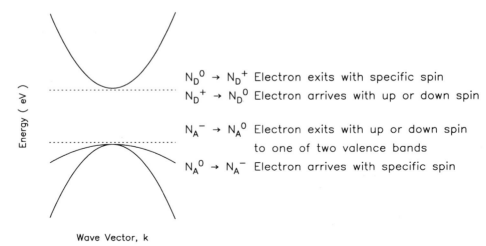

Figure 2.7: A schematic describing processes that occur during the change in ionization state for most common donors and acceptors in a semiconductor.

electron can have a spin in one of two orientations. So,

$$Q = C_{N_D^0}^{N_D} = 2^{N_D^0} \frac{N_D!}{N_D^0! (N_D - N_D^0)!} \tag{2.69}$$

gives the number of arrangements of N_D^0 electrons on N_D sites. The total number of electrons in the system is

$$n = N_D^0 + \sum_j n_j, \tag{2.70}$$

and the total energy is

$$E = N_D^0 E_D + \sum_j n_j E_j. \tag{2.71}$$

The total number of arrangements for the electrons, both in the bands and at the discrete levels, is

$$P = Q \times R = 2^{N_D^0} \frac{N_D!}{N_D^0! (N_D - N_D^0)!} \prod_j \frac{g_j!}{n_j! (g_j - n_j)!}. \tag{2.72}$$

Our procedure, as before, follows with the use of Lagrangian multipliers and Sterling's approximation to maximize P. We need to evaluate the differential:

$$
\begin{aligned}
&d[\ln(P) + \mathcal{A}(n - N_D^0 - \sum_j n_j) + \mathcal{B}(E - N_D^0 E_D - \sum_j n_j E_j)] \\
&= d\left\{ N_D^0 \ln(2) + \ln(N_D!) - \ln(N_D^0!) - \ln[(N_D - N_D^0)!] + \right. \\
&\quad \sum_j [\ln(g_j!) - \ln(n_j!) - \ln[(g_j - n_j)!]] + \\
&\quad \left. \mathcal{A}n - \mathcal{A}N_D^0 - \mathcal{A}\sum_j n_j + \mathcal{B}E - \mathcal{B}N_D^0 E_D - \mathcal{B}\sum_j n_j E_j \right\}. \quad (2.73)
\end{aligned}
$$

Maximizing with respect to (w.r.t.) N_D^0 leads to

$$
\ln 2 - \ln N_D^0 + \ln\left(N_D - N_D^0\right) - (\mathcal{A} + \mathcal{B}E_D) = 0, \quad (2.74)
$$

and

$$
\frac{N_D^0}{N_D} = \frac{1}{1 + \frac{1}{2}\exp\left(\mathcal{A} + \mathcal{B}E_D\right)}. \quad (2.75)
$$

Similarly, maximizing w.r.t. to n_j leads to

$$
\frac{n_j}{g_j} = \frac{1}{1 + \exp\left(\mathcal{A} + \mathcal{B}E_j\right)}, \quad (2.76)
$$

for the occupation of conduction band states. This we had derived before, but the distribution function for occupation of electrons at the donor levels is the new result. Substituting for the thermodynamic variables as before, we may write the distribution function for electron occupation at the donor level as

$$
f\left(E_D\right) = \frac{N_D^0}{N_D} = \frac{1}{1 + \frac{1}{2}\exp\left[(E_D - \xi_f)/kT\right]}, \quad (2.77)
$$

and since the total number of donors are divided between those that are neutral and those that are ionized, i.e., $N_D = N_D^0 + N_D^+$, we obtain

$$
\frac{N_D^+}{N_D} = \frac{1}{1 + 2\exp\left[(\xi_f - E_D)/kT\right]}. \quad (2.78)
$$

Occupation statistics for acceptors can be found in a similar manner. Most common acceptors occupy substitutional sites in the lattice and are deficient by one electron in their outer shell. The electron can go in the outer shell of a neutral acceptor in only one way without violating Pauli's exclusion principle. This electron may, however, come out with either spin and go to one of the two degenerate valence bands that commonly occur for most semiconductors of interest. This leads to (see Problem 3):

$$f(E_A) = \frac{N_A^-}{N_A} = \frac{1}{1 + 4\exp\left[(E_A - \xi_f)/kT\right]}, \qquad (2.79)$$

and

$$\frac{N_A^0}{N_A} = \frac{1}{1 + \frac{1}{4}\exp\left[(\xi_f - E_A)/kT\right]}. \qquad (2.80)$$

Most trap or deep levels can be treated in a similar manner as above. The degeneracy of the levels, a term associated with the factors 2 and 4 above (strictly speaking, 2 and 1/4), may be different for a trap level because these need not be from the adjacent columns of the periodic table. An additional feature of the deep level, as a result, is that they may have more than one ionized state within the bandgap. The impurity can be in any of these states, and the equilibrium statistics depend on all the other possible states of the system. Problem 23 of Chapter 3 is an advanced exercise related to this subject.

2.3.2 Band Occupation in Semiconductors

We have now discussed the occupation probabilities of carriers in states that arise from many different causes. In a device, the parameter that is of more interest is the number of carriers that are available for conduction. To determine this, in addition to the probability of occupation of a state, we need to know the number of such states. In a three-dimensional crystal, there is a distribution of such states, a quasi-continuum, since the number of states is large. This may be lumped together in a density of state function $D(E)$ which is the number of states in a unit energy range at any energy E. Thus in an energy range dE at the energy E, the total number of states is $D(E)dE$. The number of carriers, knowing the density of states as a function of energy, can now be determined as a function of the Fermi energy. If E_c represents the energy of the bottom of the conduction band, then, at thermal equilibrium,

$$n = \int_{E_c}^{E_{max}} f(E)D(E)dE, \qquad (2.81)$$

where $D(E)$ is the density of states available for occupation, and E_{max} is the maximum energy available in the conduction band. Since the distribution function decreases exponentially, we can replace E_{max} by ∞ without any significant error.

Consider only the simplest approximation of bands, the parabolic band. The density of states $(D(E))$, i.e., the number of states per unit volume of the crystal per unit energy, can be found by considering arguments based on allowed wave vectors of the crystal. For a one-dimensional crystal, the permitted values of the wave vector k of the wave function $\exp(jkr)$ in the periodic lattice are $\pm 2\pi l/L$, where l is an integer and L is the linear dimension. This follows from the requirement that the ends of the crystal be electron nodes. By extension, for a cube of dimension L, the permitted values of the wave vector are $\pm 2\pi l_x/L$, $\pm 2\pi l_y/L$, and $\pm 2\pi l_z/L$, corresponding to the three independent directions and with l_x, l_y, and l_z as positive integers. Thus, the magnitude of wave vector $\boldsymbol{k}$ follows as

$$k^2 = \frac{4\pi^2 \left(l_x^2 + l_y^2 + l_z^2\right)}{L^2} = \frac{4\pi^2 l^2}{L^2} = \frac{4\pi^2 l^2}{V^{2/3}}. \tag{2.82}$$

Each permitted value of $\boldsymbol{k}$ corresponds to various combinations of l_x, l_y, and l_z which result in the same l given by

$$l = \frac{L}{2\pi} k = \frac{V^{1/3}}{2\pi} k. \tag{2.83}$$

All states on a sphere, whose radius is l, result in identical energy, although the carrier may be moving in different directions. For energy varying from 0 to E, or the wave vector varying from 0 to k, the total number of modes allowed is the volume of sphere, i.e., $4\pi l^3/3$ or $Vk^3/6\pi^2$. Thus, the density of states per unit volume and unit energy, taking into account two possible spin orientations, is given as

$$D(E) = 2\frac{d}{dE}\left(\frac{k^3}{6\pi^2}\right) = \frac{1}{\pi^2}k^2\frac{dk}{dE} = \frac{1}{2\pi^2}\left(\frac{2m^*}{\hbar^2}\right)^{3/2}(E - E_c)^{1/2}. \tag{2.84}$$

In general, there may be more than one equivalent valley and the masses along the three axes may be unequal. The effective mass in the expression above, by convention, is the density of states effective mass that corresponds to the real density of states. Thus, for unequal, longitudinal (m_l^*) and transverse $(m_{t1}^*$ and $m_{t2}^*)$ masses, and a degeneracy g_v of the valleys, the density of states function $D(E)$ is given by the expression

$$D(E) = \frac{1}{2\pi^2}\left(\frac{2m_d^*}{\hbar^2}\right)^{3/2}(E - E_c)^{1/2}, \tag{2.85}$$

where

$$m_d^* = \left(g_v^2 m_l^* m_{t1}^* m_{t2}^*\right)^{1/3}. \tag{2.86}$$

Note the differences between the conduction effective mass and density of states effective mass. Given this density of states, the integral relationship for electron density now becomes

$$n = \frac{1}{2\pi^2} \left(\frac{2m_d^*}{\hbar^2} \right)^{3/2} \int_{E_c}^{\infty} \frac{(E - E_c)^{1/2} dE}{1 + \exp\left[(E - \xi_f)/kT \right]}. \tag{2.87}$$

Changing to variable $\eta = (E - E_c)/kT$, and denoting $\eta_{fc} = (\xi_f - E_c)/kT$, the integral relationship takes the form

$$n = \frac{1}{2\pi^2} \left(\frac{2m_d^*}{\hbar^2} \right)^{3/2} (kT)^{3/2} \int_0^{\infty} \frac{\eta^{1/2} d\eta}{1 + \exp(\eta - \eta_{fc})} \equiv N_C F_{1/2}(\eta_{fc}), \tag{2.88}$$

where $F_{1/2}$ is the Fermi integral of the order one-half.[9]

The parameter N_C thus has the value

$$N_C = \frac{1}{2\pi^2} \left(\frac{2m_d^*}{\hbar^2} \right)^{3/2} (kT)^{3/2} \frac{\pi^{1/2}}{2} = 2 \left(\frac{m_d^* kT}{2\pi\hbar^2} \right)^{3/2}. \tag{2.93}$$

N_C is an "effective" conduction band density of states. It may be viewed as an energy integrated density of state which approximates the effect of total density of states in the conduction band, but which appears at the edge of the conduction band. For values of of η_{fc} smaller than -3, i.e., with non-degenerate statistics dominating, the Fermi integral takes an exponential form, and the expression for electron density reduces to

$$n = N_C \exp(\eta_{fc}) = N_C \exp\left(\frac{\xi_f - E_c}{kT} \right), \tag{2.94}$$

---

[9]Fermi integrals are related to Γ functions by

$$F_\nu(\eta_f) = \frac{1}{\Gamma(\nu + 1)} \int_0^{\infty} \frac{\eta^\nu d\eta}{1 + \exp(\eta - \eta_f)}, \tag{2.89}$$

where η_f is the function η_{fc} in this instance. It can also be the corresponding valence band function η_{fv} for the calculations of hole density. Elsewhere, we have also used η_{fn} and η_{fp} synonymously for these for convenience. The derivative of $F_{1/2}$ function w.r.t. η_f is the $F_{-1/2}$ Fermi function. The Γ functions have the characteristics:

$$\Gamma(\nu + 1) = \int_0^{\infty} \eta^\nu \exp(-\eta) d\eta, \quad \Gamma(\nu + 1) = \nu\Gamma(\nu), \tag{2.90}$$

$$\Gamma\left(\frac{1}{2} \right) = \pi^{1/2}, \quad \text{and} \quad \Gamma(1) = 1. \tag{2.91}$$

For integer variables,

$$\Gamma(\nu + 1) = \nu!. \tag{2.92}$$

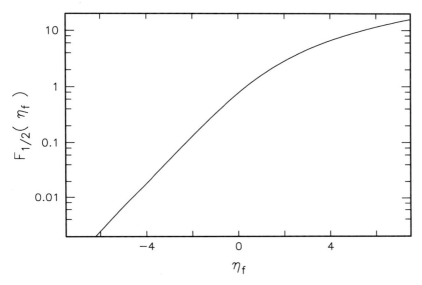

Figure 2.8: The Fermi integral $(F_{1/2}(\eta_f))$ as a function of the parameter η_f.

which is the classical Maxwell–Boltzmann relationship.

A derivation similar to the above leads to the relationship for holes (see Problem 2) as

$$p = N_V F_{1/2}(\eta_{fv}), \qquad (2.95)$$

where $\eta_{fv} = (E_v - \xi_f)/kT$, and the effective density of states for the valence band is

$$N_V = \frac{1}{2\pi^2}\left(\frac{2m_d^*}{\hbar^2}\right)^{3/2}(kT)^{3/2}\frac{\pi^{1/2}}{2} = 2\left(\frac{m_d^* kT}{2\pi\hbar^2}\right)^{3/2}, \qquad (2.96)$$

with m_d^* as the effective density of states mass for the valence band. A plot of the Fermi integral as a function of the parameter η_f (η_{fc} or η_{fv}) is shown in Figure 2.8. The product of carrier concentrations in the Maxwell–Boltzmann limit is

$$np = n_i^2 = N_C N_V \exp\left(-\frac{E_c - E_v}{kT}\right) = N_C N_V \exp\left(-\frac{E_g}{kT}\right), \qquad (2.97)$$

where n_i is the intrinsic carrier density.

Table 2.1: Symmetry points in the first Brillouin zone of face-centered cubic lattice.

Symmetry Point	Wave Vector $k =$	Degeneracy
Γ	0	1
L	$\pm(\pi/a)\langle 111\rangle$, $\pm(\pi/a)\langle\bar{1}11\rangle$, $\pm(\pi/a)\langle 1\bar{1}1\rangle$, and $\pm(\pi/a)\langle 11\bar{1}\rangle$	8
X	$\pm(2\pi/a)\langle 100\rangle$, $\pm(2\pi/a)\langle 010\rangle$, and $\pm(2\pi/a)\langle 001\rangle$	6
K	$\pm(3\pi/2a)\langle 110\rangle$, $\pm(3\pi/2a)\langle\bar{1}10\rangle$, $\pm(3\pi/2a)\langle 011\rangle$, $\pm(3\pi/2a)\langle 0\bar{1}1\rangle$, $\pm(3\pi/2a)\langle 101\rangle$, and $\pm(3\pi/2a)\langle 10\bar{1}\rangle$	12
W	$\pm(\pi/a)\langle 210\rangle$, $\pm(\pi/a)\langle 021\rangle$, $\pm(\pi/a)\langle 102\rangle$	6

2.4 Band Structure

Our discussion, so far, has already dealt with the central role of band structure in the occupation of carriers in a band and hence their behavior in thermal equilibrium. Band structure is also central to determining the transport characteristics of carriers and many other phenomena because it determines the relationship between the energy and the momentum of the carrier. In this section, we will review the band structure relationships in compound semiconductors. The consequences of the band structure in some major and general phenomena will be the subject of subsequent sections.

We have noted that the reciprocal space, also called phase space, k-space, Fourier space, and momentum space, is a convenient tool to describe the behavior of both electronic states and vibrational states. The coordinate axes of the reciprocal lattice are the wave vectors of the plane waves corresponding to the Bloch states or the vibration modes. The Wigner–Seitz unit cell in the reciprocal space is the first Brillouin zone. This was shown in Figure 2.1. This is also the first Brillouin zone for single element semiconductors Si and Ge, i.e., for diamond lattice. Let us consider the conventional definitions of the coordinate system and the important points of symmetry in the first Brillouin zone. These positions in the Brillouin zone are of particular importance because of their symmetry, and hence the relation to other characteristics that this entails. Table 2.1 summarizes the wave vectors of some of these symmetry points. Since there is symmetry of rotation around the three axes, and of mirror reflection in the three

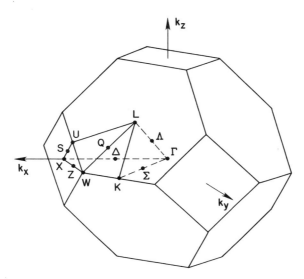

Figure 2.9: The irreducible part of the first Brillouin zone of zinc blende crystals is shown as the 1/48th part of the truncated octahedron. The major symmetry points are identified in this irreducible part.

planes, the irreducible part of this first Brillouin zone is only 1/48th in size. This irreducible part is shown in Figure 2.9. The reduced part shown in Figure 2.9 is the minimum description of wave vectors of carriers and vibrational modes in the zinc blende crystal in order to characterize their effect. In numerical calculations, it is quite often convenient to use the description of band structure only in this irreducible zone and then perform symmetry operations to generate the information as it becomes necessary during the calculations.

The task of determining the energy–momentum relationship for a three-dimensional lattice is of course a difficult one. It involves the determination of the solutions to Schrödinger equation using the crystal Hamiltonian. This crystal Hamiltonian contains energy terms due to kinetic, potential, and magnetic contributions arising from electrons in a lattice structure of the nuclei. The solution can be accomplished in a number of ways using perturbation theory: using tight binding approximation, i.e., using functions that are linear combinations of either atomic or molecular orbitals; using pseudopotentials; and using various other functional approaches such as Green's functions techniques. Perturbation theory allows the determination of final energy solution, by using the smaller Hamiltonian terms as

perturbing functions. The basis states for using this perturbation are determined from the major term associated with Coulombic interactions with
the nuclei. Since the basis functions, involving calculations based on using
atomic orbitals near the cores, are $|s>$ and $|p>$ like, the resulting functions
are combinations of these at various important points of the Brillouin zone.

Figures 2.10–2.13 show the band structures[10] of Si, Ge, GaAs, InP,
InAs, AlAs and Ga$_{.47}$In$_{.53}$As, over a large energy range, and Figure 2.14
shows the band structure of both conduction and valence band for GaAs
in larger detail.

We can make some general observations regarding these band structures.
GaAs is a direct bandgap semiconductor with the ordering of minima from
low to high energies in the conduction band as Γ, L, and X. All compound
semiconductors exhibit a universal feature of a degenerate light and heavy
hole band at zone center, and a split-off band. The split-off band comes
about because of the perturbation from the magnetic energy associated
with the spin and orbit of the electron. This coupling is generally referred
to as spin-orbit coupling or L.S coupling, and the split-off band arising
from it is 0.34 eV below the valence band maxima in GaAs. The bandgap
of intrinsic GaAs, at 300 K, is 1.42 eV, the L valley is 0.29 eV, and the X
valley about 0.48 eV above the Γ minimum.

Referring to the earlier band structure figure of silicon (Figure 2.10), Si
is an indirect bandgap semiconductor with an intrinsic bandgap of 1.10 eV;
the minimum in conduction band energy occurs $\approx$ 85% of the way to the
X point from the Γ point. Constant low energy surfaces around this valley,
loosely referred to as the X valley, are 6-fold degenerate and aligned along
the Cartesian coordinates. The next lower valley is between the line joining
X with the U point, followed by the L point. Again there is the degenerate
light and heavy hole band at zone center which is the valence band maximum. Unlike compound semiconductors, silicon does not exhibit a large
separation of the split-off band. The origin of this smaller separation is in
the spin-orbit energy that gives rise to the formation of the split-off band.
Light atoms exhibit small spin-orbit interaction. In silicon, spin-orbit splitting is only 0.044 eV. Spin-orbit energy, however, rises rapidly with atomic
mass. Germanium has a spin-orbit splitting of 0.29 eV. Germanium is also
an indirect bandgap semiconductor with a bandgap of 0.72 eV for intrinsic
material at 300 K. The ordering of the conduction band minima is L, X,
and Γ. The constant energy surfaces in the lowest conduction valley, the L

---

[10]These are based on the band structure calculations of M. Fischetti. Some of this
information is published in M. Fischetti, "Monte Carlo Simulation of Transport in Technologically Significant Semiconductors of the Diamond and Zinc Blende Structures, Part
I: Homogeneous Transport," *IEEE Trans. on Electron Devices*, **ED-38**, No. 3, p. 634,
©Mar. 1991 IEEE.

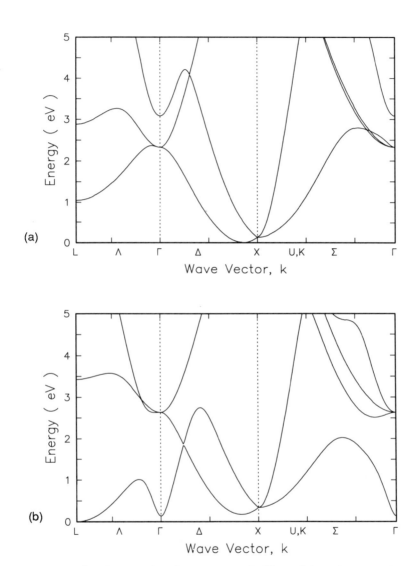

Figure 2.10: Conduction band structure of silicon (a) and germanium (b) over major points and directions in the first Brillouin zone. After M. Fischetti, "Monte Carlo Simulation of Transport in Technologically Significant Semiconductors of the Diamond and Zinc Blende Structures, Part I: Homogeneous Transport," *IEEE Trans. on Electron Devices*, **ED-38**, No. 3, p. 634, ©Mar. 1991 IEEE.

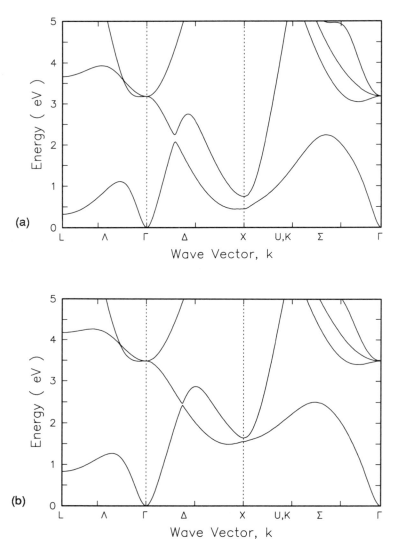

Figure 2.11: Conduction band structure of GaAs (a) and InP (b) over major
points and directions in the first Brillouin zone. After M. Fischetti, "Monte
Carlo Simulation of Transport in Technologically Significant Semiconduc-
tors of the Diamond and Zinc Blende Structures, Part I: Homogeneous
Transport," *IEEE Trans. on Electron Devices*, **ED-38**, No. 3, p. 634,
©Mar. 1991 IEEE.

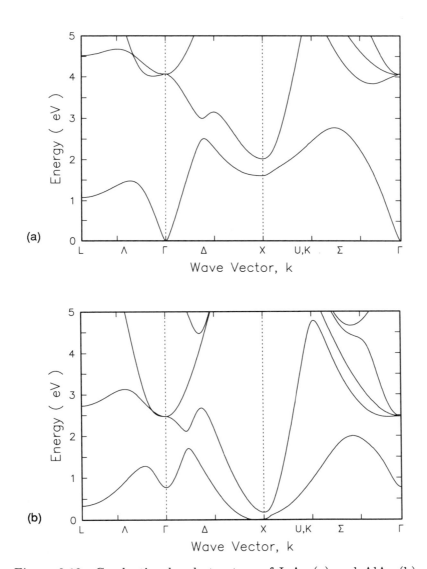

Figure 2.12: Conduction band structure of InAs (a) and AlAs (b) over major points and directions in the first Brillouin zone. After M. Fischetti, "Monte Carlo Simulation of Transport in Technologically Significant Semiconductors of the Diamond and Zinc Blende Structures, Part I: Homogeneous Transport," *IEEE Trans. on Electron Devices*, **ED-38**, No. 3, p. 634, ©Mar. 1991 IEEE.

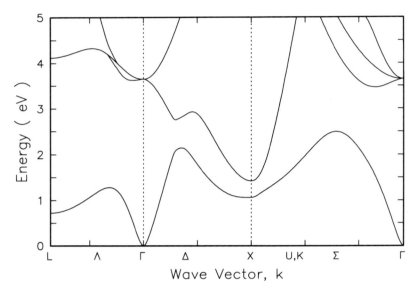

Figure 2.13: Conduction band structure of $Ga_{.47}In_{.53}As$ over major points and directions in the first Brillouin zone. After S. Tiwari, M. Fischetti, and S. E. Laux, "Transient and Steady-State Overshoot in GaAs, InP, $Ga_{1-x}In_xAs$, and InAs Bipolar Transistors," *Tech. Dig. of International Electron Devices Meeting*, p. 435, Dec. 9–12, ©1990 IEEE.

valley, are ellipsoids aligned along the $\langle 111 \rangle$ direction. These are eight-fold degenerate while the ones in silicon were six-fold degenerate. Both in silicon and germanium, the lowest conduction bands have multiple valleys associated with them. A consequence of this is the availability of a large number of states to scatter into. A large change in momentum is required for this process, and is available through zone edge phonons. The result of this is that, at low energies, both silicon and germanium experience higher scattering than direct bandgap compound semiconductors. Another interesting feature from the band structures is the observation of the lower effective mass for holes of germanium. This gives rise to nearly identical electron and hole mobilities in germanium.

InP is a direct bandgap compound semiconductor with similar ordering of conduction band minima as GaAs. Like GaAs, the light and heavy hole bands are degenerate and a split-off band occurs ≈ 0.78 eV below, at the zone center. The bandgap of intrinsic InP is 1.35 eV at 300 K. InAs is a direct bandgap compound semiconductor with a small energy gap of

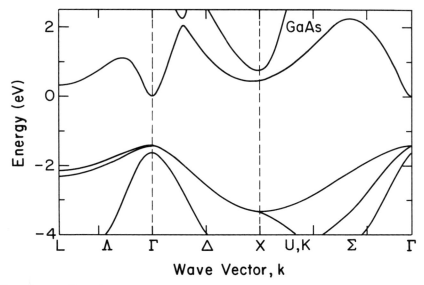

Figure 2.14: Details in medium energy region of the band structure of GaAs over major points and directions of the Brillouin zone.

≈ 0.36 eV at 300 K. Its ordering of conduction band minima is also Γ, L and X. Most compound semiconductor are direct bandgap materials. AlAs and GaP are two of the few compound semiconductors with indirect energy bandgap. In AlAs, the X valley is the minimum conduction energy valley followed by the L valley. It exhibits a behavior similar to silicon, and some of these characteristics may be traced to aluminum, which is a light atom like silicon.

2.5 Phonon Dispersion in Semiconductors

We have considered the conceptual basis for treatment of phonons as part of our discussion of one-dimensional models. To recapitulate, the Bloch function formalism considers the electron in a periodic lattice whose host atoms are stationary. At finite temperatures, atoms vibrate around this mean position. The periodic vibration leads to perturbation of the stationary periodic potential and causes the changes in the electron wave function in time. The lattice vibrations can be described in a framework that is analogous to the description of electrons. Phonons, representing these vibrations occurring at a frequency ω_q, have an energy $\hbar\omega_q$, where $\hbar$ is the reduced

Planck's constant, and have a momentum $\hbar q$. Phonons describe vibrations just as photons describe electromagnetic waves. Two phonon branches result in this idealized one-dimensional system, a high energy branch for optical phonons, and a lower energy branch for acoustic phonons. Scattering resulting from optical and acoustic phonons are commonly referred to as optical phonon scattering and acoustic phonon scattering. Since the energy of the acoustic phonon mode is smaller it produces less perturbation than the optical phonons.

Three-dimensional crystals bring with them additional complexity in phonon dispersion. In the diatomic one-dimensional crystal, the atoms could vibrate only along one axis, i.e., along the longitudinal direction. In a three-dimensional crystal, they may have longitudinal oscillations, as well as transverse oscillations in two orthogonal directions. In the primitive cell, if there are N different types of atoms either of differing mass or ordering in space, $3N$ vibration modes will result. In general, three of these branches, the acoustic branches, will disappear at zone center. The remaining $3N - 3$ branches will be optical branches. For the cubic crystals, GaAs or Si, e.g., $N = 2$, and hence there are three acoustic and three optical branches, each three comprising of one longitudinal and two transverse modes. However, along $\langle 100 \rangle$, $\langle 110 \rangle$, and $\langle 111 \rangle$ directions, the two transverse directions are indistinguishable. So, here, in the cubic crystal, the two transverse oscillations are degenerate .

Examples of phonon dispersion curves that describe the energy or frequency dependence as a function of wave vector are shown in Figure 2.15 and 2.16 for Si, Ge, GaAs, and InAs.

The order of magnitude of the optical phonon energy is between 25 and 50 meV for most semiconductors of interest, in the same order of magnitude as the thermal energy. Heavier ions should be more sluggish; the acoustic branch then should have a lower peak energy. This is reflected in the ascending ordering of peak acoustic energies of InAs, Ge, GaAs, and Si in inverse proportion to the weight of the heaviest element. A similar trend holds for the lowest optical branch energy for these materials. An additional interesting feature of these figures is that, for compound semiconductors, the longitudinal optical phonon is slightly more energetic than the transverse optical phonon at low momenta because of the presence of ionic charge. At high momenta, in certain directions, the optical branches may even cross over because the higher momenta results in a larger reduction in energy of the longitudinal mode due to the effect of the dipole. These figures also show that at specific large momenta, in directions differing from the major crystal axis direction $\langle 100 \rangle$, $\langle 110 \rangle$, or $\langle 111 \rangle$, the degenerate transverse modes split because of the breakdown of symmetry. This occurs for both optical and acoustic modes.

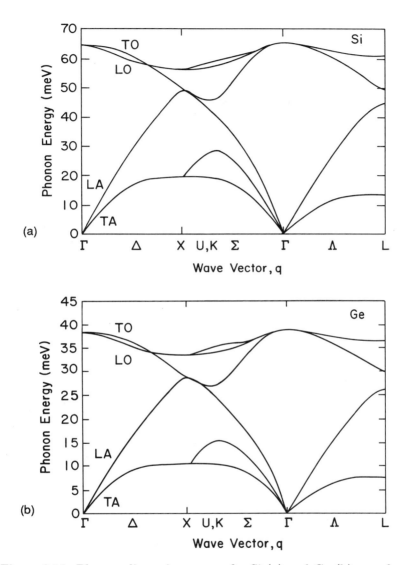

Figure 2.15: Phonon dispersion curves for Si (a) and Ge (b) as a function of the wave vector in the first Brillouin zone. After Landolt-Börnstein; O. Madelung, M. Schulz, and H. Weiss, Eds., *Semiconductors*, **V 17**, Springer-Verlag, Berlin (1982).

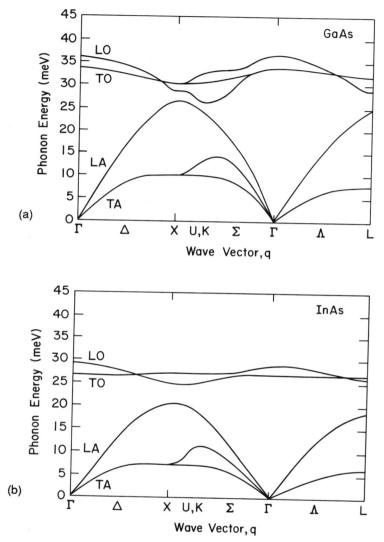

Figure 2.16: Phonon dispersion curves for GaAs (a) and InAs (b) as a function of the wave vector in the first Brillouin zone. After Landolt-Börnstein; O. Madelung, M. Schulz, and H. Weiss, Eds., *Semiconductors*, **V 17**, Springer-Verlag, Berlin (1982).

2.6 Scattering of Carriers

The carrier transport in a medium occurs under the influence of applied forces which accelerate the carriers in the preferred direction, and the carrier interacts with its environment whose general effect is to render the motion random. Scattering is a general term for the interaction of the carrier with the surroundings. Our discussion of scattering is a review, short on detail, but with a view of understanding its general consequences for compound semiconductors.

An electron in the crystal, under the influence of an external force, such as due to the electric field, picks up its energy during its acceleration, but changes energy and/or momentum by various scattering mechanisms that come about because of lattice vibrations or other carriers in its surrounding, as well as defects of the lattice. A broad classification of these scattering mechanisms is considered in Figure 2.17 based on their origin. These various scattering mechanisms are events of interaction between the carrier and its surroundings which cause it to change its momentum and/or energy. It therefore represents the individual and aggregate effect of the carrier's surroundings which influences the carrier's individual and aggregate transport behavior. The behavior of electrons in a perfect crystal can be described by a wave function. This is the Bloch function, which is the solution of the time independent Schrödinger equation. If the scattering processes were not to occur, the application of an electric field would increase the velocity linearly with time, an occurrence generally described as ballistic motion. In reality, for long sample lengths, it reaches a limiting value. This limiting value is proportional to the field at low fields ($v = \mu\mathcal{E}$, where v is the velocity, μ the mobility, and $\mathcal{E}$ the electric field, a constant mobility region), and saturates to the value v_s at high fields (the saturation velocity limit). The electron wave function can be described by the same Bloch function until it scatters, i.e., it interacts with an imperfection. When it interacts with the imperfection, the wave function is affected because of perturbation in the periodic potential. The electron has a new wave function at the end of the interaction, i.e., at the end of interaction the electron has a new wave vector and a new energy (in some scattering processes energy may remain unchanged; such scatterings are called elastic scatterings, an example is scattering by impurities). This new wave function describes the behavior of the electron until another interaction. If the perturbation is small, the effect of this interaction can be analyzed by perturbation theory. Our division of these scattering mechanisms is in three general classes: (a) scatterings that come about due to imperfections in the periodicity of the lattice, which we call defect scattering, (b) carrier-related scattering due to interactions in between the carriers, a process that becomes important at

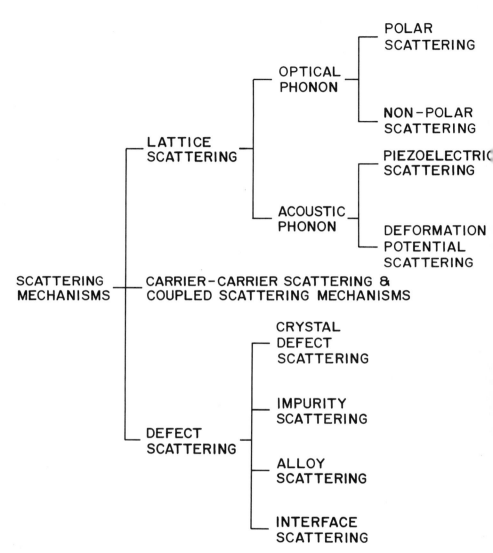

Figure 2.17: A schematic of classification of various scattering mechanisms based on their origin.

high carrier concentrations, and that can couple with other modes of scattering, and (c) scattering due to interaction of carriers with the perturbing potential produced by lattice vibrations.

2.6.1 Defect Scattering

Imperfections in the periodic potential of a crystal come about due to crystal defects such as vacancies, dislocations, interstitials, etc. However, this is quite uncommon because, except in highly lattice mismatched structures, devices of interest employ material where the density of these is less than unintentional substitutional impurities. Defect scattering also arises from impurity atoms themselves that are used to dope the crystal—this is generally referred to as impurity scattering—and originates in the deviation of the local potential around an impurity from the local and periodic potential of the atoms of the host crystal. If the impurity atom is ionized (as is almost always the case) it is called ionized impurity scattering and this particular scattering is strong. If the impurity atom is not ionized, i.e., it is neutral, it is referred to as neutral impurity scattering. Neutral impurity scattering is quite weak because of the small perturbation in potential from a neutral atom on a lattice site. Consider the scattering due to impurity atoms in a crystal as a function of temperature. We assume that impurity density is not so large so as to cause formation of an impurity band and its coalescence with the conduction band. Such a material, therefore, shows freeze-out of carriers, i.e., at low temperatures, the carriers do not have sufficient thermal energy to make the transition to the conduction or valence bands. At temperatures below the freeze-out temperature, the impurities are largely neutral, and hence there is a weak impurity scattering. As the temperature is raised, impurities ionize and impurity scattering becomes stronger and hence the carrier mobility decreases. As the temperature is continued to be raised, the thermal energy of the carriers continues to increase, their velocities are larger, and the impurity scattering process becomes weak again. An analogy may be drawn between this and the Rutherford scattering used to analyze the scattering of light atoms by heavier nuclei. A high velocity, and energy of particle, causes both a smaller time of interaction and smaller effect on the velocity of the carrier. Because of the large disparity in masses between the atoms and the moving carriers, this scattering conserves energy and is an example of elastic scattering. At high temperatures, other scattering processes become important and impurity scattering plays an increasingly smaller role. It is generally important in the 77 K (liquid nitrogen temperature) to 300 K (room temperature) range.

The third defect scattering that is important occurs in mixed crystals such as $Ga_{.47}In_{.53}As$ and $Ga_{.7}Al_{.3}As$. Due to the random variation of the

Figure 2.18: A schematic of interface scattering resulting from steps on the crystal surface.

atoms either on the cation or the anion sites of the lattice, a non-periodic perturbation occurs naturally in such material even when no defect has been added. The perturbations are sufficiently small that alloy scattering generally becomes important only below 300 K.

The last of the defect scattering processes important to devices is related to interactions at the surfaces and interfaces of the lattice. This is generally referred to as interface scattering (see Figure 2.18). This crystal defect scattering process is certainly very important at insulator–semiconductor interfaces. It can be important if interfaces are rough and have steps such as mis-oriented compound semiconductor growth. Since carriers in two-dimensional systems are confined very close to the interface, any surface potential perturbation, e.g., through surface atomic steps leads to a decrease in the mobility. In the velocity–field curves shown in Figure 2.19, e.g., the silicon surface mobility is almost half as much as the corresponding bulk mobility.

Steps are not the only way this scattering can cause a net energy or momentum change of the carriers. If there are a number of surface states,[11] they may cause perturbation effects of their own even for a perfectly smooth surface. For an ideal interface, an incident carrier goes through a reflection without losing its energy. Its momentum parallel to the interface is maintained while its momentum perpendicular to the interface is reversed. An example of this is the $Ga_{1-x}Al_xAs/GaAs$ interface which can be made atomically smooth with negligible interface state density. A low energy electron that can not penetrate the $Ga_{1-x}Al_xAs/GaAs$ barrier loses no energy and maintains the momentum parallel to the interface. From the scattering perspective, a non-ideal interface is the SiO_2/Si interface. Quite a large fraction of incident carriers lose energy and momentum at the interface because of interface states and atomic roughness. The result of such scattering, called diffuse scattering, is a thermalization of carrier energy distribution and decrease of mobility.

---

[11]Surface states result from a number of causes, primarily surface reconstruction and non-terminated bonds at the surface.

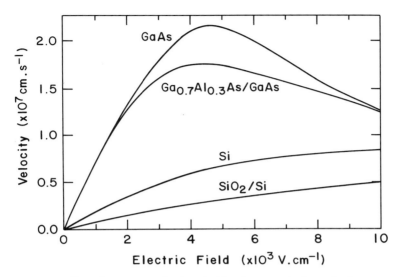

Figure 2.19: Steady-state velocity–field characteristics of electrons, in lightly doped bulk Si at the SiO_2/Si interface, in lightly doped GaAs, and at the $Ga_{.7}Al_{.3}As$/GaAs interface at 300 K.

2.6.2 Carrier–Carrier Scattering

Carrier-carrier scattering conserves the total energy and momentum of the interacting particles. This scattering, also sometimes referred to as plasma scattering, does cause redistribution of the energy and momentum among the interacting particles, and hence its primary effect is in influencing the carrier distribution as a function of energy and momentum. Thus, the scattering influences those parameters that depend on the shape of the distribution of carriers with energy and momentum. Relaxation rate of energy and transfers to other valleys or bands that depend strongly on energy tails do therefore get influenced directly by carrier–carrier interactions.

An additional important way that carrier–carrier scattering influences transport is through its coupling with other scattering mechanisms—the coupling of plasma scattering together with phonons, referred to as plasmon scattering, is an important scattering mechanism in hot electron devices.

2.6.3   Lattice Scattering

Lattice scattering is a broad term used for those scattering mechanisms that arise from the existence of the lattice itself. The lattice is a periodic structure, whose periodicity of the perturbation is incorporated in the nearly-free electron model being employed throughout this text. In this periodic lattice, at any finite temperature, the nuclei vibrate around their central positions, a behavior embodied in the dispersion behavior of the semiconductor. Lattice vibration modes that occur at high frequencies and have associated with them a relatively higher energy of 10s of meV are referred to as optical phonon modes. Those occurring at lower frequency over a wider spectrum of frequencies are known as acoustic modes. Optical modes arise from atoms vibrating out of phase with each other, resulting in their higher frequency. Acoustic modes are long range vibrations where the atoms move in phase. These vibrational modes interact with the free electron in a variety of ways, giving rise to the scattering that we call optical and acoustic phonon scattering.

Before reviewing further the phonon scattering processes, we summarize a few other terms that are often employed in the description of scattering. These terms arise from the location of the carrier in the reciprocal space as a result of the scattering process. Referring to the phonon scattering, an electron that scatters via optical phonon scattering is likely to have a significant change in momentum since optical phonons can have large momentum. Large momentum changes most often occur with changes in the valleys or bands that the carriers occupy. Holes, with their degenerate bands, can easily change bands, e.g., between light hole and heavy hole bands; thus corresponding scattering processes, termed inter-band scattering processes, can occur even at low energies. These inter-band scattering processes are unlikely for electrons at low energies. High energy electrons, such as during an avalanche process, may change bands. All such processes involving changes in bands are called inter-band scattering processes, while if a process occurs within a band it is called intra-band scattering process. Inter-band scattering may occur so that the carrier such as an electron scatters within the same valley, e.g., within the Γ valley for electrons. This is referred to as intra-valley scattering. Scattering may occur between different valleys, e.g., using an optical phonon for the large momentum change; this scattering is then referred to as inter-valley scattering. This discussion becomes semantically more elusive, because many valleys can be degenerate. The L valleys are eight-fold degenerate and the X valleys are six-fold degenerate. These are all examples of inter-valley scattering occurring with a large momentum change and hence quite likely involving optical phonons. These valleys are degenerate in energy; however, the amount of momentum change required to transfer to the different valley locations is different. This leads to a further sub-division of intra-valley scattering to f-scattering and g-scattering. f-scattering is the scattering occurring with the nearest degenerate valley locations, and g-scattering is the farthest.

2.6.4 Phonon Scattering Behavior

The optical phonon energy is considerably larger in Si than in Ge or GaAs. This may be traced to the larger average weight of the nuclei in Ge and GaAs and smaller bonding forces. Electrons need approximately this energy ≈ 37 meV in Ge, ≈ 36 meV in GaAs, and ≈ 63 meV in Si before they can emit an optical phonon, i.e., lose energy in an optical scattering process. Also, with lowering of temperature, there is a decrease in the occupation probability as described by Equation 2.56. This is referred to as freeze-out of optical phonons. Corresponding to this smaller likelihood of occurrence of optical phonons, there is a smaller likelihood of absorption of an optical phonon. Hence, at low temperatures there is a reduction in optical phonon scattering. Emission process can still occur, and continues to be leading cause for loss of energy at high electric fields. Both of these factors are important to the observed velocity–field behavior and operation of high mobility and hot electron structures as a function of temperature and bias.

Now consider the nature of the perturbing potential in acoustic mode. A change in atomic spacing, i.e., deformation of the lattice, leads to a disturbance in the potential called the deformation potential. The magnitude of the deformation potential is proportional to the strain induced by vibrations, and the corresponding scattering is called deformation potential scattering. In addition to this spacing effect, if the atoms are slightly ionic (as in all of the compound semiconductors), then the displacement of the atoms leads to a potential perturbation due to the charge on the ions. This is called the piezoelectric potential, and the resulting scattering process is called piezoelectric scattering. It is important in compound semiconductors at low temperatures in relatively pure materials. Recall that optical scattering is weak at low temperatures and low fields due to optical phonon freeze-out, etc., and impurity scattering is weak because of purity and carrier freeze-out, leaving acoustic phonon scattering as the dominant process. The deformation potential scattering is weaker than the piezoelectric scattering, and hence piezoelectric scattering dominates.

Optical vibrations also lead to potential perturbation in two ways similar to the case of acoustic vibrations. The first, a non-polar perturbation effect of strain, leads to a scattering referred to as as non-polar optical scattering. The second, due to the polarization from the ionic charge, is referred to as polar optical scattering. In GaAs, AlAs, etc., both these scattering processes are important. In general, for compound semiconductors, the polar optical phonon scattering is the dominant optical scattering mechanism.

2.7   Carrier Transport

We will now discuss the relationship of the various scattering mechanisms reviewed above with the steady-state transport behavior of carriers. The operation of a device is interlinked with the transport behavior of carriers. The velocity–field curves for carriers in a semiconductor describe the steady-state behavior of the carriers for both low fields, where perhaps an equally useful parameter is the mobility, and of high fields. The transport of the carrier is related to the scattering of the carriers. In the discussion above, we implicitly assumed a majority carrier. Minority carrier transport is important in the base of a bipolar transistor. It can be significantly different from a majority carrier device at comparable background doping and identical temperatures. This comes about because of changes in scattering behavior that result from changes in screening effects of the perturbations. Scattering can also change when the band occupation behavior is changed, such as through the quantum-mechanical effects of two-dimensional carrier gas systems. Since scattering involves transition from one state to another, it involves a transition probability embodied in a coupling coefficient that describes the likelihood of such a change taking place—a measure of the overlap function of the two states, and the number of such states that are available. It is proportional to both of these.

Consider some examples. The overlap function between valence and conduction bands is small in indirect bandgap semiconductors. There is therefore a very small likelihood of band to band transitions at low and moderate fields. Lifetimes of such materials are therefore large. In direct semiconductors, the overlap is larger, so the lifetime is generally worse. Overlap functions can of course be made large in certain device situations and are the reason for tunneling such as in the Zener effect. Zener tunneling is as likely in indirect semiconductors as in direct semiconductors. The number of available states is also directly related to the scattering. More available states increases the likelihood of the scattering event. Two-dimensional effects change the density of state distribution as well as screening and hence may influence the scattering behavior. This section reviews the relationship of the scattering with the transport properties of the carriers.

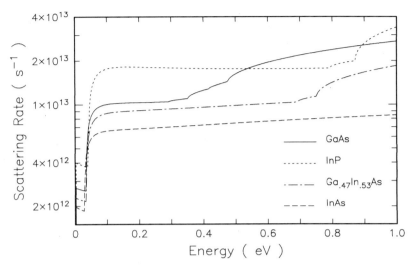

Figure 2.20: Steady-state phonon scattering rates as a function of energy, at low energies, in GaAs, InP, Ga.$_{47}$In.$_{53}$As, and InAs at 300 K. After S. Tiwari, M. Fischetti, and S. E. Laux, "Transient and Steady-State Overshoot in GaAs, InP, Ga$_{1-x}$In$_x$As, and InAs Bipolar Transistors," *Tech. Dig. of International Electron Devices Meeting*, p. 435, Dec. 9–12, ©1990 IEEE.

2.7.1 Majority Carrier Transport

We now consider scattering rates of some common compound semiconductors in order to relate them to the transport characteristics of interest in the operation of devices. Figure 2.20 shows some of the scattering rates due to phonon processes, at 300 K and low energies, for GaAs, InP, Ga.$_{47}$In.$_{53}$As, and InAs. The minimum in the scattering rates, occurring at very low energies, is due to acoustic scattering. At the low energy threshold for the optical phonon modes, scattering rises rapidly due to the large rate of optical phonon scattering at 300 K. At 77 K and lower temperatures this rate would be lower. At low energies the scatterings are intra-valley events. At higher energies, comparable to or higher than the secondary valley separation, a large density of states is available for inter-valley scattering. So, these scattering rates show another threshold where the secondary valley occurs in GaAs, InP, and Ga.$_{47}$In.$_{53}$As. InAs has a secondary valley at higher than 1 eV and does not, therefore, show such a threshold. Note that InP has the highest low energy scattering; it has both a higher effective mass that leads to higher density of states to scatter into, and a stronger coupling coefficient for scattering. Both Ga$_{.47}$In$_{.53}$As and InAs show favorable low scattering rate characteristics over the energy range, a behavior partly due to the larger secondary valley separation which inhibits inter-valley scattering over a significant energy range.

The two divisions of lattice scattering that we have considered are inter-valley and intra-valley scattering. In intra-valley scattering, low momentum phonons are involved. At low fields, this intra-valley scattering is important in materials where the conduction band minima occurs at the Γ point, e.g., GaAs, InP and most other compound semiconductors. Inter-valley scattering involves large momentum phonons and is important at low fields in materials with conduction band minimum at non-zero crystal momentum, i.e., non-Γ minimum semiconductors. Examples are Ge, which exhibits a minimum at $\langle 111 \rangle$ (i.e., the L point), and Si where it occurs close to $\langle 100 \rangle$ i.e., the X point).

The phonons that take part in inter-valley scattering are sometimes referred to as inter-valley phonons even though they may be of either the acoustic or optic type. When fields are high, electrons pick up enough energy in between scattering events, and hence even where the central valley is the minimum energy valley inter-valley scattering becomes important. The negative differential mobility of GaAs, e.g., arises from inter-valley scattering from Γ to L and X valleys. Figure 2.21 shows the velocity–field characteristics for GaAs ($\approx$ mid-10^{16} cm^{-3} doping) for various temperatures. This figure serves to compare the temperature dependence of the most dominant scattering processes in GaAs.

The decrease in mobility at low fields is due to intra-valley scattering, the decrease in velocity at higher fields is due to inter-valley scattering, and the saturation in velocity at very high fields is due to a balance between the energy gain in between scattering events and the energy loss due to high inter-valley scattering. Both of these processes and the corresponding energy and momentum loss rate are shown as a function of the field at room temperature for GaAs in Figure 2.22.

In thermal equilibrium, the energy and momentum exchange in between electrons and the lattice is balanced out by phonon absorption and emission. With an applied electric field, the energy and momentum are lost at a higher rate to the lattice, and at moderate fields the inter-valley optical phonon scattering processes dominate. The momentum loss rate $\langle dp/dt \rangle$, where p is the magnitude of the momentum, causes randomization of the electron velocity distribution in momentum space and consequently results in a finite mobility. Note that the momentum loss rate is higher than the energy loss rate (also recall that optical phonons involve a loss of $\approx$ 36 meV of energy per emission event). We will discuss subsequently the fitting of both these loss rates, phenomenologically, by a time constant called the relaxation

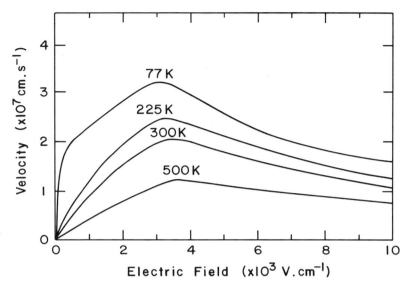

Figure 2.21: Velocity–field characteristics for electrons in lightly doped GaAs as a function of electric field for various temperatures.

time. The relaxation times for momentum (τ_p) and energy (τ_w) are in general approximations—phenomenological fitting parameters. These time constants characterize the driving force of return to equilibrium.

$$-\left\langle \frac{dp}{dt} \right\rangle\Bigg|_{loss} = \frac{\langle p \rangle}{\tau_p},$$

$$\text{and} \qquad -\left\langle \frac{dW}{dt} \right\rangle\Bigg|_{loss} = \frac{\langle W \rangle}{\tau_w}. \qquad (2.98)$$

Here, W is the energy of mean energy of the carriers and the angled brackets indicate an averaging over all the carriers.[12]

We may now describe some general features of the relative importance of different scattering mechanisms in the transport of carriers. Since ionized impurity scattering is inefficient at higher carrier energies, it becomes important in the 77 K to 300 K temperature range. Alloy scattering is impor-

---

[12]The energy relaxation time τ_w can not be exact for any scattering process. The exactness requires the final states of the scattering process to yield a Maxwell–Boltzmann distribution. So, this equation needs to be used with great caution and critical reservations. More on the relaxation time approximation and the appropriateness of the approach in Chapter 3.

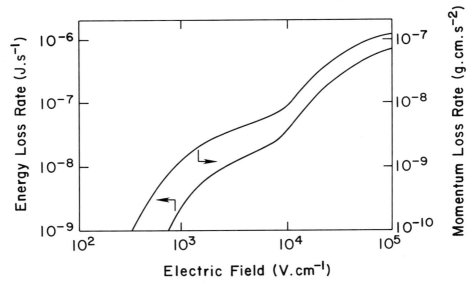

Figure 2.22: Momentum and energy loss rate in steady-state as a function of electric field for GaAs at 300 K due to inter-valley and intra-valley scattering involving polar optical phonons.

tant for transport in mixed crystals, in the temperature range below 300 K. Carrier–carrier scattering preserves both the energy and the momentum if it is between carriers of the same type. It therefore causes a substantial difference in between the transport of a majority carrier and a minority carrier. Lattice scattering becomes dominant at high temperatures and high carrier energies. Optical phonons, with their large momentum and energy compared to acoustic phonons, are dominant when carriers have large energies and are dominant in inter-valley transfers. Acoustic phonons establish a lower limit in scattering since they can exchange very low energies also. At very low temperatures, therefore, while optical phonon absorption can be frozen out because of their reduced density, acoustic phonons continue to be important. At low temperatures and low fields, both acoustic phonons and the impurity scattering are important. At high fields and very low temperatures, impurity scattering becomes weak because of the higher carrier energy, but optical phonon emission becomes important. Acoustic phonon scattering should also be considered. At high temperatures phonon scattering prevails, and both absorption and emission processes become dominant.

Now, let us consider the temperature dependence of the low field and high field behavior in the velocity–field curves of Figure 2.21. At high fields, as we have already discussed, this velocity is dominated by inter-valley scattering processes occurring largely through optical phonons, which are polar in compound semiconductors and non-polar in silicon and germanium. At low fields, other mostly inter-valley scattering processes dominate. The low field behavior is equally well characterized by the mobility or the drift mobility, which relates the velocity with the low electric field. While drift mobility is the real mobility of interest in most electronic devices, one quite often also uses Hall mobility which is a far easier transport parameter to measure accurately.

Drift Mobility

At room temperature, ionized impurity scattering and polar phonon scattering dominate the scattering processes in doped samples. As the temperature is lowered, phonon scattering becomes less efficient and the mobility is limited more by ionized impurity and deformation potential acoustic scattering. As temperature is decreased further, the ionized impurity scattering dominates. Thus, for low field regions in doped GaAs, both the ionized impurity scattering and phonon scattering are important over a broad temperature range. The dependence of mobility on carrier and impurity density in GaAs, for 300 K and 77 K, is shown in Figure 2.23.

For pure material, impurity scattering is reduced, but most of the other processes remain. Figure 2.24 shows the contribution of the various scattering processes to limiting the mobility of lightly doped GaAs, and their temperature dependence.

Drift mobility, as discussed thus far, depends on the carrier scattering embodied in the phenomenological carrier relaxation time. To calculate the mobility, we need to determine this relaxation time, including its functional dependence on energy. The average value of this relaxation time then is obtained by averaging over the energy states that the carrier occupies. For electrons, using a parabolic band approximation, we obtain

$$\tau_\mu = \overline{\tau}. \tag{2.99}$$

The averaging is over energy, which is related to the momentum in the parabolic approximation through a square law. The relaxation time, therefore, is also often written as

$$\tau_\mu = \frac{\langle (E - E_c) \tau \rangle}{\langle E - E_c \rangle} = \frac{\langle v^2 \tau \rangle}{\langle v^2 \rangle}. \tag{2.100}$$

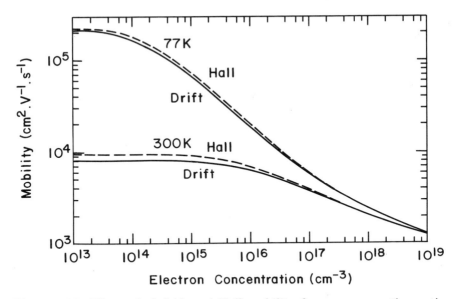

Figure 2.23: Theoretical drift and Hall mobility for a compensation ratio of two as a function of background electron density at 300 K and 77 K for GaAs.

The time constant for mobility can be calculated knowing what scattering process is dominant. Alternately, if more than one is important, they all have to be evaluated and resultant determined as a geometric mean. The geometric mean comes about because scattering rates add, and the time constant is inversely proportional to this rate. Thus, the time constant τ_μ is related to the time relating the inverse of scattering rates $\tau_1, \tau_2 \cdots$

$$\frac{1}{\tau_\mu} = \frac{1}{\tau_1} + \frac{1}{\tau_2} + \cdots, \qquad (2.101)$$

and the drift mobility is related to this time constant through[13]

$$\mu_d = \frac{q\tau_\mu}{m^*} = \frac{q}{m^*} \frac{\langle v^2 \tau \rangle}{\langle v^2 \rangle}. \qquad (2.102)$$

If one associates $\mu_1, \mu_2, \cdots$ as the mobilities with the various scattering processes corresponding to $\tau_1, \tau_2, \cdots$, then

$$\frac{1}{\mu} = \frac{1}{\mu_1} + \frac{1}{\mu_2} + \cdots \qquad (2.103)$$

---

[13]This relationship, Mathiessen's rule, has limited, not universal, validity since it is based on relaxation time approximation.

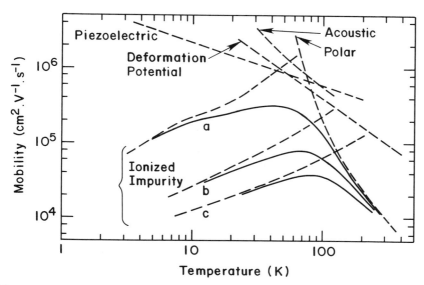

Figure 2.24: Electron mobility in n-type lightly doped GaAs as a function of temperature together with contributions from the various sources of scattering. Curve (a) is for mid-10^{13} cm^{-3} doping, curve (b) is for $\approx 10^{15}$ cm^{-3} doping, and curve (c) is for mid-10^{15} cm^{-3} doping. After J. S. Blakemore, "Semiconducting and Other Major Properties of Gallium Arsenide," *J. of Appl. Phys.*, **53**, p. R123, Oct. 1982.

For a given momentum, the velocity of carriers is inversely related to the effective mass. A larger effective mass results in smaller velocity for identical forces on the carrier. Since the velocity of carriers and their mobility is intimately related to the effective mass, we also need to consider the band structure of the crystal. The conduction band minima in both silicon and germanium do not occur at a crystal momentum of zero, i.e., the materials have an indirect bandgap. The group velocity at the minima, which are either near the X-point (about 15 percent away from it) in silicon, or the L-point in germanium, is zero, but the crystal momentum is non-zero. Crystal momentum and particle momentum are two different quantities. Perhaps we should always refer to the reduced wave vector corresponding to the crystal momentum as k_c. Usually though, the same notation is used for both reduced wave vectors—k.[14]

---

[14]i.e., we use $p = \hbar k$ where the k is the particle reduced wave vector corresponding to the particle momentum. This is related to the band structure and the crystal reduced

Since in general the effective mass is a tensor, it has a directional dependence. This dependence is quite clear in the differing curvature in different directions of Figure 2.10 for Si and Ge. Let us now consider how the occupation of the bands, the motion of carriers due to anisotropy of masses, and the occupation of more than one degenerate bands are related. In thermal equilibrium, the carriers occupy bands following Pauli's exclusion principle. Constant energy surfaces are used to show those momentum values that have constant energy and hence an equal likelihood of occupation in thermal equilibrium. The shapes of such surfaces are dependent on the band anisotropy. In cubic crystals, at thermal equilibrium, three types of constant energy surfaces are usually encountered. These are shown in Figure 2.25 with arbitrary origins of the momentum. A parabolic band has an isotropic mass; its constant energy surface is spherical. The L and X valley minima, however, have a rotational symmetry around the crystallographic directions and show a differing effective mass in the transverse direction to it than in the longitudinal direction. The constant energy surface is an ellipsoid. Valence bands are usually double degenerate at zone center. For any energy in excess of this zone center energy, the constant energy surface is warped.

Thus, even for moderate doping, in cases where the band structure is not a simple parabola, where the band minima do not occur at the zone center, and where degenerate bands exist, the effective mass for use in conduction calculations has to be derived as a suitable average that reflects the net effect. We first consider the conduction band. For the lowest conduction band minima in the Γ band, the bands are quite isotropic and we may use the usual mass. At higher energy, however, e.g., when a large field is applied and inter-valley transfer occurs, we may not use this constant

---

wave vector through the relationship

$$p = m^* v = m^* \frac{1}{\hbar} \frac{dE}{dk_c}, \tag{2.104}$$

where k_c is the crystal momentum. In this text, we use k to denote the reduced wave vector for both the particle and the crystal. The effective mass m^* is in general a tensor. The m^* tensor is not direction dependent while the $E(k)$ relationship is. The form

$$
\begin{aligned}
\frac{1}{m^*} &= \frac{1}{\hbar^2} \frac{\partial^2 E}{\partial k_i \partial k_j} \\
&= \begin{pmatrix} 1/m^*_{xx} & 1/m^*_{xy} & 1/m^*_{xz} \\ 1/m^*_{yx} & 1/m^*_{yy} & 1/m^*_{yz} \\ 1/m^*_{zx} & 1/m^*_{zy} & 1/m^*_{zz} \end{pmatrix}
\end{aligned} \tag{2.105}
$$

is suitable at the band minima where it can be directly related to the first expansion term of the $E(k)$ relationship. During a gradual change in k as the carrier travels, its effective mass undergoes a change.

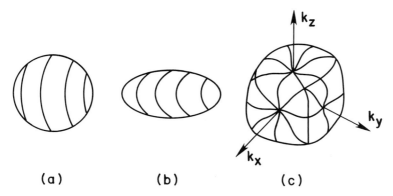

(a)   (b)   (c)

Figure 2.25: Constant low energy surfaces commonly encountered in cubic crystals. Part (a) shows a spherical surface encountered for Γ valley minima in the conduction band, part (b) shows an ellipsoidal surface encountered for L and X valley minima in the conduction band, and part (c) shows a warped surface encountered for the two degenerate valence bands. After C. Jacobani and L. Reggiani, "The Monte Carlo Method for the Solution of Charge Transport in Semiconductors with Applications to Covalent Materials," *Review of Modern Physics*, **55**, No. 3, July 1983.

isotropic effective mass. Transfer of carriers in GaAs occurs first to the L-valley and then to the X-valley with increasing energy. An L-valley has eight equivalent minima in the eight different directions. However, since the minima always occur at the Brillouin zone edge, only four equivalent complete valleys need to be considered. For X-valleys, the minima occur slightly inside the Brillouin zone, and hence all six equivalent minima and six equivalent valleys should be considered. The L and X minima are ellipsoidal surfaces of constant energy, i.e., a cut along the axis leads to an ellipse represented by

$$E - E_c = \frac{\hbar^2}{2m_l^*}(k_l - k_{ol})^2 + \frac{\hbar^2}{2m_t^*}(k_t - k_{ot})^2, \qquad (2.106)$$

where the subscripts l and t denote longitudinal and transverse directions. This is simply a translation of the axes from our original Cartesian coordinate in the reciprocal space to a new Cartesian coordinate that is aligned along the longitudinal axis, and because of symmetry the transverse variations are identical. In terms of the effective mass, if the transformed x-axis is now aligned with the longitudinal axis, then the new effective mass tensor

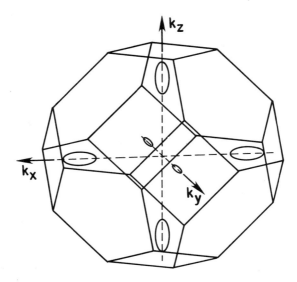

Figure 2.26: The six equi-energy ellipsoidal surfaces for the X minima shown in the first Brillouin zone. For a field aligned with one of the axes, four ellipsoid surfaces form the first equivalent set and the other two ellipsoid surfaces form the second equivalent set.

is

$$\frac{1}{m^*} = \begin{pmatrix} 1/m_l^* & 0 & 0 \\ 0 & 1/m_t^* & 0 \\ 0 & 0 & 1/m_t^* \end{pmatrix}. \tag{2.107}$$

If the electric field is applied in the $\langle 100 \rangle$ direction, as in Figure 2.26, there are four ellipsoid surfaces for the X minima for whom the transverse effective mass is the relevant mass in the determination of the response of the carrier to the field. Similarly, there are two ellipsoidal surfaces for which the longitudinal mass is relevant. Hence, in the process of approximating into the parabolic band picture, i.e., an expression of $E - E_c = \hbar^2 k^2 / 2m_c^*$, we must use the mass

$$m_c^* = \frac{1}{6} \times \left[2m_t^* + 4(m_l^* m_t^*)^{1/2} \right]. \tag{2.108}$$

For the L-valley minima (Figure 2.27), with a field in $\langle 100 \rangle$ direction,

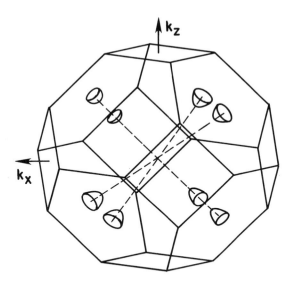

Figure 2.27: The eight equi-energy half-ellipsoidal surfaces for the L minima shown in the first Brillouin zone. For a field aligned with one of the axes all of these surfaces are equivalent.

all the valleys are equivalent, with the effective conduction mass being

$$m_c^* = \frac{1}{4} \times \left(\frac{1}{3} m_t^{*2} + \frac{2}{3} m_l^* m_t^* \right)^{1/2}. \tag{2.109}$$

The valence bands are centered in the Brillouin zone and the light and heavy hole bands are degenerate. The equivalent mass can be derived as

$$\frac{1}{m_c^*} = \left(\frac{g_{hh}(E)}{m_{hh}^*} + \frac{g_{lh}(E)}{m_{lh}^*} \right) \frac{1}{g_{hh}(E) + g_{lh}(E)}. \tag{2.110}$$

where $g_{hh}(E)$ and $g_{lh}(E)$ are the heavy hole and light hole density of states. The ratio of densities of states in the expression above represents the probability of finding the hole in a state with that mass. Since the density of states varies as $3/2$ power of the mass, the conduction mass for holes is

$$\frac{1}{m_c^*} = \frac{m_{hh}^{*\,1/2} + m_{lh}^{*\,1/2}}{m_{hh}^{*\,3/2} + m_{lh}^{*\,3/2}}. \tag{2.111}$$

The longitudinal and transverse effective masses in silicon and germanium for some of the bands are given in Table 2.2.

Table 2.2: Effective masses in Silicon and Germanium.

Material	Valence Band		Conduction Band	
	Light Hole Effective Mass	Heavy Hole Effective Mass	Longitudinal Effective Mass	Transverse Effective Mass
Silicon	$0.16m_0$	$0.5m_0$	$0.97m_0$	$0.19m_0$
Germanium	$0.04m_0$	$0.3m_0$	$1.6m_0$	$0.082m_0$

Hall Mobility

Hall mobility is a mobility parameter that results from Hall effect measurement. It is a simple and hence very common measurement to gauge the low field transport characteristic of semiconductors. Our objective here is to emphasize its relationship with the drift mobility which is of more immediate interest from a device perspective. The Hall effect is a phenomenon that results from the application of a magnetic field on a sample. Consider the n-type rectangular semiconductor sample of Figure 2.28 with a magnetic field applied perpendicular to the sample, a voltage bias applied to force a current to flow in plane of the sample, and a voltage being measured perpendicular to this flow of current. A voltage, called the Hall voltage, develops in the perpendicular direction because of the Lorentzian force acting on the charged particle. For the electron, this force is given by

$$\mathcal{F} = -q\left(v \times \mathcal{B}\right). \tag{2.112}$$

In steady-state, carriers accumulate and deplete in the transverse direction, causing a transverse electric field which opposes and cancels the field due to the Lorentzian force, to appear. This allows us to write the electric field in the transverse direction, related to the Hall voltage developed across the transverse direction, in terms of the magnetic field, the current and a parameter called the Hall coefficient. This field is related as

$$\mathcal{E}_z = R_H J_x \mathcal{B}_y, \tag{2.113}$$

where R_H is the Hall constant which is related to the characteristics of the motion and the carrier density and is given by

$$R_H = \frac{r}{q} \frac{p\mu_{dp}^2 - n\mu_{dn}^2}{\left(n\mu_{dn} + p\mu_{dp}\right)^2}, \tag{2.114}$$

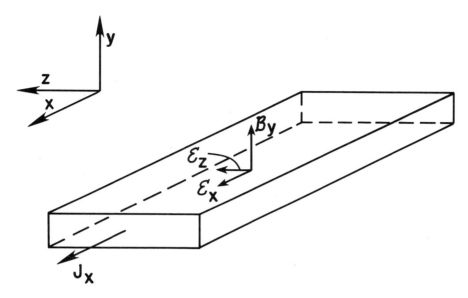

Figure 2.28: A schematic defining directions of current flow, electric field, and magnetic field for Hall effect in an n-type sample.

for a sample with both electron and hole conduction. The pre-factor r (also called Hall factor) in the Hall constant usually varies between one and two for most common scattering processes. This factor is related to the relaxation time for Hall mobility determination τ_H (analogous to τ_μ, the relaxation time for drift mobility determination) given by

$$\tau_H = \frac{\langle \tau^2 \rangle}{\langle \tau \rangle}. \tag{2.115}$$

For one carrier conduction, the Hall constant is given by

$$R_H = -\frac{1}{qn} \frac{\langle \tau^2 \rangle}{\langle \tau \rangle^2} \tag{2.116}$$

for n-type material, and

$$R_H = \frac{1}{qp} \frac{\langle \tau^2 \rangle}{\langle \tau \rangle^2} \tag{2.117}$$

for p-type material. We will consider the physical basis of these relationships during our discussion of mathematical treatments in Chapter 3.

Note the difference between this and the relaxation time associated with the drift mobility. If only acoustic deformation potential scattering dominates, and the conduction and valence bands can be assumed to be parabolic, then $r = 3\pi/8$. The motion of carriers in the transverse direction occurs due to the applied magnetic field, and is determined by the velocity in the longitudinal direction. The proportionality constant in the longitudinal direction is the drift mobility, however the proportionality in the transverse direction is a mobility term that depends on the same pre-factor term as does the Hall constant. Hence, the Hall mobility in samples with unipolar conductivity is

$$\mu_H = r\mu_d. \tag{2.118}$$

More generally,

$$\mu_H = |R_H \sigma| \tag{2.119}$$

where σ, the conductivity, is given by

$$\sigma = qn\mu_{dn} + qp\mu_{dp}. \tag{2.120}$$

Because the ratio r varies between one and two, the Hall mobility is always larger than the drift mobility. Its significance and ubiquity is largely a result of the ease of its measurement.

The Hall and drift mobility are, of course, related to each other through the characteristics of the dominant scattering processes, i.e. through the ratio r above. Figure 2.23 shows the Hall mobility in GaAs as a function of background electron concentration for 300 K and 77 K. This figure also serves to show the difference between drift and Hall mobility as a function of electron concentration.

Saturated Velocity

The mobility of carriers decreases as a function of particle energy because the scattering rate continues to increase as a function of the energy. Thus, drift and Hall mobilities become negligible at high energies because carriers move short distances in between scattering events.

It is the velocity of carriers that is of direct interest to us. At high energies, even though the mobility becomes negligible, the velocity for a high applied field is still high. At low energies, both the velocity and the mobility are useful parameters because they are directly related to each other. However, mobility is used more commonly, both because of the ease of measurement and because of simplifications in many steady-state drift-diffusion calculations in devices.

The saturation of velocity at high fields occurs for the following reason. Carriers drift in the high field in between scattering. Any increase in energy resulting from this transport in high field causes an increase in the scattering rate also. An increase in the electric field, therefore, does not

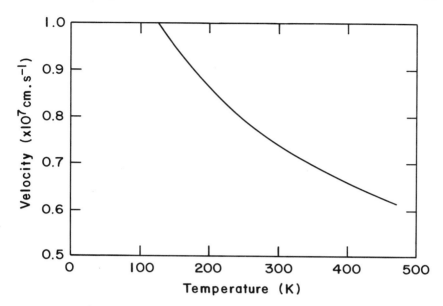

Figure 2.29: Drift velocity in GaAs as a function of temperature at a field of 50 kV.cm^{-1}.

result in a significant additional increase in the velocity, and the mobility becomes negligible. An alternative way of looking at this behavior in high fields, where the carrier is in equilibrium with its local conditions, is to note that the mobility is a function of the carrier energy. In low fields, the average carrier energy increases proportionally with the electric field, the scattering rate increases proportionally, and hence the carrier mobility remains constant and independent of the energy and the electric field. In high fields, however, the carrier energy increase causes the mobility to decrease inversely with the electric field and hence results in velocity saturation.

Optical phonon scattering is the principal source of scattering in the high field region. Since, as shown by Equation 2.56, the occupation of the optical modes decreases strongly as the temperature decreases, energy relaxation is not as strong, and hence absorption processes are less likely. The drift velocity at high fields also shows a temperature dependence. An example of this is shown in Figure 2.29 where there is nearly a 20% change in the drift velocity at high fields (approximately the saturated velocity) between 300 K and 77 K.

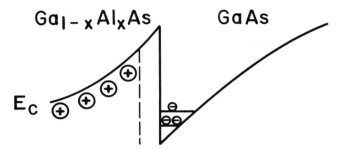

Figure 2.30: Separation of two-dimensional electron gas from the ionized donors in a two-dimensional electron gas structure. The larger bandgap material ($Ga_{1-x}Al_xAs$), from which the electron transfer occurs, is shown on the left. The right hand side shows the smaller bandgap material (GaAs) to which the transfer occurs. An undoped spacer layer, used to decrease remote Coulombic scattering, is also shown in the figure.

2.7.2 Two-Dimensional Effects on Transport

Now, compare the mobility behavior to that of two-dimensional electron gases in GaAs. Here, scattering, due to ionized impurities, occurs through a remote process because donors and carriers are separated. The perturbation in potential due to ionized donors (sometimes this is referred to as Coulombic scattering also) is separated from the electron gas as shown in Figure 2.30. Decrease of the ionized impurity scattering, and increase of the carrier screening of the ionized impurities because of large carrier densities in the two-dimensional channel, result in a significantly higher mobility. Now, scattering processes such as the piezoelectric scattering become important, at least in part of the temperature range, as shown in Figure 2.31. At the lowest of the temperatures, the residual background doping at the high mobility interface is still important, although because this is low, the mobility is very high. It is interesting to compare this figure with Figure 2.24 which shows the mobility versus temperature behavior for low doped GaAs. Note that over a fairly broad range of temperature, near and above 77 K, the behavior is very similar. Below that temperature, the improvements due to carrier screening make the mobilities of the two-dimensional electron gas system significantly larger.

The added complication in the two-dimensional electron gas structures is that one should also consider the quantization and its effects on scattering, as well as the effect of carrier–carrier interaction, and carrier–dopant interaction. The former leads to a decrease in mobility, the latter tends to

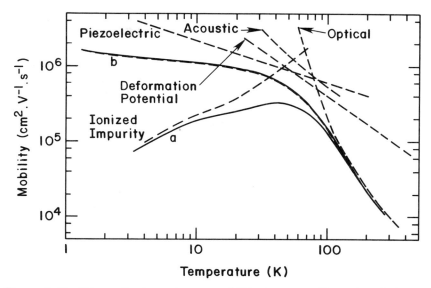

Figure 2.31: Theoretical maximum mobility in a two-dimensional electron gas at the Ga.$_7$Al.$_3$As/GaAs interface as a function of temperature is shown as the curve marked b. Curve a shows the mobility in very lightly doped GaAs. The figure also shows components of mobility arising from different scattering mechanisms in bulk GaAs.

screen the perturbation potential associated with the dopant. Thus, there is actually an optimum in carrier concentration in the channel where the two balance and allow the highest mobilities at low temperatures. Note the large increases in low field mobility that can be achieved by a decrease or removal of the ionized impurity scattering. At 300 K, since ionized impurity scattering is only one of the two dominant scattering processes, the increase is not very large. At 77 K, mobilities of greater than 70,000 cm^2.V^{-1}.s^{-1} are quite common. The behavior of two-dimensional electron gases is quite complicated; it has increased carrier–carrier scattering whose effect, as we have discussed, is to change the distribution function but to not affect the energy and momentum.

We have not discussed the interaction of carrier–carrier scattering processes with other processes, which are also important in these structures. An example of this is given in Figure 2.32, which shows the coupling in between optical phonons and the carrier plasma, called plasmon, as a function of carrier density in GaAs. Significant deviation, and large energy effects, begin to occur at mid-10^{17} cm^{-3} carrier density. Carrier densities higher

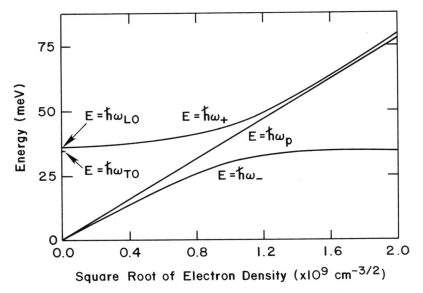

Figure 2.32: Energy of the phonon modes as a function of the square-root of electron density in GaAs. Carrier–phonon coupling results in behavior shown at high carrier densities.

than this exist in two-dimensional electron gases in device structures and therefore mobility is fairly sensitive too carrier scattering. Too low a carrier density leads to reduction of screening of Coulombic scattering from remote and local donors, and too high a density leads to increased carrier–carrier scattering effects.

2.7.3 Minority Carrier Transport

The examples, until now, have been those of majority carriers. The electron, as a majority carrier, suffers Coulombic scattering from the donors, other electrons, and any other residual impurities. Usually, the latter two are small; only at high carrier densities, such as at high dopings or in two-dimensional electron gas, do the carrier–carrier interactions also become strong. The electron, as a minority carrier, suffers Coulombic scattering from the acceptors, the holes, other electrons, and any other residual impurities. The last two, as in the case of the electron as a majority carrier, are usually small. Holes are the new twist to the environment of the electron.

The behavior of the electron as a minority carrier depends on the semi-

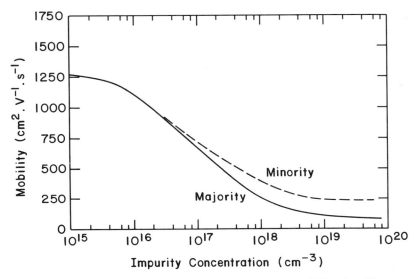

Figure 2.33: Minority and majority carrier electron mobility in silicon as a function of impurity concentration for negligible electric field at 300 K.

conductor and on any residual field in the semiconductor device. It depends on the semiconductor because any interaction between electron and hole— usually the heavy hole since it is far more common—involving exchange of momentum and energy is a function of their relative effective masses. The screening of acceptors by holes (i.e., reduction of Coulombic perturbation by holes in the vicinity of the ionized acceptor) may allow for an increase in the carrier mobility. Holes, if they are considerably heavier than the electron, may cause scattering very similar to that due to ionized accep- tors. A slight residual field in the structure causes motion of carriers in the opposite direction, which may result in a drag effect on each carrier by the other. The magnitude of the effect on each carrier is a function of their effective mass, and the disparity in the effective mass. Scattering is a strong function of temperature, hence, different considerations may apply to different temperatures.

Figure 2.33 and Figure 2.34 show plots for minority and majority carrier mobilities for Si and GaAs as a function of background donor and acceptor density. These figures are for mobility at 300 K. Silicon has a comparable mobility for electron as a majority and minority carrier, while GaAs shows a minority carrier mobility that is nearly half the value of the majority carrier mobility. In silicon, the effective masses of electron and hole are

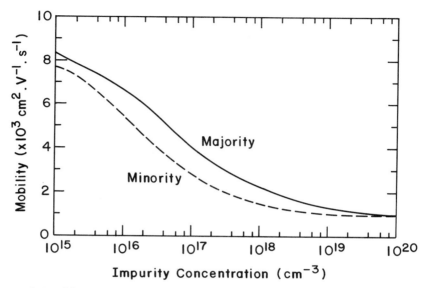

Figure 2.34: Minority and majority carrier electron mobility in GaAs as a function of impurity concentration for negligible electric field at 300 K. The majority mobility is at a compensation ratio of two. After S. Tiwari and S. L. Wright, "Material Properties of p-type GaAs at Large Dopings," *Appl. Phys. Lett.*, **56**, No. 6, p. 563, 1990.

relatively similar. Exchanges involving electrons and holes, and screening of impurities by holes, together contribute to this similarity of mobilities. In GaAs, electrons and heavy holes have widely differing masses. Heavy holes can be considered static for scattering purposes, and lead to Coulombic scattering of a magnitude similar to the ionized acceptor. This behavior may change at a different temperature. At 77 K, e.g., for GaAs in the 10^{15} cm^{-3} doping range, the minority carrier mobility for electrons is higher than as a majority carrier.

Minority carrier devices usually have fields in regions where the electron transits. In the base of the bipolar, e.g., a small field exists even for uniformly doped base structures. This field is the result of the gradient in the hole distribution established to compensate for the electron distribution. It causes a rapid decrease in mobility with the electric field, even though the mobility at zero field is quite comparable to the majority carrier mobility. The majority carrier mobility, of course, does not vary substantially. This difference in behavior is usually ascribed to the electron-hole drag effects which are quite significant for Si because of comparable effective masses.

The carrier–carrier scattering conserves momentum. In the absence of an electric field, the average drift momentum is zero. In the presence of drift field, electrons and holes move in opposite directions. The interacting electrons and holes that are in the vicinity of each other and are, on an average, moving in opposite directions, exert a Coulombic attractive force on each other. This reduces the effect of the drift field, and externally, appears as the lowering of mobility. The effect is strong, and shows up at a low electric field. At higher fields, the drag effect does not remain as important because the Coulombic scattering rate decreases and the electron now has a higher energy.

The consequence of such an effect can be strong in devices such as bipolar transistors, even if no electric field is built into the structure. In the presence of current, there exists a gradient of electron density in the base. In order to maintain charge neutrality, hole density also changes as a function of position to compensate for the excess charge due to electrons. The excess hole population is equal to the electron population under quasi-static conditions. The diffusion current due to the gradient in hole density is nearly compensated by the drift current resulting from a small electric field that is established in the base. The magnitude of this electric field can be found approximately, by setting the hole transport current to zero (it is substantially smaller than the electron current for the n–p–n bipolar transistor). The current density equation in the drift-diffusion approximation can be used to model this condition. Here we use the current density equation without proof; it will be the subject of discussion in Chapter 3. Thus,

$$J_p = q\mu_n \mathcal{E}p - q\mathcal{D}_p \frac{dp}{dz} \approx 0, \tag{2.121}$$

implying

$$\mathcal{E} \approx \frac{kT}{q} \frac{1}{p} \frac{dp}{dz}. \tag{2.122}$$

Electric field is pointed in the direction of increasing p. To maintain charge neutrality, the hole concentration is higher towards the electron injecting electrode. Hence the electric field is pointed towards the electron injecting electrode.

This is a very substantial simplification of a considerably complex many-body problem. It has been argued that under certain conditions, this drag effect should be balanced by the effect of majority carrier distribution. Hole current being small, the holes are nearly stationary. Thus, they should not transfer much momentum to electrons. Electrons, however, do transfer momentum to holes, resulting in the space charge and electric field to prevent the resulting hole current. This field aids the motion of electrons, just as in the high injection effect, and hence should compensate.

2.7.4 Diffusion of Carriers

The phenomenon of diffusion is one of the very direct consequences of scattering. Net diffusion takes place from regions of high carrier concentrations to regions of low carrier concentrations. Regions of high concentrations have a higher carrier population resulting in a higher number of scattering events. Some of these lead to carrier motion towards the low carrier concentration region. However, it is not completely compensated for by scattering events that lead to the motion of carriers from the low concentration region to the high concentration region. The net result is a flow, or diffusion, of carriers determined by the gradient of the carrier concentration, and characterized by the diffusion coefficient ($\mathcal{D}_n$ for electrons and $\mathcal{D}_p$ for holes).

Diffusion coefficients are related to mobilities because their underlying basis is in scattering. This relationship, the Einstein relationship, was derived for Brownian motion in the Maxwell–Boltzmann distribution limit of classical gases. We will show it in the Fermi–Dirac distribution limit using the drift-diffusion equation. Consider thermal equilibrium in an n-type sample, thus

$$J_n = q n \mu_n \mathcal{E} + q \mathcal{D}_n \frac{dn}{dz} = 0. \qquad (2.123)$$

Since the carrier concentration is given by

$$n = N_C F_{1/2}\left(\eta_{fc}\right), \qquad (2.124)$$

$$\frac{dn}{dz} = N_C F_{-1/2}\left(\eta_{fc}\right) \frac{d\eta_{fc}}{dz}. \qquad (2.125)$$

In thermal equilibrium (ξ_f = constant) we have

$$\frac{d\eta_{fc}}{dz} = -\frac{1}{kT} \frac{dE_c}{dz} = -\frac{q\mathcal{E}}{kT}. \qquad (2.126)$$

Therefore, we may relate the diffusion coefficient and the mobility as

$$\frac{\mathcal{D}_n}{\mu_n} = \frac{kT}{q} \frac{F_{1/2}(\eta_{fc})}{F_{-1/2}(\eta_{fc})}, \qquad (2.127)$$

the Einstein relationship for electrons; and, following similar arguments,

$$\frac{\mathcal{D}_p}{\mu_p} = \frac{kT}{q} \frac{F_{1/2}(\eta_{fv})}{F_{-1/2}(\eta_{fv})}, \qquad (2.128)$$

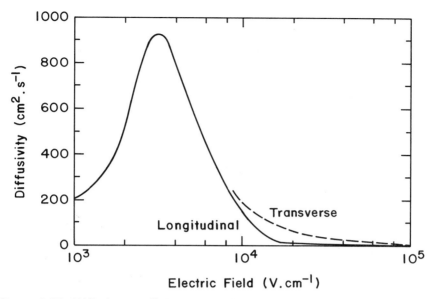

Figure 2.35: Diffusion coefficient in the longitudinal and transverse directions as a function of electric field for low-doped GaAs at 300 K.

the Einstein relationship for holes. In the Maxwell–Boltzmann limit, these can be simplified to

$$\frac{\mathcal{D}_n}{\mu_n} = \frac{\mathcal{D}_p}{\mu_p} = \frac{kT}{q}. \tag{2.129}$$

The diffusion coefficients of electrons in GaAs and other compound semiconductors are larger than in silicon for the same reason as for mobility—largely due to a smaller effective mass. The high diffusion coefficient leads to a shorter base transit time in bipolar transistors in GaAs and other compound semiconductors for comparable base widths. It also has implications for noise behavior: a larger diffusion coefficient usually results in smaller magnitude of noise. Since, in general, conduction band structure can be anisotropic, scattering processes are anisotropic, and since fields exist along specific directions, the diffusion coefficient shows anisotropy. Examples of magnitude of diffusion coefficient with field are given in Figure 2.35 for the longitudinal and transverse directions of GaAs. The anisotropic nature of polar scattering is a particularly strong contributor to this anisotropy. Since polar scattering occurs across the field range so long as it leads to carrier energies above the optical phonon threshold, the diffusion coefficient also exhibits similar behavior. Anisotropy exists both at low and high fields. At high fields, the L and X valleys are mostly occupied, the longitudinal effective mass is large, and hence the longitudinal diffusion coefficient is suppressed. It decreases to very low values compared to the zero field value, just as in the case of mobility. The pinch-off regions of a field effect transistor are regions of high fields where the carrier gradient can become large for large gate length devices. The low diffusion coefficient reduces the diffusive current in such regions even if carriers move with finite velocity (the saturated velocity) due to drift effects.

2.8   Some Effects Related to Energy Bands

There are a number of device related phenomena that can be intuitively related to the band structures. One example, that we briefly alluded to is the relationship between scattering rate and density of states. Scattering is proportional to the coupling between the two states connected by a scattering process, and the number of states available for scattering into. Thus, at the onset of secondary valley transfer, i.e., when carriers have kinetic energy approximately equal to the inter-valley separation, a rapid rise in the scattering rate occurs, as seen in Figure 2.20 for GaAs, InP, and $Ga_{.47}In_{.53}As$. Also note that at low energies, InP has the highest scattering rate. InP has the highest effective mass of the four semiconductors in the figure, and therefore has the highest density of states.

Effective mass, which is inversely proportional to the second derivative of the energy with respect to the wave vector, i.e., related to the band structure, is central to a lot of transport parameters of interest. Mobility is inversely proportional to the effective mass—a lower effective mass usually implies faster low field transport characteristics. A lower effective mass also results in improvement in the diffusion coefficient and higher tunneling probability. A smaller effective mass means lower density of states, carriers have a higher velocity for the same energy, and so long as the scattering rate is lower, the off-equilibrium effects are stronger. The highest velocity that the carriers can attain under the most ideal of circumstances (the maximum group velocity) is also limited by the band structure.

Another good example of a strong band structure relationship exists in the phenomenon of avalanche breakdown due to impact ionization. This is a process in which a carrier picks up enough energy in between scattering events to cause an electron to jump from the valence band into the conduction band during a scattering event. The process is both band structure- and scattering process-dependent.

In this section we will describe some of these relationships in more detail, and develop a comparative understanding of some of these parameters for compound semiconductors.

2.8.1 Avalanche Breakdown

The breakdown phenomenon of most interest occurs in depletion regions of devices. In bipolar transistors, it is the breakdown in the base-collector junction region. In field effect transistors, it is the breakdown in the drain to substrate depletion region. These occur because a high field exists, due to the reverse bias at the junction, and this high electric field is capable of accelerating the carrier and imparting to it a large kinetic energy. Indeed, this kinetic energy can become sufficiently large to cause an electron, or even more than one electron under the right conditions, to be removed from the valence band and transferred to the conduction band. Take the case of GaAs, where examples of the possible electron and hole initiated transitions are described in the Figure 2.36. The process, involving a single electron–hole pair generation, consists of transfer of energy from the hot carrier to an electron in the valence band, which jumps into the conduction band leaving a hole behind. For small bandgap semiconductors, even multiple such pairs may be generated.

With the complicated band structure, electrons can be from different bands, holes can be from different bands, phonons (particularly optical phonons with their large momentum) may be involved, etc. Indeed, since the band structure is orientation dependent, and impact ionization occurs due to large energy of the carrier from rapid acceleration in a field, in the event of insufficient randomization, it may even be orientation dependent.[15]

We will discuss, briefly, a very simplified model, to see how the role of band structure enters in the impact ionization process. We consider a parabolic model for the conduction band with effective mass m_e^*, and a parabolic model for the valence band with effective mass m_h^*. Under thermal equilibrium conditions, there may exist carriers that have sufficient energy for impact ionization. These carriers are in the tail of the carrier distribution function with energy, and by detailed balance, impact ionization is balanced by its reverse process—Auger recombination. With an increase in electric field, carriers pick more energy, and hence there may now exist carriers with sufficient energy for electron–hole pair or pairs generation. We consider only one electron–hole pair, one band each for the electrons and

---

[15]Impact ionization has continued to be the subject of research in semiconductors since 1960s, since it is a very complicated subject involving hot carriers. Orientation dependence is only one of many examples of this. See, e.g., T. P. Pearsall, F. Capasso, R. E. Nahory, M. A. Pollack, and J. R. Chelikowsky, "The Band Structure Dependence of Impact Ionization by Hot Carriers in Semiconductors: GaAs," *Solid-State Electronics*, **21**, p. 297, 1978, and H. Shichijo and K. Hess, "Band-Structure-Dependent Transport and Impact Ionization in GaAs," *Phys. Rev. B*, **23**, No. 8, p. 4197, 15 Apr. 1981.

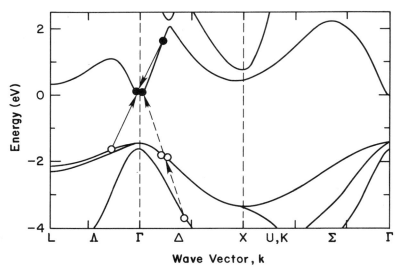

Figure 2.36: Examples of possible impact ionization processes for electron and hole initiated processes for GaAs. The ionization process initiated by a hot electron is shown using solid lines; it results in the transfer of an electron from the heavy hole band to the conduction band. The ionization process initiated by a hot hole from the split-off band is shown using dashed lines. After T. P. Pearsall, F. Capasso, R. E. Nahory, M. A. Pollack, and J. R. Chelikowsky, "The Band Structure Dependence of Impact Ionization by Hot Carriers in Semiconductors: GaAs," *Solid-State Electronics*, **21**, p. 297, ©1978 Pergamon Press plc.

holes, no phonon involvement, and a process initiated by electrons. The minimum energy that an electron must have is the bandgap E_g to cause the transfer. If i and f signify initial and final state subscripts for the electron causing the impact ionization, the creation of an electron-hole pair of momentum p_e and p_h in an electron initiated process (see Figure 2.37) requires, by conservation of energy and momentum,

$$\frac{p_i^2}{2m_e^*} = E_g + \frac{p_f^2}{2m_e^*} + \frac{p_e^2}{2m_e^*} + \frac{p_h^2}{2m_h^*},\qquad(2.130)$$

and

$$p_i = p_f + p_e + p_h.\qquad(2.131)$$

Figure 2.37: A simple one dimensional parabolic band schematic of electron-hole pair generation by impact ionization involving a hot electron.

Using the method of the Lagrangian multiplier to minimize the carrier energy under the constraint of momentum conservation,

$$d \left(E_g + \frac{p_f^2}{2m_e^*} + \frac{p_e^2}{2m_e^*} + \frac{p_h^2}{2m_h^*} \right) + \mathcal{A} d \left(p_f + p_e + p_h \right) = 0. \qquad (2.132)$$

Using partial derivatives with respect to p_e, p_h, and p_f gives

$$\frac{p_e}{m_e^*} = \frac{p_h}{m_h^*} = \frac{p_f}{m_e^*} = -\mathcal{A}, \qquad (2.133)$$

giving for the initial momentum

$$p_i = \left(2 + \frac{m_h^*}{m_e^*} \right) p_f, \qquad (2.134)$$

and for the threshold energy of the initial electron E_T

We can see the tremendous simplification in applying this relationship to real band structures. GaAs has $m_e^* = 0.067m_0$ and $m_{hh}^* = 0.5m_0$, and hence $E_T = E_g$. However, the bands are not parabolic an energy E_g into the conduction band as can be seen from Figure 2.11. So, the situation in real crystals is considerably more complicated; this threshold relationship may overestimate because it ignores phonons in momentum conservation, and the processes may even be orientation dependent.

This ionization process is characterized in a semiconductor by the parameter ionization rate. The ionization rate is the relative increase in carrier density per unit length of carrier travel. Thus, the electron ionization rate (α_n) is the rate of increase in electrons per unit length of the travel of an electron:

$$\alpha_n = \frac{1}{n}\frac{dn}{dz}. \tag{2.136}$$

For holes, the equivalent expression is

$$\alpha_p = \frac{1}{p}\frac{dp}{dz}. \tag{2.137}$$

For GaAs, the impact ionization occurs largely from carriers that are not in the lowest bands. Electrons cause ionization from higher bands, and holes generally cause it from the split-off band. At low fields, the ionization rate is also orientation dependent because of lack of randomization. Examples of ionization rates as a function of electric field are given in Figures 2.38 and 2.39.

For most compound semiconductors with greater than an eV of bandgap, the hole ionization rate is larger than the electron ionization rate. Hole effective mass being larger, for similar velocities in the high field regions it has significantly higher kinetic energy. At lower fields, electrons are more likely to have a higher energy since they usually have a smaller scattering rate than holes. Thus at low fields the electron ionization rate tends to be larger. For smaller bandgap semiconductors, e.g., InAs and $Ga_{.47}In_{.53}As$, the hole ionization rate is always low because the threshold energy for ionization is lower.

Impact ionization does not necessarily need an electric field to show a large generation rate for electrons and holes. In compound semiconductor heterostructures, there exist many smaller bandgap compounds that can be lattice matched to other semiconductors with a bandgap discontinuity for a conduction or valence band that is larger than the bandgap. Even in the absence of an electric field, high energy carriers injected into the smaller bandgap semiconductor can cause impact ionization. This process is often referred to as Auger generation.

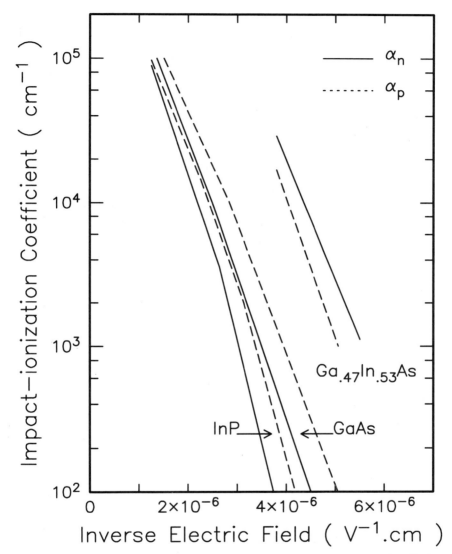

Figure 2.38: Impact ionization coefficient for electrons and holes for GaAs, InP, and Ga$_{.47}$In$_{.53}$As as a function of the inverse of the electric field.

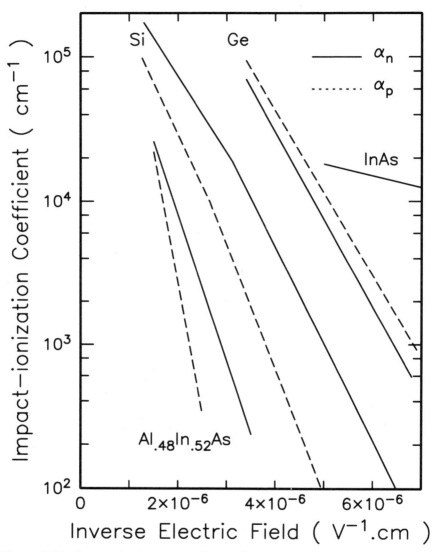

Figure 2.39: Impact ionization coefficient for electrons and holes for Si, Ge, $Al_{.48}In_{.52}As$, and InAs.

When impact ionization occurs in the presence of an electric field, such as in a p–n junction, both electrons and holes get accelerated in opposite directions, and both cause further impact ionization. The process of electron–hole pair creation, acceleration of carriers, and creation of more electron–hole pairs sets up an avalanche, and hence the generation of electron-hole pairs in an electric field is also known as an avalanche process, and the related breakdown of a p–n junction an avalanche breakdown. This is an important limiting phenomenon in the usefulness of the device and hence we will investigate it further.

The increase in electron current resulting from the ionization caused by this electron and hole current flow is

$$\frac{dJ_n}{dz} = \alpha_n J_n + \alpha_p J_p, \tag{2.138}$$

where J_n is the electron current density and J_p is the hole current density. Similarly,

$$\frac{dJ_p}{dz} = -\left(\alpha_n J_n + \alpha_p J_p\right), \tag{2.139}$$

and the total current is

$$J = J_n + J_p. \tag{2.140}$$

Excess current flows due to the presence of electron-hole pair generation. The electron carries the current to one part of the device while the hole carries the current to another part of the device. We may derive the multiplication factor in the terminal currents, due to electron-hole pair generation, as follows.

We will consider an electron initiated avalanche process. Let $J_n(-w_p)$ be the electron current at the p-edge and let $J_n(w_n)$ be the current at the n-edge of the depletion region. The hole current is predominant at the p-edge and the electron current is predominant at the n-edge, i.e. $J \approx J_n(w_n) \approx J_p(-w_p)$. In steady-state,

$$\frac{dn}{dt} = \mathcal{G}_n - \mathcal{R}_n + \frac{1}{q}\frac{dJ_n}{dz} = 0. \tag{2.141}$$

We will ignore recombination since generation processes dominate in the avalanche region. As a result, excess current due to avalanche process dominates any recombination current, and can be ignored, thus simplifying our analysis. Denoting the ratio of electron current at $z = w_n$ and $z = -w_p$ by the electron multiplication factor $\mathcal{M}_n$, we obtain

$$J = J_n(w_n) = \mathcal{M}_n J_n(-w_p). \tag{2.142}$$

We also have

$$\frac{dJ_n}{dz} = \alpha_n J_n + \alpha_p J_p = (\alpha_n - \alpha_p) J_n + \alpha_p J. \tag{2.143}$$

The solution of the linear differential equation in J_n as a function of position z is

$$J_n(z) = \frac{\int_0^z \alpha_p J \exp\left[\int_0^\zeta (\alpha_p - \alpha_n)\, d\eta\right] d\zeta + \mathcal{A}}{\exp\left[\int_0^z (\alpha_p - \alpha_n)\, d\zeta\right]}, \tag{2.144}$$

where η and ζ are dummy variables for the spatial coordinates.

The constant $\mathcal{A}$ can be determined from our expression for current continuity as

$$\mathcal{A} = J\left\{\frac{1}{\mathcal{M}_n} \exp\left[\int_0^{-w_p} (\alpha_p - \alpha_n)\, d\zeta\right] - \int_0^{-w_p} \alpha_p \exp\left[\int_0^\zeta (\alpha_n - \alpha_p)\, d\eta\right] d\zeta\right\}. \tag{2.145}$$

And hence, the electron current is given by

$$J_n(z) = J\frac{(1/\mathcal{M}_n) + \int_{-w_p}^z \alpha_p \exp\left[\int_{-w_p}^\zeta (\alpha_p - \alpha_n)\, d\eta\right] d\zeta}{\exp\left[\int_{-w_p}^z (\alpha_p - \alpha_n)\, d\zeta\right]}. \tag{2.146}$$

Evaluating this relationship at $z = w_n$ allows us to write the equality:

$$\exp\left[\int_{-w_p}^{w_n} (\alpha_p - \alpha_n)\, d\zeta\right] = \frac{1}{\mathcal{M}_n} + \int_{-w_p}^{w_n} \alpha_p \exp\left[\int_{-w_p}^\zeta (\alpha_p - \alpha_n)\, d\eta\right] d\zeta. \tag{2.147}$$

Although this expression appears to be complicated, it actually can be simplified considerably. Introducing $\mathcal{I}$ as the magnitude of the integral on the right hand side, we show

$$\mathcal{I} = \int_{-w_p}^{w_n} \alpha_p \exp\left[\int_{-w_p}^\zeta (\alpha_p - \alpha_n)\, d\eta\right] d\zeta$$

$$= \int_{-w_p}^{w_n} \alpha_n \exp\left[\int_{-w_p}^\zeta (\alpha_p - \alpha_n)\, d\eta\right] d\zeta -$$

$$\int_{-w_p}^{w_n} (\alpha_n - \alpha_p) \exp\left[\int_{-w_p}^{\zeta} (\alpha_p - \alpha_n)\, d\eta\right] d\zeta$$

$$= \int_{-w_p}^{w_n} \alpha_n \exp\left[\int_{-w_p}^{\zeta} (\alpha_p - \alpha_n)\, d\eta\right] d\zeta +$$

$$\exp\left[\int_{-w_p}^{w_n} (\alpha_p - \alpha_n)\, d\eta\right] - 1. \tag{2.148}$$

A simple expression for the multiplication factor results from this in the form

$$1 - \frac{1}{\mathcal{M}_n} = \int_{-w_p}^{w_n} \alpha_n \exp\left[-\int_{-w_p}^{\zeta} (\alpha_n - \alpha_p)\, d\eta\right] d\zeta. \tag{2.149}$$

The condition for breakdown caused by electron initiated avalanche is the condition where $\mathcal{M}_n$ goes to infinity. This occurs when the integral in this equality goes to unity. Avalanche breakdown, therefore, occurs when

$$\int_{-w_p}^{w_n} \alpha_n \exp\left[-\int_{-w_p}^{\zeta} (\alpha_n - \alpha_p)\, d\eta\right] d\zeta = 1. \tag{2.150}$$

The condition for breakdown with a hole initiated process can be found following a similar treatment (see Problem 4). The expression for the hole multiplication factor $\mathcal{M}_p$ is given as

$$1 - \frac{1}{\mathcal{M}_p} = \int_{-w_p}^{w_n} \alpha_p \exp\left[-\int_{\zeta}^{w_n} (\alpha_p - \alpha_n)\, d\eta\right] d\zeta, \tag{2.151}$$

and hence breakdown is said to have occured when

$$\int_{-w_p}^{w_n} \alpha_p \exp\left[-\int_{\zeta}^{w_n} (\alpha_p - \alpha_n)\, d\eta\right] d\zeta = 1. \tag{2.152}$$

Figure 2.40 shows the avalanche breakdown voltages as a function of doping for some of the common semiconductors for an abrupt p^+–n junction using the above criterion and the ionization rates described previously in Figure 2.38 and 2.39. While generally, at the high fields that dominate avalanche process, the hole ionization coefficient dominates, this is not true for small bandgap semiconductors such as Ga$_{.47}$In$_{.53}$As and InAs. Here, hole initiated processes are less likely, so even though they are more likely

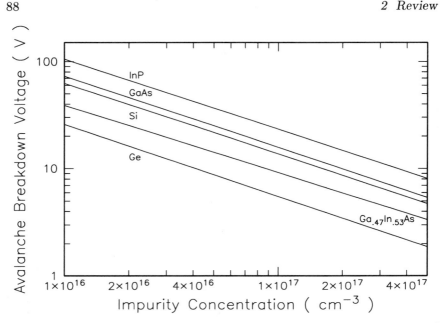

Figure 2.40: Breakdown voltages limited by avalanche process for some of the common compound and elemental semiconductors.

to avalanche by virtue of a smaller bandgap, due to suppression of hole avalanching, the decrease in breakdown voltage is not as pronounced.

Quite often, the integral on the right of the previous equations for multiplication factors is approximated by the relationship $(V/BV)^{\nu}$ where BV is the breakdown voltage, i.e.,

$$\mathcal{M} = \frac{1}{1 - (V/BV)^{\nu}}. \tag{2.153}$$

This relationship, depending on the magnitude of the power ν, shows a soft breakdown or a hard breakdown. Soft breakdown is meant to imply a slow and gradual increase in the multiplication factor with increasing bias. It occurs when the ionization rates are slowly varying functions of the electric field. A hard breakdown, with $\nu \geq 4$, leads to a rapid change in the multiplication factor and comes about when the ionization rate changes rapidly with electric field. Small bandgap materials usually have a higher magnitude of the ionization rate, and the ionization rate changes slowly with the electric field. They therefore exhibit relatively softer breakdown characteristics.

A relative comparison of breakdown voltage due to avalanche processes

Table 2.3: Bandgap and the breakdown voltage of a one-sided junction at a doping of 1×10^{17} cm^{-3} for some semiconductors.

Material	Bandgap (eV)	Breakdown Voltage (V)
InAs	0.36	2.0
Ge	0.66	5.0
Ga$_{.47}$In$_{.53}$As	0.75	6.0
Si	1.12	11.0
InP	1.35	14.5
GaAs	1.42	15.8
GaP	2.27	31.4
SiC	2.99	47.8

and its relationship with the bandgap is shown in Table 2.3. These breakdown voltages are at a donor doping of 1×10^{17} cm^{-3} in a p$^+$–n abrupt junction. This doping is quite typical of the dopings encountered in the collector region of bipolar transistors at the base-collector junction. The doping of the small bandgap semiconductors is too small to be useful at this doping or this temperature. Small bandgap semiconductors can be safely used only at low temperatures, or if the device allows a sufficiently low doping to employed to have an acceptable breakdown voltage.

The length scale over which the carriers accelerate to pick up the energy from the electric field is a few mean free path (λ). The carrier can usually continue to pick up a substantial amount of energy at these high fields because polar optical phonon scattering—the dominant scattering process— leads to only a small loss of the energy (E_p), and because it causes a small angle scattering, thus maintaining its direction of drift. We have commented that our simplified parabolic model is quite inadequate for predicting the threshold energy of impact ionization. To obtain it more accurately, the complete momentum-energy relationship and phonon effects should be included. The mean free path, the phonon energy, and the threshold energy are all important parameters in the avalanche process. Table 2.4 summarizes some approximate values of these related data. The mean free path is that for a hot carrier.

Table 2.4: Mean free path of energetic carriers, optical phonon energy, and
the threshold energy for impact ionization for some semiconductors.

Material	Mean Free Path λ (Å)	Phonon Energy E_p (eV)	Threshold Energy E_T (eV)
AlP	71.0	0.0621	3.14
AlAs	54.4	0.0501	2.80
AlSb	45.3	0.0421	2.00
GaP	32.5	0.0500	2.94
GaAs	39.3	0.0354	1.72
GaSb	48.5	0.0242	0.0848
Ge	78.0	0.037	0.67
InP	38.9	0.0428	1.69
InAs	62.9	0.0296	0.392
InSb	100.2	0.0223	0.187
Si	68.0	0.063	1.15

2.8.2   Zener Breakdown

In the context of breakdown mechanisms, it is also instructive to discuss
Zener breakdown. Zener breakdown is not really a breakdown in the traditional sense of breakdown's meaning. Zener breakdown is a form of internal
emission which takes place by tunneling. We will discuss tunneling and the
mechanism by which the Zener process occurs in the Chapter 4. Here we
describe it to show the relationship between tunneling effects and band
structure.

Figure 2.41 shows schematically the band transition involved in Zener
tunneling and breakdown. Tunneling is a manifestation of the wave nature
of electrons. If an overlap between the states in the valence band and the
conduction band is large, then electron can tunnel from the valence band
to the conduction band in the junction. This is simply to state that one
may treat the forbidden gap as a potential barrier of the classical barrier
tunneling problem.

Zener tunneling will be substantial whenever the potential barrier height
is small, the potential barrier width is short, or the effective mass is small.
Any of these allow for a large overlap of the electron and hole wave function. For a given band structure, the potential barrier height and width,
related as the electric field, are conducive to tunneling at large dopings.
Small bandgap semiconductors have intrinsically a smaller potential bar-

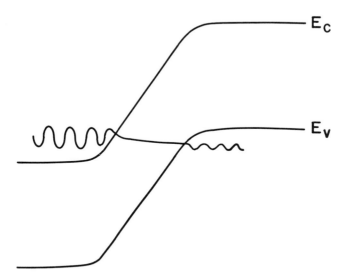

Figure 2.41: Schematic figure representing the emission of a carrier involved in Zener tunneling in p–n junction. The process results in tunneling between the conduction and the valence bands.

rier, leading to larger Zener tunneling probabilities, even at low doping. Thus, the process, has an effect on breakdown voltages for both highly doped semiconductors and small bandgap semiconductors. The breakdown voltages can be obtained by determination of the transmission function and the source function, the two characterizing the probability of tunneling and the availability of carriers for tunneling.

Like processes leading to avalanche breakdown, Zener processes may involve phonon processes. In the classical problem of tunneling through a barrier, we ignore it, i.e., we consider only energy conserving elastic processes. However, phonon assisted tunneling, an inelastic process, can be quite strong. During such a process, a longitudinal or transverse optical or acoustic phonon is emitted or absorbed in order to conserve momentum. In fact, for indirect semiconductors this is the dominant form of tunneling because it can accommodate the large differences in momenta at the bottom of conduction band and the top of valence band. Figure 2.42 shows an example of phonon assisted tunneling process in real and k-space. Tunneling, as we will see later, is a complex phenomenon. Like phonons, photons may also assist in tunneling. This phenomenon is usually referred to as the Franz–Keldysh effect.

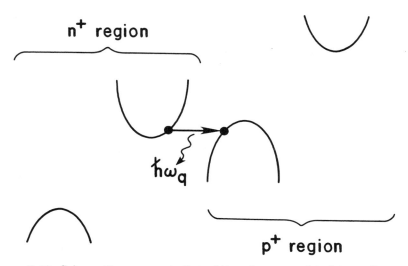

Figure 2.42: Schematic representation of the phonon-assisted tunneling process in real and k-space. This process may, e.g., occur during the Zener tunneling process discussed in the previous figure. Tunneling occurs between states with differing wave vectors and involves emission or absorption of a phonon.

2.8.3 Density of States and Related Considerations

We have discussed the role density of states plays in scattering processes. It also is intimately involved in several other ways, the most important of which is its effect on occupation of bands, and the role of degeneracy at high doping. A larger density of states implies larger carrier density before the Boltzmann approximation breaks down and the conduction electrons begin occupying higher and higher energy states in the band. Figure 2.43 shows electron density in GaAs at 300 K and 77 K as a function of the Fermi energy position with respect to the conduction band edge. This includes the changes in the density of states as a function of energy. Note that at negative Fermi energy with respect to the conduction band edge energy, the Boltzmann approximation is valid, i.e., an exponential variation occurs. However, for positive values, considerably complicated density of states effects occur.

The Fermi energy of the carriers is increased by introducing more donors in the semiconductor. Large doping implies smaller spacing between the donors. In the section on occupation statistics, we treated the donor levels as discrete and degenerate levels whose density was equal to the donor

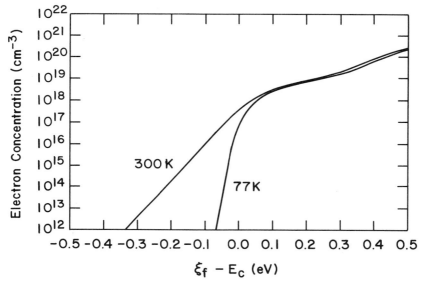

Figure 2.43: Electron concentration as a function of Fermi energy with respect to the bottom of the conduction band for GaAs at 300 K and 77 K. After M. Fischetti, "Monte Carlo Simulation of Transport in Technologically Significant Semiconductors of the Diamond and Zinc Blende Structures, Part I: Homogeneous Transport," *IEEE Trans. on Electron Devices*, **ED-38**, No. 3, p. 634, ©Mar. 1991 IEEE.

density. Close proximity of the donors causes overlap interaction between the carriers confined to them, leading to energy splitting and formation of impurity bands. One may get an estimate of the doping condition for this by determining the Bohr radius (a) in the crystal

$$a = \frac{\epsilon_s}{\epsilon_0} \frac{m_0}{m^*} a_0, \qquad (2.154)$$

where ϵ_s is the permittivity of the semiconductor, ϵ_0 is the permittivity of vacuum, and a_0 is the free space Bohr radius whose magnitude is .529 Å. For semiconductors with a large effective mass, such as Si, this occurs at larger dopings, while for semiconductors with a low effective mass this occurs at a lower doping (see Problem 6).

If the doping is continued to be increased, impurity bands coalesce with the conduction band and conduction may now take place through the impurity states. Consequently the semiconductor does not exhibit carrier freeze-out. The doping at which this transition takes place, leading to

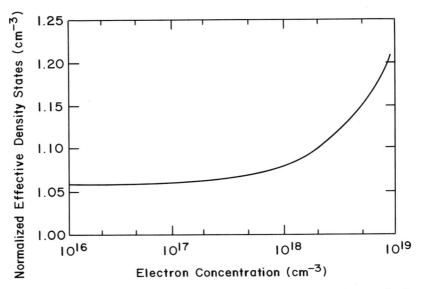

Figure 2.44: Normalized effective density of states, including heavy doping effects, as a function of electron density for GaAs at 300 K. The normalization constant is 3.99×10^{17} cm^{-3}. After J. S. Blakemore, "Semiconducting and Other Major Properties of Gallium Arsenide," *J. of Appl. Phys.*, **53**, p. R123, Oct. 1982.

metallic conduction through the impurity levels, is called the Mott transition. Similar to the degeneracy condition above, Mott transition is small in GaAs (low-10^{16} cm^{-3}, and large in Si (mid-10^{18} cm^{-3}).

These impurity level effects, related to the band structure, should be considered in the treatment of occupation statistics. Generally, though, we use the simple relationship using effective density of states, which is a delta function of the density of states at the band edge, and in our derivation we assume a parabolic band. This ad hoc treatment can be extended and we may treat the effective density of states as a function of carrier density and use the Fermi–Dirac integral. Figure 2.44 shows the variation of the effective density of states for the electrons with carrier density as a function of temperature for GaAs.

The effect of heavy doping, or the lack thereof, can be very significant in device operation at low temperatures. Mott transition and freeze-out effects are central to the design of field effect transistors and bipolar transistors in silicon for 77 K operation. Low doping in the substrate, the base, and the collector, and consequent freeze-out, can lead to large threshold voltage changes and high resistances. On the other hand, in GaAs this is not usually a consideration since freeze-out occurs at low dopings.

Increase in density of states, merging of impurity bands, etc., occurs with a decrease in effective bandgap and an increase in effective intrinsic carrier concentration. Bandgap shrinkage, e.g., in the base of the hetero-junction bipolar transistor, decreases the barrier to the conducting carrier, e.g., the electron in an n–p–n device. Devices with heavy doping in the base, thus, have a higher current density for identical bias. If this bandgap shrinkage takes place in the emitter because the emitter is higher doped (e.g., in the conventional homojunction bipolar transistor), then the injection efficiency degrades.

2.8.4 Limiting and Operational Velocities of Transport

The occupation of higher energy states in the conduction band by the electrons also results in an increase in the magnitude of the average velocity with which the individual carriers move from the thermal velocity (v_θ) to the Fermi velocity (v_F). These are really average speeds of the carriers since the average velocities, at thermal equilibrium, are zero, but the average speeds, which ignore the directional dependence, are not. But, by convention, we call them velocities. For low doping conditions, where Maxwell–Boltzmann statistics are valid, the thermal velocity may be approximately obtained from equal partition of energy, i.e.,

$$\frac{1}{2}m^*v_\theta^2 = \frac{3}{2}kT, \tag{2.155}$$

giving

$$v_\theta = \sqrt{3kT/m^*}. \tag{2.156}$$

We can also find it from the distribution of the carrier velocity from the Maxwell–Boltzmann distribution function. The average velocity is defined from the average value of the square of velocity as

$$v_\theta = (\langle v^2 \rangle)^{1/2} = \left[\frac{\int_0^\infty v^2 \exp\left(-m^*v^2/2kT\right) v^2 dv}{\int_0^\infty \exp\left(-m^*v^2/2kT\right) v^2 dv} \right]^{1/2} = \left(\frac{8kT}{\pi m^*} \right)^{1/2}, \tag{2.157}$$

slightly different from the value found from arguments based on a non-degenerate classical gas which assigns equal weight to all particles independent of the probability of occurrence.

When doping is large, so that the Fermi energy is in the conduction band, the average energy of the carriers can be larger, and they also move with larger speed. Now, we need to include the effect of the exclusion principle in determining the averages.

A consequence of Pauli's exclusion principle is that the number of independent states in a unit volume in the momentum space for a unit volume in real space is given by $1/(2\pi)^3$. Let k_F be the wave vector at the Fermi energy. Since n is the electron concentration in real space,

$$
\begin{aligned}
n &= 2\frac{1}{(2\pi)^3}\int_0^{k_F} 4\pi k^2 dk \\
  &= \frac{k_F^3}{3\pi^2},
\end{aligned}
\tag{2.158}
$$

where the first factor of 2 is from spin.[16] This gives

$$
k_F^3 = \left(\frac{m^* v_F}{\hbar}\right)^3 = 3\pi^2 n,
\tag{2.159}
$$

and hence the Fermi velocity v_F

$$
v_F = \frac{\hbar}{m^*}\left(3\pi^2 n\right)^{1/3}.
\tag{2.160}
$$

Here, again, m^* is the density of states effective mass in view of our parabolic band approximations. This velocity, called the Fermi velocity, characterizes the average speed with which the carriers move under degenerate conditions. For the example of GaAs ($m^* = 0.067m_0$ at the bottom of the conduction band), Table 2.5 summarizes the relevant velocities for two different doping levels at various temperatures.

Fermi velocity is relatively insensitive to temperature, and hence diffusion constants, etc., do not not decrease significantly with temperature, unlike in non-degenerate semiconductors. So heavy doping and lower density of states allow some of the devices to function better at lower temperatures.

Another band structure related carrier velocity effect that should be considered is the highest group velocity that the carrier can have given the band structure relationship of energy and momentum. The group velocity (v_g) can be found from the energy-momentum relationship,

$$
v_g = \frac{1}{\hbar}\frac{\partial E}{\partial k},
\tag{2.161}
$$

as discussed earlier. The maximum group velocities for some semiconductors, determined from their band structures, are given in Table 2.6.

---

[16]This is exact at $T = 0$ K. At any other temperature, it is approximate since carriers will occupy states with $k > k_F$.

Table 2.5: Thermal and Fermi velocity in GaAs at various temperatures for non-degenerate and degenerate dopings.

Temperature (K)	v_θ at $n = 1 \times 10^{16}$ cm^{-3} (cm.s^{-1})	v_F at $n = 1 \times 10^{18}$ cm^{-3} (cm.s^{-1})
300	4.5×10^7	$\approx 5.4 \times 10^7$
77	2.3×10^7	$\approx 5.4 \times 10^7$
4	5.2×10^6	$\approx 5.4 \times 10^7$

Table 2.6: Maximum group velocities in some semiconductors.

Material	Group Velocity (cm.s^{-1})
GaAs	1×10^8
Si	6×10^7
Ge	1.3×10^8
InP	9.5×10^7
InAs	1.3×10^8
AlAs	6.2×10^7

In this table, the maximum group velocity for Ge occurs in the Γ valley towards $\langle 100 \rangle$. The Γ valley is very sparsely populated—L valley is the one most occupied—and hence it is highly unlikely that structures taking advantage of this could be designed. The maximum group velocities are high in the compound semiconductors, almost twice those of silicon. AlAs is one of the exceptions in these compound semiconductors. It band structure is quite similar to that of silicon. Also, these are theoretically predicted maximum velocities, which could conceptually occur only in the absence of scattering and without inter-valley transfer, both of which will occur.

When scattering dominates, we have discussed various velocity–field behaviors for different materials. A generalized plot of maximum steady state velocities, based on simplistic theoretical arguments,[17] is shown in

---

[17]See D. K. Ferry, in W. Paul, Ed., *Handbook of Semiconductors*, Vol. 1, North-Holland, Amsterdam, p. 584, 1982.

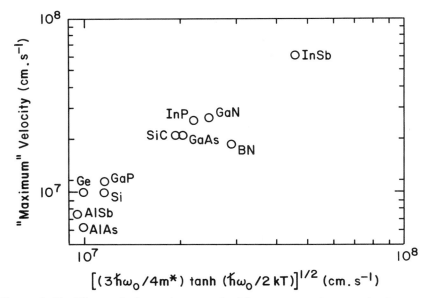

Figure 2.45: Theoretical maximum velocities, saturated or peak, for various semiconductors with respect to an energy function involving optical phonons. After D. K. Ferry, in W. Paul, Ed., *Handbook of Semiconductors*, Vol. 1, North-Holland, Amsterdam, p. 584, 1982.

Figure 2.45 for various pure semiconductors. Many of the data in this figure are experimental. The horizontal scale in this figure is a good approximation for a velocity when it is limited by non-polar optical phonons (the case with elemental semiconductors but not with compound semiconductors). Other scattering mechanisms do not change the functional form substantially and hence the approximate straight line relationship of this figure.

Many of these velocity-related features can be seen to be inversely related to the effective mass. The larger the mass, smaller the representative velocity. This is true for group velocity, thermal velocity, Fermi velocity, peak velocity for GaAs, InP, $Ga_{.47}In_{.53}As$, InAs, Si, Ge, etc. The effective mass also correlates with the bandgap in compound semiconductors. Lower electron effective mass occurs with smaller bandgaps in compound semiconductors as shown in Figure 2.46.

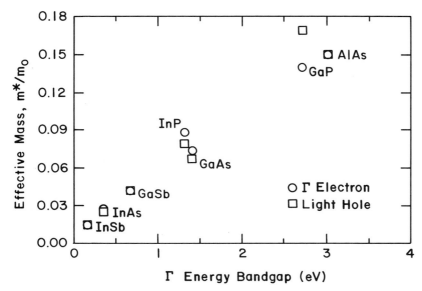

Figure 2.46: The variation of effective mass at Γ minimum and light hole band maximum as a function of the direct bandgap for some compound semiconductors. After H. L. Grubin, in H. L. Grubin, D. K. Ferry, and C. Jacobani, Eds., "The Physics of Submicron Semiconductor Devices," *NATO ASI Series*, **B180**, ©1988 Plenum Publishing Corp., New York.

2.8.5 Tunneling Effects

We have discussed tunneling, in a phenomenological way, in our discussion of Zener breakdown. Tunneling is the important example of the wave-particle duality. An estimate of the importance of wave aspect can be obtained by comparing the de Broglie wavelength with the critical device dimension. At device dimensions smaller than or comparable to the de Broglie wavelength, it is more appropriate to discuss electrons in crystals as a wave, and the classical picture that assumes negligible time scale of scattering, drift and diffusion averaged from a number of scattering events, etc., is increasingly poor. Tunneling, e.g., cannot be explained in the particle description, it requires the wave description. The de Broglie wavelength is related to the effective mass, and is shown in Figure 2.47. Smaller effective mass leads to large de Broglie wavelength and increasing wave effects, tunneling being the most prominent of them.

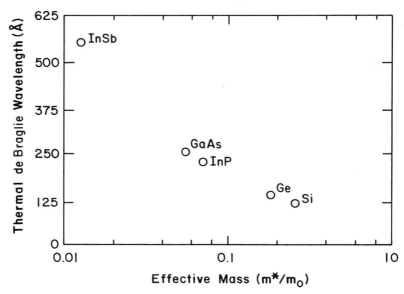

Figure 2.47: Thermal de Broglie wavelength for an electron at the bottom of the conduction band for various semiconductors. After H. L. Grubin, in H. L. Grubin, D. K. Ferry, and C. Jacobani, Eds., "The Physics of Submicron Semiconductor Devices," *NATO ASI Series*, **B180**, ©1988 Plenum Publishing Corp., New York.

2.9 Summary

In this chapter we have reviewed the underlying concepts of quantum mechanics and considered their implications for semiconductor physics by considering one-dimensional models. The ideas include the wave–particle duality, the use of plane waves in describing wave packets representing localized particles, and the derivation of quantized states in energy and momentum for carriers or phonons caused by confinement or periodic boundary conditions. We showed that a large crystal, and therefore confinement in a large box, leads to quasi-continuum of these states which has a finite density of states. The electrons in a crystal of periodic potential can, at their simplest, be modelled as having a differing effective mass. The periodicity, a condition where Bloch's theorem and the concept of reciprocal lattice are particularly useful, leads to the formation of forbidden bands, called the bandgap, and multiple bands; the valence bands are filled and the conduction bands are empty at absolute zero. At finite temperature valence bands are partially empty, and conduction of electrons in valence bands can be modelled by "particles" that we call holes, while conduction in the conduction band can be modelled by electrons. Electrons and holes obey Fermi–Dirac statistics, while phonons, the modes of lattice vibrations, obey Bose-Einstein statistics.

Subsequently, we analyzed the consequences of perturbation from our idealized behavior that leads to scattering. Using the band structure and phonon dispersion characteristics of the three-dimensional crystals of interest to us, we discussed some of the band to band processes involving electrons, holes, and phonons, and studied the general consequences of the behavior of these materials to the operation of some of the semiconductor devices of interest. The transport of carriers in the three-dimensional lattice with the constraints of band structure and scattering processes was considered in a general way to understand the behavior of carriers at low energies and at high energies. We also reviewed the behavior of this transport for minority carriers where a gradient in carrier concentration can lead to large diffusive effects, and compared the behavior of minority carriers with majority carriers. Finally, we reviewed effects that are intimately related to band structure: impact ionization and breakdown, limiting velocities, and tunneling.

General References

1. B. K. Ridley, *Quantum Processes in Semiconductors*, Clarendon Press, Oxford (1988).

2. B. R. Nag, *Electron Transport in Compound Semiconductors*, Springer-Verlag, Berlin (1980).

3. C. M. Wolfe, N. Holonyak Jr., and G. E. Stillman, *Physical Properties of Semiconductors*, Prentice-Hall, Englewood Cliffs, N.J. (1989).

4. K. Seeger, *Semiconductor Physics: An Introduction*, Springer-Verlag, Berlin (1985).

5. N. W. Ashcroft and N. D. Mermin, *Solid State Physics*, Holt, Rinehart and Winston, Philadelphia, PA (1976).

6. C. Kittel, *Introduction to Solid State Physics*, John Wiley, N.Y., (1968).

7. W. A. Harrison, *Solid State Theory*, Dover, N.Y. (1979).

8. J. S. Blakemore, *Solid State Physics*, Cambridge University Press, Cambridge (1985).

9. J. S. Blakemore, "Semiconducting and Other Major Properties of Gallium Arsenide," *J. of Appl. Phys.*, **53**, No. 10, Oct. R123 (1982).

10. I. M. Tsidilkovsky, *Band Structure of Semiconductors*, Pergamon Press, Oxford (1982).

11. M. Jaros, "Electronic Properties of Semiconductor Alloy Systems," *Rep. Prog. Phys.*, **48**, p. 1091, 1985.

12. F. Capasso, "Physics of Avalanche Photodiodes," in W. T. Tsang, Ed., *Semiconductors and Semimetals*, **22**, Part D, Academic Press, San Diego, CA (1985).

Problems

1. Consider a system of independent and distinguishable particles. Show that for this system, the ratio of the number of occupied particles in the state j to its degeneracy is the Maxwell–Boltzmann distribution function given by

$$f(E) = \frac{n_j}{g_j} = \exp\left(-\frac{E_j}{kT}\right) = \exp\left(-\frac{E - \xi_f}{kT}\right). \qquad (2.162)$$

2. Show that, as in the case of electrons, the hole occupation following Fermi–Dirac statistics leads to

$$p = N_V F_{1/2}\left(\eta_{fv}\right), \qquad (2.163)$$

where $\eta_{fv} = (E_v - \xi_f)/kT$. For GaAs, draw the electron and hole density, at 300 K, for the functions η_{fc} and η_{fv} within $5kT$ of the conduction and valence band.

3. For the most common type of acceptors in semiconductors, show that the degeneracy is 4, and also show that

$$f(E_A) = \frac{N_A^-}{N_A} = \frac{1}{1 + 4\exp\left[(E_A - \xi_f)/kT\right]}, \qquad (2.164)$$

and

$$\frac{N_A^0}{N_A} = \frac{1}{1 + \frac{1}{4}\exp\left[(\xi_f - E_A)/kT\right]}. \qquad (2.165)$$

4. Show that when holes initiate and dominate the avalanche process, the avalanche integral of interest is related by

$$1 - \frac{1}{M_p} = \int_{-w_p}^{w_n} \alpha_p \exp\left[-\int_{\zeta}^{w_n} (\alpha_p - \alpha_n)\, d\eta\right] d\zeta, \qquad (2.166)$$

following arguments based on hole generation.

5. For bipolar transport, i.e., with both electrons and holes contributing to conduction, show that the Hall coefficient R_H is related to the carrier densities and drift mobilities by

$$R_H = \frac{r}{q}\left[\frac{p\mu_{dp}^2 - n\mu_{dn}^2}{(n\mu_{dn} + p\mu_{dp})^2}\right]. \qquad (2.167)$$

6. Estimate the Bohr radius, and the impurity density corresponding to an average distance between impurities equal to the Bohr radius, for n-type GaAs, Si, Ge, InP, $Ga_{.47}In_{.53}As$, InAs, GaSb, and InAs. The Bohr radius also gives an indication of the binding energy of the donor, assuming a hydrogenic model for binding of the electron to the donor. Estimate this for these semiconductors, and compare with the observed energies.

7. What is the shape of the constant-energy surface for a semiconductor whose conduction band edge energy changes as the fourth power of the reduced wave vector? Also find the density of states with energy and the effective mass with energy assuming that the coefficient of the E–k relationship is α.

8. What is the relationship, in the previous problem, between the carrier concentration and the Fermi energy?

9. Plot the Maxwell–Boltzmann distribution function and Fermi–Dirac distribution function (assuming $\xi_f - E_c = 3kT$) at 300 K and 77 K as a function of the square of the velocity of carriers assuming $m^* = 0.067m_0$ for the effective mass of GaAs. What is the width in energy at half of the maximum for the two distributions? At what velocity do the distributions drop to 10% of the magnitude at zero velocity?

10. We will consider, simplistically, how one may obtain the phonon behavior in a three-dimensional lattice such as that of GaAs—a zinc blende lattice. To keep the problem tractable, we will consider the oscillations in ⟨111⟩ direction for (111) planes. Alternating planes

contain atoms of Ga and As. If a is the linear dimension of the unit cell, the planes alternate with a spacing of $a/4$ and $a/12$. One bond to the atoms extends perpendicular to these planes and three bonds extend at an angle of 70.53 degrees in the opposite half. The resulting strength of either is the same.

(a) Derive and plot the longitudinal energy-momentum relationship by considering the differential equation that governs the displacement and the force for the two species of atoms.

(b) Now consider the same problem for a diamond lattice, i.e., with only one type of atom. How is this result different from the previous? What is its significance?

11. Why are there differences in the variation with wave vector, including a reversal of order, for the energies of the LO and TO phonons?

12. For a semiconductor with isotropic spherical constant energy surfaces in conduction, and two valence bands, find the Hall coefficient. Let τ_n, τ_{p1}, and τ_{p2} be the momentum relaxation times for the carrier whose densities are n, p_1, and p_2 in the respective bands.

Chapter 3

Mathematical Treatments

3.1 Introduction

In using semiconductor devices, our principal interest is in knowing the behavior of currents and voltages at the device terminals due to an applied stimulus, a voltage, or a current, as a function of time. To obtain this knowledge, we need to describe the behavior of the carriers in the device. We need information related to how the carriers move and how they stay stationary. The ensemble of carriers moves in most semiconductor devices as a result of an electric field (we are not considering any magnetic field-based devices in this text), and it stays stationary because there exists no net external force on the ensemble. A stationary ensemble has carriers moving randomly as a result of scattering, a Brownian motion, while a moving ensemble exhibits a net response to the applied force. Both of these are important in studying device behavior. Another important aspect, interrelated with these in the device, is the effect of a rapidly changing electric field during a time transient.

The current in the device, at any cross-section, results from moving charged particles (the particle current) and from changing displacement field (the displacement current, $d(\epsilon_s \mathcal{E})/dt$). These currents have to be related to the applied currents or voltages. We may also use different levels of sophistication in modelling the different regions of the device, depending on how critical they are to the overall device behavior, and the level of accuracy and hence sophistication that is acceptable. These sections of the device, their interfacing with each other, and their interfacing with the device terminals (the contacts that we access) and device surfaces where there are no contacts, have to be modelled accurately to reflect the overall behavior.

So, to analyze the device, our mathematical models must include behavior of carriers in semiconductors, boundary conditions between semiconductor sections, time-dependent and time-independent behavior, and carrier behavior at semiconductor contacts and surfaces. Our aim in this chapter is to review and discuss some of the more important general approaches for the analysis so that we may develop the tools necessary to understand the operation of devices.

In trying to describe the behavior of transport in semiconductors, the principal problem to be tackled is that of determining current in the presence of the electric field that moves the carriers, and scattering that limits the unlimited increase in momentum and energy of the particles as a result of the applied field.

The most common formulations for describing this behavior are classical. They treat electrons and holes as particles, of certain position, momentum, and energy. The energy that these carrier can possess can be treated as being continuous, except in the bandgap, because the energy level separations are much less than kT for the macroscopic sample sizes employed. We call this methodology the kinetic approach since it uses the kinetics of an average particle with a simple treatment of collisions (scattering) to describe its equation of motion. The approach is quite successful in describing macroscopic properties of the semiconductors in which off-equilibrium effects are not important.

A more common approach than the kinetic approach is the use of the Boltzmann Transport Equation (BTE) for describing carrier kinetics. The formulation considers the equation of motion for the distribution function of the particles $f(\boldsymbol{r}, \boldsymbol{p}; t)$. This distribution function f is the probability of finding a particle with the momentum $\boldsymbol{p}$ at the position $\boldsymbol{r}$ at the time t. We will discuss this and simplifications resulting from it (drift-diffusion follows from the the BTE following many approximations) since they have become increasingly popular in describing the hot carrier phenomenon. One additional advantage of BTE is that it leads to a methodology which is not very computationally intensive.

Instead of considering the distribution function, we may consider individual particles themselves, selected at random, and collect enough information about them to determine the average ensemble behavior. So, one follows individual particles under the influence of the governing equations and allows them to scatter with probability of scattering determined by the randomness and the relative frequency of occurrence of the specific scattering event. By calculating over several scattering events and/or over behavior of several particles (usually many thousands), one can determine the statistical average behavior of the transport. The Monte Carlo approach is not restricted to many of the assumptions inherent in the simplified BTE approaches, such as simplified treatments of the band structure and distribution functions, in order to obtain the solution. However, these advantages may be offset by a lack of information or simplifications in implementations. An example of the former are the impact ionization processes, whose microscopic features and hence means of implementation are still open to debate. An example of the latter are coupling coefficients for scattering where not enough may be known of the underlying characteristics of the semiconductor itself. The Monte Carlo approach, in spite of these inadequacies, has been highly successful. The reason for this has been that the simplifying assumptions of BTE become questionable well before the inadequacies of a well-implemented Monte Carlo code become suspect. Examples of these successes include description of high field transport and transport involving situations where simplified distribution functions do not exist, such as in very short dimension structures.

The next level in sophistication has been broadly covered in the Quantum Transport approach, which describes behavior in situations where the quantum nature of states and kinetics is included. This becomes necessary in problems where spatial and temporal extents are similar to the de Broglie wavelength and the duration of scattering events, etc. The kinetic, BTE, and Monte Carlo approaches now become unacceptably inaccurate. Examples of problems where this is true are those involving size quantization, e.g., calculations related to tunneling in coupled barrier structures where the momentum is quantized along one direction. A common Quantum Transport approach is the density matrix approach based on the Liouville equation. The Liouville equation employs the statistical density operator or the density matrix, representing the ensemble of the states of the system, together with the Hamiltonian to describe the state of the system. This approach has shown some success in modelling the coupled barrier problem. All variations of the quantum transport approaches are very complex, and are presently needed in only some of the devices of interest. Hence, in the following few sections, we will restrict ourselves to the discussion of the kinetic approach, the BTE approach, and the Monte Carlo approach. Our objective is to appreciate the methodology, to understand the limits of these approaches, and to relate these to the physical meaning of the results derived.

3.2 Kinetic Approach

The simplest of the classical formulations is the particle kinetic approach, which models the kinetics of the motion of an average particle in the presence of a force with a simple treatment of collisions. It provides useful insights and is quite intuitive for many of the standard problems of semiconductor transport. To describe the behavior of the system we need the equation of motion and the relationship between this motion and the current. These, following classical mechanics, can be written as the following for an applied electric field ($\mathcal{E}$), magnetic field ($\mathcal{B}$), and current density ($\boldsymbol{J}$):

$$\frac{d\boldsymbol{p}}{dt} = q\left(\mathcal{E} + \boldsymbol{v} \times \mathcal{B}\right) - \frac{\boldsymbol{p}}{\tau},$$
$$\boldsymbol{v} = g(\boldsymbol{p}),$$
$$\text{and} \quad \boldsymbol{J} = nq\langle\boldsymbol{v}\rangle, \tag{3.1}$$

where q is the charge of the carriers, n is the carrier density, and $\langle\boldsymbol{v}\rangle$ the average velocity of the carriers. Here, τ is a time constant representing a linear damping term representing the loss of momentum due to scattering. In this set of equations, an ad hoc model is employed to describe the effect of applied fields with an energy dependent time constant τ, but the quantum nature of the problem is accounted for by an effective mass in the functional relationship of the second equation. Quite often groups of particles may be encountered. Some examples of this are different charge of particles such as in ambipolar transport consisting of both electrons and holes, and two valley transport such as with light holes and heavy holes in p-type material or electrons in two valleys under high field conditions. In these situations, the groups of particles may be represented by suitable modifications of the previous equations to a set of equations representing each group of particles.

Let us now consider some simple examples of application of the kinetic approach, as well as of its limitations. In its simplest form, the semiconductor band structure can be considered parabolic, leading to a constant effective mass; and particles can be considered non-interacting at low concentrations, leading to a Maxwell–Boltzmann distribution of carriers with most particles at the bottom of the band. This leads to a closed-form solution which is useful in the calculation of average properties at low concentrations and negligible fields. For this semiconductor with the spherical constant energy surface, assuming an energy independent scattering, and only electrons as the carriers, Equation 3.1 reduces to

$$\frac{d\boldsymbol{p}}{dt} = q\left(\mathcal{E} + \boldsymbol{v} \times \mathcal{B}\right) - \frac{\boldsymbol{p}}{\tau},$$
$$\boldsymbol{v} = \frac{\boldsymbol{p}}{m^*},$$
$$\text{and} \quad \boldsymbol{J} = nq\boldsymbol{v}, \tag{3.2}$$

giving

$$\tau\frac{d\boldsymbol{J}}{dt} + \boldsymbol{J} - \mu\boldsymbol{J} \times \mathcal{B} = nq\mu\mathcal{E}, \tag{3.3}$$

where

$$\mu = \frac{q\tau}{m^*} \tag{3.4}$$

is the mobility of the particle. In our discussion of drift mobility in Chapter 2, this relationship was used implicitly. It shows the approximations related to band structure, and the degeneracy involved. There is an additional approximation of steady-state implicit in this too. We can estimate the time constant—consider GaAs with a mobility of 4000 cm^2.V^{-1}.s^{-1}, typical for a donor density of 1×10^{17} cm^{-3} at 300 K. The time constant is 0.15 ps. Transient effects related to the relaxation of momentum take place during a time period extending over many time constants. This is not included in the description, and hence may not be expected to be reproduced. Thus, the set of equations should be used for time-varying parameters whose time variation is parameterized at higher than this time constant.

Let us consider another example, the response to a sinusoidally varying field with radial frequency ω in an isotropic material with the field along the coordinate axis. The electric field has a magnitude which is the real part of

$$\mathcal{E} = \hat{\mathcal{E}} \exp{(j\omega t)}, \tag{3.5}$$

and whose response is a particle current,

$$J = \hat{J} \exp{(j\omega t)}. \tag{3.6}$$

We will discuss the methodology of small-signal analysis later in this chapter. The phasor $\hat{J}$ follows from the earlier analysis as

$$\hat{J} = \frac{nq\mu}{1 + \omega^2\tau^2}\hat{\mathcal{E}} - j\frac{nq\mu\omega\tau}{1 + \omega^2\tau^2}\hat{\mathcal{E}}. \tag{3.7}$$

The term $nq\mu$ is the static conductivity of the material $\bar{\sigma}$. From the Maxwell equations,[1] we can derive the wave equation as

$$\nabla^2\mathcal{E} = -\frac{\epsilon_s\omega^2}{\epsilon_0 c^2}\mathcal{E} + \frac{\omega}{\epsilon_0 c^2}J, \tag{3.8}$$

where c is the speed of light in free space, and hence $1/\epsilon_0 c^2$ is the permeability of free space.

Since we are looking at the frequency-dependent response, whose time period is still larger than the relaxation time, we use the frequency dependence of the semiconductor permittivity ϵ_s:

$$\epsilon_s = \epsilon_{s1} - j\omega\epsilon_{s2}, \tag{3.9}$$

---

[1]Discussed in Chapter 9.

and obtain

$$\nabla^2 \hat{\mathcal{E}} = \frac{1}{\epsilon_0 c^2} \left[-\omega^2 \left(\epsilon_{s1} - \frac{\tau \overline{\sigma}}{1 + \omega^2 \tau^2} \right) + j\omega \left(\omega \epsilon_{s2} + \frac{\overline{\sigma}}{1 + \omega^2 \tau^2} \right) \right] \hat{\mathcal{E}}. \quad (3.10)$$

The real part of this response, the response to a sinusoidal signal, goes through a phase change when the effective permittivity

$$\epsilon_{eff} = \epsilon_{s1} - \frac{\tau \overline{\sigma}}{1 + \omega^2 \tau^2} = 0. \quad (3.11)$$

The effective permittivity is dependent on the carrier density and the low frequency permittivity, which can be approximated to ϵ_{s1}. We also introduce a frequency parameter parameter ω_p, the plasma radial frequency; the interrelationships of these are:

$$\epsilon_{eff} = \epsilon_{s1} - \frac{q n \mu \tau}{1 + \omega^2 \tau^2} = \epsilon_{s1} \left(1 - \frac{\omega_p^2 \tau^2}{1 + \omega^2 \tau^2} \right), \quad (3.12)$$

and

$$\omega_p^2 = \frac{1}{\tau \tau_d}, \quad (3.13)$$

where τ_d is the dielectric relaxation time

$$\tau_d = \frac{\epsilon_{s1}}{\overline{\sigma}}. \quad (3.14)$$

The significance of the real part of the wave equation going negative is that the sinusoidal signal is attenuated, and the frequency at which this occurs is defined by

$$\omega = \left(\omega_p^2 - \frac{1}{\tau^2} \right)^{1/2} = \left(\frac{1}{\tau \tau_d} - \frac{1}{\tau^2} \right)^{1/2}. \quad (3.15)$$

For our example of GaAs, of mobility 4000 $cm^2.V^{-1}.s^{-1}$, donor density of 1×10^{17} cm^{-3}, time constant 0.15 ps, and $\epsilon_{s1} = 12.9\epsilon_0$, the plasma frequency $\omega_p/2\pi$ is 3.1 THz, dielectric relaxation time τ_d is 1.79 fs, and the frequency at which this reversal occurs is 9.7 THz.

We can extend the mobility relationship from the case of a single carrier and single time constant to that of two carriers and two time constants (see Problem 1), and show a similar relationship for current for the ambipolar situation (see Problem 2). We can also extend it to multiple time constants in the same general way. Here we give two examples of the calculation of the time constants for electrons, by relating them to the scattering mechanisms for a density of states $g(E)$ in the conduction band and an occupation probability related by the distribution function f. For the average of any transport parameter $\varphi(E)$, following translation of the energy axis to the bottom of the conduction band, changing the primary variable from velocity to energy, and using approximations similar to that employed for occupation statistics in Chapter 2, we obtain (see Problem 3),

$$\langle \varphi \rangle = -\frac{2 \int_0^\infty \varphi(E) g(E) \left(\partial f(E)/\partial E \right) E \, dE}{3 \int_0^\infty g(E) f(E) dE}. \tag{3.16}$$

For parabolic band density of states, the density of states has a square-root dependence on energy, and for non-degenerate conditions the Maxwell–Boltzmann relationship gives an exponential energy dependence for the distribution function, hence,

$$\begin{aligned} \langle \varphi \rangle &= \frac{2 \int_0^\infty \varphi(E) E^{3/2} \exp\left(-E/kT\right) dE}{3kT \int_0^\infty E^{1/2} \exp\left(-E/kT\right) dE} \\ &= \frac{\int_0^\infty \varphi(E) E^{3/2} \exp\left(-E/kT\right) dE}{\int_0^\infty E^{3/2} \exp\left(-E/kT\right) dE}. \end{aligned} \tag{3.17}$$

If the variable $\varphi(E)$ has a power law dependence on the energy E, i.e., $\varphi(E) = AE^\nu$, then the integral can be related through the Γ functions

$$\langle \varphi \rangle = A(kT)^\nu \frac{\Gamma(\nu + 2.5)}{\Gamma(2.5)}, \tag{3.18}$$

where the Γ functions satisfy the properties described in our discussion of the Fermi integrals.

The average for relaxation time for the mobility, with a momentum relaxation time τ_p, is

$$\tau_\mu = \langle \tau_p \rangle = -\frac{2 \int_0^\infty \tau_p(E) g(E) \left(\partial f(E)/\partial E \right) E \, dE}{3 \int_0^\infty g(E) f(E) dE} = \frac{\langle v^2 \tau \rangle}{\langle v^2 \rangle}. \tag{3.19}$$

Here, the integral was transformed over velocity instead of energy in order to express it in the previously expressed form of Chapter 2. For speed, the relationship of the variable is $\varphi = v = (2/m^*)^{1/2} E^{1/2}$, and hence

$$v_\theta = \left(\frac{2}{m^*}\right)^{1/2} (kT)^{1/2} \frac{\Gamma(3)}{\Gamma(2.5)} = \left(\frac{2kT}{m^*}\right)^{1/2} \frac{2 \times 1}{1.5 \times \pi^{1/2}} = \left(\frac{8kT}{\pi m^*}\right)^{1/2}. \tag{3.20}$$

For scattering, by ionized impurity scattering, the Conwell-Weisskopf relationship is

$$\tau_{II} = AE^{3/2}, \tag{3.21}$$

where A is a coupling constant. Hence, the momentum relaxation time for ionized impurity scattering follows as

$$\tau_\mu = A(kT)^{3/2} \frac{\Gamma(4)}{\Gamma(2.5)} = 4A \frac{(kT)^{3/2}}{\pi^{1/2}}. \tag{3.22}$$

For acoustic scattering,

$$\tau_{ac} = BE^{-1/2}, \tag{3.23}$$

and hence the momentum relaxation time is given by

$$\tau_\mu = B(kT)^{-1/2} \frac{\Gamma(1.5)}{\Gamma(2.5)} = \frac{2B}{3(kT)^{1/2}}. \tag{3.24}$$

Now let us consider the relaxation time of interest in Hall effect measurements. When a magnetic field $\mathcal{B}$ is present, the current, in steady-state may be derived as (see Problem 4)

$$
\begin{aligned}
\mathbf{J} = & \left\langle \frac{q^2 n\tau}{m^* (1 + \omega_c^2 \tau^2)} \right\rangle \mathcal{E} - \left\langle \frac{q^3 n\tau^2}{m^{*2} (1 + \omega_c^2 \tau^2)} \right\rangle \mathcal{E} \times \mathcal{B} + \\
& \left\langle \frac{q^4 n\tau^3}{m^{*3} (1 + \omega_c^2 \tau^2)} \right\rangle (\mathcal{E}.\mathcal{B})\, \mathcal{B},
\end{aligned}
\tag{3.25}
$$

where $\omega_c = q\mathcal{B}/m^*$ is the cyclotron resonance frequency. If the magnetic field is small, then

$$\mathbf{J} = \left\langle \frac{q^2 n\tau}{m^*} \right\rangle \mathcal{E} - \left\langle \frac{q^3 n\tau^2}{m^{*2}} \right\rangle \mathcal{E} \times \mathcal{B}. \tag{3.26}$$

Hall effect measurements involve, as discussed before, a magnetic field orthogonal to the plane of the semiconductor, i.e., orthogonal to the electric field. Let z be the orthogonal direction of the magnetic field, and let x be the direction of the applied field. In steady-state, no current flows perpendicular to this either in the y direction or the z direction. Then,

$$
\begin{aligned}
J_x &= \left\langle \frac{q^2 n\tau}{m^*} \right\rangle \mathcal{E}_x - \left\langle \frac{q^3 n\tau^2}{m^{*2}} \right\rangle \mathcal{E}_y \mathcal{B}_z, \\
J_y &= \left\langle \frac{q^2 n\tau}{m^*} \right\rangle \mathcal{E}_y + \left\langle \frac{q^3 n\tau^2}{m^{*2}} \right\rangle \mathcal{E}_x \mathcal{B}_z = 0,
\end{aligned}
$$

and
$$J_z = \left\langle \frac{q^2 n\tau}{m^*} \right\rangle \mathcal{E}_z = 0. \tag{3.27}$$

The current flow and the the in-plane orthogonal electric fields, ignoring second order terms, are related by

$$\mathcal{E}_y = -\frac{qB_z}{m^*}\frac{\langle\tau^2\rangle}{\langle\tau\rangle}\mathcal{E}_x$$

and
$$J_x = -\frac{qn}{B_z}\frac{\langle\tau\rangle^2}{\langle\tau^2\rangle}\mathcal{E}_y. \qquad (3.28)$$

This shows that the Hall constant and Hall factors that we applied earlier are related as:

$$R_H = \frac{\mathcal{E}_y}{J_xB_z} = -\frac{1}{qn}\frac{\langle\tau^2\rangle}{\langle\tau\rangle^2},$$

and
$$r = \frac{\langle\tau^2\rangle}{\langle\tau\rangle^2}\mathcal{E}_y. \qquad (3.29)$$

We can now show that the Hall pre-factor r, when acoustic deformation potential scattering dominates (see Problem 5), is

$$r = \frac{3\pi}{8}, \qquad (3.30)$$

and that the Hall mobility is associated with the Hall relaxation time

$$\tau_H = \frac{\langle\tau^2\rangle}{\langle\tau\rangle}. \qquad (3.31)$$

This treatment can thus be extended over several complicated situations, all leading to complicated mathematical relationships even with the assumption of the Maxwell–Boltzmann relationship and explicit energy dependence. Many of the restrictions are violated in practical situations, e.g., degenerate semiconductors, presence of hot carriers, etc. Indeed, the problem rapidly becomes intractable and more sophisticated approaches have to be utilized.

3.3 Boltzmann Transport Approach

In the kinetic approach, we introduced classical kinetics to describe the motion of a group of particles with common characteristics described by the relaxation time, charge, and the relationship between momentum and velocity. The BTE[2] describes the kinetics using the equation of motion for the distribution function of the particles $f(\boldsymbol{r}, \boldsymbol{p}; t)$. The distribution function f is the probability of finding a particle with a crystal momentum $\boldsymbol{p}$ at a position $\boldsymbol{r}$ at a time t. In thermal equilibrium, the probability of finding a carrier in a state with energy $E(\boldsymbol{p})$ is given by the Fermi–Dirac distribution function f_0

$$f_0\left[E\left(\boldsymbol{p}\right)\right] = \frac{1}{1 + \exp\left\{\left[E\left(\boldsymbol{p}\right) - \xi_f\right]/kT\right\}}, \tag{3.32}$$

where ξ_f is the Fermi energy. When the energy of the electrons is larger than the Fermi energy by a few kTs, the distribution function can be approximated by

$$f_0\left[E\left(\boldsymbol{p}\right)\right] \approx \exp\left\{\left[\xi_f - E\left(\boldsymbol{p}\right)\right]/kT\right\}, \tag{3.33}$$

the Maxwell–Boltzmann distribution function, or Boltzmann distribution function in short. For heavy doping, however, the Fermi–Dirac distribution function should be considered.

The effect of the band structure on the distribution function is included in the relationship between energy and momentum. For a spherical constant energy surface with parabolic bands whose minimum is at the zone center, this relationship takes the form

$$E = \frac{p^2}{2m^*}, \tag{3.34}$$

while for more complex forms of parabolic bands, with different energy-momentum relationships in different directions, it still has the square dependence on the momentum but in a more complex form,

$$E = \frac{p_x^2}{2m_x^*} + \frac{p_y^2}{2m_y^*} + \frac{p_z^2}{2m_z^*}. \tag{3.35}$$

The constant energy surface is now an ellipsoid. The momentum $\boldsymbol{p}$ has an equivalence with the crystal wave vector $\boldsymbol{k}$ through the reduced Planck's constant. In the rest of our discussion of the BTE, we will use the momentum as $\boldsymbol{p}$, although using $\boldsymbol{k}$ is just as common and is even employed elsewhere in this book. Degenerate minima, such as for the light hole and heavy hole valence bands, may be included by using their energy–momentum relationship and their degeneracy. Close to thermal equilibrium, for many problems, it may be sufficient to treat these as two independent constant

---

[2]For an extended discussion of the BTE and its application in semiconductor transport theory, see E. M. Conwell, "Transport: The Boltzmann Equation," in W. Paul, Ed., *Handbook on Semiconductors*, Vol. 1, North-Holland, Amsterdam (1982).

energy surfaces of equal energy (e.g., in unstrained materials spheres at very low energies).

The force $\mathcal{F}$ on the particle in an electric field and magnetic field causes a change in drift momentum related by

$$\mathcal{F} = \frac{d\boldsymbol{p}}{dt} = q(\boldsymbol{\mathcal{E}} + \boldsymbol{v} \times \boldsymbol{\mathcal{B}}), \tag{3.36}$$

between scattering events.

Let the distribution in real and phase space at any instant of time under general conditions be given by $f(\boldsymbol{r}, \boldsymbol{p}; t)$. For the changes that occur between time instants t and $t + \Delta t$, consider the effect of drift. The carriers which had the momentum $\boldsymbol{p} - \Delta t d\boldsymbol{p}/dt$ Δt time interval earlier, drift to have the momentum $\boldsymbol{p}$ after the elapsing of time Δt. Therefore, the effect of drift in time interval Δt is to cause a change in the distribution function of $f(\boldsymbol{r}, \boldsymbol{p} - \Delta t d\boldsymbol{p}/dt) - f(\boldsymbol{r}, \boldsymbol{p})$. Hence, the rate of change of the distribution function with time due to drift is given by

$$\left. \frac{\partial f}{\partial t} \right|_{drift} \Delta t = - \left. \frac{\partial f}{\partial \boldsymbol{p}} \right|_{\boldsymbol{r}, t} \cdot \frac{d\boldsymbol{p}}{dt} \Delta t, \tag{3.37}$$

i.e.,

$$\left. \frac{\partial f}{\partial t} \right|_{drift} = -\dot{\boldsymbol{p}} \cdot \boldsymbol{\nabla}_p f. \tag{3.38}$$

The rate of change due to diffusion, e.g., resulting from concentration or temperature gradients, can be found similarly. The carriers which were at the position $\boldsymbol{r} - \Delta t d\boldsymbol{r}/dt$ Δt time interval earlier, diffuse to the position $\boldsymbol{r}$ after the elapsing of time Δt. Therefore, the effect of diffusion in time interval Δt is to cause a change in the distribution function of $f(\boldsymbol{r} - \Delta t d\boldsymbol{r}/dt, \boldsymbol{p}) - f(\boldsymbol{r}, \boldsymbol{p})$. Hence, in a similar manner as above, the rate of change of the distribution function with time due to diffusion is given by

$$\left. \frac{\partial f}{\partial t} \right|_{diff} = -\boldsymbol{v} \cdot \boldsymbol{\nabla}_r f. \tag{3.39}$$

These are the changes in the distribution function due to drift and diffusion. Changes in the distribution function also occur due to scattering. Carriers may be scattered into and out of the elemental volume at $(\boldsymbol{r}, \boldsymbol{p}; t)$. Treating these as independent processes, the total rate of change in the distribution function f can be written as

$$\begin{aligned} \frac{\partial f}{\partial t} &= \left. \frac{\partial f}{\partial t} \right|_{drift} + \left. \frac{\partial f}{\partial t} \right|_{diff} + \left. \frac{\partial f}{\partial t} \right|_{scatt} \\ &= -\dot{\boldsymbol{p}} \cdot \boldsymbol{\nabla}_p f - \boldsymbol{v} \cdot \boldsymbol{\nabla}_r f + \left. \frac{\partial f}{\partial t} \right|_{scatt}. \end{aligned} \tag{3.40}$$

This long argument establishing the change of distribution function in the six-dimensional (r, p) space can be derived directly by application of Liouville's theorem for semi-classical equation of motion. A consequence of this theorem is the conservation of (r, p)-space volume.[3] Thus,

$$\left.\frac{\partial f}{\partial t}\right|_{scatt} = \lim_{\Delta r, \Delta p; \Delta t \to 0} \frac{f(r + \Delta r, p + \Delta p; t + \Delta t) - f(r, p; t)}{\Delta t}$$

$$= \dot{p}.\nabla_p f + v.\nabla_r f + \frac{\partial f}{\partial t}, \tag{3.41}$$

an equivalent result.

The BTE incorporates effects taking place in time. Thus, should it be possible to make appropriate analytic expressions, the BTE will describe the time transient effects too if the scattering relationship can be characterized. The expression is also valid for steady-state; the distribution function does not change in steady-state, giving

$$\left.\frac{\partial f}{\partial t}\right|_{scatt} = q(\mathcal{E} + v \times \mathcal{B}).\nabla_p f + v.\nabla_r f. \tag{3.42}$$

Both of the above are convenient forms for the BTE. For multiple bands (e.g., two valleys) we have two distribution functions and two equations as a generalization of the one band treatment. The same is true for degenerate bands such as those due to holes. The scattering term in these cases must include effects on the distribution function due to scattering events taking place within the band, into the band, and out of the band. A similar generalization procedure could be used for more valleys.

3.3.1   Relaxation Time Approximation

One of the most important parameters to be introduced as part of the discussion on scattering is the relaxation time. In the kinetic approach, we introduced it in an ad hoc manner to incorporate the loss term in the momentum. In our following discussion, we will establish it on a firmer basis, and also establish limits of the validity of using a relaxation time. It is a very justifiable time constant for certain scattering processes, where it can be established as a single valued function of momentum or energy. However, scattering processes do exist where this may be unjustified, e.g., in polar optical scattering. In such cases it is an engineering approximation, or a fitting parameter, used to allow simplification of calculation using the

---

[3]For arguments leading to this, see N. W. Ashcroft and N. D. Mermin, *Solid State Physics*, Holt, Rinehart and Winston, Philadelphia, PA (1976).

distribution function approach. In interpreting results of such a calculation, these approximations should be considered in understanding the limits of validity.

Let $\mathcal{S}(\boldsymbol{p}, \boldsymbol{p}')$ be the probability per unit time of a transition from an initial state $\boldsymbol{p}$ to a final state $\boldsymbol{p}'$ in unit volume of the momentum space at $\boldsymbol{p}'$. The term $\mathcal{S}$, thus, incorporates the effect of selection rules of transitions, etc. The distribution function at $\boldsymbol{p}'$ describes the probability of finding the electron in that momentum state. The probability of a scattering occurring into this state is proportional to the likelihood of this state being empty, which is given by $1 - f(\boldsymbol{p}')$. Thus, the number of carriers scattered from $\boldsymbol{p}$ per unit time is $\int f(\boldsymbol{p})\mathcal{S}(\boldsymbol{p}, \boldsymbol{p}')(1 - f(\boldsymbol{p}'))d^3\boldsymbol{p}'$ with the integral over the entire $\boldsymbol{p}'$ volume to cover all such scattering events. Similarly, the number scattered into $\boldsymbol{p}$ from the entire $\boldsymbol{p}'$ volume is proportional to the distribution function at $\boldsymbol{p}'$, the scattering probability, and the probability of there being an empty state at $\boldsymbol{p}$. This is given by the integral $\int f(\boldsymbol{p}')\mathcal{S}(\boldsymbol{p}', \boldsymbol{p})(1 - f(\boldsymbol{p}))d^3\boldsymbol{p}'$.

The total rate of change of the distribution function f due to scattering can be found from the difference of these two integrals as:

$$\left.\frac{\partial f}{\partial t}\right|_{scatt} = \int \left[f(\boldsymbol{p}')\mathcal{S}(\boldsymbol{p}', \boldsymbol{p}) - f(\boldsymbol{p})\mathcal{S}(\boldsymbol{p}, \boldsymbol{p}') \right] d^3\boldsymbol{p}', \tag{3.43}$$

provided we assume non-degenerate conditions and hence neglect the effect of exclusion principle. This condition also implies that $f \ll 1$.

The relaxation time approximation implies that this integral can be simplified, under certain conditions, to the expression

$$\left.\frac{\partial f}{\partial t}\right|_{scatt} = -\frac{f(\boldsymbol{p}) - f_0(\boldsymbol{p})}{\tau(\boldsymbol{p})}, \tag{3.44}$$

where $f_0(\boldsymbol{p})$ is the unperturbed Maxwell–Boltzmann distribution function. We now consider the conditions and the scattering processes for which this is clearly justified.

Let us consider the effect of a small perturbation in the thermal equilibrium distribution function. Such a perturbation may occur from small changes in field, temperature, or external optical excitation. Let this perturbation cause a new distribution function

$$f = f_0 + \Delta f, \tag{3.45}$$

where the perturbation Δf in the distribution function is much smaller than the thermal distribution function f_0. Since the distribution function remains unchanged in thermal equilibrium, we have

$$\frac{\partial f_0}{\partial t}\bigg|_{scatt} = \int \left[f_0(\boldsymbol{p}')\mathcal{S}(\boldsymbol{p}',\boldsymbol{p}) - f_0(\boldsymbol{p})\mathcal{S}(\boldsymbol{p},\boldsymbol{p}') \right] d^3\boldsymbol{p}' = 0. \qquad (3.46)$$

Then, in the presence of the perturbation, our scattering equation reduces to

$$\frac{\partial f}{\partial t}\bigg|_{scatt} = \int \Delta f(\boldsymbol{p}')\mathcal{S}(\boldsymbol{p}',\boldsymbol{p})d^3\boldsymbol{p}' - \Delta f(\boldsymbol{p}) \int \mathcal{S}(\boldsymbol{p},\boldsymbol{p}')d^3\boldsymbol{p}'. \qquad (3.47)$$

This perturbation causes a current flow because any disturbance in equilibrium causes motion of the charged particles. Since this current is proportional to the velocity, it is an odd function of the velocity. Hence, the perturbation in the distribution function Δf is also an odd function of the velocity. Clearly, if the first term in the scattering equation vanishes, then

$$\begin{aligned} \frac{\partial f}{\partial t}\bigg|_{scatt} &= -\Delta f(\boldsymbol{p}) \int \mathcal{S}(\boldsymbol{p},\boldsymbol{p}')d^3\boldsymbol{p}' \\ &= -\frac{\Delta f(\boldsymbol{p})}{\tau(\boldsymbol{p})}. \end{aligned} \qquad (3.48)$$

where

$$\tau(\boldsymbol{p}) = \frac{1}{\int \mathcal{S}(\boldsymbol{p},\boldsymbol{p}')d^3\boldsymbol{p}'}. \qquad (3.49)$$

Thus, a sufficient condition for the validity of relaxation time approximation is that the first term in the scattering equation above vanish, i.e.,

$$\int \Delta f(\boldsymbol{p}')\mathcal{S}(\boldsymbol{p}',\boldsymbol{p})d^3\boldsymbol{p}' = 0. \qquad (3.50)$$

We now discuss the conditions and the scattering processes for which this is true.

If $\Delta f(\boldsymbol{p}')$ is an odd function of momentum (or velocity) then this integral vanishes if $\mathcal{S}(\boldsymbol{p}',\boldsymbol{p})$ is an even function of momentum (or velocity). For the scattering probability to be an even function, the scattering probability is equal for all combinations involving the momenta and their conjugates, i.e.,

$$\mathcal{S}\left(\boldsymbol{p},\boldsymbol{p}'\right) = \mathcal{S}(\boldsymbol{p},\boldsymbol{p}'^*) = \mathcal{S}(\boldsymbol{p}^*,\boldsymbol{p}') = \mathcal{S}(\boldsymbol{p}^*,\boldsymbol{p}'^*). \qquad (3.51)$$

The momentum $\boldsymbol{p}^*$ is in exactly the opposite direction of the momentum $\boldsymbol{p}$. Scattering processes that randomize the momentum (or velocity) fulfill this condition. Intuitively, this makes sense since random processes can not result in any net scattering into either $\boldsymbol{p}$ or $\boldsymbol{p}'$. Let us consider the

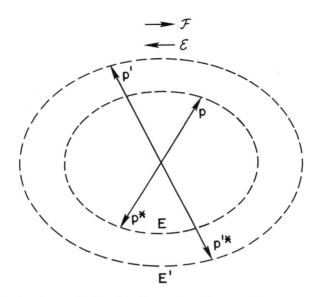

Figure 3.1: A schematic for the discussion of scattering processes in momentum space under the influence of a force $\mathcal{F}$ for a semiconductor with ellipsoidal constant energy surfaces. After E. Conwell, in W. Paul, Ed., *Handbook of Semiconductors*, Vol. 1, North-Holland, Amsterdam (1982).

ellipsoidal surfaces, i.e., consider the anisotropic mass of energy E and E' as shown in Figure 3.1.

In the presence of the force $\mathcal{F}$, the states at $\boldsymbol{p}'^*$ have an increase in population of the carriers, and the ones at $\boldsymbol{p}'$, its complex conjugate, have a decrease in the population of the carriers. Thus, we have the change in distribution functions, $\Delta f(\boldsymbol{p}'^*) > 0$ and $\Delta f(\boldsymbol{p}') = -\Delta f(\boldsymbol{p}'^*)$. If the momentum $\boldsymbol{p}'^*$ has more carriers, it results in an increase in transition of carriers from the momentum $\boldsymbol{p}'^*$ to the momentum $\boldsymbol{p}$, with an equal decrease in transition from momentum $\boldsymbol{p}'$ and momentum $\boldsymbol{p}$. This occurs because the scattering probability to the momentum $\boldsymbol{p}$ from the momentum $\boldsymbol{p}'$ and its conjugate $\boldsymbol{p}'^*$ are equal, i.e., $\mathcal{S}(\boldsymbol{p}',\boldsymbol{p}) = \mathcal{S}(\boldsymbol{p}'^*,\boldsymbol{p})$. So, the sum of the two transition rates from momentum $\boldsymbol{p}'$ and momentum $\boldsymbol{p}'^*$ into $\boldsymbol{p}$ remains unchanged in presence of small applied force $\mathcal{F}$. In this same manner, we may pair all the states in the same valley. So the total rate of transition from the momentum $\boldsymbol{p}'$ to the momentum $\boldsymbol{p}$ remains the same as it was before (i.e., when it was in thermal equilibrium with distribution function f_0), and is balanced by the rate of transition from the states of momenta $\boldsymbol{p}$, again with the distribution function f_0. The remaining small change $\Delta f(\boldsymbol{p})$ is caused only by the scattering out of the momentum $\boldsymbol{p}$, independent of the momentum $\boldsymbol{p}'$, and hence the change in scattering can be defined in terms of the initial state in accordance with the relaxation time.

So, for scattering processes that involve randomization of the momenta or velocity,

$$\frac{\partial f}{\partial t}\bigg|_{scatt} = -\frac{f(\boldsymbol{p}) - f_0(\boldsymbol{p})}{\tau(\boldsymbol{p})} = \frac{\Delta f(\boldsymbol{p})}{\tau(\boldsymbol{p})}, \tag{3.52}$$

where

$$\frac{1}{\tau(\boldsymbol{p})} = \int \mathcal{S}(\boldsymbol{p}, \boldsymbol{p}') d^3\boldsymbol{p}'. \tag{3.53}$$

In addition to scatterings that randomize the momentum, the relaxation time is also justified if (a) scattering is elastic (i.e., scattering involves a negligible loss of energy) and (b) the scattering probability depends on the momentum $\boldsymbol{p}$ and the momentum $\boldsymbol{p}'$ only through the angle θ' between them. The consequence of the first, with no loss of energy, is that $\mathcal{S}_A(\boldsymbol{p}', \boldsymbol{p}) = \mathcal{S}_A(\boldsymbol{p}, \boldsymbol{p}')$ where the additional subscript A refers to the scattering probability normalized to unit area of the constant energy surface at $\boldsymbol{p}'$. These terms are proportional to the power crossing the cross-section which should be the same for elastic processes. If this is the case, then

$$\frac{\partial f_0}{\partial t}\bigg|_{scatt} = 0 = \int \left[f_0(\boldsymbol{p}')\mathcal{S}_A(\boldsymbol{p}', \boldsymbol{p}) - f_0(\boldsymbol{p})\mathcal{S}_A(\boldsymbol{p}, \boldsymbol{p}') \right] d^2\boldsymbol{p}', \tag{3.54}$$

and because this integral is evaluated on a constant energy surface,

$$f_0(\boldsymbol{p}') = f_0(\boldsymbol{p}). \tag{3.55}$$

The appropriateness of the use of relaxation time follows.

Now consider the presence of electric field $\boldsymbol{\mathcal{E}}$. Since the change in distribution is proportional to the energy supplied by the field,

$$f = f_0 + g\boldsymbol{\mathcal{E}}.\boldsymbol{v}, \tag{3.56}$$

where g is a constant which is a function of the carrier mass, etc. From this, we can write the rate of change in the distribution function due to scattering as

$$\frac{\partial f}{\partial t}\bigg|_{scatt} = \int \left(g\boldsymbol{\mathcal{E}}.\boldsymbol{v}' - g\boldsymbol{\mathcal{E}}.\boldsymbol{v} \right) \mathcal{S}_A(\theta') d^2\boldsymbol{p}'. \tag{3.57}$$

We choose the z-axis in the direction of the velocity v and the cylindrical coordinate system. The elemental area at the momentum p' is given by

$$d^2p' = (p'\,d\theta').(p'\sin\theta'\,d\varphi') = p'^2\sin(\theta')d\theta'\,d\varphi', \qquad (3.58)$$

and hence

$$\left.\frac{\partial f}{\partial t}\right|_{scatt} = g\mathcal{E}.v \int \left[1 - \cos(\theta')\right] \mathcal{S}_A(\theta')d^2p'. \qquad (3.59)$$

This equation follows from the fact that the only the z-component of the velocity v' contributes to the above integral, i.e., the term $\left|v'\right|\cos(\theta')$ represents the difference contribution. The rest integrate to zero because they are functions of φ'. The relaxation time follows, then, as

$$\frac{1}{\tau(p)} = \int \left[1 - \cos(\theta')\right] \mathcal{S}_A(\theta')d^2p'. \qquad (3.60)$$

Since the mean free time between collisions (τ_c) is

$$\frac{1}{\tau_c} = \int \mathcal{S}_A(\theta')d^2p', \qquad (3.61)$$

the relaxation time is

$$\tau = \frac{\tau_c}{\left\langle 1 - \cos(\theta')\right\rangle}, \qquad (3.62)$$

where the denominator is averaged over all the collisions. If all final directions are equally likely, i.e., they are random, $\langle\cos(\theta')\rangle = 0$ and $\tau = \tau_c$. If they involve small angle scattering, then $\langle\cos(\theta')\rangle$ would be closer to unity and $\tau \gg \tau_c$, and the time constant will only be a function of energy.

Consider now if it is meaningful to talk about relaxation time in its strict sense for various scattering processes. For spherical constant energy surfaces, relaxation time, which is a function of energy only, exists both for acoustic and ionized impurity scattering. For non-polar semiconductors, optical scattering is randomizing, so it also can be treated by a relaxation time which is a function of energy. Multiple valleys, inter-valley scattering, and optical intra-valley scattering also allow the relaxation time approximation in non-polar semiconductors; however, acoustic mode relaxation time approximation is justified only for highly isotropic energy surfaces. For non-polar semiconductors, only for ionized impurity scattering for elliposidal energy surfaces does the use of relaxation time approximation turn out to be poor.

For polar semiconductors, polar optical scattering is among the more important scattering mechanisms. The relaxation time approximation is incorrect in this case. However, in practice, a proper Monte Carlo calculation is performed to generate a relevant engineering fitting parameter for the relaxation time. This technique allows, then, the use of a simpler and less intensive BTE approach to analyze a problem, while still maintaining accuracy because it uses Monte Carlo results for the fitting parameter of relaxation time.

3.3.2   Conservation Equations

We have now discussed procedures that allow us to determine the distribution function, provided the scattering can be described mathematically either through the relaxation time approximation or the transition probability $\mathcal{S}$. The distribution function, for some problems, can be determined in an analytical form, and for some by iterative techniques. However, application of these procedures is still limited to the simplest of the problems of interest. We will discuss, in the next section, one such problem in order to compare the BTE approach with the Monte Carlo approach. A more common technique than these is to use numerical procedures to compute the distribution functions by methods similar to those used in the Monte Carlo approach. This technique also removes the restrictions on using the simplest of distribution functions or simplified dependence on energy alone by restricting the spatial extent of the analysis. By relaxing these requirements on the distribution function, it also allows application of the BTE approach to problems with more complex distribution functions and scattering processes. An example of the latter is the problem of hot carrier transport.

But, in the study of devices, we are interested in the carrier density, the current carried by the carriers, the energy of the carriers, etc.; not the probability of finding a carrier at a certain position, with a certain momentum, at a certain time. A set of conservation equations can be generated from the BTE that directly determine these parameters, without having to determine the distribution function. The principle is very similar to, and a simple extension of, the method of determining current in the kinetic approach. For a multi-valley semiconductor, the current, in the kinetic approach, can be written as

$$J = \sum_{valleys} \sum_{p} q f(p) v(p). \tag{3.63}$$

To derive the conservation equations, consider the BTE:

$$\frac{\partial f}{\partial t} = -\dot{p}.\nabla_p f - v.\nabla_r f + \frac{\partial f}{\partial t}\bigg|_{scatt}. \tag{3.64}$$

If we wish to determine the expectation of the variable φ, we multiply the BTE by φ and integrate over the momentum space. This allows us to determine the following conservation equation for φ:

$$\frac{\partial}{\partial t} < \varphi >= -\langle \dot{p}.\nabla_p \varphi \rangle - \nabla_r.\langle v\varphi \rangle + \left\langle \frac{\partial}{\partial t}\varphi\bigg|_{scatt}\right\rangle, \tag{3.65}$$

where we have used

$$\begin{aligned}\nabla_r.v &= 0 \\ \text{and} \quad v.\nabla_r f &= \nabla_r.(vf),\end{aligned} \tag{3.66}$$

because the velocity v is a function of momentum p alone. We now have an equation that describes the dependence of $\langle\varphi\rangle$ on position and time. Different choices of $\varphi(p)$ lead to different equations of conservation for the parameter. These are referred to as moment equations; the method, method of moments; because of the way they are derived from the BTE.

The parameters that are of interest to us, such as the carrier concentration and the current, are related to the distribution function $f(r,p;t)$ and the choice of the variable $\varphi(p)$. As an example, consider $\langle\varphi\rangle$ for $\varphi(p) = 1$, then

$$\langle\varphi\rangle = \int f(r,p;t)d^3p, \tag{3.67}$$

which is the carrier concentration $n(r,t)$. If we choose the variable $\varphi(p) = -qp/m^*$, i.e., charge times the velocity of a carrier: current due to a carrier, then

$$\langle\varphi\rangle = \int \left(-\frac{qp}{m^*}\right) f(r,p;t)d^3p. \tag{3.68}$$

This is the equation of charge flux, i.e., the current density $J(r,t)$. If we choose $\varphi(p) = p^2/2m^*$, i.e., the kinetic energy of the individual particle, then

$$\langle\varphi\rangle = \int \left(\frac{p^2}{2m^*}\right) f(r,p;t)d^3p, \tag{3.69}$$

the kinetic energy $W(r,t)$ of the carrier ensemble. When divided by the number of carriers that this represents, it is the average kinetic energy per carrier (w) in the specific ensemble. With this as the meaning of the various moments, Equation 3.65 assumes the conventional meaning of conservation of particle and conservation of current, and extends it further to energy, etc. Note that if recombination effects are important, scattering between bands representing the recombination processes should also be included in the BTE. The meaning of the various moment equations is now quite explicit. These moment equations are to be seen as describing the balancing of the averaged quantities they represent.

If one could employ the scattering relationship,

$$\left. \frac{\partial f}{\partial t} \right|_{scatt} = -\frac{f - f_0}{\tau}, \qquad (3.70)$$

an approximation for some of the scattering processes, but exact for others, then we can write

$$\frac{\partial}{\partial t} \langle \varphi \rangle = -\left\langle \dot{\boldsymbol{p}}.\frac{\partial \varphi}{\partial \boldsymbol{p}} \right\rangle - \frac{\langle \varphi \rangle - \langle \varphi \rangle_0}{\tau_\varphi} - \frac{\partial}{\partial \boldsymbol{r}}.\langle \boldsymbol{v}\varphi \rangle. \qquad (3.71)$$

This equation is similar in its functional form to the continuity equation. The left hand side is the rate of change of $\langle \varphi \rangle$, the first term is a generation term of $\langle \varphi \rangle$ by forces that cause changes in the particle momentum $\boldsymbol{p}$, the second term is a loss term such as a recombination term of $\langle \varphi \rangle$ due to scattering, and the last term is a gradient of the flux of $\langle \varphi \rangle$. When $\varphi = 1$, the last term is proportional to the gradient of the current carried by particles. For $\varphi = 1$, $\tau_\varphi \to \infty$ (no particles are lost in scattering, unless we consider the inter-band scattering term represented by generation and recombination), and

$$\frac{\partial}{\partial \boldsymbol{r}}.\langle \boldsymbol{v}\varphi \rangle = \frac{1}{q}\frac{\partial}{\partial \boldsymbol{r}}.\boldsymbol{J}. \qquad (3.72)$$

So, in the absence of generation and recombination, this yields

$$\frac{\partial n}{\partial t} = \frac{1}{q}\boldsymbol{\nabla}.\boldsymbol{J}. \qquad (3.73)$$

With the interband terms included, it is

$$\frac{\partial n}{\partial t} = \mathcal{G} - \mathcal{R} + \frac{1}{q}\boldsymbol{\nabla}.\boldsymbol{J}. \qquad (3.74)$$

where $\mathcal{G}$ and $\mathcal{R}$ are the generation and recombination terms due to interaction between holes and electrons. This is the first moment equation, the particle conservation equation, and has the same form as used in drift diffusion analysis.

For $\varphi = -q\boldsymbol{p}/m^*$, we have

$$\langle\varphi\rangle = \boldsymbol{J},$$
$$\langle\varphi\rangle_0 = 0,$$
$$\frac{\partial}{\partial\boldsymbol{r}}\langle\varphi\rangle = -\frac{2q}{m^*}\frac{\partial}{\partial\boldsymbol{r}}W,$$

and $\qquad \left\langle \dot{\boldsymbol{p}}.\frac{\partial\varphi}{\partial\boldsymbol{p}} \right\rangle = -\frac{qn}{m^*}\boldsymbol{\mathcal{F}},$ \hfill (3.75)

giving

$$\frac{\partial\boldsymbol{J}}{\partial t} = \frac{qn}{m^*}\boldsymbol{\mathcal{F}} - \frac{\boldsymbol{J}}{\langle\tau_p\rangle} + \frac{2q}{m^*}\boldsymbol{\nabla}_r W. \tag{3.76}$$

This is the second moment equation, the current conservation equation. In this equation, on the right hand side, the first term represents the effect of the applied force, the second represents the effect of scattering, and the third represents the effect on change in energy of the particles which results in a change in their velocity.

For $\varphi = p^2/2m^*$, the kinetic energy,

$$\langle\varphi\rangle = W,$$
$$\langle\varphi\rangle_0 = W_0,$$
$$\left\langle\frac{\partial\varphi}{\partial\boldsymbol{p}}\right\rangle = \frac{\boldsymbol{p}}{m^*},$$

and $\qquad \left\langle \dot{\boldsymbol{p}}.\frac{\partial\varphi}{\partial\boldsymbol{p}} \right\rangle = \frac{\boldsymbol{\mathcal{F}}}{m^*}.\frac{-m^*}{q}\boldsymbol{J} = -\frac{1}{q}\boldsymbol{\mathcal{F}}.\boldsymbol{J},$ \hfill (3.77)

giving

$$\frac{\partial W}{\partial t} = \frac{1}{q}\boldsymbol{\mathcal{F}}.\boldsymbol{J} - \frac{W - W_0}{\tau_w} - \boldsymbol{\nabla}_r.\left\langle v\frac{p^2}{2m^*}\right\rangle. \tag{3.78}$$

This is the third moment equation, the energy conservation equation. In this equation, on the right hand side, the first term represents energy gain from the effect of force, the second term represents the energy loss due to scattering, and the last term is the equivalent in energy of the last term of the current conservation equation. The equation is based on averages; this term represents the ensemble effect of carrier energy change due to ensemble motion. For symmetric distribution functions, such as the displaced Maxwell–Boltzmann distribution, this last term vanishes. Note that we put in a relaxation time constant approximation in these relations, and we have not included heat conductivity of electron gas $\boldsymbol{q}$.[4] In the absence

---

[4]This should be distinguished from fundamental charge q.

of the time constant approximation, this set of equations is

$$\frac{\partial n}{\partial t} = \frac{1}{q}\boldsymbol{\nabla}.\boldsymbol{J} + \left(\frac{\partial n}{\partial t}\right)_{scatt},$$

$$\frac{\partial \boldsymbol{J}}{\partial t} = \frac{q}{m^*}n\boldsymbol{\mathcal{F}} + \frac{2q}{m^*}\boldsymbol{\nabla} W + \left(\frac{\partial \boldsymbol{J}}{\partial t}\right)_{scatt},$$

and $$\frac{\partial W}{\partial t} = \frac{1}{q}\boldsymbol{\mathcal{F}}.\boldsymbol{J} - \boldsymbol{\nabla}_r.(\boldsymbol{v}W) - \boldsymbol{\nabla}_r.\boldsymbol{q} + \left(\frac{\partial W}{\partial t}\right)_{scatt}. \quad (3.79)$$

Let us now see how these equations may be extended to a more complex situation of two valleys, a problem that occurs when considering the effect of the Γ and L valleys in hot carrier transport in GaAs. We will also consider additional simplifications to make the problem more tractable. We will have two sets of equations corresponding to the distribution functions in the two valleys. Identifying each of the individual valleys by the subscripts j and k, and the averages by an overline (this was included in the averaging in our earlier equations), we can write mean momentum and energies as

$$\overline{\boldsymbol{p}}_j = m_j^* n_j \overline{\boldsymbol{v}}_j,$$

and $$\overline{W}_j = \frac{3}{2}n_j kT_j + \frac{1}{2}m_j^* n_j v_j{}^2$$

$$(3.80)$$

for each of the valleys. We introduce the lower case w for the energy of a single carrier, i.e.,

$$\overline{w}_j = \frac{\overline{W}_j}{n_j}. \quad (3.81)$$

In the Equation 3.80 we have included an additional term of $3n_j kT_j/2$, which is the mean energy in the Boltzmann approximation even in thermal equilibrium when $\overline{\boldsymbol{v}}_j = 0$. The temperature of the carriers in the valleys may be different, and is either T_j or T_k. Writing for the jth valley,

$$\frac{\partial n_j}{\partial t} = -\boldsymbol{\nabla}_r.(n_j \overline{\boldsymbol{v}}_j) + \mathcal{G}_j - \mathcal{R}_j + \left.\frac{\partial n_j}{\partial t}\right|_{scatt},$$

$$\frac{\partial \overline{\boldsymbol{v}}_j}{\partial t} = -\frac{\boldsymbol{\mathcal{F}}}{m_j^*} - \overline{\boldsymbol{v}}_j.\boldsymbol{\nabla}\overline{\boldsymbol{v}}_j - \frac{2}{3}\frac{1}{m_j^* n_j}\boldsymbol{\nabla} n_j \overline{w}_j + \frac{1}{3}\frac{1}{n_j}\boldsymbol{\nabla} n_j \overline{\boldsymbol{v}}_j{}^2 +$$

$$\left.\frac{\partial \overline{\boldsymbol{v}}_j}{\partial t}\right|_{scatt},$$

and $$\frac{\partial \overline{w}_j}{\partial t} = -\overline{\boldsymbol{v}}_j.\boldsymbol{\mathcal{F}} - \overline{\boldsymbol{v}}_j.\boldsymbol{\nabla}\overline{w}_j -$$

$$\frac{2}{3}\frac{1}{n_j}\boldsymbol{\nabla}\cdot\left\{\left(n_j\overline{\boldsymbol{v}}_j - \frac{\kappa_j}{k}\boldsymbol{\nabla}\right)\left(\overline{w}_j - \frac{1}{2}m_j^*v_j^2\right)\right\} +$$

$$\left.\frac{\partial\overline{w}_j}{\partial t}\right|_{scatt}. \tag{3.82}$$

Here, the terms $\mathcal{G}_j - \mathcal{R}_j$ represent generation and recombination terms in the jth valley. Energy can also be transferred in the form of heat, i.e., through phonons. Such a term due to heat flux, $(\boldsymbol{\nabla}\boldsymbol{q}_j)$ has been included in these equations using the thermal conductivity κ_j and the carrier temperature T_j,

$$\boldsymbol{q}_j = \kappa_j\boldsymbol{\nabla}T_j = \frac{2}{3}\frac{\kappa_j}{k}\boldsymbol{\nabla}\left(\overline{w}_j - \frac{1}{2}m_j^*v_j^2\right). \tag{3.83}$$

For two valleys, inter-valley processes may now be included in the scattering term. Consider with relaxation time approximation, one can write for two valleys[5] (this can be generalized for more than two valleys in a similar way)

$$
\begin{aligned}
\left.\frac{\partial n_j}{\partial t}\right|_{scatt} &= -\frac{n_j}{\tau_{n,jk}} + \frac{n_k}{\tau_{n,kj}}, \\
\left.\frac{\partial\overline{\boldsymbol{v}}_j}{\partial t}\right|_{scatt} &= -\overline{\boldsymbol{v}}_j\left(\frac{1}{\tau_{p,jj}} + \frac{1}{\tau_{p,jk}} - \frac{1}{\tau_{n,jk}} + \frac{n_k}{n_j}\frac{1}{\tau_{n,kj}}\right), \\
\left.\frac{\partial\overline{w}_j}{\partial t}\right|_{scatt} &= \frac{\overline{w}_j - \overline{w}_{j0}}{\tau_{w,jj}} - \overline{w}_j\left(\frac{1}{\tau_{w,jk}} - \frac{1}{\tau_{n,jk}}\right) + \\
&\quad \frac{n_k}{n_j}\left(\frac{\overline{w}_k}{\tau_{w',kj}} - \frac{\overline{w}_j}{\tau_{n,kj}}\right).
\end{aligned}
\tag{3.84}
$$

where in the relaxation times, following our earlier notation in the BTE, the n subscript denotes carrier relaxation, the p subscript denotes momentum relaxation (not holes, which are not included here), the w subscript denotes energy relaxation, and we distinguish the relaxation time constants $\tau_{w,jk}$ and $\tau_{w',kj}$ due to energy exchange with phonons and differences in the references for kinetic energy in the two valleys. They differ by the potential energy of the conduction band minima.

The various terms of Equation 3.84 can be given specific meaning in line with the earlier description. In the particle conservation equation, the

---

[5]See the discussions in K. Bløtekjær, "Transport Equations for Electrons in Two-Valley Semiconductors," *IEEE Trans. on Electron Devices*, ED -17, No. 1, p. 38, Jan. 1970 and P. A. Sandborn, A. Rao, and P. A. Blakey, "An Assessment of Approximate Non-Stationary Charge Transport Models Used for GaAs Device Modelling," *IEEE Trans. on Electron Devices*, **ED-36**, p. 1244, Jul., 1989.

two terms describe the transfer of particles from one valley to the other. In the current conservation equation, the four terms describe the transfer of momentum within a valley and from one valley to the other. In the energy conservation equation, the first term describes the relaxation towards lattice temperature in intra-valley and inter-valley scattering, the second term describes the energy loss from the jth valley to the kth valley by inter-valley scattering, and the last term describes the energy gained by the jth valley from the kth valley. We distinguish the relaxation time constants $\tau_{w,jk}$ and $\tau_{w',kj}$ due to the differences described above.

3.3.3 Limitations

In view of the way we derived this formulation, we should recognize some of its limitations. It assumes Maxwell–Boltzmann or displaced Maxwell–Boltzmann distribution. By the latter we mean that the distribution in energy of electrons is still of the same form that the Maxwell–Boltzmann distribution has in thermal equilibrium (i.e., the effect of the exclusion principle is not important), but that it shifts in energy to have excess mean finite energy and velocity. It continues to be a symmetric function in momentum. In the process of using the relaxation times, polar optical scattering is not well modelled because it is neither elastic nor randomizing nor dependent on the cosine of the angle between the initial and final momentum. The relaxation time constants are, therefore, fitting parameters.

Polar optical scattering favors smaller angles as shown in Figure 3.2. This figure shows the relative probability of scattering of an electron at various angles in GaAs due to polar optical scattering. Since polar scattering favors small angle scattering, the momentum of electrons is not destroyed. The higher the energy of the electron, the more favorable is the tendency of the electron to continue in close to the same direction. While this tendency leads to favoring of carrier transit at higher velocities in specific situations, this lack of randomization and the approximation involved in the use of Maxwell–Boltzmann-like statistics lead to errors in specific modelling situations. We make an engineering approximation for the polar-optical scattering by forcing a relaxation time fitted to a more accurate Monte Carlo calculation, but the use of the displaced Maxwell–Boltzmann approximation is not correctable because of the way it is implemented.

Figures 3.3 shows the scattering rate due to phonons and for impact ionization in GaAs and Si. The scattering rate variations are somewhat complicated due to the nature of the band structures. We discuss this for phonon processes in silicon (see Figure 3.4). Consider the inter-valley scattering rates from an X valley to another X valley as in silicon. The constant energy surfaces are ellipsoids with two different degeneracies. There

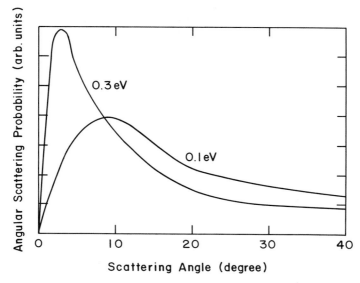

Figure 3.2: Angular scattering probability as a function of scattering angle for polar optical scattering in GaAs at 300 K. The two curves are for electron energies of 0.1 eV and 0.3 eV.

are four equivalent ellipsoids (see Figure 2.26) for scattering whose major axis is in the plane perpendicular to the major axis of the ellipsoid that the scattering occurs from, and there is one ellipsoid along the same major axis. So, inter-valley scattering, in silicon, is of two different types, one to the four equivalent ellipsoids that are closest (called f-scattering), and one to the ellipsoid that is farthest away (called g-scattering). The scattering process requires either the emission or absorption of a phonon to conserve the large change in the momentum. Note that the g-emission involving release of a phonon is the strongest. While this distinction is not of concern for an electron in the Γ band of GaAs, it is related to the inter-valley scattering effect of a hot electron in the L valley. An electron undergoing a scattering from one ellipsoid to another ellipsoid sees six equivalent ellipsoids in the perpendicular plane and one ellipsoid along the axis. Like f- and g-scattering, the probability of absorption of the phonons will be different since different momentum changes are involved.

The low and moderate energy scatterings are lower in GaAs than in silicon. There are specific energies where the scattering rates show the onset of additional processes, such as the X-L scattering process in silicon,

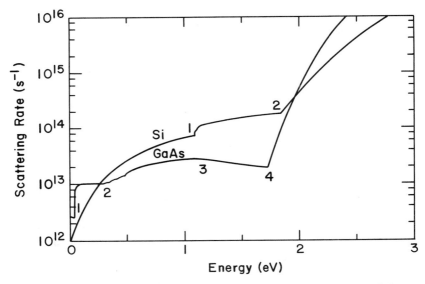

Figure 3.3: Scattering rate in GaAs and Si at 300 K as a function of electron energy due to phonon scattering and impact ionization. On the GaAs curve, 1 identifies the onset of optical phonon scattering. The last sections of the curves correspond to impact ionization processes. The rest of the identified points of rapid changes occur due to effects related to density of states.

or the Γ-L or Γ-X processes in GaAs. At the lowest energy, the scattering in GaAs is limited by the acoustical mode scattering and optical phonon scattering involving absorption of optical phonons as shown in Figure 3.5. At an energy equal to the optical phonon energy ($\approx$ 36 meV for GaAs), optical phonon emission also occurs, resulting in a sudden increase in the scattering rate at this energy.

We commented that at the lowest energies, ionized impurity scattering processes are important for useful dopings. Figure 3.6 shows the scattering rate for GaAs, noting the effect of compensation. The scattering rate is comparable to the polar optical phonon scattering rate at low energies and decreases with energy, as expected.

While discussing these scattering mechanisms, we should comment on carrier–carrier scattering. In bipolar transistors, e.g., the electron is a minority carrier in the midst of a lot of holes in the base. In the case of a two-dimensional electron gas there exists a very high local density of electrons. In the former, electron–hole scattering will be important, while in

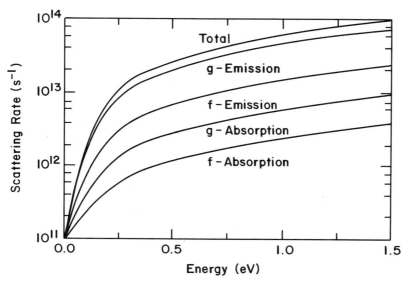

Figure 3.4: Scattering rate at 300 K as a function of energy due to the two different types of inter-valley scattering in Si from the X valley. For a carrier in any valley, f-scattering denotes scattering involving the nearest 4 valleys, and g-scattering denotes scattering involving the opposite valley. The total for both emission and absorption processes is also shown.

the latter, electron–electron scattering will be important. This scattering, which we have also referred to as plasma scattering, can occur coupled with other scattering modes (e.g., together with optical phonons), and its effect can be very strong on optical phonon scattering at high hole density (see Figure 3.7). The phonon scattering rate is reduced because the holes screen the potential perturbations that are caused by the optical mode vibrations. However, now electron–hole scattering becomes important and has to be included in scattering calculations.

We have looked at these various processes to show that one can at least fit the time constant τ_w to a scattering process as a function of energy. This time constant, then, can be used phenomenologically in the moment equations approach. The last set of equations, Equation 3.84, were non-stationary equations. So, knowing all the necessary parameters involved (i.e., the τs) and assuming a displaced Maxwell–Boltzmann distribution, allow us to determine the ns, the $\overline{v}_j$s, and the $\overline{w}_j$s, and hence to characterize the temporal and spatial response of a semiconductor device. These

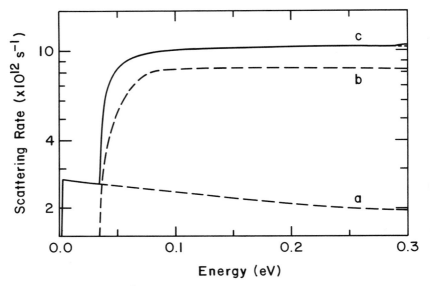

Figure 3.5: Scattering rate as a function of energy in GaAs at 300 K showing the dependence on acoustical processes and optical absorption (marked a), and optical emission (marked b). The curve marked c shows the total scattering rate due to phonons.

equations, in spite of the approximations, are still quite complicated. However, knowing the τs, they are computationally much less intensive than the Monte Carlo approach. In addition, they can be simplified for certain situations even from this form.

An example of this simplification, for the transit in a spatially homogenous sample under the influence of a force $\mathcal{F}$ in a single valley semiconductor, ignoring recombination effects, is the set of equations

$$\frac{\partial \overline{n}_j}{\partial t} = \left. \frac{\partial \overline{n}_j}{\partial t} \right|_{scatt} = 0,$$

$$\frac{\partial \overline{v}_j}{\partial t} = -\frac{\mathcal{F}}{m_j^*} + \left. \frac{\partial \overline{v}_j}{\partial t} \right|_{scatt},$$

and $$\frac{\partial \overline{w}_j}{\partial t} = -\overline{v}_j . \mathcal{F} + \left. \frac{\partial \overline{w}_j}{\partial t} \right|_{scatt}. \qquad (3.85)$$

Because of the assumption of spatial homogeneity, these equations do not account for diffusive effects which are included in the spatial gradient

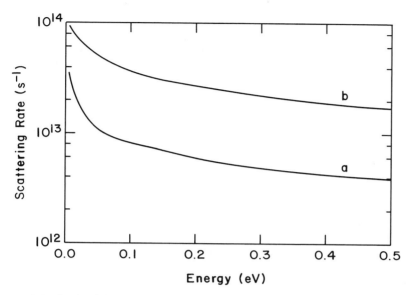

Figure 3.6: Ionized impurity scattering rate for GaAs at 300 K, for compensated (marked a) and uncompensated (marked b, compensation ratio of 2) 1×10^{17} cm^{-3} doping levels.

terms of the BTE. For a two-valley problem (e.g., in GaAs considering the Γ and the L valleys), we may define, with the subscript l for the lower valley and u for the upper valley, the following for carrier density, average velocity, and average energy per carrier,

$$
\begin{aligned}
n &= n_l + n_u, \\
\overline{v} &= \frac{n_l}{n_l + n_u}\overline{v}_l + \frac{n_u}{n_l + n_u}\overline{v}_u, \\
\overline{w} &= \frac{n_l}{n_l + n_u}\overline{w}_l + \frac{n_u}{n_l + n_u}\overline{w}_u, \\
\text{and} \quad \overline{m}^* &= \frac{n_l}{n_l + n_u}\overline{m}_l^* + \frac{n_u}{n_l + n_u}\overline{m}_u^*, \quad (3.86)
\end{aligned}
$$

leading to the following equations in relaxation time approximation for a spatially homogenous but two-valley semiconductor:

$$
\begin{aligned}
\frac{\partial n}{\partial t} &= 0, \\
\frac{\partial \overline{v}}{\partial t} &= -\frac{\mathcal{F}}{\overline{m}^*} - \frac{\overline{v}}{\tau_p},
\end{aligned}
$$

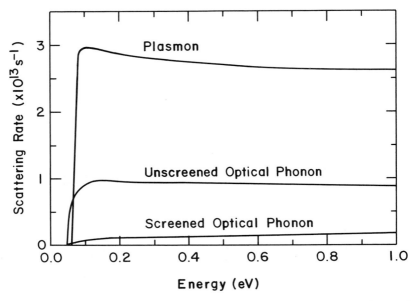

Figure 3.7: Electron scattering rate due to optical phonon and coupled mode scattering in the presence of holes in GaAs at 300 K. The background doping is 5×10^{18} cm^{-3} typical of bipolar transistors. Note the large effect of screening.

$$\text{and} \qquad \frac{\partial \overline{w}}{\partial t} = -\overline{v}.\mathcal{F} - \frac{\overline{w} - \overline{w_0}}{\tau_w}, \qquad (3.87)$$

where τ_p and τ_w lump the effects of related time constants. These time constants could be fitted from a steady-state solution of the Monte Carlo calculation of this same problem of transit in the presence of the force $\mathcal{F}$. The time constants, from the steady-state solution, could be found as

$$\tau_p = -\frac{\overline{m^*v}}{\mathcal{F}}$$
$$\text{and} \qquad \tau_w = -\frac{\overline{w} - \overline{w_0}}{\overline{v}.\mathcal{F}}. \qquad (3.88)$$

Figure 3.8 shows, for 300 K and 77 K, these relaxation times for undoped GaAs.

The temporal evolution follows as a solution of our control equations. For example, Figure 3.9 shows the average energy $\overline{w}$ and the drift velocity $\overline{v}$ for GaAs at 300 K due to an electric field of 10 kV/cm applied at $t = 0$ s. These curves are from a Monte Carlo solution. The behavior shows

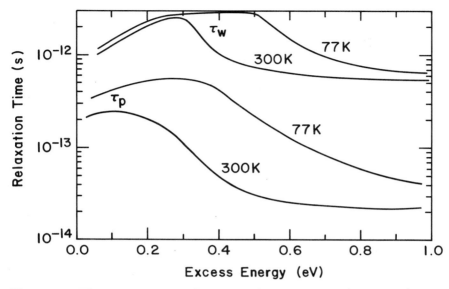

Figure 3.8: The momentum and energy relaxation times (τ_p and τ_w) as a function of energy fitted for undoped GaAs at 300 K and 77 K.

a substantial overshoot in velocity and energy in the first few ps. The response eventually settles to the steady-state characteristics. The overshoot occurs because the momentum relaxation time is lower than the energy relaxation time, leading to a rapid response in the momentum with little initial effect on the energy. It is the non-randomizing nature of the polar-optical scattering that makes this BTE solution a large approximation for GaAs. Figure 3.10 shows the angular distribution in energy associated with the parallel and perpendicular velocity to the field. In its formulation, the BTE implicitly assumed that the distribution was more like what it is at 2.5 ps where it is quite well randomized. So, during the transient, it does not determine the solution accurately. These approximations can be circumvented quite explicitly with the Monte Carlo approach since all the transitional probability relationships can be explicitly included as one follows the response of individual carriers.

3.4 Monte Carlo Transport Approach

The Monte Carlo approach is a probabilistic approach based on sampling of the parameter distributions of the semiconductor transport problem us-

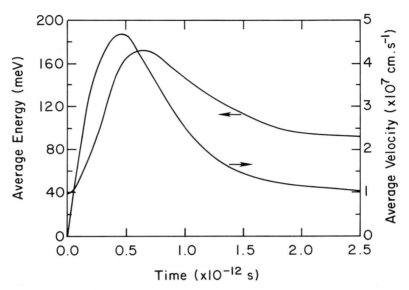

Figure 3.9: Evolution of kinetic energy and velocity as a function of time following a step change of 10 kV/cm in electric field applied at $t = 0$ in GaAs.

ing a random distribution. Parameters of the problem are governed by probability distributions. Whether a carrier undergoes lattice scattering or defect scattering, and, extending that, the particular types of these scatterings depends on possible occurrences of these events as represented by their probability. Therefore, if a random distribution is used to sample these events in a transport problem and sufficient statistics are collected, one should obtain realistic distributions of parameters of interest. Examples of these parameters may be carrier distribution as a function of energy, momentum, and position in a sample (i.e., the function f of the BTE approach). In general, this function need not be a Maxwell–Boltzmann distribution or displaced Maxwell–Boltzmann distribution in energy and the Monte Carlo approach can determine what it is. Additionally, the significance of Monte Carlo techniques goes beyond this. Small-signal analysis as derived from the drift-diffusion equation (itself a simplification of the BTE) uses superposition in the linear differential equation. This is an additional approximation for both transient and large analog signal operating conditions. A representative solution to such conditions is quite unlikely using either of the previous approaches. Monte Carlo, as a statistical method, is

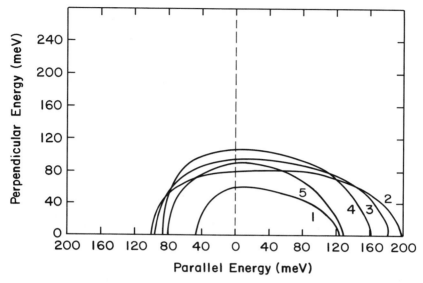

Figure 3.10: Energy associated with parallel and perpendicular components of velocity following application of an electric field of 10 kV/cm along the abscissa. Different time snapshots are shown.

a natural for obtaining solutions to such problems. As with any technique, inaccurate solutions can be obtained if the problem is described incompletely. Our discussion here is a general introduction of how problems are handled in the Monte Carlo approach.[6] It should become clear that it is highly numerically intensive; requires extensive computation; a large accumulation of data on energy, momentum, space, and time; and, as in other approaches, a proper description of boundary conditions.

The technique involves evaluation of the trajectories of a sufficient number of carriers inside the semiconductor to evaluate a statistically meaningful description of the transport. These trajectories occur under the influence of applied and built-in forces in the semiconductor, and due to scattering events; all of which can lead to a change in energy and momentum of the carrier. Examples of applied and built-in forces are those due to an applied or built-in electric field of a p–n junction, and examples of scattering

---

[6]For an intuitive discussion of the application of Monte Carlo techniques to analyzing semiconductor problems, see A. D. Boardman, "Computer Simulation of Hot Electron Behavior in Semiconductors using Monte Carlo Methods," in A. D. Boardman, Ed., *Physics Programs: Applied Physics*, John Wiley, Chichester (1980).

events are the numerous mechanisms of defect and lattice scattering. The technique evaluates both the scattering event and the time between scattering events, a period during which the carrier moves ballistically. These events are accounted for stochastically according to the relative probability of their occurrence.

Consider what this methodology means to the evaluation of the time between scattering events. If $p(\varphi)$ is the probability density of the occurrence of a variable φ (e.g., energy of a carrier) and if $p(\eta)$ is the probability density of the occurrence in a pseudo-random distribution (e.g., a uniform distribution), then

$$\int_0^\varphi p(\varphi')d\varphi' = \int_0^\eta p(\eta')d\eta'. \tag{3.89}$$

For a uniform distribution, this means that

$$\eta = \int_0^\varphi p(\varphi')d\varphi'. \tag{3.90}$$

Inversion of this equation gives a random value of φ as a function of the random distribution function η. As an example, consider scattering with a constant total scattering rate $S = 1/\tau$. The probability of the scattering event occuring after time interval t, following a scattering event at $t = 0$, is the exponential function $p(t) = S \exp(-St)$, and hence, following integration,

$$\eta = 1 - \exp(-St), \tag{3.91}$$

whose inversion gives the corresponding time between scattering as

$$t = -\frac{1}{S}\ln(1 - \eta) \equiv t = -\frac{1}{S}\ln(\eta), \tag{3.92}$$

since η has a uniform distribution. Thus, physically, in the process of selection of time for uniform scattering rate, we calculate the cumulative probability function with respect to time, and then use a uniform random number generator to select the corresponding time. This calculation and mapping has to be performed for all the processes to be accounted for. These include further complications, e.g., angle of motion for the carrier following the scattering has to be included since it determines the new direction of the particle. The selection of this angle involves an additional procedure involving probability distribution.

Determination of the time during which free flight occurs is actually a fairly difficult problem for the more general case. For a particle of wave vector $\boldsymbol{k}$ corresponding to the momentum $\boldsymbol{p}$, the probability that a scattering occurs in the time interval dt at time t is $p\left[\mathbf{k}\left(t\right)\right]dt$. Thus, the probability of no scattering in the time interval 0 to t following a scattering at $t = 0$ is

$$\exp\left\{-\int_0^t p\left[\mathbf{k}\left(\tau\right)\right]d\tau\right\}. \qquad (3.93)$$

Therefore, the probability that this particle, which has not suffered a scattering in the time interval 0 to t, will suffer a scattering during the time dt at time t is

$$P(t)dt = p\left[\mathbf{k}(t)\right]\exp\left\{-\int_0^t p\left[\mathbf{k}\left(\tau\right)\right]d\tau\right\}dt. \qquad (3.94)$$

This is not straightforward in a general problem where $p(\varphi)$ is not a constant. In such cases, the calculations are greatly simplified by introducing an additional scattering process, a virtual scattering process, that leaves the particle in the same state that it was before but which makes $p(\varphi)$ a constant and equal to its maximum value, thus allowing us to continue the procedure described. This fictitious process is referred to as "self-scattering." Now that $p(\varphi)$ is a constant, the uniform random number generation can be used, and a "self-scattering" event is treated to leave the particle state unchanged.

To summarize, the procedure at its simplest consists of starting from a random position, allowing free flight (also called ballistic flight) of the carrier, until a scattering process is identified to occur by a Monte Carlo procedure. In addition to the energy, momentum determination requires an additional Monte Carlo procedure to evaluate the angle, and then the carrier can proceed in free flight again. Following several such scattering events the final carrier parameters lose their dependence on the initial conditions.

For a steady-state homogenous problem, the behavior of the motion of the particle in this large scattering events limit would be representative of the behavior of the particle gas. This conclusion also follows from ergodicity. Let us clarify by considering the example shown in Figure 3.11. This example is for a two-dimensional system, and we study the motion of an electron in real space and $\mathbf{k}$-space (or equivalently momentum space) as it undergoes scattering and drift in time in an electric field in the $-z$ direction. Free flight is shown as a solid line and scattering as a dashed line. Corresponding to each scattering event, a change in momentum is shown via the dashed line in the $\mathbf{k}$-space, and the particle moves again under the influence of the electric field in the $-z$ direction. If this simulation was continued for many more scattering events and hence in time, the final z-position of the particle as a function of time would vary nearly linearly with time for constant effective mass, i.e., the velocity would be a constant.

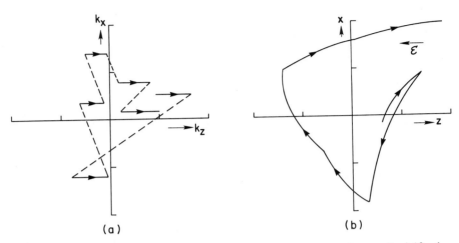

Figure 3.11: An example of a carrier motion and scattering as it drifts in an electric field ($-z$ direction) in momentum (a) and real (b) space.

The equation of motion of this particle follows from our description of particle motion. The momentum of the particle is $\hbar k$ and its energy is $\hbar^2 k^2 / 2m^*$ for the simple free particle. Under the influence of electric field $\mathcal{E}$, the equation of motion of the particle is

$$\frac{\partial (\hbar k)}{\partial t} = -q\mathcal{E}. \tag{3.95}$$

If its initial momentum corresponds to a wave vector k_i, the momentum at an instant in time t during free flight corresponds to a wave vector k, where

$$k(t) = k_i - \frac{q\mathcal{E}}{\hbar}t. \tag{3.96}$$

Since the motion is by drift, momentum only changes in the direction of the electric field, i.e., in the z direction.

For a transient problem, where one is interested in the behavior of a particle gas on a time scale during which too few scattering events occur, as well as for a spatially non-homogenous problem such as transport in the channel of a short gate-length FET or the base-collector space charge region of a bipolar transistor, many more particles have to be considered under the influence of the forces that exist in the channel or the base-collector space charge region in order to obtain meaningful results. The test of the latter is the standard deviation of the parameter of interest in order to determine its uncertainty. This test is usually accomplished by dividing the accumulated information into sufficiently large equal time intervals and determining the parameter of interest (e.g., the velocity). The average of this information is the mean of interest and its standard deviation is a measure of the uncertainty.

Real semiconductor problems are usually of significantly higher complexity. Poisson's equation, which relates the charge density and the displacement vector, should be satisfied together with the charged carrier contributions as they move in a finite sized sample. Additional complexity arises because of changing effective mass, numerous scattering processes, avalanche processes that create additional electrons and holes, momentum quantization in two-dimensional structures, the multiple bands and valleys of real band structures, and the significant deviation of these bands from parabolic approximation. For electrons in many compound semiconductors, e.g., at the least two valleys—the Γ and the L valley—should be considered for problems involving any significant heating of carriers. Quite often, X should be considered too, and neither the X nor the L valleys can be approximated as being parabolic for more than a few fractions of eV. One can make some adjustments by introducing the non-parabolicity factor α which relates the energy and momentum through

$$E\left(1 + \alpha E\right) = \frac{\hbar^2 k^2}{2m^*}, \tag{3.97}$$

so that the energy E is

$$E\left(\mathbf{k}\right) = \frac{1}{2\alpha}\left[-1 + \left(1 + 4\alpha\frac{\hbar^2 k^2}{2m^*}\right)^{1/2}\right], \tag{3.98}$$

the velocity of the particle $\mathbf{v}\left(\mathbf{k}\right)$ is

$$\mathbf{v}\left(\mathbf{k}\right) = \frac{1}{\hbar}\frac{\partial E}{\partial \mathbf{k}} = \frac{\hbar \mathbf{k}}{m^*\left(1 + 2\alpha E\right)}, \tag{3.99}$$

and the conductivity effective mass m_c is

$$m_c = m^*\left(1 + 2\alpha E\right). \tag{3.100}$$

These approximations do allow extension of the technique of using equations to describe some of the parameters of the problem. However, for small bandgap semiconductors which are both highly non-parabolic and have significant off-equilibrium effects, this inaccuracy of analytic description becomes a major limitation. Also, quite often, problems of interest are related to heterostructures with alloy grading and hence position-dependent band structure. For such problems, information could be stored as a look-up table. Monte Carlo methods thus become both numerical and storage intensive.

The boundary influences the behavior of a system. The boundary conditions, therefore, should ideally represent the true behavior of the boundary, and failing that should be selected in such a way that the final result of interest is intuitively independent of it. Thus, the initial position of the particle should be irrelevant to the final result of interest above, which was the velocity. This is an initial boundary value problem.

Spatially, the device sizes and regions to be analyzed are limited, and the boundaries have to be suitably described. In a surface oriented device, carriers incident at the surface either get reflected or recombine. If it is a majority carrier device, then the second is irrelevant if the population of the minority carrier is low. It may be irrelevant for a minority carrier device if other effects in the system are larger. In both devices, however, careful consideration must be given to what happens at the surface to the carrier as it is reflected spatially. Electron surface mobility in silicon is nearly half of its bulk value at similar carrier densities. This lowering of mobility must arise due to the partial randomization of carrier momentum of particles incident at the surface. This effect may be included by considering the scattering at the surface as partially randomizing. In heterostructures, where crystallinity is maintained between semiconductors (e.g., $Ga_{1-x}Al_xAs/GaAs$), there is a lack of surface states as well as non-planarity. These, if present, lead to potential perturbations and hence to scattering. Therefore, in a heterostructure system with low surface states and a planar interface, mobilities are maintained and true reflections occur. Thus, carriers, incident from GaAs into $Ga_{1-x}Al_xAs$ at the surface, only go through a sign reversal in the momentum normal to the surface if they go through a reflection. The carrier does not always have to reflect back. It may have the requisite momentum and energy to transfer into $Ga_{1-x}Al_xAs$. The Monte Carlo procedure must check and allow for such an event to take place.

Two boundaries of import that are often encountered are an ohmic contact and a metal–semiconductor rectifying contact. An ideal ohmic contact, unless it is intentionally a part of the intrinsic device devised to affect the operation of the device, is meant to be only a means for providing carriers to the semiconductor without a voltage drop across it. Thus, a Monte Carlo procedure may simulate this process by trying to maintain the particle density at its equilibrium value at the contact in accord with Fermi–Dirac distribution. A lost carrier may thus be replaced by a carrier whose momentum is chosen randomly for a Fermi–Dirac distribution. A metal–semiconductor contact, a rectifying contact, is considerably more complex.

Particles that do not have sufficient momentum or energy to cross into the metal get reflected, and, because of the non-planarity and interface states, may be thermalized. Carriers that have sufficient momentum and energy will traverse into the metal. Since for rectifying metal–semiconductor contacts the barrier for injection from the metal to the semiconductor is very large and scattering lengths very short, very few carriers traverse from the metal to the semiconductor, and we may ignore them in the procedure. This question is discussed in further detail in Chapter 4.

Let us now consider areas where the complexities associated with the Monte Carlo technique are unavoidable if one is to obtain a reasoned understanding. In our discussion of devices we will look at this question in significantly more detail. Here, we discuss two examples where the utility of the technique becomes quite obvious.

When carriers traverse a region where there is a significant gradient in the electric field, they undergo a velocity overshoot in parts of the region. Velocity overshoot is a term that implies that the local velocity v at a local field $\mathcal{E}$ becomes higher than what it would reach for a slowly varying field $\mathcal{E}$ of the same magnitude. A local off-equilibrium occurs. It arises because the carrier energy and the local electric field are no longer in equilibrium with each other. An example of this local equilibrium at high fields is the existence of velocity saturation. The mobility of carriers is a function of energy, as our earlier treatment shows. Because of the parallel between the energy and the electric field, the mobility can be written as a function of electric field. When the electric field becomes large, higher carrier energy results. Higher carrier energy causes larger amount of scattering, which reduces the mobility. Thus, velocity, a product of mobility and electric field with mobility varying inversely with field, becomes a constant. When a large gradient of electric field exists, the carrier does not immediately achieve the energy that corresponds to the local field if the field is rapidly varying. It actually has a lower energy and hence suffers less scattering, and yet it is accelerated by the high field. The carrier, therefore, is said to have a velocity overshoot. This large velocity occurs in spite of the fact that it has a lower energy. Another way of saying this same thing is that the relaxation rate for momentum is larger than that for energy.

A second example is related to distributions that can be determined using the Monte Carlo approach. A very instructive problem in transport is that of transport across p–n junctions. The drift-diffusion theory incorporates a mobility, and a diffusion coefficient that is proportional to the mobility, both of which are a function of the local electric field. In constant mobility approximation, this electric field dependence is ignored, and a solution such as the triode equation of the field effect transistor is found. In a p–n junction and many other junctions, large built-in fields occur because of their built-in voltages. The local electric field is high. Should the mobility for such a problem be the mobility corresponding to the local electric field? We have noted above that mobility is a parameter that can be directly related to the energy of the particle. So, while the local electric field, due to built-in considerations, is large, is the energy of the particles also correspondingly large? It is not. Particles, e.g., holes and electrons in a p–n junction, diffuse against the local field. The field can not increase the energy of these particles as it does when the particle is accelerated by the field during drift. Note that the diffusion that occurs in this problem, as elsewhere, is due to scattering that favors the direction towards regions of lower carrier density. Drift, on the other hand, causes the particles to move away from the junction. Intuition suggests that thermal-equilibrium distributions should exist everywhere, i.e., the distribution function should be a Fermi–Dirac distribution, and since this is a non-degenerate case it simplifies to Maxwell–Boltzmann distribution. Now, if we apply a forward bias to the junction, the local electric field is perturbed. Electrons and holes in the process of carrying the current receive excess energy between scatterings from the acceleration in the local field.

Can one predict the electron distribution, its energy, and, if it is a Maxwell–Boltzmann type distribution, its temperature; and would that tell us what mobility to employ? The Monte Carlo procedure is suitable for analyzing this problem since it can simulate all the necessary processes that lead to the transport. Figure 3.12 shows at thermal equilibrium and under forward bias, for a silicon p–n junction, the spatial position of the conduction and valence band edges and the distribution of carriers as a function of the kinetic energy. The average energy and the tail for the distribution for both holes and electrons in the region of quasi-neutrality still show thermal distribution corresponding to the 300 K lattice temperature. However, in the space charge region, electrons show higher average energies corresponding to a higher temperature, and extended tails corresponding to a higher temperature hot carrier tail. The distributions have been normalized to show a proper comparison of energy dependence. This heating occurred in spite of a decrease in the electric field in the junction. The distributions can be quite reasonably fitted with the exponential tail, so at least in this case, a displaced Maxwell–Boltzmann distribution (but with a changed temperature) is a good approximation to the behavior of the carriers. Since the carriers do have different temperatures, a different mobility can be argued to be necessary in the drift-diffusion approach. In this example, the electron temperature rises by approximately 15 K in the space charge region; the mobility should correspond to that energy. In reality, this mobility is smaller, but much larger than that corresponding to the high local electric field, which certainly would give erroneous results.

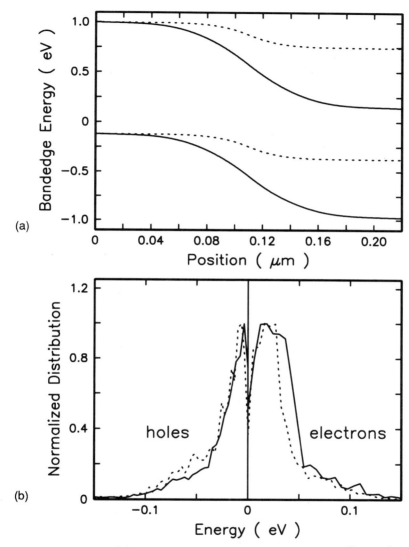

Figure 3.12: Part (a) shows the band edges for a 2×10^{17} cm^{-3} doped silicon p–n junction under thermal equilibrium (solid lines) and a forward bias of 0.87 V (dashed lines) at 300 K. Part (b) shows the electron and hole distribution for the forward bias as a function of kinetic energy for a spatial position outside (dashed lines) and inside (solid lines) the space charge region of the p–n junction.

Quite often, what helps in obtaining reasonably accurate results by the drift-diffusion approach is that errors in mobility and diffusion, which track each other, give rise to an opposing polarity of errors in the drift and diffusion currents, and hence the result is significantly more accurate. What this example does point out, however, is that applied and built-in electric fields must be treated differently.

It is also useful to consider this problem using the BTE approach.[7] Consider a problem with a built-in field of $\mathcal{E}_{bi}$ due to impurity gradient. If f_0 is the distribution function in thermal equilibrium corresponding to this condition, then from the BTE,

$$q\mathcal{E}_{bi}.\nabla_p f_0 + v.\nabla_r f_0 = 0. \tag{3.101}$$

With a field $\mathcal{E}_{bi} + \Delta\mathcal{E}$, the distribution function changes to $f_0 + \Delta f$, and the BTE gives

$$q\mathcal{E}_{bi}.\nabla_p f_0 + v.\nabla_r f_0 + q\Delta\mathcal{E}.\nabla_p \Delta f + v.\nabla_r \Delta f +$$
$$q\mathcal{E}_{bi}.\nabla_p \Delta f + q\Delta\mathcal{E}.\nabla_p f_0 + \left.\frac{\partial(f_0 + \Delta f)}{\partial t}\right|_{scatt} = 0. \tag{3.102}$$

The last term represents the effect of scattering and is the cause of diffusion; the second to last term is negligible, being the product of two perturbations; and the first two terms have a vanishing sum according to the BTE at thermal equilibrium. Current conservation requires that

$$\nabla_r . \int v\Delta f d^3 p = 0. \tag{3.103}$$

We consider only the case when the electric field $\mathcal{E}_{bi} + \Delta\mathcal{E}$ does not change with position, implying that $\nabla_r \Delta f$ does not change with position. Then, the BTE can be written as

$$q\mathcal{E}_{bi}.\nabla_p \Delta f + \left.\frac{\partial(f_0 + \Delta f)}{\partial t}\right|_{scatt} = -q\Delta\mathcal{E}.\nabla_p f_0. \tag{3.104}$$

For a built-in field of magnitude $\mathcal{E}_o$, this gives

$$q\mathcal{E}_o.\nabla_p \Delta f + \left.\frac{\partial(f_0 + \Delta f)}{\partial t}\right|_{scatt} = -q\Delta\mathcal{E}.\nabla_p f_0. \tag{3.105}$$

---

[7]For detailed arguments on this subject see J. B. Gunn, "Transport of Electrons in a Strong Built-in Electric Field," *J. of Appl. Phys.*, **39**, No. 10, p. 4602, 1968.

To understand the difference between the effects of built-in versus applied fields, consider the case of the same semiconductor with constant doping and identical distribution function f_0 at any point chosen in the previous situation. If a field of $\mathcal{E}_{ex} = \mathcal{E}_{bi} = \mathcal{E}_o$ is applied to it by external means, then the change in the distribution function $\Delta f'$ satisfies

$$q\mathcal{E}_o.\nabla_p\Delta f' + \left.\frac{\partial\left(f_0 + \Delta f'\right)}{\partial t}\right|_{scatt} = -q\mathcal{E}_o.\nabla_p f_0. \tag{3.106}$$

If the transition probability is linear and independent of the distribution function (implying negligible carrier induced scattering effects and perturbations), then using S as the scattering rate, we obtain for the built-in and the applied field cases

$$q\mathcal{E}_o.\nabla_p\Delta f + S\Delta f = -q\Delta\mathcal{E}.\nabla_p f_0$$
$$\text{and} \quad q\mathcal{E}_o.\nabla_p\Delta f' + S\Delta f' = -q\mathcal{E}_o.\nabla_p f_0. \tag{3.107}$$

For a change in electric field $\Delta\mathcal{E}$ co-directional with $\mathcal{E}_0$, the distribution functions in the two cases are related through

$$\Delta f = \frac{\Delta\mathcal{E}}{\mathcal{E}_0}\Delta f'. \tag{3.108}$$

For bias changes from thermal equilibrium for a non-homogenous semiconductor with a built-in field, the change in the distribution function is a product of a fraction equal to the relative perturbation of the field multiplied by what the distribution function change would have been had the built-in field been an externally applied field. The influence of fields in the case of a built-in field is small, and at thermal equilibrium there is no effect. Monte Carlo calculations confirm this, as follows from the distribution functions in Figure 3.12.

Since the drift velocity is given by

$$\langle v \rangle = \frac{\int v f dp}{\int f dp}, \tag{3.109}$$

we obtain in the presence of the perturbation $\Delta\mathcal{E}$ velocities in the two cases, v_d and v'_d, that are related by the same ratios (this follows from the above integrals, see Problem 9),

$$v_d = \frac{\Delta\mathcal{E}}{\mathcal{E}_0}v'_d. \tag{3.110}$$

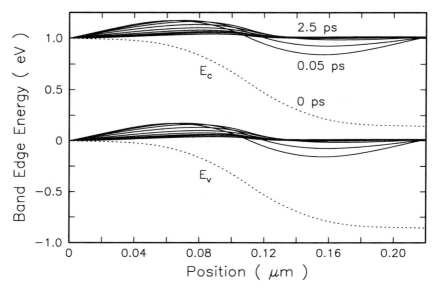

Figure 3.13: The changes in conduction and valence band edges of 2×10^{17} cm^{-3} doped silicon p–n junction when forward biased from thermal equilibrium (dashed lines) at 300 K. The band edges are shown at various instances in time.

So, the effective mobility is also scaled by this factor. This derivation made several approximations for simplifying our effort. In particular, the semiconductor was assumed to be non-degenerate and isotropic, and carrier induced perturbation of scattering was neglected. A Monte Carlo procedure can account for these in a straightforward manner.

We had remarked about the particular utility of the Monte Carlo approach towards modelling of transient phenomena. We exemplify this by showing how the transient in the p–n junction takes place when the forward bias is applied. Figure 3.13 shows the turn-on transient of the junction. Note that upon application of the bias, a field is established in the quasi-neutral region. The drift field in the quasi-neutral region vanishes rapidly as the junction charges. The junction depletion region edge can only move as rapidly as the carriers can, which is limited by the maximum group velocity of the carriers. This figure serves to demonstrate that even a simple problem such as that of a p–n junction is, in detail, quite complicated, if we are interested in understanding phenomena that take place on time scales of a few scattering events to a few pico-seconds. The steady-state result from such a calculation would be quite similar to results derived from a quasi-static analysis using the drift-diffusion equation, except where overshoot occurs due to large gradients in electric fields. The transient results, if one is only interested in time scales by which time the fields in the quasi-neutral region have died down, will also be accurate. However, in any of the cases, very short times, of the order of few pico-seconds, or in situations where velocity overshoot dominates, erroneous results will occur in the drift-diffusion approach. We will discuss this further in the chapters on devices.

Analytic formulations are appealing for two reasons: they allow a simple and rapid understanding of a problem, and they provide an appealing intuitive understanding of the problem. The drift-diffusion approach, derived from the BTE, is the most common analytical tool with these characteristics. It will be used extensively in the rest of the text, and in context we will also discuss its limitations. Here, we summarize some of its characteristics.

3.5 Drift-Diffusion Transport

The current transport equation, sometimes called the drift-diffusion equation, can be derived from the BTE equations under several assumptions. This section, therefore, actually belongs with our discussion of the BTE; it has been separated for convenience and because it is going to be the principal approach in our analysis of devices.

The second moment of the BTE yielded an equation which we may write as

$$\frac{\partial \boldsymbol{J}}{\partial t} = \frac{q}{m^*} n \boldsymbol{\mathcal{F}} - \frac{\boldsymbol{J}}{\langle \tau \rangle} + \frac{2q}{3m^*} \boldsymbol{\nabla}_r \frac{3}{2} nkT, \tag{3.111}$$

for electrons and holes. Among the assumptions inherent in this derivation are all those related to the BTE, an isotropic and parabolic single band, relaxation time approximation, and the assumption that the energy of carriers is due to thermal contributions only (non-degeneracy and no hot carrier effects). The relaxation time, $\tau_p = \langle \tau \rangle$, is related to the effective carrier mobility by

$$\mu = \frac{q\tau_p}{m^*}. \tag{3.112}$$

For transport in an electric field $\boldsymbol{\mathcal{E}}$, this gives the transport equation as:

$$\tau_p \frac{\partial \boldsymbol{J}}{\partial t} = q\mu n \boldsymbol{\mathcal{E}} - \boldsymbol{J} + \mu \boldsymbol{\nabla}_r nkT. \tag{3.113}$$

The scattering time τ_p is significantly smaller than the time scales of interest. The perturbation term embodied in the left hand side of the equation is negligible for most of the situations of interest where the time dependent change of interest occurs over many scattering events. Assuming this, and the Einstein relationship, we obtain the current transport equation

$$J = q\mu n \mathcal{E} + q\mathcal{D}\nabla_r n, \qquad (3.114)$$

where we now drop the subscript r since only spatial gradients will be considered. We derived this equation by considering the BTE for electrons; we may do the same for holes with their opposite charge. The subscripted result for electrons and holes of the drift-diffusion equations are

$$
\begin{aligned}
J_n &= qn\mu_n\mathcal{E} + q\mathcal{D}_n\nabla n \\
\text{and} \quad J_p &= qp\mu_p\mathcal{E} - q\mathcal{D}_p\nabla p.
\end{aligned}
\qquad (3.115)
$$

The major assumptions that allowed the derivation of the above are those related to relaxation time approximation, single parabolic band structure with homogenous collision time, non-degenerate material, constant temperature of lattice and carriers, several collision events and hence neglect of any off-equilibrium effects, and that the length scale for spatial variation of field, collision time, impurity concentrations, etc. are longer than the mean free path.

We introduce quasi-Fermi levels, artificial energy and potential levels akin to the Fermi level used in thermal equilibrium, to define the electron and hole densities as

$$n = n_i \exp\left(\beta\psi - \beta\phi_n\right) = n_i \exp\left(\frac{\xi_n - E_i}{kT}\right) \qquad (3.116)$$

and

$$p = n_i \exp\left(\beta\phi_p - \beta\psi\right) = n_i \exp\left(\frac{E_i - \xi_p}{kT}\right). \qquad (3.117)$$

Here, $\beta = q/kT$, ψ is the potential of the intrinsic level, and ϕ_n is the potential of the electron quasi-Fermi level. ϕ_p is defined as the potential of the hole quasi-Fermi level. The reference for these can be arbitrary: the intrinsic level, the vacuum level, etc., at some point in the crystal. The last term in these equalities is written using energies.[8] Hence,

$$
\begin{aligned}
J_n &= -qn\mu_n\nabla\phi_n = n\mu_n\nabla\xi_n \\
\text{and} \quad J_p &= qp\mu_p\nabla\phi_p = -p\mu_p\nabla\xi_p.
\end{aligned}
\qquad (3.118)
$$

---

[8]Our convention is to write potentials and energies w.r.t. electrons. Potential is higher downwards while energy is higher upwards. The reader should carefully look at these conventions and their reference levels, here and in the rest of the text. The reference levels are changed depending on the problem being analyzed in order to ease the appearance of the mathematics.

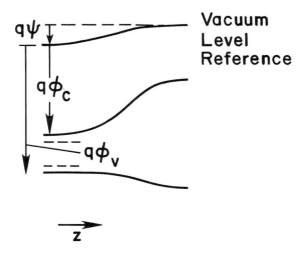

Figure 3.14: Choice of the reference coordinate system for calculations for an arbitrary slow grading.

We can also extend this equation phenomenologically for a material whose composition is slowly changing, i.e., a material with a position dependent band structure.[9] Similar treatment implies the existence of an additional force on the carrier resulting from the varying affinity of the carrier. For a slowly varying composition, this may be incorporated in quite a straightforward manner. We can write the carrier concentration as a function of the quasi-Fermi level with the coordinate system chosen as in Figure 3.14.

$$n = N_C \exp\left(\beta\psi + \beta\phi_C - \beta\phi_n\right) \qquad (3.119)$$

where ϕ_C is the conduction band edge potential with respect to the vacuum level reference and N_C is the effective density of states in the conduction

[9]Due to heavy doping effects the treatment of position dependent band structure is important in homostructure devices also. Early treatment, suitable for both heterostructures and homostructures, was related to this. See J. E. Sutherland and J. R. Hauser, "A Computer Analysis of Heterojunction and Graded Composition Solar Cells," *IEEE Trans. on Electron Devices*, **ED-24**, p. 363, 1972; A. H. Marshak and K. M. Van Vliet, "Carrier Densities and Emitter Efficiency in Degenerate Materials with Position-Dependent Band Structure," *Solid-State Electronics*, **21**, p. 429, 1978; M. S. Lundstrom and R. J. Schuelke, "Numerical Analysis of Heterostructure Semiconductor Devices," *IEEE Trans. on Electron Devices*, **ED-30**, No. 9, p. 1151, Sep. 1983; and A. H. Marshak and C. M. Van Vliet, "Electrical Current and Carrier Density in Degenerate Materials with Non-Uniform Band Structure," *Proc. of IEEE*, **72**, No. 2, p. 148, 1984.

band. Then,

$$\begin{aligned}
\boldsymbol{J}_n &= -qn\mu_n\boldsymbol{\nabla}\phi_n \\
&= -qn\mu_n\boldsymbol{\nabla}\left\{\psi + \phi_C - \frac{1}{\beta}\ln\left(\frac{n}{N_C}\right)\right\} \\
&= qn\mu_n\left\{-\boldsymbol{\nabla}\psi - \boldsymbol{\nabla}\left[\phi_C + \frac{1}{\beta}\ln(N_C)\right]\right\} + q\mathcal{D}_n\boldsymbol{\nabla}n, \quad (3.120)
\end{aligned}$$

and similarly,

$$\boldsymbol{J}_p = qp\mu_p\left\{-\boldsymbol{\nabla}\psi - \boldsymbol{\nabla}\left[\phi_V - \frac{1}{\beta}\ln(N_V)\right]\right\} - q\mathcal{D}_p\boldsymbol{\nabla}p. \qquad (3.121)$$

By defining

$$\begin{aligned}
\phi_{Cn} &= \phi_C + \frac{1}{\beta}\ln(N_C) \\
\text{and} \quad \phi_{Vp} &= \phi_V - \frac{1}{\beta}\ln(N_V), \qquad (3.122)
\end{aligned}$$

we account for variations in electron/hole affinity, effective mass, etc. This term can also include degeneracy effects related to high doping by addition of a term of $(kT/q)\ln\left[F_{1/2}(\eta_{fn})/\exp(\eta_{fn})\right]$ for n-type or a term of $(kT/q)\ln\left[F_{1/2}(\eta_{fp})/\exp(\eta_{fp})\right]$ for p-type, the terms following directly from the differences in band occupation statistics in Fermi–Dirac and Maxwell–Boltzmann distribution functions. So, for degenerate conditions

$$\begin{aligned}
\phi_{Cn} &= \phi_C + \frac{1}{\beta}\ln(N_C) + \frac{1}{\beta}\ln\left[F_{1/2}(\eta_{fn})/\exp(\eta_{fn})\right] \\
\text{and} \quad \phi_{Vp} &= \phi_V - \frac{1}{\beta}\ln(N_V) - \frac{1}{\beta}\ln\left[F_{1/2}(\eta_{fp})/\exp(\eta_{fp})\right].
\end{aligned}$$
$$(3.123)$$

From this, the current transport equation for slowly graded heterostructures is

$$\begin{aligned}
\boldsymbol{J}_n &= qn\mu_n\left\{-\boldsymbol{\nabla}\psi - \boldsymbol{\nabla}\phi_{Cn}\right\} + q\mathcal{D}_n\boldsymbol{\nabla}n \\
\text{and} \quad \boldsymbol{J}_p &= qp\mu_p\left\{-\boldsymbol{\nabla}\psi - \boldsymbol{\nabla}\phi_{Vp}\right\} - q\mathcal{D}_p\boldsymbol{\nabla}p, \qquad (3.124)
\end{aligned}$$

which, along with Poisson's equation and the current continuity equation, allows us to solve many interesting problems. Quite often, we actually ignore the ∇N_C and the ∇N_V terms of $\nabla\phi_{Cn}$ and $\nabla\phi_{Vp}$ because they tend to be small.

3.5.1 Quasi-Static Analysis

To gain physical understanding of a device, we are initially interested in the behavior of the device and its response in near-static conditions. We are interested in how the device behaves given a certain bias applied at the device terminals. What we want to know is how the current, e.g., in the device changes if we change the bias of the device under very slowly varying conditions, so that frequency-dependent effects per se are not important. This requires us to find the quasi-static solution to the problem, the simplest and the most popular example of device analysis.

Quasi-static solutions can be found by all the approaches and their approximations that we have described. A time-independent Monte Carlo solution, or results from the static BTE equation, or its approximation the drift-diffusion equation, will all give a quasi-static solution. In the p–n junction problem that we discussed, the carrier distributions were under quasi-static conditions. We assumed steady-state (i.e., the device response had completely settled) and no further time dependence in the applied bias. The carrier distributions, etc., during the time the switching occurred could be considerably different than during the quasi-static condition, as it would appear from Figure 3.13 since the device has large time-dependent displacement and particle current effects. All the techniques that we have discussed can also give transient, small-signal, and quasi-static solutions since many of the time-dependent phenomena are incorporated in them.

Often, the quasi-static solution is valid up to quite high frequencies. This allows simple predictive modelling to be applied in computer aided design of circuits. However, for both very high frequency and high speed, quite often, this solution is not sufficient. We will discuss the differences, in the context of devices, by finding the quasi-static solution as well as the small-signal frequency dependent solution within the limitations of the techniques used.

3.5.2 Quasi-Neutrality

We now apply the drift-diffusion equation, derived above, to establish some procedures that will be used repeatedly in the rest of the book. This also gives us an opportunity to describe some of the assumptions and conditions under which some of the analysis becomes invalid. First, consider the question of when a semiconductor should be considered space charge "neutral" or quasi-neutral. In the non-uniformly doped semiconductor shown in Figure 3.15, carriers separate and generate an electric field. At thermal equilibrium, one may derive this field using the current continuity equation

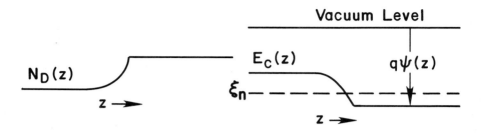

Figure 3.15: Non-uniformly doped semiconductor (a) and associated band bending (b). (b) also shows the electrostatic potential as a function of position.

as

$$\mathcal{E} = -\frac{kT}{q}\frac{\nabla n}{n} = \frac{kT}{q}\frac{\nabla p}{p}. \tag{3.125}$$

This implies the existence of a space charge given by

$$\rho = \epsilon_s \nabla.\mathcal{E}. \tag{3.126}$$

The question we wish to answer is when we should include the effect of this space charge. A semiconductor is quasi-neutral, and the space charge effect can be ignored when the space charge density is small compared to dopants placed in the material, i.e.,

$$\rho \ll q\,|N_D - N_A|. \tag{3.127}$$

Consider a compositionally homogenous semiconductor. Employing the Maxwell–Boltzmann approximation,

$$\frac{d\psi}{dz} = -\mathcal{E} = \frac{kT}{q}\frac{1}{n}\frac{dn}{dz}, \tag{3.128}$$

and hence

$$-\frac{\rho}{\epsilon_s} = \frac{d^2\psi}{dz^2} = \frac{kT}{q}\left[-\frac{1}{n^2}\left(\frac{dn}{dz}\right)^2 + \frac{1}{n}\frac{d^2n}{dz^2}\right]. \tag{3.129}$$

Table 3.1: Debye lengths in GaAs at 300 K.

Doping (cm^{-3})	Debye Length (cm)
Intrinsic	2.3
1×10^{15}	1.35×10^{-4}
1×10^{16}	0.43×10^{-4}
1×10^{17}	0.14×10^{-4}
1×10^{18}	0.04×10^{-4}

For a quasi-neutral n-type material, $n \approx N_D$, and hence

$$\rho \approx -\frac{\epsilon_s kT}{q} \left[-\frac{1}{N^2} \left(\frac{dN}{dz} \right)^2 + \frac{1}{N} \frac{d^2 n}{dz^2} \right]. \tag{3.130}$$

The condition for quasi-neutrality, then, is

$$\frac{\epsilon_s kT}{q} \left| -\frac{1}{N^2} \left(\frac{dN}{dz} \right)^2 + \frac{1}{N} \frac{d^2 n}{dz^2} \right| \ll qN, \tag{3.131}$$

i.e.,

$$\frac{\epsilon_s kT}{q^2 N} = \lambda_D{}^2 \gg \left[\left| -\frac{1}{N^2} \left(\frac{dN}{dz} \right)^2 + \frac{1}{N} \frac{d^2 n}{dz^2} \right| \right]^{-1}. \tag{3.132}$$

The parameter λ_D, the extrinsic Debye length, is a characteristic length for space charge screening. The mobile charge screens the disturbance created by the fixed charge over the Debye length scale. An interesting example of this is shown in Figure 3.16 which shows the movement of mobile charge resulting from the placement of a plane of doping in the semiconductor. The characteristic length of the electron tail is approximately the Debye length. The figure points out the severe consequences of rapid change in doping. So long as the rate of change in doping is slow enough on the Debye length scale, the material can be considered quasi-neutral. Debye lengths for various dopings in GaAs are summarized in Table 3.1. A rapid change in doping can break down the condition of charge neutrality in a semiconductor, and associated space charge effects must be considered. Only when the Debye length scale is larger than the length scale of doping change, as described in Equation 3.132, should one consider the material charge neutral. A lower doping makes the Debye length large, and hence maintaining

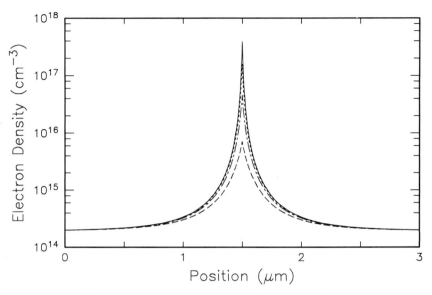

Figure 3.16: Electron concentration as a function of position resulting from the placement of planes of doping varying in sheet density from 1×10^{11} cm^{-2} to 3.3×10^{12} cm^{-2} in lightly doped GaAs.

charge neutrality easier. Likewise, high doping with a high rate of change in it has the opposite effect. An interesting aside is that exponential profiles are always quasi-neutral; for $N(z) = N_0 \exp(-z/\lambda)$ because

$$\left| -\frac{1}{N^2}\left(\frac{dN}{dz}\right)^2 + \frac{1}{N}\frac{d^2n}{dz^2} \right| = -\frac{1}{N_0^2}\frac{N_0^2}{\lambda^2} + \frac{1}{N_0}\frac{N_0}{\lambda^2} = 0. \qquad (3.133)$$

We can also derive the condition for a varying composition structure. Intuitively, a rapid rate of change in composition should lead to breakdown of quasi-neutrality. An abrupt interface, e.g., should have a large interface charge and a significant accumulation or depletion of charge at it. On the other hand, a linearly varying ϕ_{Cn} should leave the material quasi-neutral, with the only effect being that of a quasi-electric field.

Application of the same approach of compositionally homogenous material to this compositionally heterogeneous case leads to the following inequality for maintaining quasi-neutrality (see Problem 10):

$$\lambda_D{}^2 \gg \left[\left| -\frac{1}{N^2}\left(\frac{dN}{dz}\right)^2 + \frac{1}{N}\frac{d^2n}{dz^2} + \left(\frac{q}{kT}\right)^2\frac{d^2\phi_{Cn}}{dz^2} \right| \right]^{-1}. \qquad (3.134)$$

As expected, a linearly varying ϕ_{Cn}, while satisfying the homogenous case conditions, leads to quasi-neutrality. The effect here is the same as an exponentially changing doping. An interesting aside is that for any condition under which a compositionally homogenous material is not quasi-neutral, there exists, theoretically, a specific compositional profile where it would be quasi-neutral in a compositionally heterogeneous material.

3.5.3 High Frequency Small-Signal Analysis

While a detailed understanding of the quasi-static behavior of a device is extremely useful, both in understanding the specifics of the functioning of the device and in its correlation and verification using practical and easier quasi-static measurements, it is not, and can not be, a complete predictor of the behavior of the devices at high frequencies. High frequencies bring with them dispersive effects. The transport and storage of carriers is affected by the frequency more strongly than is predicted by quasi-static modelling, and the signal attenuates more strongly due to losses associated with frequency-dependent phase and amplitude effects. A reason for this is that rapid changes in signal, associated with high frequencies or fast transients, cause a large time-dependent displacement current. The effect of this is only partly included in quasi-static models where capacitors are included to model charge modulation and storage effects. Current transport is an aggregate of particle current and displacement current. Thus, current response at high frequencies is different from that under quasi-static conditions and hence the device behavior is too. Therefore, for high frequency and high speed devices, in order to understand their behavior at those frequencies, we need to model the physics of the devices at those frequencies.

Specific examples, pertaining to both the field effect transistor and the bipolar transistor, clarify this importance of modelling at the frequency conditions of interest. Consider the generic field effect transistor first. When a fast time-varying signal is applied at the gate of the transistor, as the control region underneath the gate also varies, the time-varying channel current responds to these changes. In the case of a MESFET or a junction FET (JFET), the control mechanism is the movement of the edge of the gate depletion region which occurs together with the charging and discharging

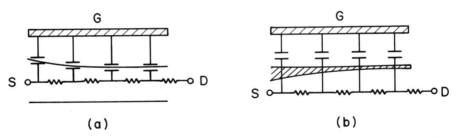

Figure 3.17: A schematic distributed transmission-line representation for modelling the time-varying dependence of channel control by the gate. (a) shows the effect in a MESFET and (b) in an HFET.

at region's edge; in the case of insulating gate field effect transistors (examples: MOSFET, MISFET, SISFET), this is build-up or build-down of the charge in the channel. This charging and discharging occurs through both the displacement and particle current flowing in the channel, and mostly or entirely displacement current flowing in the space charge or the insulator region. The conductance in the channel region is dependent both on time and position while the control region responds. The response of the control region is, therefore, not identical at the different positions along the gate, and a simple description of this phenomenon would be to model it as a distributed transmission line of conductances and capacitances as shown in Figure 3.17. The response, in amplitude and phase, of such distributed transmission lines is frequency-dependent.

In a bipolar transistor, displacement currents are particularly important in the base-collector space charge regions. For a moving charge packet in a depleted region, the termination of the electric field (an opposite polarity charge) associated with the charge moves from one edge of the depleted transit region to the other. The transit of the charge causes a displacement current since the electric field changes during transit of this charge. Quasi-static arguments would ignore this current, but, at high frequencies, its effect is substantial, and the effect of transport delay can not be ignored. Simplistically thinking, one would expect the effect of this carrier transit to correspond to a time delay equal to the time it takes to transit this space charge region. The delay in current response, however, does not correspond to the transit time, and about half its value is a better approximation if the carriers transit at a constant velocity throughout this space charge region. We will discuss this further in our treatment of bipolar transistors; it is a consequence of both displacement and particle current being present. We generally ignore the amount of time it takes for a signal to cause a change

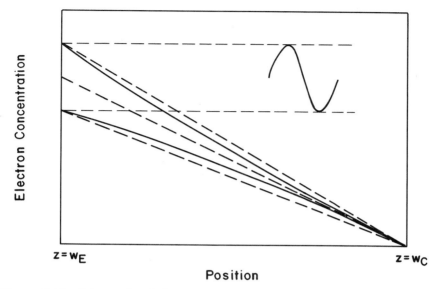

Figure 3.18: Schematic of the response in minority carrier density in the base of a n–p–n bipolar transistor with a signal of cycle time equal to twice the base transit time. The dashed lines represent the electron concentration distribution under quasi-static conditions.

in the carrier concentration at the base-emitter junction. It is taken to be instantaneous. Similar to the delay in the base-collector space charge region, it should be incorporated if it becomes a significant factor in the operation of the high frequency device. Fortunately, for useful bias conditions of the bipolar transistor, the time it takes to store the charge in the base-emitter space charge region, which occurs at the same time as this transit of carriers, is significantly larger and hence we may ignore the latter.

A third example in the bipolar transistor that clarifies the frequency dispersion is shown in Figure 3.18. Assume that the particle density at the base-emitter junction instantaneously responds to an applied signal, i.e., transit through base-emitter space charge region and the storage in the space charge region is not limiting. If, now, one applies a high frequency signal of a cycle-time twice the amount of time it takes for the carrier to transit the quasi-neutral base region (no displacement current here since it is quasi-neutral), how would the particle density respond? The input signal goes through half a cycle in the amount of time it takes for carriers injected at the base-emitter junction to reach the base-collector junction.

So, carriers reaching the base-collector junction are out of phase with the signal at the injecting junction. Quasi-static analysis assumes that they are in phase. Dissipation occurs as a result of this phase lag, and is important to the high frequency operation, and especially operation at a fraction of the ultimate figures of merit that are of interest to us.

So, a small-signal analysis attempts to incorporate such effects. For such problems, within approximations of the drift-diffusion approach, this may be done by taking the current continuity equation and incorporating the frequency-dependent terms in the equation. The current continuity equation is a linear differential equation, and hence, for small-signals, linear superposition techniques can be applied. For the problem, this would mean that as a function of position z, time t, and a sinusoidal signal of frequency ω (implicitly understood as the real part a complex signal varying as $\exp(j\omega t)$ for the sake of simplicity) applied, the current densities and voltages may be written as

$$J(z,\omega,t) = \overline{J}(z) + \tilde{J}(z,\omega,t) = \overline{J}(z) + \hat{J}(z,\omega)\exp\left(j\omega t\right), \qquad (3.135)$$

and

$$V(z,\omega,t) = \overline{V}(z) + \tilde{V}(z,\omega,t) = \overline{V}(z) + \hat{V}(z,\omega)\exp\left(j\omega t\right). \qquad (3.136)$$

The overline indicates the quasi-static current; the hat indicates the phasor of the complex signal. The sinusoidal signal is the real part of the complex signal.

Electron densities, hole densities, and other parameters vary with frequency in a similar fashion, because of the superposition principle. The one-dimensional current density, and current continuity equations, with these substitutions, are

$$
\begin{aligned}
\overline{J}_n(z) + \hat{J}_n(z,\omega)\exp\left(j\omega t\right) = {}& -q\left[\overline{n}(z) + \hat{n}(z,\omega)\exp\left(j\omega t\right)\right]\mu_n \times \\
& \left[\frac{\partial}{\partial z}\overline{V}(z) + \frac{\partial}{\partial z}\hat{V}(z,\omega)\exp\left(j\omega t\right)\right] + \\
& q\mathcal{D}_n\left[\frac{\partial}{\partial z}\overline{n}(z) + \frac{\partial}{\partial z}\hat{n}(z,\omega)\exp\left(j\omega t\right)\right],
\end{aligned}
$$

$$(3.137)$$

$$
\begin{aligned}
\overline{J}_p(z) + \hat{J}_p(z,\omega)\exp\left(j\omega t\right) = {}& -q\left[\overline{p}(z) + \hat{p}(z,\omega)\exp\left(j\omega t\right)\right]\mu_p \times \\
& \left[\frac{\partial}{\partial z}\overline{V}(z) + \frac{\partial}{\partial z}\hat{V}(z,\omega)\exp\left(j\omega t\right)\right] -
\end{aligned}
$$

$$q\mathcal{D}_p \left[\frac{\partial}{\partial z}\overline{p}(z) + \frac{\partial}{\partial z}\hat{p}(z,\omega) \exp\left(j\omega t\right) \right],$$

(3.138)

$$\frac{\partial}{\partial t}\left[\overline{n}(z) + \hat{n}(z,\omega) \exp\left(j\omega t\right)\right] = \overline{\mathcal{G}}_n + \hat{\mathcal{G}}_n \exp\left(j\omega t\right) -$$
$$\overline{\mathcal{R}}_n - \hat{\mathcal{R}}_n \exp\left(j\omega t\right) +$$
$$\frac{1}{q}\frac{\partial}{\partial z}\left[\overline{J}_n(z) + \hat{J}_n(z,\omega) \exp\left(j\omega t\right)\right],$$

(3.139)

and

$$\frac{\partial}{\partial t}\left[\overline{p}(z) + \hat{p}(z,\omega) \exp\left(j\omega t\right)\right] = \overline{\mathcal{G}}_p + \hat{\mathcal{G}}_p \exp\left(j\omega t\right) -$$
$$\overline{\mathcal{R}}_p - \hat{\mathcal{R}}_p \exp\left(j\omega t\right) -$$
$$\frac{1}{q}\frac{\partial}{\partial z}\left[\overline{J}_p(z) + \hat{J}_p(z,\omega) \exp\left(j\omega t\right)\right].$$

(3.140)

These set of equations can be rewritten separately for the time independent and and time-dependent part (varying with $\exp\left(j\omega t\right)$, i.e., only the fundamental terms) as

$$\overline{J}_n(z) = -q\overline{n}(z)\mu_n \frac{\partial}{\partial z}\overline{V}(z) + q\mathcal{D}_n\frac{\partial}{\partial z}\overline{n}(z),$$

(3.141)

$$\overline{J}_p(z) = -q\overline{p}(z)\mu_p \frac{\partial}{\partial z}\overline{V}(z) - q\mathcal{D}_p\frac{\partial}{\partial z}\overline{p}(z),$$

(3.142)

$$0 = \overline{\mathcal{G}}_n - \overline{\mathcal{R}}_n + \frac{1}{q}\frac{\partial}{\partial z}\overline{J}_n(z),$$

(3.143)

and

$$0 = \overline{\mathcal{G}}_p - \overline{\mathcal{R}}_p - \frac{1}{q}\frac{\partial}{\partial z}\overline{J}_p(z),$$

(3.144)

for the steady-state terms, and

$$\hat{J}_n(z,\omega) = -q\left[\overline{n}(z)\frac{\partial}{\partial z}\hat{V}(z,\omega) + \hat{n}(z,\omega)\frac{\partial}{\partial z}\overline{V}(z)\right] + q\mathcal{D}_n\frac{\partial}{\partial z}\hat{n}(z,\omega),$$

(3.145)

$$\hat{J}_p(z,\omega) = -q \left[\overline{p}(z)\frac{\partial}{\partial z}\hat{V}(z,\omega) + \hat{p}(z,\omega)\frac{\partial}{\partial z}\overline{V}(z) \right] - q\mathcal{D}_p\frac{\partial}{\partial z}\hat{p}(z,\omega), \quad (3.146)$$

$$\frac{\partial}{\partial t}\hat{n}(z,\omega) = \hat{\mathcal{G}}_n - \hat{\mathcal{R}}_n + \frac{1}{q}\frac{\partial}{\partial z}\hat{J}_n(z,\omega), \quad\quad (3.147)$$

and

$$\frac{\partial}{\partial t}\hat{p}(z,\omega) = \hat{\mathcal{G}}_p - \hat{\mathcal{R}}_p - \frac{1}{q}\frac{\partial}{\partial z}\hat{J}_p(z,\omega), \quad\quad (3.148)$$

for the time-varying terms.

Network parameters can be determined from these, the most common one in theoretical analysis being being y-parameters, because we directly determine current as a function of voltage. From experiments, S-parameters are more common since they are directly available from the apparatus, while equivalent circuits are best described using h-parameters because of the high output conductance and hence current source nature of both field effect transistors and bipolar transistors. Network parameters are treated separately in the Appendix A. These network parameters serve several useful purposes. Microwave circuit design and small-signal equivalent circuit modelling can be conveniently performed using them. Device stability can be evaluated because stability factors (e.g., the Linvill stability factor) can be easily determined. Similarly, the ultimate limits of the device both in gain (such as the unilateral gain or the maximum available gain) and frequency figures of merit (such as short-circuit unity current gain frequency and maximum frequency of oscillation) can be directly determined. The network parameters, thus, serve several useful purposes.

This analysis is still within the limits of drift-diffusion approximation. It does not include the off-equilibrium effects that occur at small dimensions and which allow achievement of the next level of accuracy. In the case of our FET example, the movement of the depletion region occurs with velocities not limited by the steady-state mobility-field relationship that is incorporated in the transport equation, but at higher velocities, because significant off-equilibrium effects occur due to the gradient in the electric field at the edge of the space charge region. This was shown in the example of p-n junction in our discussion of Monte Carlo approach.

Transient and high frequency large signal conditions are two other places where the quasi-static modelling is inadequate. We will discuss transient operation as part of the treatment on devices; large signal operation will not be covered in the text.

3.6 Boundary Conditions

Device analysis requires us to describe the operation in a chosen part. This part defines the section that we may be capable of describing mathematically, and that is of interest to us. However, we must be able to describe the couplings of this section at all its boundaries. These couplings, the boundary conditions, must be physically representative and must not cause a change in the result from what it would have been in the absence of this sectioning. For example, in a bipolar transistor, we may wish to describe the behavior in the semiconductor region through some representative equations. But the coupling of the current supplied from the power supplies, and carried in the device by the charged carriers, electrons and holes, requires us to suitably interface the two in the form of some mathematical model. And this holds true for voltages, etc., too. We may wish to describe the transport as a one-dimensional simplification in the quasi-neutral part of a base, in which case we must be able to mathematically model the two ends, the base-emitter junction and the base-collector junction, and these must account for the behavior of the emitter and the collector. Thus, boundary conditions, if chosen properly, allow us to simplify the analysis of a problem and give us a better physical insight by isolating the problem. In this section, we will discuss some of the commonly employed boundary conditions in device analysis by a variety of techniques. As a consequence of the discussion, we should also get insight into the simplifications and the underlying behavior that leads to the choice of these boundary conditions.

We discuss the more commonly used boundary conditions for drift-diffusion analysis first.[10] Most devices of interest to us incorporate p–n junctions: in bipolar transistors for injection and collection, and in n-channel field effect transistors for isolation of the device from the substrate. Many of these devices also incorporate ohmic contacts, metal–semiconductor contacts, or heterojunctions to contact and modulate the carriers. Our discussion is related to these device structures, and we will create mathematical models that can be easily introduced for drift-diffusion analysis first.

3.6.1 Shockley Boundary Conditions

Let us first look at the common approximation of the constant quasi-Fermi level in the space charge region of a homogenous p–n junction (see Figure 3.19). Since the current carried by each carrier can be expressed in

---

[10]For an extended discussion of these, see R. M. Warner and B. L. Grung, *Transistors: Fundamentals for the Integrated-Circuit Engineer*, John Wiley, N.Y. (1983).

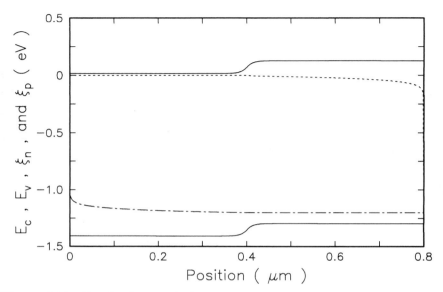

Figure 3.19: Band edges and the quasi-Fermi levels of a symmetrically doped n–p junction of GaAs in forward bias.

terms of the quasi-Fermi levels for the carrier, we can derive the quasi-Fermi level change from the edge of the depletion region of p and n sides as

$$\exp\left(\frac{\xi_n}{kT}\right)\bigg|_{z=w_n} - \exp\left(\frac{\xi_n}{kT}\right)\bigg|_{z=-w_p} = \int_{-w_p}^{w_n} \frac{J_n \exp\left(E_i/kT\right)}{qn_i kT \mu_n} dz, \quad (3.149)$$

for electrons. Hence the change in the electron quasi-Fermi level is

$$\begin{aligned}
\Delta\xi_n &= \xi_n|_{z=w_n} - \xi_n|_{z=-w_p} \\
&= kT\ln\left\{1 + \frac{1}{qn_i kT \mu_n}\int_{-w_n}^{w_p} |J_n| \exp\left[\left(E_i - \xi_n|_{z=-w_p}\right)/kT\right]\right\}.
\end{aligned}$$
$$(3.150)$$

The change in the quasi-Fermi level is logarithmically related to the integral; a large current flow may cause a significant drop in the quasi-Fermi levels. To check the validity, let us assume that a significant current density is flowing through the structure to cause the variation in quasi-Fermi level position. Since a large current flows, a near flat band condition occurs. The width of the space charge region, under these conditions, is small—certainly less than 1000 Å for typical junction conditions involving current densities of 10^4 A.cm^{-2}. Therefore, we may calculate the approximate magnitude of this integral; for GaAs, an approximate magnitude of this is 10^{-2} eV assuming a mobility of 4000 cm^2.V^{-1}.s^{-1}. Thus, our assumption of flat quasi-Fermi level is well justified for this example. Also, given this, the amount of current across this junction is determined by the amount of current being extracted from it by the other regions of the device, e.g., the base of the bipolar or the base of the p–n junction (the quasi-neutral region). This does not always have to be so. If the mobility of the material is extremely poor, or a very high current density is attempted to be extracted, or the temperature is low, the quasi-Fermi levels may change by substantially more than the thermal voltage, since it directly affects the integral. Devices based on p–n junctions, operating with small changes in the quasi-Fermi levels in any region, are said to be operating with low level injection in those regions.

Analysis of transport in the quasi-neutral regions can be simplified in drift-diffusion analysis because of the above. In a p–n junction, under these low level injection conditions, where the quasi-Fermi level is flat, the carrier concentrations on the n-side and p-side are related through the difference in the electrostatic potential. Under thermal equilibrium, this electrostatic potential is the junction built-in voltage ψ_{j0}, the contact potential of the junction. This is given by:

$$\psi_{j0} = \frac{kT}{q} \ln \left(\frac{N_D N_A}{n_i{}^2} \right) = \frac{kT}{q} \ln \left(\frac{n_{n0} p_{p0}}{n_i{}^2} \right), \tag{3.151}$$

where N_D and N_A are the donor and acceptor densities, and n_{n0} and p_{p0} are the majority carrier densities in the n-side and p-side at thermal equilibrium. From the law of mass action, the minority carrier densities in either regions, p_{n0} and n_{p0}, also satisfy

$$p_{n0} n_{n0} = n_{p0} p_{p0} = n_i{}^2, \tag{3.152}$$

for homogenous junctions.

With application of a bias of V, the electrostatic potential change at the junction is $\Delta \psi = \psi_{j0} - V$, so that

$$n_p(-w_p) = n_{n0} \exp \left(-q \frac{\psi_{j0} - V}{kT} \right)$$

$$\text{and} \quad p_n(w_n) = p_{p0} \exp \left(-q \frac{\psi_{j0} - V}{kT} \right). \tag{3.153}$$

Under bias, the majority and minority carrier concentrations both change. In low level injection, the percent change in majority carrier concentration is insignificant compared to that in minority carrier concentration. Consequently electron and hole majority carrier population change can be ignored and Equation 3.153 still remains valid. This allows us to write the equations

$$n_p(-w_p) = n_{p0} \exp\left(\frac{qV}{kT}\right)$$

$$\text{and} \quad p_n(w_n) = p_{n0} \exp\left(\frac{qV}{kT}\right). \tag{3.154}$$

These equations are commonly referred to as the law of the junction or Shockley boundary conditions. In the rest of the book, we will refer to these as Shockley boundary conditions. The use of the term "law of the junction" implies a broad validity; these equations are valid for only low level injection conditions in homogenous junctions. In compound semiconductor devices, quite often, we deal with heterogeneous junctions, where this relationship is quite often inappropriate.

Shockley boundary conditions describe an exponential relationship between carrier densities at both edges of the depletion region. This exponential factor is sometimes referred to as the Boltzmann factor; its derivation depends on low level injection and Boltzmann approximation being valid. In the derivation of the Shockley boundary conditions, we made two principal assumptions. The first is that we extract a very insignificant fraction of the current that the junction is capable of supplying, i.e., the current is much smaller than either the diffusion or drift current carried by either of the carriers at the junction. We may estimate the latter. If the junction has a doping density of 1×10^{18} cm^{-3} and the carriers moved at a velocity of 10^7 cm.s^{-1}, then the maximum current density can be nearly 1.6×10^6 A.cm^{-2}. The second assumption in the derivation was related to there being an insignificant change in the majority carrier concentration at either side of the junction.

3.6.2   Fletcher Boundary Conditions

Quite often, the latter assumption is the first one to breakdown under bias in actual devices, and it happens in the lower doped side of the junction, e.g., in the base of a p–n junction or a bipolar transistor. Under these conditions, the quasi-Fermi level is still be flat within the junction region, but the majority carrier density in one of the regions is affected by the minority carrier injection. Boundary conditions formulated by Fletcher and Misawa are more useful under these conditions. We consider the Fletcher boundary conditions first. We have derived the relationship between the carrier concentrations at the edges of the depletion region (Equation 3.153).

Rewriting these equations using V_j for the voltage across the junction,

$$p(w_n) = p(-w_p) \exp\left(-\frac{q\psi_{jo}}{kT}\right) \exp\left(\frac{qV_j}{kT}\right)$$

$$\text{and} \quad n(-w_p) = n(w_n) \exp\left(-\frac{q\psi_{jo}}{kT}\right) \exp\left(\frac{qV_j}{kT}\right). \quad (3.155)$$

Since

$$\exp\left(-\frac{q\psi_{jo}}{kT}\right) = \frac{p_{n0}}{p_{p0}} = \frac{n_{p0}}{n_{n0}} \quad (3.156)$$

from the thermal equilibrium, we may rewrite these as

$$p(w_n) = p(-w_p)\frac{p_{n0}}{p_{p0}} \exp\left(\frac{qV_j}{kT}\right),$$

$$\text{and} \quad n(-w_p) = n(w_n)\frac{n_{p0}}{n_{n0}} \exp\left(\frac{qV_j}{kT}\right). \quad (3.157)$$

The increase in majority carrier concentration at the edge of the depletion region, which we ignored in the Shockley boundary condition, occurs in order to maintain charge neutrality. Its increase therefore is equal to the increase in the minority carrier charge at the edge of the depletion region, i.e.,

$$p(-w_p) - p_{p0} = n(-w_p) - n_{p0},$$

$$\text{and} \quad n(w_n) - n_{n0} = n(w_n) - n_{n0}. \quad (3.158)$$

These four equations describe the carrier concentrations at the edge of the depletion region under high injection conditions but still assume flat quasi-Fermi levels in the junction region (we ascribe the potential V_j to the change in the junction potential). By elimination, they can be written in a more appropriate form relating to the thermal equilibrium concentrations (see Problem 11) as

$$p(w_n) = \left[n_{n0}p_{p0} - n_{p0}p_{n0} \exp\left(\frac{2qV_j}{kT}\right)\right]^{-1} \times$$

$$\left[(p_{p0} - n_{p0})\, n_{n0}p_{n0} \exp\left(\frac{qV_j}{kT}\right) + \right.$$

$$\left. (n_{n0} - p_{n0})\, n_{p0}p_{n0} \exp\left(\frac{2qV_j}{kT}\right)\right]$$

$$n(-w_p) = \left[n_{n0}p_{p0} - n_{p0}p_{n0} \exp\left(\frac{2qV_j}{kT}\right)\right]^{-1} \times$$

$$\left[(n_{n0} - p_{n0})\, n_{p0} p_{p0} \exp\left(\frac{qV_j}{kT}\right) + \right.$$

$$\left. (p_{p0} - n_{p0})\, n_{p0} p_{n0} \exp\left(\frac{2qV_j}{kT}\right) \right]$$

$$p(-w_p) = \left[n_{n0} p_{p0} - n_{p0} p_{n0} \exp\left(\frac{2qV_j}{kT}\right) \right]^{-1} \times$$

$$\left[(p_{p0} - n_{p0})\, n_{n0} p_{p0} + (n_{n0} - p_{n0})\, n_{p0} p_{p0} \exp\left(\frac{qV_j}{kT}\right) \right]$$

$$n(w_n) = \left[n_{n0} p_{p0} - n_{p0} p_{n0} \exp\left(\frac{2qV_j}{kT}\right) \right]^{-1} \times$$

$$\left[(n_{n0} - p_{n0})\, n_{n0} p_{p0} + (p_{p0} - n_{p0})\, n_{n0} p_{n0} \exp\left(\frac{qV_j}{kT}\right) \right].$$

$$(3.159)$$

These are the Fletcher boundary conditions. They relate the electrostatic potential change at the junction V_j, assuming no change in the quasi-Fermi levels in the junction depletion region, with the carrier densities at the edge of the depletion region. In the above relationships, changes in the quasi-Fermi levels may take place in the rest of the quasi-neutral region; their effect is included.

3.6.3 Misawa Boundary Conditions

Figure 3.20 shows the quasi-Fermi levels and the electrostatic potential at a p–n diode under high level injection conditions. Any applied bias at the ohmic contact can drop at the ohmic contacts, in the quasi-neutral regions, and at the junctions. The potential V_j used in the Fletcher boundary condition equations above is the part of the applied bias voltage that dropped at the junction. We considered only the junction part in our analysis, and need to relate it to the electrostatic potential changes across the device itself. As Figure 3.20 shows, this involves knowing the changes that occur at the ohmic contacts and in the quasi-neutral regions. We can find the other potential drops as a continuation of our analysis, and another reason for doing this is that the Fletcher boundary conditions are also somewhat cumbersome in appearance. A simpler form results if we write these as products of carrier densities, and as functions of the quasi-Fermi level splitting. This leads to a form known as the Misawa boundary conditions.

As an extension of the discussion of Figure 3.20 consider the quasi-Fermi level splitting and voltage drops across the junction. Excess potential drop occurs in both the n-doped and the p-doped side. Of the applied

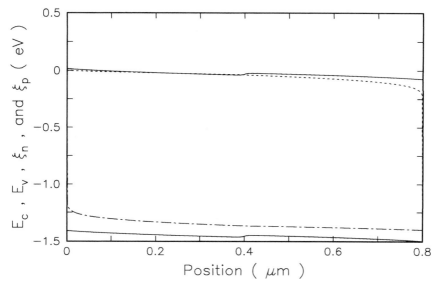

Figure 3.20: Band edges and the quasi-Fermi levels at a GaAs n–p junction in high level injection.

terminal voltage V, only part of it (V_j) drops across the junction, and the rest of it drops across in the quasi-neutral regions. The quasi-Fermi level splitting at the junction is less than the applied voltage. The relationship in the voltage across the junction and the quasi-Fermi level splitting, again assuming Maxwell–Boltzmann relationship, is

$$V_j = \phi_p - \phi_n + \frac{kT}{q} \ln \left[\frac{p(-w_p)}{p_{p0}} \right] + \frac{kT}{q} \ln \left[\frac{n(w_n)}{n_{n0}} \right]. \qquad (3.160)$$

This relationship follows from the definition of an ohmic contact as a junction where the quasi-Fermi levels merge and the carrier concentrations are equal to their thermal equilibrium values. Misawa conditions can then be derived, and these state (see Problem 12)

$$[n(-w_p) + N_A] \, n(-w_p) = p_{p0} n_{p0} \exp \left(q \frac{\phi_p - \phi_n}{kT} \right)$$

$$\text{and} \quad p(w_n) \, [p(w_n) + N_D] = p_{n0} n_{n0} \exp \left(q \frac{\phi_p - \phi_n}{kT} \right). \quad (3.161)$$

The significance of the Misawa boundary conditions is in their simplicity. They are a restatement of the Fletcher boundary conditions, having been derived from them together with the charge neutrality equations. The approximations made in deriving Fletcher boundary conditions are also the approximations for Misawa boundary conditions.

The equations can also be recast in another form in terms of only the carrier concentrations. The product of the carrier concentrations at the edges of the depletion region can be written from above using the charge neutrality conditions as (see Problem 13)

$$n(-w_p)p(-w_p) = p_{p0}n_{p0} \exp\left(q\frac{\phi_n - \phi_p}{kT}\right)$$

$$\text{and} \quad n(w_n)p(w_n) = p_{n0}n_{n0} \exp\left(q\frac{\phi_n - \phi_p}{kT}\right). \quad (3.162)$$

3.6.4 Dirichlet Boundary Conditions

We stated earlier that we may model the ohmic contact as a junction where the carrier concentrations are equal to their values at the thermal equilibrium. This has the physical meaning that the surface recombination velocity at the contact is so large that any excess carrier concentration disappears due to recombination. In reality, of course, ohmic contacts may have smaller recombination velocities. It is also possible to have a method of ohmic contact formation that allows efficient ohmic flow of carriers of only one type as in the graded bandgap contacts. $Ga_{1-x}In_xAs$ (n-type) and $Ga_{1-x}Sb_xAs$ (p-type) are two examples of such contacts to n-type and p-type GaAs, and we will discuss these examples in Chapter 4. The former has a barrier to hole flow and the latter a barrier to electron flow. So, ohmic contacts may be based on different principles, but they serve one purpose: efficient supply and gathering of carriers. A good ohmic contact provides any requisite amount of carriers with negligible voltage drop across it, so as to not limit the device operation.

The most common types of ohmic contacts are based on the use of heavy doping at an interface between a metal and a semiconductor with tunneling as the physical basis for ohmicity. Most allow one of the carrier types to tunnel. Near the surface region, carrier concentrations are determined by the characteristics of this flow which we may embody in the surface recombination velocity S of the contact. For a minority carrier in p-type material, the current is

$$J_n = qS(n - n_0). \quad (3.163)$$

For a p–n junction problem, this will now have to be solved self-consistently with the the other transport parameters (see Problem 29 of Chapter 4). If the surface recombination were infinite in the above, the carrier concentration would be its thermal equilibrium value and the contact would be ohmic for the minority carrier. If it is low, it may represent an emission velocity for minority carrier into the metal, which may, for some problems, be an adequate representation for metal–semiconductor contact. These various conditions at interfaces bring out some common features employed in drift-diffusion modelling in accord with the behavior of these interfaces. Mathematically, these features can be represented in the form of an identity on the electrostatic potential or the carrier concentration, and is referred to as a Dirichlet boundary condition.

As an example, consider the ohmic contact. There could actually be excess immobile charge qN at the interface. The behavior of the parameters would follow the following mathematical equalities:

$$\psi - \psi_{j0} - V = 0 \tag{3.164}$$

for the electrostatic potential in a voltage controlled contact, and

$$\int \mathbf{J}.d\mathbf{A} - I = 0 \tag{3.165}$$

for current. Here, $d\mathbf{A}$ is the vector normal to the elemental area with the magnitude of the elemental area. At thermal equilibrium, the carrier densities can be found by using the law of mass action and the equation for charge neutrality. Let the excess immobile charge at the interface be qN. The carrier density equations at the contact are:

$$np = n_i{}^2$$
$$\text{and} \quad n - p = N. \tag{3.166}$$

The last two equations for carrier densities at the contact may also be written as

$$n = \frac{1}{2}\left(N + \sqrt{N^2 + 4n_i{}^2}\right)$$
$$\text{and} \quad p = \frac{1}{2}\left(-N + \sqrt{N^2 + 4n_i{}^2}\right). \tag{3.167}$$

These constitute the Dirichlet boundary conditions for ohmic contacts.

Similar boundary conditions may also be written for rectifying metal–semiconductor contacts, although many of the diverse behaviors of various metal–semiconductor contacts can not be written as simply. If the barrier height for injection from the metal to the semiconductor is ϕ_B, the contact potential is ψ_{j0}, the surface recombination velocity for the contact is S_n

and S_p for electrons and holes, then the Dirichlet boundary conditions can be derived as follows in a somewhat simplified form. The potential follows the relationship

$$\psi - \psi_{j0} + \phi_B - V = 0. \tag{3.168}$$

Using the definition of recombination velocity and the charge relationships, we obtain the current and carrier densities are

$$J_n = qS_n \left[n - \frac{1}{2} \left(N + \sqrt{N^2 + 4n_i{}^2} \right) \right]$$

$$\text{and} \quad J_p = qS_p \left[p - \frac{1}{2} \left(-N + \sqrt{N^2 + 4n_i{}^2} \right) \right]. \tag{3.169}$$

These equations constitute one form of Dirichlet boundary conditions for a metal and semiconductor interface with surface recombination velocities of S_n and S_p.

3.6.5   Neumann Boundary Conditions

Besides ohmic and rectifying interfaces, boundaries between semiconductors and insulators are also encountered in practice. No or negligible current flows through the insulators. Gauss's law requires that for the electric fields normal to the interface $\mathcal{E}_n$ in the semiconductor and in the insulator,

$$\epsilon_s \mathcal{E}_n|_{sem.} - \epsilon_{ins} \mathcal{E}_n|_{ins.} = Q_{surf}, \tag{3.170}$$

where Q_{surf} is the interface charge density. If the insulator can be assumed to be infinitely thick, the field in the insulator can be ignored and one has

$$\epsilon_s \mathcal{E}_n|_{sem.} = Q_{surf}. \tag{3.171}$$

If one further assumes that the surface is charge-free, then

$$\mathcal{E}_n|_{sem.} = 0, \tag{3.172}$$

which is often referred to as the Neumann boundary condition. In compound semiconductors, there is significant surface state density and hence quite often a significant surface charge. The Neumann boundary condition is rarely valid. Fermi level pinning at the surface, which results from this surface charge, may be incorporated in two ways. We may make the ad hoc assumption that the surface potential is pinned due to this Fermi level pinning; this is akin to an electrostatic potential Dirichlet boundary condition. Alternately, we may consider a large surface state density of donor and acceptor traps that do not allow Fermi level excursion because of charge imbalance. In this case we may use charge neutrality including these interface charges, as well as Gauss's law at the interface.

3.7 Generation and Recombination

In our discussion of the BTE, we emphasized the scattering events that lead to carriers changing momentum and energy within the conduction bands or the valence bands. In our derivation of the drift-diffusion equations from the BTE, processes in between the conduction bands and the valence bands, described as generation and recombination processes since they involved creation or annihilation of electron–hole pairs, were introduced as another form of scattering event.

Such interactions between carriers of opposite types are particularly important in compound semiconductors because most of them are direct bandgap semiconductors. Additional interactions between electrons and holes through levels within the bandgap, due to traps, are also important in compound semiconductors. So, several forms of excitation processes can result in either the recombination of electron and hole pairs or their generation. Radiative recombination, usually involving band to band transitions with the energy released as photons, is the basis of light-emitting diodes and lasers. Radiative recombination can also occur sometimes through discrete levels, e.g., in the Zn-O center combination in GaP. This is a unique example where one Zn-O pair forms an isoelectronic recombination center by replacing a Ga-P pair with an energy level in the bandgap. Recombination can also occur without the involvement of radiative processes, such as through traps, through other processes where momentum conservation involves phonons, and through Auger processes where excess energy of annihilation or creation of an electron–hole pair is associated with a third carrier. All these processes are important under different conditions in devices.

In order to understand the statistics of these processes and their significance, we use the principle of detailed balance which states that, for a system in thermal equilibrium, the rate of any process and its inverse balance each other. The adjective "detailed" emphasizes that this balance occurs in all details of the process. Consider radiative recombination in thermal equilibrium. The detailed balance requires that, in thermal equilibrium, the same fraction of electron–hole recombination be radiative as is the electron–hole pair generation, and with the same spectral characteristics as the radiation that occurs with electron–hole pair recombination.

The principle of detailed balance is a statistical principle, whose manifestations include the principle of microscopic reversibility for probabilities instead of rates, or Kirchoff's law. The principle of detailed balance was used implicitly in setting the net of scattering rates for all the scattering processes to zero, in thermal equilibrium, in our discussion of the BTE.

3.7.1   Radiative Recombination

Let ν be the photon frequency; the principle of detailed balance allows us to write the radiation rate in the band of $d\nu$, at the frequency ν, as

$$\mathcal{R}(\nu)d\nu = \mathcal{P}(\nu)\varrho(\nu)d\nu, \tag{3.173}$$

where $\mathcal{R}$ is the emission rate, $\mathcal{P}$ the net probability per unit time of absorbing a photon, and ϱ is the photon density at the frequency ν given by

$$\varrho(\nu)d\nu = \frac{8\pi\nu^2\vartheta^3}{c^3}\frac{d\nu}{\exp{(h\nu/kT)} - 1}, \tag{3.174}$$

where ϑ is the index of refraction. The total number of recombinations can now be derived by integrating over the frequency as

$$\mathcal{R} = \frac{8\pi\vartheta^2(kT)^3}{c^3h^3}\int_0^\infty \frac{1}{\tau(\nu)}\frac{\eta^2}{\exp{(\eta)} - 1}d\eta, \tag{3.175}$$

where $\tau(\nu)$ is the mean lifetime of the photon in the semiconductor and $\eta = h\nu/kT$. This is the van Roosbroeck-Shockley relationship for radiative recombination rate, valid in thermal equilibrium, and useful in calculating the shape of the luminescence spectrum given by Equation 3.173 if the photon lifetime or alternately the absorption coefficient $\alpha(\nu) = \vartheta/c\tau(\nu)$ is known.

The radiative recombination rate is proportional to electron and hole densities that are recombining, i.e.,

$$\mathcal{R}_r = c_r np, \tag{3.176}$$

and in thermal equilibrium,

$$\overline{\mathcal{R}}_r = c_r n_0 p_0 = c_r n_i^2, \tag{3.177}$$

so that

$$\mathcal{R}_r = \overline{\mathcal{R}}_r \frac{np}{n_i^2}. \tag{3.178}$$

Away from thermal equilibrium, the change in recombination rate is

$$\mathcal{R}'_e = \mathcal{R}_e - \overline{\mathcal{R}}_e = \frac{\left(n_0 + n'\right)\left(p_0 + p'\right)}{n_0 p_0}\overline{\mathcal{R}}_e - \overline{\mathcal{R}}_e, \tag{3.179}$$

where p' and n' are incremental deviations in carrier concentrations. For small deviations, this gives

$$\mathcal{R}'_e = \left(\frac{n'}{n_0} + \frac{p'}{p_0}\right)\overline{\mathcal{R}}_e. \tag{3.180}$$

The radiative lifetime then follows, since excess populations of either carriers are the same and given by

$$\tau_r = \frac{n'}{\mathcal{R}'_e} = \frac{n_i{}^2}{n_0 + p_0} \frac{1}{\mathcal{R}'_e}. \tag{3.181}$$

The capture rate c_r is usually denoted by the factor $\mathcal{B}$, the probability of radiative recombination, and given by

$$\mathcal{B} = \frac{\overline{\mathcal{R}}_e}{n_i{}^2}, \tag{3.182}$$

using the thermal equilibrium condition. The radiative lifetime follows as

$$\tau_r = \frac{1}{\mathcal{B}(n_0 + p_0)}. \tag{3.183}$$

The constant $\mathcal{B}$ for various compound semiconductors is included in the Appendix B. This description is quite pertinent for moderate doping conditions. It breaks down at very large doping conditions—both experiments and detailed theory indicate that τ_r saturates in the limit of very high doping.

3.7.2 Hall–Shockley–Read Recombination

Traps, because they can exist in more than one charge state, have the ability to act as recombination-generation centers. This ability comes about because these centers can capture an electron and/or a hole from the bands as well as emit one to the bands. Thus, electrons and holes can recombine or lead to a generation process at these traps. The statistics for this were first analyzed by Hall, Shockley, and Read, and are known as Hall–Shockley–Read (HSR) recombination.

We will consider a donor-like trap, of density N_T, in an n-type crystal. Being a donor-like trap, it exists either in a positively charged state, where we consider a donor that exists only as a singly ionized donor of density N_T^+, or in a neutral state with density N_T^0. The singly ionized state of the trap exists in the bandgap at an energy E_T. In thermal equilibrium, the distribution of these densities is

$$N_T = \overline{N_T^+} + \overline{N_T^0}, \tag{3.184}$$

and away from thermal equilibrium conditions,

$$N_T = N_T^+ + N_T^0. \tag{3.185}$$

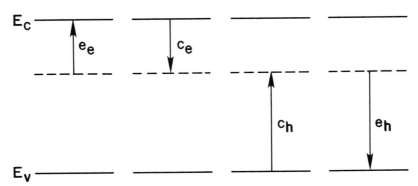

Figure 3.21: Capture and emission processes for electrons and holes and involving a donor-like trap in an n-type semiconductor.

The different processes taking place between the trap and the bands are shown in Figure 3.21. Mathematically, these processes may be represented by

$$N_T^0 \rightleftharpoons N_T^+ + e^-$$
$$\text{and} \quad N_T^0 + h^+ \rightleftharpoons N_T^+, \tag{3.186}$$

where the first is the emission and capture process between the conduction band and the trap, and the second describes the capture and emission process between the valence band and the trap.

The rate of capture of electrons from the conduction band is proportional to the number of positively-charged centers N_T^0 and the number of carriers available for capture n. Thus,

$$\mathcal{R}_{ce} = c_e n N_T^+. \tag{3.187}$$

The rate of emission of electrons to the conduction band is proportional to the number of neutral centers and the available states for capture in the conduction band, i.e.,

$$\mathcal{G}_{ce} = e_e N_T^0 (N_C - n). \tag{3.188}$$

In thermal equilibrium, the rate of capture and emission are the same so,

$$\overline{\mathcal{R}}_{ce} = \overline{\mathcal{G}}_{ce}$$
$$c_e n_0 \overline{N_T^+} = e_e \overline{N_T^0} (N_C - n_0). \tag{3.189}$$

Similarly, the two processes associated with the valence band yield

$$\mathcal{R}_{vh} = c_h p N_T^0 \qquad (3.190)$$

for capture rate of holes, and

$$\mathcal{G}_{vh} = e_h N_T^+ (N_V - p) \qquad (3.191)$$

for emission rate of holes, and in thermal equilibrium

$$\overline{\mathcal{R}}_{vh} = \overline{\mathcal{G}}_{vh}$$
$$c_h p_0 \overline{N}_T^0 = e_h \overline{N}_T^+ (N_V - p_0). \qquad (3.192)$$

The capture rates of the traps are equal to the product of the capture cross-section of the trap, the thermal velocity, and the density of such traps that can capture. The capture cross-section multiplied by the thermal velocity is a volume associated with each trap per unit time. Any carrier—and carriers move with thermal velocities—incident within this volume gets captured. The capture cross-section is thus a parameter which defines an area in which the trap captures efficiently. The total volume associated with the number of traps, then, determines the capture rate. Thus capture rate of electrons by the ionized trap level gives

$$c_e = \sigma v_\theta N_T^+. \qquad (3.193)$$

We evaluate the statistics under conditions in which the Boltzmann approximation is valid. This also implies that the effective density of states is much larger than the carrier concentrations in the band. Our thermal equilibrium equations, using the principle of detailed balance, allow us to evaluate some of the constants. The constants for emission are related to the constants for capture as above by

$$e_e = c_e \frac{n_0}{N_C} \frac{\overline{N}_T^+}{\overline{N}_T^0}, \qquad (3.194)$$

and

$$e_h = c_h \frac{p_0}{N_V} \frac{\overline{N}_T^0}{\overline{N}_T^+}. \qquad (3.195)$$

Here, we have assumed that the carrier concentration is less than the corresponding effective density of states. In thermal equilibrium, we have

$$n_0 = N_C \exp\left(\frac{\xi_f - E_c}{kT}\right)$$

and

$$p_0 = N_V \exp\left(\frac{E_v - \xi_f}{kT}\right) = \frac{n_i^2}{n_0}. \qquad (3.196)$$

The rate of decrease of an excess carrier concentration n' is given by the difference between capture and emission processes

$$-\frac{dn'}{dt} = \mathcal{R}_{ce} - \mathcal{G}_{ce}$$

$$= c_e \left(nN_T^+ - n_0 N_T^0 \frac{\overline{N_T^+}}{\overline{N_T^0}} \right)$$

$$= c_e \left[nN_T^+ - N_C \exp\left(-\frac{E_c - E_T}{kT} \right) N_T^0 \right]. \qquad (3.197)$$

Similarly, from recombination and emission processes for holes, we have:

$$-\frac{dp'}{dt} = \mathcal{R}_{vh} - \mathcal{G}_{vh}$$

$$= c_h \left(pN_T^0 - p_0 N_T^+ \frac{\overline{N_T^0}}{\overline{N_T^+}} \right)$$

$$= c_h \left[pN_T^0 - N_V \exp\left(-\frac{E_T - E_v}{kT} \right) N_T^+ \right]. \qquad (3.198)$$

In steady-state conditions away from thermal equilibrium conditions, the rate of change of excess electron and hole densities are identical since they only interact with each other during generation and recombination, i.e.,

$$\frac{dn'}{dt} = \frac{dp'}{dt}, \qquad (3.199)$$

allowing us to write the positively and neutrally charged trap densities as

$$N_T^+ = N_T \left[c_e N_C \exp\left(-\frac{E_c - E_T}{kT} \right) + c_h p \right] \times$$
$$\left\{ \left[n + N_C \exp\left(-\frac{E_c - E_T}{kT} \right) \right] c_e + \right.$$
$$\left. \left[p + N_V \exp\left(-\frac{E_T - E_v}{kT} \right) \right] c_h \right\}^{-1}, \qquad (3.200)$$

and

$$N_T^0 = N_T \left[c_e n + c_h N_V \exp\left(-\frac{E_T - E_v}{kT} \right) \right] \times$$
$$\left\{ \left[n + N_C \exp\left(-\frac{E_c - E_T}{kT} \right) \right] c_e + \right.$$
$$\left. \left[p + N_V \exp\left(-\frac{E_T - E_v}{kT} \right) \right] c_h \right\}^{-1}. \qquad (3.201)$$

This allows us to write the net rate of decrease of excess density as

$$-\frac{dp'}{dt} = -\frac{dn'}{dt} = \left(np - n_i^2\right) \times$$

$$\left\{\left[n + N_C \exp\left(-\frac{E_c - E_T}{kT}\right)\right]\frac{1}{c_h N_T} + \left[p + N_V \exp\left(-\frac{E_T - E_v}{kT}\right)\right]\frac{1}{c_e N_T}\right\}^{-1}.$$

$$(3.202)$$

This expression relates the net rate of change in carrier concentration with time, i.e., $-\mathcal{U} = \mathcal{G} - \mathcal{R} = dp'/dt = dn'/dt$.

Usually, one defines parameters τ_{p0} and τ_{n0} as

$$\tau_{p0} = \frac{1}{c_h N_T}$$

$$\text{and} \quad \tau_{n0} = \frac{1}{c_e N_T}. \qquad (3.203)$$

and hence the minority carrier lifetime for the n-type semiconductor, for a donor-like trap with a single state, is

$$\tau_p = -\frac{p'}{dp'/dt}$$

$$= \frac{p'}{np - n_i^2} \times \left\{\left[n + N_C \exp\left(-\frac{E_c - E_T}{kT}\right)\right]\tau_{p0} + \left[p + N_V \exp\left(-\frac{E_T - E_v}{kT}\right)\right]\tau_{n0}\right\}, \qquad (3.204)$$

and majority carrier lifetime is

$$\tau_n = -\frac{n'}{dn'/dt}$$

$$= \frac{n'}{np - n_i^2} \times$$

$$\left\{\left[n + N_C \exp\left(-\frac{E_c - E_T}{kT}\right)\right]\tau_{p0} + \left[p + N_V \exp\left(-\frac{E_T - E_v}{kT}\right)\right]\tau_{n0}\right\}. \qquad (3.205)$$

In deriving the above equations we considered a single level donor-like trap in the limit of Maxwell–Boltzmann approximation. Traps can be acceptor-like; the equations have the same form as above. A considerably more complex situation occurs when a trap has multiple levels in the band, e.g., transition metals with incomplete d-orbitals usually have several levels associated with different states of ionization. Chromium, e.g., has three levels within the bandgap of GaAs, and it is among the impurities used to obtain semi-insulating GaAs. Statistics in such situations can be quite complicated, because the presence of one state excludes the other. The method of grand partition function to obtain the statistics, summarized in the previous chapter for the statistics of discrete levels, is particularly useful in such situations. Problem 23 discusses such a complicated problem, and its implication on lifetimes.

3.7.3   Auger Recombination

Electrons and holes can recombine, without the HSR type interaction at a trap or the involvement of photon emission, by releasing the excess energy to another carrier. Such processes involving more than two carriers are usually referred to as Auger processes. Impact ionization, discussed earlier, is actually one example of an Auger process, a generation process that involves the creation of an additional electron–hole pair due to the excess energy of a hot carrier.

The number of Auger recombination processes is large. Auger transitions may take place through band-to-band, band-to-shallow levels, shallow levels-to-shallow levels transitions involving excitons, etc. In all these cases the excess energy released in the transition is picked up by a mobile carrier. Because of the involvement of carriers, shallow levels, and transition across much of the bandgap, the process becomes stronger both with increase in doping and by reduction in bandgap. In small bandgap materials, it is also strongly temperature-dependent. Electrons are also more easily likely to get energetic since they usually have a lighter mass. Light holes are also likely to pick up excess energy; however, most holes exist as heavy holes because of the larger density of states for heavy hole band. Band structure is therefore very central to how and which Auger recombination processes are important, as it is for its inverse, the Auger generation process.

We can derive the statistics for this process in a similar manner to those for HSR recombination. When the excess energy is received by an electron in the conduction band, we have interaction taking place between two electrons and one hole. Thus,

$$\mathcal{R}_{Ae} \propto n^2 p, \qquad\qquad (3.206)$$

and in thermal equilibrium

$$\mathcal{R}_{Ae} \propto n_0^2 p_0. \tag{3.207}$$

The generation process depends only on the electron concentration

$$\mathcal{G}_{Ae} \propto n, \tag{3.208}$$

and in thermal equilibrium,

$$\overline{\mathcal{G}}_{Ae} \propto n_0. \tag{3.209}$$

Relating the rates using the principle of detailed balance, in thermal equilibrium,

$$\overline{\mathcal{G}}_{Ae} = \overline{\mathcal{R}}_{Ae}, \tag{3.210}$$

so, away from thermal equilibrium,

$$\mathcal{R}_{Ae} = \overline{\mathcal{R}}_{Ae} \frac{n^2 p}{n_0^2 p_0}, \tag{3.211}$$

and

$$\mathcal{G}_{Ae} = \overline{\mathcal{G}}_{Ae} \frac{n}{n_0}, \tag{3.212}$$

and

$$\mathcal{R}_{Ae} - \mathcal{G}_{Ae} = \overline{\mathcal{G}}_{Ae} \left(\frac{n^2 p}{n_0^2 p_0} - \frac{n}{n_0} \right) = \gamma_n n \left(np - n_i^2 \right), \tag{3.213}$$

where

$$\gamma_n = \overline{\mathcal{G}}_{Ae} \frac{1}{n_0^2 p_0} \tag{3.214}$$

is an Auger constant. An equivalent expression for holes is

$$\mathcal{R}_{Ah} - \mathcal{G}_{Ah} = \gamma_p p \left(np - n_i^2 \right). \tag{3.215}$$

Consequently, the net Auger recombination rate due to electrons and holes is given by:

$$\mathcal{U}_A = \mathcal{R}_{Ae} - \mathcal{G}_{Ae} + \mathcal{R}_{Ah} - \mathcal{G}_{Ah} = \left(\gamma_n n + \gamma_p p \right) \left(np - n_i^2 \right). \tag{3.216}$$

We can show that under low level injection these expressions reduce to a form where the net recombination rate is proportional to the excess carrier

concentration (see Problem 14). For p-type material, such as the base of a
bipolar transistor, the net recombination rate is

$$\mathcal{U}_A = \gamma_p N_A{}^2 n', \tag{3.217}$$

where n' is the excess carrier concentration. Using this, in a way similar
to that of HSR recombination, an Auger lifetime can be defined whose
magnitude is

$$\tau_A = \frac{1}{\gamma_p N_A{}^2}. \tag{3.218}$$

For most semiconductors, where the band to band Auger processes domi-
inate, in the limit of heavy doping the lifetime does appear to follow an
inverse dependence with the square root of the doping.

3.7.4   Surface Recombination

In this discussion of the behavior of excess carriers, a particularly important
one both at surfaces and in treatments of certain boundary conditions, is
that of recombination at surfaces[11] and also by extension at interfaces.
Compound semiconductor surfaces usually occur with considerable numbers
of states in the forbidden gap. Generally, there is a distribution of states
and the surface is quite often pinned because of the large number of states.
Sometimes this pinning may occur in the bands, e.g., in InAs it occurs
in conduction band while for GaSb it occurs in the valence band. When
these states occur in the forbidden gap, recombination transitions occur
through the non-radiative HSR process at the surface. Carriers within a
few diffusion lengths can readily recombine, leading to a net flow of current
to the surface that we will call surface recombination current.

The treatment of surface recombination is actually quite complex. Simple relations, however, can be derived for low level injection conditions.
When excess carriers exist in the bulk, we can derive the diffusive current
towards the surface for a uniformly doped sample. It is this diffusive flux
that supplies the surface recombination current under low level injection
conditions. We will treat this first in a simple way, to show the simplified
equations that are used in the Dirichlet boundary conditions and that readily lead to the concept of surface recombination velocity. Following that,
we will discuss where this simplified treatment will break down.

Let ϱ_s be the reflection coefficient at the surface representing the probability that a particle returns to the bulk without recombining at the surface.
Similarly, let ϱ_b be the reflection coefficient representing the probability

---

[11]Our comments here closely follow the discussion in J. P. McKelvey, *Solid State and
Semiconductor Physics*, Krieger, Malabar, FL (1982).

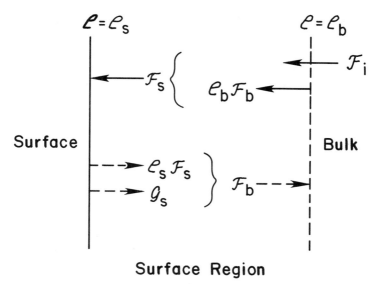

Surface Region

Figure 3.22: Fluxes of carriers representing transport processes taking place in the surface region of semiconductors during surface recombination. After J. P. McKelvey, *Solid State and Semiconductor Physics*, Krieger, Malabar, FL (1982).

that a carrier headed towards the bulk will show up at the surface. We assume, for the simple analysis, that these are independent of current density and carrier concentrations. Following Figure 3.22, the total flux $\mathcal{F}_s$ to the surface from the bulk is given by

$$\mathcal{F}_s = \mathcal{F}_i + \varrho_b \mathcal{F}_b, \tag{3.219}$$

where $\mathcal{F}_i$ is the incident flux and $\mathcal{F}_b$ is the total reverse flux. Likewise, we may write the total flux from the surface to the bulk as

$$\mathcal{F}_b = \mathcal{G}_s + \varrho_s \mathcal{F}_s, \tag{3.220}$$

where $\mathcal{G}_s$ is the generation rate at the surface. The total fluxes, in terms of the reflection parameters, generation rate, and the incident flux, are

$$\mathcal{F}_s = \frac{\mathcal{F}_i + \varrho_b \mathcal{G}_s}{1 - \varrho_s \varrho_b}$$

$$\text{and} \quad \mathcal{F}_b = \frac{\mathcal{G}_s + \varrho_s \mathcal{F}_i}{1 - \varrho_s \varrho_b}. \tag{3.221}$$

For a classical gas distribution, i.e., for non-degenerate materials, the fluxes $\mathcal{F}_s$ and $\mathcal{F}_b$ are related to the thermal velocity and the free carrier concentration as $\approx n v_\theta / 4$, which reduces to $\approx \overline{n}_0 v_\theta / 4$ in thermal equilibrium. This follows from the argument that, at any given instant of time, half of the carriers are directed towards the surface, and since their direction is random they have an average velocity of $\approx v_\theta / 2$ in the orthogonal direction. This allows us to write the generation rate and the incident flux $\overline{\mathcal{F}}_i$, at thermal equilibrium, at the surface, as

$$\overline{\mathcal{G}}_s = \frac{\overline{n}_0 v_\theta}{4}(1 - \varrho_s)$$

$$\text{and} \quad \overline{\mathcal{F}}_i = \frac{\overline{n}_0 v_\theta}{4}(1 - \varrho_b). \tag{3.222}$$

Note that the generation rate remains constant; this follows using similar arguments as with the HSR recombination mechanism. Departure from equilibrium results in a disparity between the net flux to and from the surface, the difference of which is the recombination flux. The method applied here is similar to that sometimes used in discussion of metal–semiconductor junctions. The flux directed towards the surface originates from a distance on average equal to the mean free path. Thus, the fluxes to and from the surface at any position are

$$\mathcal{F}_s = \frac{v_\theta}{4}\left(n - \gamma \frac{dn}{dz}\right)$$

$$\text{and} \quad \mathcal{F}_b = \frac{v_\theta}{4}\left(n + \gamma \frac{dn}{dz}\right), \tag{3.223}$$

where γ is a proportionality constant. The sum of the surface and bulk fluxes is $n_s v_\theta / 4$ at the surface. Our equations may now be used to determine the incident flux as

$$\mathcal{F}_i = \frac{n_s v_\theta}{2}\frac{1 - \varrho_s \varrho_b}{1 + \varrho_s} - \frac{\overline{n}_0 v_\theta}{4}(1 + \varrho_b)\frac{1 - \varrho_s}{1 + \varrho_s}, \tag{3.224}$$

and hence the net flux of carriers, assuming only diffusive transport of carriers is

$$\mathcal{F}|_{surf} = \mathcal{F}_s - \mathcal{F}_b = \frac{n_s - \overline{n}_0}{2}v_\theta \frac{1 - \varrho_s}{1 + \varrho_s} = -\mathcal{D}_n \frac{dn'}{dz}\bigg|_{surf}. \tag{3.225}$$

This is simply written as

$$-\mathcal{D}_n \frac{dn'}{dz}\bigg|_s = S\,\Delta n|_s, \tag{3.226}$$

with S, the surface recombination velocity, as

$$S = \frac{v_\theta}{2} \frac{1 - \varrho_s}{1 + \varrho_s}. \tag{3.227}$$

Note that if all carriers incident at the surface recombine, then the surface recombination velocity is $v_\theta/2$, its largest value. Assuming an infinite surface recombination velocity, as is common for many boundaries encountered in device modelling, is tantamount to assuming that the excess carrier concentration at the surface is zero. For many practical cases, this assumption is justified. The function ϱ_s represents the statistics of recombination at the surface, because it represents the probability that a carrier will return. Thus, $1 - \varrho_s$ is proportional to the recombination rate at the surface. We have discussed the statistics of HSR recombination; these also hold for most surfaces, because the recombination occurs through deep traps. The recombination rate at the surface can be represented by similar expressions to those used for a single level. Assuming that there exists a single dominating trap level, the surface recombination rate is

$$\mathcal{R} = \sigma_n \sigma_p v_\theta N_{Ts} \frac{n_s p_s - n_i^2}{(n_s + n_{Ts}) \sigma_n + (p_s + p_{Ts}) \sigma_p}, \tag{3.228}$$

where n_s and p_s are the surface carrier concentrations, and n_{Ts} and p_{Ts} are the surface carrier concentrations if the Fermi level were at the trap level. Consider an example that is a simplification of this. For a trap with equal hole and electron capture cross-section, if the electron is a minority carrier hence if hole concentrations are large, then the recombination rate is

$$\mathcal{R}_s \approx \sigma v_\theta N_{Ts} n_s. \tag{3.229}$$

The rate of recombination of electrons, per unit time, per unit area, is given by the above expression. The reflection coefficient is related to this rate since it expresses the probability of not recombining (see Problem 15). The recombination rate is proportional to the carrier concentration; it is provided by the difference of incident and reflected flux, and the constant of proportionality $\sigma v_\theta N_{Ts}$ is the surface recombination velocity in the absence of any surface space charge, with identical conditions at the surface as in the bulk. For low interface state density oxide-silicon interfaces, $\sigma \approx 10^{-16}$ cm^2 and $N_{Ts} \approx 10^{10}$ cm^{-2}, resulting in a surface recombination velocity on the order of ≈ 10 cm.s^{-1}. In compound semiconductors such as GaAs, even if we ignore the effect of surface space charge and other effects which we will soon discuss, $N_{Ts} \approx 10^{14}$ cm^{-2}, $\sigma \approx 10^{-15}$ cm^{-2}, and hence surface recombination velocity is of the order of 10^6 cm.s^{-1}. Strictly speaking, recombination rate characterizes the surface effectively for most purposes. Surface recombination velocity is a concept introduced because it characterizes a meaningful constant in some situations. If, in an ad hoc manner, we defined it as a parameter that related the recombination rate to the excess carrier concentration, then it would vary as a function of biasing condition, etc., because the above is only a simple derivation from the more complicated HSR expression.

3.7.5   Surface Recombination with Fermi Level Pinning

In the presence of a charge region at the surface, the surface recombination velocity takes an even more complicated form. It now depends on the surface state density, the characteristics of the surface states, surface charge, etc.—parameters which determine the surface carrier concentrations. We can see from the relationships above that it will actually go through a maximum when $\sigma_n (n_s + n_{Ts}) + \sigma_p (p_s + p_{Ts})$ goes through a minimum. This will occur when both electron capture and hole capture processes are equally active, i.e., when the quasi-Fermi levels straddle the trap level.

Since the surface space charge situation is important to most compound semiconductors, we will look at it in more detail. We consider variations introduced on our simple model due to the presence of band bending from Fermi level pinning. For low level injection (see Figure 3.23), we may again derive a relationship that relates surface carrier concentrations to the bulk. The presence of surface depletion changes the thermal equilibrium concentration at the surface—in the case of Figure 3.23 this would mean changing from the bulk thermal equilibrium magnitude by the Boltzmann exponential related to the total band bending at the surface. The total recombination rate is a more general case of the above, where we again ignore the n_{Ts} and p_{Ts} terms and the intrinsic carrier density terms, as

$$\mathcal{R}_s = \sigma_n \sigma_p v_{\theta n} v_{\theta p} N_{Ts} \frac{n_s p_s}{n_s \sigma_n v_{\theta n} + p_s \sigma_p v_{\theta p}}. \tag{3.230}$$

Let the bulk value of carrier concentrations be n_0 and p_0 under thermal equilibrium, and let n and p be the bulk values under the low level injection condition. If the recombination is the rate-limiting step, then the low level injection condition implies flat quasi-Fermi levels between the bulk and the surface. If the total band bending is $\Delta\psi_S$ from the bulk, the carrier concentration at the surface is

$$n_s \;=\; n \exp\left(-\frac{q\Delta\psi_S}{kT}\right)$$

$$\text{and} \quad p_s \;=\; p \exp\left(\frac{q\Delta\psi_S}{kT}\right), \tag{3.231}$$

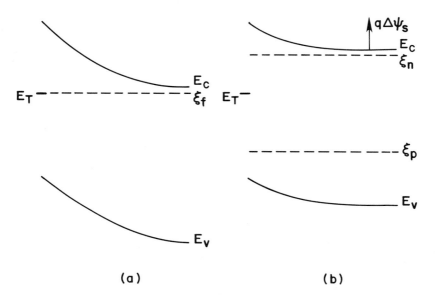

Figure 3.23: Band diagram for estimating surface recombination at a trap, in the presence of Fermi level pinning, at (a) thermal equilibrium, and (b) low level injection conditions.

with

$$np = n_i{}^2 \exp\left(\frac{\xi_n - \xi_p}{kT}\right),\qquad(3.232)$$

at both the surface and in the bulk. The surface recombination rate can, then, be written as

$$\mathcal{R}_s = \frac{(\sigma_n \sigma_p v_{\theta n} v_{\theta p})^{1/2} N_{Ts}(n_s p_s)^{1/2}}{(n_s \sigma_n v_{\theta n}/p_s \sigma_p v_{\theta p})^{1/2} + (p_s \sigma_p v_{\theta p}/n_s \sigma_n v_{\theta n})^{1/2}}.\qquad(3.233)$$

The occupation probability f of the trap N_{Ts} at the surface is given by

$$f = \frac{\sigma_n v_{\theta n} n_s + \sigma_p v_{\theta p} p_{Ts}}{\sigma_n v_{\theta n}\,(n_s + n_{Ts}) + \sigma_p v_{\theta p}\,(p + p_{Ts})}.\qquad(3.234)$$

Since, at thermal equilibrium, the Fermi level is pinned at the trap level E_T, the occupation probability is very close to $1/2$ (see Problem 16). Away from thermal equilibrium, the resultant surface charge density is given by

$$\mathcal{Q}_s = qN_{Ts}\left(f - \frac{1}{2}\right) = \frac{qN_{Ts}}{2}\frac{\sigma_n v_{\theta n} n_s - \sigma_p v_{\theta p} p_s}{\sigma_n v_{\theta n} n_s + \sigma_p v_{\theta p} p_s},\qquad(3.235)$$

where we again ignore the insignificant terms away from thermal equilibrium. For large trap densities that lead to Fermi level pinning, this charge is insignificant compared to the charge qN_{Ts} if all the traps were ionized, or equivalently only a very small deviation from the $1/2$ occupation probability occurs. This, however, implies that

$$\frac{n_s}{p_s} = \frac{\sigma_p v_{\theta p}}{\sigma_n v_{\theta n}}. \tag{3.236}$$

The ratio of carrier concentrations at the surface is a constant, and in a large trap density, and low level injection limit, the recombination rate is

$$\mathcal{R}_s = \frac{(\sigma_n \sigma_p v_{\theta n} v_{\theta p})^{1/2} N_{Ts}}{2} (n_s p_s)^{1/2}. \tag{3.237}$$

The recombination rate is not proportional to the minority carrier concentration any more; it is proportional to the square root of the electron and hole concentration at the surface, and since the quasi-Fermi levels are flat, it is also proportional to the square root of the electron and hole concentration in the bulk. A consequence of this is an $\exp(qV/2kT)$ dependence of surface recombination current for low level injection bias conditions where this analysis applies. This square root dependence also appears for recombination in a p-n junction space charge region, and is responsible for similar $\exp(qV/2kT)$ dependence there.

Surface recombination velocity, defined as a pre-factor to excess minority concentration during recombination calculations, is no longer a constant but a function of bias conditions. Arguments have been forwarded that we should define surface recombination velocity in an alternate form that preserves a constancy.[12] In the above, one may introduce s_0 as an intrinsic surface recombination velocity that follows the recombination relation

$$\mathcal{R}_s = S_0 \sqrt{n_s p_s}. \tag{3.238}$$

This intrinsic recombination velocity is related to the conventional definition of surface recombination velocity in terms of excess carrier concentration (e.g., one useful in Dirichlet boundary conditions) via

$$S = S_0 \sqrt{\frac{n_s}{p_s}}. \tag{3.239}$$

---

[12]See the discussion in G. J. Rees, "Surface Recombination Velocity—a Useful Concept?" *Solid-State Electronics*, **28**, No. 5, p. 517, 1985, and P. De Visschere, "Comment on G. J. Rees 'Surface Recombination Velocity—a Useful Concept?' " *Solid-State Electronics*, **29**, No. 11, p. 1161, 1986.

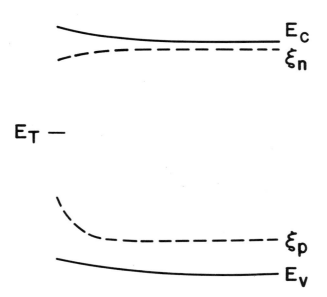

Figure 3.24: The quasi-Fermi levels and conduction and valence band edges in the presence of a high surface recombination rate.

Intrinsic surface recombination velocity is the surface recombination velocity that occurs when the electron and hole densities at the surface are equal. Rigorous calculations of these parameters show that significant deviations occur at high level injection conditions, where our theory does not appear valid. Differences also occur when the surface recombination rate is the rate-limiting step. Under these conditions, the band diagram and the quasi-Fermi levels look as in Figure 3.24, indicating that the carrier flux needed for recombination occurs via both drift and diffusion at the surface.

We will continue discussion for the more complicated cases of this subject in Chapter 7 on bipolar transistors. Here, we have only considered the one-dimensional situation and have assumed the quasi-Fermi levels to be constant. The trap recombination statistics have to be self-consistently analyzed with the ability of the semiconductor to provide the carriers for recombination at the surface. A changing quasi-Fermi level implies that supply of carriers becomes a rate-limiting step. The recombination rate is now limited by supply of carriers from the bulk and does not necessarily follow this analysis. Additionally, in real devices, two-dimensional considerations predominate under certain bias conditions, i.e., carriers injected

along the channel occurring due to Fermi level pinning must also be considered in our analysis.

3.8   Summary

This chapter reviewed and developed methods of analysis that will be used
in the rest of the book. In particular, we have emphasized the approximations of various approaches, and how one may isolate sections of a larger
problem in order to analyze related behavior. To do this we use appropriate boundary conditions, which mimic the behavior at the interface where a
section is isolated, and the appropriate model to describe the events taking
place within the section.

The first method of analysis considered is based on a simple treatment
of the kinetics of particles treated as an ensemble. This was the particle
kinetic approach and allowed us to develop meaningful methods for analysis of information related to low-field mobility, etc. The second method is
the Boltzmann Transport Equation, which considers the use of distribution
functions describing the phase space behavior of particles. The development
of this distribution function, or the moments of it (statistical averages corresponding to macroscopic properties) in time allows us to describe the
behavior of the system. However, in order to obtain results conveniently,
while still maintaining the basis in distribution functions, we had to limit
the methodology to displaced Maxwell–Boltzmann distributions and we had
to approximate scattering effects by relaxation time approximations. The
method, thus, deserves to be used with extreme caution. These limitations
are avoided, at the expense of demanding numerical computation, in the
Monte Carlo approach where statistics are developed either for an individual particle or an ensemble of particles. The limitations of the Monte Carlo
approach are related to the adequacy of the description of the processes taking place in the semiconductor. The collection of a vast amount of statistics
on particles avoids the need to assume certain distribution functions for the
particles as required in the Boltzmann Transport Approach.

Conventional semiconductor devices are usually analyzed within the
drift-diffusion approximation. We derived the relevant semiconductor transport equations from the Boltzmann Transport Equation and discussed the
limitations and approximations in making the simplification to drift-diffusion equation. We also showed how one may derive small-signal solutions
to transport problems of interest based on this approach by considering
the frequency dependence in the drift-diffusion equation. In the rest of
the book, although this approach is stressed the most, we will also resort
to Monte Carlo approach where the application of drift-diffusion is clearly

inadequate.

In analyzing device problems with any of these approaches, we still need to consider models for the boundaries of the system. Many p–n junction problems can be analyzed adequately with the use of Shockley and Fletcher or Misawa boundary conditions. These boundary conditions allow one to determine the carrier densities over a range of biases in device structures where such junctions exist, one example being the bipolar transistor. We also considered Dirichlet boundary conditions that can be applied at many semiconductor interfaces—metal–semiconductor junctions, ohmic contacts, and high recombination surfaces being some examples. Finally we considered the use of Neumann boundary conditions at semiconductor surfaces. Recombination is a particularly strong and important attribute of compound semiconductors; they are largely direct bandgap materials and hence exhibit radiative recombination. In addition, they have a large trap density and therefore a large non-radiative recombination. In addition, the smaller bandgap semiconductors can have significant Auger recombination. The necessary underlying mathematical basis of all these processes was discussed together with that for surface recombination.

General References

1. B. K. Ridley, *Quantum Processes in Semiconductors*, Clarendon Press, Oxford (1988).

2. P. J. Price, "Monte Carlo Calculation of Electron Transport in Solids," in R. K. Willardson and A. C. Beer, Eds., *Semiconductors and Semimetals*, **14**, Academic Press, N.Y. (1979).

3. C. Jacobani and L. Reggiani, "The Monte Carlo Method for the Solution of Charge Transport in Semiconductors with Applications to Covalent Materials," *Review of Modern Physics*, **55**, No. 3, p. 645, July 1983.

4. P. J. Price, "Calculation of Hot Electron Phenomena," *Solid-State Electronics*, **21**, p. 9, 1978.

5. S. Datta, *Quantum Phenomena*, Addison-Wesley, MA (1989).

6. E. M. Conwell, *High Field Transport in Semiconductors*, in *Solid State Physics*, Suppl. **9**, Academic Press, N.Y. (1967).

7. A. C. Smith, J. F. Janak, and R. B. Adler, *Electronic Conduction in Solids*, McGraw-Hill, N.Y. (1967).

8. R. W. Hockney and J. W. Eastwood, *Computer Simulation Using Particles*, McGraw-Hill, N.Y. (1981).

9. C. Jacobani and P. Lugli, *The Monte Carlo Method for Semiconductor Device Simulation*, Springer-Verlag, Wien (1989).

10. W. Paul, Ed., *Handbook on Semiconductors*, Vol. 1, North-Holland, Amsterdam (1982).

11. J. P. McKelvey, *Solid State and Semiconductor Physics*, Krieger, Malabar, FL (1982).

12. S. Selberherr, *Analysis and Simulation of Semiconductor Devices*, Springer-Verlag, Wien (1984).

13. K. Hess, *Advanced Theory of Semiconductor Devices*, Prentice Hall, Englewood Cliffs, N.J. (1988).

14. J. S. Blakemore, *Semiconductor Statistics*, Dover, N.Y. (1986).

15. J. Pankove, *Optical Processes in Semiconductors*, Dover, N.Y. (1971).

16. M. V. Fischetti and S. E. Laux, "Monte Carlo Analysis of Electron Transport in Small Semiconductor Devices Including Band Structures and Space Charge Effects," *Phys. Rev. B*, **38**, No. 14, p. 9721, 1988.

17. L. Reggiani, Ed., *Hot-Electron Transport in Semiconductors*, Springer-Verlag, Berlin (1985).

18. M. Kurata, *Numerical Analysis of Semiconductor Devices*, D. C. Heath, MA (1982).

19. C. M. Snowden, *Semiconductor Device Modeling*, Peter Peregrinus, London (1988).

20. T. H. Glisson, C. K. Williams, J. R. Hauser, and M. A. Littlejohn, "Transient Response of Electron Transport in GaAs Using the Monte Carlo Method," in *VLSI Electronics: Microstructure Science*, **4**, Academic Press, FL (1982).

21. K. Hess, "Aspects of High-field Transport in Semiconductor Heterolayers and Semiconductor Devices," in *Advances in Electronics and Electron Physics*, **59**, Academic Press, N.Y. (1982).

Problems

1. We have derived the low field mobility relationship as a function of a single time constant; this is an idealization of a single valley semiconductor. At higher fields, though, more than one valley may contribute to conduction. GaAs, e.g., at fields of $\approx 3.0 \times 10^3$ V.cm^{-1}, also exhibits conduction in the L valley and to a lesser extent in the X valley. Consider the problem of conduction in two valleys and derive the mobility relationship assuming that a single time constant characterizes momentum relaxation in each of these valleys.

2. An important problem encountered in minority carrier devices using low dopings or high injection—such as p–i–n diodes, thyristors, and bipolar transistors—is the problem of conduction when both minority carriers and majority carriers are present in sufficient numbers to be strong contributors to current. The problem is usually encountered only at small electric fields. Derive the mobility relationship important in such conditions, expressing it in terms of single time constant approximations for both electrons and holes. This mobility is commonly referred to as the ambipolar mobility.

3. For a parameter $\varphi(E)$ related to transport, show that the weighted average over all carriers is given by the relationship

$$\langle \varphi \rangle = -\frac{2 \int_0^\infty \varphi(E) g(E) \left(\partial f(E)/\partial E \right) E dE}{3 \int_0^\infty g(E) f(E) dE} \tag{3.240}$$

which reduces to

$$\langle \varphi \rangle = \frac{\int_0^\infty \varphi(E) E^{3/2} \exp\left(-E/kT\right) dE}{\int_0^\infty E^{3/2} \exp\left(-E/kT\right) dE} \tag{3.241}$$

in the Boltzmann approximation for semiconductors with parabolic bands.

4. Show, using the kinetic approach, that the current obeys the relationship in Equation 3.25 in the presence of a small electric and magnetic field.

5. Show that the Hall pre-factor r for acoustic deformation potential scattering is given as

$$r = \frac{3\pi}{8}. \tag{3.242}$$

6. The thermal velocity $v_\theta = \sqrt{8kT/\pi m^*}$ is the root mean square speed or the root mean square of the magnitude of velocity. What is the most probable speed or magnitude of velocity in the Boltzmann approximation?

7. For a carrier density of n, show that the flux of carriers in any direction is given by $nv_\theta/4$.

8. The internal energy per unit volume, following our discussion of thermal velocity, etc., is

$$W = \int_0^\infty Ef(E)D(E)dE. \tag{3.243}$$

Show that, for the Maxwell–Boltzmann approximation, this leads to $W = 3nkT/2$, where n is the carrier density. At absolute zero, show that the energy is $W = 3n\xi_f/5$.

9. Consider the problem of transport in external and built-in fields. Using the relationship between the perturbation in the distribution function and the isotropy of the semiconductor, and assuming randomizing scattering processes, show that the drift velocities are related as

$$v_d = \frac{\Delta\mathcal{E}}{\mathcal{E}_0}v_d'. \tag{3.244}$$

10. Show that a non-homogenous semiconductor can be treated as quasi-neutral provided

$$\lambda_D{}^2 \gg \left[\left|-\frac{1}{N^2}\left(\frac{dN}{dz}\right)^2 + \frac{1}{N}\frac{d^2n}{dz^2} + \left(\frac{q}{kT}\right)^2\frac{d^2\phi_{Cn}}{dz^2}\right|\right]^{-1}. \tag{3.245}$$

11. Derive the Fletcher boundary conditions (Equation 3.159) from the relationships between carrier concentrations at the edge of the depletion region and the condition of charge neutrality.

12. Show, mathematically, that the Fletcher boundary conditions (Equation 3.159) and the Misawa boundary conditions (Equation 3.161) are equivalent.

13. Show that the Misawa boundary conditions stated in Equation 3.161 and Equation 3.162 are equivalent.

14. Show that under low level injection, the expression for the Auger recombination rate reduces to a direct proportionality to the excess carrier concentration with the proportionality constant dependent on the square of dopant concentration.

15. Argue how the recombination rate at the surface is related to the reflection coefficient at the surface. Cast the relationship in a mathematical form.

16. Show that if the Fermi level is pinned at the surface, i.e., if the sheet density of traps substantially exceeds the sheet charge density in the surface depletion region, then the occupation probability of a trap level is close to 1/2. Under what conditions may this break down?

17. Is it possible for the momentum relaxation time to be larger than the energy relaxation time? Explain how, and if true what would the consequences be?

18. Assuming Boltzmann and relaxation time approximation, and small departure from thermal equilibrium, show, using the BTE approach, that a gradient of carrier density leads to diffusion whose diffusion coefficient is $\lambda v_\theta / 3$ where λ is the mean free path.

19. Consider a semi-infinite n-type sample of GaAs doped to 10^{16} cm^{-3} and with a mid-gap trap that has identical electron and hole capture cross-sections of 10^{-15} cm^{-2}. Derive, using justifiable simplifications, the expression for decay of excess carriers in both the low injection limit and the high injection limit. What are the short time and long time limits if 10^{15} cm^{-3} carriers were created? What are the short and long time limits if 10^{17} cm^{-3} carriers were created?

20. If $\tau_{n0} = \tau_{p0} = \tau_0$, show that the maximum lifetime occurs when the intrinsic and Fermi energy coincide. Show that this lifetime is

$$\tau = \tau_0 \left[1 + \cosh \left(\frac{E_T - E_i}{kT} \right) \right]. \qquad (3.246)$$

21. If the flux of carriers in any direction is given as $n v_\theta / 4$, how can the surface recombination velocity exceed $v_\theta / 4$?

22. In this problem, we will evaluate statistics of trap energy levels arising from independent defects. Consider a semiconductor with two defects associated with energy levels E_1 and E_2.

(a) What is the occupation probability of the two levels?

(b) If the Fermi energy is between the two levels with $E_2 - \xi_f \gg kT$, what do the occupation probabilities reduce to?

23. We now extend the previous problem to analysis of recombination in a semiconductor when the energy levels arise from the same impurity. A technologically relevant example of this is gold, a transition element, as a substitutional impurity in silicon. In GaAs, multiple levels occur due to chromium, and will be discussed in Chapter 5. For gold, a closed d-shell is inert. A positively charged state can accept an electron, a negatively charged state can donate an electron. The higher negatively charged states occur in the valence band. So, the substitutional impurity in Si causes a lower energy level E_D, a donor level, and a higher energy level E_A, an acceptor level.

(a) Calculate the equilibrium density of Au atoms in the positively charged state (N_T^+), the density of Au atoms in the neutral charge state (N_T^0), the density of Au atoms in the negatively charged state (N_T^-), and the total charge density on Au. Consider the degeneracies as g_a, and g_d.

(b) What are the rate equations? Let σ_n^+ represent the capture cross-section for the electron capture process on the donor site leading to a capture rate of c_n^+. Let σ_n^0 represent the capture cross-section for the capture of an electron on an acceptor leading to the capture rate of c_n^0. Let σ_p^0 represent the capture cross-section for the hole capture process on the donor site leading to a capture rate of c_p^0. Let σ_p^- represent the capture cross-section for the capture of a hole on an acceptor leading to the capture rate of c_p^-. Let e_n^0, e_n^-, e_p^+, and e_p^0 identify the electron and hole emission rates from donor and acceptor sites.

(c) Invoke the principle of detailed balance to evaluate the emission rates in terms of the capture rates at thermal equilibrium.

(d) What is the low level neutral region recombination rate?

(e) Calculate the low level neutral region minority carrier lifetime for trap concentrations significantly smaller than the shallow donor concentration.

(f) In high level injection conditions, what is the the limiting form of the lifetime?

(g) Capture cross-sections for Au in Si, at 300 K, are as follows: $\sigma_n^+ = 3.5 \times 10^{-15}$ cm^2, $\sigma_n^0 = 5.0 \times 10^{-16}$ cm^2, $\sigma_p^- = 1.0 \times 10^{-15}$ cm^2, and $\sigma_p^0 = 3.0 \times 10^{-16}$ cm^2. Assuming that recombination occurs only due to Au in Si, what are the low-level and majority carrier lifetimes for 10^{16} cm^{-3} Au-doped Si at 300 K for intrinsic material, 10^{16} cm^{-3} n-type material, and 10^{16} cm^{-3} p-type material?

24. Let E_D and g_d be the donor energy in a semiconductor sample that is n-type but compensated. Show that the carrier concentration is related as

$$n \left(n + N_A - \frac{n_i^2}{n} \right) = \frac{N_C}{g_d} \left(N_D - N_A - n + \frac{n_i^2}{n} \right) \exp \left(-\frac{E_c - E_D}{kT} \right).$$
$$(3.247)$$

How does the activation energy behave in a plot of carrier concentration with temperature in different temperature ranges?

Chapter 4

Transport Across Junctions

4.1 Introduction

We begin our analysis of devices by first studying two-terminal structures. The theory and understanding of these are important to the understanding of transistors, since junctions are natural parts of these devices. For example, metal–semiconductor junctions are used to control the channel of a metal–semiconductor field effect transistor, p–n junctions are the basis for both injection and isolation in bipolar transistors, and heterojunctions are applicable to both field effect transistors and bipolar transistors. This discussion serves another useful purpose: it allows us to discuss applications of the methods of analysis of the previous chapter in simpler structures, even as they have the richness and diversity that clearly exemplify the approximations of our analysis techniques.

This chapter discusses transport at junctions formed between metals and semiconductors, and semiconductors and semiconductors. Drift-diffusion, thermionic emission, and thermionic field emission theories are developed for the metal–semiconductor junctions exemplifying assumptions of equilibrium between the electrons in metal and semiconductor in the drift-diffusion theory, or the lack thereof with significant hot carrier effects in the other theories. The drift-diffusion and the thermionic emission theories are then serially combined to show a theory that still assumes a classical drift-diffusion equation and hence its assumptions of distribution functions. We have discussed transport in built-in and applied fields in p–n junctions when discussing the relative merits of the Monte Carlo and the BTE approaches and have pointed out their effects on the distribution function. That discussion, however, did not bring out a particularly severe approximation in the distribution function at the metal–semiconductor junction itself because of the negligible flow of carriers from the metal into the semiconductor and because almost all carriers entering the metal are lost. Thus, this discussion of metal–semiconductor transport theories brings out the breadth of the problem quite well.

The discussion of metal–semiconductor junctions is naturally amenable to extensions involving transport across heterojunction discontinuities. In transport across heterojunctions, processes encountered in metal–semiconductor junctions are still equally important, but with some additional complications. Now, the transport across the abrupt barrier from one semiconductor to the other can not be dealt with as simplistically as in the metal–semiconductor case, because the barriers are smaller and the scattering rates and energy losses in the semiconductors are much smaller than in a metal. The latter additionally leads to a fair fraction of injected carriers being capable of returning, should their momentum change direction. The lower barrier heights of heterojunctions also results in an increased importance of tunneling, with energy conservation (elastic) and without energy conservation (inelastic), and a particularly interesting example of resonance transmission in the form of resonant Fowler–Nordheim tunneling. This is therefore discussed in the context of heterojunctions.

The p–n junction transport in quasi-static approximation is discussed next, to develop examples in high injection, to point out some of its assumptions, and to develop a general charge control approach that is commonly known as the Gummel–Poon analysis after its originators. This analysis will be extended, like the earlier discussions of other junctions, in our discussion of three-terminal device structures. The p–n junctions can also be formed with heterojunctions and are the basis of heterostructure bipolar transistors; the discussion ends with this subject.

The chapter concludes with a discussion of ohmic contacts, the most general forms of which are based on tunneling between metals and semiconductors.

4.2   Metal–Semiconductor Junctions

We start with a preliminary discussion of contacts. Consider two metals far away. The metals are neutral, they have differing Fermi energies, which characterize the average energy of carriers with respect to the vacuum. When these metals are brought together with a return path for the flow of carriers, as shown in Figure 4.1, carriers move from one to the other

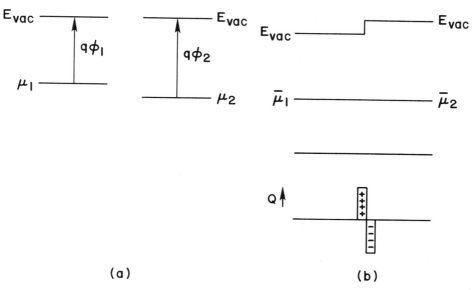

(a)   (b)

Figure 4.1: The vacuum level and the Fermi level of two metals separated are shown in (a), and in (b) the levels are shown when the two metals are brought together in contact with a path for current flow. The lower part of (b) shows the electric dipole formed at the interface due to charge transfer.

until the electrochemical potential is identical. Carriers move from higher energy to lower energy, so that, in thermal equilibrium, the equilibrium Fermi energy lies in between. The charge transfer results in formation of an electric dipole at the interface that also allows the discontinuous change in the vacuum level.

The electrostatic potential differences in the bulk, when the metals are separated, are

$$\mu_1 = E_{vac} - q\phi_1$$

$$\text{and} \quad \mu_2 = E_{vac} - q\phi_2, \tag{4.1}$$

and in contact, in thermal equilibrium,

$$\overline{\mu_1} = E_{vac} - q\phi_1 - q\psi_{10}$$

$$\text{and} \quad \overline{\mu_2} = E_{vac} - q\phi_2 - q\psi_{20}, \tag{4.2}$$

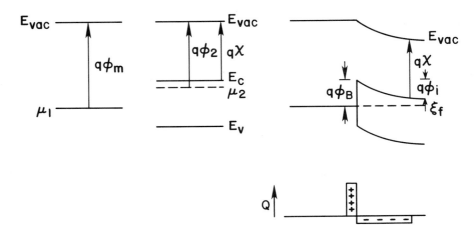

Figure 4.2: Formation of an idealized metal–semiconductor diode; (a) shows the vacuum levels, band edges, work functions, and electron affinities, and (b) shows these in contact in thermal equilibrium with a current path for electron transfer.

and the two chemical potentials are equal,

$$\overline{\mu_1} = \overline{\mu_2}, \tag{4.3}$$

so that

$$\psi_{20} - \psi_{10} = \phi_1 - \phi_2. \tag{4.4}$$

The differences in electrostatic potential balance differences in chemical potential when the two metals are brought in contact and are in thermal equilibrium.

Now let us extend this picture to an idealized metal–semiconductor junction as shown in Figure 4.2. Here, we have assumed that interface effects play no role, an assumption that is often incorrect. For now, this serves as a simplified example. When the contact is formed, electrons at the Fermi level in the metal see a barrier of height $q\phi_B$,[1] and the electrons at the conduction band edge in the semiconductor see a barrier of height $q\psi_{j0}$. In thermal equilibrium, for this idealized situation, these barrier heights from the metal to the semiconductor and from the semiconductor to the metal are related by

$$
\begin{aligned}
q\phi_B &= q\left(\phi_m - \chi\right) \\
\text{and} \quad q\psi_{j0} &= q\left(\phi_m - \chi\right) - qV_n,
\end{aligned} \tag{4.5}
$$

---

[1]We also use $q\phi_M$ to denote this barrier height elsewhere in this text.

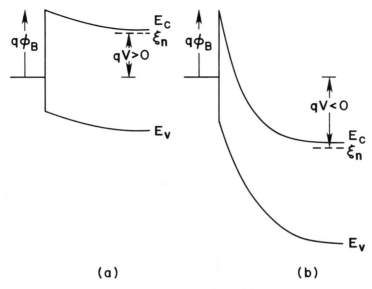

(a)   **(b)**

Figure 4.3: Metal-semiconductor diode with (a) a forward bias and (b) a reverse bias applied to the diode.

where $qV_n = E_c - \xi_f$ is evaluated in the bulk. The approximation in the above is the neglect of interface effects. Usually, semiconductors have a significant number of interface states, quite often in the bandgap. Interface states also occur due to surface reconstruction when a metal is placed in intimate contact on a nascent semiconductor surface. The metal–semiconductor barrier height, therefore, is never quite in accord with the above. Silicon comes close, but most compound semiconductors do not. In practice, therefore, we replace the metal–semiconductor barrier height by the experimentally measured values.

When we forward or reverse bias this junction (see Figure 4.3), the barrier for injection from the semiconductor to the metal is either reduced or increased. In the forward direction, a reduced barrier results in an increase in injection into the metal of electrons from the semiconductor, while in the reverse direction, an increased barrier results in a decrease in injection into the metal of electrons from the semiconductor. The injection of electrons from the metal to the semiconductor remains essentially the same since the metal is highly conducting and, in this simple picture, the barrier is constant. Therefore the diode conducts a large current in the forward direction and a small current in the reverse direction.

The interesting aspects of rectification were recognized very early, and some of the earliest work in hot electron transport was actually related to rectification properties. Many of these theories of the 1930's and Bethe's thermionic emission model of the 1940's have gone through significant modifications as our understanding of the hot electron phenomenon has improved. Before discussing the transport, though, we will discuss some of the modifications that we must include in the above description to make it consistent with real structures.

Barrier heights in compound semiconductors are strongly influenced by chemical interactions and other interface phenomena. Even for ideal systems such as vacuum cleaved surfaces, surfaces behave in different ways, in different orientations, at different temperatures, with different metals, because surfaces do not maintain a physical structure that is identical to that of the bulk. This is a result of the drive to lower energy, and different metals influence it differently, causing different interface states to appear in the bandgap. The surface recombination velocity, which described the surface recombination and was used in modelling the Dirichlet boundary conditions, was one of the consequences of these interface states. These interface states, if they exist in sufficient numbers, cause Fermi level pinning, because a large charge dipole can exist at the interface without any deviation in the Fermi level. In practice, therefore, barrier heights in compound semiconductors are best determined experimentally. This data is sufficiently reproducible, and it adequately describes the required parameter of the interface modelling problem.

There are effects on this barrier height which are unrelated to the existence or absence of interface states. One such effect is caused by image force, and it results in a lowering of barrier height that is called image force lowering. A charge in close proximity to a highly conducting surface has a lowering of its energy. This follows directly from the solution of Laplace's equation describing this problem, with the boundary condition that the conducting plane be treated as an equi-potential surface. Intuitively, this can be described via an image charge of equal and opposite magnitude, that the charge sees behind the conducting plane, with the plane being the mirror. An electron and the metal follow this description with the metal as the equi-potential conducting plane. The electron feels an image charge of the opposite type in the metal at an equal distance from the interface. The force on the electron in the semiconductor is given by

$$\mathcal{F} = \frac{q^2}{4\pi\epsilon_s (2z)^2}, \tag{4.6}$$

where z is its distance from the interface, which corresponds to a potential energy of $-q^2/16\pi\epsilon_s z$. In the presence of the electric field $\mathcal{E}$, at the interface,

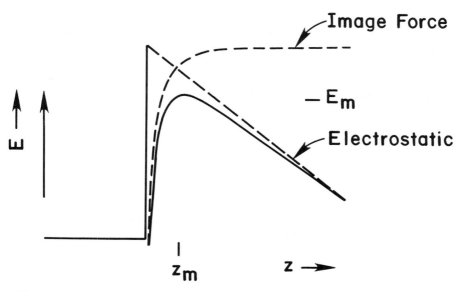

Figure 4.4: Energy of an electron (solid line) as a function of position in the presence of image charge effect near the metal–semiconductor interface.

the potential energy due to Coulombic interactions (following Figure 4.4) is

$$E = -\frac{q^2}{16\pi\epsilon_s z} - q\mathcal{E}z. \tag{4.7}$$

Several assumptions have clearly been made in the act of writing this equation. A primary one is that we have assumed that our band structure model holds all the way to the interface. The electric field is assumed constant in the above; in a depleted semiconductor with a uniform charge concentration, it varies linearly with distance. However, close to the interface in a very short region where we utilize this equation, it can be treated as a constant. The maximum in this energy occurs at

$$z = z_m = \sqrt{\frac{q}{16\pi\epsilon_s\mathcal{E}}}, \tag{4.8}$$

with a maximum in energy of

$$E = E_m = -\frac{q}{2}\sqrt{\frac{q\mathcal{E}}{\pi\epsilon_s}}. \tag{4.9}$$

So, actual metal–semiconductor barrier height is a function of the electric field at the interface and hence of bias applied; it differs from the low field value by the amount of the maximum energy given by Equation 4.9. The effect is most important in reverse bias where $\mathcal{E}$ is large because of a larger amount of charge in the semiconductor. Typical barrier heights for GaAs vary between 0.65 eV and 0.85 eV for n-semiconductors and 0.5 eV and 0.6 eV for p-semiconductors. For GaAs again, if an electric field of 100 kV.cm^{-1} were present, it would lead to a lowering of barrier height by 0.36 eV. Such large fields usually occur either in large reverse bias for moderately doped junctions, or near zero bias conditions for highly doped junctions. Effective barrier height therefore can decrease substantially with doping.

In our discussion of the different approaches to modelling transport, we concluded that, for a p–n junction, the drift-diffusion approach models transport adequately because the distribution function remains a displaced-Maxwell–Boltzmann function, and because any errors of drift or diffusion compensate. Away from the metal–semiconductor interface by a few scattering lengths, this should still be an adequate description of the transport in the semiconductor. Near the metal–semiconductor interface, the situation is unlike the p–n junction problem because carriers that get injected into the metal are quite unlikely to return because of a rapid loss in carrier energy in a very short distance. Thus, the random orientation of velocities of the Maxwell–Boltzmann distribution is lost. Within a scattering length of the interface, due to the lack of any returning electrons from the metal, the velocities are all pointed in a hemisphere directed towards the metal, and the distribution function is actually more like a hemi-Maxwell–Boltzmann distribution function. But away from the interface by a number of scattering lengths, transport in the semiconductor region is by drift-diffusion, as usual. The transport at the interface is governed by two considerations. Carriers that have sufficient energy to surmount the barrier, that have velocity pointed towards the metal, and that do not lose these properties during a collision as they reach the metal–semiconductor interface will emit into the metal. This is the thermionic emission process. Due to the small barrier that has to be traversed by the carriers that are close to the interface, and due to the short distance, some carriers may tunnel through into the metal. This is known as field emission, and when thermionic emission is combined together with this, the process is called thermionic field emission. All of these transport processes take place in the metal–semiconductor diode, with current conservation being maintained, and the current magnitude being determined by whichever limits the current to a lower value. In case they are comparable, one may have to include all if one is interested in knowing the magnitude. Often, when combining all

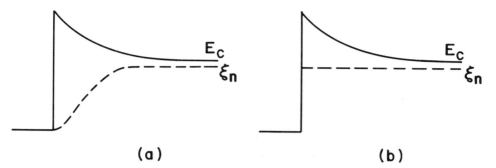

Figure 4.5: Energy band diagram under an applied bias at a metal–semiconductor junction for (a) drift-diffusion and (b) thermionic and thermionic field emission theories.

these processes, field emission is ignored, and the thermionic–drift-diffusion formulation adopted.

Early theories assumed that the transport was being limited by the rate at which carriers became available at the metal–semiconductor interfaces, i.e., these theories considered drift-diffusion as the rate-limiting step. The carriers in the semiconductor and the metal at the interface are, then, in equilibrium. For the moment, we ignore the consequences of hemi-Maxwell–Boltzmann distribution to the concept of quasi-Fermi level defined with respect to Maxwell–Boltzmann distribution and modified appropriately for the Fermi–Dirac distribution, since the problem of dealing with this question arises over distances of the order of a few scattering lengths. This assumption of drift-diffusion limited transport means that the quasi-Fermi level is no longer flat in the semiconductor, since it limits the transport, and since carriers are in equilibrium with each other on either side of the interface (see Figure 4.5). In fact, since the transport to the metal–semiconductor interface is the rate-limiting step, the density of carriers at the interface is unaltered by bias, i.e., the quasi-Fermi level at the interface continues to coincide with the metal Fermi level.

Electrons are emitted into the metal at the interface at an energy in excess of the metal Fermi energy. These electrons rapidly lose their energy over few 10s of Ås. In thermionic emission theory and its variations, the drift-diffusion in the depletion region is not the rate limiting step, i.e., like the p–n junction in the low-level injection condition, the quasi-Fermi level remains flat, and the current in the structure is substantially smaller than either the drift or the diffusion current flowing at any cross-section in the semiconductor depletion region. The quasi-Fermi level, therefore, remains flat up to the interface, and then discontinuously drops to the metal Fermi level. Strictly speaking, the quasi-Fermi level concept was introduced for the Boltzmann distribution, which is not necessarily valid in the metal. The existence of thermionically limited transport implies that the electrostatic potential change in the semiconductor over a region whose width is smaller than the mean free path of the carrier should be larger than a couple of thermal voltages. This is known as the Bethe condition, and we will show its validity in our derivations.

We can find metal–semiconductor combinations where either of these theories are better approximations. Materials with extremely poor mobilities, such as some of the p-type structures or some of the n-type II-VI materials, show transport limited by drift-diffusion in the semiconductor depletion region. Differing temperatures also affect the limiting mechanism.

In this section, we will first derive the simple form of the drift-diffusion theory, then the thermionic emission theory and its modification including field emission, and finally we will unify these two theories to extend the validity to both limits of transport.

The dominating component of the drift-diffusion mechanism, drift or diffusion, is dependent on the bias applied and is the cause of the rectification ability. Drift and diffusion at the rectifying junction lead to current in opposite directions. In forward bias, as seen in Figure 4.6, the drift component decreases because of a decrease in drift field, and hence the diffusion component predominates. In reverse bias, the drift current increases because of an increase in the field, and therefore, it determines the reverse current, unless generation effects predominate.

4.2.1   Drift-Diffusion

We first consider transport assuming that the net current is limited by drift-diffusion transport in the depletion region. We will also assume that diffusivity and mobility are related by the Einstein relationship, i.e., the field strength is low enough for the non-degenerate junction. Consider a forward bias of V on a metal–semiconductor junction whose resultant depletion width is z_d. Let J be the current flowing in the metal–semiconductor diode, and let us ignore generation–recombination effects. In terms of ψ, the electrostatic potential, integration of the current continuity equation following multiplication by the exponential factor $\exp\left(-q\psi/kT\right)$ yields

$$\int_0^{z_d} J \exp\left(-\frac{q\psi}{kT}\right) dz =$$

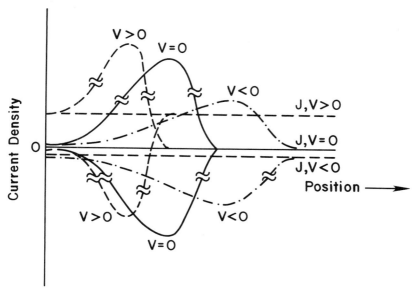

Figure 4.6: A schematic of drift (bottom half), diffusion (top half), and net currents as a function of position in the depletion region in thermal equilibrium ($V = 0$), forward bias ($V > 0$), and reverse bias ($V < 0$) of a metal–semiconductor junction. After H. K. Henisch, *Semiconductor Contacts—An Approach to Ideas and Models*, Clarendon Press, Oxford (1984).

$$q\mathcal{D}_n \int_0^{z_d} \left[-\frac{qn}{kT} \exp\left(-\frac{q\psi}{kT}\right) d\psi + \right.$$
$$\left. \exp\left(-\frac{q\psi}{kT}\right) dn \right]$$
$$= q\mathcal{D}_n \int_0^{z_d} \left\{ nd\left[\exp\left(-\frac{q\psi}{kT}\right)\right] + \exp\left(-\frac{q\psi}{kT}\right) dn \right\}$$
$$= q\mathcal{D}_n n \exp\left(-\frac{q\psi}{kT}\right)\Big|_0^{z_d}. \tag{4.10}$$

If recombination current is absent, current is constant, and we obtain

$$J = q\mathcal{D}_n \frac{n \exp\left(-q\psi/kT\right)\big|_{z=0}^{z=z_d}}{\int_0^{z_d} \exp\left(-q\psi/kT\right) dz}. \tag{4.11}$$

If we assume for convention that the electrostatic potential at the junction is the reference, we have

$$\psi(0) = 0$$
$$\text{and} \quad \psi(z_d) = \phi_B - V_n - V$$
$$\text{where} \quad qV_n = E_c - \xi_f \tag{4.12}$$

in the bulk. The carrier concentration at the junction and the junction depletion region edge are

$$n(0) = N_C \exp\left(-\frac{q\phi_B}{kT}\right),$$
$$\text{and} \quad n(z_d) = N_C \exp\left(-\frac{qV_n}{kT}\right), \tag{4.13}$$

and hence

$$J = q\frac{\mathcal{D}_n}{\int_0^{z_d} \exp\left(-q\psi/kT\right) dz} N_C \exp\left(-\frac{q\phi_B}{kT}\right) \left[\exp\left(\frac{qV}{kT}\right) - 1\right]. \tag{4.14}$$

The exponential factor in bias voltage gives rectification character to the current. The second factor in this equation has units of velocity, and is sometimes called a drift-diffusion velocity because it indicates the average velocity with which carriers move via drift-diffusion in the depletion region. The use of the specific carrier concentration relationships above assumes the form of behavior of the quasi-Fermi level shown in part (a) of Figure 4.5. We may simplify this result in reverse and moderate forward bias, where depletion approximation holds. Specifically, the electrostatic potential then follows

$$\frac{q\psi}{kT} = \frac{q^2 N_D}{2\epsilon_s kT} z_d^2 - \frac{q^2 N_D}{2\epsilon_s kT}(z_d - z)^2. \tag{4.15}$$

Denoting the first term in the above as a^2 and the second as η^2,

$$\int_0^{z_d} \exp\left(-\frac{q\psi}{kT}\right) dz = \exp\left(-a^2\right) \int_0^{z_d} \exp\left(\eta^2\right) dz$$
$$= \sqrt{\frac{2\epsilon_s kT}{q^2 N_D}} \exp\left(-a^2\right) \int_0^a \exp\left(\eta^2\right) d\eta$$
$$\approx \frac{\epsilon_s kT}{q^2 N_D z_d}, \tag{4.16}$$

where we have used the maximum magnitude of $1/2a$ of the error-function integral.

This lets us write, for moderate bias conditions,

$$J = \frac{q^2 \mathcal{D}_n N_C}{kT} \sqrt{\frac{2q\left(\psi_{j0} - V\right) N_D}{\epsilon_s}} \exp\left(-\frac{q\phi_B}{kT}\right) \left[\exp\left(\frac{qV}{kT}\right) - 1\right]. \quad (4.17)$$

The pre-factor of the current–voltage relationship has a small voltage dependence. It can be ignored in the forward direction where the exponential term dominates, but in the reverse direction, where the exponential is small, it is this term, and the image force barrier lowering, that give rise to bias dependence. Also recall that we ignored generation and recombination in the derivation. In the reverse direction, where drift-diffusion current is small, the generation processes can predominate for materials with low lifetime.

4.2.2 Thermionic Emission

In order for drift-diffusion to be a limiting step, the transport of carriers has to be considerably slow. The drift-diffusion approach assumes that the transport to the barrier is a rate-limiting step. More often, though, the transport is limited by injection over the barrier. A simple schematic of this is shown in Figure 4.7, which shows the Bethe condition for transport to be limited by the emission process at the metal–semiconductor junction. The Bethe condition states that for the transport at the metal–semiconductor junction to be limited by the interface, the width of the layer at the interface in the semiconductor across which a kT change in electrostatic energy occurs should be significantly less than the mean free path of the semiconductor.[2] Let z_c be the width of this region over which the kT change in energy occurs.

$$z_c \ll \lambda \qquad (4.18)$$

for emission at the metal–semiconductor interface to dominate. We will call this width the collision-less width to indicate its intent. The basis of this condition is that by ensuring it, one assures that the carrier distribution function at $z = z_c$ is independent of the current.[3] If the carrier distribution function at z_c and beyond is independent of the current, then it follows that the quasi-Fermi level in the region beyond is constant and the current is limited by the interface transport and not by the drift-diffusion transport in

---

[2]Our discussion here follows the arguments forwarded in F. Berz, "The Bethe Condition for Thermionic Emission Near an Absorbing Boundary," *Solid-State Electronics*, **28**, No. 10, p. 1007, Oct. 1985. Also, see discussion in G. Baccarani, "Current Transport in Schottky-barrier Diodes," *J. of Appl. Phys.*, **47**, No. 9, p. 4122, Sep. 1976.

[3]Strictly speaking, the electrostatic energy change should be more than kT to ensure sufficient change; more on this condition in the thermionic–drift-diffusion section.

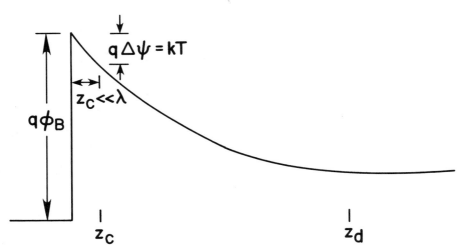

Figure 4.7: Schematic showing the Bethe condition for transport to be limited by thermionic emission. The width of the region over which a kT drop in potential energy occurs at the barrier (z_c) should be much smaller than the mean free path λ. z_d is the depletion region width of the junction.

the bulk. Clearly, an implication of the above condition is that under large forward biases, where the region z_c gets broad, the condition should be expected to break down. Thus, there are limitations to the bias conditions up to which we may apply the following analysis. At some high bias point, most semiconductors would require a unified thermionic emission and drift-diffusion description.

From the semiconductor, the carriers emitted into the metal are the ones with sufficient energy and appropriate velocity in the collision-less region. Likewise from the metal, except that the large barrier height and large scattering rate of metals makes such a current small.

Our analysis here follows a standard and intuitive approach. Later, in our discussion of thermionic–drift-diffusion transport, we will follow a more exact analysis that relaxes the approximation adopted here related to the distribution function being Maxwell–Boltzmann even at the metal-semiconductor interface. The total current density is the difference in the current flow from the semiconductor into the metal J_{sm} and the current flow from the metal into the semiconductor J_{ms}. The former is the integral

$$J_{sm} = \int_{\xi_{n,sem}+q(\phi_B-V)}^{\infty} qv_z dn \qquad (4.19)$$

at the interface. This equation states that only those particles with sufficient energy to cross the barrier, and moving in the $-z$ direction, actually cross the barrier. Assuming Boltzmann and parabolic band approximation,

$$dn = \frac{(2m^*)^{3/2}}{2\pi^2\hbar^3}\sqrt{E - E_c}\exp\left(\frac{E - \xi_{n,sem}}{kT}\right)dE. \qquad (4.20)$$

If we further assume that any excess energy is kinetic in nature, i.e.,

$$E - E_c = \frac{1}{2}m^*v^2, \qquad (4.21)$$

we get

$$dn = 2\left(\frac{m^*}{2\pi\hbar}\right)^3\exp\left(-\frac{E_c - \xi_{n,sem}}{kT}\right)\exp\left(-\frac{m^*v^2}{2kT}\right)4\pi v^2 dv, \qquad (4.22)$$

and

$$
\begin{aligned}
J_{sm} &= 2q\left(\frac{m^*}{2\pi\hbar}\right)^3\exp\left(-\frac{E_c - \xi_{n,sem}}{kT}\right)\int_{-\infty}^{\infty}\exp\left(-\frac{m^*v_x^2}{kT}\right)dv_x \times \\
&\quad \int_{-\infty}^{\infty}\exp\left(-\frac{m^*v_y^2}{kT}\right)dv_y \times \\
&\quad \int_{-\sqrt{2q(\psi_{j0}-V)/m^*}}^{\infty}v_z\exp\left(-\frac{m^*v_z^2}{kT}\right)dv_z, \\
&= \mathcal{A}^*T^2\exp\left(-\frac{q\phi_B}{kT}\right)\exp\left(\frac{qV}{kT}\right), \qquad (4.23)
\end{aligned}
$$

where

$$\mathcal{A}^* = \frac{qm^*k^2}{2\pi^2\hbar^3} \qquad (4.24)$$

is known as the effective Richardson's constant (120 A.cm^{-2}.K^{-2} for free electrons). The carriers in the metal in the very tail of the energy should be expected to have an exponential tail, even though they do not obey Maxwell–Boltzmann statistics. So, one should expect a similar exponential relationship

$$J_{ms} = \mathcal{A}^{*'}T^2\exp\left(-\frac{q\phi_B}{kT}\right), \qquad (4.25)$$

where the constant $\mathcal{A}^{*'}$ does not necessarily have the same meaning or interpretation as in a semiconductor because of the differences in statistics of the two materials. However, at zero bias, since the two currents are equal, we obtain $\mathcal{A}^* = \mathcal{A}^{*'}$ in this model. The resulting thermionic current can be written, as a function of applied bias of V, as the difference of these two currents

$$J = \mathcal{A}^* T^2 \exp\left(-\frac{q\phi_B}{kT}\right)\left[\exp\left(\frac{qV}{kT}\right) - 1\right]. \tag{4.26}$$

Note that the functional form of this equation is the same as in the drift-diffusion theory.

We know the effective mass for an isotropic (spherical constant energy surface) single valley semiconductor. However, what should it be for other semiconductors? For anisotropic constant energy surfaces, such as at L or X minima, or for degenerate bands such as light and heavy hole bands, the appropriate mass is obtained by summing over the relevant masses in the direction of injection, weighted by the probability of the occurrence of a carrier with that mass, i.e., the masses are weighted by the density of states. Problem 1 deals with this question. The effective mass, for arbitrary orientations, appropriate to the calculation of Richardson's constant is

$$m^* = \left(m_x m_y l_z^2 + m_y m_z l_x^2 + m_z m_x l_y^2\right)^{1/2}, \tag{4.27}$$

where l_x, l_y, and l_z are the direction cosines. For X-like minimas, this reduces to

$$m^* = 2m_t^* + 4(m_l^* m_t^*)^{1/2} \tag{4.28}$$

for injection in $\langle 100 \rangle$ direction, and

$$m^* = 2\sqrt{3}\left(m_t^{*2} + 2m_l^* m_t^*\right)^{1/2} \tag{4.29}$$

for injection in $\langle 111 \rangle$ direction. Similarly, for L-valleys,

$$m^* = \frac{4}{\sqrt{3}}\left(m_t^{*2} + 2m_l^* m_t^*\right)^{1/2} \tag{4.30}$$

for injection in $\langle 100 \rangle$ direction, and

$$m^* = m_t^* + \left(m_t^{*2} + 8m_l^* m_t^*\right)^{1/2} \tag{4.31}$$

for injection in $\langle 111 \rangle$ direction.

4.2.3 Field Emission and Thermionic Field Emission

Our discussion above has ignored quantum mechanical effects. Carriers that have the velocity and the energy to inject over the barrier may suffer

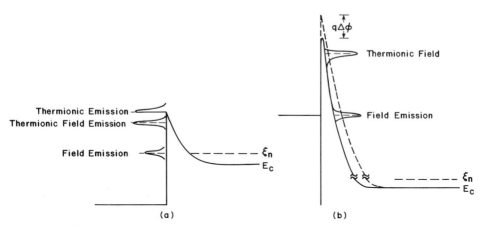

Figure 4.8: The different components of injection at a metal–semiconductor barrier shown in forward (a) and reverse (b) bias.

quantum-mechanical reflections because of the abrupt barrier. Carriers that do not have the requisite energy or velocity to inject over the barrier may still cross into the metal by tunneling through the thin barrier. This finite non-zero probability of penetration can make a substantial difference in the magnitude of current at low temperatures where the thermionic current reduces due to lower density of carriers in the tails, as well as high dopings where the barriers are thinner. This emission behavior is often referred to as field emission, and its inclusion in the thermionic emission model leads to the thermionic field emission model.

The field emission probability (see Figure 4.8) increases with temperature because electrons at higher energies see a thinner barrier with a lower barrier height. Since the field emission current over any energy range is proportional to the source function, the carrier concentration, and the characteristics of the barrier, at high energies it decreases again because of the decrease of carrier density at those energies. Thus, it peaks at a certain energy. For very high temperatures, the carrier density is high with high enough energy for crossing the barrier and most current is thermionic. Thermionic field emission current is thus important at low and intermediate temperatures, and high dopings.

Simple calculations of tunneling probability usually invoke the WKB approximation (so named after Wentzel, Kramers, and Brillouin)[4] that we

---

[4]The WKB approximation allows us to determine the transmission probability accurately when the change in barrier is continuous. For discontinuous changes, the pre-factor changes and hence one can only determine the proportional dependence on the exponen-

will discuss later in the context of tunneling in heterostructures. It relates the elastic tunneling transmission probability T_t with the carrier momentum (or, equivalently, the reduced wave vector) through

$$T_t \propto \exp\left(-2\int |k_z|\,dz\right), \tag{4.32}$$

where tunneling is assumed to take place in the direction z. The larger the evanescent effect of the carrier momentum in a barrier or the longer the tunneling distance, the smaller the probability. The WKB approximation, for a triangular barrier, gives the result of probability for elastic tunneling transition as

$$T_t \propto \exp\left[-\frac{4\sqrt{2m^*}}{3\hbar q\mathcal{E}}(q\phi_B)^{3/2}\right]. \tag{4.33}$$

The maximum field at a metal–semiconductor junction is related to the junction potential (the diffusion potential), doping, and the permittivity through

$$\mathcal{E}_{max} = \sqrt{\frac{2qN_D\psi_{j0}}{\epsilon_s}}, \tag{4.34}$$

and hence we may write

$$T_t \propto \exp\left[-\frac{2(q\phi_B)^{3/2}}{3E_{00}\sqrt{q\psi_{j0}}}\right], \tag{4.35}$$

where

$$E_{00} = \frac{q\hbar}{2}\sqrt{\frac{N_D}{m^*\epsilon_s}}. \tag{4.36}$$

E_{00} is a characteristic energy parameter which determines, by virtue of its magnitude vis-a-vis the thermal energy, the temperature below which field emission becomes important.[5] Figure 4.9 shows this parameter for some of the compound semiconductors.

The lower density of states of most of the compound semiconductors and the smaller electron effective mass lead to smaller barriers that are easier to penetrate. Thus, field emission is relatively more important in compound semiconductors than in silicon. Thus, an accurate description of transport in metal–semiconductor junctions in compound semiconductors requires inclusion of both thermionic and field emission. The thermionic field current is given by a relation quite similar to the thermionic emission

---

tial term of WKB approximation.

[5]See F. A. Padovani and R. Stratton, "Field and Thermionic Field Emission in Schottky Barriers," *Solid-State Electronics*, **9**, p. 695, 1966.

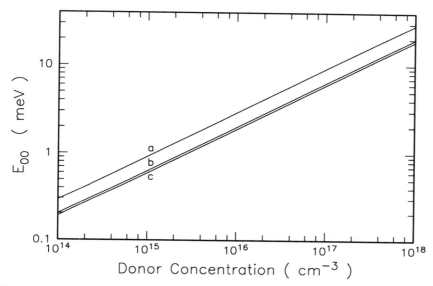

Figure 4.9: The characteristic energy E_{00} as a function of donor concentration for some of the semiconductors. The curve a identifies InAs, the curve b identifies GaAs, and the curve c identifies InP.

relationship, although it is relatively more complex to derive (see Problem 2). The current due to thermionic field emission is given by

$$J = J_s \exp\left[\frac{qV}{E_{00} \coth\left(E_{00}/kT\right)}\right]\left[1 - \exp\left(-\frac{qV}{kT}\right)\right], \qquad (4.37)$$

where J_s is related through

$$J_s = J_m \frac{q\sqrt{\pi E_{00} q\left(\phi_B - V - V_n\right)}}{kT \cosh\left(E_{00}/kT\right)} \exp\left[\frac{-q(\phi_B + V_n)}{E_{00} \coth\left(E_{00}/kT\right)}\right\}, \qquad (4.38)$$

and J_m is the current density under flat band conditions, i.e.,

$$J_m = \mathcal{A}^* T^2 \exp\left(-\frac{qV_n}{kT}\right). \qquad (4.39)$$

Here qV_n is the energy difference between the conduction band edge and the quasi-Fermi level in the bulk.

In this relationship, with increasing temperature, the transition from field emission-dominated characteristics to thermionic emission-dominated characteristics occurs naturally through the property of the coth function. Note that the denominator of the voltage exponential asymptotically reaches either E_{00} in the low temperature limit or kT in the high temperature limit, again consistent with field emission-dominated behavior in the low temperature limit, and thermionic emission-dominated behavior in the high temperature limit.

We had alluded to the importance of image force lowering, as well as quantum-mechanical reflections, to the transport. Image force lowering can be directly included by an appropriate change in the barrier heights. Since image force lowering also affects the shape of the barrier at the interface, it also subtly influences the field emission current of the above. Image force lowering also leads to a broadening of the barrier at the interface and hence leads to lower reflection problems, all of which increase the magnitude of the current.

The image force effect and the effect of emission from the metal become stronger in reverse bias, which leads to thinner barriers for injection from the metal. In reverse bias, as shown in Figure 4.8, the current from the metal into the semiconductor increases while the current component from the semiconductor into the metal continues to decrease unless effects due to inversion occur at the interface.

Increasing image force lowering, due to the increased field at the interfaces, continuously changes the apparent barrier height of the metal–semiconductor junction. Additionally, field emission and tunneling also increase with reverse bias, as does generation current. Hence, there is a continuous increase of the current with reverse bias which is much stronger than in p–n junctions. The current equation for thermionic field emission, where image force lowering was not included explicitly, continues to hold for reverse bias, except for the neglect of generation current.

An approximation of it, for reverse bias V larger than a few thermal voltages, is

$$J = -J_s \exp\left\{ \frac{qV}{E_{00}} \left[\frac{E_{00}}{kT} - \tanh\left(\frac{E_{00}}{kT} \right) \right] \right\}, \qquad (4.40)$$

where the approximation for J_s is

$$J_s = \mathcal{A}^* T \frac{\sqrt{\pi E_{00}}}{k} \left[q\left(V - V_n\right) + \frac{q\phi_B}{\cosh^2\left(E_{00}/kT\right)} \right]^{1/2} \times$$
$$\exp\left[-\frac{q\phi_B}{E_{00}\coth\left(E_{00}/kT\right)} \right]. \qquad (4.41)$$

In reverse bias, field emission occurs from the metal into the semiconductor through an approximately triangular barrier. Its low temperature limit is given by (see Problem 3)

$$J = \mathcal{A}^* \frac{E_{00}{}^2}{k^2} \frac{\phi_B - V - V_n}{\phi_B} \exp\left(-\frac{(2q\phi_B)^{3/2}}{3E_{00}\sqrt{q\phi_B + qV - qV_n}}\right). \quad (4.42)$$

We have discussed the implication of this field emission from the metal in increasing the reverse bias current; it is also the reason for ohmicity for the large number of ohmic contacts based on heavily doped metal–semiconductor interfaces.

4.2.4 Thermionic Emission-Diffusion theory

We have now developed a description for transport across a metal-semiconductor junction based on thermionic emission and with tunneling effects incorporated in it to describe thermionic field emission. We have also developed the description based on drift-diffusion limited transport in the semiconductor depletion region. The objective of a more general theory, that encompasses both thermionic emission and drift-diffusion, is to be applicable to metal–semiconductor junctions in both the limits, over wider bias range, for any semiconductor. A general theory also allows us to understand the relative importance of the two components of transport for any given semiconductor. Thus, a high mobility semiconductor would show a flat quasi-Fermi level to the metal–semiconductor interface, indicating that the drift-diffusion transport in the semiconductor is not rate-limiting. For a low mobility semiconductor, on the other hand, the quasi-Fermi level should change with position in the depletion region and reach the metal Fermi level at the metal–semiconductor interface, indicating that it is the rate-limiting step.

To develop such an analysis, we need to express these currents—drift-diffusion in the bulk part of the depletion region, and thermionic emission in the interface part of the depletion region—in terms of the semiconductor parameters and the bias, and then we need to force them to be continuous, thus maintaining the required self-consistency. This requires us to obtain the quasi-Fermi level in the bulk, or equivalently the electron distribution function in the bulk as well as at the interface.

The analysis follows, assuming as before a perfectly absorbing boundary condition at the metal–semiconductor junction, i.e., no quantum mechanical reflections, etc. Figure 4.7 showed a schematic description of the problem, broken up into the collision-less region of width z_c at the interface responsible for thermionic emission, and the rest of the depletion region from z_c to z_d responsible for the drift-diffusion. First, we wish to obtain the carrier distribution as a function of the velocity distribution at the interface. All the carriers at the interface $z = 0$ with velocity in the direction of the metal cross the interface. Hence, at $z = 0$ the distribution function is hemi-Maxwellian, while at $z = z_c$ the distribution is Maxwellian for carriers. The current is given by

$$J = q \int_{-\infty}^{\infty} v f(v, z) dv \tag{4.43}$$

at any position z in the depletion region, where v is the velocity. Here we are expressing the distribution function as a function of velocity, which is equivalent to describing it in terms of momentum. Consider the forms of the distribution functions. For carriers beginning with velocity v_c pointing towards the metal at the edge of the collision-less region z_c, the velocity in the collision-less region evolves as

$$v(z)^2 = v_c^2 - \frac{2q}{m^*} \left[\psi(z_c) - \psi(z) \right] \qquad \text{for} \qquad 0 \le z \le z_c. \tag{4.44}$$

At the interface, carriers with positive $v(0)$ return back while those that are still negative cross the barrier. The carriers that return back to $z = z_c$ have velocity identical to when they started but of opposite sign. Therefore, at $z = 0$ carriers have velocities that are less than zero only, and at $z = z_c$ the carriers have velocities less than $\sqrt{2q\Delta\psi/m^*}$, where $\Delta\psi$ is the change in electrostatic potential in the collision-less region. In the absence of collision, the distribution function $f(v, z)$ remains the same in the collision-less region, i.e.,

$$f(v, z) = f(v_c, z_c) \qquad \text{for} \qquad 0 \le z \le z_c. \tag{4.45}$$

The distribution of carriers of interest at $z = z_c$ is

$$f(v, z_c) = \mathcal{B} \exp\left(-\frac{m^* v^2}{2kT} \right) \tag{4.46}$$

for $v < \sqrt{2q\Delta\psi/m^*}$ and zero otherwise. Likewise, at $z = 0$,

$$f(v, 0) = \mathcal{B} \exp\left(-\frac{m^* v^2}{2kT} - \frac{q\Delta\psi}{kT} \right) \tag{4.47}$$

for $v < 0$ and zero otherwise. This distribution function is usually referred to as a hemi-Maxwellian distribution function since it is one of the hemispheres of the constant energy spheres in momentum space. By integrating the distribution function over all possible velocities, the carrier density at $z = z_c$ and $z = 0$ follow as

$$n(z_c) = B\sqrt{\frac{\pi kT}{2m^*}}\left[1 + \mathrm{erf}\left(\frac{q\Delta\psi}{kT}\right)\right],$$

$$n(0) = B\sqrt{\frac{\pi kT}{2m^*}}\exp\left(-\frac{q\Delta\psi}{kT}\right), \qquad (4.48)$$

and hence

$$n(0) = \frac{n(z_c)}{1 + \mathrm{erf}\,(q\Delta\psi/kT)}\exp\left(-\frac{q\Delta\psi}{kT}\right). \qquad (4.49)$$

Recall the Bethe condition, which stems from the requirement that the carrier distribution function at $z = z_c$ be unaffected by the current, or equivalently that the carrier concentration at $z = 0$ be significantly smaller than that at $z = z_c$. We may assure this by requiring the electrostatic energy change in the collision-less region to be $2kT$ or more. This will assure a large change in carrier concentration (see Problem 4). Since $\Delta\psi$ is factor of two or larger than the thermal voltage, the error function term is approximately unity and we obtain

$$n(0) \approx \frac{n(z_c)}{2}\exp\left(-\frac{q\Delta\psi}{kT}\right). \qquad (4.50)$$

Note that the carrier concentration change is larger than what a flat quasi-Fermi level would have implied in this region (see Problem 5). The current follows from Equation 4.43 as

$$J = qn(0)\sqrt{\frac{2kT}{\pi m^*}}, \qquad (4.51)$$

and one can define an effective velocity v_{eff} of transport across the interface as

$$v_{eff} = \sqrt{\frac{2kT}{\pi m^*}}. \qquad (4.52)$$

Recall our discussion of thermionic emission; this is half of the thermal velocity of the carriers. It averages the velocity in the hemisphere.

In terms of carrier density at $z = z_c$, the current can be expressed as

$$J = qn(z_c)\sqrt{\frac{kT}{2\pi m^*}}\exp\left(-\frac{q\Delta\psi}{kT}\right). \qquad (4.53)$$

We can now invoke the drift-diffusion effect on transport by considering the transport in the rest of the depletion region. If we assume a flat quasi-Fermi level, i.e., no drift-diffusion limitation on transport, then

$$n(z_c) = n(z_d) \exp\left(q\frac{\psi(z_c) - \psi(z_d)}{kT}\right) \qquad (4.54)$$

describes the relationship from the edge of the depletion region to the edge of the collision-less region. Hence, in the absence of drift-diffusion transport limitations,

$$J = q\sqrt{\frac{kT}{2\pi m^*}} \exp\left(-\frac{q\Delta\psi}{kT}\right) \exp\left(q\frac{\psi(z_c) - \psi(z_d)}{kT}\right), \qquad (4.55)$$

which is

$$
\begin{aligned}
J &= qN_C\sqrt{\frac{kT}{2\pi m^*}} \exp\left(q\frac{\psi(0) - \psi(z_d)}{kT}\right) \\
  &= \mathcal{A}^* T^2 \exp\left(-\frac{q\phi_B}{kT}\right) \exp\left(\frac{qV}{kT}\right).
\end{aligned} \qquad (4.56)
$$

This is the expression for pure thermionic emission transport.

However, if the quasi-Fermi level is not flat, i.e., drift diffusion to the interface is also important, then we must utilize the form similar to Equation 4.14 to relate the band bending, i.e., we relate the carrier concentration at $z = z_c$ with $z = z_d$ by considering the drift-diffusion current, and then force this current to be the same as the thermionic emission current in order to maintain current continuity. This allows us to write (see Problem 6)

$$J = \frac{qN_C(kT/2\pi m^*)^{1/2} \exp\left(-q\phi_B/kT\right)\left[\exp\left(qV/kT\right) - 1\right]}{1 + (q/\mu kT)\,(kT/2\pi m^*)^{1/2} \exp\left(-q\phi_B/kT\right) \int_0^{z_d} \exp\left(q\psi/kT\right) dz}. \qquad (4.57)$$

This equation takes into account the hemi-Maxwellian distribution of carriers at the metal–semiconductor interface, the lack of collisions in a region of width $z_c \ll \lambda$ over which the electrostatic energy changes by $2kT$, as well as the supply of carriers by drift-diffusion across the depletion region. Problem 7 considers the thermionic emission and drift-diffusion limit of this expression.

4.3 Heterojunctions

Heterojunctions are formed when two dissimilar materials are joined together at a junction using one of the materials growth techniques. Any arbitrary combination of semiconductors or semiconductors and insulators, or growth technique, does not necessarily result in a interface that can be of benefit in devices. Unique combinations, such as SiO_2/Si, or many of the lattice matched semiconductors (to be discussed in the context of heterostructure field effect transistors), result in interfaces with low interface density, and low interface scattering. These heterojunctions are the basis for field effect and bipolar devices. It is the transport properties, both perpendicular and parallel to the interface, that we utilize to advantage in these devices. The effect of low interface states is to allow us to control, in a very reproducible way, both perpendicular and parallel transport. For insulators, such as SiO_2, the perpendicular transport is of little concern since it is negligible (displacement current may still flow) except in breakdown conditions. For semiconductors, both are of interest. Semiconductor heterojunctions are the most popular form for compound semiconductors. This section follows a similar description as before in considering perpendicular transport through the junctions; parallel transport will be considered in our discussion of heterostructure field effect transistors.

An additional complication, unlike the SiO_2/Si interface, for semiconductor heterojunctions is the effect of alloy grading, i.e., a changing of the composition of the semiconductor at the interface over a short distance. Such an arrangement is practiced in bipolar transport, its use in field effect transistors is largely to allow ohmic access to the channel of the device, since abrupt discontinuities lead to rectification. The abrupt discontinuities in these heterojunctions can appear both in the conduction band edge energy (ΔE_c), and the valence band (ΔE_v), whose origin is the strong interatomic Coulombic effects that take place at the interface. Figure 4.10 shows three common forms (Type I, II, and III) of heterojunctions encountered. There is thus a variety of behavior that occurs in realistic semiconductor heterostructures; type I being the form that has been most extensively exploited in devices. Our discussion, throughout this text is principally with this type of heterostructure. This figure showed the discontinuities at an abrupt heterojunction. In general, these discontinuities are not related by differences in electron affinity, and hence throughout this text, following our earlier discussion of this subject, we use these as experimentally obtained parameters.

Clearly, since compositional mixing can be achieved in semiconductor heterojunctions over short length scales, variations in discontinuities can also be achieved. Figure 4.11 shows, for an n-type $Ga_{.7}Al_{.3}As/GaAs$ heterojunction, the conduction band edge and electron density for abrupt and linearly graded junctions. This is an idealization. We assume that one can obtain an abrupt change to linearly varying composition. However, the conclusion that the barrier height and the discontinuity can be varied in

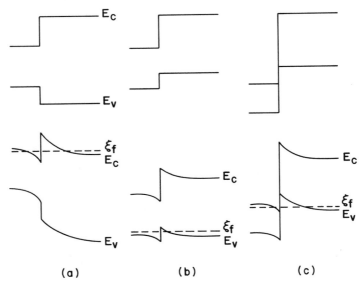

Figure 4.10: Cases (a), (b), and (c) show schematically the band edges for the three common forms (Type I, II, and III) of semiconductor heterojunctions encountered. The lower halves of these cases show the band diagrams with doping conditions commonly encountered. An example for case (a) is $GaAs/Ga_{1-x}Al_xAs$, for case (b) is $InAs/AlSb$, and for case (c) is $InAs/GaSb$.

short grading regions is correct. Recalling our discussion of quasi-neutrality, the appearance of a discontinuity in band edges at a heterojunction clearly leads to a breakdown in quasi-neutrality as seen in Figure 4.11. However, following this earlier discussion, a long grading distance at a heterojunction leads to a decrease of the alloy field, an effect of the changing material characteristics, and the quasi-neutrality may be restored provided the Debye length criterion is satisfied. An effect of changing composition over a distance is that in this region the quasi-field for an electron may be different than the quasi-field for a hole, since the change in bandgap as a function of position results in differing changes in the conduction band edge and the valence band edge. We will consider this further in the discussion on heterostructure bipolar transistors.

In this section, we consider transport across isotype interfaces, i.e., interfaces of the same polarity. Anisotype interfaces will be considered in our discussion of p–n junctions in the next section.

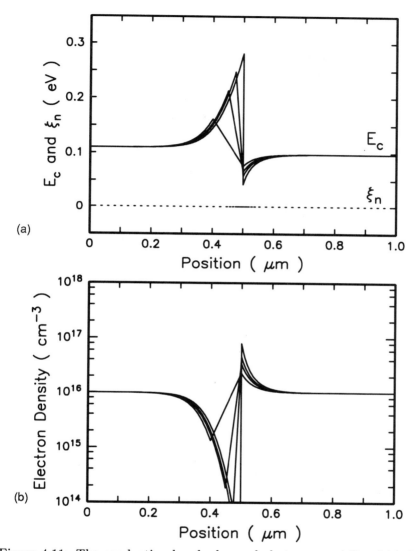

Figure 4.11: The conduction band edge and electron quasi-Fermi level at abrupt and graded $Ga_{1-x}Al_xAs/GaAs$ heterojunctions. The background doping is assumed to be 1×10^{16} cm^{-3} and the grading distances for linear grading are 250 Å, 500 Å, and 1000 Å.

4.3.1 Thermionic Emission

Conduction in heterojunctions should be expected to be similar in nature to the metal–semiconductor junction, except for some additional complexities. The behavior of electrons in the metal was characterized by a free electron mass, and a negligibly short distance over which the carrier injected from the semiconductor thermalized. The injection of carriers from the metal depended on the barrier formed with the semiconductor; no changes occurred in the injection as function of bias, since the barrier and the metal were left relatively undisturbed by the bias. When considering a semiconductor–semiconductor heterojunction, we have to consider changes in the distribution function on both sides of the junction. As an example, any applied bias is partitioned between the two semiconductors in a semiconductor–semiconductor heterojunction, while in a metal–semiconductor junction, the high conductivity of the metal forces the potential drop to occur in the semiconductor alone. We will consider the p–n heterojunction later; here we will consider an isotype heterojunction. Problem 8 considers the transport across the interface using the same procedure as in the thermionic–drift-diffusion theory of a metal–semiconductor junction. As we have discussed, usually, for electrons in the compound semiconductors, thermionic emission theory is adequate for a variety of conditions. Semiconductor heterojunctions, unlike the metal–semiconductor junctions, exhibit much larger field emission and other tunneling effects. This is a result of their smaller barrier height. Our analysis is, therefore, a simplified treatment of the problem of emission using the Maxwell–Boltzmann distribution approximation at the interface. Following Figure 4.12, we may determine the current components if we assume a Boltzmann distribution.

Let N_{D1} and ϵ_1 be the doping and the dielectric constant of the large bandgap semiconductor and let N_{D2} and ϵ_2 be the corresponding values for the small bandgap semiconductor. The current flowing from 1 to 2, evaluated at the interface in the first semiconductor, is the product of the charge of the carriers, the number of carriers available for transmission to the second semiconductor, a transmission probability T_{12} for transmission into the second semiconductor, and the carrier velocity in the z-direction. The transmission probability T_{12} depends on the availability of states in the second semiconductor for the transmission. Since most of the states are empty in the high energy tail region, the transmission probability is close to unity for transmission in both directions. However, it is not identical. In our discussion of Richardson's constant and maintaining zero net current at thermal equilibrium in the derivation of the thermionic current in the metal–semiconductor junction, we invoked an argument based on equality of flux in the two directions to obtain the desired result. A Richardson's

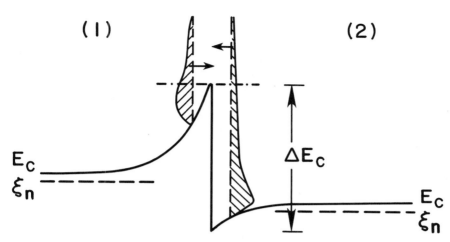

Figure 4.12: An isotype n–n heterojunction between a large bandgap and a small bandgap material. The figure shows those carriers that can cross over the barrier if they have the minimum velocity required to cross the barrier. Note the potential drops on each side of the junction in this heterojunction case, in contrast to the case of metal–semiconductor junction.

constant–based approach will naturally fall out of a discussion of thermionic emission at a semiconductor heterojunction from both sides of the interface. Detailed balance requires that, at thermal equilibrium, these be identical. We will show based on arguments of momentum conservation that the two are related and result in the desired zero net current at thermal equilibrium.

This current from semiconductor 1 to 2 (J_{12}), due to thermionic emission, obtained by integrating over all possible velocities, and assuming a Maxwell–Boltzmann distribution function at the interface, is given by the integral

$$\frac{\int_{v_{zmin}}^{\infty} \int_{-\infty}^{\infty} \int_{-\infty}^{\infty} qT_{12}N_{D1}v_z \exp\left[-m_1^* \left(v_x^2 + v_y^2 + v_z^2\right)/2kT\right] dv_x dv_y dv_z}{\int_{-\infty}^{\infty} \int_{-\infty}^{\infty} \int_{-\infty}^{\infty} \exp\left[-m_1^*/2kT \left(v_x^2 + v_y^2 + v_z^2\right)/2kT\right] dv_x dv_y dv_z}.$$

(4.58)

This integral has the following basis. If we assume drift-diffusion as not being a limitation on transport and assume a Maxwell–Boltzmann distribution function, then the carrier distribution function at the metal–semiconductor interface can be related to that at the edge of the depletion region. This argument is similar to that used in reducing to the thermionic emission limit in our thermionic–drift-diffusion analysis of a metal–semiconductor junction. The limit of integration is appropriately shifted to account for this, and, as we shall show later, to account for conservation of momentum and energy during the emission. The expression above is written as a function of N_{D1}, the donor and carrier concentration, at the edge of the depletion region. v_{zmin} is related to the emission barrier, which is related to the barrier height and the electrostatic potential V_1. V_1 is the part of the applied potential that appears across the junction depletion region in semiconductor 1. Following Problem 9, the current may be written, in terms of the effective conduction band density of states N_{C1} and a momentum-space averaged emission transmission probability T_{12}, as

$$J_{12} = N_{C1}\overline{T}_{12}\sqrt{\frac{kT}{2\pi m_1^*}}\,\exp\left(-\frac{q\phi_{B0}}{kT}\right)\exp\left(\frac{qV_1}{kT}\right). \qquad (4.59)$$

Similarly, for the conduction from material 2 the small bandgap material to material 1 (the large bandgap material), this current (J_{21}) is given by the integral

$$\frac{\int_{-v_{zmin}}^{\infty}\int_{-\infty}^{\infty}\int_{-\infty}^{\infty} qT_{21}N_{D2}v_z\exp\left[-m_2^*\left(v_x^2+v_y^2+v_z^2\right)/2kT\right]dv_x dv_y dv_z}{\int_{-\infty}^{\infty}\int_{-\infty}^{\infty}\int_{-\infty}^{\infty}\exp\left[-m_2^*/2kT\left(v_x^2+v_y^2+v_z^2\right)\right]dv_x dv_y dv_z} \qquad (4.60)$$

where the minimum velocity, in a way similar to the previous case, is determined by the condition

$$\frac{1}{2}m_2^* v_{zmin}^2 = q\phi_{B2} - qV_2, \qquad (4.61)$$

where V_2 is the part of the applied potential V that appears across semiconductor 2. This may be simplified to

$$J_{21} = N_{C2}\overline{T}_{21}\sqrt{\frac{kT}{2\pi m_2^*}}\,\exp\left(-\frac{q\phi_{B0}}{kT}\right)\exp\left(\frac{qV_2}{kT}\right). \qquad (4.62)$$

The net current, which is the difference between these two components across the junction, is

$$J = q\exp\left(-\frac{q\phi_{B0}}{kT}\right)\left[\overline{T}_{12}N_{C1}\sqrt{\frac{kT}{2\pi m_1^*}}\,\exp\left(\frac{qV_1}{kT}\right) - \right.$$
$$\left. \overline{T}_{21}N_{C2}\sqrt{\frac{kT}{2\pi m_2^*}}\,\exp\left(\frac{qV_2}{kT}\right)\right]. \qquad (4.63)$$

In thermal equilibrium, detailed balance in the above implies the condition

$$\frac{\overline{T}_{12}N_{C1}}{\sqrt{m_1^*}} = \frac{\overline{T}_{21}N_{C2}}{\sqrt{m_2^*}}. \tag{4.64}$$

This condition is, within our approximations, a mathematical abstraction of the requirement of zero current in thermal equilibrium conditions. It should be viewed only as such, that the coefficient in the transmission of the hot carriers across a heterojunction discontinuity adjusts itself to allow the balance at thermal equilibrium. The consequences of the approximation of a Maxwell–Boltzmann distribution function as opposed to the hemi-Maxwellian distribution that may actually be more appropriate at the junction, and the self-adjustment of carrier distribution in order to account for current continuity, are all lumped into this phenomenological coefficient. Note that the current varies exponentially with bias and its magnitude depends on how the division of applied voltage occurs on the two sides of the junction. The division of voltage is related by Maxwell's first equation, which relates the displacement vector with the interface charge. Problems 8 and 9 consider this example with its associated bias voltage partitioning.

As can be seen, this behavior for a semiconductor–semiconductor heterojunction is considerably more complicated than that of the metal–semiconductor junction. Fortunately, many heterojunctions of interest are highly asymmetric in doping, and in cases where there exists a sufficiently large barrier, the problem reduces to the metal–semiconductor junction–like situation. An example is a larger barrier with high doping in the smaller bandgap semiconductor. For this problem, the smaller bandgap semiconductor behaves quite similar to the metal, and a thermionic equation based on the Richardson's constant determined from the large bandgap semiconductor parameters suffices. Quite often, though, the dopings on both sides of junction are high; these are the basis for some of the ohmic contacts. In these cases, however, tunneling effects dominate.

We now return to the question of thermal equilibrium in this problem and the proper form of Richardson's constant.[6] This question can be resolved by considering conservation laws of energy and momentum for the carrier emission at the interface. When an electron transits from one semiconductor to another, across the band discontinuity, it occupies different conduction band states determined by the band structure of the new semiconductor. In the process of doing so, assuming that this transit occurs without involvement of any scattering mechanisms that lead to energy loss

---

[6]This question has been discussed in detail in A. A. Grinberg, "Thermionic Emission in Heterosystems with Different Effective Electronic Masses," *Phys. Rev. B*, **33**, No. 10, p. 7256, 1986. Our arguments follow this analysis.

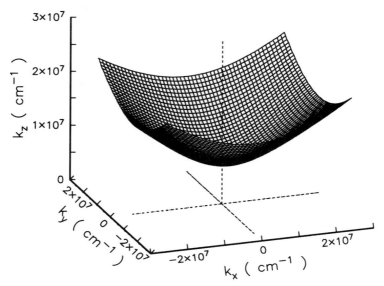

Figure 4.13: A surface plot showing the allowed wave vectors that satisfy the conditions for conservation of thermionic emission energy and momentum during injection from a low mass m_1^* material across a discontinuity of ΔE_c into a larger mass m_2^* material.

or momentum change, energy and momentum must be conserved. Continuity of the wave function parallel to the interface actually requires a stricter condition of conservation of momentum parallel to the interface $(k_\parallel)$. The equation for conservation of energy, considering the transit across a discontinuity of ΔE_c, implies

$$\frac{\hbar^2 k_{\perp 1}^2}{2m_1^*} + \frac{\hbar^2 k_\parallel^2}{2m_1^*} = \Delta E_c + \frac{\hbar^2 k_{\perp 2}^2}{2m_2^*} + \frac{\hbar^2 k_\parallel^2}{2m_2^*}. \tag{4.65}$$

For $m_1^* < m_2^*$, the constant energy surface for allowed wave vector components is an ellipsoid given by

$$k_{\perp 1}^2 + \left(1 - \frac{m_1^*}{m_2^*}\right) k_\parallel^2 = \frac{2m_1^* \Delta E_c}{\hbar^2} + \frac{m_1^*}{m_2^*} k_{\perp 2}^2. \tag{4.66}$$

The selection criterion described by this conservation equation is shown in Figure 4.13. This matching implies that for thermionic injection from material 1 to 2, there exists a non-zero minimum in energy E_{min}, which is different from the value of energy discontinuity. This minimum is obtained from the matching condition for minimum energy

$$E_{min} = \Delta E_c - \frac{(m_2^* - m_1^*)E_{\parallel}}{m_1^*}. \tag{4.67}$$

With this minimum, the thermionic current is given by the integral

$$J = \frac{qm_1^*}{2\pi^2\hbar^3} \exp\left(-\frac{qV_n}{kT}\right) \int_0^\infty \exp\left(-\frac{E_{\parallel}}{kT}\right) dE_{\parallel} \int_{E_{min}}^\infty \exp\left(-\frac{E_{\perp}}{kT}\right) dE_{\perp}. \tag{4.68}$$

The integral for emission in the other direction is identical, and hence the apparent discrepancy is resolved. There exists one unique Richardson's constant that describes the thermionic transport at the heterojunction. The ambiguity arises because of a simplified choice of minima for evaluation of the integral. Hence, the thermionic injection current for flow of carriers in the two directions is equal in thermal equilibrium. Also, it follows that the Richardson's constant is determined by the smaller effective mass.

4.3.2 Tunneling

We ignored tunneling in the above analysis. In the discussion of metal–semiconductor junctions, we included tunneling effects in the thermionic field emission theory for the junction. For transport at metal–semiconductor junctions, we treated thermionic emission as the dominant process of current flow, and field emission as a perturbation to the thermionic emission that increases the current. In our treatment of tunneling effects in heterojunctions, we will discuss the tunneling current by itself, because very often it is the dominating means of current flow in the heterojunction, owing to their smaller barriers and barrier widths.

The discussion to this point has assumed that the electron energy is conserved during the tunneling transition, although we alluded to the possibility of inelastic processes also. Inelastic interactions can occur between the electrons and phonons, plasmons, etc. Tunneling in p–n junctions, with non-Γ conduction minima, such as in silicon and germanium, occur predominantly with phonon-assisted processes. This is necessary to allow for the change in the momentum. Optical phonon interaction occurs with a change in energy of $\hbar\omega_q$. At low temperatures, phonon emission may still occur, but phonon absorption processes disappear because of the freeze-out of phonons. Also, for biases smaller than this threshold ($\hbar\omega_q/q$), neither emission nor absorption can occur at very low temperatures. Thus at these low biases, phonon-assisted tunneling should be very temperature sensitive. This is unlike the temperature insensitivity of most tunneling processes in semiconductors because bandgaps and effective masses are very weakly temperature dependent.

We briefly discuss tunneling in an indirect semiconductor versus a direct semiconductor to show the considerable variety one observes in tunneling in compound semiconductors of various thicknesses and barrier compositions. Consider the prototypical GaAs/Ga$_{1-x}$Al$_x$As/GaAs structure of Figure 4.14. This structure occurs in various forms in many devices both in field effect transistors (e.g., in SISFETs) and in coupled barriers. Tunneling may take place through processes based on Γ electrons alone, together with conservation of the energy of the electron. This would be most likely if the Ga$_{1-x}$Al$_x$As had a composition in the direct bandgap region. Even in the direct bandgap region, an electron in the Γ valley in GaAs may tunnel via the L or X valley in Ga$_{1-x}$Al$_x$As, together with the involvement of an inelastic scattering process. This means that the Γ valley electron may tunnel into the GaAs collector by coupling through the Γ barrier, the L barrier, or the X barrier. Up to a mole-fraction close to 0.4, the Γ barrier is the lowest, and most likely it would dominate the tunneling. But beyond that, the inelastic processes should dominate, since the material becomes indirect. For indirect bandgap Ga$_{1-x}$Al$_x$As, direct transitions would require the cold electrons in the Γ valley of GaAs to tunnel through a higher barrier unlike the indirect tunneling which occurs through a lower barrier. This makes it far less likely than the other inelastic processes. For increasing mole-fraction of aluminum arsenide, initially the L valley is lower, and finally the X valley is the lowest. Thus, at the highest mole-fractions, the X barrier may be the dominant cause of tunneling, barrier width permitting, even though the X valley is the third-highest valley in GaAs, and very few electrons start out from it. The larger the barrier width, the higher the likelihood of scattering, and hence the higher the likelihood of inelastic scattering in such structures. This is of importance to SISFETs as well as to coupled tunneling structures which we will discuss later in this book.

Tunneling in indirect parts of the structure can be quite complicated because equi-energy surfaces are anisotropic, and depending on the orientation, they may not even be equivalent. Thus carriers occupying different energy surfaces may behave differently. As an example, consider Ga$_{1-x}$Al$_x$As at high AlAs mole-fractions, similar to silicon, for tunneling in the $\langle 100 \rangle$ direction. X valleys are the conduction band minima and the constant energy surfaces are as shown in Figure 2.26. A set of two ellipsoids and a set of four ellipsoids are equivalent for the purposes of tunneling, and the four-ellipsoid set is favored in tunneling because it has a lower transverse mass. Tunneling from the GaAs primary valley to these surfaces will still require the involvement of transfer of energy through scattering, and may occur with phonons, plasmons, interface scattering, etc. This problem of tunnel-

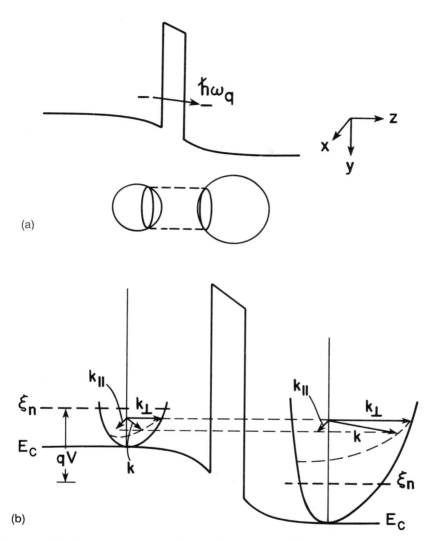

Figure 4.14: Tunneling of a Γ electron by inelastic (a) and elastic (b) process in a GaAs/Ga$_{1-x}$Al$_x$As/GaAs structure. The spheres shown are in the constant energy surfaces for the tunneling electron; the differences in the radius represent the change in potential energy during tunneling. In (b), the parallel momentum remains constant while the perpendicular momentum changes.

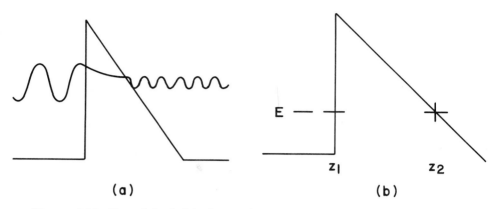

(a)   (b)

Figure 4.15: Part (a) of this figure shows tunneling through a triangular barrier for an electron with energy less than the barrier height. Part (b) shows the classical turning points for this electron.

ing in semiconductor heterostructures for such situations will be revisited during our discussion of coupled barrier structures.

We now consider the methods of analysis for determining the current densities in tunneling processes.[7] We will consider the problem of a triangular barrier, which has already occurred as part of field emission in the metal–semiconductor junction, first. This is also among the earliest of tunneling problems to have been analyzed. It is related to emission from metal surfaces into vacuum, a reverse of the field emission problem from the semiconductor into the metal. An early theory for tunneling was developed by Fowler and Nordheim assuming a triangular potential barrier for the emission into vacuum. This theory, in a modified form, is relevant to semiconductor structures also. First, we will consider the exact solution to the problem of tunneling transmission through a triangular barrier; then we will use a simpler technique that can be the basis for more general solutions.

Consider the triangular potential barrier shown in Figure 4.15. The Schrödinger equation for this single-body problem is

$$-\frac{\hbar^2}{2m^*}\nabla^2\varphi + (V - E)\,\varphi = 0, \qquad (4.69)$$

[7]Tunneling is a very rich and diverse phenomenon. Two books, C. B. Duke, *Tunneling in Solids* in *Solid State Physics—Advances in Research and Applications*, Academic Press, N.Y. (1969) and E. Burstein and S. Lundqvist, Ed., *Tunneling Phenomena in Solids*, Plenum Press, N.Y. (1969), which are also general references for this chapter and Chapter 8, give a detailed account of this. Some of our discussion here and in Chapter 8 draws from these.

where φ is the eigenfunction and E the eigenenergy. For $z < 0$, $V = 0$, and for the triangular barrier, the barrier V is described by

$$V = \xi_f + q\phi_B - q\mathcal{E}z \tag{4.70}$$

for $z > 0$, where $\mathcal{E}$ is the electric field, ξ_f the Fermi energy, and $q\phi_B$ the barrier height. The eigenfunction and the derivative of the eigenfunction are continuous. This is the statement of continuity of the probability of finding a carrier as a function of position and the continuity of the carrier's momentum as a function of position. Mathematically, this means that at the classical turning points $z = 0$ and $z = z_T$,

$$
\begin{aligned}
\varphi(z = 0^-) &= \varphi(z = 0^+), \\
\left.\frac{\partial \varphi}{\partial z}\right|_{z=0^-} &= \left.\frac{\partial \varphi}{\partial z}\right|_{z=0^+}, \\
\varphi(z = z_T^-) &= \varphi(z = z_T^+), \\
\text{and} \qquad \left.\frac{\partial \varphi}{\partial z}\right|_{z=z_T^-} &= \left.\frac{\partial \varphi}{\partial z}\right|_{z=z_T^+}. \tag{4.71}
\end{aligned}
$$

The linear dependence of potential on z appears in a number of semiconductor problems; the one here is an example of a barrier, it also shows up as a well at a heterostructure interface in the small bandgap material. This differential equation, with a linear dependence on z for the term with the zero'th order derivative, is an example of Airy equation. The eigenfunction solution is

$$\varphi(z) = \mathcal{A}\exp\left(jk_x x + jk_y y\right)\left[\exp\left(jk_z z\right) + \varrho\exp\left(-jk_z z\right)\right] \tag{4.72}$$

for $z < 0$, and

$$\varphi(z) = \mathcal{B}\left[\mathcal{B}i(\eta) + j\mathcal{A}i(\eta)\right] \tag{4.73}$$

for $z > 0$. Here, the normalized factor η is

$$\eta = -\left(\frac{2m^* q\mathcal{E}}{\hbar^2}\right)^{1/3}\left(z - \frac{q\phi_B + \xi_f - E_z}{q\mathcal{E}}\right), \tag{4.74}$$

and $Ai(\eta)$ and $Bi(\eta)$ are Airy functions. For $z < 0$, there is no barrier, and the eigenfunction describes a plane wave. For $z > 0$, the eigenfunction is described as a linear combination of Airy functions. In these equations, $\mathcal{A}$ is an amplitude normalization factor for the plane wave eigenfunction, ϱ is the reflection coefficient at the first classical turning point, and $\mathcal{B}$ is the amplitude normalization factor for the eigenfunction in the region $z > 0$.

For values of z sufficiently farther than the first classical turning point, the Airy function solution has an asymptotic form,

$$\varphi(z) = \left(\frac{\pi^2}{|\eta|}\right)^{-1/4} \mathcal{B} \exp\left(j\frac{\pi}{4}\right) \exp\left(j\frac{2}{3}|\eta|^{3/2}\right). \tag{4.75}$$

This is the transmitted wave at large distances. The probability current associated with this is

$$J = j\frac{\hbar}{2m^*}\left(\varphi\frac{d\varphi^*}{dz} - \varphi^*\frac{d\varphi}{dz}\right) = \frac{\hbar}{\pi m^*}\left(\frac{2m^*q\mathcal{E}}{\hbar^2}\right)^{1/3}|\mathcal{B}|^2. \tag{4.76}$$

The transmission probability T_t can now be found because the incident wave was a plane wave of probability current $\hbar k/m^*$. For this barrier,

$$T_t = \frac{4}{\pi}\frac{1}{k}\left(\frac{2m^*q\mathcal{E}}{\hbar^2}\right)^{1/3}\left\{\left[\mathcal{B}i(\eta_0) - \frac{1}{k}\left(\frac{2m^*q\mathcal{E}}{\hbar^2}\right)^{1/3}\dot{\mathcal{A}}i(\eta_0)\right]^2 + \left[\mathcal{A}i(\eta_0) + \frac{1}{k}\left(\frac{2m^*q\mathcal{E}}{\hbar^2}\right)^{1/3}\dot{\mathcal{B}}i(\eta_0)\right]^2\right\}, \tag{4.77}$$

where

$$\eta_0 = \frac{1}{q\mathcal{E}}\left(\frac{2m^*q\mathcal{E}}{\hbar^2}\right)^{1/3}(q\phi_B + \xi_f - E_z). \tag{4.78}$$

This is a general result for the tunneling transmission probability in a triangular barrier, assuming a constant effective mass, and only elastic processes.

We can, however, derive a simpler form based on expansion of the Airy functions, discarding the smaller terms of the series expansion. The asymptotic form of the barrier transmission probability, derived in this way (see Problem 10), is

$$T_t = 4\frac{(\xi_f + q\phi_B - E_z)^{1/2}E_z^{1/2}}{\xi_f + q\phi_B} \times \exp\left[-\frac{4}{3}\sqrt{\frac{2m^*q\mathcal{E}}{\hbar^2}}\left(\frac{|q\phi_B + \xi_f - E_z|}{q\mathcal{E}}\right)^{3/2}\right]. \tag{4.79}$$

Knowing the tunneling transmission probability T_t, we can now derive the tunneling current. The current density in a small energy range dE, at the energy E, due to tunneling, is a function of the number of carriers available for tunneling described by a source function $S(E_z)$ which is related to the distribution function and the density of states, the tunneling transmission probability T_t, the charge, and the velocity of the carriers. Thus, the tunneling current density is given by

$$J = q \int_0^\infty \frac{\hbar k_z}{m^*} T_t(E) S(E_z) dE_z. \tag{4.80}$$

For an arbitrary problem, current density due to tunneling can be determined knowing the source function, which depends on the distribution function and the density of states, and integrating over the appropriate energy range. This integral could also be written in the momentum space as

$$J = \frac{2q}{(2\pi)^3 \hbar} \int f(E) T_t(E_z) \frac{\partial E}{\partial k_z} dk_z d^2 k_\parallel, \tag{4.81}$$

where the explicit dependence of the tunneling transmission probability on the perpendicular momentum is included.

This calculation can now be performed exactly for our problem. First we determine the source function. The source function is directly related to the occupation probability related to Fermi–Dirac statistics and can be written as

$$
\begin{aligned}
S(E_z) &= \frac{2}{(2\pi)^3 \hbar} \int_{-\infty}^\infty dk_x \int_{-\infty}^\infty dk_y f(E) \\
&= \frac{4\pi m^* kT}{h^3} \ln \left[1 + \exp \left(\frac{\xi_f - E_z}{kT} \right) \right].
\end{aligned} \tag{4.82}
$$

The tunneling transmission probability is given following Equation 4.79, and hence the tunnel current follows by integration of the product. For a triangular barrier,

$$J = \frac{q^2 m_0 \mathcal{E}^2}{16\pi^2 m^* \phi_B} \exp \left[-\frac{4\sqrt{2m^*}(q\phi_B)^{3/2}}{3q\hbar\mathcal{E}} \right]. \tag{4.83}$$

Perhaps the most complicated step in this calculation was the determination of tunneling transmission probability. The source function or supply function can be quite easily determined because the occupation statistics and the density of states are known for the problems of interest. The characteristic of the barrier is what we are usually interested in determining since we do not know it as well. An approach similar to the above would be difficult for arbitrary cases. A simple example of this is a parabolic barrier. Metal-semiconductor junctions, p–n junctions, etc., all exhibit a parabolic barrier at the interface, if depletion approximation is applicable. The differential equation related to that is even more difficult to handle analytically, and perhaps a faster way would be to solve it numerically. This is at the expense of an intuitive insight that most analytical techniques provide.

A common procedure to still handle these problems in closed form is to resort to the WKB approximation. Tunneling through the barrier results in an evanescent wave characterized by an imaginary momentum $\hbar k$. The wave function during transmission through the barrier, therefore, has an exponential decay of the form $\exp(-jk_z z)$, where the term within the brackets is real negative. The probability of finding the particle at any position z is proportional to the square of the accumulated decay of the eigenfunction. Thus, it stands to reason that the tunneling transmission probability, through a barrier whose classical turning points are z_{T1} and z_{T2}, is approximately given by

$$
\begin{aligned}
T_t(E_z) &\approx \exp\left(-2\int_{z_{T1}}^{z_{T2}} k_z \, dz\right) \\
&= \exp\left[-2\int_{z_{T1}}^{z_{T2}} \sqrt{\frac{2m^*\left(V(z) - E_z\right)}{\hbar^2}} \, dz\right].
\end{aligned}
\tag{4.84}
$$

This is the WKB approximation that we had used earlier in the calculation of field emission current in a metal–semiconductor junction. The application of this to a triangular barrier where V decreases linearly with distance is straightforward. Indeed, the tunneling transmission probability is identical to the asymptotic approximation determined before (see Problem 10), and likewise the tunneling current density is determined as

$$
\begin{aligned}
J &= \frac{q^2 m_0 \mathcal{E}^2}{16\pi^2 m^* \phi_B} \exp\left[-\frac{4\sqrt{2m^*}(q\phi_B)^{3/2}}{3q\hbar\mathcal{E}}\right] \\
&= 1.54 \times 10^{-6} \left(\frac{m\mathcal{E}^2}{m^* \phi_B}\right) \exp\left[-\frac{4\sqrt{2m^*}(q\phi_B)^{3/2}}{3q\hbar\mathcal{E}}\right] \\
&= \alpha \mathcal{E}^2 \exp\left(-\frac{\mathcal{E}_0}{\mathcal{E}}\right),
\end{aligned}
\tag{4.85}
$$

where

$$
\mathcal{E}_0 = \frac{4\sqrt{2m^*}(q\phi_B)^{3/2}}{3q\hbar}.
\tag{4.86}
$$

This can also be rewritten as

$$
\ln\frac{J}{\mathcal{E}^2} = \ln\alpha - \frac{\mathcal{E}_0}{\mathcal{E}}.
\tag{4.87}
$$

Table 4.1: Characteristic field for tunneling for triangular barriers of various parameters.

	Barrier Height= 0.8 eV		Barrier Height= 0.3 eV	
	$m^* = 0.067m_0$	$m^* = 0.5m_0$	$m^* = 0.067m_0$	$m^* = 0.5m_0$
$\mathcal{E}_0$ (V.cm^{-1})	1.26×10^7	3.5×10^7	2.9×10^6	7.9×10^6

This form of the tunneling current equation is called the Fowler–Nordheim equation. Tunneling current occurs in this form in a variety of problems where energy losses during tunneling are small and a triangular barrier is a valid approximation. Recall that we arrived at this form assuming elastic processes. Small losses in energy, e.g., through a single phonon process for carriers that are otherwise high in energy, also follow this behavior. Note that tunneling of higher energy carriers is favored because they encounter a smaller barrier. This functional relationship of current and field is encountered as an approximation in a variety of problems in semiconductor devices. Tunneling through insulators such as SiO_2 is one common example, tunneling through larger bandgap barriers at heterojunctions is another example, and finally Zener tunneling where the barrier at a p–n junction can be idealized as a triangular barrier is another example. Note the lack of strong temperature dependence in all these cases because the effective mass is the strongest source for it.

Tunneling current is a sensitive function of the field parameter $\mathcal{E}_0$. Let us consider some simple examples to show this sensitivity, assuming a barrier height of 0.8 eV to represent metal–semiconductor junctions and 0.3 eV to represent heterojunctions. We consider two effective masses, $0.067m_0$ to represent electron processes, and $0.5m_0$ to represent hole processes. Table 4.1 shows the magnitude of the field $\mathcal{E}_0$ for these conditions. The smaller the characteristic field the larger the tunneling current density. Thus, factors-of-10 differences in electron tunneling current result from changes in barrier height typical of heterojunctions. This emphasizes the importance of tunneling currents in heterostructures.

4.3.3 Resonant Fowler–Nordheim Tunneling

We now use an example where the WKB approximation is not a good predictor because it is a simplification of the description of wave propagation.

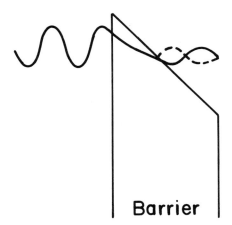

Figure 4.16: A schematic example of resonant Fowler–Nordheim tunneling due to interference of the propagating waves.

Note that the exponential relationship only considers propagating waves in the direction of tunneling. Wave nature, however, can result in effects related to the interferences from reflections that occur at all the semiconductor discontinuities. Resonant Fowler–Nordheim tunneling is such an example resulting from the interference of the waves. Figure 4.16 shows a schematic of this in the triangular well formed at a triangular barrier when it is biased appropriately. Electrons that tunnel into the conduction band of the barrier can suffer quantum-mechanical reflection at the discontinuity, and hence can interfere constructively or destructively depending on the wave propagation properties. These interferences are directly related to the tunneling transmission probability, and appear as a modulation in the current–voltage behavior at low temperatures where elastic processes dominate and other current components can be suppressed. Such interferences can be observed in many of the heterojunction tunneling structures at low temperatures. Such resonance structures have also been observed in SiO_2/Si systems. This also serves as an example of elastic hot carrier tunneling in that system, because interference requires that the energy be identical. Since the barrier height at a SiO_2/Si interface is large, a large field is required to allow for the occurrence of the resonance condition. Large barrier structures thus exhibit this form of tunneling only at the highest biases.

This phenomenon actually was among the earliest unambiguous demonstrations of scattering-free electron transport in semiconductor devices. As elastic processes become less likely, the occurrence of resonant processes also becomes less likely because of its dependence on coherent interference. Even for indirect materials, so long as elastic tunneling can take place, such as in SiO_2/Si junctions, it can be observed.

4.4 Ohmic Contacts

Our discussion of transport across junctions formed using metals with semiconductors and heterojunctions naturally leads us to a discussion of ohmic contacts. Ohmic contacts are needed in all semiconductor devices in order to allow a link through which current can flow and bias can be applied. Ohmic contacts are generally formed using heavy doping in structures to obtain large current flow without any significant bias drop across the ohmic contact regions. Current flow in these structures may be by any of the transport mechanisms that we have discussed. Ohmicity implies that the current is proportional to voltage, both in sign and magnitude. This can only occur over a limited current or bias range, because of the limitations on linearity. Contacts based on thermionic emission or thermionic field emission exhibit linearity over a smaller range because these processes are strongly non-linear. Tunneling processes exhibit linearity over a larger range. Heavy doping in all these cases serves to increase the currents and hence the current range of ohmicity; it also serves to minimize the potential drop in semiconductors. Heavy doping, therefore, also allows for reducing the effects of non-linearity from transport in the semiconductor itself by keeping the fields low.

The ohmic access to a lightly doped region of a semiconductor device from a heavily-doped region, therefore, occurs quite commonly. We considered consequences of quasi-neutrality at such junctions in our discussion of drift-diffusion transport in Chapter 3. Doping changes should occur over length scales that allow for screening of the background charge, and hence suppression of any barriers caused by the background charge. Exponential changes in doping, such as those that occur in tails of Gaussian profiles or error function profiles, naturally lead to maintaining of quasi-neutrality. If quasi-neutrality is maintained, the current density resulting from majority carrier transport, in steady-state, is a product of charge density and carrier velocity. Since fields are small, and voltage drops across an ohmic region minimal, the time-dependent electric field variations during a transient are also small, and hence the steady-state calculation is also representative of the transient calculation.

In compound semiconductor ohmic contacts, these large doping regions are usually implemented together with isotype heterojunctions or with metals. The implementation of the former specifically takes advantage of the lower barrier heights and hence higher tunneling transmission probability of heterojunctions; it is therefore a tunneling-based contact. The implementation of the latter occurs usually with larger barriers, and hence either thermionic field emission emphasizing tunneling near the top of the barrier—a temperature-sensitive process since it depends on electron energy distribution—or tunneling emphasizing processes at the bottom of the barrier—a temperature-insensitive process. Figure 4.17 shows some examples of such contacts. Parts (e) and (f) of this figure demonstrate an interesting variation on the common methods of forming ohmic contact. Fermi level pinning, which usually occurs in the bandgap, can also occur in the conduction or valence band. InAs, e.g., pins in the conduction band, and GaSb pins in the valence band. Such materials, therefore, show direct electronic conduction between the metal and the semiconductor. A suitable grading, following the requirement of quasi-neutrality, or an interface where an additional tunneling process has to be considered, allows one to link this contact region to the semiconductor of interest. Ohmic contacts are usually characterized by the parameters specific contact conductance (σ_c), or specific contact resistance (ρ_c), and defined by

$$\sigma_c = \frac{1}{\rho_c} = \frac{\partial J}{\partial V} \tag{4.88}$$

where V is the applied bias voltage.

We consider these parameters, for metal–semiconductor junctions, for the transport processes that we have analyzed in this chapter. The derivative of the expression for current can be used to show that the specific contact resistances (see Problem 11) are proportional to the exponential factors as

$$\rho_c \propto \exp\left(\frac{q\phi_B}{kT}\right) \tag{4.89}$$

for thermionic emission,

$$\rho_c \propto \exp\left[\frac{4\pi\sqrt{m^*\epsilon_s}}{h} \frac{\phi_B}{\sqrt{N_D}} \tanh\left(\frac{hq}{4\pi kT}\sqrt{\frac{N_D}{m^*\epsilon_s}}\right)\right] \tag{4.90}$$

for thermionic field emission, and

$$\rho_c \propto \exp\left[\frac{4\pi\sqrt{m^*\epsilon_s}}{h} \frac{\phi_B}{\sqrt{N_D}}\right] \tag{4.91}$$

for field emission in a triangular barrier.

Note the general features of these expressions. Thermionic emission processes are temperature-sensitive because they describe current transport due to emission from the hot electron tail of the distribution function.

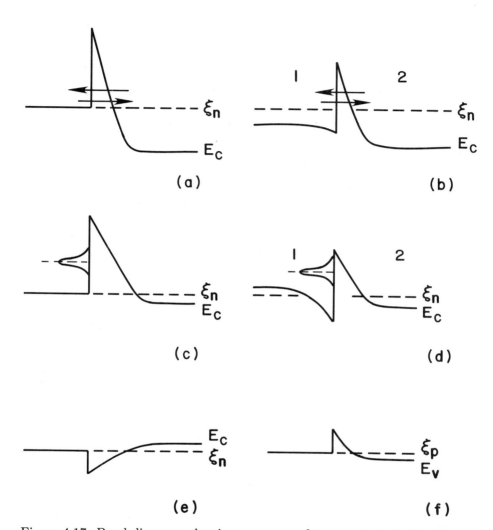

Figure 4.17: Band diagrams showing processes of transport at ohmic contacts based on tunneling-dominated transport ((a) and (b)), thermionic field emission-dominated transport ((c) and (d)), and utilizing Fermi level pinning in the bands ((e) and (f)). Parts (a) and (c) utilize metal–semiconductor junctions, (b) and (d) utilize heterojunctions, and (e) and (f) utilize Fermi level pinning in the conduction or valence band.

Table 4.2: Specific contact resistances based on depletion approximation and tunneling through a triangular barrier.

	Barrier Height= 0.8 eV		Barrier Height= 0.3 eV	
	$m^* = 0.067m_0$	$m^* = 0.5m_0$	$m^* = 0.067m_0$	$m^* = 0.5m_0$
σ_c (S.cm^{-2})	1.26×10^7	3.5×10^7	2.9×10^6	7.9×10^6

The magnitude of this resistance should be large. The thermionic field emission due to additional tunneling occurring in the higher energy tail of the carrier distribution function is less temperature-sensitive than the thermionic emission, but it is still significant. The tunneling for very thin barriers, assumed to be triangular, leads to least the temperature sensitivity because the thin barriers allow for tunneling of low-energy electrons. The last of these is the basis for many of the metal–semiconductor ohmic contacts.

We can use our expression for Fowler–Nordheim tunneling, which is the dominant tunneling process, to derive the complete form of this conductance. Using Equation 4.85,

$$\sigma_c = \frac{dJ}{dV} = \alpha \left(\mathcal{E}_0 + 2\mathcal{E} \right) \exp \left(-\frac{\mathcal{E}_0}{\mathcal{E}} \right) \frac{d\mathcal{E}}{dV}. \tag{4.92}$$

The derivative of field with potential follows from the Poisson equation as

$$\frac{d\mathcal{E}}{dV} = -\left(\frac{2qN_D(\phi_B - V)}{\epsilon_s} \right)^{1/2}, \tag{4.93}$$

and hence the doping dependence. It is instructive to evaluate the magnitude of this conductance for some examples. We consider barrier heights of 0.6 eV and 0.3 eV, and a doping of 1×10^{20} cm^{-3}. The specific contact conductance is shown in Table 4.2. The calculation, although simplistic, is instructive. A smaller barrier, such as in a heterojunction, allows for low specific contact resistance if the semiconductor used in its formation can be accessed with a low specific contact resistance also. Holes, represented here by a larger effective mass, lead to a higher specific contact resistance for similar barrier heights. Light holes exist too, but the density of the light hole states is small and hence tunneling to these is less likely. In practice, barrier heights for holes is smaller, and the specific contact resistances for both electrons and holes is similar.

The major approximations in the above relate to the exact determination of the fields, interface roughness that significantly enhances the field and hence the tunneling current, image force lowering which reduces the barrier height, heavy doping effects which introduce both Coulombic fluctuations and other band structure effects, interface oxides, etc. Clearly, depletion approximation is quite inappropriate and others only contribute to making the calculation more inexact. A more general expression of current can be derived for field emission where some of the approximations are relaxed (see Problem 12) as

$$
J = \frac{\mathcal{A}^*}{\alpha^2 k^2} \exp\left(-q\frac{\phi_B - V}{E_{00}}\right) \times
$$
$$
\left\{\frac{\pi \alpha k T}{\sin(\pi \alpha k T)}\left[1 - \exp\left(-q\alpha V\right)\right] - q\alpha V \exp\left[-\alpha\left(\xi_f - E_c\right)\right]\right\},
$$
$$(4.94)$$

where $\alpha = \ln\left[4\left(q\phi_B - qV\right)/\xi_f\right]/2E_{00}$. In deriving this expression, we have assumed that

$$
1 - \alpha k T > k T\left(\frac{1}{2E_{00}\xi_f}\right)^{1/2}.
\tag{4.95}
$$

The expression for specific contact resistance, assuming further that $\xi_f - E_c - qV > 3kT$ and $\alpha V \gg 1$ (see Problem 12), is

$$
\rho_c = \frac{(k/q\mathcal{A}^*)\exp\left(q\phi_B/E_{00}\right)}{(\pi T/\sin(\pi\alpha^* k T)) - (1/\alpha^* T)\exp\left[-\alpha^*\left(\xi_f - E_c\right)\right]}.
\tag{4.96}
$$

where α^* is the magnitude of α at zero bias. The functional form is similar to the earlier field emission description in a triangular barrier.

Isotype heterojunctions with large dopings on both sides of the junction and a metal–semiconductor junction with large doping in the semiconductor are two of the more common methods of forming ohmic contacts. Due to smaller barrier heights or barrier widths, in some of these cases even for moderate doping conditions, the barriers do not substantively impede the flow of current. One such example is the access to the two-dimensional electron gas at a heterojunction interface of a heterostructure field effect transistor. A low specific contact resistance can be obtained to the two-dimensional gas even if it is formed in a low-doped substrate. This low contact resistance mainly arises because of the low barrier height and width for carrier tunneling from the large bandgap material to the small bandgap material. The contact is still limited by the features of the barrier, and further lowering of the contact resistance may be achieved by grading of the barrier. Grading of the barrier lowers the barrier height and allows for more field emission.

The contacts formed from a semiconductor through a heterojunction to a metal that exploit lower barriers in different semiconductors and their interfaces are particularly interesting in compound semiconductor devices. A set of examples where this may be relevant is shown Figure 4.18. The choice for interfacial heterojunction is a semiconductor with favorable properties for formation of an ohmic contact. Examples for n-type GaAs are InAs and large InAs mole-fraction $Ga_{1-x}In_xAs$ which show negative barrier heights for conduction band, and Ge, which shows a low positive barrier height. Both of these contacts show minimal resistances (the sum of the specific contact resistance of the metal–semiconductor and the heterojunction interface). Because of the lower barrier height, heterojunctions may be kept abrupt. Abrupt heterojunctions between semiconductors of low barrier heights can be treated in a manner that we have already described. In these cases, the junction current as described by the tunneling current is simply

$$J = \frac{4\pi m^* qkT}{h^3} \int_{E_c}^{\infty} T_t S(E) dE, \qquad (4.97)$$

where the source function is

$$S(E) = \ln \left\{ \frac{1 + \exp\left[(\xi_f - E)/kT\right]}{1 + \exp\left[(\xi_f - E - qV)/kT\right]} \right\}. \qquad (4.98)$$

Ohmic contacts can also be formed using other techniques. Disorder at the interface, e.g., can lead to states in the band that allow hopping conduction leading to bias-symmetric conduction. One of the more common methods for forming ohmic contacts in GaAs utilizes a eutectic of gold and germanium. Thermal processing of this structure is commonly believed to lead to islands where an inter-metallic forms an ohmic contact to highly n^+ doped regions of compound semiconductor, Ge providing the doping by substituting on the Group III element site. These islands provide conduction by lateral spreading of the current; the contact is non-uniform because it has spreading effects from the islands. Very small area contacts will not follow large area behavior. More importantly, the square root dependence on doping that one expects from the tunneling behavior is not followed. Instead, one sees an inverse dependence (see Problem 13) of the specific contact resistance.

(a) Metal – Ge – GaAs

(b) Metal – InAs – GaAs

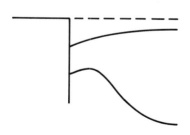

(c) Metal – InAs → GaInAs → GaAs

Figure 4.18: Metal–semiconductor junctions formed with heterostructures using abrupt interface with Ge (a), abrupt interface with InAs (b), and graded interface with InAs (c). (a) and (b) are two examples where the tunneling theory is suitably applicable. The grading in the InAs contact is shown to occur pseudomorphically to a larger bandgap compound such as $Ga_{1-x}In_xAs$ or GaAs.

4.5   p–n Junctions

We discussed in chapter 3, transport in p–n junctions using the BTE approach, and we also discussed boundary conditions for p–n junctions in both low injection conditions and high injection conditions. Our analysis and discussion here will develop the necessary junction parameters, emphasize limitations of depletion approximation, discuss voltage drop across the quasi-neutral region in high level injection conditions, and develop a general charge control approach to analyze transport in p–n junctions. The latter approach, called Gummel–Poon model after its originators, is applied to a variety of conditions including recombination, a variety of doping profiles, and heavy doping conditions, to demonstrate the transport effects in p–n junctions.

4.5.1   Depletion Approximation

We first consider, within depletion approximation, junction parameters of interest, and then we will consider limitations of the assumption of complete carrier depletion in the junction space charge region. Figure 4.19 shows band edges and carrier densities for a GaAs p–n junction.[8] In the quasi-neutral part of either the p-type or the n-type material,

$$\frac{d^2\psi}{dz^2} = -q\frac{N}{\epsilon_s} = 0, \tag{4.99}$$

where ψ is the electrostatic potential. The sign of the charge used in the above equation corresponds to that of donors. In the quasi-neutral region, there exists no net charge density, and hence the electric field is constant. Since the electric field is negligible at the ohmic contact, electric field is negligible in the quasi-neutral region, and hence the electrostatic potential ψ is constant. In depletion approximation, i.e., ignoring the mobile carrier charge, in the depletion region on the n side, the electric field is given by

$$\mathcal{E} = -\frac{d\psi}{dz} = \frac{qN_D}{\epsilon_s}z + \mathcal{A}. \tag{4.100}$$

Electric field is zero at the edge of the depletion region ($z = w_n$) of the n-type semiconductor, hence the constant

$$\mathcal{A} = -\frac{qN_D w_n}{\epsilon_s}, \tag{4.101}$$

---

[8]For the calculation of the depletion region widths, etc., we will assume origin to be at the metallurgical junction with the depletion region edge in the p-type material at $z = -w_p$ and in the n-type material at $z = w_n$.

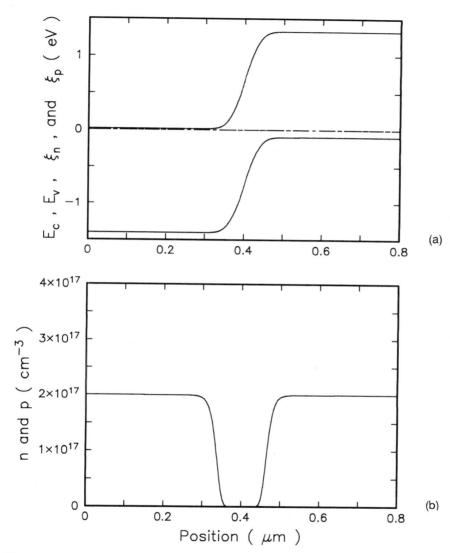

Figure 4.19: Band bending and quasi-Fermi levels (a), and electron and hole densities (b) at a p–n homojunction in GaAs in thermal equilibrium.

and

$$\mathcal{E} = -q\frac{N_D}{\epsilon_s}(w_n - z) \tag{4.102}$$

in the depletion region. This expression is similar to that for a metal–semiconductor diode. Integrating again,

$$\psi = -\int \mathcal{E}dz = q\frac{N_D}{\epsilon_s}\left(w_n z - \frac{z^2}{2}\right) + \mathcal{B}. \tag{4.103}$$

Using the boundary condition for the electrostatic potential in the n-type material, $\psi = \psi_n$ at $z = w_n$, we obtain

$$\psi = \psi_n - \frac{qN_D}{2\epsilon_s}(w_n - z)^2. \tag{4.104}$$

Similarly, on the p-side we have

$$\psi = \psi_p + \frac{qN_A}{2\epsilon_s}(z + w_p)^2, \tag{4.105}$$

where depletion region edge is assumed to occur at $z = -w_p$.

For a homogeneous material, with no interface charge, both the electric field and the electrostatic potential are continuous at the junction ($z = 0$), i.e.,

$$\psi_p + \frac{qN_A}{2\epsilon_s}w_p^2 = \psi_n - \frac{2N_D}{2\epsilon_s}w_n^2$$

$$\text{and}\quad \frac{qN_A}{\epsilon_s}w_p = \frac{qN_D}{\epsilon_s}w_n. \tag{4.106}$$

This allows us to write the charge on each side of the depletion region as

$$Q = qN_A w_p = qN_D w_n = \left[2q\epsilon_s\frac{N_D N_A}{N_D + N_A}(\psi_n - \psi_p)\right]^{1/2}, \tag{4.107}$$

where $(\psi_n - \psi_p)$ is the total band bending of the junction which is the built-in potential ψ_{j0},

$$\psi_{j0} = \frac{kT}{q}\ln\left(\frac{N_D N_A}{n_i^2}\right) \tag{4.108}$$

in the Boltzmann approximation. We can also write the two depletion widths as

$$w_n = \left[\frac{2\epsilon_s \psi_{j0} N_A}{q(N_D + N_A)N_D}\right]^{1/2} = \sqrt{2}\lambda_{Dn}\left[\frac{q\psi_{j0}N_A}{kT(N_A + N_D)}\right]^{1/2}, \tag{4.109}$$

and

$$w_p = \left[\frac{2\epsilon_s \psi_{j0} N_D}{q(N_D + N_A) N_A} \right]^{1/2} = \sqrt{2} \lambda_{Dp} \left[\frac{q\psi_{j0} N_D}{kT(N_A + N_D)} \right]^{1/2}, \qquad (4.110)$$

where λ_{Dn} and λ_{Dp} are the Debye lengths in the n-type and p-type regions. Also, the total depletion region width is

$$W = w_n + w_p = \left[\frac{2\epsilon_s \psi_{j0}}{q N_D N_A / (N_D + N_A)} \right]^{1/2}. \qquad (4.111)$$

These expressions simplify for one-sided junctions because the effective doping is the geometric mean and it becomes the smaller doping for one-sided junctions.

In reality, a mobile charge exists at the junction edges and under larger forward bias, and a significant mobile charge may exist within the depletion region. Close to the flat-band conditions of the junction, this charge can be considerable. Depletion region, therefore, is a misnomer for such conditions. The term space charge region is more appropriate. In this text, both are used interchangeably. Consider the inclusion of this charge at the edges of the junction depletion region,

$$\frac{d^2\psi}{dz^2} = -\frac{q(N_D - n)}{\epsilon_s} = -\frac{q}{\epsilon_s}\left[N_D - n_i \exp\left(\frac{q\psi}{kT}\right) \right]. \qquad (4.112)$$

If we consider the deviation of electrostatic potential from ψ_n given by $\Delta\psi = \psi_n - \psi$,

$$\begin{aligned} \frac{d^2\Delta\psi}{dz^2} &= \frac{q}{\epsilon_s}\left[N_D - n_i \exp\left(q\frac{\psi_n - \Delta\psi}{kT} \right) \right] \\ &= \frac{q N_D}{\epsilon_s}\left[1 - \exp\left(-\frac{q\Delta\psi}{kT} \right) \right]. \end{aligned} \qquad (4.113)$$

In the small region at the junction edge, with $\Delta\psi \ll kT/q$, we may expand the exponential in a Taylor series expansion, and if we include only the first perturbation term, we obtain

$$\frac{d^2\Delta\psi}{dz^2} = \frac{q^2 N_D}{\epsilon_s kT}\Delta\psi = \frac{\Delta\psi}{\lambda_D{}^2}. \qquad (4.114)$$

So, electrostatic potential actually deviates exponentially with a characteristic length given by the Debye length at the edge of the depletion region. Consequently the mobile charge density falls off in significance exponentially.

The significance of this exponential fall-off is that we may use its characteristic length, the Debye length, to judge the appropriateness of the use of depletion approximation. If the depletion region width on either side is comparable to the Debye length, then the depletion approximation is clearly invalid. We have

$$\frac{w_n}{\lambda_{Dn}} = \left[\frac{2q\psi_{j0}N_D}{kT\,(N_A + N_D)} \right]^{1/2},$$

$$\text{and} \quad \frac{w_p}{\lambda_{Dp}} = \left[\frac{2q\psi_{j0}N_A}{kT\,(N_A + N_D)} \right]^{1/2}. \tag{4.115}$$

Consider an abrupt one-sided GaAs n^+–p junction with dopings of 5×10^{17} cm^{-3} and 5×10^{14} cm^{-3}. The built-in voltage for this junction

$$\psi_{j0} = \frac{kT}{q} \ln \left(\frac{N_D N_A}{n_i{}^2} \right) = 1.19 \text{ V}, \tag{4.116}$$

and hence, $w_n/\lambda_{Dn} = .3$, and $w_p/\lambda_{Dp} = 9.6$. On the higher-doped side, the depletion approximation is poor, because a substantial electron density occurs throughout the depletion region in this n-type region. On the other hand, for a symmetric junction of 1×10^{16} cm^{-3} doping, the built-in voltage of the junction is 1.16 V, and w_n/λ_{Dn} and w_p/λ_{Dp} are ≈ 6.7, making the depletion approximation a good one on both sides of the junction. Similarly, for a large doping symmetric junction of 5×10^{17} cm^{-3}, $\psi_{j0} = 1.36$ V, and w_n/λ_{Dn} and w_p/λ_{Dp} are ≈ 7.3, a good depletion approximation. Thus, one-sided junctions lead to breakdown of the depletion approximation in the higher-doped side. The approximation is poorest because potential drop across the heavily doped layer is lower due to asymmetry and hence the depletion width in the higher-doped side is smaller.

4.5.2   High-Level Injection

We considered the low-level injection behavior of p–n junctions quite extensively during the discussion of Shockley, Fletcher, and Misawa boundary conditions. We will analyze the problem further by considering the behavior of p–n junction in high-level injection. In particular, we are interested in determining the voltage drops across the n-side and the p-side. Let J be the total current through the device. Our transport equation for the n-side is

$$J_n = q\mu_n n_n \mathcal{E} + q\mathcal{D}_n \frac{\partial n_n}{\partial z}, \tag{4.117}$$

and hence

$$\mathcal{E} = \frac{J_n}{q\mu_n n_n} - \frac{\mathcal{D}_n}{\mu_n n_n} \frac{\partial n_n}{\partial z}. \tag{4.118}$$

Particle current flows both as electron and hole current, therefore,

$$J_n(z) = J - J_p(z), \tag{4.119}$$

and hence we get the expression for the electric field as

$$\mathcal{E}(z) = \frac{J}{q\mu_n n_n(z)} - \frac{J_p(z)}{q\mu_n n_n(z)} - \frac{kT}{qn_n} \frac{\partial n_n}{\partial z}. \tag{4.120}$$

On the n-side again, we may express the hole current directly, and substitute the above for the electric field. This allows us to express the hole current density as

$$\begin{aligned} J_p(z) &= q\mu_p p_n \mathcal{E}(z) - q\mathcal{D}_p \frac{\partial p_n}{\partial z} \\ &= q\mu_p p_n(z) \left[\frac{J}{q\mu_n n_n(z)} - \frac{J_p(z)}{q\mu_n n_n(z)} - \frac{kT}{qn_n} \frac{\partial n_n}{\partial z} \right] - \\ &\quad q\mathcal{D}_p \frac{\partial p_n(z)}{\partial z}. \end{aligned} \tag{4.121}$$

Since charge neutrality implies

$$\frac{\partial p_n(z)}{\partial z} = \frac{\partial n_n(z)}{\partial z}, \tag{4.122}$$

we can simplify this to yield

$$J_p(z) = \frac{1}{\left[1 + \frac{\mu_p p_n(z)}{\mu_n n_n(z)} \right]} \left\{ \frac{\mu_p p_n(z)}{\mu_n n_n(z)} J - q\mathcal{D}_p \left[1 + \frac{p_n(z)}{n_n(z)} \right] \frac{\partial p_n(z)}{\partial z} \right\} \tag{4.123}$$

as the expression for hole current, and, substituting back in the electric field expression,

$$\mathcal{E}(z) = \frac{J}{q(\mu_n n_n(z) + \mu_p p_n(z))} - \frac{\mathcal{D}_n - \mathcal{D}_p}{(\mu_n n_n(z) + \mu_p p_n(z))} \frac{\partial p_n(z)}{\partial z} \tag{4.124}$$

as the expression for the electric field.

Note that this actually follows from the the total current relation

$$J = \sigma \mathcal{E}(z) + q\mathcal{D}_n \frac{dp_n(z)}{dz} - q\mathcal{D}_p \frac{dp_n(z)}{dz}, \tag{4.125}$$

where σ is the conductivity of the material which also includes high-level injection effects.

We can now determine the voltage drop across either of the semiconductor regions. Let V_n be the change in electrostatic potential on the n-side; it is given by

$$V_n = -\int_{w_n}^{z_n} \mathcal{E}(z)dz, \tag{4.126}$$

where z_n is the position of the ohmic contact to the n-type material. Charge neutrality and the junction boundary conditions allow us to write

$$p_n(z) = p_{n0} + \Delta p_n \exp\left(-\frac{z}{\mathcal{L}_p}\right), \tag{4.127}$$

$$n_n(z) = n_{n0} + \Delta n_n \exp\left(-\frac{z}{\mathcal{L}_p}\right), \tag{4.128}$$

$$\Delta p_n = \Delta n_n, \tag{4.129}$$

$$\text{and} \quad \Delta p_n = p_{n0}\left[\exp\left(\frac{qV_j}{kT}\right) - 1\right], \tag{4.130}$$

and hence

$$V_n = \frac{J}{q}\int_{w_n}^{z_n} \frac{1}{(\mu_n n_{n0} + \mu_p p_{n0}) + (\mu_n + \mu_p)\Delta p_n \exp(-z/\mathcal{L}_p)}dz - \int_{p_n(w_n)}^{p_{n0}} \frac{(\mathcal{D}_n - \mathcal{D}_p)}{(\mu_n n_{n0} + \mu_p p_{n0}) + (\mu_n + \mu_p)\Delta p_n \exp(-z/\mathcal{L}_p)}dp_n. \tag{4.131}$$

This may be written in the form of the integrable expression

$$V_n = \frac{J}{q}\int_{w_n}^{z_n} \frac{1}{(\mu_n n_{n0} + \mu_p p_{n0}) + (\mu_n + \mu_p)\Delta p_n \exp(-z/\mathcal{L}_p)}dz - (\mathcal{D}_n - \mathcal{D}_p)\int_{p_n(w_n)}^{p_{n0}} \frac{1}{(\mu_n n_{n0} - \mu_n p_{n0}) + (\mu_n + \mu_p)p_n(z)}dp_n, \tag{4.132}$$

and finally, following integration,

$$V_n = \frac{J(z_n - w_n)}{q(\mu_n n_{n0} + \mu_p p_{n0})} + \frac{J\mathcal{L}_p}{q(\mu_n n_{n0} + \mu_p p_{n0})} \times$$
$$\log\left\{\frac{(\mu_n n_{n0} + \mu_p p_{n0}) + (\mu_n + \mu_p)\Delta p_n \exp\left[-(w_n - z_n)/\mathcal{L}_p\right]}{(\mu_n n_{n0} + \mu_p p_{n0}) + (\mu_n + \mu_p)\Delta p_n}\right\} - $$
$$\frac{\mathcal{D}_n - \mathcal{D}_p}{\mu_n + \mu_p}\log\left[\frac{(\mu_n n_{n0} + \mu_p p_{n0})}{(\mu_n n_{n0} + \mu_p p_{n0}) + (\mu_n + \mu_p)\Delta p_n}\right]. \tag{4.133}$$

This describes completely, in terms of known prameters, the total voltage drop across the n-side of the junction. Its simplification, at low level injection (i.e., $n_{n0} >> p_{n0}, \Delta p_n$) where the logarithmic terms are negligible, is simply

$$V_n = \frac{J(z_n - w_n)}{q\mu_n n_{n0}}, \qquad (4.134)$$

or

$$\begin{aligned} \mathcal{E} &= \frac{V_n}{z_n - w_n} \\ &= \frac{J}{q\mu_n n_{n0}}, \end{aligned} \qquad (4.135)$$

which is Ohm's Law because the denominator is the conductivity of the semiconductor.

A similar expression may be found for the p-side. Usually, though, junctions are one-sided and the high-level injection effects need to be included on only one side of the junction. These relations relate the current density to the voltage drop in the quasi-neutral region. The relationship between the current and electrostatic potential at the junction is known from Chapter 3. The applied voltage is partitioned between the junction and the quasi-neutral region at high-level injection. These form a complete set of equations to determine the current for the applied bias at the device terminals.

Figure 4.20 shows the characteristics of a symmetric p–n junction under conditions of medium- and high-level injection. The diode in this example is a short diode, where the diffusion length is larger than the base width—the width of the quasi-neutral region. Depletion approximation is quite valid at low-level injection conditions, and the exponential drop in the carrier concentration is also demonstrated under these conditions. The potential drop is negligible in the low-level injection conditions, and all the applied potential drops at the junction. Figure 4.20 shows the changes from this at medium- and high-level injection.

At medium-level injection conditions, some deviation in majority carrier concentration occurs because the minority carrier concentration is becoming significant, and the change maintains quasi-neutrality away from the junction. Almost all of the voltage still drops across the junction.

Under high-level injection conditions, a significant injection of minority carriers occurs, and an equal change in majority carrier concentration occurs to re-establish quasi-neutrality. The electric field needed to maintain this gradient in the majority carriers is now clear in the variation of the conduction band edge as a function of position. Also note from this

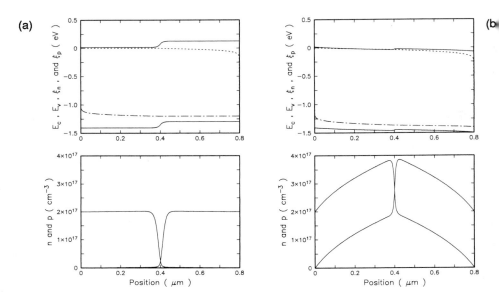

Figure 4.20: Band edges and electron and hole concentration in a symmetric GaAs p–n junction under medium- (a), and high-level (b) injection. The thermal equilibrium conditions for this junction are shown in the previous figure.

variation of the conduction band edge, and the variation of the quasi-Fermi level, that both diffusion and drift current flow now occurs in the base of the p–n junction. The low and medium injection conditions show current transport only through diffusion.

Figure 4.21 shows the current–voltage behavior of a GaAs diode; this example assumes a large lifetime and diffusion length. At the lowest biases the current increases exponentially, at the rate of 60 mV/decade in current. This increase is characteristic of the exponential current–voltage relationship that characterizes transport in an ideal p–n junction. It occurs because the flat quasi-Fermi level at the low-level injection condition relates the carrier concentration at the edges of the depletion region by the Boltzmann factor. Therefore, in the presence of diffusion-dominated transport in the low-level injection condition, and with low recombination effects so that electron and hole current densities are constant in the depletion region, the current also varies exponentially. Deviation from this behavior at the higher-level injection condition comes about because now the junction voltage is smaller than the applied voltage, and because of the

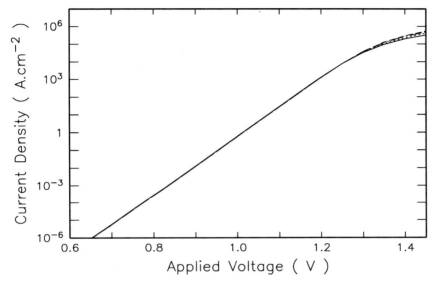

Figure 4.21: Current-voltage characteristics of an idealized one-sided n^+–p GaAs junction as a function of applied bias. The n^+-type doping is 5×10^{18} cm^{-3}, and the p-type doping is varied between 10^{17} cm^{-3} and 10^{18} cm^{-3}.

high injection, both diffusive and drift transport occur. Note also that in the exponential region the current is only very weakly dependent on the doping of the lighter doped side. It is the strongest function of the built-in voltage which is only a weak function of doping.

This discussion assumed transport with perfect ohmic contacts, i.e., contacts where the excess carrier concentration was zero. We have discussed modelling of contacts by a surface recombination velocity that relates this excess carrier concentration above its value at thermal equilibrium. Problem 29 considers incorporation of such boundary conditions in analyzing the behavior of p–n junction diodes.

4.5.3 Gummel–Poon Quasi-Static Model

The Gummel–Poon model implementation[9] for p–n junctions and bipolar transistors is a useful approach for predicting the the quasi-static trans-

---

[9]H. K. Gummel and H. C. Poon, "An Integral Charge Control Model of Bipolar Transistors," *The Bell System Technical Journal*, p. 827, May-June 1960, discusses the application of the approach in bipolar transistors.

port behavior of an arbitrary device. In our discussion of p–n junctions to this point, we have only considered uniform dopings. It may be desirable in certain structures to have varying doping profiles. p–n junctions with large dopings occur with bandgap grading due to bandgap shrinkage. Alloy grading of heterostructures occurs with bandgap grading also. These devices can, thus, have varying quasi-fields in the base either with a doping gradient or an alloy gradient. Gummel–Poon modelling is a convenient mechanism to precisely predict the quasi-static behavior for such a variety of structures. It is also useful for complex recombination situations. Our model is a simple extension of the homojunction model to include the heterostructure and the bandgap grading in the base. It is useful at both low and medium current densities.

Consider the references and charge parameters for an arbitrary p–n junction as described in Figure 4.22. Q_p and Q_n describe the integrated charge due to majority carriers in the quasi-neutral region on both sides of the junction, while $\mathcal{Q}_n^p$ and $\mathcal{Q}_p^n$ are the corresponding integrated minority carrier charge densities. The ohmic contacts are located at $z = -z_p$ and $z = z_n$ for the p-type and n-type regions, and $z = -w_n$ and $z = w_p$ are the positions of the edges of the depletion region.

The continuity equations for steady-state are

$$\frac{\partial J_n}{\partial z} = q \left[\mathcal{R}(z) - \mathcal{G}(z) \right] \tag{4.136}$$

for the electron current and

$$\frac{\partial J_p}{\partial z} = -q \left[\mathcal{R}(z) - \mathcal{G}(z) \right] \tag{4.137}$$

for the hole current. We will employ these equations, which describe the positional change of the current density, to write equations for the current density as a function of its magnitude at the contacts. Since current continuity requires the current at the two contacts to be the same, this will allow us, by referencing to either of the contacts, to write an equation for the current in terms of the changes in the quasi-Fermi level which may then be related to the voltage for different injection conditions, just as we did in the boundary condition analysis. In doing this, several well-defined integrals occur, which are variously referred to as Gummel or modified Gummel numbers. The origin of these integrals can be traced to the charge in the quasi-neutral regions, modified by factors such as alloy grading, bandgap changes, and differing diffusivities or mobilities, all of which affect current transport. In low-level injection, for uniformly-doped short diodes, the minority carrier current is proportional to the minority carrier density at the edges of the depletion region, and inversely proportional to the

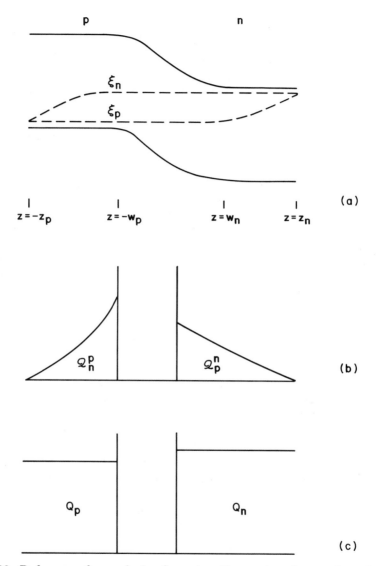

Figure 4.22: References for analysis of p–n junctions using the quasi-static Gummel–Poon approach; Q_p and Q_n are the integrated majority carrier charge densities in the p-type and n-type quasi-neutral regions, $\mathcal{Q}_n^p$ and $\mathcal{Q}_n^n$ are the integrated minority carrier charge densities in the quasi-neutral regions.

width because these two parameters define the gradient of the minority carrier charge density. The minority carrier charge density is directly related to the exponential of the junction bias—the Boltzmann factor—and the minority carrier density at thermal equilibrium. The minority carrier density, thus, is inversely related to the majority carrier density, which is a constant for a uniformly-doped and uniform-alloy composition junction in low-level injection. Under these conditions, the current density is inversely proportional to the majority carrier charge density and the width of the quasi-neutral region, i.e., the integrated majority carrier charge densities Q_n and Q_p. These are the simplest forms of Gummel numbers applicable if alloy grading, bandgap changes, diffusivity or mobility changes as well as other injection effects are negligible, and the p–n junction is a short junction. Our derivation allows for the inclusion of these, and hence we generate modified Gummel numbers,[10] which are more complex in form, but more generally applicable. Our major limitation will be that current transport is considered by drift-diffusion alone. Thus, it is not applicable to devices with any rapid changes in alloy composition that lead to band edge discontinuities and thermionic transport.

In the absence of recombination, current density for electrons and holes is constant in the junction region, the sum of which is the total current in the device. This constant current density in the junction region for both carriers is used as the basis for the calculation of total current density in the simpler diode problem. It allows us to determine the minority carrier current density at the junction edges and hence the current density of the diode. These minority currents are diffusive, i.e., there exists a gradient in the minority carrier density in the quasi-neutral region. The current supports this carrier distribution. The converse is also true: the carrier distribution supports this current.

If we integrate the current continuity equation from the contacts, the electron current density at any position z may be written w.r.t. its magnitude at one of the contacts as

$$J_n(z) = J_n(z_n) - q \int_z^{z_n} [\mathcal{R}(z) - \mathcal{G}(z)] \, dz, \qquad (4.138)$$

and

$$J_n(z) = J_n(-z_p) + q \int_{-z_p}^z [\mathcal{R}(z) - \mathcal{G}(z)] \, dz. \qquad (4.139)$$

---

[10]For a basis of this modification, see A. H. Marshak and K. M. Van Vliet, "Carrier Densities and Emitter Efficiency in Degenerate Materials with Position-Dependent Band Structure," *Solid-State Electronics*, **21**, p. 429 (1978), and A. H. Marshak and C. M. Van Vliet, "Electrical Current and Carrier Density in Degenerate Materials with Non-Uniform Band Structure," *Proc. of IEEE*, **72**, No. 2, p. 148 (1984).

The first term in the former equation is related to the injected or extracted electron current density and the second is related to generation–recombination current. The second term relates the exchange between the minority and majority carrier current density. A decrease in minority carrier current occurs with an increase in majority carrier current, and in the process, the total current remains constant. A similar set of equations for the hole current density are

$$J_p(z) = J_p(-z_p) - q \int_{-z_p}^{z} [\mathcal{R}(z) - \mathcal{G}(z)]\, dz, \qquad (4.140)$$

and

$$J_p(z) = J_p(z_n) + q \int_{z}^{z_n} [\mathcal{R}(z) - \mathcal{G}(z)]\, dz. \qquad (4.141)$$

Electron current is injected through the n-type region, and hole current through the p-type region, thus, the terms $J_n(z_n)$ and $J_p(-z_p)$ are the injected electron and hole current densities. The total current density remains constant in this one-dimensional analysis, and at any cross-section, the total current density is given by

$$
\begin{aligned}
J(z) &= J_n(z) + J_p(z) \\
&= J_n(z_n) - q \int_{z}^{z_n} [\mathcal{R}(z) - \mathcal{G}(z)]\, dz + J_p(-z_p) - \\
&\quad q \int_{-z_p}^{z} [\mathcal{R}(z) - \mathcal{G}(z)]\, dz \\
&= J_n(z_n) + J_p(-z_p) - J_{gr}(-z_p). \qquad (4.142)
\end{aligned}
$$

The current density $J_{gr}(z)$ is the integral of the charge generation–recombination rate from any position z in the diode to the end of the device defined by the n-side ohmic contact,

$$J_{gr}(z) = q \int_{z}^{z_n} [\mathcal{R}(z) - \mathcal{G}(z)]\, dz. \qquad (4.143)$$

This definition allows us to write more succinct forms of the spatial dependence of the current density as

$$J_n(z) = J_n(z_n) - J_{gr}(z) \qquad (4.144)$$

and

$$J_p(z) = J_p(z_n) + J_{gr}(z). \qquad (4.145)$$

Both electron and hole current densities are the largest at the ohmic contacts of their respective majority carrier regions. The hole current density decreases from the p-contact to the n-contact, while a corresponding increase occurs in the electron current density. The decrease in hole current density appears as an increase in the electron current density and vice-versa. An equation, similar to Equation 4.142, can be written by using the opposite references. This equation is therefore written in terms of the minority current densities at the contacts instead of the majority current densities.

$$J(z) = J_n(-z_p) + J_p(z_n) + J_{gr}(-z_p). \tag{4.146}$$

This equation may be viewed as being in terms of collected currents since the minority carriers are injected by majority carrier regions. The total current is the sum of the minority carrier current at the contacts and the $J_{gr}(-z_p)$ term due to generation–recombination effects that characterizes the exchange current of carriers.

The carrier transport can now be written using drift-diffusion formalism. We write the currents in terms of the quasi-Fermi potentials ϕ_n and ϕ_p for electrons and holes, and the electrostatic potential ψ. We have

$$J_n = -q\mu_n n \frac{\partial \phi_n}{\partial z}, \tag{4.147}$$

where

$$n = n_i \exp\left[\frac{q(\psi - \phi_n)}{kT}\right]. \tag{4.148}$$

For hole current density, we have

$$J_p = q\mu_p p \frac{\partial \phi_p}{\partial z}, \tag{4.149}$$

where

$$p = n_i \exp\left[\frac{q(\phi_p - \psi)}{kT}\right]. \tag{4.150}$$

Using the Einstein relation, and quasi-Fermi potentials instead of carrier densities as the variable, current density can be written as

$$
\begin{aligned}
J_n &= q\mu_n n_i \exp\left[\frac{q(\psi - \phi_n)}{kT}\right] \frac{\partial \phi_n}{\partial z} \\
&= q\mu_n n_i \frac{kT}{q} \exp\left(\frac{q\psi}{kT}\right) \frac{\partial}{\partial z}\left[\exp\left(-\frac{q\phi_n}{kT}\right)\right] \\
&= q\mathcal{D}_n n_i \exp\left(\frac{q\psi}{kT}\right) \frac{\partial}{\partial z}\left[\exp\left(-\frac{q\phi_n}{kT}\right)\right].
\end{aligned}
\tag{4.151}
$$

This relates the change of quasi-Fermi levels for electrons to the electron current density as

$$\frac{\partial}{\partial z}\left[\exp\left(-\frac{q\phi_n}{kT}\right)\right] = \frac{J_n(z)}{q\mathcal{D}_n n_i \exp\left(q\psi/kT\right)}. \tag{4.152}$$

We may now relate the changes in quasi-Fermi potential to its magnitude at the contacts. Integrating from the p-type ohmic contact to the n-type ohmic contact $(-z_p$ to $z_n)$,

$$\exp\left[-\frac{q\phi_n\left(z_n\right)}{kT}\right] - \exp\left[-\frac{q\phi_n\left(-z_p\right)}{kT}\right] = \int_{-z_p}^{z_n}\frac{J_n(z)}{q\mathcal{D}_n n_i \exp\left(q\psi/kT\right)}dz. \tag{4.153}$$

The current density at any position is related to the current density at the contact and the generation–recombination term through the identity $J_n(z) = J_n(z_n) - J_{gr}(z)$. Thus,

$$\exp\left[-\frac{q\phi_n\left(z_n\right)}{kT}\right] - \exp\left[-\frac{q\phi_n\left(-z_p\right)}{kT}\right] =$$
$$J_n\left(z_n\right)\int_{-z_p}^{z_n}\frac{1}{q\mathcal{D}_n n_i \exp\left(q\psi/kT\right)}dz - \int_{-z_p}^{z_n}\frac{J_{gr}}{q\mathcal{D}_n n_i \exp\left(q\psi/kT\right)}dz, \tag{4.154}$$

and hence, the majority carrier current at the contact at $z = z_n$ can be written as

$$J_n\left(z_n\right) = \left[\int_{-z_p}^{z_n}\frac{1}{q\mathcal{D}_n n_i \exp\left(q\psi/kT\right)}dz\right]^{-1} \times$$
$$\left\{\exp\left[-\frac{q\phi_n\left(z_n\right)}{kT}\right] - \exp\left[-\frac{q\phi_n\left(-z_p\right)}{kT}\right] +\right.$$
$$\left.\int_{-z_p}^{z_n}\frac{J_{gr}(z)}{q\mathcal{D}_n n_i \exp\left(q\psi/kT\right)}dz\right\}. \tag{4.155}$$

The total change in the quasi-Fermi potential between the two ohmic contacts is the applied voltage V across the diode, i.e.,

$$V = \phi_n\left(-z_p\right) - \phi_n\left(z_n\right). \tag{4.156}$$

This assumes a perfect ohmic contact with no voltage drop across it; more complicated conditions, e.g., using finite surface recombination velocity, can also be incorporated. For now, we consider ideal contacts. We multiply and divide by the factor $\exp\left[q\phi_n(-z_p)/kT\right]$, giving

$$
J_n(z_n) = \left\{ \int_{-z_p}^{z_n} \frac{\exp\left[q\left(\phi_n(-z_p)-\psi\right)/kT\right]}{q\mathcal{D}_n n_i}\,dz \right\}^{-1} \times
$$
$$
\left\{ \left[\exp\left(\frac{qV}{kT}\right)-1\right] + \right.
$$
$$
\left. \int_{-z_p}^{z_n} \frac{J_{gr}(z)}{q\mathcal{D}_n n_i}\exp\left[q\frac{\phi_n(-z_p)-\psi}{kT}\right]\,dz \right\}. \qquad (4.157)
$$

This is a general form of the expression for electron current density for arbitrary situations with doping, alloy composition, and bias variations in a one-dimensional geometry. It is a complicated form, which relates the currents through integrals of terms involving elementary charge, the diffusion coefficient, and the majority charge density. These integrals are the various forms of modified Gummel numbers. Equation 4.157 is the basic expression that we will refer to in complicated problems, but to demonstrate its utility even in simpler forms, we will consider homojunction structures. A convenient form, then, is

$$
J_n(z_n) = \frac{qn_i^{2}}{\Gamma_n}\left[\exp\left(\frac{qV}{kT}\right)-1\right] +
$$
$$
\frac{1}{\Gamma_n}\int_{-z_p}^{z_n}\frac{J_{gr}(z)n_i}{\mathcal{D}_n}\exp\left[q\frac{\phi_n\left(-z_p\right)-\psi}{kT}\right]\,dz, \quad (4.158)
$$

where

$$
\Gamma_n = \int_{-z_p}^{z_n}\frac{n_i}{\mathcal{D}_n}\exp\left[q\frac{\phi_n(-z_p)-\psi}{kT}\right]\,dz \qquad (4.159)
$$

is one form of the Gummel number for electrons. Following similar procedures, an expression may be written for hole current (see Problem 14) for the general case as

$$
J_p(z_n) = \left[\int_{-z_p}^{z_n}\frac{1}{q\mathcal{D}_p n_i \exp\left(-q\psi/kT\right)}\,dz\right]^{-1} \times
$$
$$
\left\{ \exp\left[\frac{q\phi_p(z_n)}{kT}\right] - \exp\left[\frac{q\phi_p(-z_p)}{kT}\right] - \right.
$$
$$
\left. \int_{-z_p}^{z_n}\frac{J_{gr}(z)}{q\mathcal{D}_n n_i \exp\left(-q\psi/kT\right)}\,dz \right\}, \qquad (4.160)
$$

which, under the restricted condition of no alloy grading, reduces to

$$
\begin{aligned}
J_p\left(z_n\right) \;=\;& \frac{qn_i{}^2}{\Gamma_p}\left[\exp\left(\frac{qV}{kT}\right)-1\right]- \\
& \frac{1}{\Gamma_p}\int_{-z_p}^{z_n}\frac{J_{gr}(z)n_i}{\mathcal{D}_p}\exp\left[q\frac{\psi-\phi_p(z_n)}{kT}\right]dz, \quad (4.161)
\end{aligned}
$$

where

$$
\Gamma_p=\int_{-z_p}^{z_n}\frac{n_i}{\mathcal{D}_p}\exp\left[q\frac{\psi-\phi_p(z_n)}{kT}\right]dz \quad (4.162)
$$

is a Gummel number for holes. These expressions for Gummel numbers have been written in terms of the quasi-Fermi potentials, the electrostatic potential, and the intrinsic carrier concentration. They are in a general form that is applicable to high-level injection or long base conditions. Under these arbitrary conditions, the total current, a constant at any bias, flowing in the diode can be written as

$$
\begin{aligned}
J(z) \;=\;& J_n(z_n)+J_p(z_n) \\
\;=\;& qn_i{}^2\left(\frac{1}{\Gamma_n}+\frac{1}{\Gamma_p}\right)\left[\exp\left(\frac{qV}{kT}\right)-1\right]+ \\
& \frac{1}{\Gamma_n}\int_{-z_p}^{z_n}\frac{J_{gr}(z)n_i}{\mathcal{D}_n}\exp\left[q\frac{\phi_n\left(-z_p\right)-\psi}{kT}\right]dz- \\
& \frac{1}{\Gamma_p}\int_{-z_p}^{z_n}\frac{J_{gr}(z)n_i}{\mathcal{D}_p}\exp\left[q\frac{\psi-\phi_n\left(-z_p\right)}{kT}\right]dz. \quad (4.163)
\end{aligned}
$$

We continue to assume perfect ohmic contacts with quasi-Fermi level position defined by the majority carrier, the excess minority carrier concentration going to zero, and the voltage drop at the contact negligible. The assumption of perfect ohmic contacts implies

$$
\phi_p\left(-z_p\right)=\phi_n\left(-z_p\right) \quad (4.164)
$$

and

$$
\phi_p\left(z_n\right)=\phi_n\left(z_n\right). \quad (4.165)
$$

We now restrict our discussion to applications involving low-level injection. Under these conditions, the quasi-Fermi potentials ϕ_n and ϕ_p remain approximately constant in the quasi-neutral regions and the junction space charge region. The Gummel number for electrons can then be written as

$$
\Gamma_n=\int_{-z_p}^{z_n}\frac{n_i}{\mathcal{D}_n}\exp\left[q\frac{\phi_n(-z_p)-\psi}{kT}\right]dz
$$

$$\approx \int_{-z_p}^{z_n} \frac{n_i}{\mathcal{D}_n} \exp\left[q\frac{\phi_p(-z_p) - \psi}{kT}\right] dz. \tag{4.166}$$

The Gummel number can be evaluated by integrating over the device depth. The carriers are majority carriers in part of the device; the integrand in this region is large. In the space charge region, and in the region where they are minority carriers, this magnitude is usually small in low-level injection. Using references from Figure 4.22, we have

$$\Gamma_n = \int_{-z_p}^{-w_p} \frac{p}{\mathcal{D}_n} dz + \int_{-w_p}^{z_n} \frac{n_i}{\mathcal{D}_n} \exp\left[q\frac{\phi_p(-z_p) - \psi}{kT}\right] dz, \tag{4.167}$$

where the integral is related to majority carrier density (holes) in the p-type region, and the last term is in a region where the integrand is small. Usually, therefore, in low-level injection, the last term is small, and we obtain

$$\Gamma_n \approx \int_{-z_p}^{-w_p} \frac{p}{\mathcal{D}_n} dz. \tag{4.168}$$

Similarly, we may find the hole Gummel number, whose major term is due to the electron charge density in the n-type region, i.e.,

$$\Gamma_p \approx \int_{w_n}^{z_n} \frac{n}{\mathcal{D}_p} dz. \tag{4.169}$$

In their original formulation, Gummel and Poon considered the integrated charge as the basis for the integral equations. These unmodified Gummel numbers were defined as

$$\mathcal{GN}_n = \frac{Q_n}{q} = \int_{w_n}^{z_n} n \, dz \tag{4.170}$$

and

$$\mathcal{GN}_p = \frac{Q_p}{q} = q \int_{-z_p}^{-w_p} p \, dz. \tag{4.171}$$

If the diffusivity is constant, intrinsic carrier concentration is constant, and low-level injection conditions exist, then the Gummel numbers of Equation 4.159 and 4.162 are equivalent to these. The effect of varying diffusivity of carriers is effectively incorporated in the unmodified Gummel–Poon approach by using the weighted diffusion coefficients $\overline{\mathcal{D}}_n$ and $\overline{\mathcal{D}}_p$. We may write these relationships as

$$\frac{1}{\Gamma_n} = \frac{q\overline{\mathcal{D}}_n}{Q_p}$$

$$\text{and} \quad \frac{1}{\Gamma_p} = \frac{q\overline{\mathcal{D}}_p}{Q_n}, \tag{4.172}$$

where

$$
\overline{\mathcal{D}}_n = \frac{\int_{-z_p}^{-w_p} p\,dz}{\int_{-z_p}^{-w_p} (p/\mathcal{D}_n)\,dz}
$$

$$
\text{and} \quad \overline{\mathcal{D}}_p = \frac{\int_{w_n}^{z_n} n\,dz}{\int_{w_n}^{z_n} (n/\mathcal{D}_p)\,dz}. \tag{4.173}
$$

Our expressions for electron and hole current densities at $z = z_n$ can be rewritten in terms of these integrated charge numbers as

$$
J_n(z_n) = \frac{q^2 n_i{}^2 \overline{\mathcal{D}}_n}{Q_p} \left[\exp\left(\frac{qV}{kT}\right) - 1 \right] +
$$
$$
\frac{q\overline{\mathcal{D}}_n}{Q_p} \int_{-z_p}^{z_n} \frac{J_{gr}(z)n_i}{\mathcal{D}_n} \exp\left[q\frac{\phi_n(-z_p) - \psi}{kT} \right] dz \tag{4.174}
$$

and

$$
J_p(z_n) = \frac{q^2 n_i{}^2 \overline{\mathcal{D}}_p}{Q_n} \left[\exp\left(\frac{qV}{kT}\right) - 1 \right] -
$$
$$
\frac{q\overline{\mathcal{D}}_p}{Q_n} \int_{-z_p}^{z_n} \frac{J_{gr}(z)n_i}{\mathcal{D}_p} \exp\left[q\frac{\psi - \phi_p(z_n)}{kT} \right] dz. \tag{4.175}
$$

The total current density for this homojunction, under conditions of low-level injection and with ideal ohmic contacts, is

$$
J(z) = qn_i{}^2 \left(\frac{q\overline{\mathcal{D}}_n}{Q_p} + \frac{q\overline{\mathcal{D}}_p}{Q_n} \right) \left[\exp\left(\frac{qV}{kT}\right) - 1 \right] +
$$
$$
\frac{q\overline{\mathcal{D}}_n}{Q_p} \int_{-z_p}^{z_n} \frac{J_{gr}(z)n_i}{\mathcal{D}_n} \exp\left[q\frac{\phi_n(-z_p) - \psi}{kT} \right] dz -
$$
$$
\frac{q\overline{\mathcal{D}}_p}{Q_n} \int_{-z_p}^{z_n} \frac{J_{gr}(z)n_i}{\mathcal{D}_p} \exp\left[q\frac{\psi - \phi_p(z_n)}{kT} \right] dz. \tag{4.176}
$$

This was derived from first principles including both drift and diffusion terms for the minority carriers since the quasi-Fermi potentials were used as a basis for the derivation. Thus, it does apply to conditions where the doping may be a function of position and where drift effects may be also be important. The geometrically weighted diffusivities $\overline{\mathcal{D}}_n$ and $\overline{\mathcal{D}}_p$ contain the contribution due to drift current.

The effect of motion of minority carriers is related to their motion parameters of effective diffusivity. The expressions indicate that this parameter is dependent on the material properties related to diffusivity and intrinsic carrier concentration, and to the integrated majority carrier charge density. This dependence was discussed on the basis of intuitive arguments at the beginning of this section. Thus, the contribution of electrons is related to n_i, $\overline{\mathcal{D}}_n$, and $\mathcal{Q}_p$, and the generation–recombination effects. The voltage V in these equations is the total potential drop in the semiconductor from one ohmic contact edge to the other, i.e., $V = V_p + V_j + V_n$, where V_p is the voltage drop across the p-side, V_j the voltage drop at the junction, and V_n the voltage drop across the n-side.

Let us now apply this approach to the simple situation of low-level injection in a short diode with generation–recombination considered negligible and with a uniform doping profile. We have

$$
\begin{aligned}
Q_p &= q(z_p - w_p)p_{p0}, \\
Q_n &= q(z_n - w_n)n_{n0}, \\
J_{gr}(z) &= 0, \\
\overline{\mathcal{D}}_p &= \mathcal{D}_p, \\
\text{and} \quad \overline{\mathcal{D}}_n &= \mathcal{D}_n.
\end{aligned}
\tag{4.177}
$$

This results in the following:

$$
\begin{aligned}
J_n(z_n) &= \frac{q^2 n_i^2 \overline{\mathcal{D}}_n}{Q_p}\left[\exp\left(\frac{qV}{kT}\right) - 1\right] \\
&= \frac{q n_i^2 \mathcal{D}_n}{(z_p - w_p)p_{p0}}\left[\exp\left(\frac{qV}{kT}\right) - 1\right] \\
&= \frac{q n_{p0}\mathcal{D}_n}{z_p - w_p}\left[\exp\left(\frac{qV}{kT}\right) - 1\right],
\end{aligned}
\tag{4.178}
$$

and

$$
\begin{aligned}
J_p(z_n) &= \frac{q^2 n_i^2 \overline{\mathcal{D}}_p}{Q_n}\left[\exp\left(\frac{qV}{kT}\right) - 1\right] \\
&= \frac{q n_i^2 \mathcal{D}_p}{(z_n - w_n)n_{n0}}\left[\exp\left(\frac{qV}{kT}\right) - 1\right] \\
&= \frac{q p_{n0}\mathcal{D}_p}{z_n - w_n}\left[\exp\left(\frac{qV}{kT}\right) - 1\right].
\end{aligned}
\tag{4.179}
$$

The total current, constant and independent of position in the device structure is

$$
J(z) = \left(\frac{q n_{p0}\mathcal{D}_n}{z_p - w_p} + \frac{q p_{n0}\mathcal{D}_p}{z_n - w_n}\right)\left[\exp\left(\frac{qV}{kT}\right) - 1\right].
\tag{4.180}
$$

This is the common expression for a short, uniformly-doped, homojunction diode in low-level injection, ignoring generation–recombination contributions. It is the sum of diffusion currents on either side of the junction, which is related to the gradients of the minority carrier densities which are constant in the absence of generation–recombination.

Now consider the same example with a varying doping profile to show the more general applicability in cases where the generation–recombination effects may still be ignored (by definition, the generation–recombination effects may not be ignored in a long diode). We consider a short diode with an exponential doping profile in one of the base regions.

Consider the example of Figure 4.23 . The diode has a uniformly doped p-type region with a doping N_A on one side of an abrupt junction and an exponentially increasing doping $N_D(z)$ on the other side of the junction given by

$$N_D(z) = N_{D0} \exp\left(\frac{z - z_n}{\ell}\right). \tag{4.181}$$

Such a retrograde doping profile commonly occurs in devices where constant capacitance is desired. An example is the varactor diode used for frequency multiplication in microwave applications. Because of the non-uniform doping, a built-in field exists in the n-type region, and both diffusion and drift currents occur in this region. These are included in the weighted diffusion coefficient $\overline{\mathcal{D}}_p$. For the doping profile shown, we have

$$
\begin{aligned}
Q_p &= q(z_p - w_p)N_A, \\
Q_n &= q\int_{w_n}^{z_n} N_D(z)\,dz \\
&= q\ell N_{D0}\left[1 - \exp\left(-\frac{z_n - w_n}{\ell}\right)\right], \\
J_{gr} &= 0, \\
\overline{\mathcal{D}}_p &= \frac{\int_{w_n}^{z_n} n\,dz}{\int_{w_n}^{z_n} (n/\mathcal{D}_p)\,dz}, \\
\text{and} \quad \overline{\mathcal{D}}_n &= \mathcal{D}_n.
\end{aligned}
\tag{4.182}
$$

The expression for the diffusivity $\overline{\mathcal{D}}_p$, thus, weights the contribution of the highest doping that occurs at the ohmic contact, and not at the edge of the junction depletion region. The drift effect is incorporated in this effective coefficient. Using the expression for current, under these conditions,

$$
J = \left\{\frac{qn_i^2 \mathcal{D}_n}{(z_p - w_p)N_A} + \frac{qn_i^2 \overline{\mathcal{D}}_p}{\ell N_{D0}\left\{1 - \exp\left[-\left(z_n - w_n\right)/\ell\right]\right\}}\right\} \times
$$

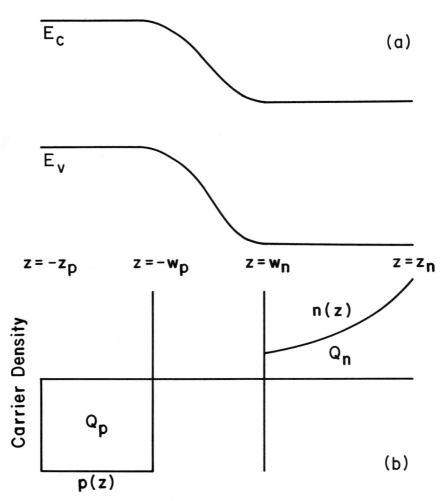

Figure 4.23: Example of a short p–n diode with an exponentially varying profile in one of the base regions. The top half of the figure shows the band edges and the bottom half shows the carrier density as a function of position.

$$\left[\exp\left(\frac{qV_j}{kT}\right) - 1 \right],$$

(4.183)

a considerably more complex expression, but obtained in a fairly straight-forward manner. The first part of this expression, due to electron transport in the short uniformly-doped p-type region, is the same as in the previous example. The second part of the expression is the variation introduced by non-uniform doping, and the drift and diffusion current contribution of the hole current.

These examples ignore recombination effects, usually a poor assumption in materials with short lifetimes such as compound semiconductors. We will therefore show the applicability of our derivation of the Gummel–Poon model to these by including simpler forms of generation–recombination effects in examples. Consider HSR-dominated generation–recombination processes, in a uniformly-doped long abrupt diode. The generation–recombination rates in the quasi-neutral regions are characterized by a lifetime, τ_n for electrons, τ_p for holes; and let $-\mathcal{U}_{scr}$ represent the net generation–recombination rate in the space charge region. Following our discussion of recombination processes, the former would be equally applicable for Auger processes due to heavy doping effects, the latter, however, is a complicated quantity because it exhibits a position-dependent variation within the space charge region of the junction. These net generation–recombination rates, for uniform doping in various parts of the diode, are given by

$$\mathcal{R}(z) - \mathcal{G}(z) = \frac{n_{p0}}{\tau_n} \left[\exp\left(\frac{qV_j}{kT}\right) - 1 \right] \exp\left(\frac{z + w_p}{L_n}\right)$$

(4.184)

in the p-type region $(-z_p \leq z \leq -w_p)$,

$$\mathcal{R}(z) - \mathcal{G}(z) = \mathcal{U}_{scr}$$

(4.185)

due to HSR recombination in the space charge region $(-w_p \leq z \leq w_n)$, and

$$\mathcal{R}(z) - \mathcal{G}(z) = \frac{p_{n0}}{\tau_p} \left[\exp\left(\frac{qV_j}{kT}\right) - 1 \right] \exp\left(-\frac{z - w_n}{L_p}\right)$$

(4.186)

in the n-type region $(w_n \leq z \leq z_n)$.

We wish to evaluate the integral containing the generation–recombination current term where the expression for the generation–recombination current is

$$J_{gr}(z) = \int_z^{z_n} [\mathcal{R}(z) - \mathcal{G}(z)]\, dz.$$

(4.187)

Let J_{scr} represent the contribution of the generation–recombination in the space charge region, i.e.,

$$J_{scr} = \int_{-w_p}^{w_n} q\mathcal{U}_{scr}(z)dz. \tag{4.188}$$

Using our simplified equation for electron current at the n-region contact,

$$
\begin{aligned}
J_n(z_n) &= \frac{q n_i^2}{\Gamma_n}\left[\exp\left(\frac{qV}{kT}\right) - 1\right] + \\
&\quad \frac{1}{\Gamma_n}\int_{-z_p}^{z_n}\frac{J_{gr}(z)n_i}{\mathcal{D}_n}\exp\left[q\frac{\phi_n(-z_p) - \psi}{kT}\right]dz \\
&\approx \frac{q n_i^2}{\Gamma_n}\left[\exp\left(\frac{qV_j}{kT}\right) - 1\right] + \frac{1}{\Gamma_n}\int_{-z_p}^{-w_p}\frac{J_{gr}(z)p_{p0}}{\mathcal{D}_n}dz.
\end{aligned}
$$
$$\tag{4.189}$$

The latter follows from the assumption of perfect ohmic contacts, which allows us to substitute $\phi_n(-z_p) = \phi_p(-z_p)$; the integral is therefore dominated by the contribution of the p-type region. We can evaluate this resulting integral involving recombination current by considering the three sections (p-type region, space charge region, and n-type region) contributing to it, and, in view of the above, we will consider the contribution of $J_{gr}(z)$ in the p-type region only.

The effect of the generation–recombination term, due to $J_{gr}(z)$, is the most complicated in form in the p-type region since the limits of integration for the generation–recombination rate are from this to z_n, which includes contributions from both the space charge region and the quasi-neutral n-type region. Let us consider this term first, i.e., the contribution of the quasi-neutral p-region of the device $(-z_p \leq z \leq -w_p)$

$$
\begin{aligned}
J_{gr}(z) &= J_{scr} + \\
&\quad \frac{q p_{n0}\mathcal{L}_p}{\tau_p}\left[\exp\left(\frac{qV}{kT}\right) - 1\right]\left[1 - \exp\left(-\frac{z_n - w_n}{\mathcal{L}_p}\right)\right] + \\
&\quad \frac{q n_{p0}\mathcal{L}_n}{\tau_n}\left[\exp\left(\frac{qV}{kT}\right) - 1\right]\left[1 - \exp\left(\frac{z + w_p}{\mathcal{L}_n}\right)\right]. \tag{4.190}
\end{aligned}
$$

In a long diode,

$$
\begin{aligned}
z_n - w_n &\ll \mathcal{L}_p \\
\text{and}\qquad z_p - w_p &\ll \mathcal{L}_n, \tag{4.191}
\end{aligned}
$$

and, therefore, the exponential in the second term is negligible. Also, by definition,

$$\frac{\mathcal{L}_p}{\tau_p} = \frac{\mathcal{D}_p}{\mathcal{L}_p}$$

$$\text{and} \quad \frac{\mathcal{L}_n}{\tau_n} = \frac{\mathcal{D}_n}{\mathcal{L}_n}, \tag{4.192}$$

yielding

$$
\begin{aligned}
J_n(z_n) &= \frac{q n_i^2}{\Gamma_n}\left[\exp\left(\frac{qV}{kT}\right) - 1\right] + \\
&\quad \frac{1}{\Gamma_n}\left\{\left(\frac{q p_{n0}\mathcal{D}_p}{\mathcal{L}_p} + \frac{q n_{p0}\mathcal{D}_n}{\mathcal{L}_n}\right)\left[\exp\left(\frac{qV}{kT}\right) - 1\right] + J_{scr}\right\} \times \\
&\quad \left[\frac{(z_p - w_p)p_{p0}}{\mathcal{D}_n}\right] - \\
&\quad \frac{2}{\Gamma_n}\frac{n_{p0}\mathcal{D}_n}{\mathcal{L}_n}\mathcal{L}_n\left[\exp\left(\frac{qV}{kT}\right) - 1\right]\left[1 - \exp\left(-\frac{z_p - w_p}{\mathcal{L}_n}\right)\right].
\end{aligned}
\tag{4.193}
$$

Again, the device being a long diode, the exponent in the last equation is negligible, and since

$$\Gamma_n \approx \left[\frac{(z_p - w_p)p_{p0}}{\mathcal{D}_n}\right] \tag{4.194}$$

for a uniformly doped device, we obtain

$$J_n(z_n) = \left(\frac{q p_{n0}\mathcal{D}_p}{\mathcal{L}_p} + \frac{q n_{p0}\mathcal{D}_n}{\mathcal{L}_n}\right)\left[\exp\left(\frac{qV_j}{kT}\right) - 1\right] + J_{scr}. \tag{4.195}$$

This is the electron current at the n-type contact in the long diode. In deriving this we consider the recombination perturbation as coming about only due to the recombination current integral in the p-type region. The basis of this is that this integral is a product of the hole density and the recombination current integral; therefore, the space charge region and the n-type region contribute little to the perturbation. Intuitively, also, since this is a long diode, all the current at the n-type contact must be electron current, since all excess holes injected in the n-type region must recombine before reaching the contact. The above expression, therefore, must also give the total current for a long diode. Let us show this by considering the hole current at the n-type contact at $z = z_n$.

We may evaluate this hole current at z_n

$$
\begin{aligned}
J_p(z_n) &= \frac{qn_i^2}{\Gamma_p}\left[\exp\left(\frac{qV}{kT}\right) - 1\right] - \\
&\quad \frac{1}{\Gamma_p}\int_{-z_p}^{z_n}\frac{J_{gr}(z)n_i}{\mathcal{D}_p}\exp\left[\frac{\psi - \phi_n(z_n)}{kT}\right]dz \\
&\approx \frac{qn_i^2}{\Gamma_p}\left[\exp\left(\frac{qV}{kT}\right) - 1\right] - \frac{1}{\Gamma_p}\int_{w_n}^{z_n}\frac{J_{gr}(z)n_{n0}}{\mathcal{D}_p}dz.
\end{aligned}
$$

$$(4.196)$$

Since the recombination perturbation is an integral involving the product of the hole density and the generation–recombination integral, it is dominated by the contribution from the quasi-neutral n-type region. This contribution is given by

$$
\begin{aligned}
J_{gr}(z) &= \int_z^{z_n}\frac{qp_{n0}}{\tau_p}\left[\exp\left(\frac{qV}{kT}\right) - 1\right]\exp\left(-\frac{z - w_n}{\mathcal{L}_p}\right)dz \\
&= \frac{qp_{n0}}{\tau_p}\mathcal{L}_p\left[\exp\left(\frac{qV}{kT}\right) - 1\right] \times \\
&\quad \left[\exp\left(-\frac{z - w_n}{\mathcal{L}_p}\right) - \exp\left(-\frac{z_n - w_n}{\mathcal{L}_p}\right)\right] \\
&\approx \frac{qp_{n0}}{\tau_p}\mathcal{L}_p\left[\exp\left(\frac{qV}{kT}\right) - 1\right]\exp\left(-\frac{z - w_n}{\mathcal{L}_p}\right).
\end{aligned}
$$

$$(4.197)$$

We have again used the inequality $(z_n - w_n) << \mathcal{L}_p$, and hence dropped the last exponential. We can now evaluate the hole current at z_n,

$$
\begin{aligned}
J_p(z_n) &\approx \frac{qn_i^2}{\Gamma_p}\left[\exp\left(\frac{qV}{kT}\right) - 1\right] - \\
&\quad \frac{1}{\Gamma_p}\int_{-w_n}^{z_n}\frac{qp_{n0}\mathcal{L}_p n_{n0}}{\tau_p\mathcal{D}_p}\left[\exp\left(\frac{qV}{kT}\right) - 1\right]\exp\left(-\frac{z - w_n}{\mathcal{L}_p}\right)dz \\
&\approx \frac{qn_i^2}{\Gamma_p}\left[\exp\left(\frac{qV}{kT}\right) - 1\right] - \\
&\quad \frac{qn_i^2}{\Gamma_p}\left[\exp\left(\frac{qV}{kT}\right) - 1\right]\left[1 - \exp\left(-\frac{z_n - w_n}{\mathcal{L}_p}\right)\right] \\
&\approx 0.
\end{aligned}
$$

$$(4.198)$$

As expected, the hole current is negligible at the n-region contact for this example of long diode.

The total diode current density, therefore, is

$$
\begin{aligned}
J(z_n) &= J_n(z_n) + J_p(z_n) \\
&= J_n(z_n) \\
&= \left(\frac{q p_{n0} \mathcal{D}_p}{\mathcal{L}_p} + \frac{q n_{p0} \mathcal{D}_n}{\mathcal{L}_n} \right) \left[\exp\left(\frac{qV}{kT} \right) - 1 \right] + J_{scr}.
\end{aligned}
\tag{4.199}
$$

The first term of this expression is the familiar form of the long diode expression for uniform doping. The second term in this expression of J_{scr} is the contribution of the space charge layer. The conventional expression ignores the space charge region contribution to generation–recombination, while this expression was derived considering contributions from all the regions. The contribution J_{scr}, therefore, arrived naturally in our expression instead of through perturbative iterations of the derivation. Since our control equations are linear differential equations, it occurs as a superposition term, as expected.

The evaluation of J_{scr} follows directly from the evaluation of the integral in Equation 4.188. Space charge recombination usually arises from HSR recombination. Hence, following our discussion in Chapter 3, we employ the simplified equation

$$
\mathcal{U} = \frac{np - n_i^2}{\tau_0 \left\{ n + p + 2n_i \cosh\left[(E_T - E_i)/kT \right] \right\}},
\tag{4.200}
$$

where we have assumed the trap to have identical τ_{n0} and τ_{p0} given by τ_0. Such traps are efficient recombination centers. However, differing values are the norm and can be treated numerically. Given this assumption, another characteristic of the net recombination rate is that a maximum in it occurs when $E_T = E_i$. A differing τ_{n0} and τ_{p0} changes this conclusion only slightly. So, deep traps are responsible for the generation-recombination effects. The np product is known in the space charge region since quasi-Fermi levels are assumed constant in low-level injection. We will evaluate this integral for these most efficient traps, i.e., for $E_T = E_i$. The contribution of the space charge region recombination current is

$$
J_{scr} = \int_{-w_p}^{w_n} \frac{n_i^2 \left[\exp\left(qV/kT \right) - 1 \right]}{\tau_0 \left(n + p + 2n_i \right)} dz.
\tag{4.201}
$$

Since the maximum contribution to this integral occurs from the region of maximum net recombination rate, we can approximate this integral as a product of the width of the space charge region $W = (w_n - w_n)$ and the maximum of the net recombination rate. n and p are know as a function of the displacement between the quasi-Fermi levels and intrinsic energy level. Hence, a maximum in $\mathcal{U}_{scr}$ can be shown to occur when

$$E_i = \frac{\xi_n + \xi_p}{2} \tag{4.202}$$

in the space charge region. This occurs where

$$n = p = n_i \exp\left(\frac{qV}{2kT}\right), \tag{4.203}$$

yielding the approximate relationship for space region current as

$$J_{scr} = \frac{qWn_i}{2\tau_0} \frac{\exp\left(qV/kT\right) - 1}{\exp\left(qV/2kT\right) + 1}. \tag{4.204}$$

In forward bias, this reduces to

$$J_{scr} \approx \frac{qWn_i}{2\tau_0} \exp\left(\frac{qV}{2kT}\right). \tag{4.205}$$

In reverse bias, this reduces to

$$J_{scr} \approx -\frac{qWn_i}{2\tau_0}. \tag{4.206}$$

The derivation for current in a long diode assumed a uniform doping density. Our Gummel–Poon equations are more general than that. Problem 15 discusses the long diode case with a non-uniform doping, similar to our short diode example.

In these examples of Gummel–Poon modelling, we made a simplifying assumption by restricting the exposition to homojunctions. This allowed us to move the term n_i outside the integral. In general, it may be position-dependent, such as when heavy doping effects or alloy grading effects predominate. We will now establish the expressions for these conditions. The derivation only needs to include the fact that the intrinsic carrier concentration may be position-dependent. Our derivation will be in terms of the current density at the edges of the depletion region, and hence, in general, use of this procedure will require the determination of both the hole current density and the electron current density.

The general expression for current density at the n-type contact ($z = z_n$), as a simple extension of the earlier derivation is, (see Problem 16)

$$J_n(-w_p) = \left\{ \int_{-z_p}^{-w_p} \frac{1}{q \mathcal{D}_n n_i \exp\left\{q\left[\psi - \phi_n(-z_p)\right]/kT\right\}} dz \right\}^{-1} \times$$

$$\left\{ \left[\exp\left(q\frac{V_p + V_j}{kT} \right) - 1 \right] + \int_{-z_p}^{-w_p} \frac{J_{gr}^p(z)}{q \mathcal{D}_n n_i \exp\left\{q\left[\psi - \phi_n(z_p)\right]/kT\right\}} dz \right\}, \qquad (4.207)$$

where

$$J_{gr}^p(z) = \int_z^{-w_p} q\left[\mathcal{R}(z) - \mathcal{G}(z)\right] dz. \qquad (4.208)$$

This equation is in a general form, applicable also for high-level injection, since we have only equated the differences in the quasi-Fermi potentials to the change in the electrostatic potential that they correspond to.

A similar expression may be written for the hole current density, and is given as

$$J_p(w_n) = \left\{ \int_{w_n}^{z_n} \frac{1}{q \mathcal{D}_n n_i \exp\left\{q\left[\phi_p(z_n) - \psi\right]/kT\right\}} dz \right\}^{-1} \times$$

$$\left\{ \left[\exp\left(q\frac{V_n + V_j}{kT} \right) - 1 \right] + \int_{w_n}^{z_n} \frac{J_{gr}^n(z)}{q \mathcal{D}_p n_i \exp\left\{q\left[\phi_p(z_n) - \psi\right]/kT\right\}} dz \right\}, \qquad (4.209)$$

where

$$J_{gr}^n(z) = \int_{w_n}^z q\left[\mathcal{R}(z) - \mathcal{G}(z)\right] dz. \qquad (4.210)$$

Assuming perfect ohmic contacts under low-level injection, we obtain

$$J_n(-w_p) = \left(\int_{-z_p}^{-w_p} \frac{p}{q \mathcal{D}_n n_i^2} dz \right)^{-1} \times$$

$$\left\{ \left[\exp\left(\frac{qV}{kT} \right) - 1 \right] + \int_{-z_p}^{-w_p} \frac{p J_{gr}^p(z)}{q \mathcal{D}_n n_i^2} dz \right\} \qquad (4.211)$$

for the electron current. This equation may be written in a form similar to the earlier equation for diodes using an effective Gummel number,

$$J_n(-w_p) = \frac{q \overline{\mathcal{D}}_n n_i^2}{\mathcal{G} \mathcal{N}_p'} \left[\exp\left(\frac{qV}{kT} \right) - 1 \right], \qquad (4.212)$$

where the effective Gummel number is

$$\mathcal{GN}_p' = \frac{Q_p}{q} \left\{ 1 + \frac{\int_{-z_p}^{-w_p} \left[pJ_{gr}^p(z)/q\mathcal{D}_n n_i^2 \right] dz}{\exp{(qV/kT)} - 1} \right\}^{-1}. \tag{4.213}$$

Likewise, the expression for hole current at the n-type contact $(z = z_n)$ is

$$J_p(w_n) = \left(\int_{w_n}^{z_n} \frac{n}{q\mathcal{D}_p n_i^2} dz \right)^{-1} \times$$
$$\left\{ \left[\exp\left(\frac{qV}{kT} \right) - 1 \right] + \int_{w_n}^{z_n} \frac{nJ_{gr}^n(z)}{q\mathcal{D}_p n_i^2} dz \right\}. \tag{4.214}$$

Writing this equation in a form similar to the expression for the electron current density, we obtain

$$J_p(w_n) = \frac{q\overline{\mathcal{D}}_p n_i^2}{\mathcal{GN}_n'} \left[\exp\left(\frac{qV}{kT} \right) - 1 \right], \tag{4.215}$$

where the effective electron Gummel number is given by

$$\mathcal{GN}_n' = \frac{Q_n}{q} \left\{ 1 + \frac{\int_{w_n}^{z_n} \left[nJ_{gr}^n(z)/q\mathcal{D}_p n_i^2 \right] dz}{\exp{(qV/kT)} - 1} \right\}^{-1}. \tag{4.216}$$

These expressions are cast in a form similar to our expressions for the long diode and short diode expressions. They reduce directly to the short diode expression since the effective Gummel number is then the majority charge divided by the electronic charge. The form of these expressions is chosen to directly show this relationship, and the perturbative effect of alloy compositional changes, large doping effects, and recombination effects. Note that the majority charge densities Q_n and Q_p and the effective diffusivities $\overline{\mathcal{D}}_n$ and $\overline{\mathcal{D}}_p$ continue to directly influence the current transport. The perturbation due to the recombination effect is generally small because the dividing factor is exponentially proportional to the applied voltage. The expressions also directly show that the recombination current density increases the current through the device, and the corresponding Gummel numbers $\mathcal{GN}_n'$ and $\mathcal{GN}_p'$ are decreased. The effect of changes in bandgap, such as through the alloy composition variation, occur through the effect on the intrinsic carrier concentration, an exponential effect. The effect on current under these conditions can thus become large, and is considered in the next sub-section.

We now use these expressions in determining the current, with recombination, for the example of varying doping profile. We will assume a long diode, unlike our earlier example that was for a varying doping profile in a short diode and with no recombination. The treatment of the p-side is the same as before for the case of a long diode with uniform doping profile. For the n-side, which is non-uniformly doped, the treatment is different and the expressions need to be reevaluated. The excess minority carrier density on the n-side is

$$p'(z) = p_{n0}(w_n) \left[\exp\left(\frac{qV}{kT}\right) - 1 \right] \exp\left(\frac{w_n - z}{\mathcal{L}_p}\right), \qquad (4.217)$$

assuming the diffusion coefficient to be a constant. The generation–recombination rate in the quasi-neutral n-type region is

$$\mathcal{R}(z) - \mathcal{G}(z) = \frac{p'(z)}{\tau_p} \qquad (4.218)$$

for $w_n < z < z_n$. To determine the effective Gummel number, we need the magnitude of the recombination current integral $J_{gr}^n(z)$ only in this interval. Hence,

$$
\begin{aligned}
J_{gr}^n(z) &= \frac{q p_{n0}(w_n)}{\tau_p} \left[\exp\left(\frac{qV}{kT}\right) - 1 \right] \int_{w_n}^{z} \exp\left(\frac{w_n - z}{\mathcal{L}_p}\right) dz \\
&= \frac{q p_{n0} \mathcal{L}_p}{\tau_p} \left[\exp\left(\frac{qV}{kT}\right) - 1 \right] \left[1 - \exp\left(\frac{w_n - z}{\mathcal{L}_p}\right) \right] \\
&= \frac{q \mathcal{D}_p p_{n0}}{\mathcal{L}_p} \left[\exp\left(\frac{qV}{kT}\right) - 1 \right] \left[1 - \exp\left(\frac{w_n - z}{\mathcal{L}_p}\right) \right]. \quad (4.219)
\end{aligned}
$$

To determine the effective Gummel number we need to determine the integral using this $J_{gr}^n(z)$, i.e., the effective Gummel number for calculation of the hole current is

$$
\begin{aligned}
\mathcal{GN}_n' &= \frac{Q_n}{q} \left\{ 1 + \frac{\int_{w_n}^{z_n} \left[n J_{gr}^n(z) / q \mathcal{D}_p n_i^2 \right] dz}{\exp(qV/kT) - 1} \right\}^{-1} \\
&\approx \frac{Q_n}{q} \left\{ 1 + \frac{p_{n0}(w_n)}{n_i^2 \mathcal{L}_p} \times \right. \\
&\qquad \left. \int_{w_n}^{z_n} \left[1 - \exp\left(-\frac{w_n - z}{\mathcal{L}_p}\right) \right] N_D(0) \exp\left(\frac{z - z_n}{\ell}\right) dz \right\}^{-1} \\
&\approx \frac{Q_n}{q} \left\{ 1 + \frac{p_{n0}(w_n)}{n_i^2 \mathcal{L}_p} \ell N_D(0) \left[1 - \exp\left(-\frac{w_n - z_n}{\ell}\right) \right] + \right.
\end{aligned}
$$

$$\frac{p_{n0}(w_n)}{n_i^2 \mathcal{L}_p} \frac{\ell \mathcal{L}_p}{\ell - \mathcal{L}_p} N_D(0)$$

$$\left[\exp\left(\frac{w_n - z_n}{\mathcal{L}_p}\right) - \exp\left(\frac{w_n - z_n}{\ell}\right)\right]^{-1}\Bigg\}.$$

$$(4.220)$$

Since $(w_n - z_n) \ll \mathcal{L}_p$, the exponential involving this factor can be ignored, and the effective Gummel number is

$$\mathcal{GN}_n' = \frac{Q_n}{q}\left\{1 + \frac{p_{n0}(w_n)N_D(0)}{n_i^2 \mathcal{L}_p}\left\{\ell\left[1 - \exp\left(-\frac{w_n - z_n}{\ell}\right)\right] - \frac{\ell \mathcal{L}_p}{\ell - \mathcal{L}_p}\exp\left(\frac{w_n - z_n}{\ell}\right)\right\}\right\}^{-1}.$$

$$(4.221)$$

Since the effective Gummel number is known, the hole current density at the edge of the depletion region follows. The electron current can be similarly derived, indeed it is the analog of this with $\ell \rightarrow \infty$. In a long diode, the total width of the depletion region is relatively short compared to the two bases of the diode. Therefore, we may ignore it, and hence determine the total current density (see Problem 17).

4.5.4   Abrupt and Graded Heterojunction p–n Diodes

Use of p–n heterojunctions is central to the operation of both the heterostructure bipolar transistor and the heterostructure laser. In these structures, the injection of carriers occurs through a heterojunction with the heterojunction aiding in the selective suppression of the injection process of one of the carriers. In abrupt heterojunctions, it occurs due to the existence of the discontinuity in the conduction and valence bands, and in the graded heterojunction diode, it occurs because the difference in bandgap appears predominately as a barrier to injection of one of the carrier types. Examples of some variations of the heterojunctions due to abruptness and grading of the alloy composition at one-sided junctions are shown in Figure 4.24. This example considers the Ga$_{.7}$Al$_{.3}$As/GaAs junction; the doping for the larger bandgap Ga$_{.7}$Al$_{.3}$As is identified by the capital N or P for n-type and p-type doping. The graded structures are formed using a grading length of 200 Å evenly distributed at the metallurgical junction. The effect of this grading is to distribute the bandgap change at the junction between the conduction and the valence band edges, whose extreme example is the case of the abrupt heterojunction under similar doping conditions, i.e., the band edges in this graded case show a smaller deviation then that caused by abrupt heterojunctions.

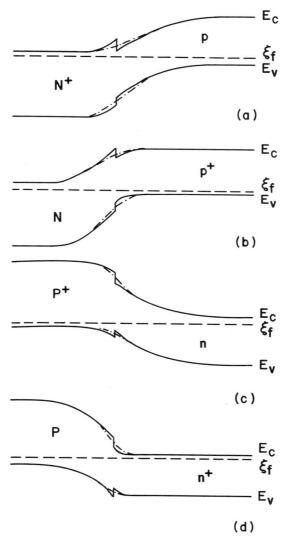

Figure 4.24: Band edges in thermal equilibrium for one-sided abrupt and graded heterojunctions of Ga$_{.7}$Al$_{.3}$As/GaAs, drawn to scale for various combinations of doping on either side of the junction. The chained lines show the band edges for the graded heterojunction with a linear alloy grading over 200 Å.

We need to establish how this band variation description comes about. Maxwell's equations continue to be valid, and to this point, we have only added the requirement that abrupt heterojunctions discontinuities in the conduction and valence band edges have to be established experimentally. Thus, Poisson's equation, which is a restatement of Maxwell's second equation, continues to be valid, and we have to incorporate variation in the dielectric constant. Gauss's law, which is a restatement of Maxwell's second equation applied at the interface, states that the difference in the displacement vector normal to the interface must differ by the interface charge density. For heterojunctions of interest, abrupt and graded, this is negligible. Thus, the displacement vector normal to the interface is continuous.

We have thus established three rules for describing the electronic behavior of abrupt heterojunctions: Poisson's equation, Gauss's Law, and the band edge discontinuities at the interface. These, together with the boundary conditions at the edges of the depletion region and the constancy of the Fermi level in thermal equilibrium, are sufficient to determine the band edge for abrupt heterojunctions in thermal equilibrium. These rules were used in determining the abrupt heterojunction band edges in Figure 4.24. As we remarked earlier, in the discussion of isotype heterojunctions, there will exist in these structures a discontinuity in the vacuum level and the electrostatic potential at the interface. The discontinuity in the vacuum level indicates that electrons at the band edge on either side of the interface require different energy to be removed into vacuum. The materials have different characteristics and this discontinuity is not unexpected. The difference in electrostatic potential has the same cause—it comes about because the electrochemical potential includes the effect of the material in the term that we have referred to as alloy potential.

The band edge variation changes when a bias is applied to the abrupt heterojunction. Under these conditions, the transport process has to be included together with the rules of Poisson's equation, Gauss's Law, and band edge discontinuity to determine the new variation. If, e.g., transport is by drift-diffusion, then we may continue to use the concept of quasi-Fermi energy as applied to a displaced Maxwell–Boltzmann distribution, and the quasi-Fermi energies stay continuous at the interface. The argument here is similar to the discussion of isotype heterojunctions. The carrier distribution at either side of the interface is very similar because of the small difference between drift and diffusion flux. With the large mobility of electrons in most compound semiconductors of interest, however, the transport of electrons at the interface is limited by thermionic emission and the higher-energy field emission current. Thus, similar to our discussion of isotype heterojunctions, the concept of quasi-Fermi energy is not strictly valid within a few kT's of changes in band edge energy at the interface. These must therefore be treated as fitting parameters in this region. Knowing the transport mathematics, however, we may include the other rules from the above and again derive the band edge variation in the abrupt heterojunction.

We have discussed the transport equations in slowly graded heterostructures, where the grading lengths are such that any band edge discontinuities can be ignored and all compositional variations can be directly included in the bandgap changes, in Chapter 3. Using this treatment of transport in graded heterostructures, Poisson's equation, and the boundary conditions at the edges of the depletion region, the band edge variation can be derived for the slowly graded heterojunctions. So, the transport through such graded heterojunctions, is determined using the modified drift-diffusion equation. The last of our Gummel–Poon quasi-static model derivation can be applied to such graded structures because it included variations in n_i, variations which result from alloy composition variation. Thus, the slowly graded heterojunction case is fairly straightforward and follows our discussion of homojunctions quite closely. We will return to analysis of slowly graded heterojunctions later in this section.

First, we consider the transport through the anisotype abrupt heterojunctions. In our discussion of isotype abrupt heterojunctions, we discussed the thermionic emission behavior. This analysis can be extended for the case of anisotype heterojunction diodes also. Consider the example of abrupt N–p$^+$ junction shown in case (b) of Figure 4.24. It is chosen because it occurs in the emitter-base junction of an abrupt heterostructure bipolar transistor. The base in these structures is more heavily doped and takes advantage of the suppression of hole injection current. Our discussion follows a similar argument as in the case of abrupt isotype heterojunctions. We will therefore be brief, with the intent of considering the problem for practical application.

The injected carriers thermalize over few mean free paths in the base and then drift-diffuse to the contact. In the analysis of which conduction mechanism is dominant, we must consider the ability of the junction to thermionically emit the carriers over the barrier and the ability of the semiconductor bulk in the emitting and collecting regions to supply and collect these carriers. Due to the use of a heterojunction, the ability to inject carriers has decreased by the increase in the built-in junction voltage for that carrier without affecting other transport characteristics. Therefore, it is this barrier transport that limits the current across the junction. In our analysis of the isotype heterojunction, we derived this thermionic current in a form where we can now ignore the component from the smaller bandgap material. A significant barrier continues to exist for the injection from the small bandgap material to the large bandgap material throughout the bias range. In a first-order analysis, this injection from the small bandgap material may be ignored. We can determine it, if needed, by considering the transport via thermionic–drift-diffusion theory as well as via field emission. This thermionically injected current is given by

$$
|J_{th}| = A\frac{m^*}{m_0}T^2\exp\left[-\frac{q}{kT}\frac{E_g + \Delta E_c + (\epsilon_1 N_D/\epsilon_2 N_A)\,V_n - V_p}{1 + \epsilon_1 N_D/\epsilon_2 N_A}\right] \times
$$
$$
\left[\exp\left(\frac{q}{kT}\frac{V}{1 + \epsilon_1 N_D/\epsilon_2 N_A}\right) - 1\right] \quad \text{for} \quad V \gg \frac{kT}{q},
$$
(4.222)

where we have parameterized the pre-factor. In this equation, J_{th} is the thermionic current density, A is the Richardson's constant for a free electron whose magnitude is 120 A.cm^{-2}.K^{-2}, m_e^* and m_0 are the effective and free electron masses, ΔE_C is the conduction band edge discontinuity, ϵ_1 and N_D and ϵ_2 and N_A are the permittivity and doping of the two semiconductors, V_n and V_p are the differences in potential between the Fermi levels and the conduction and the valence band edge in the two semiconductors, and V is the applied potential at the junction. The ratio $\epsilon_1 N_D/\epsilon_2 N_A$ in this equation accounts for the partitioning of the electrostatic potential across the depletion region. This does not have to be considered in metal–semiconductor junctions where the thermionic emission theory is usually applied.

This is a simplified equation; more rigorous equations similar in form to the metal–semiconductor thermionic field emission and thermionic–drift-diffusion equations can also be applied. Indeed, since those equations were derived by considering the transport only in the semiconductor, they are well applicable here for the specific transport directions if the correct meaning are ascribed to the parameters. As an example of the thermionic–drift-diffusion equation applied to this problem, the effective mass is for the correct orientation in the large bandgap material, the effective density of states is the effective density of states of the large bandgap material, the barrier height is the difference between the Fermi level and the top of the conduction band edge, and the potential V is the value of the partitioned quantity that occurs across the large bandgap material. Similar comments apply to the modification of the thermionic field emission equation for this problem.

Figure 4.25 shows behavior of the conduction band edge under forward bias conditions for a graded junction at two different doping levels in the n-type injector/emitter region. The higher doped emitter is well behaved in the band edge while the lower doped emitter shows a barrier at the higher injection conditions. The formation of the barrier, due to variation in composition and associated alloy potential, will be subject of further

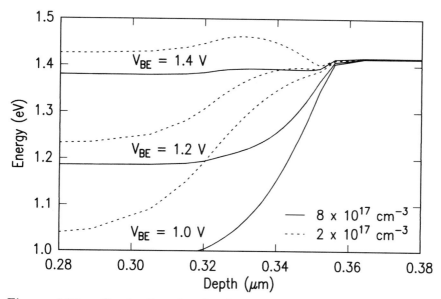

Figure 4.25: Conduction band edge energy diagram of an N–p$^+$ Ga$_{.7}$Al$_{.3}$As/GaAs heterojunction diode under low ($V = 1.0$ V), medium ($V = 1.2$ V), and high ($V = 1.4$ V) forward bias. Two n-type dopings are shown; the solid lines are for 8×10^{17} cm^{-3} doping and the dashed lines are for 2×10^{17} cm^{-3} doping. The p$^+$ doping is 5×10^{18} cm^{-3}.

discussion later in this chapter and in Chapter 7. Our treatment here assumes that the conditions of monotonic variation in band edge hold. This is usually the case for low-level injection.

Let us now consider the determination of the current in such graded anisotype heterojunctions. The application of the modified drift-diffusion equation follows in a similar manner as the simpler form of the drift-diffusion equation for transport. We will use a simple example of a heterojunction formed by compositional grading within the junction depletion region, and apply the modified drift-diffusion equation and the extended Gummel–Poon model to evaluate the currents.

Consider the analysis in one dimension of a short N–p heterojunction in low-level injection, whose band edge schematic and some parameters are shown in Figure 4.26. The composition in the quasi-neutral region is a constant in this example, and like the homojunction in low-level injection, the quasi-Fermi levels remain constant through the bulk of the device and the current transport is limited by diffusion of minority carriers in the

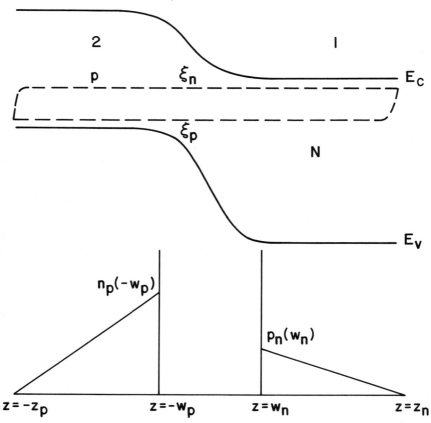

Figure 4.26: A graded p–N heterojunction formed using a large bandgap material (identified by 1), and a small bandgap material (identified by 2). The alloy grading occurs within the junction depletion region, and like the homojunction analysis in low-level injection, the transport in the junction region is not a limiting transport mechanism.

quasi-neutral regions. The device is a short device, and we also consider recombination as being absent. Thus, to determine the current, we need minority carrier current at the edge of the depletion regions. In the absence of recombination, this current density stays constant in the depletion region, and the total current is the sum of these minority carrier current densities.

The Shockley boundary conditions describe the minority carrier densities at the edges of the depletion region as

$$n_p(-w_p) = n_{p0} \exp\left(\frac{qV}{kT}\right)$$

$$\text{and} \quad p_n(w_n) = p_{n0} \exp\left(\frac{qV}{kT}\right), \tag{4.223}$$

where n_{p0} and p_{n0} are the concentrations of minority carriers in thermal equilibrium in the two different materials. The current density due to electron and hole transport in the junction quasi-neutral regions is described by the modified drift-diffusion equations

$$J_n = qn\mu_n\left(-\frac{d\psi}{dz} - \frac{\phi_{Cn}}{dz}\right) + q\mathcal{D}_n\frac{dn}{dz}$$

$$\text{and} \quad J_p = qp\mu_p\left(-\frac{d\psi}{dz} - \frac{\phi_{Vp}}{dz}\right) - q\mathcal{D}_p\frac{dp}{dz}. \tag{4.224}$$

From these,

$$J_n(-w_p) = \frac{qn_{p0}\mathcal{D}_n}{z_p - w_p}\left[\exp\left(\frac{qV}{kT}\right) - 1\right]$$

$$\text{and} \quad J_p(w_n) = \frac{qp_{n0}\mathcal{D}_p}{z_n - w_n}\left[\exp\left(\frac{qV}{kT}\right) - 1\right], \tag{4.225}$$

and hence the total current is

$$J = J_n(-w_p) + J_p(w_n) = \left(\frac{qp_{n0}\mathcal{D}_p}{z_n - w_n} + \frac{qn_{p0}\mathcal{D}_n}{z_p - w_p}\right)\left[\exp\left(\frac{qV}{kT}\right) - 1\right]. \tag{4.226}$$

The equation has the same form as our earlier example of a homojunction, but with the major difference that the thermal equilibrium minority carrier concentration is different in a heterojunction. Thus, if N_D and N_A represented the donor and acceptor doping densities, E_{g1} and E_{g2} the two bandgaps ($E_{g1} > E_{g2}$ in this example), N_{C1} and N_{C2} the effective conduction band density of states, N_{V1} and N_{V2} the effective valence band density of states, and n_{i1} and n_{i2} the intrinsic carrier concentrations in the two materials, then these are related in the Boltzmann approximation as

$$n_{p0} = \frac{n_{i2}^2}{N_A},$$

$$p_{n0} = \frac{n_{i1}^2}{N_D},$$

$$n_{i1}^2 = N_{C1}N_{V1}\exp\left(-\frac{E_{g1}}{kT}\right),$$

$$\text{and} \quad n_{i2}^2 = N_{C2}N_{V2}\exp\left(-\frac{E_{g2}}{kT}\right). \tag{4.227}$$

The hole concentration in thermal equilibrium in the larger bandgap material is significantly smaller compared to the electron concentration in thermal equilibrium in the smaller bandgap material. This ratio is

$$\frac{p_{n0}}{n_{p0}} = \frac{N_A}{N_D}\frac{n_{i1}^2}{n_{i2}^2} = \frac{N_A}{N_D}\frac{N_{C2}N_{V2}}{N_{C1}N_{V2}}\exp\left(-\frac{E_{g1}-E_{g2}}{kT}\right). \tag{4.228}$$

The difference in bandgap decreases this minority carrier concentration exponentially with the bandgap change. The corresponding hole current density also decreases by the same exponential factor. On the other hand, the electron current and the electron transport are not influenced by the presence of the heterojunction. Another way of looking at this follows arguments based on built-in voltages. In Figure 4.26, the flat quasi-Fermi levels and bandgap grading leave the barrier for injection of electrons unchanged from the homojunction case except for the second order effects related to the effective density of states and its influence on position of the Fermi level in the quasi-neutral region. However, the hole barrier increases by this increase in the bandgap, and hence hole injection is suppressed by a corresponding amount.

Thus, the presence of a heterojunction causes the built-in junction voltages for the electrons and the holes to be different. The built-in voltage now represents the sum of two effects, the electrostatic potential, which we may still define in terms of the variation of the vacuum level, and the alloy potential. The latter changes at the junction due the compositional changes.

These built-in voltages follow from the drift-diffusion equations, applied at thermal equilibrium where each of the current densities is zero and the Fermi level is flat. For electrons, the junction built-in voltage is

$$\psi_{jn} = \frac{kT}{q}\int_{n(-w_p)}^{n(w_n)}\frac{dn}{n} = \frac{kT}{q}\ln\left(\frac{N_D N_A}{n_{i2}^2}\right), \tag{4.229}$$

and for holes, the junction built-in voltage is

$$\psi_{jp} = \frac{kT}{q} \int_{p(w_n)}^{p(-w_p)} \frac{dp}{p} = \frac{kT}{q} \ln \left(\frac{N_D N_A}{n_{i1}^2} \right). \tag{4.230}$$

The junction built-in voltage for holes increased by the difference in the bandgap.

Here, only low-level injection was considered. A high current density across heterojunctions presents a particularly interesting problem because it is concomitant with a smaller space charge region (due both to the larger doping that a larger current capability requires and to the forward biasing of the junction). The smaller space charge region and decreased electrostatic field lead naturally to alloy potential effects appearing in the conduction band. One may see this as an example of the mixing of case (b) and case (d) for the graded case of Figure 4.24. A small electrostatic field at the junction occurs near flat band conditions. When this happens, the difference in alloy potential appears partially in the conduction band and partially in the valence band. Looking at the graded heterojunction examples of Figure 4.24, the difference in the alloy potential appears in the valence band edge in case (a) and case (b), and in the conduction band edge for case (c) and case (d). The magnitude and direction of the electrostatic field force the alloy potential effects to appear in the conduction or the valence band, and when the electrostatic field becomes small, such as in large forward bias, they cause the alloy effects to appear in both the band edges. Appearance of the resulting barrier limits the current density in the structure, and the junctions need to be designed to prevent this. The alloy grading length needed to prevent barrier-limited transport effects in the depletion approximation limit can be derived by consideration of the electrostatic and alloy potential variation. If the electrostatic field remains larger than the alloy field, the band bending remains monotonic, and no barrier exists. Mathematically, this requirement can be summarized as

$$\frac{d\psi_{all}}{dz} < E_{es}, \tag{4.231}$$

where E_{es} is the local electrostatic field and ψ_{all} is the alloy potential. For a one sided p–n junction, this results in (see Problem 18),

$$z_g > \frac{\Delta E_g}{\sqrt{2kT(E_g - qV - 2kT)}} \gg \lambda_D, \tag{4.232}$$

where z_g is the grading length, ΔE_g the total bandgap change, q the electronic charge, E_g the bandgap in the narrow bandgap material, V the

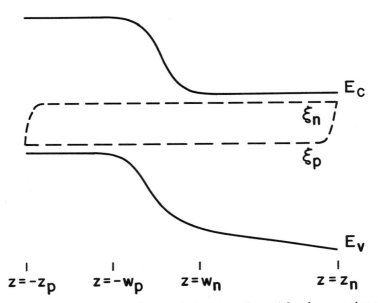

Figure 4.27: Band diagram of a p–n heterojunction with a base region with exponentially decreasing intrinsic carrier concentration.

applied bias, and λ_D the extrinsic Debye length. The expression is derived assuming the depletion approximation, and hence is inaccurate at high forward bias conditions, where the mobile charge should be taken into consideration.

Graded heterojunctions are convenient to analyze using the extended Gummel–Poon model. The derivation for current in the heterojunction example follows directly; it is the same as the expression derived in our homojunction short diode example using the Gummel–Poon model. We consider another short diode example with a change in the alloy composition in one of the bases of the diode to show the breadth of this approach. This example is similar to the example of varying doping in the base of a short diode, except here doping is constant and the intrinsic carrier concentration is allowed to change exponentially as

$$n_i^2 = n_{i0}^2 \exp\left(\frac{z - w_n}{\ell}\right) \qquad \text{for} \qquad z > w_n$$

$$\text{and} \quad n_i^2 = n_{i0}^2 \qquad \text{for} \qquad -z_p < z \le w_n. \qquad (4.233)$$

Figure 4.27 shows a schematic of this problem.

The general expressions for the electron and hole currents at the edges of the junction are given by

$$J_n(-w_p) = \left(\int_{-z_p}^{-w_p} \frac{p}{q\mathcal{D}_n n_i^2} dz \right)^{-1} \times$$

$$\left\{ \left[\exp\left(\frac{qV}{kT}\right) - 1 \right] + \int_{-z_p}^{-w_p} \frac{pJ_{gr}^p(z)}{q\mathcal{D}_n n_i^2} dz \right\}, \quad (4.234)$$

and

$$J_p(w_n) = \left(\int_{w_n}^{z_n} \frac{n}{q\mathcal{D}_p n_i^2} dz \right)^{-1} \times$$

$$\left\{ \left[\exp\left(\frac{qV}{kT}\right) - 1 \right] + \int_{w_n}^{z_n} \frac{nJ_{gr}^n(z)}{q\mathcal{D}_p n_i^2} dz \right\}. \quad (4.235)$$

Ignoring recombination in the short diode, and assuming low-level injection and a hole diffusion coefficient that is a weighted constant $\overline{\mathcal{D}}_p$, we can evaluate these currents. The minority carrier electron current density is

$$J_n(-w_p) = \left[\frac{N_A(z_p - w_p)}{q\mathcal{D}_n n_{i0}^2} \right]^{-1} \left[\exp\left(\frac{qV}{kT}\right) - 1 \right], \quad (4.236)$$

an expression identical to the earlier expression. The minority carrier hole current in the bandgap grading region is modified, and given by

$$J_p(w_n) = \left\{ \frac{N_D \ell}{q\overline{\mathcal{D}}_p n_{i0}^2} \left[\exp\left(\frac{z_n - w_n}{\ell}\right) - 1 \right] \right\}^{-1} \left[\exp\left(\frac{qV}{kT}\right) - 1 \right], \quad (4.237)$$

and the total current, which is the sum of the above two expressions, is given by

$$J = \left\{ \frac{q\mathcal{D}_n n_{i0}^2}{N_A(z_p - w_p)} + \frac{q\overline{\mathcal{D}}_p n_{i0}^2}{N_D \ell \left[\exp\left[(z_n - w_n)/\ell\right] - 1 \right]} \right\} \left[\exp\left(\frac{qV}{kT}\right) - 1 \right].$$
$$(4.238)$$

The drift current due to the quasi-field for holes in the n-type bandgap graded region is included in the above expressions. This drift field opposes the motion of holes and hence suppresses the hole current as drawn in Figure 4.27. Note that the hole concentration and the electron concentration at the edges of the junction depletion region are still identical to the homojunction example since we chose the bandgap change to occur only in the quasi-neutral n-type region. Also, in the limit of $\ell \to \infty$, i.e., no bandgap and intrinsic carrier concentration variation, this expression reduces to the expression for a short uniformly-doped homojunction diode. More examples of the application of the extended Gummel–Poon model are considered in the Problems, and we will also consider it again in our discussion of heterostructure bipolar transistors.

4.6   Summary

This chapter developed the concepts and mathematical framework for treatment of transport across junctions. We considered junctions between different materials, i.e., junctions formed between metals and semiconductors or two different semiconductors, as well as junctions formed by a change in the type of doping, i.e., p–n junctions. For the metal–semiconductor and heterojunctions, we considered the contributions due to thermionic emission and field emission in the region at the interface as well as due to other restrictions placed by drift-diffusion transport in the bulk. We showed how, in the case of thermionic emission, the consequences of Bethe condition relating the collision-less emission of carriers at the interface and the momentum matching condition lead to an unambiguous effective mass for the Richardson's constant. In considering the specialized case of heterojunctions, we also reviewed tunneling: the conventional Fowler–Nordheim tunneling that considers wave function decay in a trapezoidal barrier, and resonant Fowler–Nordheim tunneling, important at low temperatures, and involving additional wave interference effects in the barrier region. These processes are important in the near-ideal heterojunctions of compound semiconductors for a variety of devices. As an extension of the analysis we considered the formation of ohmic contacts, contacts that allow low resistance bi-directional current flow in device structures.

We also considered in this chapter the approximations of p–n junction analysis and described the bulk effects occurring at high level injection conditions. For graded p–n junctions, i.e., those lacking an abrupt barrier at the interfaces, we developed an appropriate criterion for grading. For predicting the current, we developed a formal method of quasi-static analysis, the Gummel–Poon method, that allows a description of diode behavior in a variety of practical circumstances. This was applied to cases involving doping variation as well as composition variation.

General References

1. E. H. Rhoderick and R. H. Williams, *Metal-Semiconductor Contacts*,

Clarendon Press, Oxford (1988).

2. H. K. Henisch, *Semiconductor Contacts—An Approach to Ideas and Models*, Clarendon Press, Oxford (1984).

3. F. A. Padovani and R. Stratton, in R. K. Willardson and A. C. Beer, Eds., *Semiconductor and Semimetals*, **7A**, Academic Press, N.Y. (1971).

4. R. Stratton, "Diffusion of Hot and Cold Electrons in Semiconductor Barriers," *Phys. Rev.*, **126**, No. 6, p. 2002, 1962.

5. A. G. Milnes and D. L. Feucht, *Heterojunctions and Metal–semiconductor Junctions*, Academic Press, N.Y. (1972).

6. S. M. Sze, *Physics of Semiconductor Devices*, John Wiley, N.Y. (1981).

7. S. S. Cohen and G. Sh. Gildenblat, *Metal–Semiconductor Contacts and Devices*, in N. G. Einspruch, Ed., *VLSI Electronics and Microstructure Science*, **V 13**, Academic Press, FL (1986).

8. C. B. Duke, *Tunneling in Solids*, in F. Seitz, D. Turnbull, and H. Ehrenreich, Eds., *Solid State Physics*, Suppl. **10**, Academic Press, N.Y. (1969).

9. E. Burstein and S. Lundqvist, Ed., *Tunneling Phenomena in Solids*, Plenum Press, N.Y. (1969).

Problems

1. Problems of thermionic injection encountered in metal–semiconductor junctions involve injection from many complicated situations. Injection in metal/n-$Ga_{1-x}Al_xAs$ junctions, e.g., involves electrons in the anisotropic L and X valleys. Injection in metal–p-$Ga_{1-x}Al_xAs$ junctions, e.g., occurs from valence bands with light and heavy hole mass. Consider the electron and hole injection problem and determine the mass appropriate to the determination of Richardson's constant for

   (a) injection from the L valley on (100), (110), and (111) surfaces,

   (b) injection from the X valley on (100), (110), and (111) surfaces, and

   (c) injection from valence bands.

2. We have considered the thermionic injection component and the tunneling component of injection at a metal–semiconductor junction separately. In a metal–semiconductor junction, thermionic injection occurs together with tunneling of carriers. The tunneling of higher-energy carriers is favored because these encounter smaller barrier energies and width. Show that when both of these processes dominate, the current can be expressed in the form shown in Equation 4.37.

3. Show that the low temperature limit of thermionic field emission in a metal–semiconductor junction can be expressed as

$$J = \mathcal{A}^* \frac{E_{00}{}^2}{k^2} \frac{q\phi_B - qV - qV_n}{q\phi_B} \exp\left[-\frac{(2q\phi_B)^{3/2}}{3E_{00}\sqrt{q\phi_B + qV - qV_n}} \right].$$

$$(4.239)$$

4. Plot the carrier concentration and the velocity as a function of position between $z = 0$ and $z = z_c$ for a GaAs metal–semiconductor junction on 1×10^{16} cm^{-3} doped material at thermal equilibrium at 300 K. The barrier height from the metal to the semiconductor is 0.8 eV.

5. What is the ratio of the carrier concentration in the semiconductor at the metal–semiconductor junction, if the Bethe condition is satisfied, and if the quasi-Fermi level is assumed to be flat?

6. Show that Equation 4.57 follows from the application of current continuity to the boundary of regions dominated by drift-diffusion transport and thermionic emission transport.

7. Consider the application of Equation 4.57 to the metal–semiconductor junction formed on 1×10^{16} cm^{-3} n-type GaAs. Consider a metal barrier height of 0.8 eV, and an operating temperature of 300 K. Plot the current density of this thermionic–drift-diffusion expression, as well as its thermionic and drift-diffusion limit. Which component dominates? Now consider the same structure at 77 K, and plot the three current densities. Which component dominates?

8. Consider the application of the thermionic–drift-diffusion theory to an n–n heterojunction. Compared to the metal–semiconductor junction the differences include the partitioning of the electrostatic potential on the two sides of the heterojunction and the injection of carriers of differing effective mass, and requiring appropriate momentum matching as they traverse the junction. Taking into account these differences,

find an expression similar to Equation 4.57, and derive the current voltage relationship for thermionic–drift-diffusion transport.

9. An alternate, but simplistic, way of analyzing Problem 8 is to consider transport as emission of a Maxwell–Boltzmann distribution across a barrier with transmission coefficients T_{12} and T_{21} which are functions of energy. Using detailed balance, derive an expression for current transport and compare it with the thermionic emission limit of Problem 8.

10. Using the complete solution of the triangular barrier tunneling transmission probability (Equation 4.77), derive the asymptotic form of Equation 4.79. Show that this asymptotic form follows directly from the use of the WKB approximation.

11. Using the current voltage relationships due to thermionic emission and thermionic field emission of a metal–semiconductor junction, and field emission in a triangular barrier, derive the expressions for the contact resistance together with their pre-factors. Apply the appropriate relationship to an n-type metal–semiconductor junction on GaAs. Assume a temperature of 300 K, a barrier height of 0.8 eV, and n-type dopings of 1×10^{16} cm^{-3}, 1×10^{17} cm^{-3}, 1×10^{18} cm^{-3}, 1×10^{19} cm^{-3}, and 5×10^{19} cm^{-3}.

12. For a field emission-dominated region of transport across a metal–semiconductor junction, show that the current can be expressed as shown by Equation 4.94 and that the contact resistance follows as Equation 4.96. Estimate the contact resistance resulting from this for Problem 11 at the highest doping.

13. Consider the practical contact as a distribution of point contacts on the semiconductor surface. Show that, if the spreading resistance of these point contacts dominates, then the contact resistance must vary inversely with doping.

14. Derive the expression for hole current density at z_n (Equation 4.160) using the hole quasi-Fermi level as basis for the calculation of current.

15. Consider a n$^+$–p diode with a long p-type region where the doping varies as $\Delta N_A \exp(z/\ell) + N_{A0}$. Derive the expression for electron current in such a structure using the Gummel–Poon model.

16. Derive Equation 4.207 for electron current at the p-type contact using the Gummel–Poon model.

17. Derive the current in a long diode by considering both the electron and the hole current and ignoring the recombination in the depletion region.

18. Show that the requirement that the alloy field be maintained smaller than the electrostatic field results in the following condition

$$z_g > \frac{\Delta E_g}{\sqrt{2kT(E_g - qV - 2kT)}} \gg \lambda_D \qquad (4.240)$$

to obtain an alloy-barrier free transport in a p–n junction.

19. Consider a uniformly-doped GaAs p–n junction with a doping of 1×10^{17} cm^{-3} on either side. Assuming depletion approximation, what is the junction depth in thermal equilibrium? Plot the capacitance C and $1/C^2$ as functions of reverse bias for the junction. Now consider a deep donor trap density of 1×10^{16} cm^{-3} uniformly distributed throughout the semiconductor. The trap energy is 0.75 eV below the conduction band edge, and the degeneracy is 2. How does the junction depth change in thermal equilibrium and what is the behavior of capacitance C and $1/C^2$ with reverse bias? Can one determine the position and density of the donor trap using such a measurement technique?

20. Solve Laplace's equation in the presence of a conducting plane to show that the lowering of energy described by the image charge leads to an identical result for the energy of a charged carrier as a function of position from the conducting plane.

21. The Poole-Frenkel effect is the lowering of barrier energy at a defect in a dielectric, e.g., at a positively charged defect that can capture an electron and has related Coulombic effects. The effect has parallels with the image force lowering with the major difference being the absence of the conducting plane. Find an expression for the amount of barrier lowering due to the Poole-Frenkel process described in Figure 4.28.

22. Following the previous problem, what electric field will give rise to a Poole-Frenkel barrier lowering effect equal to the donor ionization energy in GaAs (typically 6 meV)?

23. Barrier heights may be changed by incorporation of a limited amount of dopants. The region of incorporated dopants is depleted under the biasing conditions applied to the semiconductor junction. This can

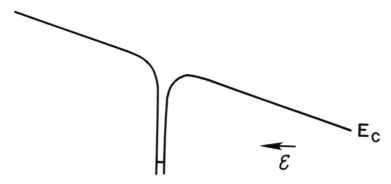

Figure 4.28: Schematic of the Poole-Frenkel process.

be employed to either increasing or forming a low-energy barrier such as in planar doped diodes, as well as to lowering barrier heights of metal–semiconductor junctions.

(a) Consider a metal–semiconductor junction where the donor density is increased to a value of N_{D2} in a region of width ℓ from the donor density of N_{D1}. Let ℓ be less than the Debye length. What is the electric field as a function of position and the magnitude of image force lowering?

(b) Consider the same metal–semiconductor junction with an acceptor layer of width ℓ and doping N_A at the junction. What is the electric field as a function of position and the magnitude of image force lowering due to the changed electric field at the interface?

24. Consider a GaAs n-type material of doping 1×10^{17} cm^{-3}. What is the barrier energy if one places a plane of acceptors of sheet density, 1×10^{10} cm^{-2}, 5×10^{10} cm^{-2}, 1×10^{11} cm^{-2}, 5×10^{11} cm^{-2}, and 1×10^{12} cm^{-2} in it?

25. Consider a GaAs p$^+$–n junction doped 2×10^{18} cm^{-3} in the p$^+$ region and 1×10^{17} cm^{-3} in the n region. For base widths of 1 μm, at what lifetime is the simplified expression that ignores recombination current suspect?

26. Consider an electric dipole layer of moment δ per unit area on either side of a junction. Show that this results in a discontinuous change in the electrostatic potential of δ.

27. A problem of particular interest in high-level injection is that of transport and storage in a p–i–n diode. The i-region is doped lightly n-type and the high injection condition exists in this lightly-doped region. Considering the electron and hole carrier concentration to be similar in the i-region, show that the carrier concentration at any position z in the i-region referenced to one of the junctions, for an i-region of width W_i, at a current density J, is

$$n(z) = \frac{\mathcal{L}J}{2q}\left\{ \frac{1}{\mathcal{D}_n}\frac{\cosh(z/\mathcal{L})}{\sinh(W_i/\mathcal{L})} + \frac{1}{\mathcal{D}_p}\frac{\cosh[(W_i - z)/\mathcal{L}]}{\sinh(W_i/\mathcal{L})} \right\}. \quad (4.241)$$

In this equation, the diffusion length $\mathcal{L} = \sqrt{\mathcal{D}\tau}$ where $\mathcal{D}$ is the ambipolar coefficient length given as

$$\mathcal{D} = \frac{2\mathcal{D}_n\mathcal{D}_p}{\mathcal{D}_n + \mathcal{D}_p}, \quad (4.242)$$

and τ is the high-level injection lifetime given by $\tau = \tau_{n0} + \tau_{p0}$.

28. What is the ratio between the diffusion capacitances of one-sided wide-base and narrow-base p–n junction diodes in terms of time constants?

29. Consider the analysis of p–n junctions, in low-level injection, but with the Dirichlet boundary conditions at the ohmic contacts. Find the minority carrier distribution for the homogeneously doped abrupt junctions. In the p-type region, the electron concentration, e.g., is given by

$$n(z) = n_{p0} + \frac{S\tau_n \sinh\left[(z + z_p)/\mathcal{L}_n\right] + \mathcal{L}_n \cosh\left[(z + z_p)/\mathcal{L}_n\right]}{S\tau_n \sinh\left[(z_p - w_p)/\mathcal{L}_n\right] + \mathcal{L}_n \cosh\left[(z_p + w_p)/\mathcal{L}_n\right]}. \quad (4.243)$$

In short devices, i.e., with $(z_p - w_p)/\mathcal{L}_n \ll 1$, show that the ohmic contacts are inefficient and lead to increased storage and smaller current if

$$S < \frac{\mathcal{D}_n}{z_p - w_p}. \quad (4.244)$$

Chapter 5

Metal–Semiconductor Field Effect Transistors

5.1 Introduction

The simplest and most common form of a field effect transistor in GaAs utilizes a metal gate to control the transport through the channel, and is called the metal–semiconductor field effect transistor (MESFET). It is a close analog of the junction field effect transistor and operationally similar in principle. The depletion region of the metal–semiconductor junction is used as a means to control the conductivity of the channel region. Since the extent of the depletion region can be modulated, the channel current can be modulated. It thus uses a capacitive coupling between the metal gate and the conducting channel for the modulation. Carriers flow from the source of the transistor to the drain of the transistor. With increasing drain bias, for a gate bias that allows for channel conduction, channel current and drain current saturate or change slowly. For large gate-length devices, with relatively low fields, this current saturation is associated with the pinching of the channel due to a gate-to-drain voltage that depletes the doped channel region. For short gate-length devices, saturation of the velocity of carriers is responsible for the current saturation. In either of the cases, a drain current that is weakly dependent on the drain bias, i.e., exhibiting a high output resistance or a low output conductance, occurs. This allows for a nearly constant drain current drive under quasi-static conditions that can be modulated by a gate voltage. Capacitive coupling allows for high input impedance, and current saturation results in a desirable current source. The device can therefore be used in analog and digital applications, with the compound semiconductors providing an improvement in the transport characteristics of the carriers, generally electrons.

In all these, the device is operationally similar to the junction field effect transistor. The metal gate, or the characteristics of the semiconductor, bring perturbations which are of significance in the behavior of the device, and ultimately in our use of the device. Many effects are common between the MESFET and the other field effect transistor of interest in compound semiconductors—the heterostructure field effect transistor (HFET). Our discussion of many of the characteristics, therefore, overlaps, and hence a few of the common phenomena have been chosen for emphasis in the discussion here, and a few of these will be discussed in the context of HFETs in Chapter 6.

Detailed differences in operational characteristics of different field effect transistors result from differences in junction characteristics, transport characteristics, and characteristics of control of mobile charge in the channel. Examples of differences in characteristics due to junctions include those due to the metal–semiconductor junction for the MESFET, p–n junction for the junction FET, or metal–insulator–semiconductor junction for the MISFET. Metal-semiconductor junctions, e.g., have a lower built-in junction voltage, and hence begin to conduct current at smaller forward biases than either the p–n junction or the metal–insulator–semiconductor junction. Examples of transport characteristics relate to differences in low field and high field behavior of carriers. Electrons in compound semiconductors generally exhibit large mobilities and, at low dopings, negative differential velocities. The former causes increased emphasis on effects related to saturation of velocity, and the latter leads to formation of accumulation and depletion regions of carriers along the channel because of the decrease in velocity with increasing longitudinal channel fields. Thus, a MESFET, which is based on a doped channel, and a MISFET, which is based on carrier transfer to a low doped channel region, exhibit differences resulting from the above. Examples of control of mobile charge are more immediate. A MESFET, e.g., is based on removal of carriers which would otherwise be present under quasi-neutral conditions in the material, while HFETs include examples of devices where this conducting charge may be induced. The effect of this basis of operation on parasitic resistances as well as on current continuity is very direct. One important consequence, e.g., is that MESFETs tend to have a saturated drain current that is closer in dependence to the square of the gate voltage while HFETs have a closer to a linear dependence.

Two additional and particularly important effects unique to compound semiconductors that will be discussed are that of sidegating (also called backgating), which is the ability of a potential distant from the device, arising from the trapping characteristics of the substrate, the substrate surface, and the interfaces, to influence the channel current; and the piezo-electric effect, which is the creation of a position-dependent charge density due to stress (e.g., from metallurgies and dielectrics) and the ionic character of compound semiconductors—an effect which again results in shifts in the electrical characteristics. We will also emphasize some of the interesting aspects and implications of the accumulation and depletion region that forms in the channel due to current continuity as well as due to negative differential mobility.

First, however, we will discuss quasi-static models, useful in predicting current–voltage characteristics and commonly applied in computer-aided design. We will discuss the principles that these are based on to understand their limitations, and we will selectively use rigorous small-signal high-frequency and two-dimensional analyses to emphasize their approximations. We will also discuss the off-equilibrium and transient behavior of devices, both in the short-channel limit, and emphasize the non–quasi-static aspects of operation in transient conditions.

The appeal of the computer-aided design models is their analytic or numerical simplicity, which brings together with it a more intuitive insight. However, these models must always be used with caution and with an understanding of their limitations.

We will discuss these models in order of increasing complexity and generality. Our discussion, therefore, extends beyond the simplest of these models to predict and evaluate the quasi-static behavior more accurately, and in the process also assimilate the limitations of simpler models. We will begin our discussion of computer-aided design models by discussing the long-channel device limit where constant mobility is an adequate approximation, then discuss the short-channel device limit where velocity is approximated as being constant along the channel, and then develop a model where both of these limits are included.

5.2   Analytic Quasi-Static Models

We have remarked that the appeal of computer-aided design models is their analytic or numerical simplicity, which brings together with it a more intuitive insight. The simplest form of such a model assumes a constant mobility operation of the transistor. Thus, for a practical semiconductor, it implies operating in a bias region where the electric fields remain low and approximately in the constant mobility region of the semiconductor velocity-field characteristics. Most semiconductors exhibit deviation from this behavior at electric fields exceeding 1×10^3 V.cm^{-1}. For operating voltages of a few volts, this requires the device gate-length to be larger than

20 μm. This electric field, the channel electric field, can of course be much smaller than the field in the depletion region that modulates the channel width and hence conductance. A particularly useful simplification in the analysis under conditions of a large difference in the electric field between the depletion and the channel region is the gradual channel approximation.

5.2.1 Gradual Channel Approximation

Since the current flow in the FET occurs due to application of drain-to-source voltage across the channel with the modulation of channel conductance by the gate-to-source voltage, one would expect the problem to be complicated with electric fields being two-dimensional in the structure. Real situations quite often lend themselves to approximations which ease the description of the physical problem. Figure 5.1 shows underneath the gate of a MESFET, in the mirrored part of a symmetric configuration, the conduction band edge and the velocity of carriers in the conducting part of the channel. The channel current flow occurs between the source and the drain, and the gate potential controls the constriction of the channel and hence the conductance. The channel current flow occurs in the z-direction, with the channel cross-section and the electric field $\mathcal{E}_z$ varying with z. In addition, there is a component of the electric field $\mathcal{E}_y$ due to the gate potential. In a symmetric structure, $\mathcal{E}_y$ goes to zero at the center (shown at the 0.25 μm in Figure 5.1(a)), but increases towards the surface. If, in the conducting part of the channel, this electric field stays very much smaller than $\mathcal{E}_z$, then the equi-potential surfaces are planes nearly orthogonal to z. The problem now reduces to one dimension, in the z-direction of the channel, and in the y-direction in the depleted region underneath the gate. An accompanying relationship of this one-dimensional framework, which is referred to as the gradual channel approximation, is that the electric field $\mathcal{E}_z$ should vary in the channel region slowly with z, and hence the carrier density should also vary slowly in the channel.

5.2.2 Constant Mobility Model

We first consider a geometry where this approximation is very accurate, the case of a device with a long gate-length, i.e., a long-channel device where the fields are sufficiently small that a constant mobility may be assumed. Figure 5.2 shows variation of the depletion region in our device. The figure also shows geometric parameters to be used in our analysis. In the long-channel model, the gradual channel approximation is valid because while the gate and drain voltages are comparable, the channel length is significantly larger than the thickness of the conducting layer. We can get

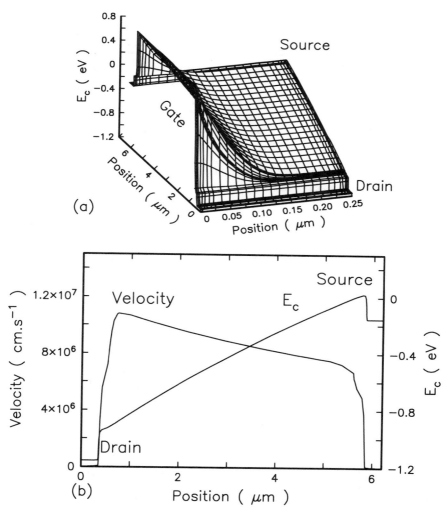

Figure 5.1: Surface plot of the conduction band edge of a 5 μm gate length GaAs MESFET is shown in (a). (b) shows the band edge and the velocity of carriers at a cross-section in the conducting part of the channel.

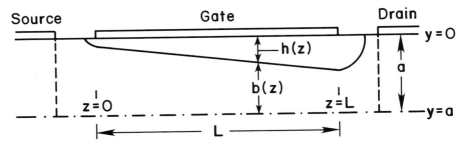

Figure 5.2: A schematic of a long-channel device with geometric parameters used in the long-channel model.

closed-form solutions if we further assume abrupt depletion and ignore diffusive current. The latter follows from a small channel electric field, which implies small deviations from quasi-neutrality and therefore a small gradient in carrier concentration. The depletion layer width $h(z)$ varies slowly along the channel. The electric field in the depletion region is along the y-axis and the electric field in the channel region along the z-axis.

In the depletion region, using depletion approximation,

$$-\frac{d^2V}{dy^2} = \frac{\rho(y)}{\epsilon_s} = \frac{qN_D}{\epsilon_s} \tag{5.1}$$

for uniform doping, and hence the depletion region width $h(z)$ at any position z along the channel is given by

$$h(z) = \left\{ \frac{2\epsilon_s\left[V(z) + V_G + \psi_{j0}\right]}{qN_D} \right\}^{1/2}, \tag{5.2}$$

where $V(z)$ is the channel voltage at z w.r.t. the source, ψ_{j0} is the built-in voltage of the metal–semiconductor junction ($\psi_{j0} = \phi_M - V_n$), and V_G is the absolute gate voltage. The square-root dependence in voltage above, or other similar parametric dependence in voltage, occurs quite often in these analytic models in field effect transistors. We will encounter a similar form in cubic root in our analysis of HFETs. These parameters form a convenient basis for deriving and writing expressions for current. These parameters characterize identifiable features of the device or their normalizations. $h(z)$ can be used as a parameter itself, or in its normalized form $u(z) = h(z)/a$, they are related to the width of the depletion region underneath the gate. We have

$$h(z = 0) = \left[\frac{2\epsilon_s(V_G + \psi_{j0})}{qN_D} \right]^{1/2} \tag{5.3}$$

at the source end, and

$$h(z = L) = \left[\frac{2\epsilon_s(V_D + V_G + \psi_{j0})}{qN_D}\right]^{1/2} \tag{5.4}$$

at the drain end. In a constant mobility long-channel model, the highest electric fields are limited. For example, 5 V across a 100 μm channel leads to an average field of 500 V.cm^{-1}, which is sufficiently small that a constant mobility may be assumed throughout the device. The saturation of the current in such a device occurs when the drain end of the gate pinches off, and is depleted of carriers. This condition, termed channel pinch-off, occurs when

$$h(z = L) = a = \left[\frac{2\epsilon(V_D + V_G + \psi_{j0})}{qN_D}\right]^{1/2}. \tag{5.5}$$

The pinch-off voltage therefore, is defined as the parameter V_p, where

$$V_p = \frac{qN_D a^2}{2\epsilon_s}. \tag{5.6}$$

This is the magnitude of the voltage between gate and source to shut off current flow in the channel, since the extent of the gate-source depletion width is now a, and hence carriers do not enter the channel. It is also the voltage between the gate and the drain when current saturation occurs, because it pinches off the drain opening in the channel. These voltages are identical due to constant mobility and gradual channel approximation. In general, such as for the models we discuss later, these voltages can be different. The gate voltage required to shut off current flow is usually referred to as the threshold voltage, and the voltage required to saturate the current is usually referred to as pinch-off voltage. We will return to the questionable practice of referring to this as a pinch-off voltage, later in our discussion of the combined constant mobility and constant velocity model.

In a first-order approximation, at any position z along the channel, we can evaluate the drain current assuming that the diffusion current is negligible. This is quite valid over much of the channel, but not at the drain end of the gate. A pinch-off of the channel maintains current continuity because a large diffusion current flows through this region. The gradient of carrier concentration is large at the drain end. We assume that this condition is met to satisfy current continuity. In other parts of the device under the gate, drift current is dominant, and hence one may write

$$I_D = qN_D\mu W\left[a - h(z)\right]\frac{dV}{dz} \tag{5.7}$$

where the drain current enters the device. Since

$$h^2(z) = \frac{2\epsilon_s(V + V_G + \psi_{j0})}{qN_D}, \tag{5.8}$$

we obtain

$$2hdh = \frac{2\epsilon_s}{qN_D}dV, \tag{5.9}$$

and hence

$$I_D = WqN_D\mu(a - h)\frac{qN_D}{\epsilon_s}h\frac{dh}{dz}. \tag{5.10}$$

We now employ the continuity of this current in the channel to eliminate the position-dependent variable $h(z)$ by integrating over the local variable. This is a common technique that is useful in problems involving current continuity, e.g., it will also be applied in HFETs and in the collector depletion region of a bipolar transistor. Application of it along a path where current continuity is maintained allows for elimination of a local variable. Continuity allows one to obtain as a result the current in terms of variables at the source or drain end in this problem. Integrating,

$$\int_0^L I_D dz = \frac{(qN_D)^2\mu W}{\epsilon_s}\int_{h(0)}^{h(L)}(a - h)hdh, \tag{5.11}$$

and

$$I_D = \frac{W\mu q^2 N_D{}^2 a^3}{6\epsilon_s L}\left\{\frac{3}{a^2}\left[h^2(L) - h^2(0)\right] - \frac{2}{a^3}\left[h^3(L) - h^3(0)\right]\right\}. \tag{5.12}$$

This is the form of the current, in constant mobility and long-channel approximation, in terms of structural variables and $h(0)$ and $h(L)$, which depend on biases applied at the terminals and parameters of the device.

The expression above can be written in terms of voltages and other normalized parameters, to make it appear more direct. Using the normalizations,

$$I_{norm} = \frac{W\mu q^2 N_D{}^2 a^3}{6\epsilon_s L}, \tag{5.13}$$

and

$$u = \frac{h(z)}{a} = \left(\frac{V + V_G + \psi_{j0}}{V_p}\right)^{1/2},$$

$$u(0) = \frac{h(0)}{a} = \left(\frac{V_G + \psi_{j0}}{V_p}\right)^{1/2},$$

$$\text{and} \quad u(L) = \frac{h(L)}{a} = \left(\frac{V_D + V_G + \psi_{j0}}{V_p}\right)^{1/2}, \tag{5.14}$$

we can express the drain current as

$$
\begin{aligned}
I_D &= I_{norm} \left\{ 3 \left[u^2(L) - u^2(0) \right] - 2 \left[u^3(L) - u^3(0) \right] \right\} \\
&= I_{norm} \left\{ 3 \frac{V_D}{V_p} - \frac{2}{V_p^{3/2}} \left[(V_D + V_G + \psi_{j0})^{3/2} - (V_G + \psi_{j0})^{3/2} \right] \right\}.
\end{aligned}
$$

(5.15)

These expressions relate the current up to the point of pinch-off of the channel. At this bias, which occurs when $h(z = L) = a$ or equivalently $u(y = L) = 1$, the drain current saturates and then remains constant independent of the drain bias. This current I_{DSS} is given by

$$
I_{DSS} = I_{norm} \left[1 - 3 \left(\frac{V_G + \psi_{j0}}{V_p} \right) + 2 \left(\frac{V_G + \psi_{j0}}{V_p} \right)^{3/2} \right],
$$

(5.16)

which occurs at a drain-to-source voltage V_{DSS} given by

$$
V_{DSS} = V_p - V_G - \psi_{j0} = \frac{q N_D a^2}{2\epsilon} - V_G - \phi_M + \frac{kT}{q} \ln \left(\frac{N_C}{N_D} \right).
$$

(5.17)

The quasi-static expression for drain current in terms of the gate and drain biases is complete, and we may determine the quasi-static device parameters using these expressions. As an example, the transconductance follows from this analysis as:

$$
g_m = \frac{\partial I_D}{\partial V_G} = \frac{2W \mu q a N_D}{L} \left[u(L) - u(0) \right],
$$

(5.18)

and the output conductance follows as

$$
g_d = \frac{\partial I_D}{\partial V_D} = \frac{2W \mu q a N_D}{L} \left[1 - u(L) \right]
$$

(5.19)

in the bias region before the channel pinches off. This region is sometimes referred to as the triode region of operation because of its poorer output conductance. For biases beyond channel pinch-off, i.e., for drain voltages exceeding V_{DSS} (V_{DSS} being dependent on the gate bias), the transconductance in this model stays constant, and the output conductance is zero. Because of the latter, the output characteristics beyond current saturation are said to be pentode-like. The transconductance can be found by finding its magnitude at V_{DSS}. So, in the triode region, the transconductance

$$
g_m = \frac{2W \mu q a N_D}{L} \left[\left(\frac{V_D + V_G + \psi_{j0}}{V_p} \right)^{1/2} - \left(\frac{V_G + \psi_{j0}}{V_p} \right)^{1/2} \right],
$$

(5.20)

and in the pentode region, the transconductance

$$g_m = \frac{2W\mu q a N_D}{L} \left[1 - \left(\frac{V_G + \psi_{j0}}{V_p} \right)^{1/2} \right]. \tag{5.21}$$

Note that in the constant mobility assumption, the derivation of the behavior is quite straightforward, and even for complicated impurity distribution functions (e.g., Gaussian, complementary error function, etc.) simple results can be derived. However, these are valid only as long as the constant mobility assumption is a good one. For GaAs devices, this requires maintaining electric fields below 1000 V/cm. So for a device with 5 V drain bias the gate-length should be maintained at $5/1000 = 50$ μm. Devices used in circuits are significantly smaller than this.

We will therefore consider the other limit of the behavior of devices— that based on a constant high field velocity throughout the channel. This follows our discussion in the introduction.

5.2.3   Constant Velocity Model

For shorter gate-lengths, with average fields exceeding 3×10^3 V.cm^{-1}, the velocity of the carriers is similar to or larger than the saturation velocity v_s, as shown in Figure 5.3. Thus, carriers transit with a velocity that is on an average higher than the saturated velocity. The simplest method of analyzing this problem would be to assume a source end velocity v_s', which is close to the saturation velocity or slightly exceeding it, because the field is the smallest at the source end and below the peak-velocity field. We may also ignore diffusive currents because the gradient in carrier concentration remains low; velocity saturation in a constant-channel cross-section requires the absence of gradient in carrier concentration. The consequence of this nearly constant carrier concentration is that even though this velocity occurs in a region where the linear velocity–field relationship is largely maintained, since it does not change appreciably any further, we may incorporate a constant velocity v_s' in evaluating the current. Thus, for gate-lengths ≈ 1 μm of interest, instead of a constant mobility assumption, we now resort to a constant velocity assumption to crudely estimate the current. This current I is

$$I \approx q n v_s' \left[a - h(0) \right] W. \tag{5.22}$$

Under quasi-static conditions this particle current is continuous, and hence this is also the drain current of the device.

$$I_{DS} \approx q W v_s' \left[a - h(0) \right] N_D$$

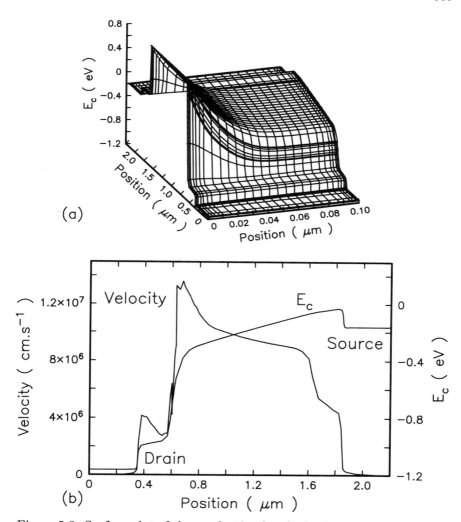

Figure 5.3: Surface plot of the conduction band edge in a 1 μm gate-length 5×10^{17} cm^{-3} doped GaAs MESFET is shown in (a). The conduction band edge and the velocity of carriers are shown in (b) in a conducting part of the channel. This figure should be compared with the 5 μm gate-length MESFET shown in Figure 5.1. Both devices are biased under identical conditions and have similar threshold voltages.

$$\propto \ I_{norm} \left[1 - \left(\frac{V}{V_p} \right)^{1/2} \right]. \qquad (5.23)$$

This expression is appealing in simplicity and extendable to obtain the transconductance of the device, since the dependence of the source-end depletion width $h(0)$ is known as a function of gate-source bias. Note that the current changes with a square-root power dependence on voltage because of the direct effect of the modulation of the width of the channel.

We have assumed that a field exists at the source end that is sufficient to cause a large velocity, comparable to the saturated velocity, even though it is largely in the constant mobility region of the device. The accuracy of this constant velocity model is limited by the extent of this constant mobility section. So long as it is short, the analysis is meaningful. However, if the extent of this region is large, the velocity in the drift current expression is an abstraction of a hypothetical average, meaningful only as a fitting parameter following an experiment, and limited in the predictive ability desired from models. So, this is an approximate relationship, and is only applicable in the region of device operation where the drain current has saturated. However, it serves well to relate the approximate variation expected in the drain current in response to a change in gate voltage for an arbitrary doping profile within the channel, or for variations of other parameters such as the velocity $v_s' \approx v_s$. For many digital circuits, there is a preference to have devices with constant but large transconductance. Hence, there has historically been an interest in doping profiles that give linear I_{DS}–V_G characteristics. From the above relationship, it can be seen that this is connected with how the depletion width $h(0)$ changes with the gate voltage. A smaller change in the depletion width with bias leads to a smaller change in the transconductance also. As we shall see later, HFETs based on inducing the conducting charge in a two-dimensional electron gas channel come quite close to having a constant transconductance because of the nearly constant depth and the linear dependence of this charge through capacitive coupling of the gate. For MESFETs, this would require a large doping of small width at the edge of the depletion region with a small doping everywhere else. This kind of doping pattern is referred to as planar doping. The device is fully turned on when this region is undepleted and turned off when it is depleted. Since the position of the depletion edge does not change substantially between these two,[1] the transconductance is

---

[1] Screening is of particular importance to this problem. It was been shown earlier that a plane of doping leads to a carrier distribution with exponential tails characterized by the Debye length. This Debye length also characterizes the shape of the carrier distribution at the edge of the depletion region. Hence the order of change in the depletion width between the on and off conditions for planar doping is of the order of Debye screening

nearly constant. Problem 2 considers the behavior of the transconductance for various channel doping profiles.

Because of the constant velocity assumption, different doping profiles in the channel should lead to a transconductance–voltage relationship which follows the behavior of capacitance–voltage relationship, which relates the controlled charge to the gate bias. Thus, the gate-to-source capacitance dependence on gate bias is related to the transconductance behavior or the dependence of drain current on the gate bias. A constant capacitance implies a linear charge control in voltage, and hence when the charge moves with a constant velocity, a linear I_{DS}-V_G characteristic is obtained. The example of planar doping pins the depletion region at the doping plane and hence leads to the nearly constant capacitance–voltage and transconductance–voltage behavior. In order to assure pinch-off of the channel of the device, only a limited amount of charge compatible with the breakdown voltage of the gate can be placed in the doped region. The drawback of such a profile is that it leads to large source and drain resistance because the undoped region cladding the doped planar region makes it difficult to form low access regions to the source and drain of the device. HFETs, which we commented on as another example of linear transconductance devices, have the desired attribute of forming this planar region using a two-dimensional carrier gas, along with a simplified low resistance implementation.

5.3 Constant Mobility with Saturated Velocity Model

We have now developed an approximate description of the behavior of the MESFET under two conditions: under a low longitudinal electric field in the channel where we applied the constant mobility model together with the gradual channel approximation, and under a high longitudinal electric field in the channel where we applied a constant velocity model. To develop a general framework we must include both of these in a unified model so that it may have a general applicability for various gate-lengths as well as for various bias conditions. Even for small gate-lengths our approximate constant velocity model is still limited to conditions of high fields and hence large drain bias conditions in the device. It does not, therefore, predict the current–voltage relationship before the onset of current saturation. At low voltages and hence low electric fields, the constant mobility model would be more appropriate. So, we now go a step further and combine the constant mobility and constant velocity assumptions in a single device. Since this

length.

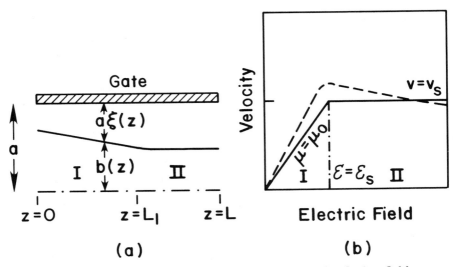

Figure 5.4: Device cross-section (a) and the assumed velocity–field curve (b) for the PHS model as an approximation to the velocity–field curve at typical channel dopings exceeding mid-10^{17} cm^{-3} in GaAs. In region I of both parts, constant mobility is assumed. In region II of both parts, a constant velocity is assumed.

would develop a model that incorporates both the low-field constant mobility feature and the high-field constant velocity feature of carrier transport, it should have more general application.

We now consider the entire velocity–field behavior of the carrier by making a constant mobility and constant velocity approximation of the two limits of the behavior. This approximation of the velocity–field curve and transport behavior in the channel is shown in Figure 5.4. Thus, we will assume the constant mobility to be an adequate approximation in the part of the device channel at the source end where the fields are low, and we will assume the constant velocity to be an adequate approximation at the drain end of the channel where the fields are high. Our discussion follows the analysis of Pucel, Haus, and Statz, and will be referred to as the PHS model.[2] The saturation in channel current occurs when the field towards

---

[2]R. A. Pucel, H. A. Haus, and H. Statz, "Signal and Noise Properties of Gallium Arsenide Microwave Field-Effect Transistors," in *Advances in Electronics and Electron Physics*, **38**, Academic Press, N.Y. (1975) also has a treatment of noise and a historic introduction to early MESFET models for computer-aided design. The PHS model is an example of quasi-static model. For a lucid discussion of the general approach that is

the drain end becomes high enough that the carriers begin moving with a constant high field limit velocity—the saturated velocity. The channel current saturation, or equivalently the drain current saturation, occurs due to the saturation in the velocity of carriers. Recall that in the constant mobility model it occurs due to the pinching of the channel. These are quite different in nature. Current saturation, with a constant cross-section for flow of carriers, implies a constant carrier distribution for current continuity in the former and a constant carrier gradient in the latter. The latter is important in long gate-length devices where the current density is small and hence the gradient in carrier concentration is feasible without violating the particle velocity limits from the characteristics of the semiconductor. In small gate-length devices, the fields at the drain end are large enough, the gradual channel approximation is no longer applicable, and the carriers move with velocities close to the saturated velocity. Once this channel current saturation occurs, the gate depletion width accommodates itself longitudinally, towards the source and the drain, to allow for the drain current to flow through the undepleted part of the channel.

We will consider the device geometry shown in part (a) of Figure 5.4. We ignore the current in the substrate of the device as well as its effect on the flow in the channel itself by considering a symmetric section of a device with the second half being a mirror reflection of the bottom edge of the device. In our earlier analysis, we had ignored the substrate boundary; implicitly, this is the geometry that we analyzed. Our velocity–field curve is a two section model (Figure 5.5) that is an approximation of the semiconductor velocity–field curve. This figure also shows the variation in channel opening, Here, at low fields a constant mobility is used, and at fields $\mathcal{E} > \mathcal{E}_s$ a constant velocity is used. We obtain continuity in the slope of the derived characteristics by matching the boundary conditions at the point at which $\mathcal{E} = \mathcal{E}_s$. This will also serve for us as a demonstration of the use of boundary conditions in matching sections with the constraint of current continuity. Among other parameters for the device, a is the channel thickness, b is the channel opening, and $\mathcal{V}$ is the a normalized potential.

In the region of length L_1, where $\mathcal{E} < \mathcal{E}_s$, gradual channel approximation is assumed to be valid; this field is low enough that it is significantly smaller than the transverse gate field and we employ a constant mobility. In the region beyond, of length L_2, the carriers are assumed to move with their limit velocity, the saturated velocity velocity v_s. The point L_1, therefore, defines the spatial point of onset of velocity saturation. We will denote the potential of this point by V_p, in conformity with the earlier analysis, since

---

the basis of PHS model see the treatment on junction field effect transistors in R. S. C. Cobbold, *Theory and Applications of Field-Effect Transistors*, Wiley-Interscience, N.Y. (1970). Cobbold's book also contains a detailed treatment of small-signal modelling.

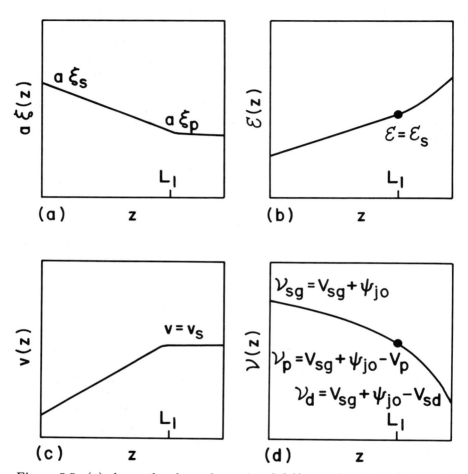

Figure 5.5: (a) shows the channel opening ($\xi(z)$) as a function of channel position, (b) shows the electric field ($\mathcal{E}(z)$) as a function of position, (c) shows the velocity of carriers ($v(z)$) as a function of position, and (d) shows the normalized potential ($\mathcal{V}(z)$) as a function of position along the channel. These figures are shown schematically for a doping of 2.5×10^7 cm.s^{-1}.

drain current saturation is intimately related to this velocity saturation.[3] We assume the source at ground potential. The potential of the channel referred to the gate is

$$\mathcal{V}(z) = V_{sg} + \phi_M - V(z). \tag{5.24}$$

This is a translated variable, whose magnitude at the source end, the drain end, and the point at which the carrier velocity saturates are

$$
\begin{aligned}
\mathcal{V}_s &= V_{sg} + \phi_M, \\
\mathcal{V}_d &= V_{sg} + \phi_M - V_{sd}, \\
\text{and} \qquad \mathcal{V}_p &= V_{sg} + \phi_M - V_p.
\end{aligned}
\tag{5.25}
$$

In the constant mobility model, we employed a variable h proportional to the square root of potential—this term represented the depletion depth. It is useful to employ a similar square-root term here. We will employ a normalization factor of the channel thickness a. These normalization factors, the reduced potentials,[4] are the following at the source end, the drain end, and at the point of velocity saturation:

$$
\begin{aligned}
\xi_s &= \left(\frac{\mathcal{V}_s}{\mathcal{V}_{00}} \right)^{1/2}, \\
\xi_d &= \left(\frac{\mathcal{V}_d}{\mathcal{V}_{00}} \right)^{1/2}, \\
\text{and} \qquad \xi_p &= \left(\frac{\mathcal{V}_p}{\mathcal{V}_{00}} \right)^{1/2},
\end{aligned}
\tag{5.26}
$$

with the reduced potential at any position z along the channel as

$$\xi(z) = \left(\frac{\mathcal{V}(z)}{\mathcal{V}_{00}} \right)^{1/2}. \tag{5.27}$$

In these expressions, the normalization potential is the potential

$$\mathcal{V}_{00} = \frac{qN_D}{2\epsilon_s} a^2, \tag{5.28}$$

---

[3]In the constant mobility model, the current saturation occurs due to a pinch-off of the channel where few carriers exist, and current transport occurs via diffusion in this pinched region. In the constant velocity model, the channel does not physically pinch off. The velocity saturates and a large carrier density exists that supports the drift current. These are distinctly different. The subscript p, whose origin is in this "pinch-off" term, should not be confused with a physical pinch-off of the channel.

[4]We employ the symbol ξ for this. It should not be confused with the symbol for quasi-Fermi energy.

which is the potential associated with full channel depletion in gradual channel approximation.

Note that because of the square root of $\mathcal{V}$, the terms ξ_s, ξ_d, ξ_p, and $\xi(z)$ correspond to the depletion region width normalized by the channel thickness, since, in the depletion region, the fields largely terminate on the gate electrode irrespective of the behavior in the channel region. The parameter ξ, therefore, corresponds to the parameter h used in our earlier constant mobility calculations. The manner in which we assure current continuity, and include velocity saturation and hence the existence of large longitudinal fields in the channel, prevents it from being identical, and allows us to include the velocity saturation effect. In the constant mobility section of the device, using gradual channel approximation,

$$\mathcal{V}(z) = \mathcal{V}_{00}\left[1 - \frac{b(z)}{a}\right]^2 \tag{5.29}$$

and

$$\xi(z) = 1 - \frac{b(z)}{a}. \tag{5.30}$$

Also, at any cross-section, the current I, which is also the drain current I_D because of current continuity, is given by

$$I = I_D = b(z)Wq\mu_0 N_D \frac{\partial \mathcal{V}}{\partial z}, \tag{5.31}$$

where W is the gate width. This gives, in the constant mobility section of the device,

$$\int_0^{L_1} I_D dz = 2aq\mu_0 N_D W \mathcal{V}_{00} \int_{\xi_s}^{\xi_p} (1 - \xi)\xi d\xi. \tag{5.32}$$

Note,

$$
\begin{aligned}
b(z) &= a\left[1 - \xi(z)\right], \\
\text{and} \quad 2\xi(z)d\xi(z) &= \frac{d\mathcal{V}(z)}{\mathcal{V}_{00}},
\end{aligned} \tag{5.33}
$$

so

$$I_D = \frac{aq\mu_0 N_D W \mathcal{V}_{00}}{L_1}\left[\xi_p{}^2 - \xi_s{}^2 - \frac{2}{3}\left(\xi_p{}^3 - \xi_s{}^3\right)\right]. \tag{5.34}$$

This relationship is identical to that derived for the constant mobility model since it is based on identical assumptions. The difference between the present model and the constant mobility model is that the latter assumes constant mobility is valid over the entire channel length. We may introduce g_o as a conductance parameter per unit width,

$$g_o = aq\mu_0 N_D, \tag{5.35}$$

giving

$$I_D = \frac{g_o W \mathcal{V}_{00}}{L_1} \left[\xi_p{}^2 - \xi_s{}^2 - \frac{2}{3} \left(\xi_p{}^3 - \xi_s{}^3 \right) \right]. \tag{5.36}$$

In the second region, where the constant velocity assumption is applied,

$$v_s = \mu_0 \mathcal{E}_s. \tag{5.37}$$

Since velocity is constant and the field in the conducting section is longitudinal, in the presence of a constant channel thickness and hence constant depletion region width, current continuity can only be maintained by a constant carrier concentration. The current at the point of velocity saturation is

$$
\begin{aligned}
I_D &= q\mu_0 \mathcal{E}_s N_D W a (1 - \xi_p) = g_o W \mathcal{E}_s (1 - \xi_p) \\
&= I_{norm}(1 - \xi_p).
\end{aligned} \tag{5.38}
$$

The normalized current $I_{norm} = q v_s N_D W a$ is the drain current if the channel is completely open and carriers move at saturated velocity. This is the maximum current that can be obtained from the device. Current continuity between the constant mobility and constant velocity sections implies

$$\frac{g_o W \mathcal{V}_{00}}{L_1} \left[\xi_p{}^2 - \xi_s{}^2 - \frac{2}{3} \left(\xi_p{}^3 - \xi_s{}^3 \right) \right] = g_o W \mathcal{E}_s (1 - \xi_p), \tag{5.39}$$

and hence the length L_1 is given by

$$L_1 = \frac{\mathcal{V}_{00}}{\mathcal{E}_s(1 - \xi_p)} \left[\xi_p{}^2 - \xi_s{}^2 - \frac{2}{3} \left(\xi_p{}^3 - \xi_s{}^3 \right) \right]. \tag{5.40}$$

We write this as

$$L_1 = L \frac{1}{\alpha(1 - \xi_p)} \left[\xi_p{}^2 - \xi_s{}^2 - \frac{2}{3} \left(\xi_p{}^3 - \xi_s{}^3 \right) \right], \tag{5.41}$$

where

$$\alpha = \frac{\mathcal{E}_s L}{\mathcal{V}_{00}} \tag{5.42}$$

is a dimensionless potential parameter that the PHS model calls saturation index. It is inversely related to the degree of importance of saturation of velocity in the determination of the device behavior. Recall that $\mathcal{V}_{00}$, for a constant mobility model, is the voltage for current saturation. For large α, the parameter $\mathcal{V}_{00} \ll \mathcal{E}_s L$; hence typically applied voltages are less than the voltages required to achieve velocity saturation across a significant fraction of the channel lengths. So, a large α results in constant mobility–dominated behavior of the device. If α is small, $\mathcal{E}_s L < \mathcal{V}_{00}$, and hence, for potentials typical of current saturation, in a substantial part of the device, the carriers travel at saturated velocities.

Let us now consider the channel potential for a given current I_D through the device. Consider the first section of the device, where the gradual channel approximation is applicable. The potential in the channel at the point of onset of velocity saturation is

$$\begin{aligned} V_p &= -(\mathcal{V}_p - \mathcal{V}_s) \\ &= -\mathcal{V}_{00}(\xi_p{}^2 - \xi_s{}^2). \end{aligned} \tag{5.43}$$

The field in the depleted region is transverse, and the field in the channel with a constant mobility conduction is longitudinal. In the second section, both in the depleted and constant velocity conduction region, the fields are two-dimensional. For this section, the potentials can not be derived in as straightforward a fashion, since gradual channel approximation is no longer applicable. We may not use the one-dimensional Poisson's equation applied in the conducting region of the first section, which followed from the existence of only a longitudinal field in the constant mobility section. In the constant velocity section, the fields are two-dimensional, and the electrostatic potential ψ is a function of both y and z. A rigorous solution would require us to solve the problem subject to the boundary conditions of continuity in potential and field at the intersection of the constant mobility and constant velocity regions for all y's and also to satisfy the appropriate approximations of boundary conditions at the gate–semiconductor interface, at the drain contact interface with the second region, and at the line of mirror symmetry at the bottom.

Our solution will be approximate. We break the problem in two parts, emphasizing the two electrostatically distinct parts. Let ψ_2 be the electrostatic potential associated with the large charge density qN_D of the depletion region lacking in mobile charge, and let ψ_1 be the electrostatic potential associated with the conducting region filled with mobile charge. The superposition of these two potentials, $\psi = \psi_1 + \psi_2$, is the electrostatic potential of the second region with ψ_1 dominating in the conducting channel part and ψ_2 in the depletion region part. The potential ψ_2 is still approximately parabolic in y because the depletion charge still largely terminates on the gate electrode. In the conducting region of the channel, carriers move at the constant saturated velocity, carrier concentration is constant to maintain current continuity, and hence the longitudinal electric field is associated with charge at the drain electrode. So, in the conducting part of the second region,

$$-\nabla^2\psi_1 = \frac{q}{\epsilon}(N_D - n) \approx 0 \qquad \text{for} \qquad L_1 < z < L, \tag{5.44}$$

i.e., ψ_1 satisfies Laplace's equation. We employ the boundary conditions of continuous longitudinal electric field and potential at $z = L_1$, and vanishing transverse field at $y = 0$, the central cross-section of the device. Mathematically,

$$\begin{aligned} \psi_1(z = L_1^-) &= \psi_1(z = L_1^+) \\ \text{and} \qquad \left.\frac{\partial\psi_1}{\partial z}\right|_{z=L_1^-} &= \left.\frac{\partial\psi_1}{\partial z}\right|_{z=L_1^+}. \end{aligned} \tag{5.45}$$

For the general case, the solution to Poisson's equation for the above problem is the sum of the solution to Laplace's equation (ψ_1) and the solution to Poisson's equation with fixed ionized charge (ψ_2). The solution to Laplace's equation is a sum of products of exponentials in the two orthogonal directions. Since, in the y-direction, the field vanishes at the surface, and since it builds up towards the drain electrode (the charge on the drain electrode sustains this field), intuitively one would expect the solution to be a harmonic function that peaks at $y = a$ and vanishes at $y = 0$, i.e., a cosine function, and an exponential function in the z direction. The solution for potential should be of the form

$$\psi_1(y,z) = \sum_{m=0}^{m=\pm\infty} A_m(-1)^{2m+1} \cos\left[\frac{m\pi(y-a)}{2a}\right] \exp\left[-\frac{m\pi(z-L_1)}{2a}\right] + B, \tag{5.46}$$

where the A_m's and B are constants.

Strictly speaking, we should have tried to force the cosine function to zero at the depletion region edge since the mobile charge nearly vanishes there. This would lead to a very complex analytic formulation. In this approximate solution we wish to emphasize the accuracy at $y = a$ so that we may calculate the potential accurately along this line of symmetry. The calculation of the potential at $y = a$ does not get severely affected by the approximation in the cosine function. So, an approximate result for the potential drop and length of the region can be found by considering the behavior at $y = a$. In the above series, the terms corresponding to $m = 0$ and $m = \pm 1$ have the largest contribution for this cross-section of symmetry. This contribution,

$$\psi_1(y, z) = -\frac{2a}{\pi} \mathcal{E}_s \cos\left[\frac{\pi(y - a)}{2a}\right] \sinh\left[\frac{\pi(z - L_1)}{2a}\right] + \psi(L_1), \qquad (5.47)$$

where $\psi(L_1)$ is the potential of the channel derived as a solution from the analysis of first region, may be seen as the part of the solution that corresponds to a point drain contact at $y = a$ and $z = L$, and the resultant variation in $\psi(y, z)$ is most accurate at the cross-section of symmetry with large deviations occurring further away. At $y = a$ and $z = L_1$, the longitudinal electric field is $\mathcal{E}_s$ and the potential is $\psi(L_1)$. Since the potential is symmetric in the y-direction around $y = a$, the transverse field vanishes at $y = a$. Also, because of the cosine term, the longitudinal field drops off towards the depletion region edge and into the depletion region. The approximation here is clear; in the conducting region, while we actually expect the potential variation in the y-direction to be negligible, this potential describes it as changing appreciably. However, at $y = a$, this potential is quite accurate. So, this potential should be considered as the approximate solution in the conducting undepleted region at $y = a$, and we will make appropriate evaluations of voltage drops, etc., at $y = a$ only.

In the depletion region, the electric field is mostly transverse, the charge is N_D^+, and the total potential (sum of contribution from the depletion region and the conducting region) increases parabolically towards the gate. Our rationale in adopting the above approach to obtain relatively simple results is that if we model behavior at $y = a$ accurately enough, maintaining current continuity, then the results will not be too different from an analysis considering more terms in the harmonic series expansion. Were we to adopt the latter approach, we could also consider the potential contribution of the charge in the depletion region too, by considering the boundary conditions in terms of equi-potential surface at the gate, no orthogonal field at $y = a$, continuity of potential and electric field at $z = L_1$, and equi-potential surface at the drain electrode. This would be unwieldy.

The longitudinal channel voltage drop in our present approximation can now be obtained as follows

$$V_{II} = -V_{00}\frac{2}{\pi}\left(\frac{a}{L}\right)\alpha\sinh\left(\frac{\pi L_2}{2a}\right), \qquad (5.48)$$

hence,

$$V_{sd} = -V_{00}(\xi_p{}^2 - \xi_s{}^2) - \frac{2}{\pi}V_{00}\frac{a}{L}\alpha\sinh\left(\frac{\pi L_2}{2a}\right)$$

$$= -\mathcal{V}_{00} \left[(\xi_p{}^2 - \xi_s{}^2) + \frac{2}{\pi} \left(\frac{a}{L} \right) \alpha \sinh \left(\frac{\pi L_2}{2a} \right) \right], \qquad (5.49)$$

where

$$L_1 = L \frac{\xi_p{}^2 - \xi_s{}^2 - \frac{2}{3}(\xi_p{}^3 - \xi_s{}^3)}{\alpha(1 - \xi_p)}, \qquad (5.50)$$

and

$$L_2 = L - L_1. \qquad (5.51)$$

Given V_{sd}, V_{sg}, and L, we know the source end opening $a(1 - \xi_s)$. Device and material parameters imply knowledge of the long-channel pinch-off normalized potential $\mathcal{V}_{00}$, the saturation index α, and the channel thickness a. This allows us to determine the drain end opening $a(1 - \xi_p)$. From this the device characteristics follow.

Figure 5.6 shows the output characteristics of GaAs MESFETs with $1\mu m$ and $10\ \mu m$ gate-lengths. The dashed lines of this figure show the output characteristics in the constant mobility limit for zero gate bias. The PHS model shows the lowest current and a finite output conductance throughout the bias range. The constant mobility model shows zero output conductance in the current saturation region, and a much higher current. Since this model ignores any velocity saturation effect, it overestimates the velocity at this short gate-length and hence overestimates the current too. Clearly, a self-consistent inclusion of constant mobility and constant velocity conduction is important in predicting the device characteristics.

Let us now look at some of the device parameters and get a feel for the degree of importance and implications of the saturation of velocity. A plot of opening in the channel at the source and drain ends ($a(1 - \xi_s)$ and $a(1 - \xi_p)$) as functions of the drain current is shown in Figure 5.7. In this figure, at a constant current, the openings are a function of the aspect ratio (L/a). Small aspect ratios and small gate-lengths mean nearly identical source and drain openings (the source opening is slightly larger because of smaller velocities) because carriers transit at saturated velocities through much of the device.

Note that because of velocity saturation, the channel can be fully open at high gate biases even when high drain biases occur. The saturated velocity forces the drain end of the channel to remain significantly open even at high drain biases in order to maintain the large drain currents.

The source opening is bigger but not significantly. At 1/10th the full channel current, it may be about 30–40% percent bigger for the longest aspect ratio device of $L/a = 10$. For a smaller aspect ratio, e.g., a 1.0 μm gate-length with a 0.3 μm channel thickness, the source opening is within 5% of the drain opening. The depletion region is essentially parallel to the

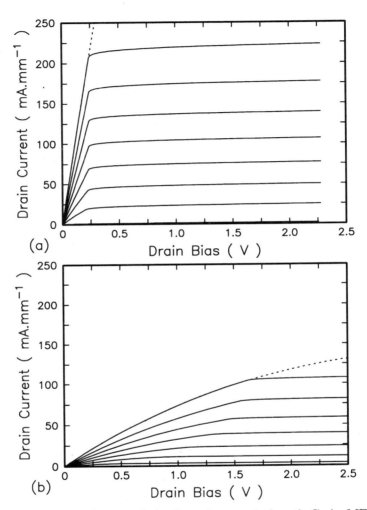

Figure 5.6: Output characteristics for a 1 μm gate-length GaAs MESFET are shown in (a) following the PHS analysis. (b) shows the output characteristics for a 10 μm gate-length GaAs MESFET following the PHS analysis. Both figures show, using dashed lines, the current for zero gate bias assuming no velocity saturation, i.e., assuming that a constant mobility is valid throughout the operation.

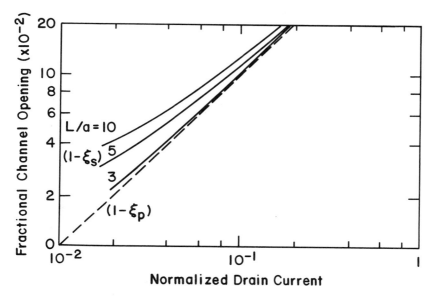

Figure 5.7: Source (solid lines) and drain (dashed line) end openings of MESFET channel derived from the PHS model at different aspect ratios with $V_{sd}/V_{00} = 1$ and $\alpha = 0.1$. The normalization parameter for current is I_{norm}. After R. A. Pucel, H. A. Haus, and H. Statz, "Signal and Noise Properties of Gallium Arsenide Microwave Field-Effect Transistors," in *Advances in Electronics and Electron Physics*, **38**, Academic Press, N.Y. (1975).

surface—which also indicates the quite good validity of gradual channel approximation.

While the differences in fractional openings at source and drain ends are insensitive to the current being driven or the drain voltage, the fractional length of the velocity saturated region is very sensitive, with L_2/L increasing rapidly with V_{sd}. Figure 5.8 shows an example of this dependence for our model device at various aspect ratios, with the smaller aspect ratio showing the larger effect. The smaller the aspect ratio of the device, the less is this region—but for shorter gate-lengths at a potential equal to the long-channel pinch-off potential V_{00}, e.g., for the device with $L/a = 3$, almost the whole longitudinal section may have carrier transport at saturated velocity. This is also seen from a plot with α as a variable shown in Figure 5.9. Notice that as the saturation index decreases. the transport effects in the saturated region become dominant.

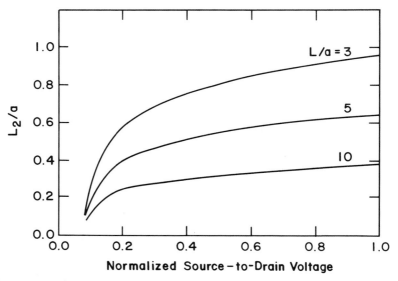

Figure 5.8: Relative length of the velocity saturated region L_2/L as a function of the normalized source-to-drain voltage $V_{sd}/\mathcal{V}_{00}$ for different aspect ratios, $I_D/I_{norm} = 0.5$, and $\alpha = 0.1$. After R. A. Pucel, H. A. Haus, and H. Statz, "Signal and Noise Properties of Gallium Arsenide Microwave Field-Effect Transistors," in *Advances in Electronics and Electron Physics*, **38**, Academic Press, N.Y. (1975).

Since we can predict the output current–voltage relationship, we can also derive secondary parameters of interest (such as transconductance, output conductance, etc.) in the quasi-static approximation. We can also derive simplified limits of the capacitances by considering the electrostatics of the problem. These are considered next.

5.3.1   Transconductance

We again use quasi-static approximation and find g_m through perturbation of the drain current.

$$
\begin{aligned}
g_m &= \left. -\frac{\partial I_D}{\partial V_{sg}} \right|_{V_{sd}} \\
&= \left. g_o \mathcal{E}_s W \frac{\partial \xi_p}{\partial V_{sg}} \right|_{V_{sd}}
\end{aligned}
$$

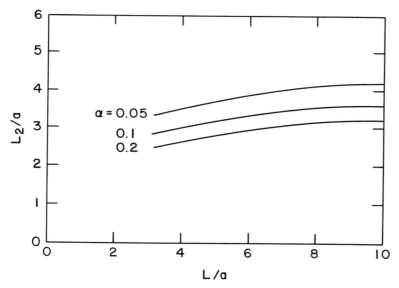

Figure 5.9: The normalized length of the saturated velocity region L_2/a, as a function of the aspect ratio L/a for varying saturation parameters. $V_{sd}/V_{00} = 1$, and $I_D/I_{norm} = 0.5$. After R. A. Pucel, H. A. Haus, and H. Statz, "Signal and Noise Properties of Gallium Arsenide Microwave Field-Effect Transistors," in *Advances in Electronics and Electron Physics*, **38**, Academic Press, N.Y. (1975).

$$= \frac{g_o \mathcal{E}_s W}{2 \xi_s \mathcal{V}_{00}} \frac{d\xi_p}{d\xi_s}. \tag{5.52}$$

This gives

$$g_m = \frac{I_{norm}}{\mathcal{V}_{00}} \frac{(1 - \xi_s) \cosh (\pi L_2/2a) - (1 - \xi_p)}{[2\xi_p(1 - \xi_p) + \alpha (L_1/L) \cosh (\pi L_2/2a) - 2\xi_p(1 - \xi_p)]}. \tag{5.53}$$

Note when $L_2 \to 0$, i.e., when there is no velocity saturated region,

$$g_m \approx \frac{g_o W}{L} (\xi_p - \xi_s), \tag{5.54}$$

which is our earlier long-channel expression. In the other limit of $L_2 \to L$, i.e., velocity saturated conduction throughout the channel, $\xi_p \approx \xi_s$, and

$$g_m \approx \frac{I_{norm}}{\mathcal{V}_{00}} \frac{1}{2\xi_p}, \tag{5.55}$$

because the cosh term dominates, and

$$g_m \approx \frac{I_{norm}}{2\mathcal{V}_{00}} \frac{1}{1 - I_D/I_{norm}}. \tag{5.56}$$

This follows from our constant velocity expression (see Problem 1). Note that these expressions predict an increasing transconductance at smaller gate-lengths.

5.3.2 Output Conductance

The output conductance is defined as

$$g_{sd} = -\left.\frac{\partial I_D}{\partial V_{sd}}\right|_{V_{sg}}. \tag{5.57}$$

Following our expression for current, this can be derived as

$$g_{ds} = \frac{I_{norm}}{\mathcal{V}_{00}} \frac{1 - \xi_p}{[2\xi_p(1 - \xi_p) + \alpha\,(L_1/L)]\cosh\,(\pi L_2/2a) - 2\xi_p(1 - \xi_p)}. \tag{5.58}$$

When, $L_2 \to 0$, i.e., no velocity saturated region,

$$\begin{aligned}
g_{ds} &= \frac{I_{norm}}{\mathcal{V}_{00}}(1 - \xi_p)\frac{\mathcal{V}_{00}}{\mathcal{E}_s L} \\
&= \frac{g_o(1 - \xi_p)W}{L}.
\end{aligned} \tag{5.59}$$

When the aspect ratio L/a gets smaller, the output conductance increases and the output characteristics of the device show higher slope. This was shown in the two separate plots for the long-channel and short-channel devices of Figure 5.6.

For short gate-length structures,

$$g_{ds} \approx \frac{\alpha}{\pi}\frac{a}{L}\frac{I_{norm}}{|V_{sd}|}\frac{1}{\xi_p}, \tag{5.60}$$

from our expressions. The most inaccurate small-signal parameter in this model for predicting characteristics of real devices is this resistance, because the idealized geometry does not include injection into the substrate and modifications resulting from that in the device—a very important consideration under short-channel conditions. We show implications of some of these effects in later sections.

5.3.3 Gate-to-Source Capacitance

Since the depletion region charge also responds to changes in bias, this calculation is quite complicated. The total gate charge $\mathcal{Q}_g$ is (see Problem 3)

$$
\mathcal{Q}_g \approx 2qN_D aW \left[\frac{\frac{2}{3}(\xi_p{}^3 - \xi_s{}^3) - \frac{1}{2}(\xi_p{}^4 - \xi_s{}^4)}{(\xi_p{}^2 - \xi_s{}^2) - \frac{2}{3}(\xi_p{}^3 - \xi_s{}^3)} + \xi_p L_2 + \right.
$$
$$
\left. \frac{\alpha}{\pi} \frac{a^2}{L} \left(\cosh \left(\frac{\pi L_2}{2a} \right) - 1 \right) \right], \tag{5.61}
$$

and the gate-to-source capacitance, following definition, is

$$
C_{gs} = \left. \frac{\partial \mathcal{Q}_g}{\partial V_{sg}} \right|_{V_{sd}}. \tag{5.62}
$$

The gate-to-source capacitance increases with increased current.

5.3.4 Gate-to-Drain Capacitance

In our analysis of device behavior at biases beyond those at which the channel current saturates, we obtained an output conductance that came about because essentially the device acts as if it has a shorter gate-length which has a higher current capability. This "shortening" of gate-length, in the case of long devices where the electric field does not reach the values necessary to achieve velocity saturation, occurs via channel pinch-off. We encountered this phenomenon in our long-channel constant mobility analysis. The effects that give rise to channel length shortening also give rise to drain-to-gate capacitance effects.

First consider the non–current-saturation region of operation. The charge underneath the gate terminates at the gate, and the depletion region edge responds to both the source and the drain bias. If the two are at identical potentials, the depletion region is symmetric in both source and drain and the gate-to-source and gate-to-drain capacitances are identical. If, now, a drain bias is applied in excess of the source bias, the depletion region at the drain extends farther, and at the source end it responds to the requirements of conserving channel current for the quasi-static conditions. As a result of application of drain bias in excess of source bias, however, the capacitance and the charge that can directly be modulated by the drain decrease. The source capacitance increases because a larger part of the charge underneath the gate can be directly modulated. Another way of viewing this is that a higher electric field that naturally forms towards the drain as a result of the application of this drain bias results in

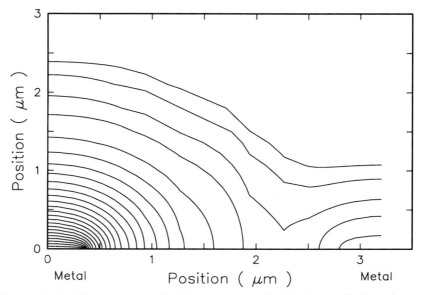

Figure 5.10: The equi-potential lines in a very lightly doped GaAs substrate due to bias applied between two metal electrodes on the surface. Electric field lines are orthogonal to these equi-potential lines.

screening of the channel and source regions. As a result, a larger and larger fraction of the depletion region underneath the gate can be associated with the source electrode and hence the gate-to-source capacitance. The net consequence, irrespective of the reason for channel current saturation, is that the drain-to-gate capacitance decreases.

If there were no conducting channel, there would still be a capacitance associated with the drain-to-gate as well as the drain-to-source electrodes because of mutual termination of fields (see Figure 5.10). This capacitance really can not be calculated independently because the presence of a conducting channel changes the field profile. However, a calculation based on a semi-insulating medium with planar electrodes on the surface gives a lower bound for the capacitance of the structure. This capacitance $C_{dg,si}$ is approximately given by

$$C_{dg,si} = (\epsilon_s + \epsilon_0) W \frac{K\left(\sqrt{1 - k^2}\right)}{K(k)}, \qquad (5.63)$$

where $K(k)$ and $K(\sqrt{1 - k^2})$ are Elliptic integrals of the first kind, and k

is given as

$$k = \left(\frac{L_{dg}}{L_{dg} + L} \right)^{1/2}. \tag{5.64}$$

In these expressions, L_{dg} is is the drain-to-gate spacing and the drain length is assumed to be the same as that of the gate.

5.3.5   Drain-to-Source Capacitance

The drain-to-source capacitance is generally the smallest of capacitances and occurs to a large extent due to the fields through the substrate. The drain is usually isolated from the source by the high field regions in the channel. Hence, the approximation above for inter-electrode capacitance gives a better approximation for this capacitance.

$$C_{ds,si} = (\epsilon_s + \epsilon_0)\, W \frac{K\left(\sqrt{1 - k^2}\right)}{K\left(k\right)}, \tag{5.65}$$

where $K(k)$ and $K(\sqrt{1 - k^2})$ are Elliptic integrals of first kind, and k is given as

$$k = \left(\frac{(2L_s + L_{ds})\, L_{ds}}{L_s + L_{ds}} \right)^{1/2}, \tag{5.66}$$

where L_s is the source length assumed to be the same as the gate and L_{ds} is the drain-to-source spacing.

We have now determined a number of elements associated with the ports of the MESFET using perturbation on the quasi-static analysis: g_m to represent the variation of drain current with gate bias, C_{dg} to represent the change in associated storage between drain and gate electrodes, C_{gs} to represent the change in storage associated with gate-to-source electrode, C_{ds} for the fringing effects between drain and source electrodes, and g_{ds} for output effects. The channel region charging the capacitance C_{gs} has associated with it a finite resistance. Later on, we will see how this naturally appears as a result of a direct small-signal analysis from the small-signal transport equations. Here we will introduce it ad hoc as a resistance R_i that represents the intrinsic resistance associated with the integrated effect of channel conductance in the gate-source region. The resulting equivalent circuit, a quasi-static equivalent circuit, is shown in Figure 5.11. This figure includes three resistances, R_s, R_d, and R_g, to represent the ohmic effects in the extrinsic regions of the device.[5]

---

[5]In equivalent circuit representation, we adopt the use the hat symbol to represent small-signal. Thus, $\hat{V}_{gs}$ is the small-signal gate-to-source voltage.

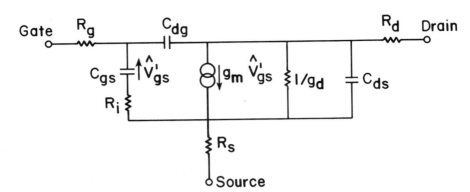

Figure 5.11: Quasi-static small-signal equivalent circuit for a MESFET based on elements derived as part of perturbation analysis on the PHS model.

5.4   Accumulation–Depletion of Carriers

In the PHS model, a nearly flat depletion region occurs as a consequence of velocity saturation, and the carrier concentration in the conducting region is nearly constant to maintain current continuity. This is not a good general description of real devices. The most important reason for this inadequacy is that the analysis entirely ignores diffusive currents, a current component that can be quite important. For example, in the long gate-length limit, this is the dominant current in the pinched region of the channel. It can be important in short gate-length devices also, depending on the design of the device and the bias conditions. Large electric fields exist in this region; they are the reason for velocity saturation, and hence these large electric fields can support and exist with deviations from quasi-neutrality.

Models incorporating diffusion current can be quite complicated. We will consider a model that includes diffusion current in the analysis of HFETs in Chapter 6. Here, a more suitable way of exemplifying the deviations from quasi-neutrality in the form of an accumulation-depletion region is by considering a full two-dimensional model based on drift-diffusion transport[6] of our 1 μm with a 5×10^{17} cm^{-3} doped channel MESFET

---

[6]i.e., the use of a two-dimensional simulator that finds the solution of the drift-diffusion equation, the continuity equation, and Poisson's equation throughout the device cross-section subject to the boundary conditions. Such simulators are usually based on finite element analysis. They solve by dividing the device into many small but finite sized elemental volumes and by assuring continuity at their interfaces and with the device surfaces. Simulators can also be based on the use of the BTE approach or the

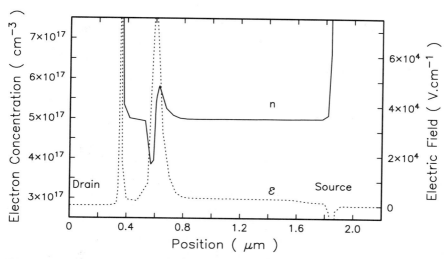

Figure 5.12: Electric field and carrier density for a 1 μm gate-length GaAs MESFET with a 5×10^{17} cm^{-3} doped channel biased into current saturation. The formation of the accumulation-depletion region is shown in proximity of the drain electrode.

example. Results of the variation in electric field and the carrier density along the channel are shown in Figure 5.12. Note that diffusion currents tend to be small at the source end but get larger in the high field region because of the rapid changes in the carrier concentration.

There are two other important features in the results associated with this phenomenon. One is that the depletion region is not entirely flat in the high field region where carriers do move with essentially a constant velocity (see Figure 5.1). This is connected to the changes in the diffusion current. Current continuity requires that the sum, drift and diffusion, be a constant following integration over the cross-section. A changing cross-section can still accommodate current continuity with diffusion currents present. The presence of diffusion current also implies the presence of a concentration gradient in the conducting region with high fields. Thus, depletion depth changes, carrier concentration changes, and diffusion currents can all be present in MESFETs, and the PHS model does not include them. However, its inclusion would be very difficult numerically within the framework of the model. Indeed, making it significantly more complex would remove its attractiveness as an intuitive model. One might as well, then, employ a

Monte Carlo approach as discussed in Chapter 3.

complete two-dimensional model.

The change in carrier concentration, shown in Figure 5.12 is of particular interest since it brings out an additional consideration. Current continuity, in the presence of a decreasing channel cross-section and high electric fields, results in an accumulation of carriers. If the change in channel cross-section takes place over a sufficiently large distance, then the associated diffusion current is small, and the accumulation of carriers is a direct consequence of current continuity alone. Eventually, when the drain is reached, the conducting channel may expand and hence a depletion of carriers may also occur. So, the large longitudinal field in the region of velocity-saturated transport occurs together with a change in the net charge density. For most short gate-length structures, employing relatively large channel dopings, current continuity is the main source of this accumulation–depletion. It occurs in all materials at appropriately short gate-lengths where velocity saturation may occur.[7]

A phenomenon similar to this, of accumulation–depletion, also occurs due to the negative differential mobility in the velocity–field characteristics of electrons in a variety of compound semiconductors. An example is GaAs. But, for this negative differential mobility to be significant, the channel doping has to be kept low—in the low-10^{17} cm^{-3} range for GaAs. This formation of an accumulation-depletion of charge in the channel region of the GaAs MESFET is shown in Figure 5.13 for a low-doped GaAs MESFET of 1 μm gate length. As carriers travel along from the source edge of the channel to the drain they encounter higher fields beyond the field corresponding to peak velocity. Following the velocity–field curve, this initially leads to a decrease in velocity and hence an increase in carrier density, the channel cross-section effects being secondary. As the carriers move further down the channel, near the drain, they encounter lower electric fields. The carriers accelerate and hence a depletion of charge occurs to maintain current continuity. Thus, along the channel, an accumulation and depletion region forms in the conducting channel region. In PHS model, a constant velocity is assumed in the high field region. This accumulation-depletion region would have formed in this part of the device underneath the drain edge of the gate, had we included the specific velocity–field curve with negative differential mobility.

---

[7]i.e., it occurs in silicon FETs also; see D. P. Kennedy and R. R. O'Brien, "Computer-Aided Two-Dimensional Analysis of the Junction Field-Effect Transistor," *IBM J. of Research and Development*, **14**, No. 2, p. 95, 1970.

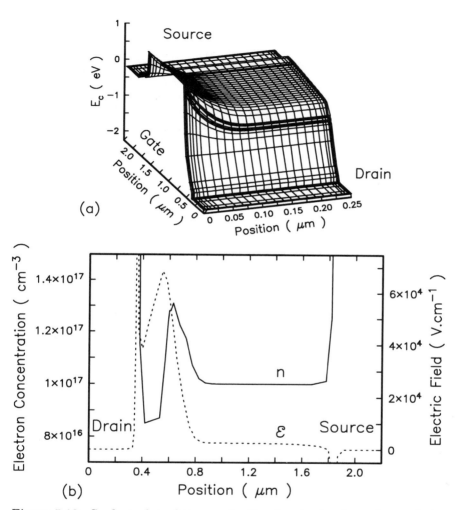

Figure 5.13: Surface plot of the conduction band edge for a 1 μm GaAs MESFET with a 1×10^{17} cm^{-3} doped channel biased into current saturation. (b) shows the electric field and the carrier concentration as a function of position along a conducting cross-section of the channel.

Short gate-length devices are usually made with channel dopings in the range of mid-10^{17} cm^{-3}; at such doping levels, the effect of negative differential mobility is weaker since peak velocity is smaller, however, the similar effect of current continuity is larger. Note that a nearly constant channel cross-section is important to arguments regarding this dipole. The PHS model considers a symmetric cross-section for the device and entirely ignores substrate injection effects, a reason why it is inadequate in modelling output conductance features. Substrate injection is also important because of its effect on the formation of the dipole. If carriers can spill over into the substrate, the cross-section can stay larger. Large electric fields from drain-to-source, both transverse and longitudinal, aid this injection into the substrate and hence mitigate substantially the formation of dipole layers. Figure 5.14 shows an example similar to Figure 5.13 in the presence of substrate injection. Note the decrease in the dipole formation. The effect is much more substantial in the case of the first type of dipole because channel cross-section is very central to the dipole's formation. Substrate and back interfaces are usually designed to suppress injection away from the gate-channel region. Any current path farther away from the gate leads to poorer control by the gate. So, for most designs of MESFETs, irrespective of the cause, some dipole effect is present in the high field region of the conducting channel. The effect is larger at smaller doping levels, and its implications are significant at very high frequencies because its principle consequence is a capacitive feedback from the output. For computer-aided design models of current–voltage characteristics, this may be neglected with an acceptable error; for quasi-static capacitance models, this may or may not be acceptable depending on the doping level and other structural features of the device.

A complete inclusion of the implications of the accumulation-depletion requires an elaborate numerical calculation, since the complex velocity–field curve does not allow for analytic or simple numerical analysis. We will, however, see how a simpler analysis may be used to incorporate first-order effects of this dipole region, by considering the negative differential mobility–induced dipole as an example. The objective of this exercise is to serve as an example of how a complex phenomenon may be analyzed within a section of a device in order to include its important consequences for device behavior. As we have remarked before, though, this particular source of dipole is relatively weak in the short gate-length heavily-doped channels.

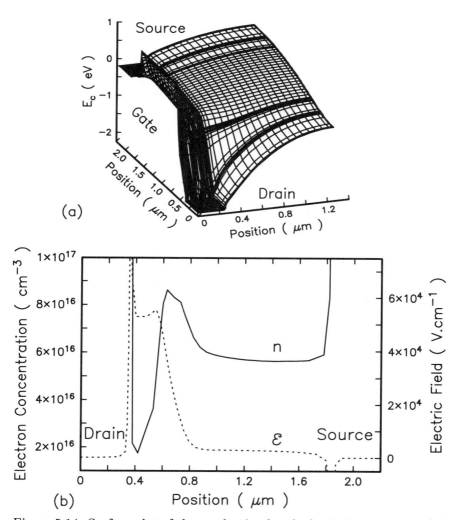

Figure 5.14: Surface plot of the conduction band edge is shown in part (a) for a 1 μm GaAs MESFET with a 1×10^{17} cm^{-3} doped channel biased into current saturation. This example allows for injection into the substrate. Part (b) shows the electric field and the carrier concentration as a function of position along a conducting cross-section of the channel. This figure should be compared with the previous figure which assumed a symmetric boundary condition at the interface between the channel and the substrate.

Mathematically, the current continuity equation for the channel carrier transport[8] is

$$J = q \left[nv(\mathcal{E}) + \frac{d\left(n\mathcal{D}(\mathcal{E})\right)}{dz} \right], \qquad (5.67)$$

where $\mathcal{D}(\mathcal{E})$ is the diffusion coefficient of electrons as a function of electric field. If N_D is the background donor concentration,

$$\frac{d\mathcal{E}}{dz} = -\frac{q}{\epsilon_s}\left(n - N_D\right), \qquad (5.68)$$

and just outside the dipole region, where the channel is quasi-neutral,

$$J = qN_D v_1, \qquad (5.69)$$

giving

$$\frac{d(n\mathcal{D})}{d\mathcal{E}} = \frac{\epsilon_s}{q} \frac{nv(\mathcal{E}) - N_D v_1}{n - N_D}. \qquad (5.70)$$

Here we have utilized the approximations of the velocities in the high field region of the device.

The solution of this differential equation is found as follows, with the perturbation of the diffusive term, a small term, ignored:

$$\int_{N_D}^{n} \frac{n' - N_D}{n'} dn' = \frac{\epsilon_s}{q} \int_{\mathcal{E}_1}^{\mathcal{E}} \frac{v - N_D v_1/n}{\mathcal{D}(\mathcal{E}')} d\mathcal{E}', \qquad (5.71)$$

or

$$\frac{n - N_D}{N_D} - \ln\left(1 + \frac{n - N_D}{N_D}\right) = \frac{\epsilon_s}{qN_D} \int_{\mathcal{E}_1}^{\mathcal{E}} \frac{v - N_D v_1/n}{\mathcal{D}(\mathcal{E}')} d\mathcal{E}'. \qquad (5.72)$$

We now consider only limited deviation from quasi-neutrality, i.e.,

$$|n - N_D|/N_D \ll 1, \qquad (5.73)$$

appropriate for low-10^{17} cm^{-3} or higher doped devices in GaAs. If the dipole forms over a region of many Debye lengths, i.e., exceeding 0.1 μm, then the limited deviation from quasi-neutrality also implies a diffusion

---

[8]Here we follow arguments discussed in detail in T. J. Fjeldy, "Analytical Modeling of the Stationary Domain in GaAs MESFETs," *IEEE Trans. on Electron Devices*, **ED-33**, No. 7, p. 874, July, 1986. We stress, however, that at doping levels of interest in practical MESFETs, this is of secondary, not primary importance. It occurs together with the other cause of dipole already discussed. It is probably of larger significance in HFETs where the background doping is much lower and hence the negative differential mobility is significantly stronger.

current density much smaller than drift current density. This justifies the prior assumption of this analysis.

$$\left(\frac{n - N_D}{N_D}\right)^2 = \frac{2\epsilon_s}{qN_D} \int_{\mathcal{E}_1}^{\mathcal{E}} \left(v - \frac{N_D}{n}v_1\right) \frac{1}{\mathcal{D}(\mathcal{E}')} d\mathcal{E}'. \tag{5.74}$$

At the peak field $\mathcal{E}_m$ in the domain,

$$\int_{\mathcal{E}_1}^{\mathcal{E}_m} \frac{v - v_1}{\mathcal{D}(\mathcal{E}')} d\mathcal{E}' = 0, \tag{5.75}$$

because $n = N_D$. This allows us to determine the maximum field $\mathcal{E}_m$—the maximum electric field in the dipole region—and hence a reduction of variables.

The total voltage drop in the dipole region V_{dip}, assuming symmetric accumulation and depletion, is

$$
\begin{aligned}
V_{dip} &= 2\int_{\mathcal{E}_1}^{\mathcal{E}_m} (\mathcal{E} - \mathcal{E}_1) dz \\
&= 2\int_{\mathcal{E}_1}^{\mathcal{E}_m} (\mathcal{E} - \mathcal{E}_1) \frac{\epsilon_s}{q(n - N_D)} d\mathcal{E}, \tag{5.76}
\end{aligned}
$$

using Poisson's Equation. The total sheet charge density in the dipole region $\mathcal{Q}_{dip}$, using Gauss's Law, is

$$\mathcal{Q}_{dip} = \epsilon_s(\mathcal{E}_m - \mathcal{E}_1), \tag{5.77}$$

and hence the capacitance associated with this charge, C_{dip}, is

$$C_{dip} = \frac{\partial \mathcal{Q}_{dip}}{\partial V_{dip}} = \epsilon_s \left[\frac{\partial V_{dip}}{\partial (\mathcal{E}_m - \mathcal{E}_1)}\right]^{-1}. \tag{5.78}$$

In the limit of large field in this dipole domain region, one can determine a parameter $\mathcal{R}$ for a given $\mathcal{D}(\mathcal{E})$ and $v(\mathcal{E})$ characteristic, whose magnitude in the high field or high voltage limit is given by

$$
\begin{aligned}
\mathcal{R} &= \int_{\mathcal{E}_1}^{\mathcal{E}_m} \frac{v(\mathcal{E}) - v_1}{\mathcal{D}(\mathcal{E})} d\mathcal{E} \\
&\approx \int_{\mathcal{E}_s}^{\infty} \frac{v(\mathcal{E}) - v_s}{\mathcal{D}(\mathcal{E})} d\mathcal{E}, \tag{5.79}
\end{aligned}
$$

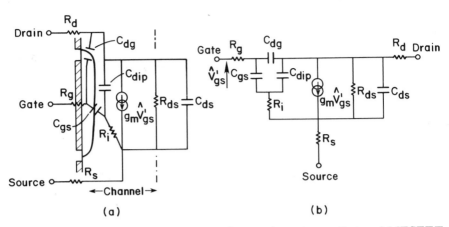

Figure 5.15: The origin of elements of a quasi-static small-signal MESFET model (a) including intrinsic and extrinsic elements derived in the quasi-static approximations. The developed model is shown in (b).

since, at the high field end ($\mathcal{E}_m \to \infty$), v_1 approaches the saturated velocity v_s. At the low field end $\mathcal{E} \to \mathcal{E}_s = v_s/\mu_0$, because a constant mobility may be assumed.

Under this assumption of a large dipole domain voltage, it can be shown that the dipole voltage reduces to (see Problem 4)

$$V_{dip} = \frac{4\sqrt{2}}{3}\lambda_D \left(\frac{\mu_0}{\mathcal{D}_0 \mathcal{R}}\right)^{1/2}(\mathcal{E}_m - \mathcal{E}_s)^2. \qquad (5.80)$$

where $\mathcal{R}$ is a constant $\approx 5.5 \times 10^7$ V.cm^{-2} for a doping $N_D = 10^{17}$ cm^{-3} in GaAs and $\mathcal{D}_0$ is the diffusion coefficient in the absence of field. The dipole capacitance follows as

$$C_{dip} = \frac{3\epsilon}{8\lambda_D}\left(\frac{\mathcal{D}_0\mathcal{R}}{2\mu_0}\right)^{1/2}\frac{1}{\mathcal{E}_m - \mathcal{E}_s}. \qquad (5.81)$$

Based on this analysis, we may now infer a quasi-static equivalent circuit model for the GaAs MESFET, by including in it the features related to current–voltage dependence, the parasitic resistances and capacitances, and the intrinsic resistances and capacitances. A model with a broad range of applicability under quasi-static conditions is shown in Figure 5.15.

The equivalence between the cross-sectional parameters and the equivalent circuit elements is quite intuitive. The intrinsic device is the part of the device used in the analysis of channel charge transport and control. Extrinsic elements are those required to link this intrinsic device to the stimulus we apply to or sample from the device. Source resistance R_s, drain resistance R_d, and gate resistance R_g (which is usually negligible due to high metal conductivity but non-negligible at very short gate-lengths due to small cross-section area) are examples of extrinsic resistances. Pads used to connect the device are also extrinsic elements. The intrinsic elements are related to the functioning of the device whose operational basis is incorporated in our model. R_i is the resistance that results from conductance effects of the channel region, and it emphasizes in a lumped element the effects of the source end of the channel. The gate-to-source capacitance C_{gs}, the drain-to-gate capacitance C_{dg}, and the drain-to-source capacitance C_{ds} have been derived before. The capacitance due to accumulation and depletion of the charge in the channel C_{dip} is incorporated in the channel path between the drain and the source. The output resistance R_{ds} is incorporated to account for effects of channel modulation by the drain bias, and the transconductance g_m is incorporated through the current source. This is a quasi-static equivalent circuit model, and we will show later that it is useful up to moderate frequencies. Significantly more complex circuits, or even network parameters, based on a direct small-signal analysis or experimental measurements at the frequencies of interest, are necessary for modelling effects at higher frequencies. This model will also be used as a basis in HFETs, the operation of the devices being similar in basis. The variation in magnitudes of the elements, may, however, be significantly different because of the nature of the charge control process itself. The model itself is quite similar and shows very similar current gain–frequency dependence and other features of interest.

5.5 Sub-Threshold and Substrate Injection Effects

The bias-dependent high fields result in injection of carriers into the substrate. This is certainly very strong in the sub-threshold region where most conduction takes place due to injection of carriers from the passive regions related to access to the channel. This injection region, with lower carrier densities, also has large changes in carrier densities. Thus, diffusive effects are large in this bias region of device operation. Since the injection is over a barrier, it has a characteristic exponential behavior with the gate bias. This sub-threshold behavior is quite similar in all FETs irrespective of how they operate in the conducting region. The reason for this is behavior is that sub-threshold conduction largely takes place in the substrate and injection and collection occur in the access regions of the source and the drains. All these regions are quite identical in FETs. We will discuss sub-threshold current in more detail in Chapter 6.

The injection of carriers into the substrate can also be very strong even when the channel is conducting. This injection is particularly strong when the device is in the saturated current mode of operation. The high fields push the conduction towards the channel–substrate interface and beyond from the doped region of the device, and the high field region related to formation of the dipole domain can make this effect stronger. Injection into the substrate occurs along longer path lengths and further away from the gate. It acts effectively as a large gate-length parasitic device and hence results in a device that is intrinsically slower and has poorer operational characteristics, such as transconductance, output conductance, etc.

Band bending in the substrate is dependent on the characteristics of the substrate material; the electric fields in the device together with this barrier result in the injection into the substrate. The problem is therefore complicated and intrinsically two-dimensional in nature. It also becomes particularly strong at small gate-lengths. Our discussion in the following section, coupled with the sub-threshold and injection behavior, addresses this in more detail, together with analytic and one-dimensional models for intuitive understanding and two-dimensional solutions for rigorous understanding of the injection problem.

5.6   Sidegating Effects

We now look at important problems in the use of compound semiconductor FETs—HFETs and MESFETs—in circuits. The first of these is the phenomenon of backgating or sidegating—a general term describing a change in device characteristics caused by changes in applied potentials at other than the device terminals themselves. An example of this is the effect of logic signals carried by metal lines in the vicinity of a device. Ideally, any signals other than those applied to the device should have no effect on the device characteristics because the substrate, such as that of GaAs, is semi-insulating. Semi-insulating substrates should be expected to effectively shielded at distances of μm's away from the line, resulting in very small effects on the characteristics. Without specific precautions in the technology, however, it can actually be quite substantial. We will call the collection of such effects sidegating effects since we are most interested in those resulting from biases on lines, etc., on the same surface. These effects are also sometimes referred to in the literature as backgating effects.

Sidegating effects can result from causes related to the semiconductor surface as well as the substrate. The direct effect of sidegating on the device occurs through changes in the depletion region at the substrate–channel interface, but effects can also occur, if device geometrical and functional considerations allow, due to changes in surface–channel interfaces. A change in the depletion region does not require any significant current flow because significant changes in device currents can occur with small changes in the depletion region, both for moderately forward biased and reverse biased substrate–channel interfaces. These small changes in the depletion region occur with small changes in the bias voltage across it. Thus, the important cause of sidegating is changes in the bias voltages across the substrate and surface junctions. These small changes in potential drops can occur without need for significant conduction, and the propagation of the voltage via the substrate is a requisite to sidegating. It should be emphasized that this propagation can also occur via surface effects because of local changes in trap concentrations in the surface region as well as at the surface itself. The subject is complicated because such propagation is two-dimensional in origin, and because this electrostatic voltage propagation in a semi-insulating substrate depends on the mechanism that leads to the semi-insulating character, this may, in reality, be position-dependent because of the processes involved in fabrication of structures.

Our discussion therefore looks at conduction mechanisms in semi-insulating material first, both without and with the deep traps and shallow levels that occur in a balance, which result in the semi-insulating character of the substrate. Most device structures are designed so that there is no source of holes, and hence unipolar considerations are sufficient in a preliminary analysis. However, hole effects can occur from small-barrier metals or the presence of p-type regions due to thermally-induced conversion of the surface, and hence may need to be included too. We will do this using only illustrative examples. In all such structures, where a source of holes may exist, bipolar transport considerations will apply. Using these, we will study the propagation of voltages in the presence of traps and relate it to the behavior of currents in devices.

5.6.1 Injection and Conduction Effects

As a preliminary exercise, we will look at conduction currents in one dimension in semi-insulating substrates for both trap-free and trap-dominated cases. The former is a case study for a very pure material such as that grown epitaxially, intrinsic GaAs, e.g., is semi-insulating. The latter is, however, more generally the case of substrate material. Quite often, here, the semi-insulating character is achieved by over-compensating the resid-

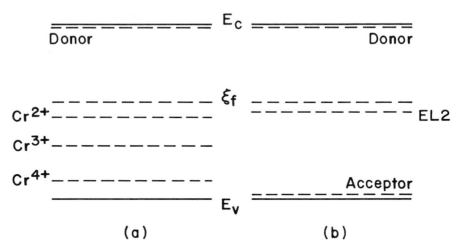

Figure 5.16: An energy-band model for obtaining semi-insulating GaAs using shallow donor and deep acceptor (Cr) is shown in (a) and using shallow acceptors, shallow donors, and deep donor (EL2) is shown in (b).

ual shallow acceptors, such as carbon, by a deep donor, such as a trap level called the EL2 level. EL2 is believed to be a native defect in GaAs. The semi-insulating material is obtained without intentionally introducing an impurity in the crystal. Shallow donors and other deep traps, both acceptor-like and donor-like, may also exist in GaAs, and these may be employed to obtain, by alternate means, semi-insulating GaAs. An example is that of obtaining semi-insulating GaAs through compensation of shallow donors and a deep acceptor such as chromium. In this latter case, the densities of traps and shallow levels are larger because these are intentionally introduced into the crystal. The energy band schematic for these various means of achieving semi-insulating GaAs is shown in Figure 5.16. The EL2 compensation scheme relies on a dominant shallow acceptor (e.g., C, Be, Mg, Mn) being present, while the chromium compensation scheme relies on a dominant shallow donor (e.g., S, Si, Se, Te, Ge) being present. We will discuss our trap model using a shallow acceptor and deep donor example; its implications can be derived using symmetry arguments for the shallow donor and deep acceptor case. The causes of sidegating can vary depending on the behavior of these traps. We will discuss some of these unusual characteristics that result due to the propagation of electrostatic potential to the device junctions in a later section. Before that, we will discuss the behavior of current in the presence of deep donor traps with larger electron capture rates than hole capture rates. These characteristics are important, since, at thermal equilibrium, the substrate is semi-insulating, and the bulk currents that flow under bias are space charge–limited.

First consider a trap-free n^+–ν–n^+ structure as shown in Figure 5.17 and its behavior with and without bias; the current transport in this structure is largely by electrons, because the contacts, with a mid-gap barrier height, can only provide electrons in significant numbers assuming that avalanching does not occur in these structures. The current in the structure, carried by the electrons injected and collected at the contacts, is largely by drift in the constant mobility regime; we are restricting applied voltages to less than $\mathcal{E}_s L$ for a sample of length L. For the 5 μm sample of Figure 5.17, this would mean that for voltages less than 1.5 V, this current is given by

$$J = q\mu n\mathcal{E} = \mu\epsilon_s\mathcal{E}\frac{d\mathcal{E}}{dz}. \tag{5.82}$$

The field is negligible at $z = 0$ at the edge of the depletion region of the injecting contact, and hence

$$\mathcal{E}(z) = \sqrt{\frac{2Jz}{\mu\epsilon_s}}, \tag{5.83}$$

the potential is

$$V(z) = -\int_0^z \mathcal{E}dz = -\sqrt{\frac{8J}{9\epsilon_s\mu}}z^{3/2}, \tag{5.84}$$

and hence the relationship in terms of terminal parameters is

$$J = -\frac{9}{8}\mu\epsilon_s\frac{V^2}{L^3}. \tag{5.85}$$

The current in trap-free material, in the constant mobility drift-dominated region, follows a direct dependence on the square of the applied voltage and an inverse dependence on the cube of the sample length. This is sometimes referred to as the trap-free square law or the Mott-Gurney law. Figure 5.17 shows the dependence on a wider voltage range than the limit of validity of this law.

Now consider this problem in the presence of shallow acceptors and deep donors as shown in Figure 5.18. This problem analyzes a geometry similar to that of Figure 5.17. Poisson's equation for this problem is

$$-\frac{d^2V}{dz^2} = -\frac{q}{\epsilon_s}\left(n + N_A - N_T^+\right). \tag{5.86}$$

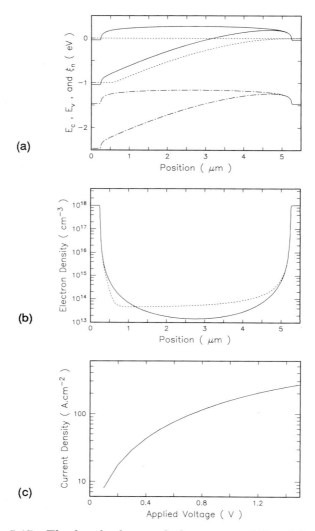

Figure 5.17: The band edges and electron quasi-Fermi level of a n^+–ν–n^+ structure in GaAs with and without bias are shown in (a). (b) shows the associated electron densities as a function of position. (c) shows the current–voltage characteristics of the structure.

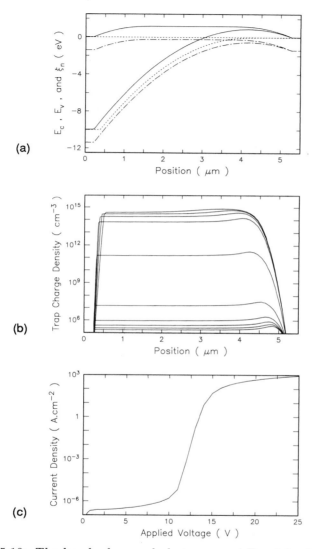

Figure 5.18: The band edges and electron quasi-Fermi levels of a $n^+-\nu-$ n^+ structure containing shallow acceptors and deep donors (10^{15} cm^{-3}) in GaAs are shown in (a) with and without bias. (b) shows the associated trap density that is charged as a function of bias. (c) shows the current–voltage characteristics of the structure. This figure should be compared with the previous figure, where no traps were present.

Using a convention similar to that in the HSR recombination problem, with n_T as an electron concentration when the quasi-Fermi level coincides with the trap level, and with the assumption that the shallow acceptor concentration is smaller than the deep donor concentration, thus imparting the semi-insulating character, we obtain

$$\frac{n_T}{n} = \frac{N_C \exp\left[-\left(E_c - E_T\right)/kT\right]}{N_C \exp\left[-\left(E_c - \xi_n\right)/kT\right]} = \exp\left(\frac{E_T - \xi_n}{kT}\right). \tag{5.87}$$

Since

$$\frac{N_T^+}{N_T} = \frac{1}{1 + \exp\left[-\left(E_T - \xi_n\right)/kT\right]}, \tag{5.88}$$

we can write the identities under the Boltzmann approximation of

$$n_T = \frac{n N_T^+}{N_T - N_T^+}, \tag{5.89}$$

and

$$N_T^+ = \frac{N_T n_T}{n + n_T}. \tag{5.90}$$

In thermal equilibrium, for semi-insulating material with low carrier concentration (i.e., neglecting the carrier concentration w.r.t. the acceptor concentration or n_T), we obtain

$$n_0 = n_T \frac{N_T - N_A}{N_A + n_T}. \tag{5.91}$$

We can now solve Poisson's equation, again assuming constant mobility transport and neglecting the n_T carrier term w.r.t. the free carrier term,

$$\frac{d^2 V}{dz^2} = \frac{q}{\epsilon_s}\left(n + N_A - \frac{n_T N_T}{n}\right). \tag{5.92}$$

The first term in this equation, which is associated with the free carrier charge, leads to the square-law behavior that we have already discussed for the trap-free case. The second term, due to the ionized acceptor charge, gives a constant term for the definite integral of voltage; this constant is given by

$$V_{TFL} = \frac{q L^2 N_A}{2\epsilon_s}, \tag{5.93}$$

a parabolic voltage term similar to that of the p–n junction problem. It corresponds to the contribution of the ionized acceptor charge of density N_A throughout the insulating region. The last term results in

$$n(z) = -\frac{J}{q\mu dV/dz} = \frac{n_0}{1 - \left[1 - n_0/n(0)\right]\exp\left(-z/L_T\right)}, \tag{5.94}$$

and

$$V(z) = -\frac{J}{qn_0\mu}\left\{z - L_T\left(1 - \frac{n_0}{n(0)}\right)\left[1 - \exp\left(-\frac{z}{L_T}\right)\right]\right\}, \quad (5.95)$$

with $n_0 \approx n_T N_T/N_A$ because $N_T \gg N_A \gg n_T$, and

$$L_T = -\frac{J\epsilon_s}{q^2\mu n_0 N_A}. \quad (5.96)$$

In these equations, the carrier density at $z = 0$ is written explicitly to account for carrier enhancement provided by the contacts beyond the thermal equilibrium value. The constant potential term due to the ionized acceptors becomes important when the other charge terms are either insignificant or closely neutralize each other. For low bias voltages, i.e., with very small injected carrier density (electrons in our example) sufficient number of deep donors ionize to compensate for the acceptor charge, the injected electron charge, and to support the applied bias. The ionized donor density is approximately the same as the acceptor charge density. The rest of the deep donor traps are neutral in much of the semi-insulating region. At a bias of V_{TFL}, sufficient electron charge density has been injected in the semi-insulating region to saturate the deep donor levels, most of which are now in the neutral state, and hence the ionized acceptors provide the net resultant charge, and the corresponding potential term is parabolic in length akin to the depletion region of a p–n junction. The significance of this potential V_{TFL} is that at higher biases the deep donors remain neutral due to the filling of the trap with electrons, and hence the term trap-filled limit used as a subscript to this voltage. At higher biases, the material can be treated as being trap-free, and should follow the square law in voltage. Figure 5.18 demonstrates the break point that this trap-filling provides, as well as the power law relationship following that.

We can make further analytic approximations in the analysis of this problem. For large current densities, i.e., a significant enhancement of carriers at $z = 0$, i.e., $n_0 \ll n(0)$ and $z/L_T \ll 1$, the applied potential V_A can be expressed using the first term of the expansion of the exponential as

$$V_A = V(z = L) = V_{TFL}\left(1 + \frac{2n_0 q\mu V_{TFL}}{3L_T J}\right), \quad (5.97)$$

and hence the current density under these approximations is

$$J = -\frac{2n_0 q\mu V_{TFL}^2}{3L_T\left(V_{TFL} - V_A\right)}. \quad (5.98)$$

This equation is very approximate; however, it serves well to indicate that a sudden rise in current should be expected at biases corresponding to the trap-filled limit voltage. In the example of Figure 5.18, the acceptor doping was 1×10^{15} cm^{-3}, and sample length 5 μm, corresponding to a V_{TFL} of 20 V. The exact number is different because of the Debye tailing effect in the low-doped region and drift approximation in the calculation. Figure 5.18 also shows that in addition to this threshold behavior at the trap-filled voltage, a significant voltage drop can occur in the space charge region and a small voltage does drop at the other junction. We would therefore expect sidegating effects to occur due this conduction mechanism.

In deriving these analytic relations, and in analyzing the space charge effect, we have considered unipolar transport even when traps were present. We have mentioned that conditions may exist where hole transport may also occur in addition to electron transport. In such situations, the capture characteristics of the traps themselves become very important, since they are important to the residual space charge in the "semi-insulating" region. Consider the characteristics of the EL2 level: since it is a deep donor it has a significant electron cross-section and a small hole cross-section. The consequence of this can be that an EL2 trap can be compensated by electrons, and a simultaneous hole conduction can lead to a very efficient propagation of electrostatic potential, as shown in Figure 5.19. We will visit this condition in greater detail later. Here, we emphasize that there is now a negligible drop in the space-charge region. Most of the voltage drop occurs at the junction, specifically at the reverse biased junction. If this reverse biased junction corresponded to the channel-substrate interface of an FET, clearly there would be a significant sidegating effect.

These two mechanisms of conduction and electrostatic propagation are quite different. The first relies on trap filling at a threshold voltage for the propagation of electrostatic potential, while the second relies on the significant differences in the capture cross-sections for the two types of carriers. The second mechanism propagates electrostatic potential extremely efficiently. It also does not have a threshold voltage associated with it. Our discussion of the bulk-dominated behavior of sidegating will be based on these two mechanisms.

5.6.2 Bulk-Dominated Behavior

We reiterate the basis of the semi-insulating character of GaAs substrates: it occurs because of the presence of compensating acceptors and donors, both shallow and deep, in the $\approx 10^{14}$ to $\approx 10^{16}$ cm^{-3} concentration range. Since the traps are larger in concentration and are deep, the Fermi level is in the middle of the semiconductor, and semi-insulating characteristics

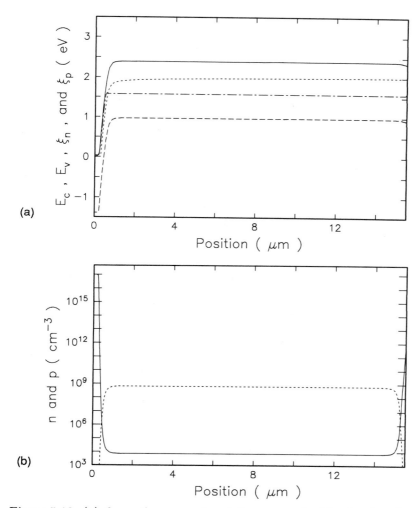

Figure 5.19: (a) shows the energy band diagram under bias for the $n^+-\nu-n^+$ example with bipolar transport. The trap is assumed to have significantly different electron and hole capture cross-sections. (b) shows the carrier densities as a function of position.

result because of the relatively large bandgap of GaAs. However, this technique of obtaining semi-insulating characteristics relies on compensation by traps that have differing electron and hole capture cross-sections. In the case of trap-filled conduction whose one-dimensional and single carrier theory we have looked at, the potential drop occurs both in the bulk and at the substrate–channel junction. This leads to one possible means of electrostatic propagation, and should have a threshold feature associated with it corresponding to the trap-filled voltage. Although we looked at a single carrier theory for this, it should also occur in ambipolar conduction, provided the capture cross-sections are similar for both the carriers. In the case of large differences in capture cross-sections of the carriers, and provided both carriers are available, propagation of potential can occur due to equilibrium between one carrier and the trap—e.g., electrons and donor traps—with the current being carried by the other carrier.[9] Since the current density is small, this allows most of the electrostatic potential drop to occur at the substrate–channel interface, hence resulting in efficient sidegating. This is the second possible mechanism for sidegating.

We discuss the first mechanism and possible means of reducing its impact first. The effect of a sidegate leading to a perturbation of electric fields in the vicinity of a device is shown in Figure 5.20. In this example, once sufficient voltage is applied across the material, so that injected carriers saturate the donor trap, ambipolar conduction determines the transport across the material. The required voltage to fill the traps is the trap-filled voltage V_{TFL}. We derived an expression for this. In general, this is proportional to the nth power of the spacing between the device and the sidegate where n is generally larger than $3/2$, depending on the characteristics of the trap. Once this trap-filled voltage is reached, conduction occurs and the depletion width at the channel substrate interface begins getting modulated, giving rise to a modulation of the channel current of the FET. The current–voltage characteristics show a decrease in current with the sidegate bias. A reverse bias is required for this conduction process to begin. Figure 5.20 shows the onset of substrate conduction and how it is affected by other potentials applied under different conditions. The two cases show the signature of a rapid rise in sidegating current when a trap-filled voltage is reached. The rapid increase in current in this example occurs at the V_{TFL} and following this rapid rise the current remains larger. It is also following this rise that the drain current changes rapidly, hence the leakage current through the substrate and associated potential propagation is directly responsible for

---

[9]For a discussion of this type of propagation, with one specific trap characteristic, see N. Goto, Y. Ohno, and H. Yano, "Two-Dimensional Numerical Simulation of Sidegating Effect in GaAs MESFETs," IEEE Trans. on Electron Devices, **ED-37**, No. 8, p. 1821, Aug. 1990.

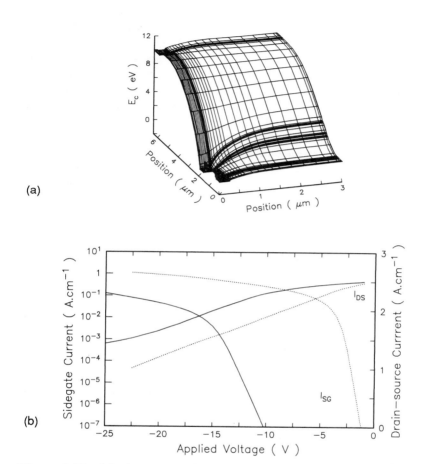

Figure 5.20: (a) shows a surface plot of the conduction band edge of a MESFET in the presence of modulation by a sidegate. The source of the device is located near the origin, followed by the drain and the sidegate. The sidegate has the large bias applied. (b) shows the current–voltage characteristics (drain-to-source current (I_{DS}) and sidegate current (I_{SG})) as a function of applied voltage at the sidegate. Results due to two different trap densities are shown; these result in different current–voltage responses.

the sidegating effect.

If sidegating were to occur through this particular mechanism, one could envision controlling it by providing a shielding of the sidegate potential and modulation of the leakage current. Such a shielding could be provided by applying an additional potential between the sidegate and the device using an ohmic contact, a n^+ region, or a metal–semiconductor junction, all of which are reverse biased. In the ohmic and n^+ region–based structures, since there is no voltage drop across the contact—the current being small—all the voltage drop occurs across the semi-insulating substrate and the channel–semi-insulating substrate interface. This leads to a decrease in the device channel saturation current due to the application of this shielding bias itself. The shield itself acts as an additional sidegate, acting as a remote gate. It reduces the current flowing in the device; since it now isolates the device from the actual sidegate potential, it also reduces the effect of the sidegate voltage. In this case, the shield at the channel interface acts just like a remote gate because of the depletion region for this applied voltage. However, it shields the other sidegate voltage applied through other metal lines. A more efficient means of providing a shielding for this sidegating mechanism is through a metal–semiconductor barrier—similar to the metal lines that caused this problem in the first place. The metal–semiconductor diode has a voltage drop across the junction and the semi-insulating regon instead of the channel-substrate interface region (see Figure 5.21, which should be compared with Figure 5.20). This capacitive shielding (ohmic and n^+ contacts provide resistive shielding) occurs without the modulation of the channel-substrate interface, and hence the channel current continues to stay large. The use of a metal–semiconductor diode does not change the device saturation current appreciably, although it shifts the onset of backgating considerably. With an ohmic shield, the saturation current changes substantially because of its conduction and proximity to the channel. Similarly, the doped region shield also reduces the channel current.

There is another means by which this sidegating mechanism can be reduced in useful circuits. This employs generation of a high concentration of traps, such as by ion implanting in the vicinity of the device, since it would raise the trap-filled voltage. The defects may be native defects, e.g., those created by implanting ionized hydrogen (i.e. protons), isoelectronic species such as boron, or inert species such as helium, argon, etc. In addition to native defects, one may generate extrinsic deep levels by implanting other impurities such as oxygen. This allows annealing away of the native defects while still maintaining the efficacy of additional deep levels. Of the species mentioned, oxygen is among the heaviest, and is apparently among the most effective because of the nature of the deep traps generated.

The trap-fill–limited sidegating phenomenon is dependent on the spac-

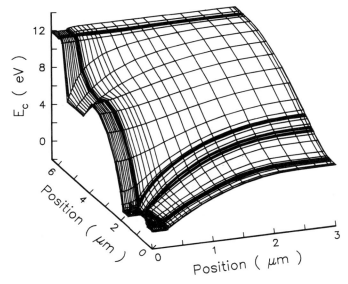

Figure 5.21: Example of shielding of sidegating effect using a metal–semiconductor junction. The figure shows a surface plot of the conduction band edge. The source is at the origin, followed by the drain, the shield electrode, and the sidegate electrode. The shield is shown to reduce the electrostatic potential propagation from a sidegate.

ing of the sidegate. None of these techniques is completely effective in suppressing sidegating at the device-to-device spacings one would like to use in circuits, which is approximately the pitch of metal lines. This dimension is too small for the above techniques to be completely effective. In practice, with implantation isolation, one employs a sufficiently large device-to-device spacing to limit the worst-case drain current variations to an acceptable range; or, equivalently, the threshold voltage and other device parameters to an acceptable range. Another possible solution of this problem is the use of a sufficiently p-doped layer to isolate devices by junction isolation. However, this also leads to an increase in capacitance to the substrate and reduces the speed of circuits.

The trap-fill–limited conduction discussed so far depends on the distance between the sidegating electrode and the device. Quite often, one encounters instances where sidegating is not dependent on the distance, and a large modulation of the channel can occur over very large distances, such as from the back of the substrate. Such cases involve sidegating through a mechanism that is obviously very different, because for substrate thickness, which is usually about 500 μm, the trap-fill–limited voltage would be very high. Here, the likelihood of the second sidegating mechanism discussed in the beginning of this section is most likely. In these cases the trap characteristics of large asymmetry between the capture cross-sections for electrons and holes plays a very important role and leads to the behavior of an otherwise semi-insulating material acting like a conductor of the potential, although very little actual current flow occurs. The reason is that electron and hole concentrations are very low for this conduction, or for contribution towards band bending. The substrate potential distribution is related to the charge on the trap, e.g., the deep donor trap due to EL2. This charge depends on the statistics of occupation and is related to the capture cross-sections. The deep donor trap has a significantly larger electron cross-section and a small electron lifetime ($\approx 10^{-9}$ s), and a small hole capture cross-section and a large hole lifetime ($\approx 2 \times 10^{-5}$ s). This leads to local neutralization of the trap throughout the substrate except at the channel-substrate interface. The potential drop occurs at the channel-substrate interface, and hence, significant sidegating that is independent of the distance occurs. Note that in this instance the semi-insulating material acts as an n–p junction.

An example of this propagation of electrostatic potential is shown in Figure 5.22 for the deep acceptor shallow donor case. An example of a deep acceptor is Cr, together with Si or other shallow donor species. So, in this case, the conduction occurs due to electrons, and the trap compensation is by holes. This mechanism shows no threshold behavior. Cr, e.g., is one of the deep acceptors in Cr-doped semi-insulating substrates, where the substrate will behave like a p-type region. The converse of this occurs with deep-donor EL2 and a shallow acceptor. Additional extensions of these occur at the surface due to temperature-induced changes in the deep and shallow level densities, and we consider an example with EL2 compensated material for this variation.

5.6.3   Surface-Dominated Behavior

During the high temperature processing for fabrication of devices, EL2 depletion can occur at the surface together with an accumulation of acceptors (such as boron and carbon). This can lead to a very lightly compensated or even a p-type layer at the surface. If the material is entirely p-type, electrostatic propagation could occur with a process that is complementary of the previous example. Now, the holes will carry the current and the electrons will equilibrate with the traps.

A surface can play a role in other ways, too. Due to the Fermi level

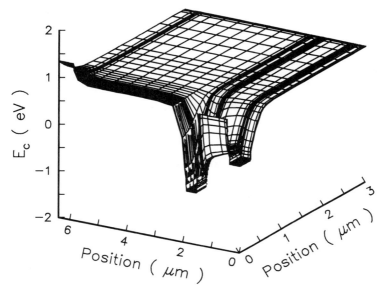

Figure 5.22: Surface plot of conduction band edge showing electrostatic potential propagation for a semi-insulating substrate consisting of a deep acceptor and shallow donor due to local neutralization of acceptors by holes. The current is carried by electrons. The figure, unlike the earlier surface plots, shows the applied sidegate bias appearing nearly entirely at the channel-substrate interface.

pinning at the surface, unoccupied trap concentration can become lower than in the bulk, leading to a lowering of the trap-filled voltage limit, and hence to the mechanism that we discussed earlier.

We have emphasized the role of traps and trap statistics in our discussion. The role of carriers and carrier injection is also important, since the equilibrium is achieved with carriers available. Avalanching at a large biased junction can, e.g., cause both electrons and holes to be available. So, results from metal electrodes, even if they are poor sources of holes at low bias, can be different at high bias. Both these cases will show different sidegating behavior. It should be clear from our discussion that sidegating can occur by a variety of means in different materials due to differing compensation and surface conditions brought about by the traps, the processing, and the geometry. The sidegating thresholds can be raised by increased trap concentrations in some of the cases where trap-filling is the dominant mechanism; these traps can be introduced using damage implantation. However, these may have little effect in other cases where electrostatic propagation occurs because of the trap characteristics and the polarity of the material. In all these cases, one possible solution is to use sufficiently p-type doped material so that junction isolation is achieved. This would isolate the sidegating voltage from the substrate–channel interface by dropping across the sidegate–substrate junction, but at the expense of an increase in capacitance.

5.7 Piezoelectric Effects

A second problem, unique to compound semiconductors, is related to their ionic character. This ionic character leads to a piezoelectric nature of the crystal, and an effect resulting from stress that is termed the piezoelectric effect. The basis of this effect lies in stress-induced polarization in the crystal because of the ionic charge and distortion of the lattice. The polarization can be viewed to result in additional charge because

$$\nabla.(\epsilon_s \mathcal{E} + \mathcal{P}) = \rho, \qquad (5.99)$$

where $\mathcal{P}$ is the polarization vector and ρ is the charge density.

The polarization results in additional charge and hence device parameter variation because of stress[10] in the material. This stress can occur in devices because of different expansion coefficients of the materials employed; it may result from the dielectric films employed on semiconductor surfaces or the metallizations of the gate and ohmic contacts which are in proximity to the device active region.

We first show in a simple way how this polarization occurs in GaAs as shown schematically in Figure 5.23. This figure is drawn in one plane. When a compressive stress is applied, the angle between the bonds for the

---

[10]The problem of determining stress and its piezoelectric consequences is a subject intimately related to finite element analysis and the theory of elasticity of solids. This section is written as an introduction without undue reliance on the mathematical details of the description of the problem. The interested reader is referred to S. P. Timoshenko and J. N. Goodier, *Theory of Elasticity*, McGraw–Hill, N.Y. (1970); P. M. Asbeck, C. P. Lee, and M. F. Chang, "Piezoelectric Effects in GaAs FETs and Their Role in Orientation-Dependent Device Characteristics," *IEEE Trans. on Electron Devices*, **ED-31**, p. 1377, 1984; J. C Ramirez, P. J. McNally, L. S. Cooper, J. J. Rosenberg, L. B. Freund, and T. N. Jackson, "Development and Experimental Verification of a Two-Dimensional Numerical Model of Piezoelectrically Induced Threshold Voltage Shifts in GaAs MESFETs," *IEEE Trans. on Electron Devices*, **ED-35**, p. 1232 1988; and T. Onodera and H. Nishi, "Theoretical Study of Piezoelectric Effect on GaAs MESFETs on (100), (011), ($\overline{111}$)Ga, and (111)As," *IEEE Trans. on Electron Devices*, **ED-36**, p. 1580, 1989, for further details.

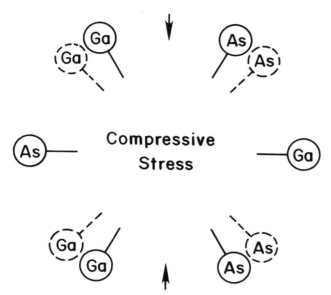

Figure 5.23: Polarization effect in GaAs as a result of compressive stress. The atoms are shown in one plane viewed from the $\langle 111 \rangle$ direction.

atoms away from the center section increases and there is a net displacement of the positive charge to the left and the negative charge to the right. Consequently, there is now a net dipole moment in the crystal pointing towards the right and the crystal is polarized. A shear stress, instead of compressive stress, would have similarly resulted in a net dipole moment pointing upwards. This polarization, caused by stress, can be determined in general by tensor analysis as

$$\mathcal{P} = \boldsymbol{d} \times \boldsymbol{\sigma}, \tag{5.100}$$

where $\boldsymbol{d}$ is the piezoelectric tensor and $\boldsymbol{\sigma}$ is the stress tensor.

There is an additional requirement for this polarization to lead to a cumulative charge effect and hence a macroscopic variation with position. This is that the crystal be non-centro-symmetric. Non-centro-symmetry leads to the accumulation of the polarization of the unit cells because the polarizations are additive and have the same direction. Centro-symmetric crystals do not accumulate the polarization of unit cells since they oppose in alternate unit cells.

Most of the elements of $\boldsymbol{d}$ are zero for compound semiconductors, except

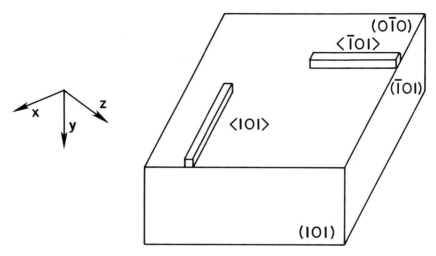

Figure 5.24: Orientation nomenclature. We have treated y axis as pointed towards the substrate. Thus, the surface of GaAs is (010). The figure also shows the gate orientations in MESFETs.

the ijth component for which $i \neq j$. This results in a polarization vector

$$
\begin{bmatrix} \mathcal{P}_x \\ \mathcal{P}_y \\ \mathcal{P}_z \end{bmatrix} = d \begin{bmatrix} \sigma_{xy} \\ \frac{1}{2}\left(\sigma_{xx} - \sigma_{zz}\right) \\ \sigma_{yz} \end{bmatrix},
\tag{5.101}
$$

where d is the magnitude of the equal non-zero components, the x subscript denotes the $\langle 101 \rangle$ direction, the y subscript denotes the $\langle 010 \rangle$ direction, and the z subscript denotes the $\langle \bar{1}01 \rangle$ direction, as shown in Figure 5.24.

Since the strongest effect occurs due to the surface dielectric and metal films in close vicinity to the active region of the device, we need to assess the effect of stress in geometries such as those shown in Figure 5.25. The effect from stress due to the gate and the dielectric films may be present simultaneously. Both the gate and the dielectric films can be under compressive or tensile stress, and hence the effect can cause electric parameter (e.g., threshold voltage) change of both signs. In addition, depending on the orientation of the gate or dielectric lines, the polarization results are different. Si_3N_4, one of the dielectric films employed, is usually under compressive stress, while amorphous SiO_2, another dielectric employed, is usually under tensile stress. The resulting polarization from the dielectric film is therefore of the opposite sign, and hence threshold voltage varia-

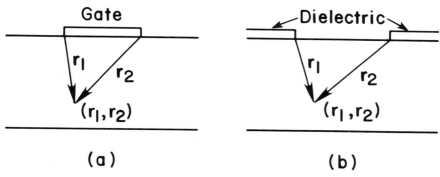

Figure 5.25: Two common sources of piezoelectric effects due to stress generated by the mismatch of materials. (a) shows the effect occuring from the gate of a MESFET, while (b) shows that due to dielectric films. Both of these effects may be present simultaneously.

tion is opposite. Using the stress tensor we can determine the polarization. Knowing the polarization,

$$\rho_{pol} = \mathbf{\nabla}.\mathcal{P}, \tag{5.102}$$

the charge density associated with the polarization can be found for the bulk.

At the surface of the crystal, a surface charge density of $\mathcal{Q}_{pol}$ is also induced,

$$\mathcal{Q}_{pol} = \hat{n}.\mathcal{P}, \tag{5.103}$$

where $\hat{n}$ is the unit vector normal to the surface. For a $\langle 010 \rangle$ surface, this results in $\mathcal{Q}_{pol} = \mathcal{P}_y$. As an example, consider the geometry of Figure 5.25. The maximum stress occurs at the corner of the substrate and the film. For gates and dielectric films, the two corners are the logical choice for a coordinate system $(\mathbf{r}_1, \mathbf{r}_2)$, where the radial distance is measured from the the points of highest stress. The resulting stress components can be written as

$$\sigma_{xx} = -\frac{2}{\pi} \nu \sigma_f t \left(\frac{z_1}{r_1^2} - \frac{z_2}{r_2^2} \right),$$

$$\sigma_{yy} = -\frac{2}{\pi} \sigma_f t \left(\frac{z_1 y^2}{r_1^4} - \frac{z_2 y^2}{r_2^4} \right),$$

$$\sigma_{zz} = -\frac{2}{\pi} \sigma_f t \left(\frac{z_1^3}{r_1^4} - \frac{z_2^3}{r_2^4} \right),$$

$$\text{and} \qquad \sigma_{zy} \;=\; -\frac{2}{\pi}\sigma_f t \left(\frac{z_1^2 y}{r_1^4} - \frac{z_2^2 y}{r_2^4}\right), \qquad (5.104)$$

where t is the film thickness, which has also been assumed to be much smaller than the substrate, ν is the Poisson ratio, and σ_f is the stress in the film. Note that the maximum stress occurs at the substrate–film interface and that it disappears for large distances into the substrate. This gives for the polarization charge density in the bulk and at the surface:

$$\rho_{pol} \;=\; \gamma_b \sigma_f t \left\{ \frac{z_1 y \left(z_1^2 - \beta y^2\right)}{r_1^6} - \frac{z_2 y \left(z_2^2 - \beta y^2\right)}{r_2^6} \right\}$$

$$\text{and} \qquad Q_{pol} \;=\; \gamma_s \sigma_f t \left(\frac{1}{z_1} - \frac{1}{z_2}\right). \qquad (5.105)$$

The magnitudes, for GaAs, are $\nu = 0.31$, $\gamma_b = 2d(4 + \nu)/\pi \approx 390$ electron charges per dyne, $\beta = (2+\nu)/(4+\nu) = 0.53$, and $\gamma_s = d(1-\nu)/\pi \approx 40$ electron charges per dyne. An example of the piezoelectric charge distribution for the $\langle \bar{1}01 \rangle$ oriented device in compressive stress is shown in Figure 5.26.

The piezo-charge gets stronger towards the center of the gate and away from the surface, and is dependent on the angle. Near the tail of the doping density, its effect can be substantial because of a reduction in dopant charge. It can, therefore, significantly disturb the low current behavior of devices, i.e., it affects parameters such as threshold voltage and conductance at low biases. The effect is strongly dependent on the orientation. For a device perpendicular to this direction, the charge sign will actually reverse, and hence the electric parameters will change in the opposite direction. This leads to the problem of changes in threshold voltage in devices that are oriented differently on the same wafer—a significant design constraint. In the example that we have looked at, the threshold voltage shifts negative, and the device is more conductive at any given bias because of the piezoelectric effect. However, devices perpendicular to this device have a threshold voltage shift that is positive, with the device being less conductive.

The threshold shift also becomes more pronounced with a decrease in dimensions, i.e., shortening of the gate length. Shortening of the gate-length naturally leads to an increase in the stress underneath the gate due to stress from dielectric films. This gate-length effect is in addition to the effect of the short-channel phenomenon. Figure 5.27 shows the threshold variation in an example of GaAs MESFETs. Note the rapid increase in the shifts in the sub-5 μm range of gate lengths. Thicker films lead to larger stress, and hence stronger shifts.

While these results are for compressive Si_3N_4 films, the results for SiO_2

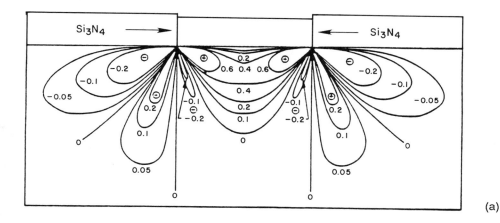

(a)

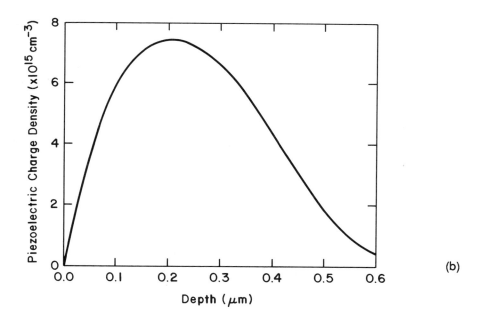

(b)

Figure 5.26: Normalized piezoelectric charge distribution due to compressive Si_3N_4 film is shown in (a). (b) shows the charge density at the center of the gate and going into the substrate. After P. M. Asbeck, C. P. Lee, and M. F. Chang, "Piezoelectric Effects in GaAs FETs and Their Role in Orientation-Dependent Device Characteristics," *IEEE Trans. on Electron Devices*, **ED-31**, p. 1377, ©1984 IEEE.

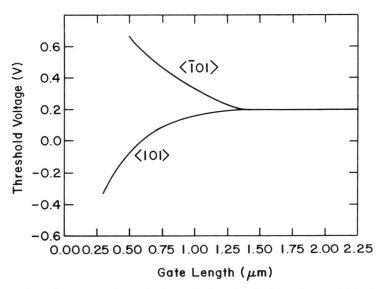

Figure 5.27: A schematic variation of the threshold voltage shift due to compressive stress in dielectric films as a function of gate-length for GaAs MESFETs. Note the differing effect on two different orientations.

are similar but of the opposite sign because those films are usually tensile in nature.

Interestingly, it turns out that variations due to such stress effects on (101) oriented substrates with $\langle 101 \rangle$ or $\langle \bar{1}01 \rangle$ oriented FETs are minimal because of the symmetry of the piezoelectric charge distribution. It changes sign at the center of the gate on either side of the plane through the substrate. Figure 5.28 shows the piezoelectric charge distribution effects due to stress in gate metallurgy placed on a (101) surface of GaAs. This results in a significantly lower threshold voltage dependence on the gate-length for this substrate orientation.

In GaAs, the (100) family of substrate surfaces is among the most difficult surfaces to work with, due to the large piezoelectric effect. All other major orientations show periodic modulation of charge, as in the above example, leading to smaller threshold voltage shifts. Figure 5.29 shows the gate orientation dependence of threshold voltage with various crystallographic surfaces due to stress caused by gate metallurgy. Effects similar to dielectric stress effects occur due to gate stress. In addition, these stress effects show up as a function of temperature, even if a device is fabricated

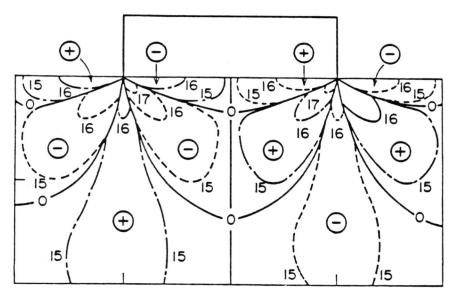

Figure 5.28: Piezoelectric charge distribution for a $\langle\overline{1}01\rangle$ oriented gate on a (101) substrate. Contours correspond to electron density of 10^{15}, 10^{16}, and 10^{17} cm^{-3}. After T. Onodera and H. Nishi, "Theoretical Study of Piezoelectric Effect on GaAs MESFETs on (100), (011), $(\overline{111})$Ga, and (111)As," *IEEE Trans. on Electron Devices*, **ED-36**, p. 1580, ©1989 IEEE.

to have a negligible stress for any specific temperature of operation.

5.8 Signal Delay along the Gate

We now return to the modelling of the devices at high frequencies. Our quasi-static analysis is strictly true only for moderate and low frequencies since it ignores all dispersive effects and treats the problem as a steady-state problem. The discussion in this and the following sections is related to understanding the device operation from a high frequency perspective.

MESFETs are high frequency devices, and we have been analyzing them, in quasi-static approximation, for phenomena underneath the gate between the source and the drain. We have assumed the gate to be an equi-potential region because it is usually made of metal. Both at high frequencies and for unusually wide gate structures, these approximations break down. A metal line, such as the gate on a substrate, is actually a distributed transmission line in the frequency range of interest to us (see

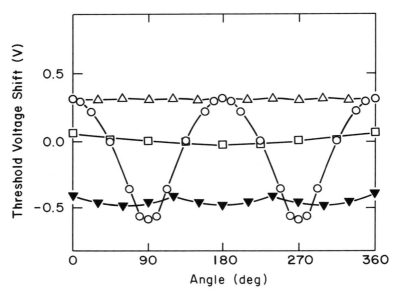

Figure 5.29: Threshold voltage shift as a function of gate orientation on various GaAs substrate surfaces. The angles are w.r.t. $\langle 10\bar{1}\rangle$ orientation. The open circles are for (010) surface, the squares are for (101) surface, the normal triangles are for $(\overline{111})$ Ga surface, and the inverted triangles are for 111) As surface. The periodic changes in the threshold voltage are the largest for the (010) substrate. After T. Onodera and H. Nishi, "Theoretical Study of Piezoelectric Effect on GaAs MESFETs on (100), (011), $(\overline{111})$Ga, and (111)As," *IEEE Trans. on Electron Devices*, **ED-36**, p. 1580, ©1989 IEEE.

Figure 5.30). Thus, a signal fed into the gate is delayed and decreases in amplitude as it transmits further down the gate. Depending on the termination at the end of this gate line, reflected waves may also be set up. We do not wish, here, to describe the complexity of this problem. But we do wish to show the limits of validity of frequency analysis due to the distributed nature of the problem. As Figure 5.30 shows, the gate itself may be modelled as a transmission line network; the per unit length inductance $\mathcal{L}_g$ and capacitance $\mathcal{C}_g$ are related through the gate's characteristic impedance Z_g. Both the capacitance and the impedance are bias-dependent. If the resistive drops (ignored in the above) were also important, the amplitude of the signal would change during the transmission along the gate. The

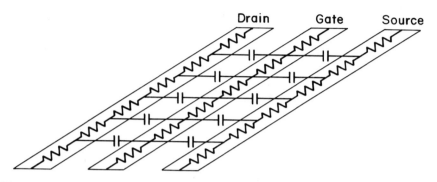

Figure 5.30: A schematic for transmission of signal along the gate for a long gate-width MESFET, together with lumped model representation of sections.

characteristic impedence Z_g is related by

$$Z_g = \left(\frac{\mathcal{L}_g}{\mathcal{C}_g}\right)^{1/2}. \tag{5.106}$$

The phase delay in the drain current as a result of this propagation, assuming that the drain electrode, of the same form as the gate, is fed from the opposite end, is:

$$\theta_g = \omega(\mathcal{L}_g\mathcal{C}_g)^{1/2}W. \tag{5.107}$$

Problem 7 analyzes an approximate way of understanding the limits that this poses to the frequency of operation. Since transmission-line effects are dominant at long gate widths, terminations at the end of the gate and drain lines can also have a significant impact on the predicted results, and a clever use of such a termination together with appropriate modification of the gate and drain transmission lines can be used to advantage in getting specific frequency-dependent gain characteristics.

5.9 Small-Signal High Frequency Models

We have estimated most of the basic elements that characterize the operation of the transistor. This has been based on a quasi-static formulation, since we did not employ the small-signal transport equations. Small-signal analysis based on solution of the static and small-signal transport equations is a difficult proposition because the practical MESFET problem is inherently a two-dimensional problem. Solutions based on a one-dimensional approximation of a two-dimensional or three-dimensional problem are inherently more inaccurate. An analogy would be that our analysis using gradual channel approximation was relatively accurate for the static device characteristics, i.e., we could make predictions of the static currents that would flow under static bias conditions. Our predictions from this quasi-static analysis of the perturbational quantities such as output conductance were relatively more inaccurate. A first-order model should not be expected to predict a second-order parameter. Even if we could make an accurate quasi-static second-order model that included two-dimensional effects, etc., and applied perturbation analysis to it, it would be valid only under quasi-static conditions, i.e., at frequencies quite lower than the limit frequencies of the device. In devices that are dominated by time constants resulting from charging of capacitances that are lumped in nature, such as the extrinsic elements, the quasi-static approximation is generally a good one even for predicting these limit frequencies. This is so because the lumped representations of these resistors and capacitors still remains valid, and hence they continue single pole roll-off beyond the frequencies where the validity of quasi-static models is questionable. Should lumped representation be questionable, such as for very wide gates where there is a phase delay effect resulting from transmission of the signal down the gate, the modelling approach becomes entirely invalid.

Our discussion here points out that a true small-signal representation can thus be derived only from small-signal equations, which would include dispersive, i.e., frequency-dependent signal transmission effects. For a two-dimensional geometry this discussion points to using network parameters as a small-signal model for the device. We may derive equivalent circuits from these network parameters by physical insight, however, the equivalent circuits are approximations that represent the modelled or measured small-signal characteristics of the device. The problems related to equivalent circuit representation based on physical insight of the storage and transport processes occurring in a device relate to issues of charge partitioning. Associating a capacitance between two terminals is to make use of linear independence of charge partitioning, which is not always valid in a multi-terminal device. Thus, equivalent circuits are not unique. However, equivalent circuits are very useful because their valid use in a range of operating conditions allows ease in the designing of circuits, and, if physically correct, an insight into the operation of the device.

While we have made these comments for frequency domain characteristics, they are equally valid for time domain characteristics. The charge partitioning problem arises in time domain as a charge non-conservation problem. Since capacitances are represented between nodes to model the charges, and the charge itself is partitioned based on physical insight from specific regions of the device at specific bias points, the calculated charges are path-dependent in a multi-terminal device. This is to say that while charge is the unique variable, use of capacitance to represent its effect results in a non-unique (although still relatively accurate) solution for a change from one state to another state. The capacitances are path dependent, and hence non-unique solutions may result.

The coupling of the small-signal response to an equivalent circuit with a range of validity is valuable, if physical in basis, and if the limitations are clearly understood. We will elaborate on an approach to the small-signal analysis of the MESFET by considering the small-signal transport equation—the transmission line equation in gradual channel approximation, and at the end of this section we will relate this to the equivalent circuit referred to in our quasi-static analysis. The network response of the two port network with the gate as the input port and the drain as the output port is

$$\left[\begin{array}{c} \tilde{I}_g \\ \tilde{I}_d \end{array} \right] = \left[\begin{array}{cc} y_{gg} & y_{gd} \\ y_{dg} & y_{dd} \end{array} \right] \times \left[\begin{array}{c} \tilde{V}_g \\ \tilde{V}_d \end{array} \right], \tag{5.108}$$

where all the small-signal currents flow into the port.

Figure 5.31 shows a model of the incremental cross-section, where we assume that gradual channel approximation and the constant mobility assumption holds. We will make several simplifying assumptions, some already commented on, regarding this analysis, which is a high frequency equivalent of the quasi-static analysis. It employs equations that are similar in basis[11] but with frequency-dependent variables. We will derive it for the bias condition under which the channel current just saturates as a result of carrier velocity saturation. This occurs at the field $\mathcal{E}_s$. So, the electric field becomes $\mathcal{E}_s$ at the position $z = L$. We can thus employ the low field constant mobility model over the entire channel and derive an equivalent circuit at the channel current saturation point, giving a solution that we may apply in the current saturation region within a limited approximation.

In the low field model, the drain current of one half of the symmetric structure is given as

$$I = qn\mu aW \left(1 - \frac{h(z)}{a} \right) \frac{\partial V}{\partial z}, \tag{5.109}$$

where $h(z)$ is the depletion width at position z along the channel and only

---

[11]The interested reader may wish to refer to J. A. Geurst, "Calculation of High-Frequency Characteristics of Field-Effect Devices," *Solid-State Electronics*, **V8**, p. 563, 1965 from where we have adopted details of this analysis.

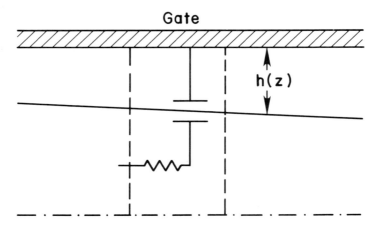

Figure 5.31: Cross-section of an incremental part of the gate-controlled channel region of the MESFET in gradual channel approximation. The capacitor and resistor model the effect of the incremental section.

drift current is considered. This equation is also written as

$$I = qn\mu aW \left[1 - \left(\frac{\mathcal{V}}{\mathcal{V}_{00}}\right)^{1/2}\right] \frac{\partial \mathcal{V}}{\partial z}. \tag{5.110}$$

Both $\mathcal{V}$ and I are time-dependent quantities in the above if a time-dependent signal is applied, $\mathcal{V}$ is given by

$$\mathcal{V} = \psi_{j0} + V_g - V, \tag{5.111}$$

where both V_g and V are time-dependent terms. This notation is the same as employed in the discussion of the PHS model; $\mathcal{V}_{00}$ is the potential associated with full channel depletion in gradual channel approximation. In order to bring the equation to a simple form, it can be written in terms of normalized quantities defined as follows:

$$\xi = \frac{z}{L},$$

$$\theta = \frac{t}{t_0},$$

$$\chi = \frac{\mathcal{V}}{\mathcal{V}_{00}},$$

$$\text{and} \quad \iota = \frac{I}{I_{norm}}, \tag{5.112}$$

where

$$t_0 = \frac{\epsilon_s}{qn\mu} \left(\frac{L}{a}\right)^2$$

$$\text{and} \quad I_{norm} = \frac{aq\mu n \mathcal{V}_{00}}{3L}. \tag{5.113}$$

Using a transformation $\kappa = 2\sqrt{\chi} - 1$, where

$$\chi = \frac{\psi_{j0} + V_g - V}{\mathcal{V}_0}, \tag{5.114}$$

one can rewrite the current equation as

$$\iota = \frac{3}{4}\left(1 - \kappa^2\right)\frac{d\kappa}{d\xi} = \frac{d\eta}{d\xi}, \tag{5.115}$$

where

$$\eta\left(\kappa\right) = \frac{3}{4}\left(\kappa - \frac{\kappa^3}{3}\right). \tag{5.116}$$

The time-dependent continuity equation, assuming no generation and recombination since there is negligible minority carrier density, is

$$\frac{\partial I}{\partial z} = -\frac{\partial \rho}{\partial t} = \frac{\partial}{\partial t}\left[qN_D aW\left(\frac{\mathcal{V}}{\mathcal{V}_{00}}\right)^{1/2}\right]. \tag{5.117}$$

This may be written as

$$\frac{\partial}{\partial \xi}\left[\left(1 - \kappa^2\right)\frac{\partial \kappa}{\partial \xi}\right] = \frac{4}{3}\frac{\partial \kappa}{\partial \theta}. \tag{5.118}$$

The use of superposition of steady-state and time-dependent parts for the normalized variables κ and ι,

$$\kappa\left(\xi, \theta\right) = \overline{\kappa}\left(\xi\right) + \tilde{\kappa}\left(\xi, \theta\right)$$

$$\text{and} \quad \iota\left(\xi, \theta\right) = \overline{\iota}\left(\xi\right) + \tilde{\iota}\left(\xi, \theta\right), \tag{5.119}$$

leads to the steady-state solution of

$$\overline{\iota} = \eta\left(\overline{\kappa}_d\right) - \eta\left(\overline{\kappa}_s\right), \tag{5.120}$$

with

$$\xi = \frac{\eta\left(\overline{\kappa}\right) - \eta\left(\overline{\kappa}_s\right)}{\eta\left(\overline{\kappa}_d\right) - \eta\left(\overline{\kappa}_s\right)} \tag{5.121}$$

in terms of the magnitude of κ at the source and the drain end. The transconductance for the steady-state term, using this, is

$$\overline{g}_m = \frac{q\mu N_D a W}{L}\left(\overline{\kappa}_d - \overline{\kappa}_s\right), \tag{5.122}$$

which agrees with the quasi-static transconductance derived before.

The time-dependent part of the equation is

$$\frac{\partial}{\partial\overline{\kappa}}\left(\frac{1}{1-\overline{\kappa}^2}\frac{\partial\tilde{v}}{\partial\overline{\kappa}}\right) = \frac{3}{4\overline{\iota}^2}\frac{\partial\tilde{v}}{\partial\theta}, \tag{5.123}$$

where

$$\tilde{v} = \frac{\partial\eta\left(\overline{\kappa}\right)}{\partial\overline{\kappa}}\tilde{\kappa},$$

$$\text{and}\qquad \tilde{\iota} = \frac{4}{3}\frac{\overline{\iota}}{1-\overline{\kappa}^2}\frac{\partial\tilde{v}}{\partial\overline{\kappa}}. \tag{5.124}$$

Using

$$\tilde{v} = \hat{v}\exp\left(j\omega t\right) = \hat{v}\exp\left(j\omega_1\theta\right)$$

$$\text{and}\qquad \tilde{\iota} = \hat{\iota}\exp\left(j\omega t\right) + \hat{\iota}\exp\left(j\omega_1\theta\right), \tag{5.125}$$

where $\omega_1 = \omega t_0$, and introducing additional variables α and ν through

$$z = \alpha\overline{\kappa}\qquad \text{where}\qquad \alpha = \left(\frac{3\omega_1}{\overline{\iota}^2}\right)^{1/4}\exp\left(j\frac{3\pi}{8}\right), \tag{5.126}$$

and

$$\nu = \frac{\alpha^2}{4} - \frac{1}{2}, \tag{5.127}$$

we can write

$$\frac{d}{dz}\left(\frac{1}{\nu + \frac{1}{2} - \frac{1}{4}z^2}\frac{d\hat{v}}{dz}\right) + \hat{v} = 0. \tag{5.128}$$

This is a simplified form of the control equation for small signals in the channel of the MESFET using normalized parameters. This equation is similar to Weber's equation, which is generally written in the form

$$\frac{d^2w}{dz^2} + \left(\nu + \frac{1}{2} - \frac{1}{4}z^2\right)w = 0. \tag{5.129}$$

Our control equation can be solved numerically, using the known solutions of Weber's equation, following further compaction. Approximate solutions can also be obtained for it using Taylor series expansion (see Problem 8). This Taylor series expansion technique could also have been applied in the original form of the time-dependent current continuity equation.

The major reason for using this technique is to show that if desired, complete solutions of the differential equations involved in one-dimensional approximations of time-dependent behavior can be found.[12] Intuitive understanding of the behavior, however, is lost because of the complexity of the form of these solutions. Thus, approaches involving iterative procedures to evolve series expansion solutions can be particularly rewarding. We will address the latter techniques later in this section. First, we establish a compact form of the solution to y-parameters based on the Weber-like equation. Let w_1 and w_2 be the two independent solutions of Weber's equation[13] (parabolic cylinder functions are particular solutions of Weber's equation). We can now write the y-parameters from the solution of these equations. The common-gate parameters are easier to write because the channel transport equation is written from source end to drain end. These can be subsequently changed to common-source parameters using the network parameter transformations described in Appendix A. The evaluation of different y-parameters and their subsequent transformation to the more desired form is a technique of convenience that will be applied to bipolar transistors also. In the common-gate mode these y-parameters, with overdots representing derivatives, are

$$y_{ss} = \bar{g}_m \frac{\alpha^3 \bar{\imath}}{6\Delta}(1 - \bar{\kappa}_s) \begin{vmatrix} w_1(\alpha \bar{\kappa}_s) & w_2(\alpha \bar{\kappa}_s) \\ \dot{w}_1(\alpha \bar{\kappa}_d) & \dot{w}_2(\alpha \bar{\kappa}_d) \end{vmatrix},$$

$$y_{sd} = -\bar{g}_m \frac{\alpha^3 \bar{\imath}}{6\Delta}(1 - \bar{\kappa}_d) \begin{vmatrix} w_1 & w_2 \\ \dot{w}_1 & \dot{w}_2 \end{vmatrix},$$

$$y_{ds} = -\bar{g}_m \frac{\alpha^3 \bar{\imath}}{6\Delta}(1 - \bar{\kappa}_s) \begin{vmatrix} w_1 & w_2 \\ \dot{w}_1 & \dot{w}_2 \end{vmatrix},$$

$$\text{and} \quad y_{dd} = \bar{g}_m \frac{\alpha^3 \bar{\imath}}{6\Delta}(1 - \bar{\kappa}_d) \begin{vmatrix} w_1(\alpha \bar{\kappa}_d) & w_2(\alpha \bar{\kappa}_d) \\ \dot{w}_1(\alpha \bar{\kappa}_s) & \dot{w}_2(\alpha \bar{\kappa}_s) \end{vmatrix}, \quad (5.130)$$

---

[12]We will employ a similar procedure to show small-signal equations and their solutions in specific instances for HFETs in Chapter 6 and HBTs in Chapter 7.

[13]See, e.g., M. Abramowitz and I. A. Stegun, *Handbook of Mathematical Functions with Formulas, Graphs, and Mathematical Tables*, U.S. Government Printing Office, Washington, D.C., p. 358 (1964) for a summary of Weber's equation and Weber functions. Weber functions are Bessel functions of the second kind.

where Δ is the determinant

$$\Delta = \begin{vmatrix} \dot{w}_1\left(\alpha\overline{\kappa}_s\right) & \dot{w}_2\left(\alpha\overline{\kappa}_s\right) \\ \dot{w}_1\left(\alpha\overline{\kappa}_d\right) & \dot{w}_2\left(\alpha\overline{\kappa}_d\right) \end{vmatrix}. \qquad (5.131)$$

The common-source y-parameters now follow using the matrix transformations described in Appendix A. This is the rigorous method of determining the y-parameters within the gradual channel approximation and using the drift equation. It is unwieldy, so attempts have been made to describe this by either breaking the channel in various cross-sections or by making a Taylor series expansion as part of the solution to the current continuity equation. The accuracy of the solution from series expansion depends, then, on the number of terms of expansion considered.

As an example of this technique, we will consider Hauser's general treatment[14] for small-signal modelling of field effect devices. This analysis is applicable to the MESFET as well as the HFET with some approximations. Here, we will describe it and use it; the use in the HFET case will be left as an exercise where we should obtain similar results as we will in the case of our small-signal treatment in Chapter 6.

Our equations for transmission of the signal along the channel will be written as a general model composed of conductances and capacitances. The conductance, at any cross-section, where there exists a channel potential V, is influenced by the applied gate potential and the channel potential. Gradual channel approximation allows us to write this as the difference between the two potentials, i.e., the conductance G per unit length is

$$G\left(V, V_g, z\right) = G\left(V - V_g, z\right). \qquad (5.132)$$

Similarly the capacitance of any control section C per unit length is given by

$$C\left(V, V_g, z\right) = C\left(V - V_g, z\right). \qquad (5.133)$$

The corresponding transmission-line equations are

$$\frac{\partial I}{\partial z} + \frac{C}{L}\frac{\partial\left(V - V_g\right)}{\partial t} = 0$$

$$\text{and} \qquad \frac{\partial\left(V - V_g\right)}{\partial z} + \frac{1}{LG}I = 0. \qquad (5.134)$$

---

[14] J. R. Hauser, "Small-Signal Properties of Field-Effect Devices," *IEEE Trans. on Electron Devices*, **ED-12**, p. 605, 1965. An extension of this into the saturated current region of operation is suggested in P. L. Hower and N. G. Bechtel, "Current Saturation and Small-Signal Characteristics of GaAs Field-Effect Transistors," *IEEE Trans. on Electron Devices*, **ED-20**, No. 3, p. 213, 1973.

In their integral forms, these equations are

$$\int_z^L \frac{C}{L} \frac{\partial (V - V_g)}{\partial t} dz' = -\int dI = -(-I_d - I) = I_d + I$$

and

$$\int_{V-V_g}^{V_d-V_g} Gd(V - V_g) = -\frac{1}{L} \int_z^L Idz', \qquad (5.135)$$

where we use the drain as a reference and the convention for the drain current is into the drain port. A similar set of integral equations with the source as a reference is

$$\int_0^z \frac{C}{L} \frac{\partial (V - V_g)}{\partial t} dz' = -\int dI = -(I - I_s) = -I + I_s$$

and

$$\int_{V_s-V_g}^{V-V_g} Gd(V - V_g) = \frac{1}{L} \int_0^z Idz'. \qquad (5.136)$$

We can derive the static and sinusoidal components of these equations in order to relate the two using the substitution for currents and voltages. First we use this method for the drain reference equations.

$$\begin{aligned} I &= \bar{I} + \tilde{I} &= \bar{I} + \hat{I} \exp(j\omega t) \\ \text{and} \quad V &= \bar{V} + \tilde{V} &= \bar{V} + \hat{V} \exp(j\omega t). \end{aligned} \qquad (5.137)$$

For the drain reference set of integral equations, we obtain

$$\bar{I} + \bar{I}_d = 0$$

and

$$\int_{\bar{V}}^{\bar{V}_d} GdV = -\frac{1}{L} \int_z^L Idz' \qquad (5.138)$$

for the static equations with all time dependences set to zero, and

$$\hat{I} + \hat{I}_d = \frac{j\omega}{L} \int_z^L C\left(\hat{V} - \hat{V}_g\right) dz',$$

and

$$\left(\hat{V}_d - \hat{V}_g\right) G_d - \left(\hat{V} - \hat{V}_g\right) G = -\frac{1}{L} \int_z^L \hat{I} dz' \qquad (5.139)$$

for the sinusoidal equations. Combining the static equations,

$$\bar{I}_d \left(1 - \frac{z}{L}\right) = \int_{\bar{V}}^{\bar{V}_d} GdV, \qquad (5.140)$$

which lets us write the position z as

$$z = L\left(1 - \frac{1}{\bar{I}_d} \int_{\bar{V}}^{\bar{V}_d} GdV'\right). \qquad (5.141)$$

Evaluating at source end $(z = 0)$, we obtain

$$\bar{I}_d = \int_{\bar{V}_s}^{\bar{V}_d} G dV, \tag{5.142}$$

and a transformation between position and potential may be obtained by using

$$dz = -\frac{L}{\bar{I}_d} G dV, \tag{5.143}$$

where we have employed the results of the static integral equations. Similarly, combining the sinusoidal equations,

$$\left(\hat{V}_d - \hat{V}_g\right) G_d - \left(\hat{V} - \hat{V}_g\right) G =$$
$$-\frac{1}{L} \int_z^L \left\{ -\hat{I}_d + \frac{j\omega}{L} \int_{z'}^L \left[\left(\hat{V} - \hat{V}_g\right) C \right] dz'' \right\} dz', \tag{5.144}$$

or

$$\left(\hat{V} - \hat{V}_g\right) G \ =$$
$$\left(\hat{V}_d - \hat{V}_g\right) G_d - \hat{I}_d \left(1 - \frac{z}{L}\right) +$$
$$\frac{j\omega}{L^2} \int_z^L \left\{ \int_{z'}^L \left[\left(\hat{V} - \hat{V}_g\right) C \right] dz'' \right\} dz'. \tag{5.145}$$

We transform position to potential as a variable using Equations 5.143 and 5.141, giving

$$\left(\hat{V} - \hat{V}_g\right) G \ = \ \left(\hat{V}_d - \hat{V}_g\right) G_d - \frac{\hat{I}_d}{\bar{I}_d} \int_{\bar{V}}^{\bar{V}_d} G dV' +$$
$$\frac{j\omega}{L^2} \int_{\bar{V}}^{\bar{V}_d} \left\{ \int_{V'}^{V} \left[\left(\hat{V} - \hat{V}_g\right) C \frac{GL}{\bar{I}_d} \right] dV'' \frac{LG}{\bar{I}_d} \right\} dV'. \tag{5.146}$$

We write the sinusoidal potential at any position in the channel as

$$\left(\hat{V} - \hat{V}_g\right) \ = \ \left(\hat{V}_d - \hat{V}_g\right) \frac{G_d}{G} - \frac{\hat{I}_d}{\bar{I}_d G} \int_{\bar{V}}^{\bar{V}_d} G dV' +$$
$$j\omega \frac{1}{\bar{I}_d^2 G} \int_{\bar{V}}^{\bar{V}_d} G \left[\int_{V'}^{V} \left(\hat{V} - \hat{V}_g\right) G C dV'' \right] dV' \tag{5.147}$$

which has the form

$$\left(\hat{V} - \hat{V}_g\right) = \Theta + jw\Omega, \tag{5.148}$$

where Θ is a function of the conductances and Ω is the operator on $\left(\hat{V} - \hat{V}_g\right)$. Our definitions of these are

$$\Theta\left(V\right) = \left(\hat{V}_d - \hat{V}_g\right)\frac{G_d}{G} - \frac{\hat{I}_d}{\overline{I}_dG}\int_V^{V_d} GdV' \tag{5.149}$$

and

$$\Omega\left[\Theta\left(V\right)\right] = \frac{1}{G\overline{I}_d^2}\int_V^{V_d} G\left[\int_{V'}^{V_d} \Theta\left(V''\right)GCdV''\right]dV'. \tag{5.150}$$

The equation has been recast in this form in order to allow an iterative series expansion of this integral equation as

$$\left(\hat{V} - \hat{V}_g\right) = \left[1 + jw\Omega + (jw\Omega)^2 + \cdots\right]\Theta. \tag{5.151}$$

Increasingly accurate approximations may be obtained using an increasing number of terms in the expansion. These terms are multiple integrals that may be evaluated, at least numerically. Evaluating this expansion at the source end results in

$$\begin{aligned}
\left(\hat{V}_s - \hat{V}_g\right) &= \left(\hat{V}_d - \hat{V}_g\right)\frac{G_d}{G_s}\left[1 + \frac{jw}{jw_{11}} + \left(\frac{jw}{w_{12}}\right)^2 + \cdots\right] \\
&\quad - \hat{I}_d\left[1 + \frac{jw}{jw_{01}} + \left(\frac{jw}{w_{02}}\right)^2 + \cdots\right],
\end{aligned} \tag{5.152}$$

where $w_{11}, w_{12}, \ldots$ and $w_{01}, w_{02}, \ldots$ are characteristic frequencies that may be evaluated from the static equations. Following the above, these characteristic frequencies can be derived as

$$\left(\frac{1}{w_{1n}}\right)^n = \mathcal{O}_{1n}\left(\overline{V}_s\right), \tag{5.153}$$

with

$$\mathcal{O}_{10}\left(V\right) = 1, \tag{5.154}$$

and

$$\mathcal{O}_{1(n+1)}\left(V\right) = \frac{1}{\overline{I}_d^2}\int_{\overline{V}}^{\overline{V}_d} G\left[\int_{V'}^{V_d} C\mathcal{O}_{1n}\left(V'\right)dV''\right]dV', \tag{5.155}$$

and similarly,

$$\left(\frac{1}{\omega_{0n}}\right)^n = \mathcal{O}_{0n}\left(\overline{V}_s\right),$$
(5.156)

with

$$\mathcal{O}_{00}\left(V\right) = \frac{1}{\overline{I}_d}\int_V^{V_d} G dV',$$
(5.157)

and

$$\mathcal{O}_{0(n+1)}\left(V\right) = \frac{1}{\overline{I}_d^2}\int_V^{\overline{V}_d} G\left[\int_{V'}^{V_d} C\mathcal{O}_{0n}\left(V''\right) dV''\right] dV'.$$
(5.158)

The small-signal drain current phasor may now be written as:

$$\hat{I}_d = -\left(\hat{V}_s - \hat{V}_g\right) G_s \frac{1}{1 + (j\omega/\omega_{01}) + (j\omega/\omega_{02})^2 + \cdots} +$$
$$\left(\hat{V}_d - \hat{V}_g\right) G_d \frac{1 + (j\omega/\omega_{11}) + (j\omega/\omega_{12})^2 + \cdots}{1 + (j\omega/\omega_{01}) + (j\omega/\omega_{02})^2 + \cdots}.$$
(5.159)

This establishes the forward and output admittance parameters for the common-gate configuration. In order to establish the other two, we need to determine the small-signal source current phasor as a linear function of $\left(\hat{V}_s - \hat{V}_g\right)$ and $\left(\hat{V}_d - \hat{V}_g\right)$. We may do that in the same way as the derivation of the small-signal drain current phasor. We derived the drain current using Equation 5.135. Similarly, using Equation 5.136, we can derive the series expansion for small-signal source phasor. This problem is left as an exercise in this chapter (see Problem 9). The small-signal source current phasor is given by

$$\hat{I}_s = \left(\hat{V}_s - \hat{V}_g\right) G_s \frac{1 + (j\omega/\omega_{21}) + (j\omega/\omega_{22})^2 + \cdots}{1 + (j\omega/\omega'_{01}) + (j\omega/\omega'_{02})^2 + \cdots} -$$
$$\left(\hat{V}_d - \hat{V}_g\right) G_d \frac{1}{1 + (j\omega/\omega'_{01}) + (j\omega/\omega'_{02})^2 + \cdots}.$$
(5.160)

Here, as in the small-signal drain current phasor case,

$$\left(\frac{1}{\omega_{2n}}\right)^n = \mathcal{O}_{2n}\left(\overline{V}_d\right),$$
(5.161)

with

$$\mathcal{O}'_{20}\left(V\right) = 1,$$
(5.162)

and

$$\mathcal{O}_{2(n+1)}\left(V\right) = \frac{1}{I_d^2} \int_{\overline{V}_s}^{\overline{V}} G\left[\int_{V_s}^{V'} C\mathcal{O}_{2n}\left(V''\right) dV''\right] dV', \qquad (5.163)$$

and similarly

$$\left(\frac{1}{\omega'_{0n}}\right)^n = \mathcal{O}_{0n}\left(\overline{V}_d\right), \qquad (5.164)$$

with

$$\mathcal{O}'_{00}\left(V\right) = \frac{1}{I_d} \int_{\overline{V}_s}^{\overline{V}} G dV', \qquad (5.165)$$

and

$$\mathcal{O}'_{0(n+1)}\left(V\right) = \frac{1}{I_d^2} \int_{\overline{V}_s}^{\overline{V}} G\left[\int_{V_s}^{V'} C\mathcal{O}'_{0n}\left(V''\right) dV''\right] dV'. \qquad (5.166)$$

Thus, the common-gate admittance parameters may be written as

$$y_{ss}^g = G_s \frac{1 + (j\omega/\omega_{21}) + (j\omega/\omega_{22})^2 + \cdots}{1 + (j\omega/\omega'_{01}) + (j\omega/\omega'_{02})^2 + \cdots},$$

$$y_{sd}^g = -G_d \frac{1}{1 + (j\omega/\omega'_{01}) + (j\omega/\omega'_{02})^2 + \cdots},$$

$$y_{ds}^g = -G_s \frac{1}{1 + (j\omega/\omega_{01}) + (j\omega/\omega_{02})^2 + \cdots},$$

and $\quad y_{dd}^g = G_d \frac{1 + (j\omega/\omega_{11}) + (j\omega/\omega_{12})^2 + \cdots}{1 + (j\omega/\omega_{01}) + (j\omega/\omega_{02})^2 + \cdots}. \qquad (5.167)$

The common-source y-parameters follow from Appendix A using matrix transformation. Obviously, this is an unwieldy method, with any intuitive interpretation lost. We may, however, obtain some understanding of the significance of the small-signal high frequency effects by looking at the small-signal behavior in an intermediate frequency range where only the first-order frequency term is important. The y-parameters for common-gate reduce to

$$y_{ss}^g = G_s \frac{1 + (j\omega/\omega_{21})}{1 + j\omega/\omega'_{01}},$$

$$y_{sd}^g = -G_d \frac{1}{1 + j\omega/\omega'_{01}},$$

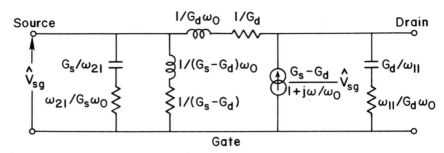

Figure 5.32: A common-gate equivalent circuit for intermediate frequencies derived as an approximation from the small-signal solution.

$$y_{ds}^g = -G_s \frac{1}{1+j\omega/\omega_{01}},$$

and
$$y_{dd}^g = G_d \frac{1+j\omega/\omega_{11}}{1+j\omega/\omega_{01}}. \qquad (5.168)$$

The above equations show a common pole at the characteristic frequency $\omega_{01}, \omega_{01}'$, which are actually the same. This can be shown by manipulation of the corresponding integrals (see Problem 10). We will refer to this characteristic frequency as ω_0; it is given by:

$$\frac{1}{\omega_0} = \frac{1}{\overline{I}_d^2} \int_{\overline{V}_s}^{\overline{V}_d} G \left[\int_{V'}^{V_d} C \left(\frac{1}{\overline{I}_d} \int_{V''}^{V_d} G dV''' \right) dV'' \right] dV' \qquad (5.169)$$

in terms of the quasi-static parameters.

These y-parameters serve as an ideal basis for deriving equivalent circuits and establishing the quasi-static equivalent circuit (and its variations) derived earlier on a rational basis. The y-parameters that we have derived, and their truncated first-order forms above, are representations of resistances, capacitances, and current sources. An equivalent circuit, modelling the truncated form of the admittance parameters for common-gate configuration, is shown in Figure 5.32. The most common use of the devices is in common-source configuration, where we may write the y-parameters following Appendix A as (see Problem 11)

$$y_{gg}^s = G_s \frac{j\omega/\omega_{21}}{1+j\omega/\omega_0} + G_d \frac{j\omega/\omega_{11}}{1+j\omega/\omega_0},$$

$$y_{gd}^s = -G_d \frac{j\omega/\omega_{11}}{1+j\omega/\omega_0},$$

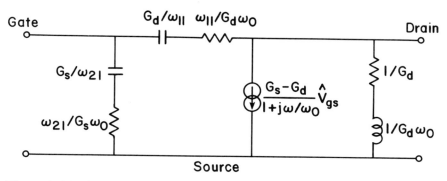

Figure 5.33: An equivalent circuit, valid at intermediate frequencies, following the y-parameter derivation, in the common-source mode.

$$y_{dg}^s = G_s \frac{1}{1 + j\omega/\omega_0} - G_d \frac{1 + j\omega/\omega_{11}}{1 + j\omega/\omega_0},$$

$$\text{and} \quad y_{dd}^s = G_d \frac{1 + j\omega/\omega_{11}}{1 + j\omega/\omega_0}. \tag{5.170}$$

An equivalent circuit corresponding to this is shown in Figure 5.33. The similarity between these equivalent circuits and the quasi-static form derived earlier is clear. The current sources have a single pole repsponse in the common-source form. This serves as a basis to explain the use of an exponential delay factor in the current source of $\exp(-j\omega\tau)$ in many common forms of equivalent circuit. The cause of this is the characteristic frequency ω_0, which models the signal delay in the channel transmission-line.

Our theory, so far, has actually ignored current saturation. We wish to include it as an extension of our analysis. First we will show how the saturated velocity forms a basis for this exponential phase-delay term. Consider a device in which carriers are moving at a constant velocity v. The source-to-drain current per unit width flowing through this device, using the source-end channel opening, is

$$J = qN_D \left\{ a - \left[\frac{2\epsilon_s (\psi_{j0} - V)}{qN_D} \right]^{1/2} \right\} v, \tag{5.171}$$

where a is the epitaxial thickness, ψ_{j0} is the built-in potential of the barrier, and v is the velocity. No current flows through the space charge region because it is assumed depleted. Similarly, the charge per unit length in any incremental cross-section is

$$\Delta Q = qN_D \left[\frac{2\epsilon_s \left(\psi_{j0} - V \right)}{qN_D} \right]^{1/2} \Delta z. \tag{5.172}$$

The continuity equation requires that in the incremental cross-section of length Δz,

$$\frac{\partial \left(\Delta Q / \Delta z \right)}{\partial t} = \frac{\partial I}{\partial z}, \tag{5.173}$$

leading to, for the time-varying terms,

$$-qN_D \frac{1}{2} \left(\frac{2\epsilon_s}{qN_D} \right)^{1/2} \frac{1}{(\psi_{j0} - V)^{1/2}} \frac{\partial \tilde{V}}{\partial \tilde{t}} = qN_D \frac{1}{2} \left(\frac{2\epsilon_s}{qN_D} \right)^{1/2} \times$$

$$\frac{1}{(\psi_{j0} - V)^{1/2}} \frac{\partial \tilde{V}}{\partial z} v, \tag{5.174}$$

i.e.,

$$\frac{\partial \tilde{V}}{\partial t} = -v \frac{\partial \tilde{V}}{\partial z}. \tag{5.175}$$

For a sinusoidal signal $\tilde{V} = \hat{V} \exp \left(-j\omega t \right)$, the solution as a function of position is

$$\hat{V} = \hat{V}_0 \exp \left(-j\omega \frac{z}{v} \right), \tag{5.176}$$

where $\hat{V}_0$ is independent of position. Corresponding to this, we may write the small-signal voltage phasor as

$$\tilde{V} = \tilde{V}_0 \exp \left(-j\omega \frac{z}{v} \right), \tag{5.177}$$

and the current phasor, by perturbative expansion, as

$$\tilde{I} = qN_D \left[\frac{2\epsilon_s \left(\psi_{j0} - \overline{V} \right)}{qN_D} \right]^{1/2} \frac{\tilde{V}_0 \exp \left(-j\omega z/v \right)}{2 \left(\psi_{j0} - \overline{V} \right)}$$

$$= \left[\frac{q\epsilon_s N_D}{2 \left(\psi_{j0} - \overline{V} \right)} \right]^{1/2} \tilde{V}_0 \exp \left(-j\omega \frac{z}{v} \right), \tag{5.178}$$

which shows a phase delay of z/v from the origin to the position z. Thus, the phase delay in the current generator, commonly implemented in the small-signal models, arises from the transmission-line delay associated with carrier transport.

We now derive some of these frequencies for the case of the gradual channel constant mobility model of MESFETs, to note how they differ from quasi-static models. Our derivation is for the intermediate frequencies with only first-order frequency terms being considered. The quasi-static drain current is

$$\overline{I}_d = \int_{\overline{V}_s}^{\overline{V}_d} G \, dV, \tag{5.179}$$

where

$$G = G_0 \left[1 - \left(\frac{\overline{V} - \overline{V}_g}{V_p} \right)^{1/2} \right], \tag{5.180}$$

where

$$G_0 = q N_D \mu a \tag{5.181}$$

is the open channel conductance and V_p is the threshold voltage of the device. The capacitance per unit length is given by

$$C = C_0 \left(\frac{\psi_{j0}}{\overline{V} - \overline{V}_g + \psi_{j0}} \right)^{1/2}, \tag{5.182}$$

where C_0 is the capacitance per unit length at thermal equilibrium. The drain current, using this, is given by

$$\overline{I}_d = G_0 V_p \left(\xi_d - \xi_s - \frac{2}{3} \xi_d^{3/2} + \frac{2}{3} \xi_s^{3/2} \right), \tag{5.183}$$

where the ξ's are normalized parameters,

$$\xi_s = \frac{\overline{V}_s - \overline{V}_g - \psi_{j0}}{V_p}$$

$$\text{and} \quad \xi_d = \frac{\overline{V}_d - \overline{V}_g - \psi_{j0}}{V_p}. \tag{5.184}$$

The characteristic frequency ω_0 is related, following integrations, as (see Problem 12)

$$\frac{1}{\omega_0} = \frac{C_0}{G_0} \left(\frac{\psi_{j0}}{V_p} \right)^{1/2} \left(\xi_d - \xi_s - \frac{2}{3} \xi_d^{3/2} + \frac{2}{3} \xi_s^{3/2} \right)^{-3} \times$$

$$\left[\frac{2}{21} \left(\xi_d^{7/2} - \xi_s^{7/2} \right) - \frac{1}{3} \left(\xi_d^3 - \xi_s^3 \right) + \frac{4}{15} \left(\xi_d^{5/2} - \xi_s^{5/2} \right) - \frac{2}{3} \xi_d^2 \xi_s^{3/2} + \frac{2}{3} \xi_d^{3/2} \xi_s^2 + \xi_d^2 \xi_s - \xi_d \xi_s^2 - \frac{4}{3} \xi_d^{3/2} \xi_s + \frac{4}{3} \xi_d \xi_s^{3/2} \right]. \tag{5.185}$$

Similarly, the two other characteristic frequencies may be derived as

$$\frac{1}{\omega_{11}} = \frac{C_0}{G_0}\left(\frac{\psi_{j0}}{V_p}\right)^{1/2}\frac{\frac{2}{3}\xi_d^{3/2} - \frac{1}{3}\xi_d^2 - 2\xi_d^{1/2}\left(\xi_s - \frac{2}{3}\xi_s^{3/2}\right) + \frac{4}{3}\xi_s^{3/2} - \xi_s^2}{\left(\xi_d - \xi_s - \frac{2}{3}\xi_d^{3/2} + \frac{2}{3}\xi_s^{3/2}\right)^2},$$

(5.186)

and

$$\frac{1}{\omega_{21}} = \frac{C_0}{G_0}\left(\frac{\psi_{j0}}{V_p}\right)^{1/2}\frac{\frac{4}{3}\xi_d^{3/2} - \xi_d^2 - 2\xi_s^{1/2}\left(\xi_d - \frac{2}{3}\xi_d^{3/2}\right) + \frac{2}{3}\xi_s^{3/2} - \frac{1}{3}\xi_s^2}{\left(\xi_d - \xi_s - \frac{2}{3}\xi_d^{3/2} + \frac{2}{3}\xi_s^{3/2}\right)^2}.$$

(5.187)

In the region where drain voltage is very small, the MESFET behaves as a linear resistor $\xi_d \approx \xi_s$ and the characteristic frequencies tend to the limit

$$\frac{1}{\omega_0}, \frac{1}{\omega_{11}}, \frac{1}{\omega_{21}} \to \approx \frac{C_0}{G_0}\left(\frac{\psi_{j0}}{V_p}\right)^{1/2} \propto \frac{\epsilon_s}{\sigma}\left(\frac{L}{a}\right)^2.$$

(5.188)

These frequencies are related to the dielectric relaxation frequency ($\propto \sigma/\epsilon_s$) together with a proportionality constant related to the device geometry. This is intuitively expected since majority carrier devices have ultimate limitations placed by this frequency. At the onset of current saturation, near which devices are usually operated, these frequencies are somewhat more compact and are

$$\frac{1}{\omega_0} = \frac{27}{35}\frac{C_0}{G_0}\left(\frac{\psi_{j0}}{V_p}\right)^{1/2}\frac{1 + 5\xi_s^{1/2} + \frac{10}{3}\xi_s}{\left(1 - \xi_s^{1/2}\right)\left(1 + 2\xi_s^{1/2}\right)^3},$$

$$\frac{1}{\omega_{11}} = 3\frac{C_0}{G_0}\left(\frac{\psi_{j0}}{V_p}\right)^{1/2}\frac{1 + 3\xi_s^{1/2}}{\left(1 - \xi_s^{1/2}\right)\left(1 + 2\xi_s^{1/2}\right)^2},$$

$$\text{and} \quad \frac{1}{\omega_{21}} = 3\frac{C_0}{G_0}\left(\frac{\psi_{j0}}{V_p}\right)^{1/2}\frac{1 + \xi_s^{1/2}}{\left(1 - \xi_s^{1/2}\right)\left(1 + 2\xi_s^{1/2}\right)^2}.$$

(5.189)

We have thus evaluated the elements of the equivalent circuit model because all the parameters have been determined. They are not precisely their values of the quasi-static model because of the dispersion effects at the higher frequencies. The method derived here, in terms of conductances, etc., is similarly valid for the case of HFETs in the same general form. Since the capacitance per unit length and conductance are also determinable as here, a similar model follows for HFETs. While our discussion of small-signal behavior of HFETs will follow an alternate approach mainly because of its simplicity, we will leave this as an exercise for the curious.

This method of analysis is general, however, the steady-state equations that we employed are strictly valid only for biases below the point of current saturation. Let us now consider how this may be extended to include effects of current saturation. Our PHS model incorporated both a low field constant mobility and a saturated velocity transport behavior. The saturation of the current occurred when the field reached a critical electric field of $\mathcal{E}_s$ in the channel. So the current saturation occurred when the electric field reached $\mathcal{E}_s$, and this is quite a good approximation of the behavior of devices when negative differential velocity effects do not dominate. This condition for saturation allowed us to write the length L_1 and the current as

$$L_1 = L \frac{1}{\alpha(1 - \xi_p)} \left[\xi_p{}^2 - \xi_s{}^2 - \frac{2}{3} \left(\xi_p{}^3 - \xi_s{}^3 \right) \right], \qquad (5.190)$$

$$I_D = \frac{aq\mu_0 N_D W \mathcal{V}_{00}}{L_1} \left[\xi_p{}^2 - \xi_s{}^2 - \frac{2}{3} \left(\xi_p{}^3 - \xi_s{}^3 \right) \right]. \qquad (5.191)$$

In both these equations, ξ is the channel opening factor. This allows us to write the current relationship as

$$\bar{I}_{dsat} = G_0 V_p \alpha \left(1 - \xi_{sat}^{1/2} \right), \qquad (5.192)$$

where α as defined earlier is the ratio $\mathcal{E}_s L / \mathcal{V}_{00}$—the saturation index. When higher voltage than this is applied, the electric gate length shrinks. We will ignore that effect here and consider only the consequences of current saturation. Thus, the current can be obtained by the simultaneous satisfaction of the above two equations, which gives us the value of ξ_{sat}. Under these assumptions, the saturation current is now a function only of the potential at the source end of the channel, i.e., of the function ξ_s. Our earlier general small-signal analysis can now be employed, with the modifications coming about due to the saturation of velocity at $\mathcal{E}_s$. Note, we do not include the effects of gate-length shrinkage, i.e., L_2 of our earlier quasi-static analysis. The drain current change due to a change in the normalized saturation potential using the above equation yields

$$\hat{I}_d = -\frac{G_0 \alpha}{2\xi_s{}^{1/2}} \left(\hat{V}_d - \hat{V}_g \right). \qquad (5.193)$$

Substituting this in the first order current equations yields the related y-parameters, e.g.,

$$\hat{I}_g = \hat{V}_g G_s \left[\frac{j\omega/\omega_{21}}{1 + j\omega/\omega_0} + \frac{\gamma}{1+\gamma} \frac{j\omega/\omega_{11}}{(1 + j\omega/\omega_\gamma)(1 + j\omega/\omega_0)} \right], \quad (5.194)$$

and

$$\hat{I}_d = \hat{V}_g \frac{G_s}{1 + 2\alpha^{-1}\xi_s^{1/2}\left(1 - \xi_s^{1/2}\right)} \frac{1}{1 + j\omega/\omega_\gamma}, \quad (5.195)$$

where

$$\frac{1}{\omega_\gamma} = \frac{1}{1+\gamma}\left(\frac{1}{\omega_0} + \frac{\gamma}{\omega_{11}}\right) \quad (5.196)$$

and

$$\gamma = \frac{2\xi_s^{1/2}}{\alpha\left(1 - \xi_s^{1/2}\right)}. \quad (5.197)$$

In the drain current expression, the ratio term of the source end conductance is the intrinsic quasi-static value of the transconductance. The saturation brings together with it modifications through the factors ξ_{sat}, α, and γ in the expressions above. The frequency ω_γ is the modification to the effect of ω_0 due to saturation. It may be seen as the 3 dB drop-off frequency for the transfer admittance in the common-source configuration. It occurs before ω_0. Since this analysis is for a one pole approximation, we can actually drop the second-order term that would come about if ω_0 were also included together with ω_γ in the expression for the gate current phasor above.

We can see some of the modifications that occur as a result of these changes due to incorporation of saturation effects. The input admittance, from the expression for the drain current phasor, can be approximated by a capacitance and series resistance as

$$C_{gs} = G_s\left(\frac{1}{\omega_{21}} + \frac{\gamma}{1+\gamma}\frac{1}{\omega_{11}}\right), \quad (5.198)$$

and

$$R_i = \frac{1}{\omega_\gamma C_{gs}}. \quad (5.199)$$

This is the basis for the intrinsic resistance introduced in our quasi-static equivalent circuit.

5.10 Limit Frequencies

Since the objective of a large body of compound semiconductor devices is to operate at high frequencies or with high switching speeds, it is pertinent to relate much of the discussion on small-signal analysis, quasi-static modelling, and equivalent circuit models with these applications. Network parameters, if known as a function of frequencies, characterize the expected behavior of a circuit incorporating the device, since any mismatches in impedances at input and output, and any feedback effects, can be accounted in the calculation. Thus, any circuit's performance can be predicted. In order to understand the ability of the device to operate and function at high frequencies, it is naturally useful to look for characteristic frequencies that are not dependent on specific circuits, and that take an idealized approach of the best possible condition for operation.

Since power gain is one natural area of use of these devices, a power gain frequency figure of merit is of interest. The most common one of interest is the frequency at which the unilateral power gain of the device goes to unity, or 0 dB. A unilateral power gain is the gain from the device in an amplifier made using only non-lossy, passive, and reciprocal matching networks. It therefore assumes the use of idealized matching conditions at the input and at the output, and a feedback circuit, all of which have the above characteristics. The resulting ratio of the power into the load to the power into the input of the device is the unilateral power gain. One can derive the unilateral power gain knowing the network parameters, and some of the expressions for various network parameters are summarized in Appendix A. The frequency at which this gain is unity is the maximum frequency at which the devices still provide a power gain. It is, therefore, also the highest frequency at which an ideal oscillator made using the device will still be expected to operate. Thus the frequency at which the unilateral power gain goes to unity is known as the maximum frequency of oscillation (f_{max}).

For practical reasons, two other power gains are also in use—one is the maximum available power gain, which occurs only if the device is stable, and the other is the maximum signal gain, which is a characteristic useful in the case of unstable devices. Some of these are summarized in Appendix A.

Another figure of merit in frequency of interest is the frequency at which the current gain in a common-source (or common-emitter in case of a bipolar transistor) configuration becomes unity for short-circuit conditions at the output. This is referred to as the unity current-gain frequency (f_T). Again, knowing network parameters, one can calculate the current gain under short-circuit conditions (it is h_{21}), and hence one can find the unity current-gain frequency for the device.

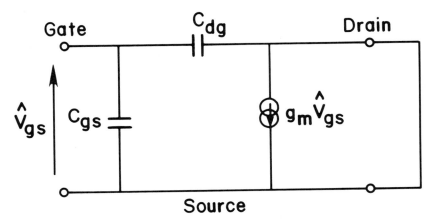

Figure 5.34: A simplified equivalent circuit with capacitance-dominated behavior under conditions of short circuit at the output.

Both power gain and current gain, once they begin to roll off as a function of frequency, usually follow a 6 dB/octave slope. Note that in dB's, the unilateral power gain (U) is given as $10\log(U)$, and the current gain (h_{21}) is $20\log(h_{21})$. Since extrinsic lumped elements usually are the dominant effects, single pole effects dominate (we will consider deviations from this in Chapters 6 and 7, and in our small-signal analysis we have shown higher order frequency terms also), due, at the least, to the role played by parasitics. The parasitics can usually be adequately modelled by simple lumped resistors and capacitors and hence they generally contribute only simple pole effects.

At its simplest, the high frequency operation of a transistor is dominated by capacitive effects. In the common-source configuration, the gate-to-source capacitance C_{gs} and the drain-to-gate capacitance C_{dg} dominate, and a simple equivalent circuit (see Figure 5.34) can be drawn. The current gain can then be written as

$$\frac{\hat{I}_d}{\hat{I}_g} = \frac{g_m \hat{V}_{gs}}{(j\omega C_{gs} + j\omega C_{dg})\,\hat{V}_{gs}}. \tag{5.200}$$

The radial frequency at which this gain becomes unity is

$$\omega_T = \frac{g_m}{C_{gs} + C_{dg}}, \tag{5.201}$$

and hence the unity current gain frquency for this simple equivalent circuit

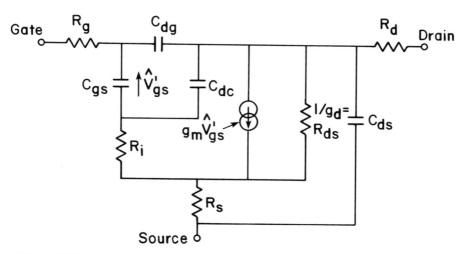

Figure 5.35: An equivalent circuit of the MESFET incorporating drain-to-source resistance, and drain and source resistances.

is

$$f_T = \frac{g_m}{2\pi \left(C_{gs} + C_{dg}\right)}. \tag{5.202}$$

This is, of course, a highly simplified analysis; output drain to source resistance, drain resistance, and source resistances are significant in small gate-length devices where capacitances begin to scale to low values. A higher-order modification to the equivalent circuit may then be described by Figure 5.35. Calculations for this problem get quite unwieldy; a simplification of this, with output shorted, is

$$\frac{\hat{I}_d}{\hat{I}_g} = \frac{1}{g_m}\left[\omega\left(C_{gs} + C_{dg}\right)\left(1 + \frac{R_d + R_s}{R_{ds}}\right) + \omega C_{dg} g_m \left(R_d + R_s\right)\right], \tag{5.203}$$

which gives an approximation for the unity current gain frequency as

$$f_T = \frac{g_m}{2\pi}\left[\left(C_{gs} + C_{dg}\right)\left(1 + \frac{R_d + R_s}{R_{ds}}\right) + C_{dg} g_m \left(R_d + R_s\right)\right]^{-1}. \tag{5.204}$$

An important term in the above is the ratio $\left(R_d + R_s\right)/R_{ds}$; its origin is the division of return current between the paths provided by R_{ds} and the parasitic resistances of the transistor R_s and R_d. Also, the effect of the feedback term is increased by the Miller feedback gain factor of $g_m \left(R_d + R_s\right)$.

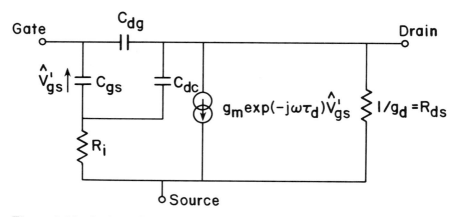

Figure 5.36: A simplified equivalent circuit including the effect of internal resistance R_i, a channel capacitance C_{dc}, and the phase delay factor τ_d of the current source.

Note therefore that parasitics do degrade the current gain of a transistor. A similar approximation, for low parasitic resistances, is

$$f_{max} = \frac{g_m}{2\pi \left(C_{gs} + C_{dg}\right)} \times \left\{ 4 \frac{R_g + R_s + R_i}{R_{ds}} \times \left[1 + 4\pi f_T R_{ds} C_{dg} \times \left(1 + \frac{R_s}{R_g + R_s + R_i} + \frac{2\pi\tau}{\left(R_g + R_s + R_i\right) C_{gs}}\right)\right] \right\}^{-1/2}, (5.205)$$

where f_T follows from Equation 5.204.

While we have incorporated the parasitic resistances in these equations one may also wish to understand the effects related to any dipole capacitances. A simplified equivalent circuit, excluding the parasitics which usually dominate, is shown in Figure 5.36. In this simplified equivalent circuit we have included the effect of internal resistance R_i, a channel capacitance C_{dc}, and the phase delay factor of the current source. All these are therefore additional perturbations of earlier analyses. The unilateral gain for this structure[15] is given as

$$U = \frac{g_m^2 R_{ds} \left(1 + \omega^2 R_i^2 C_{dc}^2\right)}{4 C_{gs} R_i \left[C_{gs} - g_m R_{ds} C_{ds} \cos\left(\omega\tau_d\right)\right] \omega^2}. \quad (5.206)$$

In the above expression, C_{dg} does not figure because it can be tuned by an ideal passive loss-less inductor. The expression with the control source

---

[15]See H.-O. Vikes, "Note on Unilateral Power Gain as Applied to Sub-Micrometer Transistors," *Electronic Letters*, **24**, No. 24, p. 1503, 24th Nov. 1988.

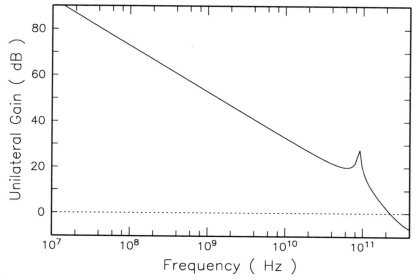

Figure 5.37: Unilateral gain as a function of frequency for the intrinsic part of 1μm gate-length MESFET, assuming a nominal phase delay.

across C_{gs} and R_i is similar in form,

$$U = \frac{g_m^2 R_{ds} \left[1 + \omega^2 R_i^2 (C_{gs} + C_{dc})^2\right]}{4 C_{gs} R_i \left[C_{gs} - g_m R_{ds} C_{ds} \cos{(\omega \tau_d)}\right] \omega^2}. \qquad (5.207)$$

So, inclusion of the resistance effectively increases the input capacitance by the amount C_{dc}; the feedback capacitance C_{dg} can still be tuned out. The inclusion of parasitics in addition to this capacitance is not straightforward; it is best dealt with by using network parameter equations—a procedure we will use in the discussion of small-signal properties of bipolar transistors.

The significance of the above is that the second term in the denominator is negative. At high frequencies, the contribution of the cosine term leads to a vanishing denominator at

$$f_r = \frac{1}{2\pi \tau_d} \arccos \left(\frac{C_{gs}}{g_m R_{ds} C_{dc}}\right), \qquad (5.208)$$

a resonance frequency. Figure 5.37 shows this resonance for an idealized small gate-length device. This behavior is, however, conditioned on negligible parasitics and high output impedance, which are practically difficult to achieve in small device structures.

5.11 Transient Analysis

Based on our small-signal analysis, some general comments can be made regarding the transient behavior of the device also. The quasi-static equivalent circuit representation serves as a poor basis for describing the transient effects due to sudden changes in voltage if these changes take place on a very fast time scale, since such a change has frequency components that are closer to limit frequencies where the quasi-static equivalent circuit has limited validity. However, if changes are slower, and logic gates driving other logic gates have this occur naturally due to the loading effect of other gates and capacitances, then the quasi-static circuits that are valid at intermediate frequencies would be applicable. The modifications that we have to make to these equivalent circuits is to represent, by diodes, conduction paths such as that between the gate and the source and drain, or equivalently a bias dependent resistance and capacitance.

The fast transient process resulting from a sudden change at the input or output is, however, of considerable interest, since it, like the limit frequencies, describes a limit of operation. Fast transients also have rapid changes in electric fields associated with them, hence, one would expect displacement effects to also play a substantial role in them. One term that was associated with displacement effects is the phase delay in the current source. It arose out of the transmission-line gate delay, the charging of the gate capacitance through the resistive channel. In a transient response, this would reflect in the output response of the device as a lagging term in the drain current versus the source current. When the gate of the device is turned on,[16] the displacement current flowing through the gate depletion region occurs together with the particle current in the source region. Some displacement current also flows in the source, but this is much smaller. Figure 5.38 shows such a transient response due to a a gate turn-on pulse.

As carrier build-up occurs in the channel, with a time-delay that is related to the phase delay in the current-source, and once the carriers reach the constricted channel region, drain current begins to build up. Following this, further delay arises mostly from the time constants associated with the gate-to-drain and the drain-to-source capacitances.

---

[16]For a systematic discussion of two-dimensional aspects of the device operation during transient conditions, see J. V. Faricelli, J. Frey, and J. P. Krusius, "Physical Basis of Short-Channel MESFET Operation II: Transient Behavior," *IEEE Trans. on Electron Dev.*, **ED-29**, No. 3, p. 377, Mar. 1982.

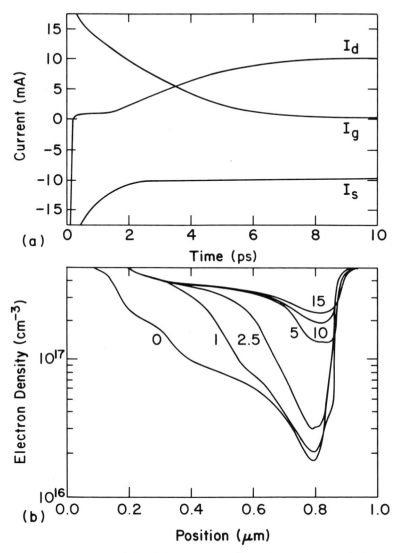

Figure 5.38: Transient of the drain current, the gate current, and the source current due to a rapid change in gate bias for a 0.5 μm gate-length GaAs MESFET is shown in (a). The device is 100 μm in width and is biased positive by 0.25 V on the gate from near-threshold condition. (b) shows the electron density along a cross-section in the conducting part of the channel at various instances of time in pico-seconds. The source is on the left, and the drain is on the right.

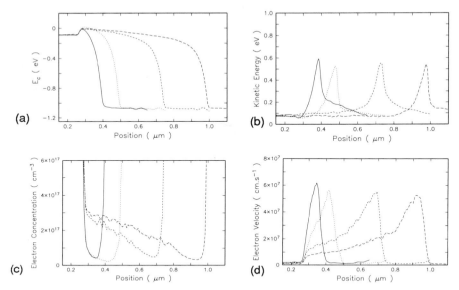

Figure 5.39: Conduction band edge (a), electron kinetic energy (b), electron density (c), and electron velocity (d) as a function of position along a cross-section in the channel for devices with gate-lengths of 0.15 μm, 0.25 μm, 0.5 μm, and 0.75 μm. After S. E. Laux, M. V. Fischetti, and D. J. Frank, "Monte Carlo Analysis of Semiconductor Devices: The DAMOCLES Program," *IBM J. of Research and Development*, **34**, No. 4, p. 466, July 1990.

5.12   Off-Equilibrium Effects

In a discussion of very short channel devices, one needs to include the effect of velocity overshoot as an example of off-equilibrium effect and its relationship with the device performance. This is also a natural place for discussion of how results of calculations which incorporate the off-equilibrium effect compare with the results from a drift-diffusion analysis.

We consider the latter question first. Figure 5.39 shows some of the parameters for various gate lengths for GaAs MESFETs that have a 250 Å channel thickness with 1.2×10^{18} cm^{-3} channel doping, a gate-to-source voltage of 0.5 V and a gate-to-drain voltage of 1.0 V.

Note that at the edge of the gate towards the drain, the electric field is the highest, and in the longer channel devices it is quite small over a fair fraction of the channel. The large gradient in electric field occurs at the drain edge, where the channel pinch-off occurs, and corresponding to which a lower carrier density is observed. The pinch-off is stronger and the carrier density in the pinch-off region correspondingly lower for the largest gate-length device. Corresponding to the large gradient in electric field in this region, in all these devices, velocity overshoot occurs, and nearly to the same velocity.

We discussed velocity overshoot in Chapter 2 as a consequence of the higher relaxation rate of momentum compared to energy. The carrier does not gain all the energy required to bring it in equilibrium with the local electric field over the short spatial scale. As a result, it has a lower scattering rate corresponding to its lower energy even though it is transiting in a high electric field region. The velocity is therefore higher than the velocity that would occur in a long sample with the same field under steady-state, or as seen in the steady-state velocity–field curves such as those of Chapter 2. Even though this velocity overshoot occurs in the 0.75 μm gate-length device, the velocity over a large fraction of this device is low, and the kinetic energy closer to a few kT, i.e., near the thermal value. The overshoot is only in a small fraction of the channel region, the average velocity corresponding to both drift and diffusion in the 1μm channel is still just a little higher than 1×10^7 cm.s^{-1}. This velocity is between the saturated velocity of $\approx 1 \times 10^7$ cm.s^{-1} and the peak velocity of $\approx 1.5 \times 10^7$ cm.s^{-1} at the corresponding carrier density. Thus, use of the assumptions inherent in the drift-diffusion approach is largely justified at 0.75 μm. At 0.5 μm the assumptions are clearly beginning to break down, and drift-diffusion is clearly not adequate in modelling the 0.15 μm device where overshoot occurs over a large fraction of the channel. This brings into question prediction of channel currents, frequency response, etc., in small gate-length devices. Note, though, that the relative changes will still occur as predicted by drift-diffusion formalism. Off-equilibrium effects bring into question quantitative accuracy, a question that exists for the Monte Carlo method also because of its reliance on numerous parameters that can not be independently experimentally measured. Note, however, that many effects of importance, such as sub-threshold effects, or output conductance effects due to substrate injection effects, are still relatively immune to the off-equilibrium phenomenon.

In the case of GaAs, the reduction in kinetic energy occurs when the carriers begin transferring to the L valley. Such a transfer can occur after carriers have acquired an energy of 0.36 eV (the Γ–L inter-valley transfer threshold energy). When such a transfer can occur, because the carrier has acquired the requisite energy, the carrier scattering rate becomes larger, and the carriers begin to lose energy. The Γ–L transfer of carriers takes place in a time scale of at most 100 fs, and hence the carriers can still transit distances corresponding to this, $\approx$ 300 Å before energy begins to reduce.

Obviously, such a transfer could be prevented by lowering the drain voltage. With the drain degenerately doped (Fermi energy in the drain for nominal 1×10^{18} cm^{-3} doping is ≈ 0.1 eV; the drain bias should therefore be nearly 0.2 V), this is a bias condition which is of no practical use. Even though the kinetic energy is lowered because of the lowering of the bias, the overshoot is nearly identical. The group velocity in GaAs in the Γ valley, at the L valley energy, is approximately 6×10^7 cm.s^{-1}. At both these biases, this maximum velocity is being nearly reached.

Considering that band structure does play an important role, as evidenced by the above, it is instructive to compare various compound semiconductor materials with their differing band structures. Figure 5.40 shows some of the parameters of interest at identical bias conditions of $V_{GS} = 0.5$ V and $V_{DS} = 1.0$ V in MESFETs of GaAs, InP, and Ga$_{.47}$In$_{.53}$As. InP has a higher scattering rate and higher effective mass at lower energies than either GaAs or Ga$_{.47}$In$_{.53}$As, and hence its velocities are lower towards source end. However, at higher carrier energies, its velocity overshoot is higher than that of GaAs, because GaAs exhibits a larger scattering rate corresponding to the inter-valley transfer. Ga$_{.47}$In$_{.53}$As has the most favored overshoot characteristic because of its lower scattering rate and large inter-valley transfer threshold.

We end by looking at distributions of which valleys the carriers are at for different positions in the device. We use the 0.75 μm and 0.15 μm GaAs MESFETs as examples for $V_{GS} = 0.5$ V and $V_{DS} = 1.0$ V bias (see Figure 5.41). At the source end, most carriers are in the Γ valley. 60% of the carriers are still in the Γ valley in the 0.75 μm device at the point of maximum velocity overshoot. In the 0.15 μm device, however, a significantly smaller number stays in the Γ valley. Thus, a significant transfer of carriers occurs to secondary valleys and this is particularly strong at shorter gate lengths.

5.13   Summary

This chapter discussed the theory of operation of MESFETs and other phenomena that are important to the use of these devices. First, we developed a simple but intuitive long-channel model where we assumed that the mobility in the device was a constant. An important underlying approximation for this analysis was the gradual channel approximation which allowed us to decouple the electric field in the conducting region of the channel from the electric field in the depletion region of the gate. Compound semiconduc-

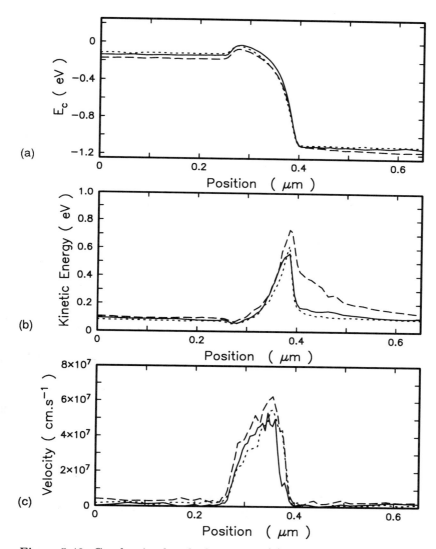

Figure 5.40: Conduction band edge energy (a), kinetic energy (b), and band velocity (c) in 0.15μm gate-length GaAs, InP, and Ga$_{.47}$In$_{.53}$As MESFETs at a bias of V_{GS} = 0.5 V and V_{DS} = 1.0 V. The solid lines are for GaAs, the short dashed lines are for InP, and the long dashed lines are for Ga$_{.47}$In$_{.53}$As.

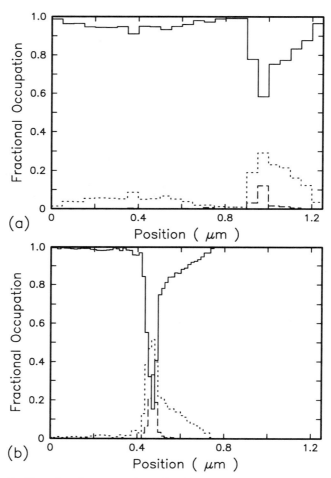

Figure 5.41: Fractional occupation of valleys along the channel in 0.75 μm (a) and 0.15μm (b) gate-length GaAs MESFETs. The Γ, L, and X valley populations are shown as solid, short dashed, and long dashed lines.

tors have large mobilities, hence, long-channel models are of limited utility since they are strictly applicable to only those devices exceeding a minimum gate-length for a specific low field mobility and saturated velocity. As another extreme we considered the constant velocity model, which assumes that the carriers enter the channel region under the gate at a high velocity, which is approximately equal to the velocity with which they traverse the channel. Finally, we combined the constant mobility and constant velocity approximations of the above two models in different parts of one device, and derived quasi-static characteristics. We also considered the effects of current continuity and the negative differential mobility to the operating characteristics and the channel capacitances of the device. Both of these cause formation of a dipole region at the drain end of the channel. The specific design of the device and the operating conditions determine which one of these effects dominates.

In using a MESFET, several parasitic phenomena become important—one is due to the effect of remote potentials in a semi-insulating substrate, and another is due to the effect of strain in a compound semiconductor piezoelectric crystal. The former results in propagation of potential in the crystal and hence influences the current flowing in the device. The latter results in additional charge, which causes a change in the threshold voltage of the device, an effect that becomes increasingly important at short gate-lengths. We also considered the effects of signal delay along a gate in a MESFET, its effect on the operating frequency of the device, and methods to prevent this from becoming important. We also analyzed, at length, methods to perform small-signal analysis without the quasi-static approximation, and applied them to deriving the small-signal behavior of MESFETs. Finally, we concluded with a study of transients in the device and of off-equilibrium effects at short gate lengths.

General References

1. P. H. Ladbrooke, *MMIC Design: GaAs FETs and HEMTs*, Artech House, Boston, MA (1989).

2. M. A. Lampert and P. Mark, *Current Injection in Solids*, Academic Press, N.Y. (1970).

3. J. V. DiLorenzo and D. D. Khandelwal, Eds., *GaAs FET Principles and Technology*, Artech House, Boston, MA (1982).

4. M. S. Shur, *GaAs Devices and Circuits*, Plenum, N.Y. (1987).

5. R. A. Pucel, H. A. Haus, and H. Statz, "Signal and Noise Properties of Gallium Arsenide Microwave Field-Effect Transistors," in *Advances in Electronics and Electron Physics*, **38**, Academic Press, N.Y. (1975).

6. R. S. Pengelly, *Microwave Field-Effect Transistors—Theory, Design, and Applications*, Research Studies, Chichester (1982).

7. S. M. Sze, *Physics of Semiconductor Devices*, John Wiley, N.Y. (1981).

Problems

1. Starting from the expression for current in the constant velocity approximation, derive the relationship for the transconductance g_m as a function of the normalizing current I_{norm}, the normalizing voltage $\mathcal{V}_{00}$, and the drain current I_D. Show that this expression, as expected, is identical to the expression for the transconductance of the PHS model in the limit of $L_2 \to L$.

2. Find the transconductance behavior, in the constant velocity approximation, for a GaAs MESFET with a threshold voltage of -1.5 V. Assume a gate barrier height of 0.8 eV. Consider a uniform doping of 1×10^{17} cm^{-3}, a Gaussian doping profile with a peak doping of 2×10^{17} cm^{-3} with the peak occurring 1000 Å below the surface, and a plane of dopants 1000 Å below the surface.

3. The gate charge is the excess charge of the gate electrode that terminates in the depletion region and the conducting region, towards the source, channel, and the drain. Gauss's law tells us that this is given as $\mathcal{Q}_g = \epsilon_s \mathcal{E}_y$ at the interface between the gate metal and the semiconductor. The electric field in region I of the PHS model follows from the partial derivative of potential in gradual channel approximation, and the electric field in region II follows from the approximation, using the first term, of the Fourier series expansion. Show that

$$
\mathcal{Q}_g \approx 2qN_D aW \left[\frac{\frac{2}{3}(\xi_p{}^3 - \xi_s{}^3) - \frac{1}{2}(\xi_p{}^4 - \xi_s{}^4)}{(\xi_p{}^2 - \xi_s{}^2) - \frac{2}{3}(\xi_p{}^3 - \xi_s{}^3)} + \xi_p L_2 + \right.
$$
$$
\left. \frac{\alpha}{\pi} \frac{a^2}{L} \left(\cosh \frac{\pi L_2}{2a} - 1 \right) \right]. \tag{5.209}
$$

4. Consider our approximate treatment of formation of a stationary dipole in lightly doped, thick-channel MESFETs operating under con-

ditions where the channel is largely open. This is the region of operation where our treatment is least approximate. Considering only a small perturbation from quasi-neutrality, show that the voltage drop across the stationary domain region can be derived as

$$V_{dip} = \frac{4\sqrt{2}}{3}\lambda_D \left(\frac{\mu_0}{\mathcal{D}_0 \mathcal{R}}\right)^{1/2} (\mathcal{E}_m - \mathcal{E}_s)^2, \qquad (5.210)$$

where $\mathcal{R}$ is a constant $\approx 5.5 \times 10^7$ V.cm^{-2} for a doping $N_D = 10^{17}$ cm^{-3} in GaAs.

5. We derived the Mott-Gurney law for space charge–limited current in semiconductors in the text. Derive the relationship for space charge–limited current in vacuum,

$$J = \frac{4}{9}\epsilon_0 \left(\frac{2q}{m_0}\right)^{1/2} \frac{V^{3/2}}{L^2}. \qquad (5.211)$$

This is the Langmuir-Childs law.

6. Consider the distortion of single-crystal quartz under shear stress, and show that it results in a dipole moment perpendicular to that of compressive stress.

7. Estimate the frequency response and the effect of phase delay due to signal propagation along a gate of a GaAs MESFET for a gate metal with resistance of 10 $\Omega/\square$ for the following conditions:

   (a) 1.0 μm gate length and 100 μm gate width,

   (b) 1.0 μm gate length and 10 μm gate width,

   (c) 0.1 μm gate length and 100 μm gate width, and

   (d) 0.1 μm gate length and 10 μm gate width.

8. Consider the modified time-dependent control equation for current continuity,

$$\frac{d}{dz}\left(\frac{1}{\nu + \frac{1}{2} - \frac{1}{4}z^2}\frac{d\hat{v}}{dz}\right) + \hat{v} = 0. \qquad (5.212)$$

Consider the Taylor series expansion of the solution for $\hat{v}$ as a function of z and find the coefficient of the first two powers of z. Show that the term of first power in frequency is identical with that following Hauser's analysis discussed in the text.

9. Our derivation of the drain current phasor, which was used as the basis for deriving output and transfer admittances, employed an iterative procedure on the integral form of the drain current equation (Equation 5.135). A similar integral form of the source current equation has been shown in Equation 5.136. Use a similar iterative procedure to show that the source current phasor can be expressed in the form

$$
\hat{I}_s = \left(\hat{V}_s - \hat{V}_g\right) G_s \frac{1 + (j\omega/\omega_{21}) + (j\omega/\omega_{22})^2 + \cdots}{1 + (j\omega/\omega_{01}') + (j\omega/\omega_{02}')^2 + \cdots} -
$$
$$
\left(\hat{V}_d - \hat{V}_g\right) G_d \frac{1}{1 + (j\omega/\omega_{01}') + (j\omega/\omega_{02}')^2 + \cdots}, \quad (5.213)
$$

where the symbols have the meaning defined in the text. Show that the input and reverse transfer admittances follow.

10. Using the integral expansions of the pole frequencies ω_{01} and ω_{01}', show by using separation of variables that the frequencies are identical and may be replaced by the frequency ω_0.

11. We derived the common-gate admittance parameters using the drain and source current phasors as a function of voltage phasors referenced to the gate. Transferring of these parameters to other common ports is an exercise in matrix manipulation. In the case of definite two-port parameters, this is relatively straightforward and discussed in Appendix A. Using results presented there, derive the common-source admittance parameters (Equation 5.170) from the common-gate parameters considering only those terms important to the first power in frequency in the final result.

12. Using the approximations of the constant mobility model of the operation of a MESFET, show that the characteristic frequency ω_0 is related as shown in Equation 5.185.

13. Explain, using simple physical arguments, why the gate-to-drain capacitance should decrease and the gate-to-source capacitance should either increase or stay constant with increasing drain-to-source bias for a conducting FET.

14. Estimate the gate width for a 1 μm gate-length GaAs MESFET beyond which transmission-line effects due to signal propagation along the width of the device should be included for 8 GHz operation.

15. Consider the effects of formation of a dipole in a MESFET. It can arise as a consequence of the negative differential mobility, as well as due to the requirement of current continuity in a channel of varying conducting thickness (even in the absence of negative differential mobility).

   (a) Should the dipole domain capacitance of a MESFET arising from negative differential mobility in velocity–field characteristics decrease with increased doping in the channel and should this result in a reduced or increased output conductance with doping?

   (b) What would be the dipole effect on output conductance arising from current continuity in the conducting channel?

   (c) Now consider design of MESFET structures for digital logic, i.e., designed for large current drive, or for high-frequency operation, i.e., designed for large limit frequencies. Which should be important for digital logic, and which should be important for high frequency operation?

   (d) What are the implications of scaling on the relative importance of the two causes of dipole domain formation?

16. Should the strain that gives rise to the piezoelectric effect in a GaAs MESFET also lead to increased piezoelectric scattering?

17. Show that for one carrier space charge–limited current, the current density, at very low voltages, can be written as

$$J = \frac{3\mu\epsilon_s kT}{L^3 q} V = \frac{3\mathcal{D}}{L^3} V. \tag{5.214}$$

18. Should the piezoelectric charge distribution also lead to a change in the local ionized impurity scattering?

19. Consider the 1 μm gate-length GaAs MESFET with a $1 \times 10^{17}\text{cm}^{-3}$ doped channel which is 0.25 μm thick. The metal–semiconductor barrier height is 0.8 eV. We will ignore the effects of signal delay along the gate. Using our small-signal analysis, find the characteristic frequencies and the equivalent circuit at the point of onset of current saturation at a gate bias of 0 V. What is the effect of velocity saturation?

Chapter 6

Insulator and Heterostructure Field Effect Transistors

6.1 Introduction

In Chapter 2, we discussed the improvements in in-plane transport behavior resulting from a variety of causes associated with the use of heterostructures, the most important being low ionized impurity scattering and screening of Coulombic scattering effects by carriers in the channels formed at heterostructure interfaces. Heterostructure field effect transistors are field effect devices that take advantage of these improvements. There are additional intrinsic advantages in the control and transport of carriers in HFETs compared to MESFETs. MESFETs employ doped channels and have a difficult problem of threshold control, and gate leakage current during forward bias of the gate due to limited metal–semiconductor barrier heights—both cause limitations in operating range of voltages and of integration. This is a subject of significant interest, since it places constraints on the use of increasingly smaller devices in digital circuits. We will discuss this question in depth in Chapter 9. HFETs, particularly those based on semiconductor–insulator–semiconductor structures, offer certain advantages from this digital perspective. They exhibit improvements in threshold voltage control; gate leakage current improvements also come about in many HFET implementations due to the smaller thermionic emission and tunneling components of the current. Additional operational improvements result from the controlled and nearly constant depth of the sheet charge at the heterostructure interface, which results in a more linear control of the current by the gate, as discussed in Chapter 5. All these features are appealing in both digital and analog usage of the devices.

In many respects, our discussion of HFETs is a continuation of the treatment of MESFETs. The details of the modelling may be different, because of the differences in control of charge, transport of charge, and parasitic conduction; however, the underlying principles embodied in gradual channel approximation, the effects of negative differential velocity and low saturation fields, and small-signal effects, such as that of gate transmission line, are still similar. These are all issues related to the intrinsic operation of the device. In our discussion of MESFETs, we emphasized sidegating effects and piezoelectric effects as two important phenomena arising from inadequacy of the substrates or consequences of strain in a non-centro-symmetric crystal. These also exist in HFETs. There are additional effects in HFETs; a particularly important one that occurs in n-type $Ga_{1-x}Al_xAs$ is due to DX centers, which have unusual emission and capture characteristics for carriers. The DX abbreviation arose, historically, to denote a donor complex, although conventional wisdom suggests that it does not have to be a complex, and may just be due to distortion of lattice. Similar to the discussion of sidegating and piezoelectric effects, in Chapter 5, we will discuss the effect of DX centers on the operation of devices.

The discussion of the intrinsic device is divided in two broad parts. First we will consider heterostructure devices where the insulator will be treated as a perfect insulator and we will assume that Boltzmann statistics describe the charge control adequately. The adequacy and inadequacy of this latter assumption, in different bias ranges, together with that of Fermi–Dirac statistics with three-dimensional and two-dimensional density of states distributions will be discussed later, and the assumptions will be compared. The Boltzmann treatment will let us discuss the basic physics of field effect transistors based on channels formed by inversion unencumbered by a variety of effects that are unique to compound semiconductors and heterostructures. In this respect, the treatment of the devices will be similar to that of silicon metal–oxide–semiconductor field effect transistors. The prominent attribute of the oxide in the silicon structures is that it provides an ideal insulator with an ideal interface for charge control with the semiconductor. While better interfaces for transport can be obtained in compound semiconductors using lattice matched and slightly mismatched compositionally different compound semiconductors, these are at best semi-insulating in character, with approximately 10^{10} Ω.cm as the maximum resistivity. Silicon dioxide, on the other hand, can be for most practical purposes treated as a perfect insulator.

Thus, our insulator field effect transistor analysis will need additional modifications to incorporate the effect of conduction in the large bandgap barrier materials. This subsequent analysis will also allow us to include a variety of effects that are unique to or become more important in compound semiconductors. This modification of the treatment will include size quantization and related band and band-occupation effects in the two-dimensional channel of carrier gas. The differences in details of channel charge control and transport result in differences in details of the small-signal operation, transient operation, and off-equilibrium effects. We will include this in our discussion in a form similar to that adopted for MESFETs. We begin with a general treatment of heterostructures, continuing from the discussion in Chapter 4.

6.2 Heterostructures

One of the unique attributes of compound semiconductors, exploited in many specialized device structures, is the rich selection of various semiconductors that form heterojunctions of the same lattice constant yet differing in bandgap, thus allowing for a broader selection and tuning of desired characteristics. Two figures pertinent to this are Figure 7.1, which describes the bandgap change of compound semiconductors as a function of the lattice constant, and Figure 7.2, which describes the conduction and valence band discontinuity of semiconductors with lattice constant. Figure 7.1 allows us to show those compound semiconductor systems that are lattice matched or slightly mismatched to allow pseudomorphic[1] growth, and hence allow the fabrication of structures where bandgap changes take place. Figure 7.2 is an approximate description, based on observed bandgap discontinuities, of the expected conduction and valence band discontinuities if one were to make an ideal interface between two semiconductors. The lines connecting have been drawn to follow the behavior of Figure 7.1 for the conduction band edge. The figure is approximate and should be used with caution.

$Ga_{1-x}Al_xAs$ remains closely lattice matched at all mole fractions of aluminum arsenide and gallium arsenide (approximate lattice mismatch remains less than 0.1% to both GaAs and AlAs). $Ga_{1-x}In_xP$, GaP_xSb_{1-x}, etc., lattice match only at specific mole fractions. For a limited thickness of the mismatched layers, however, the resulting strain can be accommodated. Thus, materials with close match only at specific mole fractions can still be

---

[1]A *pseudomorph* is a mineral that possesses the external form characteristic of another mineral. When a semiconductor is grown on another semiconductor, the grown semiconductor, if it is crystalline, takes the in-plane lattice periodicity of the basis semiconductor. It is said to be pseudomorphic.

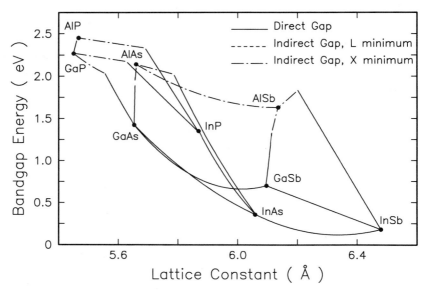

Figure 6.1: Bandgap versus lattice constant for compound semiconductors. Legend in the figure describes the key to the lowest valley in the various combinations of materials.

grown with a fair degree of reproducibility. Examples of these of interest are $Ga_{.47}In_{.53}As$ and $Al_{.48}In_{.52}As$ to InP, and $Ga_{.51}In_{.49}P$ to GaAs. The ability to grow such structures allows one to tune the characteristics of the semiconductor medium to achieve desired characteristics in the devices. A differing barrier height than that of a metal–semiconductor junction can be achieved by the use of heterostructures; low barrier heights being useful in microwave detection, large barrier heights being useful in suppressing gate leakage current in field effect transistors. The changes in bandgap can be used in suppressing injection characteristics of one particular carrier through a suitable use of grading and doping, as in heterostructure bipolar transistors. A major usage of the changes in bandgap at a spatially abrupt interface has been in obtaining two-dimensional channels of carriers for operation in field effect transistors, the subject of this chapter.

Practical limitations quite often arise from our ability to grow lattice mismatched semiconductors. Lattice strain prevents growth of semiconductors with large mismatch because of energy considerations. Thermal energy provides the requisite barrier energy for this relaxation process and the lattice regains its unstrained lattice constant. A lattice with high strain

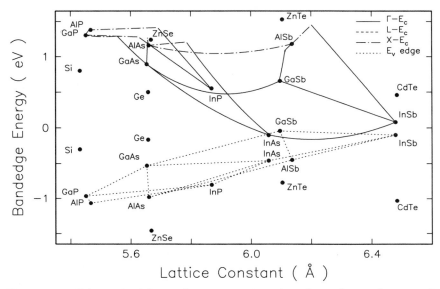

Figure 6.2: Discontinuities at heterostructure interfaces for various semi-conductor systems, plotted as a function of lattice constant. This figure is approximate and should be used with great caution. It has been assembled using available and reliable discontinuity data, most of which are for unstrained interfaces. The strained data have also been incorporated approximately, and hence the behavior described by this figure should be considered an approximate description of what one might expect if one were to make an ideal interface of two sets of materials.

relaxes by generating defects such as dislocations, etc. A low temperature of growth allows one to make structures that are metastable and have a significant strain due to large lattice mismatch, but they may relax by radiative or non-radiative means during operation of a device. In any case, there is a critical thickness beyond which, for given growth conditions, one can not grow a suitable high quality semiconductor. The thicknesses required for HFETs are usually sufficiently small that quite significant mismatch can be accommodated. An example of this is the considerable work in $Ga_{1-x}Al_xAs/Ga_{1-x}In_xAs$ and $GaAs/Ga_{1-x}In_xAs$ systems where low mole fractions of indium arsenide have been used to obtain improvements in the amount of charge and transport for HFETs. The effect of strain in such structures is both in band structure itself and in scattering due to changes in band structure.

6.3   Strained Heterostructures

During the growth of a structure, the grown epitaxial layer assumes the lattice constant of the crystal, and is under a strain of $\epsilon = \Delta a / a$, where a is the lattice constant of unstrained crystal. This strain has an effect on the band structure because it introduces a spatially periodic perturbation differing from that of the unstrained crystal. The strain introduced due to the growth, caused by lattice in the plane perpendicular to the growth orientation, is a biaxial strain. Strain can also be introduced by external means on a grown crystal. The inverse process of the piezoelectric effect, discussed in Chapter 5, should give rise to strain. This is by electrical means. Stress can be applied uniaxially by compressing a crystal in an anvil; it can be applied hydrostatically using hydrostatic pressure. A hydrostatic pressure on a crystal produces the same strain in all directions. The effect of such strains on band structure is quite complicated. In general, strain causes a change in bands–bandgap, as well as band curvature, i.e., effective masses.

Symmetry of the band structure has a principal relationship with the polarity of the effect on the bandgap. This is stressed in Figure 7.3, which shows a schematic of the effect on the band structure for a variety of conditions. Biaxially compressive strain, i.e., a larger lattice constant material grown on a smaller lattice constant material for a direct bandgap crystal, e.g., $Ga_{1-x}In_xAs$ grown on GaAs (see Figure 7.3), leads to a decrease in the bandgap. The effect of compressive strain in an indirect gap crystal is more complex. For Si with (100) surface, the effect of compressive strain is to raise a set of conduction band constant energy ellipsoidal surfaces and lower the other set of constant energy surfaces. Thus, in-plane and out-of-plane valleys behave differently.

These changes in bandgap occur together with changes in the band curvature at the band minimum of the conduction band and the band maxima of the valence bands, with the changes in the latter being particularly significant. The constant energy surface for the valence bands, as discussed in Chapter 2, is a warped surface. It exhibits differing curvature in different directions. Also, at these energies, we need to consider both the light hole and the heavy hole band. The primary effect of the strain in these, as seen in Figure 7.3, is to lift the degeneracy. For the example of $Ga_{1-x}In_xAs$ grown on GaAs, it causes, at zone center, the heavy hole band to become the lower hole energy band. Additionally, the in-plane effective mass, at zone center for the heavy hole band, becomes lower than that of the light hole band even for small incorporation of indium arsenide in the crystal. This effective mass change is orientation-dependent because the constant energy surface is warped. Significant complications have now been intro-

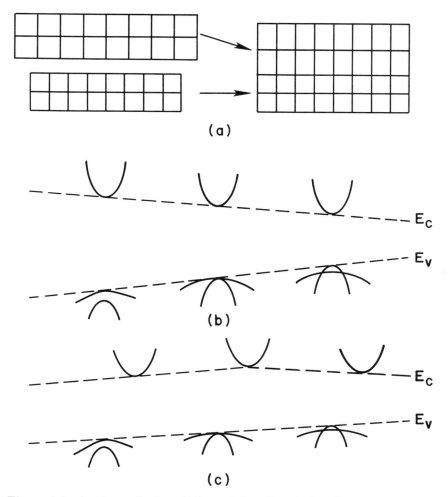

Figure 6.3: A schematic description of the effect in band structure due to biaxial strain occuring due to assembling of crystals with differing lattice constants (a). (b) shows the behavior of a direct gap semiconductor with compressive on left and tensile on right. (c) shows the effect on an indirect gap semiconductor such as Si biaxially strained in the (100) plane. This indirect semiconductor differs from the direct gap example because the four-fold and two-fold degenerate valleys of Si behave differently. The conduction band edge moves lower in energy for one of these sets of valleys. The valence band edges behave the same as in (b). Other indirect bandgap semiconductors and other orientations will behave differently.

duced in the description of the electronic structure due to introduction
of this strain. Since the masses are different in different orientations and
degeneracy has been lifted at zone center, we would expect interesting consequences both for hole transport, due to the effect on mass and scattering,
and for electron–hole interactions, such as in optical processes.

For transport, the primary effect is the lifting of the light hole and heavy
hole degeneracy, which reduces scattering, and hence improves the hole
transport properties. Strained HFETs with p-type channels are believed
to show improvements due to these band structure improvements. This
is also believed to be important for the in-plane transport of holes in p-
SiGe bases of bipolar transistors. The consequence for effective mass was
shown schematically in Figure 7.3. The example of $Ga_{1-x}In_xAs$ grown on
GaAs and $Si_{1-x}Ge_x$ grown on (100) Si approximately follow the behavior
shown in this figure, both for the bandgap and the in-plane effective mass.
The effect on distortion of conduction bands is smaller than that for hole
bands in these examples. Compressive biaxial strain does lead to a slight
improvement in the electron effective mass for the direct gap and an orientation-dependent change for the indirect gap example discussed here. The
effect on transport is more complicated because we must incorporate the
effects on scattering in this discussion. The in-plane distortion of the crystal
due to the biaxial strain is four-fold symmetric. Band edge symmetries can
therefore be destroyed by this strain. For the examples considered, for
electrons, the out-of-plane transport of electrons improves but the in-plane
transport deteriorates.

6.4   Band Discontinuities

We continue here our discussion of band edges at heterojunctions initiated
in Chapter 4. In heterojunctions, the semiconductor band structure may
be considered undisturbed right up to the interface. At the interface, in
the abrupt heterojunctions, discontinuities occur in the conduction band
(ΔE_c) and valence band (ΔE_v), both of whose magnitude is experimentally
determined.

Many ab-initio models and calculations of band discontinuities have
been presented, some of which are in fair agreement with experiments. Historically, the first of the models was Anderson's electron affinity rule, which
assumed continuity of the vacuum level. This predicts $\Delta E_c = q(\chi_2 - \chi_1)$,
where the χ's are the electron affinities in the two semiconductors. This
is in error for most compound semiconductors. For the SiO_2/Si interface,
it is relatively close, but does not predict the observed orientation dependence. Atomic orbital–based calculations predict that when anion species

are common at a heterojunction, such as As at $Ga_{1-x}Al_xAs/GaAs$, most
of the discontinuity should occur in the conduction band. This is because
valence band states are p-orbital–like and come from the anion which is
common. This is known as the common-anion rule. In practice this is not
so, for $Ga_{1-x}Al_xAs/GaAs$, the conduction band discontinuity is $\approx 0.65\Delta E_g$
for aluminum arsenide mole-fractions of less than 0.4. More success appears
to be occurring by semi-empirical techniques. One example is an approach
postulating that, similar to the Fermi level in the case of metal–metal junctions, there exists a level that should be considered primary and equi-energy
across the interface in a semiconductor. Most often this level would lie inside the energy gap, but it doesn't necessarily have to do so. It is this
level that determines barrier heights on nascent surfaces, i.e., on vacuum
cleaved and unreconstructed surfaces. Thus, metal–semiconductor barrier
heights may be a good indicator of discontinuities. The discontinuity plot
(Figure 7.2) of this chapter appears to follow this empirical relationship.

The discontinuity is, of course, with respect to specific band minimum.
Different minima have different discontinuities which follow different trends,
represented, e.g. in the bowing parameters of the alloy. At specific compositions, the different minima may be the lowest minima. At a heterojunction, which minima should be considered to be important, and hence which
discontinuity should be used, depends on the dominant physical process.
All the discontinuities—Γ–Γ, L–L, X–X, etc.—exist at the heterojunction
interface. As an example, consider tunneling through a barrier of large
bandgap material such as $Ga_{1-x}Al_xAs$ from GaAs. Electrons in GaAs are
largely in the Γ valley; if the $Ga_{1-x}Al_xAs$ barrier is relatively thin, these
Γ electrons are most likely to tunnel via Γ-like states in the barrier. However, if the barrier is thick, they may do so through X-like states, a process
that will require phonon interaction in order to allow for change in momentum. Thus, for thick barriers, the difference in energy between the Γ
valley of GaAs and the X valley of $Ga_{1-x}Al_xAs$ may be more pertinent.
Consider the same for transfer of carriers from donors in $Ga_{1-x}Al_xAs$ to
GaAs. The transfer of carriers takes place from whichever is the lowest
valley in $Ga_{1-x}Al_xAs$, irrespective of the width of the barrier. This interesting behavior of the discontinuity, with its implication to the charge
transfer process and to perpendicular current flow, is shown in Figure 7.4
for the $Ga_{1-x}Al_xAs/GaAs$ system as a function of aluminum mole-fraction.

At aluminum arsenide mole-fractions of greater than 0.45, the X band
is the minimum and forms the lower barrier. So, for $x > 0.45$, the barrier between the X valley in $Ga_{1-x}Al_xAs$ and the Γ valley in GaAs is
lower, and in fact for charge transfer, and for tunneling in larger barrier
widths, the relevant maximum discontinuity occurs at $x \approx 0.45$. For the

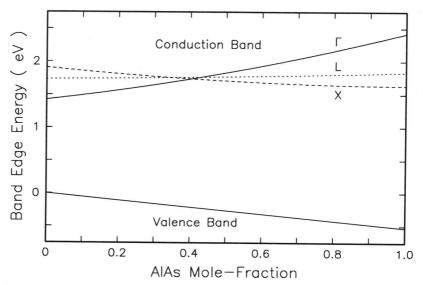

Figure 6.4: The conduction and valence band edges in the $Ga_{1-x}Al_xAs$ system as a function of the AlAs mole-fraction for the Γ, L, and X valleys. After J. Batey and S. L. Wright, "Energy Band Alignment in GaAs:(Al,Ga)As Heterostructures: The Dependence on Alloy Composition," *J. of Appl. Phys.*, **59**, No. 1, p. 200, 1986.

GaAs/$Ga_{1-x}Al_xAs$/GaAs SISFET, this results in a larger leakage current between the gate and the channel as mole-fractions go towards the AlAs end of the semiconductor composition, counter-intuitive to the simplistically expected trend. The importance of appropriate discontinuity will be considered again in Chapter 8 due to its importance in tunneling structures. Here, we stress that the barrier thickness and barrier height for each of the valleys, need to be considered together with elastic and non-elastic tunneling processes to determine the dominant barrier in the current transport mechanism.

6.5   Band Bending and Subband Formation

We have discussed the details of drawing of band edges at a heterostructure in Chapter 4. If we know the discontinuities in the conduction and the valence band, then the heterojunction band diagram follows, similar to the case of homojunction band diagram and consistent with the requirements of

Poisson's equation in the bulk (which may have two-dimensional constraints on the carrier distribution), Gauss's Law at the interface, and the additional condition of the discontinuity at the interface.

For abrupt ideal heterojunctions, band edges are parallel at the interface because there is no macroscopic interface charge; for graded heterojunctions the change in accelerating potential for electrons and holes is given by the electrostatic potential plus the alloy grading potential due to composition-induced changes in the conduction and valence band energies. The quasi-fields are the negative of the gradient of the electrostatic and alloy grading potential. These were shown in Figure 4.11 for no interface charge.

The effect of confinement at an abrupt interface has been discussed in our introduction to heterostructures in earlier chapters. Carriers confined in the narrow potential well, perpendicular to the interface of a discontinuity, do not behave classically, (i.e., according to the nearly-free electron model). When the carrier de Broglie wavelength becomes of the same order of magnitude as the classical length scale, i.e., the classical turning point $(= kT/q\mathcal{E})$, the classical treatment of the carrier motion in the direction perpendicular to the interface breaks down. In this direction, the carrier can have only specific momenta, i.e., the momentum is quantized, and this result is a direct consequence of the principles of quantum mechanics in the presence of the strong confinement. The eigenfunction solution of Schrödinger's equation, whose square represents the probability of finding a carrier at any position, decays beyond both the discontinuity and the classical turning point as a result of the barriers. The solution of Schrödinger's equation for the specific potential well and carrier mass, etc., can be interpreted as nearly-free electron behavior parallel to the interface, a direction in which no change from the classical behavior and boundary conditions has occurred. Perpendicular to the interface, however, only certain eigenfunctions are the solution; energies associated with them are quantized. These discrete energies represent the minimum in energy in a band of allowed energies. Each continuum of states attached to one of the discrete energies is called a subband. The behavior of carriers under conditions of confinement in one direction is called two-dimensional behavior.

Consider an isotropic semiconductor whose energy bands in three dimension behavior can be approximated by parabolic bands. The confinement in one dimension results in energy subbands whose allowed energies are of the form

$$E = E_n + \frac{\hbar^2}{2m^*}(k_x{}^2 + k_y{}^2). \tag{6.1}$$

Here, E_n is the quantized energy associated with the perpendicular motion referenced to the unquantized condition. It represents the set of minima in energy for carriers occupying the subbands. The various E_ns allowed, rep-

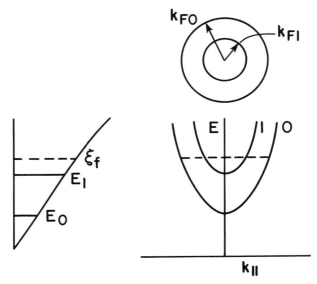

Figure 6.5: Confinement and the minima of the resulting energy subbands are shown on the left. The constant energy surfaces, at the Fermi energy, are shown on the right.

resenting the minima in subband energies, are shown in Figure 7.5, together with constant energy Fermi surfaces in various subbands.

We drew this figure in the isotropic parabolic band approximation, relevant at low energies for the Γ band of GaAs. The formation of subbands and the allowed perpendicular momenta are considerably more complicated for holes as well as for electrons in Si and Ge, etc. This is because of the presence of different hole masses (light and heavy) whose perpendicular momentum and energy are quantized differently, and because of the anisotropy of the electron mass in silicon and germanium.

As an example of this more complicated situation, consider the quantization of a silicon n-type inversion layer together with anisotropy of the mass. We leave the case of holes as an exercise (see Problem 1). Recall that the constant energy surfaces near the bottom of the conduction band are ellipsoids with a large longitudinal mass and a low transverse mass. Consider inversion or accumulation on a (100) surface using SiO_2 as an insulator. There are two equivalent ellipsoids occupied by electrons whose momentum is confined in the longitudinal direction, and four equivalent ellipsoids occupied by electrons whose momentum is confined in the trans-

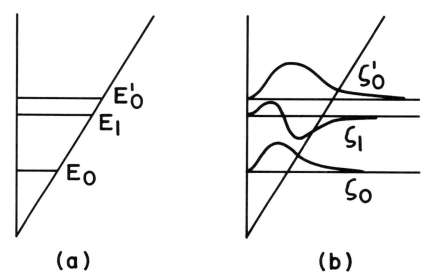

Figure 6.6: Schematic of the lowest energy subband minima (a) and associated envelope functions (b) for inversion layers on (100) Si.

verse direction. The longitudinal mass is larger than the transverse mass, and leads to a smaller energy E_n of the bottom of the corresponding two-dimensional subband because of the inverse dependence of the energy on carrier mass. Figure 7.6 shows the quantization of the two sets of subbands and their envelope functions.[2] Because of the anisotropy, different surfaces of silicon behave differently just as they behave differently for tunneling or thermionic injection. These follow in a similar manner and are considered as an exercise (see Problem 2) in this chapter.

The energy levels in the quantized two-dimensional channels must be derived by solving Poisson's equation and Schrödinger's equation for the quantum well simultaneously, and by taking into account in this solution all carrier energy–related effects. Poisson's equation, here, identifies the relationship between potential energy and the charge distribution, Schrödinger's equation identifies the relationship between the allowed total energy and the momentum, and the energy effects include the consequences of many-body effects such as exchange and correlation effects on carriers. These latter effects usually result in lowering of the energy. The effective poten-

---

[2]The envelope function together with the Bloch function factor yields the wave function of the electron.

tial energy $V(y)$ in Schrödinger's equation can be written as

$$V(y) = -q\psi(y) + \sum V_i. \tag{6.2}$$

Here $\psi(y)$ is the electrostatic potential, the V_i's are the potential energy terms associated with exchange correlation, image, grading, etc., and y, following our convention is the direction perpendicular to the interface. Schrödinger's equation for the envelope function $\varsigma_i(y)$ in subband i is

$$-\frac{\hbar^2}{2}\frac{d}{dy}\left[\frac{1}{m^*(y)}\frac{d}{dy}\varsigma_i(y)\right] + V(y)\varsigma_i(y) = E_i\varsigma_i(y), \tag{6.3}$$

where $m^*(y)$ is the position-dependent effective mass, which can depend on i and E_i in general cases, and E_i is the energy at the bottom of the ith subband. Unlike the SiO_2/Si system, which has a large discontinuity, in the compound semiconductor heterojunction systems, ΔE_c is usually of the order of 0.5 eV or less (see Figure 7.2), and the electron wave function can penetrate a significant distance into the barrier just as it does in the other direction. Compound semiconductors, with their small effective masses for electrons, exhibit strong quantization effects; the reduced confinement by the barrier layer adds to the complexity of the problem.

Poisson's equation, using the envelope function, is of the form

$$\frac{d}{dy}\left[\epsilon_s(y)\frac{d\psi(y)}{dy}\right] = q\sum N_{si}\varsigma^2(y) - \rho_I(y), \tag{6.4}$$

where

$$N_{si} = \frac{m^*kT}{\pi\hbar^2}\ln\left[1 + \exp\left(\frac{\xi_f - E_i}{kT}\right)\right], \tag{6.5}$$

and $\epsilon_s(y)$ is the position-dependent permittivity, N_{si} is the number of electrons per unit area in subband i, ξ_f the Fermi energy, E_i the minimum of the ith subband energy, m^* the density of states effective mass of the inversion layer material, and ρ_I the ionized impurity charge density, $\rho_I = -qN_A$ for an acceptor doping of N_A.

Since the carrier charge is confined in a short distance at the interface, and the ionized impurity charge density is usually smaller by a significant amount, the potential well is sometimes approximated as an infinite triangular well. This is to say that the carrier charge is assumed to be a sheet charge giving rise to a constant electric field which is not significantly disturbed by the weak acceptor charge density. For this approximation, which is not an unreasonable description of moderate and strong inversion conditions, an approximate solution to the problem is

$$\varsigma_i = \mathcal{A}i_i\left[\left(\frac{2m^*q\mathcal{E}_s}{\hbar^2}\right)^{1/3}\left(y - \frac{E_i}{q\mathcal{E}_s}\right)\right], \tag{6.6}$$

where $\mathcal{A}i$ are the Airy functions,

$$\mathcal{A}i(y) = c_1 f(y) - c_2 g(y). \tag{6.7}$$

The problem of a triangular barrier was encountered in our discussion of tunneling, where we showed how Airy functions[3] solved the problem exactly and led to a similar expression as the WKB expression in asymptotic limit. This problem is similar, of a triangular well instead of a triangular barrier (see Problem 3), and the subband minima energies are given by

$$E_i = \left(\frac{\hbar^2}{2m^*}\right)^{1/3} \left[\frac{3}{2}\pi q \mathcal{E}_s \left(i + \frac{3}{4}\right)\right]^{2/3} \tag{6.9}$$

for large values of i.[4] This is similar to the asymptotic solution of the triangular barrier problem.

In the limit of large inversion charge, or negligible acceptor charge density, Gauss's law gives $\mathcal{E}_s = qN_S/\epsilon_s$ for the electric field at the heterostructure interface. Here, N_S is the carrier sheet charge. Thus, in the limit of moderate to strong inversion, low acceptor charge, and ignoring of barrier penetration effects, the subband energy levels can be related to either the electric field at the interface or the sheet carrier charge at the interface. For $Ga_{1-x}Al_xAs/GaAs$ junctions, this results in the energy level positions as

$$
\begin{aligned}
E_0 &= 1.83 \times 10^{-6} \mathcal{E}_s^{2/3} \\
\text{and} \quad E_1 &= 3.23 \times 10^{-6} \mathcal{E}_s^{2/3},
\end{aligned} \tag{6.10}
$$

where the energies are given in units of eV and electric fields are in units of $V.m^{-1}$.

---

[3]The functions $f(y)$ and $g(y)$ are given by

$$
\begin{aligned}
f(y) &= \sum_{n=0}^{\infty} 3^n \left(\frac{1}{3}\right)_n \frac{y^{3n}}{(3n)!}, \\
\text{and} \quad g(y) &= \sum_{n=0}^{\infty} 3^n \left(\frac{2}{3}\right)_n \frac{y^{3n+1}}{(3n+1)!}.
\end{aligned} \tag{6.8}
$$

See, e.g., M. Abramowitz and I. A. Stegun, *Handbook of Mathematical Functions with Formulas, Graphs, and Mathematical Tables*, U.S. Government Printing Office, Washington, D.C., p. 446, 1964.

[4]A more exact solution suggests replacement of the bracketed index terms of $(i+3/4)$ by 0.7587, 1.7540, and 2.7525 for $i = 0, 1$, and 2. The reader would find T. Ando, A. B. Fowler, and F. Stern, "Electronic Properties of Two-Dimensional Systems," *Reviews of Modern Physics*, **54**, No. 2, p. 437, April 1982, a very thorough review. Even though this long article, suggested as a general reading, concentrates on inversion layers in silicon, the basic concepts are general. Our treatment here draws extensively from this reference.

We should also consider, in addition to these, the changes in density of states in the two-dimensional systems as a result of the formation of the subbands. Again, recall that density of states in a ν-dimensional k-space is $(2\pi)^{-\nu}$. So, for two-dimensional systems, the density of states is

$$D(E) = \frac{2g}{(2\pi)^2} 2\pi k \frac{dk}{dE}, \tag{6.11}$$

where g is the valley degeneracy and the factor 2 accounts for spin degeneracy. If we assume isotropic parabolic bands,

$$E = E_0 + \frac{\hbar^2 k_{\parallel}^2}{2m^*}, \tag{6.12}$$

with

$$
\begin{aligned}
D(E) &= \frac{gm^*}{\pi\hbar^2} \quad \text{for} \quad E > E_0 \\
&= 0 \quad \text{for} \quad E < E_0.
\end{aligned} \tag{6.13}
$$

The density of states is constant in any subband, and zero below the subband edge. As energy increases, more subbands become populated and there is a piece-wise discontinuous increase in the density of states. For an n-type inversion layer in GaAs, $g = 1$ because there is only one equivalent Γ band, and the density of state distribution for the two-dimensional system is straightforward and shown in part (a) of Figure 7.7. For (100) silicon, $g = 2$ for the two ellipsoids whose longitudinal momentum is confined and $g = 4$ for the ones whose transverse momentum is confined. The density of state distribution is slightly more complicated and is shown in part (b) of Figure 7.7. The mass to be used in Equation 7.13 is the density of state effective mass for the two orientations, i.e.,

$$m^* = m_t \tag{6.14}$$

for the 2-fold degenerate subbands and

$$m^* = (m_t m_l)^{1/2} \tag{6.15}$$

for the four-fold degenerate subbands. Holes and other orientations of Si are different and follow similar considerations (see Problem 1 and Problem 2).

At absolute zero, the sheet carrier density and the Fermi level energy (if only the zero'th band is occupied) are related by

$$N_s = D(E)(\xi_f - E_0), \tag{6.16}$$

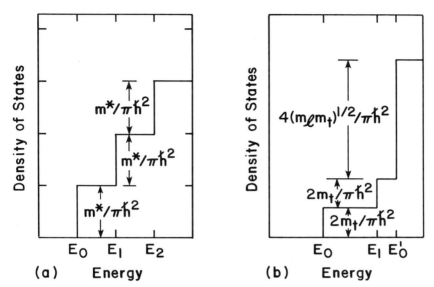

Figure 6.7: Density of states distribution in an electron inversion layer
on GaAs surface as a function of energy is shown in (a). (b) shows the
density of states distribution function as a function of energy for an electron
inversion layer on a (100) Si surface.

and, following our calculations for Fermi velocity in Chapter 2,

$$k_F = \left(\frac{2\pi N_s}{g}\right)^{1/2}$$

$$\text{and} \quad v_F = \frac{\hbar}{m^*}\left(\frac{2\pi N_s}{g}\right)^{1/2}. \tag{6.17}$$

Consider GaAs with $N_s = 6 \times 10^{11}$ cm^{-2} electron density, a common
sheet density in high mobility samples at Ga$_{1-x}$Al$_x$As/GaAs interfaces,
the Fermi velocity is 3.2×10^7 cm.s^{-1} at absolute zero. Note that this is
larger than thermal velocity in non-degenerate GaAs at 300 K. The treatment leading to these equations ignores several second-order effects related
to band structures. The parabolic band approximation is accurate only for
small energies, and higher-order subbands are influenced strongly by the
wave-function penetration into the larger bandgap material at the discontinuity. Hole bands are highly anisotropic, as are bands in smaller bandgap
materials such as Ga$_{.47}$In$_{.53}$As. p-channel devices also have complications

due to multiple bands, warped surfaces, and low barriers with larger wave-function penetration even for the lowest energy subbands. The parabolic infinite barrier picture is too simplistic for these, and we have to rely almost exclusively on numerical techniques to obtain parameters of interest.[5]

Numerical techniques, based on the quantum-mechanical treatment described above, yield results that, although accurate for specific modelling conditions only, can be employed to yield fitting equations that describe variation over a parameter space. Consider, e.g., position of the lowest subband energy E_0 as a parameter that is of interest to us, since in thermal equilibrium a major fraction of carriers occupy this subband. The energy E_0 is a function of both the sheet mobile charge and the background immobile charge. The former has a strong effect on the electric field at the interface, hence confinement at the interface, and the latter has a strong effect on the fields away from the interface, and hence higher-order subbands. The composition of the barrier, and the discontinuity at the barrier are central to the penetration of the wave function, and hence E_0 should depend on these too. A calculation over a parameter space allows us to write simple interrelationships that are convenient in modelling and sufficiently accurate for analysis. The lowest subband energy, e.g., is related as

$$E_0 = \left(\frac{3}{2}\right)^{5/3}\left(\frac{q^2\hbar}{m^{*1/2}\epsilon_s}\right)^{2/3}\frac{N_{As} + \frac{55}{96}N_s}{\left(N_{As} + \frac{11}{32}N_s\right)^{1/3}}, \qquad (6.18)$$

where N_{As} is the sheet density of the acceptors in the depletion region at the interface (two-dimensional gas being formed in a p-type GaAs medium) and N_s the sheet concentration in the two-dimensional electron gas. Application of this equation to finding the parameters of interest, e.g. the sheet electron density, requires an iterative procedure, since occupation statistics have to be considered together with the above as well as higher-order subbands and the total sheet acceptor charge consistent with the band bending. When $N_{As} \ll N_s$, the acceptor charge may be ignored, and the subband energy is proportional to the 2/3rd power of the sheet carrier density. Since the sheet carrier density is proportional to the electric field, the zero'th subband energy is proportional to the 2/3rd power of the electric field, a result shown earlier employing the Airy function solution to the triangular barrier problem. A triangular barrier can be formed only with

---

[5]The reader is referred to F. Stern and S. Das Sarma, "Electron Energy Levels in GaAs/Ga$_{1-x}$Al$_x$As Heterojunctions," *Phys. Rev. B*, **30**, No. 2, p. 840, 1984, and F. Stern "Doping Considerations for Heterojunctions," *J. of Appl. Phys.*, **43**, No. 10, p. 974, 15 Nov. 1983, for a discussion of calculations in the Ga$_{1-x}$Al$_x$As/GaAs system. Some of the results described in these papers are reproduced here as part of our discussion.

no acceptor charge to allow constant field. The two results are consistent. The pre-factors of the equation can also be shown to be consistent. This also points out the approximate basis of the triangular well approximation, and its inadequacy when N_s is low and within factors of four of the acceptor density. A 1×10^{15} cm^{-3} doped material, with a 1 μm depletion for typical bias conditions, has a sheet acceptor density of 1×10^{11} cm^{-2}, making this approximation inaccurate under conditions of 4×10^{11} cm^{-2} sheet carrier density. Clearly, if non-degenerate conditions prevail, a test that can be made *a posteriori,* application of the Boltzmann approximation may be more acceptable. We will return to the discussion of these approximations again later; Problem 5 considers the above question for a subject of some interest in HFETs.

When $N_{As} < N_s$, the subband energies are quite independent of temperature and the temperature dependence of the Fermi level position can be found by integrating the Fermi function over the subbands. The subband positions being insensitive to temperature, these can be modelled more accurately than the expressions based on triangular well approximation. An example for the Ga$_{1-x}$Al$_x$As/GaAs system is shown in Figure 7.8 at absolute zero.

Figure 7.6 showed schematically the envelope functions and the position of the first three subband levels in an inversion layer on a (100) surface in silicon. The lowest two of these subband levels arose from the lower transverse mass and the third from the higher longitudinal mass. Since the conduction band at the band minimum is isotropic in GaAs, the subbands do not exhibit the complexity of (100) silicon surface; the pattern is similar to that exhibited by the X valleys of silicon corresponding to the transverse masses and the envelope functions have a similar general shape. The average distance of electrons from the interface is related to the shape of the envelope function. Figure 7.9 shows, at various temperatures, the average distance of the two-dimensional electron gas from the interface as a function of the sheet charge density. With an increase in the carrier density in the channel, even though higher-order bands get more occupied, the average distance from the interface decreases because it leads to a stronger confinement, again through semiconductor band bending at the interface. The electrons in the lowest subband are on an average about $\approx$ 40 Å into GaAs, and those in the first excited subband $\approx$ 150 Å away from the interface at high–10^{11} cm^{-2} sheet electron density in GaAs. We will discuss this average distance later in this chapter since it can be a substantial fraction of the thickness of the large bandgap semiconductor and hence quite important to the characteristics of HFETs.

The minimum subband energy is also a function of the sheet electron density since these are interrelated with the degree of confinement. Fig-

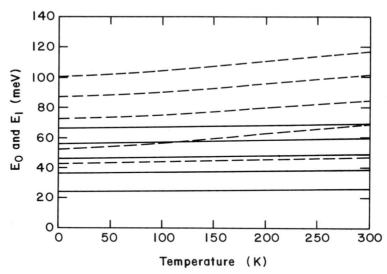

Figure 6.8: Zero'th energy level E_0 (solid lines) and the first energy level (dashed lines) for the $Ga_{1-x}Al_xAs/GaAs$ system as a function of temperature for a barrier energy of 0.3 eV, and acceptor concentration of 3×10^{14} cm^{-3}. The two-dimensional sheet electron densities are 1, 3, 5, 7.5, and 10×10^{11} cm^{-2} for the successive curves starting at the bottom. After F. Stern and S. Das Sarma, "Electron Energy Levels in GaAs/Ga$_{1-x}$Al$_x$As Heterojunctions" *Phys. Rev. B*, **30**, No. 2, p. 840, 1984.

ure 7.10 shows the movement of the energy of the zero'th energy level as a function of sheet electron density. The influence of channel carrier density in moving the energy levels is strongest for the lowest acceptor doping because it leads to the largest band bending and confinement at the interface. At higher acceptor densities, the triangular well approximation is increasingly inaccurate, and the well characteristics are strongly influenced by the acceptor charge in addition to being influenced by the carrier charge.

For the influence of thermal energy in the occupation statistics, Figure 7.11 shows the occupation of individual subbands as a function of the sheet electron density for various temperatures. The occupation of the higher bands increases with temperature, as expected, as also with larger sheet electron density. At 5×10^{11} cm^{-2} channel electron density, the ground state is $\approx 60\%$ occupied, the first excited state is $\approx 20\%$ occupied, and the second excited state is $\approx 10\%$ occupied at 300 K, showing the

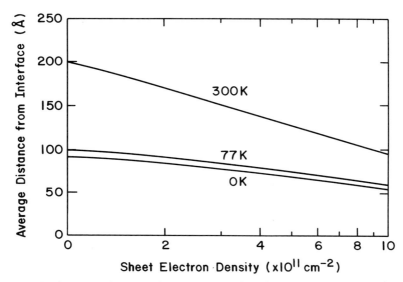

Figure 6.9: Average distance from the interface for the $Ga_{1-x}Al_xAs/GaAs$ example. The acceptor density is 3×10^{14} cm^{-3} and the different curves are for 0, 77, and 300 K. After F. Stern and S. Das Sarma, "Electron Energy Levels in GaAs/Ga$_{1-x}$Al$_x$As Heterojunctions" *Phys. Rev. B*, **30**, No. 2, p. 840, 1984.

significance of all of these in the electron statistics. The resulting Fermi energy as a function of channel electron density and temperature is shown in Figure 7.12, which shows the variation of the Fermi energy again for the various temperatures. Note the rapid movement of the Fermi energy with respect to the extrapolated bottom of the conduction band with channel electron density at 300 K.

Actual implementations of heterostructures in a transistor may be significantly different from the structure we have been discussing. We assumed the existence of the larger bandgap semiconductor $Ga_{1-x}Al_xAs$ as a heterostructure with GaAs, and concentrated on the behavior of GaAs and occupation of states in GaAs. Practical structures quite often employ doped $Ga_{1-x}Al_xAs$ with an undoped spacer layer to create the electron or hole two-dimensional carrier gas. The donor in this $Ga_{1-x}Al_xAs$ or other barrier material may be deep, thus the excursion in the Fermi energy at the interface in GaAs, and the sheet carrier density is not unlimited. As the Fermi energy increases in the smaller bandgap material, the donors in

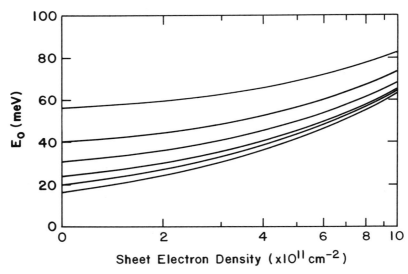

Figure 6.10: Zero'th subband position as a function of the sheet electron concentration at absolute zero for our example of the 0.3 eV barrier, $Ga_{1-x}Al_xAs/GaAs$ heterojunction with N_A varying. The sheet acceptor density N_{As} corresponding to the acceptor densities shown are 0.146, 0.46, 0.80, 1.47, 2.56, and 4.69 × 10^{11} cm^{-2}. After F. Stern and S. Das Sarma, "Electron Energy Levels in GaAs/Ga$_{1-x}$Al$_x$As Heterojunctions" *Phys. Rev. B*, **30**, No. 2, p. 840, 1984.

the larger gap material themselves begin being occupied by electrons and lose their ability to influence the electric field and the electron charge at the interface.

We now return to the question of statistics and the relevance of various approximations in a device problem. We consider this using the problem of $Ga_{.7}Al_{.3}As/GaAs$ with 3×10^{14} cm^{-3} p-type GaAs, and we consider the prediction of carrier density with Fermi energy for the quantum-mechanical calculation, the triangular well approximation, the Boltzmann approximation, and the Fermi–Dirac approximation. The results are shown in Figure 7.13. Note that at the lowest carrier densities, the Boltzmann, the Fermi–Dirac (which reduces to Boltzmann in the non-degenerate limit), and the quantum-mechanical calculation yield quite accurate results. In the modest carrier density range of low to mid 10^{11} cm^{-2}, the Fermi–Dirac approximation and the quantum-mechanical calculation agree reasonably

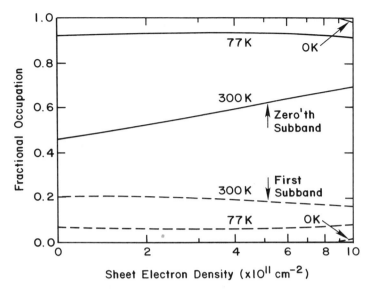

Figure 6.11: Fraction of the occupation of zero'th and first subbands as a function of sheet electron density for 0 K, 77 K, and 300 K. Barrier energy is 0.3 eV and the sheet acceptor density is 8×10^{10} cm^{-2} in the substrate. After F. Stern and S. Das Sarma, "Electron Energy Levels in GaAs/Ga$_{1-x}$Al$_x$As Heterojunctions" *Phys. Rev. B*, **30**, No. 2, p. 840, 1984.

well; the triangular approximation is in agreement at the highest carrier densities. Thus, the use of simplifications to the quantum-mechanical calculation in the form of the triangular well approximation are useful in the larger carrier density situation; in others, simplified three-dimensional approximations suffice. We will relate this to the behavior in devices, where low and high carrier densities exist simultaneously in different parts of the device, later.

Our conclusion from this calculation is restricted to 300 K temperature of operation. The thermal energy is .026 eV; it is not surprising that only when the quantization energies become sufficiently stronger than this thermal energy do we need to consider the effect of quantization on the carrier densities. At 77 K, the thermal energy is only $\approx$ .005 eV and quantization effects must be considered (see Problem 6).

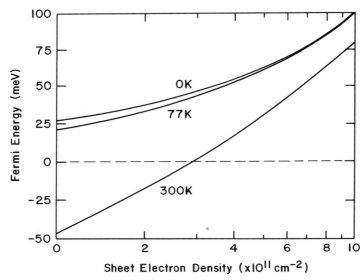

Figure 6.12: Fermi energy w.r.t. extrapolated E_c as a function of sheet electron density for $Ga_{1-x}Al_xAs/GaAs$ for 0 K, 77 K, and 300 K. Sheet density of the accepter charge in the depletion layer is 8×10^{10} cm^{-2} and barrier energy is assumed to be 0.3 eV. After F. Stern and S. Das Sarma, "Electron Energy Levels in GaAs/Ga$_{1-x}$Al$_x$As Heterojunctions" *Phys. Rev. B*, **30**, No. 2, p. 840, 1984.

6.6 Channel Control in HFETs

Doped barrier HFETs[6] are usually fabricated using a spacer layer of undoped material at the interface in the larger gap material. In addition, donors in Ga$_{1-x}$Al$_x$As have large ionization energies, a subject we will discuss later. Figure 7.14 shows the channel control under one-dimensional conditions with and without bias, showing schematically the effect of the spacer layer, deep donor layer, and bias in a modulation-doped HFET. The major consequence of a spacer layer and large ionization energy in the large bandgap material is that fewer carriers occur in the well than in the situation without the spacer or with a shallow donor. Due to the large ionization energy, the donor states in the larger gap material themselves begin being

---

[6]These have also been called MODFETs for modulation-doped field effect transistors, HEMTs for high electron mobility transistors, and SDHTs for selectively doped heterostructure transistors for historical reasons. We will refer to these as doped barrier HFETs implicitly acknowledging the common roots of the different varieties of HFETs.

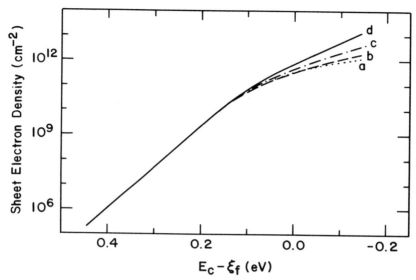

Figure 6.13: Calculated results for a Ga$_{.7}$Al$_{.3}$As/GaAs heterostructure with 3×10^{14} cm^{-3} p-type GaAs, showing the dependence of predicted sheet carrier density as a function of Fermi energy for the triangular well approximation (a), the quantum-mechanical calculation (b), the Fermi–Dirac approximation (c), and the Boltzmann approximation (d) at 300 K.

occupied at large forward voltages on the gate. This reduces the ability to influence the electric field and the electron charge at the interface. The gate then loses control of the two-dimensional channel.

Once the bias reaches these conditions, a further increase in gate voltage leads to electrons in the channel formed in Ga$_{1-x}$Al$_x$As being modulated. These electrons are either available through ohmic contacts such as in an FET, through generation process, and by injection through the barrier regions. Since the generation time constant is of the order of 1 ns, absent ohmic contacts, in a capacitor structure, frequency dispersion effects in capacitances occur around 1 GHz. In FETs, the ohmic contacts supply the necessary carriers, and the bias effects of electron capture by donors in the larger bandgap material are always present. Devices thus have a parasitic Ga$_{1-x}$Al$_x$As MESFET channel in the high gate bias region when modulation-doped barrier semiconductors are employed. The channel will form irrespective of shallow or deep levels—the problem just occurs earlier in the latter case. For Ga$_{1-x}$Al$_x$As, with $x \approx 0.4$, the donor ionization

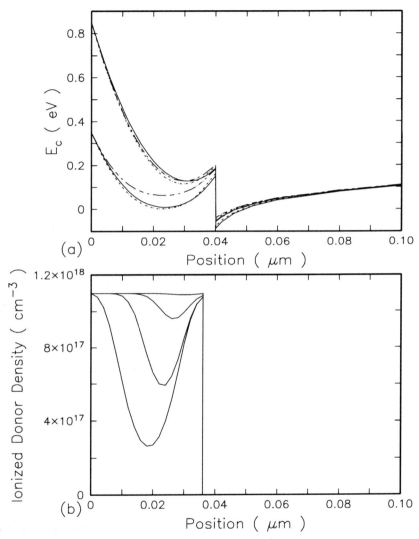

(a)

(b)

Figure 6.14: Effect of spacer layer, deep donors, etc., on the channel control without and with bias at a $Ga_{.7}Al_{.3}As/GaAs$ heterojunction. (a) shows conduction band edge assuming shallow donors (solid lines), deep donors (long dashed lines), and shallow donors with a spacer layer (short dashed lines). The lower set of lines is with a forward bias of 0.5 V. (b) shows the ionized donor density as a function of forward bias on the gate of 0 V, 0.25 V, 0.5 V, and 0.75 V for the deep donor case. It shows the significant electron capture by donors occuring in this example.

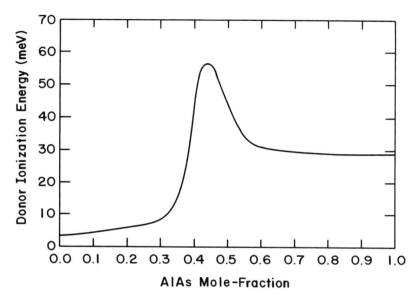

Figure 6.15: Ionization energy associated with doping due to Si in $Ga_{1-x}Al_xAs$ as a function of AlAs mole-fraction. This figure is approximate with an error bar of the order of 20 meV at the peak and has been assembled from various data in the literature.

energy is $\approx$ 60–80 meV. Later, we also discuss other peculiar phenomena in $Ga_{1-x}Al_xAs$—that of DX centers. Figure 7.15 shows the magnitude of ionization energy of a silicon donor as a function of aluminum arsenide mole-fraction, and, for a selected structure, the ratio of sheet charge in the channel versus in the $Ga_{.7}Al_{.3}As$ as a function of sheet density in the channel. In this figure, it can be seen that at $x \approx 0.4$, there is a rapid rise in the ionization energy. This rapid rise is believed to result from the lattice distortion effect as well as possible consequences of a large coupling between the Γ, L, and X bands near the band crossover point.

HFETs based on confinement of carriers can obviously be made in several different forms, for one need not confine oneself to a quasi-triangular well. We could equally well use a square well by cladding a smaller bandgap material by doped larger gap materials, or even undoped larger gap materials. The doped large gap material doesn't have to be at the surface—it could be buried underneath the channel, and it would still have a confined channel showing the enhanced mobility. This is known as the inverted structure.

These variations in structures are a consequence of the freedom afforded by morphic and pseudomorphic growth of dissimilar bandgap materials in compound semiconductors. Figure 7.16 shows many of the variations of HFET structures in the form of one-dimensional band diagrams showing the formation of the two-dimensional channel. The various device structures are differing attempts at alleviating shortcomings of other structures or improving on some of the features. These may include higher mobility or use of a different material, use of an inverted interface to obtain buried devices, better threshold control, suppression of the DX center problem, a larger amount of charge control by improving the efficiency of charge transfer, higher mobilities by introducing strain in the crystal, use of alternate confinement to obtain better short-channel effects, etc.

One can think of several other structures and material combinations, each with a trade off between advantages and disadvantages. Doping, as we have discussed, leads to formation of parasitic conducting channels. These would also happen in inverted structures. Inverted structures present a discontinuity to the conducting channel away from the charge control electrode. Ideally, this should result in a suppression of injection of warm carriers from the conducting channel to the substrate. However, in doped $Ga_{1-x}Al_xAs$, where a lower energy region exists between the discontinuity and the substrate, carriers can get injected towards the substrate, resulting in their trapping in the $Ga_{1-x}Al_xAs$. This phenomenon can lead to instability problems in switching because these thermalized carriers can not come back to the channel as easily. This same instability can occur in the capacitor structure discussed in this section. Generation–recombination times are of the order of 1 ns. If a forward bias rapid pulse is applied to cause an injection of carriers from the channel into the large bandgap material region of the capacitor, and they are captured by the donor, the process occurs on a slow time scale. Thus the capacitor structure shows time transients at these time scales. In FETs, with bias changes, etc., as during a switching transient, they will lead to larger time constant decays and rises.

An alternative to the doping of the larger gap material is to use undoped material—making the larger material a closer analogue of the SiO_2 insulator, a process that also removes the above time transient effects. These structures, such as the GaAs gate SISFET, have a threshold voltage closer to zero and result in p-channel devices with a similar close-to-zero threshold voltage. Undoped $Ga_{1-x}Al_xAs$, etc., may also be used in certain cases to increase the barrier height of a Schottky barrier, such as to $Ga_{1-x}In_xAs$. The example (b.1) of Figure 7.16 accomplishes this.

The common feature to all these devices is a conducting channel where there are usually no intentional dopants. This allows the low field mobilities

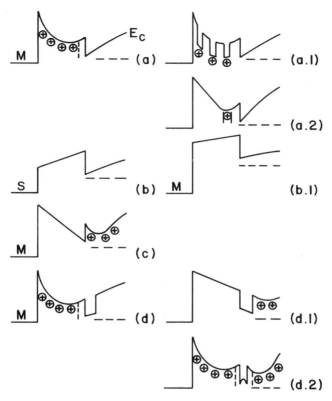

Figure 6.16: Some of the many variations of HFET structures, based on confinement of carriers in one of the dimensions using heterostructures. (a) shows the doped-barrier HFET. (a.1) and (a.2) show variations of this structure using doping in GaAs in a superlattice to avoid deep donor problems and the use of planar doping to achieve high sheet carrier densities. (b) shows the SISFET structure with (b.1) as its variation using a metal gate (also called MISFET for metal–insulator semiconductor field effect transistor and HIGFET for heterostructure–insulator gate field effect transistor). (c) shows the inverted heterostructure where an underlying layer of the large bandgap material is doped. (d), (d.1) and (d.2) show structures based on utilization of stronger confinement using square wells.

to be larger, largely because of reduction in ionized impurity scattering. The threshold voltage is controlled by appropriate thicknesses of the doped barrier layer, or to the natural value that may result from the structure. In SiO_2 MOSFETs, the threshold is controlled by doping the channel in order to accomplish the appropriate band bending. Channels have also been doped in compound semiconductor HFETs, but usually the objective is to suppress substrate injection by appropriate band shaping at the back interface.

Undoped barrier devices have been particularly appealing because they avoid deep donor related problems and have better threshold control. Being extensions of the SiO_2-based structures, they are also easier to discuss and consider as basis for further refinements. We have also discussed the appropriateness of Boltzmann statistics, Fermi–Dirac statistics, triangular-well approximation, and full quantum-mechanical calculation in the context of calculating channel charge in a two-dimensional carrier gas. Boltzmann approximation was shown to be adequate, for 300 K, for low sheet densities. This would suggest that in sub-threshold conditions, low channel densities, and in the region of the device where the channel is pinched off, i.e., near the drain for a normally operating FET, the Boltzmann approximation will be an adequate representation. Initially, we will make an additional assumption of treating the larger bandgap material, to be called an insulator, as entirely insulating. Such devices can be made in some of the compound semiconductors, and hence serve as a reference for the development of our analysis. They also serve to refresh the physics of MOSFETs, since they are similar in form.

6.7   Quasi-Static MISFET Theory Using Boltzmann Approximation

The simplest of the field effect transistor structures based on inversion or accumulation at semiconductor interfaces utilize large bandgap insulators such as silicon dioxide, and the most convenient means of analyzing the structures based on the two-dimensional carrier system is based on treating it as a sheet charge utilizing the Boltzmann approximation.[7]   Such

---

[7]Sheet charge models were originally developed for silicon MOSFETs where they have been extensively applied. Four references that describe various degrees of sophistication as well place it in a historical perspective are: H. C. Pao and C. T. Sah, "Effect of Diffusion Current on Characteristics of Metal–Oxide(Insulator)–Semiconductor Transistors," *Solid-State Electronics*, **9**, No. 9, 1966; G. Baccarani, M. Rudan, and G. Spadani, "Analytical I.G.F.E.T Model Including Drift and Diffusion Currents," *IEE J. on Solid-State and Electron Devices*, **2**, No. 2, 1978; P. A. Muls, G. J. Declerck, and R. G. Van

structures, based on an ideal insulator and statistics of non-degenerate conditions, have limited validity in compound semiconductors. Insulators such as silicon dioxide have fairly large interface state density at the surfaces of most compound semiconductors, reducing or precluding control of inversion layers, and at the least leading to substantial slow and fast states related effects. Large bandgap semiconductors conduct some current, and hence hot carrier and tunneling effects are substantially more important and part of the device operation over much of the bias range, unlike in the oxide-based structure where the tunneling and hot carrier effects are more of a concern in device reliability. Likewise, most of the large-mobility compound semiconductors have a low effective density of states, and strong inversion requires inclusion of degenerate statistics, and, because of the smaller effective mass, inclusion of quantization effects in the source end part of the device. But, both low inversion conditions (i.e., low gate biases) and drain biases leading to pinch-off at the drain end cause a substantial part of the channel to be in weak inversion; these are all regions where the Boltzmann approximation is appropriate. Quasi-static characteristics for both capacitor structures and field effect transistor structures, assuming ideal insulators and the Boltzmann approximation, are of general interest before we consider the consequences of breakdown of these approximations.

Our discussion of insulator semiconductor physics for use in field effect transistors will begin with a discussion of the semiconductor–insulator–semiconductor capacitor structures. Consider the band bending when two pieces of semiconductors or a semiconductor and metal are brought together with an intervening insulating medium, in thermal equilibrium, and in steady-state, with positive and negative bias applied as shown in Figure 7.17. Because the insulator does not support any flow of charge (an example of such an insulator is SiO_2 under most conditions), the system is in thermal equilibrium even with bias, with the only mechanism for the redistribution of charge being through the processes of generation and recombination. In the figure, case (b) is when a voltage V_{FB} is applied to the metal gate to lead to flat band conditions in the semiconductor. Flat-band is defined as the condition when the band edges are flat in the region of interest, i.e., at the interface in the semiconductor. For a structure with a changing bandgap, we may define this as when the band edge of interest corresponding to the carrier of interest is flat, e.g., conduction band edge (corresponding to electrons) for formation of inversion layer in the p-type

Overstraeten, "Characterization of the MOSFET Operating in Weak Inversion," in L. Marton, Ed., *Advances in Electronics and Electron Physics*, **47**, Academic Press, N.Y., (1978); and J. R. Brews, "Physics of the MOS Transistor," in D. Kahng, Ed., *Applied Solid State Science*, Suppl. 2A, Academic Press, San Diego, CA (1981). Our description here closely follows the last reference with changes to suit compound semiconductors.

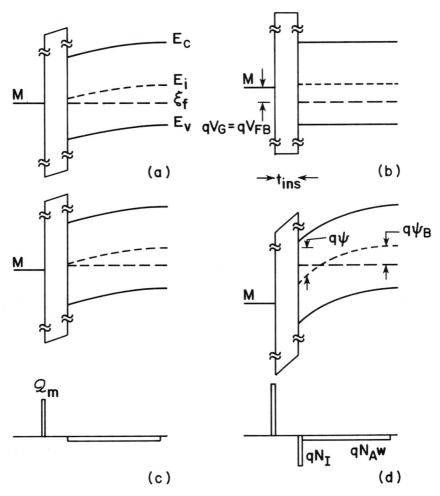

Figure 6.17: (a) shows the energy band diagram when the metal and semi-conductor are far away and (b) shows the same when they are connected together and the flat-band voltage V_{FB} is applied. (c) and (d) show the energy band diagrams in depletion and inversion, together with a schematic of charge distribution in the lower half of the figure. (d) also shows the band bending convention adopted here.

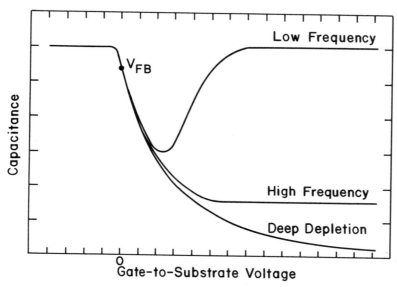

Figure 6.18: Capacitance–voltage characteristics for the metal–insulator–semiconductor structure. The accumulation region is on the left part, and the positive bias for this p-type substrate shows various possible capacitances depending on the frequency of operation.

material. In the base–emitter junction of the n–p–n bipolar transistor, it is again the conduction band edge for both the emitter and the base since the electron transport is the major transport and the feature of interest. Case (c), in this figure, shows the depletion condition at the interface when the bias causes removal of holes from near the insulator interface. When more bias is applied, electron charge results at the surface, as shown in case (d). This charge region is known as the inversion layer because in this example we used a p-type semiconductor. If a polarity opposite to this were applied, we would obtain an accumulation layer of holes at the interface. In all these cases, because no current flow can occur, not being allowed by the insulator, the carrier redistribution occurs through generation and recombination of carriers, and thermal equilibrium prevails.

The corresponding capacitance–voltage characteristics that we would see for this example are shown in Figure 7.18. The observed capacitance, which measures the change in charge to any change in applied bias, depends on how rapidly the signal used to measure the capacitance is changed. In the absence of a source of carriers, the process of addition or removal of carriers from the interface and the depletion region are limited by generation and recombination, which have a finite time constant. Recall that the insulator allows no flow of charge. In a FET, this moving charge at the interface is available through the ohmic electrodes of source and drain. But, in a capacitor structure, with only the gate and the substrate electrode for bias, it has to occur through carrier generation and recombination, either through a thermal process or through some external means of excitation. An example of the latter is light-induced generation of carriers.

Let us assume that we have an accumulation layer formed in the structure and we have a negative bias applied for the p-type substrate structure. So, we have accumulation of holes at the insulator–semiconductor interface. These holes were attracted to the interface by the negative bias voltage applied and the lower hole energy at the band edge that resulted. If we measure the capacitance of this structure at this bias by a small-signal voltage, we will essentially measure the capacitance of the insulator. Because the substrate is p-type, carriers will move to the interface or away from the interface through the conducting path. The frequency of signal that we apply can either be low or high (within limits of dielectric relaxation frequency); the accumulation region will be able to respond through the conducting path and we will continue to observe this high capacitance so long as accumulation exists, i.e., below flat-band voltage bias for the p-type substrate. We now change the bias that is applied to the substrate. As we change the bias to greater than the flat-band voltage, the capacitance response becomes a function of the measurement conditions, because these determine whether the electrons can appear at the interface or not, depending on rate of sweep of the bias, and if they do appear whether they can respond or not.

First consider the changing of bias and measurement of the capacitance using a small applied signal in the absence of any excitation such as light. If the sweep of bias, even though it is slow, does not lead to generation of carriers because the semiconductor material has good lifetime, then, independent of the frequency of measurement, no carriers appear at the interface and the depletion region continues to expand into the semiconductor. The capacitor structure is then said to go into deep depletion. On the other hand, if the bias sweep is such that it allows the carriers to appear at the interface, then the capacitance response depends on whether these carriers at the interface can be modulated by the small signal. If the frequency is high, then even though the inversion layer is formed, the carriers in the inversion layer do not respond to the applied signal, only the substrate depletion region responds, and one sees a low capacitance. This low capacitance occurs together with an inversion layer; the bias drop across the substrate region is smaller than that in the deep-depletion case where no inversion layer is formed. So, this capacitance is higher than the deep-depletion capacitance. Now, if the applied small signal is at a low frequency, the inversion layer may respond through carrier generation and recombination. So, with change in bias, at low frequencies, we gradually see the capacitance reaching back to similar values as in the accumulation case. Minor differences in the magnitude of this limit capacitance will occur because the total band bending also depends on density of states, which are different for the conduction and valence bands. So, when the bias is swept from below flat-band voltages to higher than flat-band voltages, the capacitor structure can show three distinctly different behaviors, to be referred to as deep-depletion behavior, high-frequency behavior, and low-frequency behavior. The inversion layer is formed in both the high-frequency and the low-frequency case, even though the capacitance observed is different because the inversion layer fails to respond to the high-frequency signal but does respond to the low-frequency signal.

We may also have an excitation, such as light, during or during part of the measurement of the structure. The deep-depletion type of measurement obtained by sweeping the bias from below flat-band voltage to above the flat-band voltage is quite often accomplished by applying a slowly varying bias (a ramp, e.g.), and observing the current to determine the capacitance. The current is displacement current in the insulator, and, because of the slow change in bias, conduction current in the semiconductor. If we apply a bias larger than the flat-band voltage bias and form an inversion layer (e.g., by using light excitation), and then use the slow ramp sweep to measure the capacitance as we apply a bias going towards accumulation, a capacitance–voltage behavior similar to the low frequency behavior will be observed. Such a behavior is usually referred to as quasi-static capacitance–voltage behavior. Its observation indicates the insulating character of the insulator. Together with the high-frequency behavior it forms a strong basis for first-order understanding of interface state distribution, etc. (see Problem 8).

We will define the various states of charge distribution at the insulator–semiconductor interface for the p-type substrate here. The definitions for the n-type substrate are analogous to this.

Weak inversion is defined as the state when the surface is barely n-type with the electron concentration greater than the intrinsic carrier concentration but less than the background hole concentration of the substrate, i.e.,

$$n_i \leq n(0) \leq p_0, \tag{6.19}$$

and equivalently,

$$\psi_B \leq \psi_S \leq 2\psi_B. \tag{6.20}$$

We define the potentials w.r.t. the bulk where the reference is the intrinsic level. The bulk potential is ψ_B, the hole Fermi energy level being $q\psi_B$ below the intrinsic energy level. The surface potential is ψ_S. The two equations above for carrier concentration and potential can then be seen to be equivalent.

Likewise, moderate inversion is the state when the carrier concentration at the surface exceeds the bulk hole concentration, but the total charge in the inversion layer is still less than the charge in the depletion region.

$$n(0) \geq p_0 \text{ and } Q_I \leq Q_{dep}. \tag{6.21}$$

Let ψ_H be the potential at the surface when the inversion charge becomes equal to the depletion layer charge. The above condition can be summarized to

$$2\psi_B \leq \psi_S \leq \psi_H. \tag{6.22}$$

Strong inversion is defined as the condition at which the total charge in the inversion layer exceeds the depletion layer charge. Thus, at strong inversion

$$
\begin{aligned}
n(0) &\geq p_0, \\
Q_I &\geq Q_{dep.}, \\
\text{and} \quad \psi_S &\geq \psi_H.
\end{aligned}
\tag{6.23}
$$

Part (d) of Figure 7.17 shows the convention for definition of potentials and energies. We use the bulk intrinsic level as our reference. Then, the electrostatic potential is defined as the band bending with respect to the bulk. It is positive if the bands bend downwards. The Fermi level in the bulk is at $q\psi_B$ below the intrinsic level. Let us derive a general solution to the problem, relating the band bending to the material properties. We will later extend this by relating the band bending to the applied potentials themselves. Consider the case of a uniformly doped semiconductor with

$$N_A(y) = N_A, \tag{6.24}$$

independent of the position. We consider the Boltzmann approximation for the carriers,

$$p = n_i \exp\left(q\frac{-\psi + \psi_B}{kT}\right). \tag{6.25}$$

In thermal equilibrium, and this insulator case is always in thermal equilibrium because the insulator allows no current to flow,

$$n = n_i \exp\left(q\frac{\psi - \psi_B}{kT}\right). \tag{6.26}$$

Also, again by definition,

$$N_A = n_i \exp\left(\frac{q\psi_B}{kT}\right). \tag{6.27}$$

Poisson's equation allows us to write the problem in the form

$$-\frac{d^2\psi}{dy^2} = -\frac{qN_A(y)}{\epsilon_s} + \frac{qp(y)}{\epsilon_s} + \frac{qN_D(y)}{\epsilon_s} - \frac{qn(y)}{\epsilon_s}. \tag{6.28}$$

Consider the case of no donors (p-type substrate) and uniform doping.

$$\frac{d^2\psi}{dy^2} = \frac{qN_A}{\epsilon_s} - \frac{qn_i}{\epsilon_s} \exp\left(q\frac{-\psi + \psi_B}{kT}\right) + \frac{qn_i}{\epsilon_s} \exp\left(q\frac{\psi - \psi_B}{kT}\right)$$

$$\text{for} \quad 0 < y < w$$

and $\quad \dfrac{d^2\psi}{dy^2} = 0 \quad \text{for} \quad y > w.$ $\tag{6.29}$

Multiplying by $d\psi/dy$, using the equality

$$\frac{1}{2}\frac{d}{dy}\left(\frac{d\psi}{dy}\right)^2 = \left(\frac{d\psi}{dy}\right)\left(\frac{d^2\psi}{dy^2}\right), \tag{6.30}$$

and neglecting the hole contribution, we obtain

$$\frac{1}{2}\frac{d}{dy}\left(\frac{d\psi}{dy}\right)^2 = \frac{qN_A}{\epsilon_s}\left[1 + \frac{n_i}{N_A}\exp\left(q\frac{\psi - \psi_B}{kT}\right)\right]\frac{d\psi}{dy}. \tag{6.31}$$

This is simply written as

$$\frac{1}{2}\frac{d}{dy}\left(\frac{d\psi}{dy}\right)^2 = \left(\frac{kT}{q\lambda_D}\right)^2\left[\frac{q}{kT} + \frac{q}{kT}\left(\frac{n_i}{N_A}\right)^2\exp\left(\frac{q\psi}{kT}\right)\right]\frac{d\psi}{dy}, \tag{6.32}$$

where λ_D is the Debye length given by

$$\lambda_D = \sqrt{\frac{\epsilon_s kT}{q^2 N_A}}, \tag{6.33}$$

yielding the form

$$\frac{1}{2}\frac{d}{dy}\left(\frac{d\psi}{dy}\right)^2 = \left(\frac{kT}{q\lambda_D}\right)^2\frac{d}{dy}\left[\frac{q\psi}{kT} + \left(\frac{n_i}{N_A}\right)^2\exp\left(\frac{q\psi}{kT}\right)\right]. \tag{6.34}$$

Integrating from the insulator–semiconductor interface ($y = 0$) to the edge of the depletion region ($y = w$) we obtain

$$\left(\frac{d\psi}{dy}\right)^2_{y=w} - \left(\frac{d\psi}{dy}\right)^2_{y=0} = 2\left(\frac{kT}{q\lambda_D}\right)^2\left[\frac{q}{kT}\psi + \left(\frac{n_i}{N_A}\right)^2 \exp\left(\frac{q\psi}{kT}\right)\right]\Bigg|^{y=w}_{y=0}. \tag{6.35}$$

At the edge of the depletion region both the electrostatic potential and the electrostatic field are zero, and at the insulator–semiconductor interface the potential is the surface potential. So the boundary conditions at $y = w$ are

$$\psi = 0 \qquad \text{and} \qquad \frac{d\psi}{dy} = 0, \tag{6.36}$$

and at $y = 0$ is

$$\psi = \psi_S. \tag{6.37}$$

So, we obtain

$$-\frac{d\psi}{dy}\Bigg|_{y=0} = \frac{\sqrt{2}kT}{q\lambda_D}\left\{\frac{q\psi_S}{kT} + \left(\frac{n_i}{N_A}\right)^2\left[\exp\left(\frac{q\psi_S}{kT}\right) - 1\right]\right\}^{1/2}. \tag{6.38}$$

The terms in this equation can be associated with the charge terms of Poisson's equation. The first term is due to the ionized impurity charge N_A and the second term is due to the electron charge corresponding to ψ_S of band bending. We may show this quite simply, Poisson's equation solution for band bending and field for the ionized charge solution is the simplest level of approximation in p–n junction theory. The field and the band bending from Poisson's equation as a function of the depth y are

$$-\frac{d\psi}{dy} = \frac{qN_A}{\epsilon_s}y,$$

$$\text{and} \qquad \psi = \frac{qN_A}{2\epsilon_s}y^2, \tag{6.39}$$

which, when y is eliminated, is

$$-\frac{d\psi}{dy} = \frac{qN_A}{\epsilon_s}\sqrt{\frac{2\epsilon_s\psi}{qN_A}} = \frac{\sqrt{2}}{\lambda_D}\psi. \tag{6.40}$$

This equation, derived earlier, is a more general solution of the transition region problem that includes the contribution of both mobile (electron) and immobile charge. It still ignores hole contribution which could have been included, but would have resulted in a more complicated expression. In the insulator–semiconductor problem, our interest is in the realistic modelling of the charge that gives rise to conduction, i.e., the electrons in inversion. The hole charge contribution is negligible because the inversion condition, where these devices are usually biased to operate, leads to significant enough band bending that very little hole charge exists. Most of the other charge contribution appears from the ionized acceptor charge in the depletion region. The electron charge is spread out near the insulator–semiconductor interface, and, as we will see, mostly confined in a region which is fairly thin (of the order of at most $\approx$ 100 Å's). Our further approximation, for a simple iterative but manageable solution, entails assuming that we will approximate the inversion charge of electrons to be confined in a thin sheet of charge at the interface. We will take into account all the contributions by approximating the limited spread of the electron charge into this infinitesimally thin layer. We could have used numerical integration of the electron density given by

$$n = n_i \exp\left(q\frac{\psi - \psi_B}{kT}\right) \tag{6.41}$$

and a solution of ψ for calculating this. Instead we treat it as simply a sheet charge at the interface of sheet density N_I, equal to the numerically integrated value of the above. Gauss's law applied to Equation 7.38 allows us to do this simply. The total charge in the semiconductor that gives rise to the band bending and the field is $\mathcal{Q}_{sem}$ and is

$$\mathcal{Q}_{sem} = -qN_I - qN_A w. \tag{6.42}$$

Once an inversion layer forms, and conditions allow the supply or removal of electrons for any further change in bias, most of the charge is induced in the inversion layer; w does not change substantially then. The reason for this is that the gate bias is screened by the electron inversion charge once the inversion layer has been formed. ψ_S then does not change substantially. Since neutrality prevails in the substrate beyond the depletion region, and the field is zero at the depletion region edge, $\mathcal{Q}_{sem}$ simply follows from Gauss's law using a volume whose one surface is at the interface and other is beyond the depletion region.

$$\epsilon_s \left.\frac{d\psi}{dy}\right|_{y=0} = \mathcal{Q}_{sem}, \tag{6.43}$$

and

$$\mathcal{Q}_{sem} = -\frac{\sqrt{2}\epsilon_s kT}{q\lambda_D}\left\{\frac{q\psi_S}{kT} + \left(\frac{n_i}{N_A}\right)^2\left[\exp\left(\frac{q\psi_S}{kT}\right) - 1\right]\right\}^{1/2}, \tag{6.44}$$

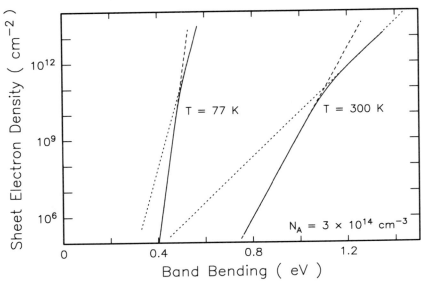

Figure 6.19: Sheet electron density in the Boltzmann approximation as a function of band bending for GaAs at 300 K and 77 K; the p-type doping in the substrate is assumed to be 3×10^{14} cm^{-3}. The dotted lines show fits to the two segments of the variations.

and

$$qN_A w = qN_A\sqrt{2}\lambda_D \left(\frac{q\psi_S}{kT}\right)^{1/2}. \tag{6.45}$$

So,

$$N_I = N_A\sqrt{2}\lambda_D \left\{ \left[\frac{q\psi_S}{kT} + \left(\frac{n_i}{N_A}\right)^2 \exp\left(\frac{q\psi_S}{kT}\right)\right]^{1/2} - \left(\frac{q\psi_S}{kT}\right)^{1/2} \right\}. \tag{6.46}$$

In this equation, we have dropped the unity term of the electron charge part of the semiconductor charge term because in moderate and strong inversion, $\exp(q\psi_S/kT)$ is large. In weak inversion the whole term by itself is small. Figure 7.19 shows the inversion carrier density of electrons in p-type substrate of 3×10^{14} cm^{-3} doping. As can be seen in this, with the Boltzmann approximation, and we elaborate on this further later in our discussion of FET characteristics, the weak inversion and strong inversion regions are characterized by exponentially-varying electron charge densities.

Mathematically, these can be related by looking at the approximations of Equation 7.46.

In weak inversion, the surface is barely n-type, $n_i \leq n(0) \leq p_0$, $\psi_B \leq \psi_S \leq 2\psi_B$, and $Q_I \leq Q_{dep.}$. Our approximation of Equation 7.46 is

$$
\begin{aligned}
N_I \quad &\approx \quad N_A \sqrt{2} \lambda_D \left\{ \left(\frac{q\psi_S}{kT} \right)^{1/2} \left[1 + \left(\frac{n_i}{N_A} \right)^2 \frac{\exp\left(q\psi_S/kT\right)}{2q\psi_S/kT} \right] - \right. \\
&\qquad \left. \left(\frac{q\psi_S}{kT} \right)^{1/2} \right\} \\
&\approx \quad N_A \lambda_D \left(\frac{n_i}{N_A} \right)^2 \sqrt{\frac{kT}{2q\psi_S}} \exp\left(\frac{q\psi_S}{kT} \right).
\end{aligned} \tag{6.47}
$$

In strong inversion, the inversion charge of electrons is significantly larger than the depletion charge due to ionized acceptors, and the exponential term dominates.

$$
\begin{aligned}
N_I \quad &\approx \quad N_A \sqrt{2} \lambda_D \left(\frac{n_i}{N_A} \right) \exp\left(\frac{q\psi_S}{2kT} \right) \\
&\approx \quad \sqrt{2} \lambda_D n_i \exp\left(\frac{q\psi_S}{2kT} \right).
\end{aligned} \tag{6.48}
$$

The slope of the sheet charge, with respect to band bending in strong inversion, changes to approximately half of the exponential dependence of the weak inversion region. The change in surface potential, necessary to cause any fraction of change in inversion carrier density, is smaller by nearly a factor of two in strong inversion compared to weak inversion. The largest changes in carrier densities occur at the insulator–semiconductor interface in a thin region; the relative change in band bending required to cause a fractional change the charge density here is very small because of the exponential relationship.

The relationship we seek is the relationship between applied gate bias, the external stimulus, and the carrier density, in order to apply it in the calculation of the current flow of the field effect transistor. The band bending occurs as an implicit parameter in this derivation since it is the primary parameter in calculations of charge densities. So, first we seek the relationship between this band bending and the gate bias of the insulator capacitor structure.

Let the V_G be the gate bias with respect to the substrate. The electrostatic potential of the gate, assuming for the moment that the flat-band voltage, a translational variable, is zero, is the applied potential V_G itself.

The electric field in the insulator, $\mathcal{E}_{ins}$, is

$$\mathcal{E}_{ins} = \frac{V_G - \psi_S}{t_{ins}}, \tag{6.49}$$

where t_{ins} is the thickness of the insulator. In the absence of any interface charge, the displacement field is continuous across the interface, i.e., $\mathcal{D}_{ins} = \mathcal{D}_{sem}$, and

$$\epsilon_{ins} \frac{V_G - \psi_S}{t_{ins}} = \frac{\sqrt{2}\epsilon_s}{\lambda_D} \frac{kT}{q} \left\{ \frac{q\psi_S}{kT} + \left(\frac{n_i}{N_A} \right)^2 \left[\exp\left(\frac{q\psi_S}{kT} \right) - 1 \right] \right\}^{1/2}, \tag{6.50}$$

or

$$C_{ins} (V_G - \psi_S) = \sqrt{2} C_{FB} \frac{kT}{q} \left\{ \frac{q\psi_S}{kT} + \left(\frac{n_i}{N_A} \right)^2 \left[\exp\left(\frac{q\psi_S}{kT} \right) - 1 \right] \right\}^{1/2}, \tag{6.51}$$

where

$$C_{ins} = \frac{\epsilon_{ins}}{t_{ins}} \tag{6.52}$$

is the capacitance associated with the insulator, and

$$C_{FB} = \frac{\epsilon_s}{\lambda_D} \tag{6.53}$$

is the capacitance associated with the semiconductor under flat-band conditions (see Problem 9).

These capacitances are defined per unit area, as are many of the earlier parameters such as the sheet charge carrier densities, etc. The flat-band capacitance C_{FB} is the semiconductor portion of the total capacitance of the structure when flat band conditions prevail. Our analysis of p–n junctions also gives rise to similar capacitance when the junction is biased to near flat-band conditions. We now have the relations between the gate bias and surface band bending and we have already derived the relationship between surface band bending and the sheet carrier density. For the insulator capacitor, we now have a relatively complete description of the parameters of interest, from which most others can be derived.

Figure 7.20 shows the variation of surface potential with applied gate bias for a GaAs substrate of background doping 3×10^{14} cm^{-3} at 300 K and 77 K , assuming an idealized insulator of undoped Ga$_{1-x}$Al$_x$As, of dielectric constant of 11.9, and a bandgap assumed to be very large. At low biases, the surface potential changes rapidly with gate bias through the depletion and weak inversion region. At larger biases, the surface potential

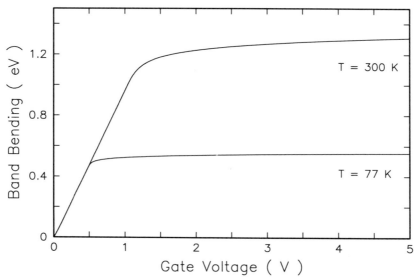

Figure 6.20: Surface band bending with applied gate at 300 K and 77 K for GaAs of background doping 3×10^{14} cm^{-3}, with an insulator of 400 Å thickness and a dielectric constant of 11.9 (an idealization of undoped Ga$_{1-x}$Al$_x$As assuming a large bandgap). This calculation assumes a flat-band voltage of zero, so the gate voltage should be translated along the ordinate axis by V_{FB}.

does not change as much with bias. The change to this lower bias dependence begins to occur at moderate inversion conditions where the inversion sheet charge is similar in magnitude to the depletion layer sheet charge. Once the semiconductor surface is in strong inversion, the surface potential changes very little, almost as if it is pinned. Together with this, the depletion region width also becomes nearly constant. The capacitance associated with the semiconductor, therefore, is nearly a constant in this bias region. The screening by inversion charge is responsible for this. The surface band bending does not have to change much to induce a large change in the sheet charge because most of the charge resides in a sheet at the insulator–semiconductor interface, and, recall, the carrier density changes exponentially with the energy difference between the Fermi energy and the conduction band edge energy.

For modest doping in the substrate, this change in behavior of the surface potential occurs close to the condition where the charge in the inversion

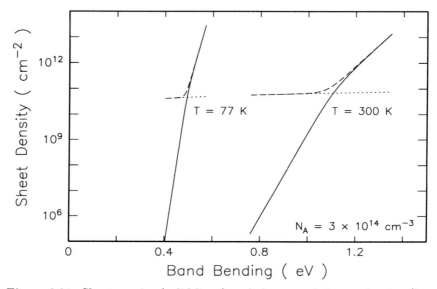

Figure 6.21: Sheet carrier (solid lines) and sheet total charge density (long dashes) as a function of band bending for 3×10^{14} cm^{-3} acceptor doping in GaAs at 300 K and 77 K. The variation of sheet density due to acceptor in the depletion region is shown using short dashes.

layer becomes comparable to the ionized dopant charge in the depletion region. It becomes equal at ψ_H; the region between a band bending of $2\psi_B$ and ψ_H is characterized by the change to saturation behavior of the surface potential. We tend to use the magnitude $2\psi_B$ for evaluation of the condition of onset of inversion at modest or large substrate dopings because it is easy to evaluate.

Figure 7.21 shows the behavior of sheet charge, at the interface, in the semiconductor region, as a function of surface potential for the 3×10^{14} cm^{-3} acceptor doping. ψ_B for this doping is ≈ 0.49 V. Here, the use of $2\psi_B$ as a criterion for inversion would be quite erroneous. However, at larger dopings in the substrate, exceeding 2×10^{16} cm^{-3}, the change over to modest and strong inversion comes close to the $2\psi_B$ criterion. The lighter-doped material has too little integrated charge due to ionized impurities; the band bending that creates a similar amount of sheet carrier density results in a small carrier density and a large difference between the conduction band edge and the Fermi energy at the interface. A larger sheet density than this is needed before the band bending exhibits the expected pinning.

The condition at which this change to rapid increase in sheet carrier density with band bending occurs is reflected in the threshold condition, since conduction can be large for biases exceeding this. For the low-10^{16} cm^{-3} doped substrate, this threshold condition can be defined in terms of when the band bending reaches the magnitude of $2\psi_B$. This is the onset of modest inversion; as the trend of sheet carrier density with band bending shows, a more accurate description may be the onset of strong inversion which occurs nearly 100 mV to 200 mV above the $2\psi_B$ band bending. For lighter-doped substrates, we may define this to occur at conditions when the inversion sheet carrier density approaches mid-10^{10} cm^{-2}, a condition when the band edge and Fermi energy are quite close and inversion charge changes rapidly without any significant change in band bending. So, for higher-doped substrates, the threshold voltage in our derivation where we ignore the exponential factor is

$$V_T \approx 2\psi_B + \frac{C_{FB}}{C_{ins}} \left(2\frac{q\psi_B}{kT} \right)^{1/2},$$ (6.54)

with a translational parameter of flat-band voltage, and more accurately, the term of $2\psi_B$ may be replaced by $2\psi_H$. In the case of low-doped substrates, we need to determine the surface band bending for mid-10^{10} cm^{-2} sheet carrier density and employ that value.

6.7.1 Capacitance of the MIS Structure

We have now determined the relationship between the band bending and the applied bias, and the relationship between the carrier density and the band bending. The capacitances (quasi-static, since this analysis uses the static equations) follow from the change in charge due a change in potential. The semiconductor capacitance C_{sem}, e.g., is

$$
\begin{aligned}
C_{sem} &= -\frac{dQ_{sem}}{d\psi_S} \\
&= \frac{C_{FB}}{\sqrt{2}} \left[1 + \left(\frac{n_i}{N_A} \right)^2 \exp\left(\frac{q\psi_S}{kT} \right) \right] \times \\
&\quad \left\{ \frac{q\psi_S}{kT} + \left(\frac{n_i}{N_A} \right)^2 \left[\exp\left(\frac{q\psi_S}{kT} \right) - 1 \right] \right\}^{-1/2}.
\end{aligned}
$$ (6.55)

In order to obtain an expression for capacitance valid from depletion to strong inversion, we must consider the charge contribution from electrons and holes; the above equation considers electron charge but ignores hole charge. Thus, the expression is invalid in accumulation and flat-band conditions since the hole contribution can be substantial in these regions. The validity of the above expression is thus restricted to $q\psi_S/kT \gg 0$. Had we included this (see Problem 9),

$$
\begin{aligned}
C_{sem} &= \frac{C_{FB}}{\sqrt{2}} \left\{ 1 - \exp\left(-\frac{q\psi_S}{kT}\right) + \left(\frac{n_i}{N_A}\right)^2 \left[\exp\left(\frac{q\psi_S}{kT}\right) - 1\right] \right\} \times \\
&\quad \left\{ \left[\exp\left(-\frac{q\psi_S}{kT}\right) + \frac{q\psi_S}{kT} - 1\right] + \right. \\
&\quad \left. \left(\frac{n_i}{N_A}\right)^2 \left[\exp\left(\frac{q\psi_S}{kT}\right) - \frac{q\psi_S}{kT} - 1\right] \right\}^{-1/2}.
\end{aligned}
\tag{6.56}
$$

This equation should be valid in all regions. This corresponds to the low-frequency behavior discussed earlier, since we have assumed quasi-static conditions to prevail in the semiconductor. Note that when $\psi_S \to 0$, $C_{sem} \to C_{FB}$, the semiconductor capacitance at flat-band conditions, related through the Debye length. The capacitance we are most interested in is the capacitance C_{MIS} of the complete structure, related as

$$
C_{MIS} = \frac{dQ_G}{dV_G} = -\frac{dQ_{sem}}{dV_G}.
\tag{6.57}
$$

We assume a perfect insulator that contains no charge, hence,

$$
C_{ins}\left(V_G - \psi_S\right) = -Q_{sem},
\tag{6.58}
$$

and

$$
\begin{aligned}
C_{ins}\left(1 - \frac{d\psi_S}{dV_G}\right) &= -\frac{dQ_{sem}}{dV_G} \\
&= C_{sem}\frac{d\psi_S}{dV_G}.
\end{aligned}
\tag{6.59}
$$

The relationship between the band bending and the applied bias, in terms of the capacitances, following the above, is

$$
\frac{d\psi_S}{dV_G} = \frac{C_{ins}}{C_{ins} + C_{sem}},
\tag{6.60}
$$

from which it follows that

$$
C_{MIS} = -\frac{dQ_{sem}}{dV_G} = -\frac{dQ_{sem}}{d\psi_S}\frac{d\psi_S}{dV_G} = \frac{C_{sem}C_{ins}}{C_{sem} + C_{ins}}.
\tag{6.61}
$$

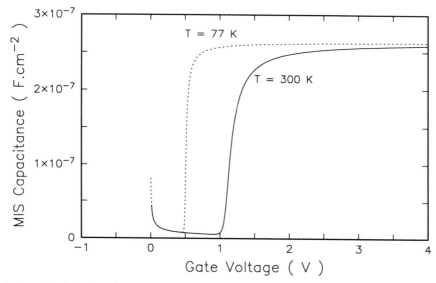

Figure 6.22: Capacitance per unit area of the MIS structure consisting of a $3 \times 10^{14} \text{cm}^{-3}$ GaAs with 400 Å $\text{Ga}_{1-x}\text{Al}_x\text{As}$ as a function of bias. Curves are shown for 300 K and 77 K.

The capacitance of the metal–insulator–semiconductor structure is a series combination of capacitors. Using the above relationships, we can also now determine the capacitance of the structure, since both the insulator and the semiconductor are known. This capacitance, for the example of $3 \times 10^{14} \text{ cm}^{-3}$ substrate of GaAs at 300 K and 77 K, with a 400 Å $\text{Ga}_{1-x}\text{Al}_x\text{As}$ insulator, is shown in Figure 7.22. This corresponds to the low-frequency capacitance of the structure, considered qualitatively earlier, since it is based on the use of static equations.

6.7.2 Flat-Band Voltage

So far, we have assumed that flat bands, i.e. $\psi_S = 0$, occured at $V_G = 0$, the rationale being that this is only a translational variable. In practice it occurs at a finite non-zero value, which we have called V_{FB} ($\psi_S = 0$ at $V_G = V_{FB}$); we will determine it now to evaluate its quantitative magnitude for structures of interest to us. This non-zero flat-band voltage is a consequence of work function differences, insulator charge, etc. Our equations derived so far will all be correct if we replace the potential V_G by $V_G - V_{FB}$. For example, the threshold voltage V_T for modest or higher substrate dopings

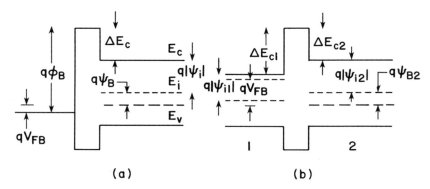

Figure 6.23: Flat-band conditions obtained by applying a gate voltage equal to the flat-band voltage to metal–insulator–semiconductor (a) and semiconductor–insulator–semiconductor (b) structures. The parameters are defined in the figure.

is

$$V_T = V_{FB} + 2\psi_B + \sqrt{2}\frac{C_{FB}}{C_{ins}}\frac{kT}{q}\left(2\frac{q\psi_B}{kT}\right)^{1/2}. \qquad (6.62)$$

We have been treating the insulator–semiconductor interface as ideal with no interface state density, and the insulator as containing no charge. The flat band will occur when a gate voltage of V_{FB} is applied that is the sum of the difference in the metal barrier height and the conduction band discontinuity and the bulk potential, i.e.,

$$V_{FB} = \phi_B - \frac{\Delta E_c}{q} - |\psi_i| - \psi_B, \qquad (6.63)$$

as shown in Figure 7.23. In reality, modifications have to be made to this. Insulators have some unintentional charge in them. Undoped $Ga_{1-x}Al_xAs$ quite often has acceptor charge, while SiO_2 has a very small positive charge. The interface may also have a non-negligible interface state density which also must be considered in the charge analysis. Problem 10 discusses some of these considerations and their effect on capacitances and flat-band voltages.

When semiconductor gate electrodes are used instead of metal electrodes, such as in a semiconductor–insulator–semiconductor structure, the flat-band voltage under the idealized conditions, assuming discontinuities of ΔE_{c1} and ΔE_{c2}, and an energy difference between conduction band edge and the Fermi level of $q\psi_{B2}$ in the control electrode (see (b) in Figure 7.23), is

$$V_{FB} = \frac{\Delta E_{c1}}{q} + |\psi_{i1}| + \psi_{B1} - \frac{\Delta E_{c2}}{q} - |\psi_{i2}| - \psi_{B2}, \qquad (6.64)$$

which reduces to the bulk potential differences when identical semiconductors are employed. Note that ψ_{B1} is negative in the example shown. In the semiconductor–semiconductor system we have to independently determine the discontinuities. Fixed charge at the interface is less likely to be substantial at the interfaces due to crystal continuity; however, charge may still exist in the insulator. V_{FB} is therefore most accurately characterized by experiments because of its sensitivity to so many different parameters.

6.7.3 MISFET Models Based on Sheet Charge Approximation

We have now established the framework for treating the field effect transistor based on this approach utilizing sheet charge approximation and band bending in the semiconductor as the parameter. We will refer to this analysis as the sheet charge model and we will follow Brews' discussion (see general references) of its application to silicon MOSFETs. No current flows in the ideal MIS structure; it is always in thermal equilibrium, and the carriers in inversion layer come about due to generation and recombination processes or external stimuli. Light is one of these. We can also introduce the carriers by introducing ohmic electrodes of source and drain. These ohmic contacts can supply and extract the carriers in the inversion channel. The carriers are now introduced through the external circuit that biases the source and drain electrodes. The carriers are now limited in availability by the dielectric relaxation time constant and any other limitations prescribed by the time-dependent operational characteristics of the device. Our idealized MISFET is, therefore, a MIS capacitor with source and drain ohmic electrodes located on opposite sides of the gate, and doped opposite to the polarity of the substrate in order to have a small barrier for injection of carriers into the inversion channel, as shown schematically in Figure 7.24.

The gate bias allows significant flow of current above the threshold voltage. The surface has an inversion layer and the carrier flow occurs between the source and the drain along the surface channel due to injection and collection of carriers at the source and the drain. As with the MESFET, gradual channel approximation allows us to reduce our analysis to one dimension and derive a useful model. The arguments and the conditions for applicability of this approximation are similar to those for the MESFET.

A major difference between the ideal MIS structure and this structure is that the structure is no longer always in thermal equilibrium. In the presence of current flow, the electrons and holes readjust to accommodate

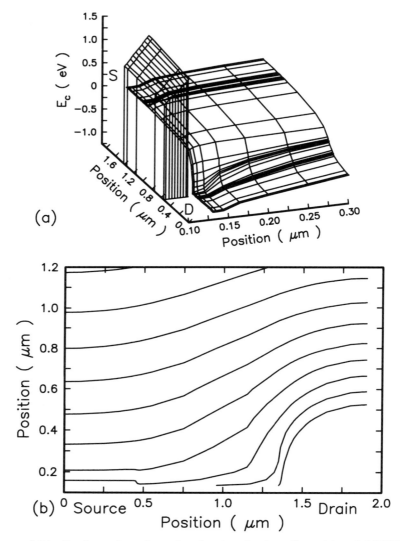

Figure 6.24: Surface plot of conduction band edges for a biased MISFET structure is shown in (a). (b) shows contours of constant potential. The source and the drain electrodes are identified in the figure. Band edge, with a constant slope, is shown for the idealized insulator.

the current flow and the bias, or equivalently in the Boltzmann approximation, the Fermi level splits into non-identical quasi-Fermi levels. Current
flow occurs partly by drift, i.e., it is driven by an electric field, and partly
by diffusion, i.e., due to carrier density gradient. We may write

$$p = n_i \exp\left(q\frac{-\psi + \phi_p}{kT}\right),$$

$$n = n_i \exp\left(q\frac{\psi - \phi_n}{kT}\right), \tag{6.65}$$

and the current density of electrons, for the n-channel device, is

$$J = -q\mu_n n\frac{d\psi}{dz} + qD_n\frac{dn}{dz}, \tag{6.66}$$

where $D_n = kT\mu_n/q$. Here the first term is the drift component of the
current and the second is the diffusion component of the current. We may
rewrite the current expression in terms of the quasi-Fermi levels as

$$J = -q\mu_n n\frac{d\phi_n}{dz} \tag{6.67}$$

for current carried by electrons in an n-channel device. Since the source
and drain electrodes supply only electrons, the holes remain at equilibrium
as before in the MIS capacitor, i.e.,

$$\phi_p = \phi_F = \psi_B = \frac{kT}{q}\ln\left(\frac{N_A}{n_i}\right). \tag{6.68}$$

Figure 7.25 shows the Fermi level in a MIS capacitor and the splitting
to quasi-Fermi levels in a MISFET due to carrier flow through source and
drain electrodes. To calculate the current by the inversion layer, we must
integrate the current continuity equation (Equation 7.67) across the depth
(the y direction) and then multiply by the channel width W, since there
exists a uniform current density along the width of the gate. The electron quasi-Fermi level, perpendicular to the surface, remains relatively flat
because there is little if any current flow between the source and drain electrodes and the substrate electrode. Recall that, even in a forward biased
p–n junction under low-level injection conditions, the quasi-Fermi levels remain flat. The electron quasi-Fermi level approaches the hole quasi-Fermi
level deep into the substrate on a length scale of the order of the diffusion length similar to that in a wide-base p–n junction. We may treat
the electron quasi-Fermi level as independent of y, which is the coordinate
perpendicular to the surface in the region of interest.

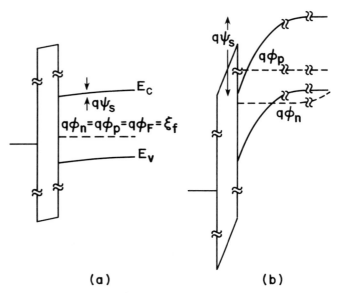

(a)   **(b)**

Figure 6.25: The Fermi level in a MIS capacitor (a) and the splitting of the quasi-Fermi level in a MISFET (b) due to carrier flow through ohmic source and drain electrodes.

The current I at any position z of the channel is given by

$$I = -W q \mu_n N_I \frac{d\phi_n}{dz}, \tag{6.69}$$

which relates the electron current in terms of the sheet carrier density in the two-dimensional carrier gas and the quasi-Fermi level. We have determined this inversion charge density for the MIS capacitor before in terms of the unsplit Fermi level. Including the splitting of the quasi-Fermi level, the sheet charge equation for the MISFET structure is

$$
\begin{aligned}
N_I = \; & N_A \sqrt{2} \lambda_D \left\{ \left[\frac{q\psi_S}{kT} + \right. \right. \\
& \left. \left(\frac{n_i}{N_A} \right)^2 \exp\left(\frac{q\psi_S}{kT} \right) \exp\left(q\frac{-\phi_n + \phi_F}{kT} \right) \right]^{1/2} - \\
& \left. \left(\frac{q\psi_S}{kT} \right)^{1/2} \right\}.
\end{aligned}
$$

$$\tag{6.70}$$

While this follows directly from the earlier MIS-related equation, the relationship is not directly obvious. Therefore, consider its derivation directly as a solution from Poisson's equation. Ignoring the hole contribution,

$$\frac{d^2\psi}{dy^2} = \frac{qN_A}{\epsilon_s} - \frac{qn_i}{\epsilon_s} \exp\left(q\frac{\psi - \phi_n}{kT}\right) \qquad \text{for} \qquad 0 < y < w. \qquad (6.71)$$

The solution follows using a similar technique of multiplying by $d\psi/dy$. The equation for band bending in terms of the quasi-Fermi level is

$$\frac{1}{2}\frac{d}{dy}\left(\frac{d\psi}{dy}\right)^2 = \left(\frac{kT}{q\lambda_D}\right)^2 \times$$
$$\frac{d}{dy}\left[\frac{q\psi}{kT} + \left(\frac{n_i}{N_A}\right)^2 \exp\left(\frac{q\psi}{kT}\right) \exp\left(q\frac{\phi_F - \phi_n}{kT}\right)\right]. \qquad (6.72)$$

The boundary condition at the edge of the depletion region ($y = w$) is given by

$$\psi = 0$$
$$\text{and} \quad \frac{d\psi}{dy} = 0. \qquad (6.73)$$

Integrating, we obtain

$$-\left|\frac{d\psi}{dy}\right|_{y=0} = \frac{\sqrt{2}kT}{q\lambda_D}\left\{\frac{q\psi_s}{kT} + \left(\frac{n_i}{N_A}\right)^2 \exp\left(q\frac{-\phi_n + \phi_F}{kT}\right) \times \right.$$
$$\left.\left[\exp\left(\frac{q\psi_s}{kT}\right) - 1\right]\right\}^{1/2}. \qquad (6.74)$$

Introducing the sheet charge approximation for identical reasons,

$$-qN_I = \mathcal{Q}_{sem} - (-qN_Aw) = \epsilon_s\left.\frac{d\psi}{dy}\right|_{y=0} + qN_A\sqrt{2}\lambda_D\left(\frac{q\psi_s}{kT}\right)^{1/2}, \qquad (6.75)$$

gives the carrier density in the inversion layer as

$$N_I = N_A\sqrt{2}\lambda_D\left\{\left[\frac{q\psi_s}{kT} + \left(\frac{n_i}{N_A}\right)^2 \exp\left(\frac{q\psi_s}{kT}\right) \exp\left(q\frac{-\phi_n + \phi_F}{kT}\right)\right]^{1/2}\right.$$
$$\left. -\left(\frac{q\psi_s}{kT}\right)^{1/2}\right\}. \qquad (6.76)$$

We wish to actually go a step further. In addition to the source and drain contacts, the substrate contact can also be used to control the channel by an externally applied stimulus. The substrate bias (V_{BS} for the applied bias between the source and the body) leads to additional changes in the splitting of the quasi-Fermi levels. For any applied bias condition, if a larger reverse bias is applied at the substrate, then a larger voltage drops across the depletion region, the depletion region extends deeper into the substrate, and the inversion and depletion region charges reconfigure. Consider the source end and the drain end of the channel at the same bias so that no current flows between them. They also have the same reverse bias with respect to the substrate. The quasi-Fermi level is the same along the channel from source to drain. Due to the application of the substrate bias, the electron quasi-Fermi level ϕ_n and the bulk Fermi level ϕ_F differ by the applied substrate bias V_{BS}. Substrate biases are usually negative ($-V_{BS}$ with $V_{BS} > 0$ for the n-channel device utilizing a p-type substrate), and are used to control substrate effects. In the presence of this reverse bias, the quasi-Fermi level splitting is established by this reverse bias, and the inversion layer charge density is given by

$$
N_I = N_A \sqrt{2} \lambda_D \left\{ \left[\frac{q\psi_S}{kT} + \left(\frac{n_i}{N_A} \right)^2 \exp\left(\frac{q\psi_S}{kT} \right) \exp\left(-\frac{qV_{BS}}{kT} \right) \right]^{1/2} - \left(\frac{q\psi_S}{kT} \right)^{1/2} \right\}.
$$

$$(6.77)$$

With the application of substrate bias, the total band bending ψ_S has also changed. Since this substrate bias V_{BS} drops across the substrate region, in strong inversion, one would expect the band bending to be approximately equal to the substrate bias V_{BS}. It deviates in part due to the reconfiguration of the charge. An increased integrated charge in the depletion region causes some changes in the inversion layer charge for identical gate-to-source bias as a consequence of the continuity of the displacement vector. In strong inversion, these are small because the sheet carrier charge is the largest component of the semiconductor charge. In this limit, from the equation of inversion charge in the presence of substrate bias (Equation 6.77),

$$
N_I \approx \sqrt{2} \lambda_D n_i \exp\left(q \frac{\psi_S - V_{BS}}{2kT} \right).
$$

$$(6.78)$$

Thus, to obtain the same sheet charge N_I with the source-to-body bias V_{BS}, the new band bending increases by V_{BS}, i.e., the depletion region charge increases to accommodate this increase in band bending from

$$Q_{dep} = qN_A\sqrt{2}\lambda_D\left(\frac{q\psi_S}{kT}\right)^{1/2} \tag{6.79}$$

to

$$Q_{dep} = qN_A\sqrt{2}\lambda_D\left(q\frac{\psi_S + V_{BS}}{kT}\right)^{1/2}. \tag{6.80}$$

As an elaboration of this argument, consider the situation where the gate-to-source bias is kept constant and the bias of source and drain, which are being maintained the same are increased with respect to the substrate by V_{BS}. Since the gate charge remains the same, the semiconductor charge remains the same and equal. Since the depletion region has widened and the charge in it increased, the inversion layer charge must decrease to maintain the same semiconductor charge density.

In the presence of the substrate bias, we may now rewrite the threshold voltage expression, Equation 7.62, with respect to the substrate as

$$V_T = V_{FB} + 2\psi_B + V_{BS} + \sqrt{2}\frac{C_{FB}}{C_{ins}}\frac{kT}{q}\left(2\frac{q\psi_B}{kT} + \frac{qV_{BS}}{kT}\right)^{1/2}. \tag{6.81}$$

The threshold voltage referenced to the source V_{TS} can be written as

$$V_{TS} = V_{FB} + 2\psi_B + \sqrt{2}\frac{C_{FB}}{C_{ins}}\frac{kT}{q}\left(2\frac{q\psi_B}{kT} + \frac{qV_{BS}}{kT}\right)^{1/2}, \tag{6.82}$$

where the substrate bias occurs as a perturbative effect through the depletion region charge.

In these equations, the effect of the substrate and the substrate bias occurs through the terms involving C_{FB}. The relative intensity of the effect of the substrate bias compared to that of the gate bias is described by the body parameter a, defined as

$$a = \sqrt{2}\frac{C_{FB}}{C_{ins}} = \sqrt{2}\frac{\epsilon_s}{\lambda_D}\frac{t_{ins}}{\epsilon_{ins}}. \tag{6.83}$$

The larger the flat-band capacitance, or the smaller the insulator capacitance, the stronger is the control of the substrate bias vis-a-vis the gate bias. Quite often, a term γ, called the body coefficient, is also employed. This is given by

$$\gamma = \frac{(2q\epsilon_s N_A)^{1/2}}{C_{ins}} = \sqrt{2}\frac{C_{FB}}{C_{ins}}\frac{kT}{q} = a\frac{kT}{q}. \tag{6.84}$$

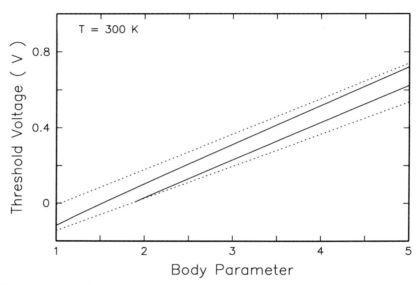

Figure 6.26: Threshold voltage with respect to the source as a function of the substrate-insulator parameter for various insulator thicknesses and acceptor doping at 300 K for GaAs/Ga$_{1-x}$Al$_x$As/GaAs. The lower and upper solid lines are for constant insulator thicknesses of 750 Å and 250 Å respectively, with acceptor doping allowed to vary. The lower and upper dashed lines are for constant acceptor doping of 5×10^{15} cm^{-3} and 5×10^{17} cm^{-3} respectively, with insulator thickness allowed to vary.

The relative insensitivity of the threshold voltage to substrate bias, for typical parameters of a GaAs/Ga$_{1-x}$Al$_x$As/GaAs is shown in Figure 7.26 for various acceptor dopings and insulator thicknesses. This figure assumes a 5×10^{18} cm^{-3} doped gate semiconductor. The substrate voltage is used to obtain a stronger control on back side injection effects; the design of device structures can be made convenient and easy only by making the body coefficient and body parameter small. Note that in the area between the inside lines, the insulator thickness may vary between 250 Å and 750 Å, and the acceptor doping may also be allowed to vary substantially without causing large changes in the threshold voltage. These calculations assume no charge in the insulator and no fixed charge at the interface.

Current–Voltage Characteristics

We have now determined the expression for the inversion layer sheet charge density as a function of band bending and substrate bias. We want to determine the current of the device, given V_{GS}, the gate bias relative to the source; V_{DS}, the drain bias relative to the source; and $-V_{BS}$, where $V_{BS} >, 0$ the body-to-source bias. We shall place several restrictions on our discussion in order to derive an analytic theory which has most of the features of the real device and a realistic physical representation. To allow the use of gradual channel approximations, we consider the geometrical condition of $L \gg t_{ins}$, so that the analysis can be separated between the transverse direction for control of the charge and the longitudinal direction for drift-diffusion transport effects related to the current. The variation in electric field along the channel, i.e., the dependence on z is very weak compared to the variation with depth, i.e., the dependence on y. We will consider length scales to be such that the depletion width in the substrate and the junction depths at the ohmic contact do not influence the device operation. Two-dimensional current flow effects, due to injection into the substrate, etc., are therefore considered to have been minimized in these structures by use of small junction depths. We will keep, for convenience purposes only, the flat-band voltage as zero. To include it, one can replace V_G by $V_G - V_{FB}$ and obtain the general relationship.

The current is

$$I = -W q \mu N_I \frac{d\phi_n}{dz}. \tag{6.85}$$

Our charge expressions have been determined in terms of the surface band bending ψ_S. We will maintain the use of this parameter because it occurs naturally in the exponentials relating charge with energies, and linearly in other energy terms, and is quite basic. Since we know the sheet carrier density, we need to find the quasi-Fermi level as a function of band bending. We know

$$
\begin{aligned}
C_{ins} \left(V_G - \psi_S \right) &= -\mathcal{Q}_{sem} \\
&= \sqrt{2} C_{FB} \frac{kT}{q} \left\{ \left[\frac{q\psi_S}{kT} + \left(\frac{n_i}{N_A} \right)^2 \exp \left(q \frac{-\phi_n + \phi_F}{kT} \right) \times \right. \right. \\
&\qquad \left. \left. \left[\exp \left(\frac{q\psi_S}{kT} \right) - 1 \right] \right\}^{1/2},
\end{aligned}
\tag{6.86}
$$

i.e.,

$$\left\{ \frac{C_{ins} \left(V_G - \psi_S \right)}{\sqrt{2} C_{FB}} \frac{q}{kT} \right\}^2 - \frac{q\psi_S}{kT} = \left(\frac{n_i}{N_A} \right)^2 \left[\exp \left(\frac{q\psi_S}{kT} \right) - 1 \right] \times$$

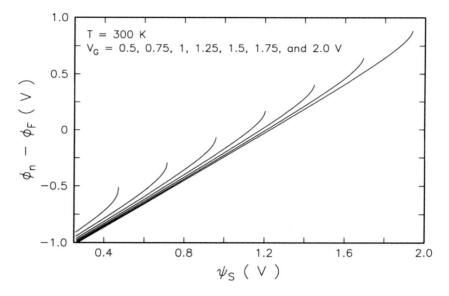

Figure 6.27: Variation of the quasi-Fermi level with respect to the bulk Fermi level ($\phi_n - \psi_S$) with the surface band bending (ψ_S) for the GaAs MIS structure being considered in the calculations (insulator thickness is 400 Å). Gate voltage V_G is the parameter for the different curves.

$$\exp\left(q\frac{-\phi_n + \phi_F}{kT}\right), \tag{6.87}$$

and hence

$$\frac{q\phi_n}{kT} = \frac{q\phi_F}{kT} - \ln\left\{\frac{[(qV_G - q\psi_S)^2/(akT)^2] - (q\psi_S/kT)}{(n_i/N_A)^2 [\exp(q\psi_S/kT) - 1]}\right\}. \tag{6.88}$$

This allows us to obtain the normalized quasi-Fermi level position, $q\phi_n/kT$, as a function of the normalized band bending $q\psi_S/kT$ at any point along the channel. The variation of the quasi-Fermi level with respect to the bulk Fermi level ($\phi_n - \phi_F$) with the band bending (ψ_S) is shown in Figure 6.27 for our GaAs/Ga$_{1-x}$Al$_x$As/GaAs example.

The electron quasi-Fermi level moves rapidly when the band bending approaches specific value for various gate biases. This is the condition delineating the tracking of the quasi-Fermi level and the band bending. The tracking entails a nearly constant separation between the band edge and the quasi-Fermi level, the condition of large charge in the channel, the inversion condition. The rapid change of the separation describes the condition where band bending in the substrate and the movement of the quasi-Fermi level are disjointed because the integrated substrate immobile charge is a significant fraction of the sheet charge density in the semiconductor. This behavior corresponds to the depletion part of the characteristics considered earlier. The breakpoint in this band bending will be called the saturation band bending ψ_{sat}. From the relationship of the Fermi level position already derived in Equation 7.88, this rapid change occurs when the numerator of the logarithmic term goes to zero, i.e., when

$$\frac{1}{a^2}\left(\frac{qV_G}{kT} - \frac{q\psi_{sat}}{kT}\right)^2 = \frac{q\psi_{sat}}{kT},$$

$$\text{or} \quad \left(\frac{qV_G}{kT} - \frac{q\psi_{sat}}{kT}\right)^2 = a^2 \frac{q\psi_{sat}}{kT}. \tag{6.89}$$

Following our discussion, this condition of saturation in band bending describes the condition when the channel has pinched off and the inversion layer has very few carriers.

We can show that the inversion layer carrier density is negligible by noting that the total sheet charge density in the semiconductor

$$-\mathcal{Q}_{sem} = C_{ins}\left(V_G - \psi_S\right), \tag{6.90}$$

at this condition of $\psi_S = \psi_{sat}$, becomes

$$
\begin{aligned}
-\mathcal{Q}_{sem} &= C_{ins} a \frac{kT}{q}\left(\frac{q\psi_S}{kT}\right)^{1/2} \\
&= C_{FB}\sqrt{2}\frac{kT}{q}\left(\frac{q\psi_S}{kT}\right)^{1/2} \\
&= \frac{\epsilon_s}{\lambda_D}\sqrt{2}\frac{kT}{q}\frac{w}{\sqrt{2}\lambda_D} \\
&= qN_Aw.
\end{aligned} \tag{6.91}
$$

Since the total semiconductor charge is close to the charge in the depletion region, the charge in the inversion layer is negligible. The proportional change in the quasi-Fermi level for electrons with surface band bending follows intuitively, since a negligible inversion charge implies that any band bending occurs due to the acceptor charge and leads directly to a movement of the quasi-Fermi level for electrons. The quasi-Fermi level is no longer near the conduction band edge. It moves rapidly further and further away with band bending.

The magnitude of this saturation voltage can also be derived from the full charge control derived earlier (see Problem 11). Explicitly, this saturation band bending is given by

$$\frac{q\psi_{sat}}{kT} = \frac{qV_G}{kT} + \frac{a^2}{2} - a\left(\frac{qV_G}{kT} + \frac{a^2}{4}\right)^{1/2}, \tag{6.92}$$

as a function of the gate voltage and the body parameter. The form of this equation in terms of the body parameter explicitly justifies the use of the body parameter, since it directly scales the applied gate bias with its substrate linkage at the condition of the disappearance of inversion.

We have now established the direct relationship between the quasi-Fermi level, the inversion charge and the band bending. This can be utilized in our current equation in order to show the current at any cross-section as a function of the band bending at the surface at that cross-section. To complete the evaluation of the current, we should consider the boundary conditions on this band bending at the source edge and the drain edge of the channel. We can determine these, since we know the quasi-Fermi levels at the source end and the drain end of the channel since they are directly influenced by the applied biases. The position of the quasi-Fermi level ϕ_n at the source end of the channel is

$$\phi_n|_{source} = \phi_F + V_{BS}. \tag{6.93}$$

No particle current flows in the circuit involving the gate electrode, and the splitting at the source end is directly related to the effect of the bias between the source and the bulk substrate. Here, we do not use the Fermi level position of the source ohmic contact region, which is in the extrinsic part of the device and enters only indirectly in the current calculation. Instead, we employ the quasi-Fermi level position at the end of the channel at the source end, which may differ from the quasi-Fermi level at the source end. For clarifying this, see Figure 7.28. This is so because the behavior of the device is assumed to be determined by the behavior in the channel and not of the injecting regions, i.e., the contact regions are assumed to be capable of supplying as many carriers as determined by the channel behavior and any changes in the ohmicity of these regions should not influence the operation of the device (the current it carries, etc.). This is quite similar to the low-level injection behavior of p–n junctions. The current under low-level injection is determined by the rate at which carriers transport in the quasi-neutral regions and not in the junction regions. The drift and diffusion currents, delicately in balance in the junction region, are capable of supplying much larger current than is extracted in the low-level injection condition. Similarly, in this problem, the ohmic contact and the

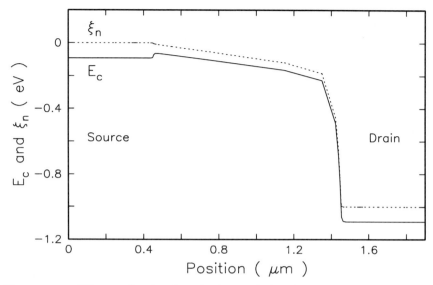

Figure 6.28: The conduction band edge and the electron quasi-Fermi level at the surface and along the channel for a GaAs MISFET. Note the nearly flat quasi-Fermi level at the source edge.

source region is assumed to be capable of supplying much larger current than is the actual magnitude limited by the channel transport. The carrier densities are very large in the doped region and hence capable of supporting large current densities—the product of the electron charge, the carrier density, and the thermal velocity ($\approx 10^7$ cm.s^{-1}) through it. Locally, at the interface of the contact and the channel region, the quasi-Fermi level is flat, and the band bending at the channel end and the quasi-Fermi levels are determined by the charge control relationship of the channel and the bias constraints that determine the quasi-Fermi level.

As the source-to-bulk bias $-V_{BS}$ increases, the band bending ψ_S increases until the pinch-off occurs, i.e., inversion ends. Then, the band bending saturates to ψ_{sat} as in Figure 7.27. For a bulk-to-source bias V_{BS} of zero, the quasi-Fermi level at the source end of the channel is the same as the bulk Fermi level ($\phi_n = \phi_F$), and Equation 7.88 gives

$$\frac{qV_G}{kT} - \frac{q\psi_S}{kT} = a\left\{\frac{q\psi_S}{kT} + \left(\frac{n_i}{N_A}\right)^2\left[\exp\left(\frac{q\psi_S}{kT}\right) - 1\right]\right\}^{1/2}, \qquad (6.94)$$

which also follows from Gauss' Law. If the source-to-bulk bias V_{BS} is zero, then the band bending at the source end of the channel is the same as the band bending that we derived for the MIS capacitor for a bias at the gate of V_G, since no current flow occurs between the source and the substrate.

Let us reemphasize that in this analysis we are assuming that the current transport is affected by what happens in the channel, not by the transition region between the source contact region and the source end of the channel. That is why we use the band bending ψ_S following Equation 7.93 and not that determined from the built-in potential of the associated contact–channel p–n junction. Since the source is assumed to not limit the supply of needed carriers, and the ohmic drops are assumed to be small, the quasi-Fermi level is constant in this transition region.

Similarly, the band bending at the drain end of the channel is determined from

$$\phi_n|_{drain} = \phi_F + V_{BS} + V_{DS}, \tag{6.95}$$

which is the explicit effect on the quasi-Fermi level as a result of the application of the bias at the drain and the substrate, and Equation 7.88, which relates the position of the quasi-Fermi level with the band bending. In the band bending at the drain, $\psi_S \rightarrow \psi_{sat}$ at large enough drain-to-source bias V_{DS} just as the behavior of the source end does as a function of the bulk-to-source bias V_{BS}. A large enough drain-to-source bias increases the depletion region in the substrate at the drain end, resulting in a decrease of the inversion charge, all other bias constraints remaining constant. The occurrence of a band bending of ψ_{sat} reflects removal of inversion conditions at the drain end, i.e., the occurrence of pinch-off of the channel at the drain end. This behavior is similar to that of the long channel constant mobility model of the MESFET, where pinch-off also represented disappearance of mobile charge. Like the MESFET, under this condition, the current flow through the channel saturates. This drain-to-source voltage, corresponding to pinch-off of the channel, will be referred to as the drain saturation voltage V_{Dsat}. To estimate it, we approximate the asymptotic and the linear increase portion with intersecting straight lines. The asymptotic line is given by

$$\psi_S = \psi_{sat}, \tag{6.96}$$

and the linear line by

$$\phi_n - \phi_F = \psi_S - \psi_{s0} \tag{6.97}$$

So the drain saturation voltage follows, by evaluation at the intersection of linear and asymptotic lines, as

$$V_{Dsat} = \phi_n - \phi_F|_{intersection} = \psi_{sat} - \psi_{s0}. \tag{6.98}$$

For drain voltages higher than the drain saturation voltage, i.e., $V_{DS} > V_{Dsat}$, even though the band bending ψ_S does not increase, a large increase in the lateral field in the drain end channel transition region can occur. The size of this transition region increases with drain bias, causing a reduction in the electrical channel length L. In small devices, this could be appreciably different from the source-to-drain separation, which is usually described as the metallurgical channel length L_m. We will discuss the consequences of this decrease in the effective electrical channel length in a discussion of the extension of this model to short channel conditions.

Having determined these boundary conditions, we can now determine the current, following similar procedures as in our analysis of MESFETs. From the equation for quasi-Fermi level (Equation 6.88), we have

$$\frac{d}{dz}\left(\frac{q\phi_n}{kT}\right) = \left[1 + \frac{2\left(qV_G - q\psi_S\right)/a^2 kT + 1}{\left(qV_G - q\psi_S\right)^2/(akT)^2 - q\psi_S/kT}\right] \times$$
$$\frac{d}{dz}\left(\frac{q\psi_S}{kT}\right), \qquad (6.99)$$

and

$$qN_I = C_{ins}\left(V_G - \psi_S\right) - qN_A\lambda_D\sqrt{2}\left(\frac{q\psi_S}{kT}\right)^{1/2}, \qquad (6.100)$$

all in terms of band bending and gate bias.

Substituting, at any vertical cross-section along the channel, the current is given by

$$I = -Wq\mu N_I \frac{d\phi_n}{dz}$$
$$= -W\mu C_{ins}\left(\frac{kT}{q}\right)^2 \left[\frac{qV_G}{kT} - \frac{q\psi_S}{kT} - a\left(\frac{q\psi_S}{kT}\right)^{1/2}\right]\frac{d}{dz}\left(\frac{q\phi_n}{kT}\right)$$
$$= -W\mu C_{ins}\left(\frac{kT}{q}\right)^2 \left[\frac{qV_G}{kT} - \frac{q\psi_S}{kT} - a\left(\frac{q\psi_S}{kT}\right)^{1/2} + \right.$$
$$\left. \frac{2\left(qV_G - q\psi_S\right)/kT + a^2}{\left(qV_G - q\psi_S\right)/kT + a(q\psi_S/kT)^{1/2}}\right]\frac{d}{dz}\left(\frac{q\psi_S}{kT}\right) \qquad (6.101)$$

Under quasi-static conditions, this current is constant throughout the channel due to current continuity. Like the MESFET analysis, we can integrate with respect to z,

$$\int I\, dz = IL$$

$$
= -W\mu C_{ins}\left(\frac{kT}{q}\right)^2\int_{source}^{drain}\left[\frac{qV_G}{kT} - \frac{q\psi_S}{kT} - a\left(\frac{q\psi_S}{kT}\right)^{1/2} + \frac{2(qV_G - q\psi_S)/kT + a^2}{(qV_G - q\psi_S)/kT + a(q\psi_S/kT)^{1/2}}\right]\frac{d}{dz}\left(\frac{q\psi_S}{kT}\right).
$$

$$
(6.102)
$$

A complete result of this can be obtained by a lengthy integration (see Problem 12) as

$$
I = \frac{W\mu C_{ins}}{L}\left(\frac{kT}{q}\right)^2\left\{\left(\frac{qV_G}{kT} + 2\right)\frac{q\psi_S}{kT} - \frac{1}{2}\left(\frac{q\psi_S}{kT}\right)^{1/2} - \right.
$$

$$
\frac{2\sqrt{2}}{3}\frac{(q\psi_S/kT - 1)^{3/2}}{C_{ins}}\exp\left(\frac{q\phi_F}{2kT}\right) +
$$

$$
4\sqrt{2}\frac{(q\psi_S/kT - 1)^{1/2}}{C_{ins}}\exp\left(\frac{q\phi_F}{2kT}\right) -
$$

$$
2\left(\frac{qV_G}{kT} - \frac{q\psi_1}{kT}\right)\ln\left[\left(\frac{q\psi_1}{kT} - 1\right)^{1/2} + \left(\frac{q\psi_S}{kT} - 1\right)^{1/2}\right] +
$$

$$
\left.2\left(-\frac{qV_G}{kT} + \frac{q\psi_2}{kT}\right)\ln\left[\left(\frac{q\psi_2}{kT} - 1\right)^{1/2} + \left(\frac{q\psi_S}{kT} - 1\right)^{1/2}\right]\right\}\Bigg|_{\psi_{s0}}^{\psi_{sL}},
$$

$$
(6.103)
$$

where

$$
\frac{q\psi_1}{kT} = \frac{qV_G}{kT} + \frac{1}{C_{ins}^2}\exp\left(\frac{q\phi_F}{kT}\right) - \frac{\sqrt{2}}{C_{ins}}\exp\left(\frac{q\phi_F}{2kT}\right) \times
$$

$$
\left[\frac{qV_G}{kT} - 1 + \frac{1}{2C_{ins}^2}\exp\left(\frac{q\phi_F}{kT}\right)\right]^{1/2}, \qquad (6.104)
$$

$$
\frac{q\psi_2}{kT} = \frac{qV_G}{kT} + \frac{1}{C_{ins}^2}\exp\left(\frac{q\phi_F}{kT}\right) + \frac{\sqrt{2}}{C_{ins}}\exp\left(\frac{q\phi_F}{2kT}\right) \times
$$

$$
\left[\frac{qV_G}{kT} - 1 + \frac{1}{2C_{ins}^2}\exp\left(\frac{q\phi_F}{kT}\right)\right]^{1/2}, \qquad (6.105)
$$

and the boundary conditions on band bending are written as

$$
\psi_S|_{z=0} = \psi_{s0} \qquad (6.106)
$$

at the source end and

$$\psi_S|_{z=L} = \psi_{sL} \tag{6.107}$$

at the drain end.

Fairly accurate approximations can be made to the complicated forms above. Brews has described a subtle approximation that makes the problem quite a bit less cumbersome. The first two terms in the integral above describe the contribution from the strong inversion charge qN_I, i.e., the contribution of $q^2 N_I / kT C_{ins}$. The last term is the perturbation term important in regions of weak inversion. It is important when the strong inversion terms are negligible. This weak inversion condition prevails near pinch-off condition, i.e., in the region of the channel where it is pinched off. It is important at drain biases close to drain saturation voltage, in the drain region of the channel, and at gate biases in the sub-threshold and threshold region, for all parts of the channel. An adequate approximation may be obtained by substituting for this fraction its magnitude at pinch-off, the condition at which it is important and hence must be accurate. This leads to simpler approximation,

$$\frac{2 \left(qV_G - q\psi_S \right) / kT + a^2}{\left(qV_G - q\psi_S \right) / kT + a(q\psi_S/kT)^{1/2}}$$
$$\approx \frac{2a(q\psi_S/kT)^{1/2} + a^2}{2a(q\psi_S/kT)^{1/2}} \approx 1 + \frac{a}{2(q\psi_S/kT)^{1/2}}, \tag{6.108}$$

and hence, Equation 7.102 takes the form

$$IL = -W\mu C_{ins} \left(\frac{kT}{q} \right)^2 \int_0^L \left[\frac{qV_G}{kT} - \frac{q\psi_S}{kT} - \right.$$
$$\left. \frac{2}{3}a \left(\frac{q\psi_S}{kT} \right)^{1/2} + 1 + \frac{a}{2} \left(\frac{q\psi_S}{kT} \right)^{-1/2} \right] \frac{d}{dz} \left(\frac{q\psi_S}{kT} \right). \tag{6.109}$$

The band bending at the source end and the drain end can be derived since we know the quasi-Fermi levels at source and drain from the applied biases, and because we have related the quasi-Fermi level to the band bending through Equation 7.88. Integration of the approximate equation yields

$$I = -\frac{W\mu C_{ins}}{L} \left(\frac{kT}{q} \right)^2 \left\{ \left(1 + \frac{qV_G}{kT} \right) \left(\frac{q\psi_{sL}}{kT} - \frac{q\psi_{s0}}{kT} \right) - \right.$$
$$\left. \frac{1}{2} \left[\left(\frac{q\psi_{sL}}{kT} \right)^2 - \left(\frac{q\psi_{s0}}{kT} \right)^2 \right] - \right.$$

$$\frac{2}{3}a\left[\left(\frac{q\psi_{sL}}{kT}\right)^{3/2} - \left(\frac{q\psi_{s0}}{kT}\right)^{3/2}\right] +$$
$$a\left[\left(\frac{q\psi_{sL}}{kT}\right)^{1/2} - \left(\frac{q\psi_{s0}}{kT}\right)^{1/2}\right]\bigg\}. \tag{6.110}$$

Here,

$$V_G = V_{GS} + V_{BS}, \tag{6.111}$$

which can be translated for a non-zero flat-band voltage using the procedure discussed with MIS capacitors. The source end band bending ψ_{s0} is found from

$$\phi_n = \phi_F + V_{BS}, \tag{6.112}$$

and the quasi-Fermi level position from Equation 6.88; and the drain end band bending ψ_{sL} is found from

$$\phi_n = \phi_F + V_{BS} + V_{DS}, \tag{6.113}$$

and the quasi-Fermi level position from Equation 6.88.

The current–voltage characteristics have now been derived as a function of the band bending parameter. The current–voltage characteristics can be calculated using several different procedures based on the above set of equations. As an example, consider the derivation of the output characteristics for a bulk-to-substrate bias of V_{BS}. If we choose the band bending at the source end ψ_{s0}, then the quasi-Fermi level position at the source end ($z = 0$), following Equation 6.88, is related by

$$\frac{q\phi_n}{kT}\bigg|_{z=0} = q\frac{\phi_F + V_{BS}}{kT} = \frac{q\phi_F}{kT} -$$
$$\ln\left\{\frac{(qV_G - q\psi_S)^2/(akT)^2 - q\psi_S/kT}{(n_i/N_A)^2\,[\exp(q\psi_S/kT) - 1]}\right\}. \tag{6.114}$$

This gives the gate bias as a function of source end band bending ψ_{s0} and bulk-to-substrate bias V_{BS},

$$\frac{qV_G}{kT} = \frac{q\psi_{s0}}{kT} + a\left\{\frac{q\psi_{s0}}{kT} + \left(\frac{n_i}{N_A}\right)^2 \exp\left(\frac{q\psi_{s0}}{kT}\right)\exp\left(-\frac{qV_{BS}}{kT}\right)\right\}^{1/2}$$
$$= \frac{\psi_{s0}}{kT} + a\left[\frac{q\psi_{s0}}{kT} + \exp\left(q\frac{\psi_{s0} - 2\phi_F - V_{BS}}{kT}\right)\right]^{1/2}. \tag{6.115}$$

Now, choosing ψ_{sL} in the range $0 \le \psi_{sL} \le \psi_{sat}$, one can derive the corresponding drain-to-source voltage, since we now know the quasi-Fermi level

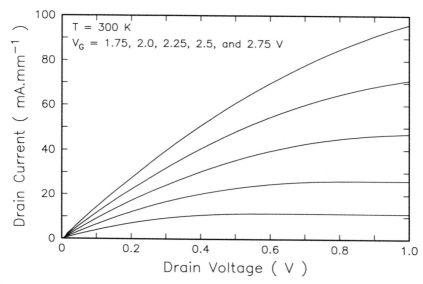

Figure 6.29:   The output characteristics for a MISFET using GaAs/Ga$_{1-x}$Al$_x$As/GaAs and using the procedure based on parameterization of the band bending. The device has a gate length of 10 μm, a mobility of 4000 cm^2.V^{-1}.s^{-1}, an acceptor doping of 3×10^{14} cm^{-3}, and an insulator thickness of 400 Å. This device will also be used as an example in subsequent figures to show the use of numerical procedures.

position at the drain end of the channel. The band bending at the drain end ψ_{sL} is limited between the value of zero, which corresponds to a drain-to-source bias of zero, and ψ_{sat}, which is in the region of channel pinch-off and which corresponds to the drain-to-source bias $V_{DS} \to \infty$. By varying the value of the drain end band bending ψ_{sL}, we generate the current-to-drain-to-source voltage behavior for constant source end band bending, i.e., constant gate voltage. This is one of the characteristics; others for other source end band bending ψ_{s0} result in characteristics for different gate voltages. We use the band bending as a parameter in these calculations because we had parameterized our problem in terms of band bending and had derived the analytic form of the current equation in terms of it. The characteristics of our MISFET example utilizing the GaAs/Ga$_{1-x}$Al$_x$As/GaAs example are shown in Figure 7.29.

Ideally, one likes to know the current for a specified gate and drain bias. This, however, will require an iterative procedure, because our equations are not written explicitly in terms of the voltages in an analytic form. Since the quasi-Fermi level position is known through Equation 7.88 and is related to the applied bias at the contacts through the boundary conditions of Equations 7.93 and 7.95, one obtains an iteration of a new approximation of the band bending in the form

$$
\frac{q\psi_{s0}{}^{i+1}}{kT} = \frac{qV_{BS}}{kT} + 2\frac{q\phi_F}{kT} - 2\ln a + \ln\left[\left(\frac{qV_G}{kT} - \frac{q\psi_{s0}{}^{i}}{kT}\right)^2 - a^2\frac{q\psi_{s0}{}^{i}}{kT}\right].
$$
$$(6.116)$$

The band bending at the drain end can be allowed to vary between zero and the asymptotic value of ψ_{sat}. From Equation 7.110, we can now derive the drain current–voltage characteristics for any gate bias.

Quite often, one wishes to find the magnitude of the saturation current with gate bias, i.e., the drain current at a band bending of $\psi_{sL} = \psi_{sat}$ as a function of the gate voltage V_G. This derivation is similar to before. For any chosen band bending at the source end of ψ_{s0}, we can compute the gate bias V_G from our equation. The saturation current can be obtained since ψ_{sat} is known, and hence the saturation drain current I_{DSS} can be obtained. This saturation drain current versus gate voltage behavior, known as transfer characteristic, is shown in Figure 7.30 for our MISFET example.

One may also be interested in drain current versus gate voltage for a given drain voltage, unlike the previous example in which it was determined for channel pinch-off conditions. Since the drain bias V_{DS} and bulk-to-substrate bias V_{BS} are known, the band bending at the drain end ψ_{sL} can be determined by the quasi-Fermi level equation (Equation 7.88) and Equation 7.95. We also know that

$$
\frac{qV_G}{kT} = \frac{q\psi_{sL}}{kT} + a\left[\frac{q\psi_{sL}}{kT} + \exp\left(q\frac{\psi_{sL} - 2\phi_F - V_{BS} - V_{DS}}{kT}\right)\right]^{1/2}, \quad (6.117)
$$

and hence the gate bias V_G is known. Knowing the gate bias V_G, the source end band bending ψ_{s0} can be determined iteratively as before and hence the current follows. An example of this is shown in Figure 7.31.

The procedure is not as accurate, and needs refining, for calculation in the sub-threshold region, i.e., the region where the entire channel is in weak and moderate inversion. The reason for this may be traced to our approximation of the perturbation term in the quasi-Fermi level expression to its magnitude at ψ_{sat}. When the entire channel region is in weak inversion, moderate inversion, or depletion, the resulting expression can not be expected to be representative. We can, however, derive a procedure, considering the weak inversion component to be the dominant component.

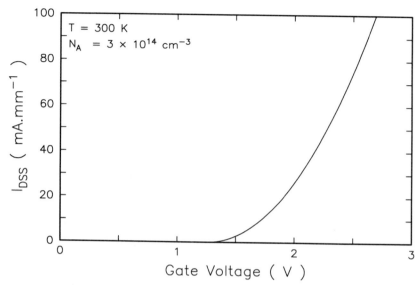

Figure 6.30: The saturation drain current I_{DSS} as a function of gate voltage V_G for the MISFET example employing GaAs/Ga$_{1-x}$Al$_x$As/GaAs.

The band bending at the source end is

$$\psi_{s0} < 2\phi_F = 2\psi_B. \tag{6.118}$$

Clearly, the exponential term in the quasi-Fermi level equation (Equation 7.88) is small, there is very little electron charge of N_I. We eliminate the gate bias V_G by using the identity

$$\left(\frac{qV_G}{kT} - \frac{q\psi_{sat}}{kT} \right)^2 = a^2 \left(\frac{q\psi_{sat}}{kT} \right), \tag{6.119}$$

in an equation similar to Equation 7.117 for the source end, hence,

$$\frac{qV_G}{kT} = \frac{q\psi_{s0}}{kT} + a \left[\frac{q\psi_{s0}}{kT} + \exp\left(q\frac{\psi_{s0} - 2\phi_F - V_{BS}}{kT} \right) \right]^{1/2}. \tag{6.120}$$

We can use this in the iteration

$$\frac{q\psi_{s0}{}^{i+1}}{kT} = \frac{q\psi_{sat}}{kT} \left[1 - \frac{\delta - a}{(q\psi_{sat}/kT)^{1/2} + \left(q\psi_{s0}{}^i/kT \right)^{1/2} + \delta} \right]^2, \tag{6.121}$$

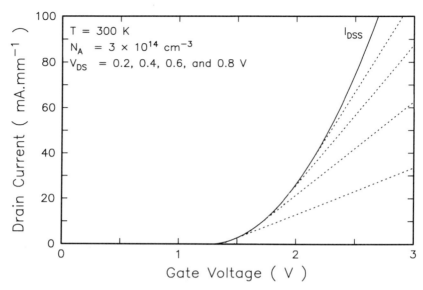

Figure 6.31: The drain current as a function of gate bias for various drain voltages near and above the threshold voltage for the GaAs/Ga$_{1-x}$Al$_x$As/GaAs MISFET example.

where

$$\delta = a \left[1 + \left(\frac{q\psi_{s0}{}^i}{kT} \right)^{-1} \exp \left(q \frac{\psi_{s0}{}^i - 2\phi_F - V_{BS}}{kT} \right) \right]^{1/2} . \qquad (6.122)$$

In the sub-threshold region, there is very little electron charge in the channel, and since ψ_{sat} approaches the value of the source end band bending ψ_{s0}, the drain end band bending ψ_{sL} approaches the value of source end band bending ψ_{s0}, and most of the current flow occurs via diffusion. Expanded characteristics of the drain current I_D with gate voltage V_G are shown in the logarithmic plot of Figure 7.32.

Figure 7.33 shows expanded output characteristics of the MISFET with a schematic of the band bending and carrier distribution along the channel for the different regions of operation. At the large gate biases considered, the surface is strongly inverted. At small drain biases V_{DS}, the drain edge is also strongly inverted. The channel behaves like a resistor with an ohmic drop of V_{DS} across it, and the majority of the current flows as drift current. The quasi-Fermi level change between the source end and the drain end

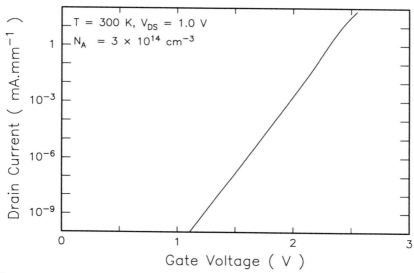

Figure 6.32: Drain current versus gate voltage characteristics for the GaAs/Ga$_{1-x}$Al$_x$As/GaAs MISFET example emphasizing the behavior in the sub-threshold region. The conventional linear region appears at high currents in this figure.

is equal to this drain-to-source voltage V_{DS}. As the drain-to-source bias V_{DS} increases, the ohmic drop increases, lowering the quasi-Fermi level ϕ_n further. As this approaches the saturation voltage, the quasi-Fermi level ϕ_n begins to drop faster than the band bending ψ_S, and diffusion current begins to become important. The carrier density at the drain end begins to reduce and hence the ohmic drop in this region gets larger and larger. More drain bias is needed to increase the current, curves begin to flatten from the linear behavior, and it becomes more parabolic. When the drain-to-source bias V_{DS} exceeds the saturation voltage V_{Dsat}, carrier density at the drain end becomes very low—a condition we described as channel pinch-off. Band bending now becomes independent of carrier density and most of the excess voltage now drops across this region.

The behavior in the sub-threshold region is largely limited by the diffusion current, because there are very few carriers to support any appreciable drift current. In the sub-threshold region, as shown in Figure 7.32, the current is exponential in nature. This is because no inversion channel exists; the carrier transport is by injection across a barrier, giving a characteristic

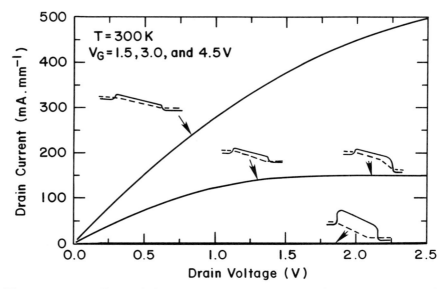

Figure 6.33:   Expanded output characteristics of the example of GaAs/Ga$_{1-x}$Al$_x$As/GaAs MISFET together with a schematic of the band bending along the channel in the different regions of operation.

exponential dependence. This is also sometimes referred to as the barrier modulated region of operation. The diffusive character of this transport should be compared with large drift dominated component in regions where strong inversion exists. Figure 7.34 serves to emphasize the relative importance of drift and diffusive currents by showing the net current resulting in our device from drift current. This figure serves to emphasize the net drift or diffusion effects in the device under various bias conditions.

In the sub-threshold region of operation, i.e., for $V_G < V_T$, or equivalently for $\psi_S < 2\phi_F + V_{BS}$, weak inversion exists. Since we may approximate

$$N_I = N_A \lambda_D \left(\frac{n_i}{N_A} \right)^2 \exp\left(q \frac{-\phi_n + \phi_F}{kT} \right) \left(2 \frac{q\psi_S}{kT} \right)^{-1/2}, \qquad (6.123)$$

and since weak inversion implies that the band bending is nearly the asymptotic magnitude of ψ_{sat} along the channel,

$$N_{I0} - N_{IL} = N_A \lambda_D \left(\frac{n_i}{N_A} \right)^2 \exp\left(\frac{q\psi_{sat}}{kT} \right) \exp\left(-\frac{qV_{BS}}{kT} \right) \times$$

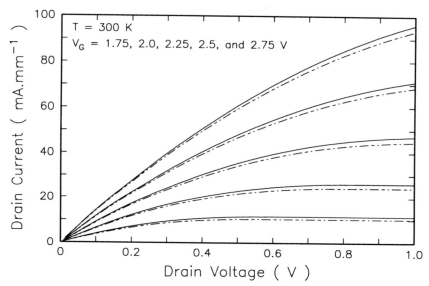

Figure 6.34: The solid lines show characteristics by including both drift and diffusive components of the current in the GaAs/Ga$_{1-x}$Al$_x$As/GaAs MISFET example. The dot-dashed lines show characteristics using only drift current.

$$\left(2\frac{q\psi_{sat}}{kT}\right)^{-1/2}\left[1 - \exp\left(-\frac{qV_{DS}}{kT}\right)\right], \qquad (6.124)$$

where N_{I0} and N_{IL} are the inversion charge density at the source end and the drain end. Since $\psi_S \approx \psi_{sat}$ over the channel, the electric field along the channel

$$\mathcal{E}_z = -\frac{\partial\psi_S}{\partial z} \approx 0, \qquad (6.125)$$

and most of the current is diffusive. We can, therefore, determine it from

$$I = q\mathcal{D}W\frac{dN_I}{dz}. \qquad (6.126)$$

The current being constant in the steady-state condition, the weak inversion charge varies linearly, and hence,

$$\frac{dN_I}{dz} = \frac{N_{IL} - N_{I0}}{L}, \qquad (6.127)$$

and hence,

$$I = -\frac{W\mu q N_A \lambda_D}{L}\frac{kT}{q}\left(\frac{n_i}{N_A}\right)^2 \exp\left(-\frac{qV_{BS}}{kT}\right)\exp\left(\frac{q\psi_{sat}}{kT}\right) \times$$
$$\left[1 - \exp\left(-\frac{qV_{DS}}{kT}\right)\right]\left(2\frac{q\psi_{sat}}{kT}\right)^{-1/2}, \tag{6.128}$$

where from our earlier approximation,

$$\frac{q\psi_{sat}}{kT} = \frac{qV_{GS}}{kT} + \frac{qV_{BS}}{kT} + \frac{a^2}{2} - a\left(\frac{qV_{GS}}{kT} + \frac{qV_{BS}}{kT} + \frac{a^2}{4}\right)^{1/2}. \tag{6.129}$$

The important attributes of this behavior are that the current is nearly exponentially dependent on the gate bias, that for drain-to-source bias a few times larger than the thermal voltage there exists little or no drain bias dependence, and that there is a rapid reduction of current with gate bias when a reverse bias is applied between the body and the source.

As voltage levels of operation decrease, the gate bias swing and the corresponding current in its conducting and non-conducting states become increasingly important. The rapidity with which the drain current is reduced by gate bias in the sub-threshold region is usually measured by the sub-threshold gate swing S given by

$$S = \frac{\ln(10)}{d\ln(I)/dV_G}. \tag{6.130}$$

We may derive the magnitude of this in terms of the parameters of our device. We have

$$S = \frac{kT}{q}\ln(10)\left[\frac{d\ln(I)}{d(q\psi_{sat}/kT)}\right]^{-1}\left[\frac{d(q\psi_{sat}/kT)}{d(qV_G/kT)}\right]^{-1} \tag{6.131}$$

and

$$\frac{d(qV_G/kT)}{d(q\psi_{sat}/kT)} = 1 + \frac{a}{2(q\psi_{sat}/kT)^{1/2}}. \tag{6.132}$$

In sub-threshold, the change in current with the saturation band bending is related by

$$\frac{d\ln(I)}{d(q\psi_{sat}/kT)} = 1 - \frac{1}{2(q\psi_{sat}/kT)}. \tag{6.133}$$

The relationship between the change in gate bias and the saturation band bending may be related by the insulator and body effects represented in the corresponding capacitances. The depletion capacitance, for band bending of ψ_{sat}, is given by

$$C_{depsat} = C_{dep}|_{\psi=\psi_{sat}} = C_{FB} \left(2 \frac{q\psi_{sat}}{kT} \right)^{-1/2}, \qquad (6.134)$$

because there is very little inversion charge. Therefore,

$$\frac{\partial (qV_G/kT)}{\partial (q\psi_{sat}/kT)} = 1 + \frac{C_{depsat}}{C_{ins}}, \qquad (6.135)$$

and

$$\frac{\partial \ln(I)}{\partial (q\psi_{sat}/kT)} = 1 - \frac{2}{a^2} \left(\frac{C_{depsat}}{C_{ins}} \right)^2. \qquad (6.136)$$

The sub-threshold gate swing S can now be explicitly written as

$$S = \frac{kT}{q} \frac{1 + C_{depsat}/C_{ins}}{1 - (2/a^2)} \left(\frac{C_{depsat}}{C_{ins}} \right)^2 \ln(10). \qquad (6.137)$$

The larger the insulator capacitance, i.e., the larger the control by the gate through the insulator, the more nearly ideal is the sub-threshold swing, the gate being still efficient in controlling the channel. The smaller the body parameter (the weaker the body effect and hence two-dimensional and punch-through effects due to substrate parameters), the more ideal the sub-threshold behavior is.

Since the asymptotic band bending affects the sub-threshold gate swing S, device comparisons for logic at small voltages must be made at similar current densities in order for them to be equally effective in the off state and the on state. A device for large integration requires careful control of the sub-threshold behavior in order to establish clear off voltages and in order to decrease power dissipations.

We have described the modelling of the MISFET in quite rigorous detail. Usually, such preciseness can be sacrificed for the purposes of simplistic understanding. Simplified treatments also lend themselves for some applications where their inaccuracy is not a major hindrance; most often such applications involve computer aided design of circuits where the device needs to be treated as functional block, reproducible enough and whose characteristics can be described in the least numerically time consuming and most robust manner to allow for circuit design. Such simplifications can be extracted quite easily, based on our derivation. Consider the triode region of operation, i.e., $V_G > V_T$ and $V_{DS} < V_{Dsat}$. Now the band bending ψ_S and the quasi-Fermi level position ϕ_n change by the same amount along the channel. This is the unity slope region of the plot showing their dependence on each other. Given that

$$\phi_n|_{z=L} - \phi_n|_{z=0} = V_{DS}, \tag{6.138}$$

it follows that

$$\psi_{sL} = \psi_{s0} + V_{DS}, \tag{6.139}$$

and our current equation (Equation 7.110) reduces to

$$I = -\frac{W\mu C_{ins}}{L} \left\{ \left(V_G - \psi_{s0} - \frac{1}{2}V_{DS} \right) V_{DS} - \frac{2}{3}a \left(\frac{kT}{q} \right)^{1/2} \left[(\psi_{s0} + V_{DS})^{3/2} - (\psi_{s0})^{3/2} \right] \right\}, \tag{6.140}$$

where the square root term has been ignored. We now take advantage, in the triode region, of $\psi_{s0} > V_{DS}$. The first two terms of the Taylor series expansion of the last term in the above yield

$$(\psi_{s0} + V_{DS})^{3/2} - (\psi_{s0})^{3/2} = \frac{3}{2}(\psi_{s0})^{1/2}V_{DS} + \frac{3}{2}\frac{1}{2}(\psi_{s0})^{-1/2}\frac{1}{2}V_{DS}^2. \tag{6.141}$$

This gives the current in the devices as

$$I = -\frac{W\mu C_{ins}}{L} \left\{ V_G - \psi_{s0} - a\frac{kT}{q} \left(\frac{q\psi_{s0}}{kT} \right)^{1/2} - \frac{1}{2}\left[1 + \frac{a}{2}\left(\frac{q\psi_{s0}}{kT} \right)^{-1/2} \right] V_{DS} \right\} V_{DS}. \tag{6.142}$$

Note that if we ignore the negative half power term of ψ_{s0}, it being smaller than unity, we obtain our familiar simplistic expression of the current–voltage characteristics of FETs,

$$I = -\frac{W\mu C_{ins}}{L} \left[(V_{GS} - V_T)V_{DS} - \frac{1}{2}V_{DS}^2 \right], \tag{6.143}$$

where we have introduced the threshold voltage as V_T.

Quasi-Static Equivalent Circuit Elements

Having derived the current–voltage characteristics, the equivalent circuit elements, based on perturbation analysis, follow in a straightforward manner.

The drain conductance g_d, e.g., following definition, is given as

$$
\begin{aligned}
g_d &= \left. \frac{\partial I}{\partial V_D} \right|_{V_G} \\
&= \mu q \frac{W}{L} N_{IL} \\
&= \mu q \frac{W}{L} N_{I0} \frac{(qV_G - q\psi_{sL})/kT - a(q\psi_{sL}/kT)^{1/2}}{(qV_G - q\psi_{s0})\,kT - a(q\psi_{s0}/kT)^{1/2}} .
\end{aligned}
\tag{6.144}
$$

In weak inversion, this reduces to

$$
g_d = \mu q \frac{W}{L} N_{I0} \exp\left(-\frac{qV_{DS}}{kT}\right) .
\tag{6.145}
$$

Similarly, the transconductance g_m is by definition

$$
\begin{aligned}
g_m &= \left. \frac{\partial I}{\partial V_G} \right|_{V_{DS}} \\
&= \frac{W \mu C_{ins}}{L} (\psi_{sL} - \psi_{s0}) .
\end{aligned}
\tag{6.146}
$$

Model Extensions

We now discuss some of the shortcomings of the theory. We have assumed constant mobility; velocity saturation was not taken into account. It is increasingly important in shorter gate-length structures because lateral fields become large. We can relax the assumption of constant mobility, still in a closed form, by assuming a hyperbolic relationship, i.e.,

$$
\mu = \frac{\mu_0}{1 - \mathcal{E}_z/\mathcal{E}_c} ,
\tag{6.147}
$$

where $\mathcal{E}_z$ is the electric field in the channel direction and $\mathcal{E}_c$ is a critical field so that saturated velocity $v_s = \mathcal{E}_c\mu_0$. Alternately, we may write

$$
v = \frac{v_s}{1 - \mathcal{E}_c/\mathcal{E}_z} .
\tag{6.148}
$$

This is quite a good approximation to the equilibrium velocity–field relationship in silicon, for holes in compound semiconductors, but poorer for electrons in compound semiconductors. We can show how this leads to a

closed form following a similar analysis since the field dependence is in the
z-direction and the integrals are similar. We have

$$\mu = \frac{\mu_0}{1 + d\psi_S/(\mathcal{E}_c dz)},\tag{6.149}$$

using the longitudinal field dependence in the channel for calculation in
gradual channel approximation.

Our current equation (Equation 6.67) yields

$$I = -Wq\frac{\mu}{1 + d\psi_S/(\mathcal{E}_c dz)}N_I\frac{d\phi_n}{dz}.\tag{6.150}$$

So,

$$\int_0^L I\left(1 + \frac{1}{\mathcal{E}_c}\frac{d\psi_S}{dz}\right)dz = -Wq\mu_0\int N_I d\phi_n.\tag{6.151}$$

This yields, from Equation 6.88, and after integration,

$$\begin{aligned}
I = {} & -\frac{W\mu_0 C_{ins}}{L}\frac{(kT/q)^2}{1 + (\psi_{sL} - \psi_{s0})/\mathcal{E}_c L}\left\{\left(1 + \frac{qV_G}{kT}\right)\left(\frac{q\psi_{sL}}{kT} - \frac{q\psi_{s0}}{kT}\right) - \right.\\
& \frac{1}{2}\left[\left(\frac{q\psi_{sL}}{kT}\right)^2 - \left(\frac{q\psi_{s0}}{kT}\right)^2\right] - \frac{2}{3}a\left[\left(\frac{q\psi_{sL}}{kT}\right)^{3/2} - \left(\frac{q\psi_{s0}}{kT}\right)^{3/2}\right] + \\
& \left. a\left[\left(\frac{q\psi_{sL}}{kT}\right)^{1/2} - \left(\frac{q\psi_{s0}}{kT}\right)^{1/2}\right]\right\}.
\end{aligned}\tag{6.152}$$

The equation has a very similar form to that of Equation 6.110, and identical procedures may be adopted to obtain the required characteristics or
equivalent circuit parameters (see Problem 13). Comparison of results,
with and without velocity saturation, for a 2.5 μm device are shown in
Figure 6.35 to emphasize the effects of velocity saturation in reducing the
current drive capability of a device.

This modelling is also poor in its prediction of output conductance. The
conductance of the channel was assumed to occur in a narrow channel along
the interface, and does not include the stronger two-dimensional effects of
conduction in the bulk, a component we identified as being very important
in MESFETs. Also, a major cause for output conductance is the shortening of the electrical length of the device. Recall that when pinch-off occurs,
a larger and larger fraction of the applied bias in excess of the pinch-off
voltage drops in the pinched region. The region of the device where inversion exists, and where our formulation continues to truly hold, becomes
shorter and shorter, as shown in Figure 6.36. With the shortening of this

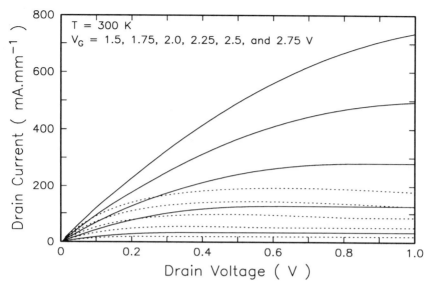

Figure 6.35: Example of characteristics with (dashed lines) and without (solid lines) velocity saturation for a 2.5 μm GaAs/Ga$_{1-x}$Al$_x$As/GaAs MIS-FET.

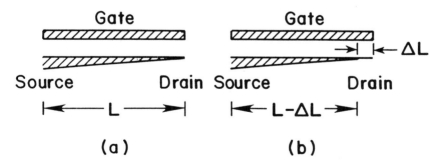

Figure 6.36: Shortening of the channel with application of bias beyond the pinch-off voltage. (a) shows schematically the channel at low bias voltages, while (b) shows the shortening of channel at high bias voltages.

effective gate length, our model predicts that the device current would increase. So, when a drain-to-source bias of V_{DS} in excess of the saturation voltage V_{Dsat} is applied, the extent of the inversion region is shortened by ΔL, where most of the excess voltage above V_{Dsat} drops. Diffusion current is the dominating current here. The device current becomes larger with the increase in the drain-to-source bias V_{DS} because the device behaves as one of electrical gate length L, shorter than L_m, the metallurgical gate length, and because the current is inversely proportional to the electrical gate length. This gives rise to finite output conductance. Approximating that all the excess voltage drop occurs across the pinched off weak inversion region, we obtain

$$\Delta L \approx \sqrt{2}\lambda_D \left\{ [\psi_{sat} + (V_{DS} - V_{Dsat})]^{1/2} - (\psi_{sat})^{1/2} \right\}, \qquad (6.153)$$

and the current is given by

$$
\begin{aligned}
I = & -\frac{W\mu C_{ins}}{L_m - \Delta L}\left(\frac{kT}{q}\right)^2 \left\{ \left(1 + \frac{qV_G}{kT}\right)\left(\frac{q\psi_{sL}}{kT} - \frac{q\psi_{s0}}{kT}\right) - \right. \\
& \frac{1}{2}\left[\left(\frac{q\psi_{sL}}{kT}\right)^2 - \left(\frac{q\psi_{s0}}{kT}\right)^2\right] - \frac{2}{3}a\left[\left(\frac{q\psi_{sL}}{kT}\right)^{3/2} - \left(\frac{q\psi_{s0}}{kT}\right)^{3/2}\right] + \\
& \left. a\left[\left(\frac{q\psi_{sL}}{kT}\right)^{1/2} - \left(\frac{q\psi_{s0}}{kT}\right)^{1/2}\right]\right\}
\end{aligned}
\qquad (6.154)
$$

for the constant mobility assumption, and

$$
\begin{aligned}
I = & -\frac{WL_m}{L_m - \Delta L}\left(\frac{kT}{q}\right)^2 \frac{\mu_0 C_{ins}}{1 + (\psi_{sL} - \psi_{s0})/\mathcal{E}_c L} \times \\
& \left\{ \left(1 + \frac{qV_G}{kT}\right)\left(\frac{q\psi_{sL}}{kT} - \frac{q\psi_{s0}}{kT}\right) - \right. \\
& \frac{1}{2}\left[\left(\frac{q\psi_{sL}}{kT}\right)^2 - \left(\frac{q\psi_{s0}}{kT}\right)^2\right] - \frac{2}{3}a\left[\left(\frac{q\psi_{sL}}{kT}\right)^{3/2} - \left(\frac{q\psi_{s0}}{kT}\right)^{3/2}\right] + \\
& \left. a\left[\left(\frac{q\psi_{sL}}{kT}\right)^{1/2} - \left(\frac{q\psi_{s0}}{kT}\right)^{1/2}\right]\right\}
\end{aligned}
\qquad (6.155)
$$

for the hyperbolic velocity–field relationship. Figure 7.37 shows the effect of shortening of the electrical gate length on the behavior of the device in the constant mobility limit.

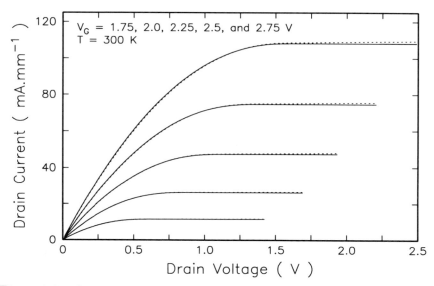

Figure 6.37: Output conductance resulting from reduction in electric gate-length of the 10 μm gate length GaAs/Ga$_{1-x}$Al$_x$As/GaAs MISFET example in the constant mobility limit. The dashed curves result due to shortening of the channel.

6.8 Quasi-Static HFET Theory Using Analytic Approximations

In the previous section, we considered the behavior of a field effect transistor, based on a two-dimensional carrier gas approximated as a sheet charge, by ignoring band occupation effects related to quantization, the exclusion principle, and parasitic conduction in Ga$_{1-x}$Al$_x$As, and assuming either a constant mobility or a hyperbolic velocity–field relationship. Our rationale for this was based in part on the observation that the Boltzmann approximation was justifiable over a significant part of the channel and bias range.

We now wish to to relax some of these assumptions and include some of the special features of confinement effects in the compound semiconductors. As a general introduction, consider how we may modify the quasi-static MISFET theory procedure to include some of these in calculating the current–voltage characteristics for arbitrary conditions, but still utilizing the sheet charge approximation. We will assume that since our specific interest is in taking advantage of the larger carrier velocities in these structures, we will use low background doping in the substrates.

Our analytic procedure will be a variation on the methodology adopted in MISFET analysis. Our intention is to complement the previous treatment by including analytically some of the effects of complications introduced in using compound semiconductors in the barrier region and elsewhere. We may use the Boltzmann approximation only where the quasi-Fermi level is $\approx kT$ or more below the conduction band edge at the interface. Here, as a result of $(\phi_n - \psi_S) > kT/q$, the sub-band levels E_0, E_1, etc., come close together and the total sheet charge in the semiconductor, which is the sum of the sheet charge in the inversion layer and the sheet charge due to acceptors in the semiconductor ($N_s = N_I + N_{As}$) is small. This is accurate in the sub-threshold region but not always in the near-threshold region. When $\phi_n - \psi_S < kT/q$, we need to begin including Fermi–Dirac statistics. As carrier concentrations reach 3×10^{11} cm^{-2} and more, we need to include confinement effects and the Fermi–Dirac statistics. At the kind of background dopings employed commonly (5×10^{14} cm^{-3}), at $N_s = N_I > 3 \times 10^{11}$ cm^{-2}, the acceptor charge may begin to be neglected and N_I begins to be large enough for the triangular well approximation to be quite accurate. But below these carrier concentrations, the Airy function solution employed in the triangular well approximation is about as accurate as the Fermi–Dirac solution assuming a continuum band, and it is poor in sub-threshold conditions because it neglects acceptor doping effects. So, depending on ψ_S and ϕ_n, we may determine $N_s(\psi_S, \phi_n)$ using the appropriate statistics and density of state distribution.

Knowing a bias V_G (again assuming the flat-band voltage of zero but can be introduced by a translation of V_{FB} in gate voltage), we can determine the total semiconductor charge. As an example, for an undoped Ga$_{1-x}$Al$_x$As insulator in a GaAs gate SISFET, assuming no interface charge and charge variation in Ga$_{1-x}$Al$_x$As, this semiconductor charge density is $C_{Al}(V_G - \psi_S)$, where C_{Al} is the capacitance ϵ/t_{Al} per unit area of the insulator. The relationship for V_G may be substantially more complicated for the variety of these structures, e.g., for the doped barrier HFET with uniformly doped Ga$_{1-x}$Al$_x$As, and no spacer in undoped region, the band bending from Figure 7.38 gives

$$-qN_s = -Q_{sem} = C_{Al}\left(V_G - \phi_M + \frac{qN_D t_{Al}^2}{2\epsilon_{Al}} + \frac{\Delta E_c}{q} + |\psi_i| + \psi_B - \psi_S\right). \quad (6.156)$$

Thus, for the general problem, the charge control equation allows us to relate the quasi-Fermi levels to ψ_S for various V_G's.

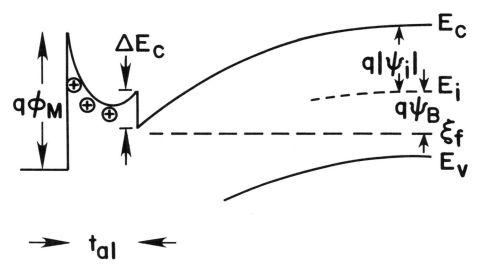

Figure 6.38: Schematic showing parameters related to the determination of sheet charge using Gauss' Law for a doped barrier HFET.

We need to identify the boundary conditions. Unlike the case of the MISFET, we generally do not employ a substrate bias (we use lightly doped p-type buffer and semi-insulating substrates). Remembering from our discussion of built-in voltages of MISFETs, the device transport is determined by drift-diffusion in the channel of the device, and not by the ability of the source and drain junctions. So, we need to find the band bending ψ_S at the source end and the drain end. Since $V_{BS} = 0$,

$$\xi_n(z = 0) = \xi_p = -q\phi_F = -q\psi_B, \tag{6.157}$$

and

$$\xi_n(z = L) = -q\phi_F - qV_{DS} = -q\psi_B - qV_{DS}. \tag{6.158}$$

These give us the magnitude of band bending ψ_S at the source end and the drain end.

Now we employ the drift-diffusion equation in steady-state,

$$I = W\mu N_I \frac{d\xi_n}{dz}. \tag{6.159}$$

Here W is the width of the device, μ is the chordal mobility $v/\mathcal{E}$, N_I is the inversion or accumulation layer mobile charge, and $d\xi_n/dz$ accounts, by using quasi-Fermi levels, for both the drift and diffusion currents.

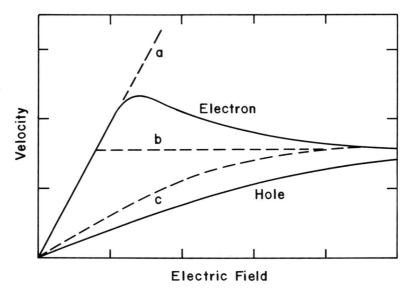

Electric Field

Figure 6.39: Stationary velocity–field curve and various approximations for GaAs employed in analytic and non-analytic device modelling. (a) is the constant mobility approximation, (b) shows the constant mobility with saturated velocity approximation adopted in the PHS model for MESFETs, and (c) is a hyperbolic approximation. These should be compared with the electron and hole velocity–field characteristics present in inversion layers of $Ga_{1-x}Al_xAs/GaAs$ heterostructures.

In the case of a MISFET, using $\mu = \mu_0$, or

$$\mu = \left| \frac{v}{\mathcal{E}} \right| = \frac{\mu_0}{1 + |\mathcal{E}/\mathcal{E}_c|}, \qquad (6.160)$$

allowed us to get a closed-form solution because N_I could be written in terms of V_G and ψ_S, and so could $d\phi_n/dz$. This led to the closed-form expression for I by integration over the device gate length L in terms of ψ_{s0} and ψ_{sL}.

Consider the additional complications of HFETs. Ideally, the velocity–field curve (neglecting off-equilibrium effects) will have a negative differential mobility region, which has been approximated in several forms, as shown in Figure 6.39.

For a general velocity–field curve, the complete solution will have to be performed iteratively according to the procedure above with the condition that

$$I = W\mu\left(d\psi_S/dz\right)N_I(\psi_S, V_G)\frac{d\xi_n}{dz} \tag{6.161}$$

is satisfied throughout the channel region, and

$$
\begin{aligned}
\text{at} &\quad y = 0, &\quad \psi_S &= \psi_{s0}, \\
\text{and at} &\quad y = L, &\quad \psi_S &= \psi_{sL}.
\end{aligned}
\tag{6.162}
$$

We now discuss some approximations. First consider the calculation of charge. In the case of MISFET analysis, we have already discussed the use of the Boltzmann approximation, and the closed-form solutions we could find with it. In the beyond-threshold region, we should include Fermi–Dirac statistics. Poisson's equation in the Boltzmann approximation is

$$
\begin{aligned}
-\frac{d^2\psi}{dy^2} &= \frac{q}{\epsilon_s}\left[N_A^- + n\right] \\
&= \frac{q}{\epsilon_s}N_A + \frac{q}{\epsilon_s}n_i\exp\left(q\frac{\psi - \phi_n}{kT}\right)
\end{aligned}
\tag{6.163}
$$

for $0 < y < w$. For $y > w$,

$$-\frac{d^2\psi}{dy^2} = 0. \tag{6.164}$$

This gives the solution, for the Boltzmann approximation,

$$
\begin{aligned}
N_I &= N_A\sqrt{2}\lambda_D\left\{\left[\frac{q\psi_S}{kT} + \left(\frac{n_i}{N_A}\right)^2\exp\left(\frac{q\psi_S}{kT}\right)\exp\left(q\frac{-\phi_n + \phi_F}{kT}\right)\right]^{1/2}\right.\\
&\quad\left. - \left(\frac{q\psi_S}{kT}\right)^{1/2}\right\}.
\end{aligned}
\tag{6.165}
$$

In this equation, the second term is the total depletion charge, while the first term contains the total semiconductor charge, the difference being the mobile charge. With Fermi–Dirac statistics, we consider the Poisson's equation in $0 < y < w$ as

$$\frac{d^2\psi}{dy^2} = \frac{q}{\epsilon_s}N_A + \frac{q}{\epsilon_s}N_C F_{1/2}\left(q\frac{\psi - \phi_n}{kT}\right). \tag{6.166}$$

An approximation of the Fermi integral of the order $1/2$ is the Ehrenreich approximation, given by

$$F_{1/2}(\eta) = \frac{4\exp(\eta)}{4 + \exp(\eta)}, \tag{6.167}$$

which is quite accurate for $\eta \leq 2$. So, the Fermi level can be ≈ 50 meV at 300 K into the conduction band while still giving a fairly accurate estimate for the sheet density of semiconductor charge. This allows quite accurate modelling of the inversion/accumulation charge at near-threshold conditions. Using the technique of multiplying by $d\psi/dy$ on both sides, we get in the Ehrenreich approximation

$$
N_I = N_A\sqrt{2}\lambda_D \left\{ \left\{ \frac{q\psi_S}{kT} + \frac{4N_C}{N_A} \ln\left[1 + \frac{1}{4}\exp\left(q\frac{\psi_S - \phi_n}{kT}\right)\right] \right\}^{1/2} - \left(\frac{q\psi_S}{kT}\right)^{1/2} \right\}.
\tag{6.168}
$$

This can be shown to reduce to the Boltzmann approximation expression for $q(\psi_S - \phi_n)/kT << 1$ by perturbative expansion of the logarithmic term.

In strong inversion, we consider the effect of occupation of the various subbands with the Fermi–Dirac distribution function determining the occupation statistics. Consider the simple example of occupation by Γ electrons. The density of states (D) for the subbands E_0, E_1, etc., is constant and equal to $qm^*/\pi\hbar^2$ (for GaAs, $\approx 3.24 \times 10^{13}$ cm^{-2}.eV^{-1}). We may then write the Fermi level ξ_f w.r.t. the conduction band edge E_c as

$$
\begin{aligned}
N_I &= D\int_{E_0}^{\infty} \frac{dE}{1 + \exp\left[(E - \xi_f)/kT\right]} + \\
&\qquad \int_{E_1}^{\infty} \frac{1}{1 + \exp\left[(E - \xi_f)/kT\right]} dE + \cdots \\
&= DkT \ln\left\{ \left[1 + \exp\left(\frac{\xi_f - E_0}{kT}\right)\right] \left[1 + \exp\left(\frac{\xi_f - E_1}{kT}\right)\right] \cdots \right\}.
\end{aligned}
\tag{6.169}
$$

The density of states of all the subbands is equal for the isotropic and parabolic Γ valley, and the sheet density of carriers in the inversion layer are related via a simple logarithmic expression. For the other, more complicated subband forms, e.g., of the hole inversion layer, the results are necessarily more complex (see Problem 14), and emphasize the importance of heavy hole bands due to both their lower energy shifts and their larger density of states.

In the carrier concentration range of interest, i.e., $\approx 5\text{-}10 \times 10^{11}$ cm^{-2}, the third subband is about 10% occupied for GaAs at 300 K according to Figure 6.11, but quite often this is ignored. The complication in analysis using the confinement effects arises in the form of the dependence of

E_0 and E_1 themselves on the inversion layer sheet density N_I because N_I determines the electric field and hence the confinement in the triangular well. To obtain accuracy in the solution one would have to use an iterative procedure at this step. However, considering that the triangular well approximation is only being adopted for the strong inversion condition, we may employ the approximations for the subband energies described earlier using the Airy function solutions,

$$E_0 = \gamma_0 N_I{}^{2/3}; \quad \gamma_0 = 2.5 \times 10^{-12} \text{ J.m}^{4/3},$$
$$E_1 = \gamma_1 N_I{}^{2/3}; \quad \gamma_1 = 4.0 \times 10^{-12} \text{ J.m}^{4/3}, \qquad (6.170)$$

for GaAs. Using for the moment our notation in terms of energies, we now obtain

$$
\exp\left(\frac{N_I}{DkT}\right)\exp\left(\frac{E_0 + E_1}{kT}\right)
$$

$$
= \quad \exp\left(\frac{E_0 + E_1}{kT}\right) + \exp\left(\frac{\xi_f + E_1}{kT}\right) +
$$

$$
\exp\left(\frac{\xi_f + E_0}{kT}\right) + \exp\left(\frac{2\xi_f}{kT}\right) +
$$

$$
\left[\exp\left(\frac{E_0}{kT}\right) + \exp\left(\frac{E_1}{kT}\right)\right]\exp\left(\frac{\xi_f}{kT}\right) +
$$

$$
\exp\left(\frac{E_0 + E_1}{kT}\right)\left[1 - \exp\left(\frac{N_I}{DkT}\right)\right]. \qquad (6.171)
$$

The solution for ξ_f, the Fermi energy, follows as

$$
\xi_f = kT \ln\left\{-\frac{1}{2}\left[\exp\left(\frac{E_0}{kT}\right) + \exp\left(\frac{E_1}{kT}\right)\right] + \mathcal{A}\right\}, \qquad (6.172)
$$

where $\mathcal{A}$ is

$$
\mathcal{A} = \frac{1}{4}\left[\exp\left(\frac{E_0}{kT}\right) + \exp\left(\frac{E_1}{kT}\right)\right]^2 -
$$

$$
\exp\left(\frac{E_0 + E_1}{kT}\right)\left[1 - \exp\left(\frac{N_I}{DkT}\right)\right]^{1/2}
$$

$$
= \left\{\frac{1}{4}\left[\exp\left(\frac{E_0}{kT}\right) - \exp\left(\frac{E_1}{kT}\right)\right]^2 +
$$

$$
\exp\left(\frac{E_0 + E_1}{kT}\right)\exp\left(\frac{N_I}{DkT}\right)\right\}^{1/2}. \qquad (6.173)
$$

This can be written, as a function of sheet inversion charge density, for strong inversion, as

$$
\begin{aligned}
\xi_f = \ & kT \ln \left\{ -\frac{1}{2} \left[\exp\left(\frac{\gamma_0 N_I^{2/3}}{kT} \right) + \exp\left(\frac{\gamma_1 N_I^{2/3}}{kT} \right) \right] + \right. \\
& \left\{ \frac{1}{4} \left[\exp\left(\frac{\gamma_0 N_I^{2/3}}{kT} \right) - \exp\left(\frac{\gamma_1 N_I^{2/3}}{kT} \right) \right]^2 + \right. \\
& \left. \left. \exp\left(\frac{\gamma_0 + \gamma_1}{kT} N_I^{2/3} \right) \exp\left(\frac{N_I}{DkT} \right) \right\}^{1/2} \right\} .
\end{aligned} \tag{6.174}
$$

Recall that this is still an approximation using a quasi-triangular well. Over the years this problem has been treated with further analytic simplifications to allow simple solutions of current–voltage characteristics. These may be treated as fitting techniques suitable only for specific conditions since acceptor doping effects are still ignored or assimilated in the parameters of the approximation, thus limiting its applicability for wider substrate parameters. Some examples of analytic approximations[8] are

$$
\xi_f = \xi_{f0}, \tag{6.175}
$$

i.e., no quasi-Fermi level variation,

$$
\xi_f = \xi_{f0} + aN_I, \tag{6.176}
$$

i.e., a linear variation of the quasi-Fermi level, and

$$
\xi_f = K_1 + K_2(N_I + K_3)^{1/2}, \tag{6.177}
$$

and

$$
\xi_f = \xi_{f0} + \gamma N_I^{2/3}, \tag{6.178}
$$

which are various forms of power law variation. Referring to our accurate calculation of the variation of the sheet carrier density with Fermi energy at the the $Ga_{1-x}Al_xAs/GaAs$ interface, Figure 6.40 shows representative examples of some of the previous approximations. The merit of these ana-

[8]Some examples of these and discussion related to them can be found in J. Yoshida, "Classical versus Quantum-Mechanical Calculation of the Electron Distribution at the n-AlGaAs/GaAs Heterointerfaces," *IEEE Trans. on Electron Devices*, **ED-33**, No. 1, p. 154, Jan. 1986; K. Park and K. D. Kwack, "Calculation of the Two-Dimensional Electron Gas Density at the $Al_xGa_{1-x}As/GaAs$ Interface," *IEEE Trans. on Electron Devices*, **33**, No. 11, p. 1831, Nov. 1986; S. Kola, J. M. Golio, and G. N. Maracas, "An

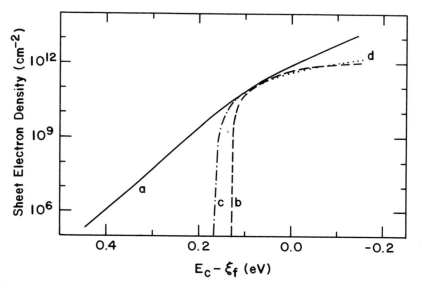

Figure 6.40:   Sheet electron density as a function of $E_c - \xi_f$ at $Ga_{1-x}Al_xAs/GaAs$ interface for 3×10^{14} cm^{-3} acceptor doping in the substrate. (a) shows the calculation in Boltzmann approximation, (b) in the linear approximation, (c) in the 2/3rd power law approximation, and (d) shows a fit with triangular well approximation. Appropriate fitting parameters were chosen to obtain most accuracy at sheet densities of interest.

lytic approximations, is in the convenience of fitting an analytic expression to a strong inversion behavior that is considerably complicated.

We could employ our procedure together with these relationships to arrive at the current–voltage characteristics even for arbitrary velocity–field curves. Some examples of velocity–field relationships were illustrated in Figure 7.39. The first of these is the hyperbolic relationship, used earlier, which allows a closed-form solution of the current equations derived above if the charge relationship allows analytic integrations. For silicon long channel devices, the hyperbolic curve is quite accurate. For compound semiconductors exhibiting large mobilities or a negative differential relationship, it is

Analytical Expression for Fermi Level versus Sheet Carrier Concentration for HEMT Modelling," *IEEE Electron Device Letters*, **9**, No. 3, p. 136, Mar. 1988; and A-J Shey and W. H. Ku, "On the Charge Control of the Two-Dimensional Electron Gas for Analytic Modelling of HEMTs," *IEEE Electron Device Letters*, **9**, No. 12, p. 624, Dec. 1988.

quite inaccurate, and a hyperbolic relationship may be treated as a fitting technique with fitting parameters chosen to best reflect the behavior at specific electric fields. Finally, there have been many attempts at modelling the compound semiconductors by fitting closely the low-field behavior and the high-field behavior. These piece-wise fits are illustrated in the next set of velocity–field curves in this figure.

One common thread in most of these fits is their inaccuracy at moderate fields even if low-field behavior is fitted inaccurately in order to obtain more accuracy at moderate fields. None can simulate negative differential mobility, none can then simulate moderate field behavior (i.e., near the peak velocity fields) without sacrificing the accuracy of high-field behavior. This moderate field behavior is certainly quite important to the carrier transport. Channel pinch-off occurs when these fields are reached, so much of the region of transistor operation in logic and in analog applications occurs under these field conditions. Quite generally, therefore, when applying these characteristics, one is forcibly fitting a behavior, and the choice of parameters such as low-field mobility and saturated velocity are dictated not by the actual material parameters but by values that provide minimum error fit to actually observed device characteristics. So, when we use these velocity–field curves, it is implicitly understood that quoted mobilities or velocities are fits and not actual mobilities or velocities.

A general analytic expression that shows negative differential mobility, but which does not allow closed-form solutions, is

$$v(\mathcal{E}) = \frac{\mu\mathcal{E} + v_s(\mathcal{E}/\mathcal{E}_c)^n}{1 + (\mathcal{E}/\mathcal{E}_c)^n}. \qquad (6.179)$$

For lightly doped GaAs at room temperature, $n \approx 4$.

We now discuss our sheet charge approximations. The rapid drop-off of Fermi energy with sheet carrier concentration at low carrier concentration is extremely difficult to model, considering the variability of N_A, which has also been assumed to be a constant. Like the velocity–field relationship, fitting can be performed only over a very narrow range, and does not lead to an accurate model over a wide variety of bias conditions extending from the sub-threshold region to high gate biases. The effect of background doping and temperature on these approximations is quite large and one should carefully obtain the parameters to achieve an acceptable accuracy in the end results of the calculations—device characteristics.

Before we derive a simplistic closed-form solution of the current–voltage and capacitance–voltage characteristics, let us emphasize some of the important differences of HFETs vis-a-vis silicon MOSFETs. HFETs use much lower p-type doping in order to maintain a high low-field mobility. So, the

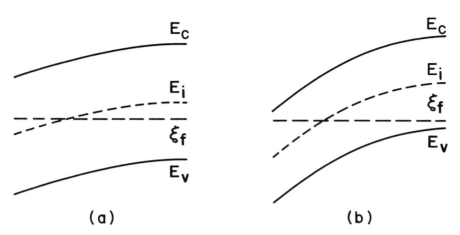

Figure 6.41: Schematic comparing the band bending in SiO_2/Si MOSFETs (a) to HFETs, which employ low background dopings (b) at the $\psi_S = 2\psi_B$ condition.

criterion $\psi_S = 2\psi_B$ would lead to a condition where there are still very few carriers for conduction. This is shown in Figure 7.41. So, unlike in SiO_2/Si MOSFETs, neither ψ_S saturation nor a rapid change in electron charge with bias would occur. Therefore, the threshold can not be defined as simply in terms of band bending for HFETs. Practically speaking, V_T is a fitting parameter that we get from measured characteristics of devices. It signifies the gate voltage, at a given drain voltage, for which a rapid onset of conduction occurs. We may, therefore, write it as a minimal N_I needed for this conduction to take place. For example, at a velocity of 1×10^7 cm.s^{-1}, an inversion charge of $Q_I = 8 \times 10^{-9}$ C.cm^{-2} (i.e., $N_I = 5 \times 10^{10}$ cm^{-2}) results in a current of 8 mA.mm^{-1}—a small current compared to the 200-400 mA.mm^{-1} current that devices are capable of. Thus, this may be used as a definition of threshold voltage, and certainly one can see from the plots of sheet carrier concentration, it is around this concentration that the rapid change in Fermi level position occurs—similar to what happens in SiO_2/Si MOSFETs. One could define it more precisely for arbitrary conditons by actually determining the region where this rapid change in N_I occurs. This would depend on the charge density in the background—a quantity dependent on the the bulk potential, i.e., ψ_B.

Note that the rapid onset region is dependent on both the background doping and the temperature. The onset condition also occurs fairly near where the conduction band edge and the quasi-Fermi level are nearly coin-

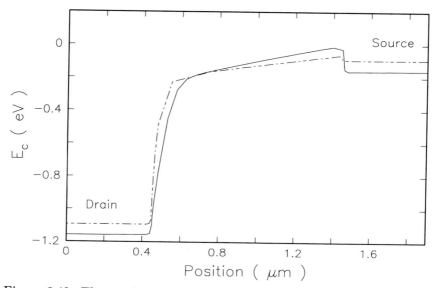

Figure 6.42: The conduction band edges in a GaAs MESFET (solid line) and $GaAs/Ga_{1-x}Al_xAs/GaAs$ MISFET (dot-dashed line) for 1 μm gate-length at a 1 V drain bias.

cident. So the amount of band bending is $\approx \psi_B + \psi_i$ instead of $2\psi_B$.

In a manner similar to the analysis of the PHS model for MESFETs, we could have broken up our problem in various sections that could be treated analytically or numerically. As in PHS model, this leads to a discontinuity in $dv/d\mathcal{E}$. We could match where the peak velocity and field occur, i.e., we could solve the current equation in the two sections and match the two sections by maintaining current continuity in a way similar to the constant mobility and constant saturated velocity analysis of MESFETs. This methodology does not lead to analytic form of sufficient accuracy; however, non-analytic procedures with confinement effects in the low-field region and sheet charge approximation in the Boltzmann limit in the velocity saturated region do lead to acceptable accuracy. Compared to the MESFET, there is one particularly interesting difference between the two devices, which is shown in Figure 7.42. In the PHS model, the region of high-field and saturated velocity transport can be very large. In a MISFET, where the channel is induced, the pinch-off region is shorter at the larger gate lengths, since the carrier pinch-off is more efficient, and the fields in this region are larger for similar drain voltages. This behavior is size-dependent, the smaller the gate length of a device, the more similar the two devices become in the behavior of the potential.

Our analytic calculation of the current–voltage characteristics[9] follow in light of the approximations discussed. We consider a smooth velocity–field curve based on the hyperbolic relationship, letting us write complete relationships instead of a procedure, but with the caveat that the mobility and saturated velocity are fitting parameters that emphasize a more accurate reflection of the velocity–field behavior in the moderate electric field region. Away from this region, in regions in which they normally would have a physical meaning, they are mere fitting parameters. We have ignored the consequences of negative velocity–field characteristics present at the low background doping, which may exhibit substantial hot carrier effects in certain HFET structures. We will return to this shortcoming by considering its two-dimensional nature later in this chapter.

We use our general relationship within the sheet charge approximation of

$$I = \mu N_I W \frac{d\xi_n}{dz}. \tag{6.180}$$

We ignore the diffusive current component—the model is incorrect in the sub-threshold region, which we treat separately. Our control equation then is

$$I_D = q\mu N_I W \frac{dV}{dz}, \tag{6.181}$$

where V is the electrostatic potential of the channel. We will refer this to the bottom of the conduction band at the source end of the channel. So, $V = 0$ at the source end. Recall, based on potential drops and Gauss's law, at a position z somewhere along the channel,

$$
\begin{aligned}
-q(N_{As} + N_I) &= C_{Al} \times \\
&\quad \left\{ V_G - V - \frac{\xi_f}{q} + \frac{\Delta E_c}{q} + \frac{qN_D}{\epsilon_{Al}}(t_{Al} - t_{sp})^2 - \phi_M \right\} \\
&= C_{Al} \left\{ V_G - V - \left[\phi_M - \frac{qN_D}{\epsilon_{Al}}(t_{Al} - t_{sp})^2 - \right. \right. \\
&\qquad \left. \left. \frac{\Delta E_c}{q} + \frac{\xi_f}{q} \right] \right\}. \tag{6.182}
\end{aligned}
$$

---

[9]The model discussed here is a simple extension of A-J Shey and W. H. Ku, "An Analytical Current–Voltage Characteristics Model for High Electron Mobility Transistors Based on Nonlinear Charge Control Formulation," *IEEE Trans. on Electron Devices*, **ED-36**, No. 10, p. 2299, Oct. 1989.

We use the non-linear approximation for

$$\xi_f = \xi_{f0} + \gamma N_I{}^{2/3}, \tag{6.183}$$

which parameterizes the confinement effects, and is accurate over a modest range of mobile charge in the two-dimensional carrier gas under conditions of low background doping with a proper selection of the fitting parameters ξ_{f0} and γ. For modest (low 10^{15} cm^{-3} or lower) doping in GaAs, a suitable fit is

$$\gamma = 1.79 \times 10^{-9} \text{ eV.cm}^{4/3}, \tag{6.184}$$

with ξ_{f0} dependent on the background doping (see Problem 15). So,

$$N_I = \frac{C_{Al}}{q}\left\{V_G - V - V_{T0} - \gamma N_I{}^{2/3}\right\}, \tag{6.185}$$

where

$$V_{T0} = \phi_M - \frac{qN_D}{\epsilon_{Al}}(t_{Al} - t_{sp})^2 - \frac{\Delta E_c}{q} - \frac{\xi_{f0}}{q} \tag{6.186}$$

is a "threshold" voltage at which charge is induced rapidly at the source end where $V = 0$. This leads to

$$N_I = \frac{C_{Al}\left(V_G - V - V_{T0}\right)}{q\left(1 + C_{Al}\gamma N_I{}^{-1/3}/q^2\right)}. \tag{6.187}$$

For an example of Ga$_{1-x}$Al$_x$As with $\epsilon_{Al} \approx 1 \times 10^{-12}$ F.cm^{-1}, $t_{Al} = 400$ Å, and $N_I \approx 1 \times 10^{12}$ cm^{-2}, the perturbation term in the sheet carrier density is

$$\frac{C_{Al}\gamma}{q^2}N_I{}^{-1/3} \approx 0.28, \tag{6.188}$$

i.e., deviations from a constant quasi-Fermi level relationship with carrier density account for nearly a quarter of the effect in the carrier density, a substantial effect.

Our velocity–field relationship is

$$v(\mathcal{E}) = \frac{\mu_0 \mathcal{E}}{1 + |\mathcal{E}/\mathcal{E}_c|}. \tag{6.189}$$

For convenience, we now introduce a parameter

$$\varpi = \frac{C_{Al}\gamma}{q^2}. \tag{6.190}$$

Using these relationships, the drain current can be expressed as

$$I_D = -\frac{WC_{Al}}{1 + \varpi N_I{}^{-1/3}}\left[V_G - V_{T0} - V\right]\frac{\mu_0 \mathcal{E}}{1 + |\mathcal{E}/\mathcal{E}_c|}, \tag{6.191}$$

and

$$N_I = \frac{C_{Al}}{q\left(1 + \varpi N_I^{-1/3}\right)} (V_G - V_{T0} - V), \qquad (6.192)$$

where $\varpi N_I^{-1/3}$ is the perturbation term in carrier occupation due to the movement of the subband levels with changing occupation. We may also interpret it as a renormalization term for the capacitance C_{Al} to $C_{Al}/\left(1 + \varpi N_I^{-1/3}\right)$. The charge could be viewed as $t_{Al} \times \varpi N_I^{-1/3}$ further away from the interface, leading to poorer modulation of the charge. For our example, this effective increase in the spacing for the sheet carrier density is $\approx 0.28 \times 400 = 112$ Å. Strictly speaking, we can determine this from the complete calculation based on the envelope wave function and the occupation in all the subbands. This excess spacing is

$$\Delta t_{Al} = \frac{\frac{2}{3} \sum_i N_{Ii} y_i}{\sum_i N_{Ii}}, \qquad (6.193)$$

where N_{Ii} is the sheet carrier density in the ith subband. Figure 6.43 shows variation of this parameter, and compares it with the fitted parameter determined above.

We can utilize our expression for N_I as a function of the bias voltages in the equation for I_D, and ignore the second-order expansion terms in voltage, leading to

$$I_D = -\frac{W C_{Al} (V_G - V_{T0} - V)}{1 + \varpi (C_{Al}/q)^{-1/3}(V_G - V_{T0} - V)^{-1/3}} \frac{\mu_0 \mathcal{E}}{1 + |\mathcal{E}/\mathcal{E}_c|}. \qquad (6.194)$$

We now define

$$\lambda = \varpi \left(\frac{C_{Al}}{q}\right)^{-1/3} = \left(\frac{\epsilon_{Al}}{t_{Al}}\right)^{2/3} \frac{1}{q^{2/3}} \frac{\gamma}{q}. \qquad (6.195)$$

For our example, $\lambda \approx 0.24$ V$^{1/3}$. Substituting for field and above,

$$I_D = -\frac{W \mu_0 C_{Al} (V_G - V_{T0} - V)}{1 + \lambda(V_G - V_{T0} - V)^{-1/3}} \frac{-\partial V/\partial z}{1 + \partial V/\mathcal{E}_c \partial z}. \qquad (6.196)$$

This is an equation that can be solved analytically; the other power relationship fits also allow this, albeit in slightly different forms. Again, like our earlier MESFET example, we introduce the normalized variable of voltage,

$$\xi = (V_G - V_{T0} - V)^{1/3}, \qquad (6.197)$$

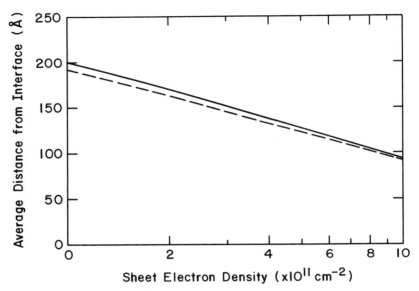

Figure 6.43: Excess distance, as calculated from accurate calculations of the two-dimensional electron gas at $Ga_{1-x}Al_xAs/GaAs$ interface (solid line), and from the fitting parameter of the power relationship (dashed line).

which has the magnitude of

$$\xi = (V_G - V_{T0})^{1/3} \tag{6.198}$$

at the source $(z = 0)$, and the magnitude of

$$\xi = (V_G - V_{T0} - V_D)^{1/3} \tag{6.199}$$

at the drain $(z = L)$. Here, V_D is the drain voltage with the source as a reference. We have

$$
\begin{aligned}
d\xi &= \frac{1}{3(V_G - V_{T0} - V_D)^{2/3}} \\
&= -\frac{1}{3\xi^2}dV.
\end{aligned}
\tag{6.200}
$$

So,

$$I_D = -W\mu_0 C_{Al}\frac{\xi^3}{1 + \lambda\xi^{-1}}\frac{3\xi^2}{1 - 3\xi^2 d\xi/\mathcal{E}_c dz}\frac{d\xi}{dz}, \tag{6.201}$$

and hence,

$$
I_D \int \left[z - \frac{3\xi^2}{\mathcal{E}_c} \frac{d\xi}{dz} \right] d\xi = -3W\mu_0 C_{Al} \int \frac{\xi^6}{\xi + \lambda} d\xi. \tag{6.202}
$$

Integrating from $z = 0, \xi = \xi_s$ to $z = L, \xi = \xi_d$, we obtain

$$
\begin{aligned}
I_D &= -\frac{3W\mu_0 C_{Al}}{L\left[1 - (\xi_d{}^3 - \xi_s{}^3)/\mathcal{E}_c L\right]} \left[\frac{\xi^6}{6} - \lambda\frac{\xi^5}{5} + \lambda^2\frac{\xi^4}{4} - \right. \\
&\qquad \left. \lambda^3\frac{\xi^3}{3} + \lambda^4\frac{\xi^2}{2} - \lambda^5\frac{\xi}{1} + \lambda^6 \ln(\xi + \lambda) \right]\Bigg|_{\xi_s}^{\xi_d},
\end{aligned} \tag{6.203}
$$

which can be expressed as

$$
I_D = -\frac{3W\mu_0 C_{Al}}{L\left(1 + V_D/\mathcal{E}_c L\right)} \left\{ \left[\sum_{i=1}^{6} \frac{(-1)^i}{i} \lambda^{6-i} \left(\xi_d{}^i - \xi_s{}^i\right) \right] + \lambda^6 \ln\left(\frac{\xi_d + \lambda}{\xi_s + \lambda} \right) \right\}, \tag{6.204}
$$

and in terms of applied voltages as

$$
\begin{aligned}
I_D &= -\frac{3W\mu_0 C_{Al}}{L\left(1 + V_D/\mathcal{E}_c L\right)} \left\{ \left[\sum_{i=1}^{6} \frac{(-1)^i}{i} \lambda^{6-i} \left((V_G - V_{T0} - V_D)^{i/3} - \right. \right. \right. \\
&\quad \left. \left. \left. (V_G - V_{T0})^{i/3} \right) \right] + \lambda^6 \ln\left[\frac{(V_G - V_{T0} - V_D)^{1/3} + \lambda}{(V_G - V_{T0})^{1/3} + \lambda} \right] \right\}, \tag{6.205}
\end{aligned}
$$

or

$$
\begin{aligned}
I_D &= \frac{3W\mu_0 C_{Al}}{L\left(1 + V_D/\mathcal{E}_c L\right)} \left\{ \left[\sum_{i=1}^{6} \frac{(-1)^i}{i} \lambda^{6-i} \left((V_G - V_{T0})^{i/3} - \right. \right. \right. \\
&\quad \left. \left. \left. (V_G - V_{T0} - V_D)^{i/3} \right) \right] + \lambda^6 \ln\left[\frac{(V_G - V_{T0})^{1/3} + \lambda}{(V_G - V_{T0} - V_D)^{1/3} + \lambda} \right] \right\}. \tag{6.206}
\end{aligned}
$$

This relation reduces to the familiar parabolic approximation in the triode region, i.e., for drain voltages below the channel pinch-off voltage. We show this by considering the approximations. The natural logarithm term is small because it is in the powers of λ, a small quantity. The sum term has as its largest contributor the term whose coefficient is λ^0, i.e.,

$i = 6$. Consider this term,

$$
\begin{aligned}
I_D &= \frac{3W\mu_0 C_{Al}}{L\left(1 + V_D/\mathcal{E}_c L\right)} \times \\
&\quad \frac{1}{6}\left[(V_G - V_{T0})^2 - (V_G - V_{T0})^2 - V_D{}^2 + 2\left(V_G - V_{T0}\right)V_D\right] \\
&= \frac{3W\mu_0 C_{Al}}{L\left(1 + V_D/\mathcal{E}_c L\right)}\left[(V_G - V_{T0})V_D - \frac{1}{2}V_D{}^2\right]. \tag{6.207}
\end{aligned}
$$

The earlier association of V_{T0} as a threshold term follows from this expression.

We must make a cautionary remark regarding our extended expression: it is still an approximation, since it does not consider diffusive current, and uses gradual channel approximation and parametric fits to the carrier density and the velocity–field relationships. However, it provides through its simplicity a convenient means to represent, adequately for computer-aided design models, the quasi-static current–voltage relationship. Its approximation in the velocity–field parameterization may be extended further by considering a PHS model–like partitioning of the velocity–field relationship: breaking it in a hyperbolic part and a constant velocity part, and matching the two field regions in the device.

A more convenient form of including this effect, within the simpler modelling procedure, is to consider the origin of the saturation of current and the evolution of the current–voltage characteristics at drain biases exceeding this. The saturation of the current, in high mobility materials, is associated with the appearance of a sufficiently high-field region at the drain end of the channel to cause saturation of velocity. Thus, a simple parameterization of this field allows us to determine the condition of saturation of velocity and current (see Problem 16). This happens at the voltage V_{Dsat} associated with the specific gate bias of the structure. Beyond this point, the current–voltage characteristics may be considered to arise due to shortening of the channel as in the MISFET extension. Since $L_e < L$, I_D, which is inversely proportional to L_e, continues to increase, leading to what is usually described as the increase in current due to the short channel effect.

A simpler, and for many instances adequate, model would be to consider the assimilation of a model similar to the constant velocity model of MESFETs. If we ignore the effects of the gradual channel approximation region, where voltage drops are small and consider the entry of the carriers in the channel at the saturated velocity v_s, then in the piece-wise approximation, this current saturation occurring at a channel voltage of V_{sat} leads to the drain current given by

$$I_D = W q \mu_0 C_{Al} \frac{(V_G - V_{T0} - V_{sat})}{1 + \lambda(V_G - V_{T0} - V_{sat})^{1/3}} v_s. \qquad (6.208)$$

The current continuity condition requires this current and the current from the gradual channel approximation expression be identical at the drain saturation voltage V_{sat}, giving the magnitude of V_{sat}. The deviations in electric length from the metallurgical gate length now follow, allowing for the determination of the current due to electric gate length shortening.

The major limitation of this and other procedures that do not consider effects of transport in the substrate region directly is that they underestimate the output conductance. Proper output conductance modelling can only be made through two-dimensional models where the substrate characteristics are adequately accounted. We may continue these refinements by including the parasitic MESFET and its screening of the charge transport in the two-dimensional carrier gas etc. (see Problem 17). These are simpler extensions and will not be covered.

6.8.1 Sub-Threshold Currents

This quasi-static model for current flow is probably least adequate in the sub-threshold region, where many of the approximations break down quite entirely. The parametric power law fit of carrier density is very inadequate in this region, and as discussed earlier, the Boltzmann approximation or Fermi–Dirac approximation of three-dimensional statistics is quite adequate. Sub-threshold current, then, follows from our discussion related to the MISFET. Here, we employ the Ehrenreich approximation for Fermi–Dirac statistics. Since the current is by diffusion, we need the gradient of the carrier distribution.

$$I_{Dsubthr} = W q \mathcal{D} \frac{dN_I}{dz}. \qquad (6.209)$$

We have at the source end

$$N_I|_{source} = N_A \sqrt{2} \lambda_D \left\{ \left[\frac{q\psi_{s0}}{kT} + 4 \frac{N_C}{N_A} \ln \left[1 + \frac{1}{4} \exp \left(q \frac{\psi_{s0} - \phi_n}{kT} \right) \right] \right]^{1/2} \right.$$
$$\left. - \left(\frac{q\psi_{s0}}{kT} \right)^{1/2} \right\}, \qquad (6.210)$$

in the degenerate limit, and

$$N_I|_{source} = \frac{\lambda_D N_C}{\sqrt{2}(q\psi_{s0}/kT)^{1/2}} \exp \left(\frac{\xi_f}{kT} \right) \qquad (6.211)$$

in the the non-degenerate limit. Since N_I is small,

$$\xi_f = -q\,(V_G - V_{T0})\,, \tag{6.212}$$

and

$$N_I|_{drain} = N_I|_{source} \exp\left(-q\frac{V_{DS}}{kT}\right). \tag{6.213}$$

So,

$$I_{Dsubthr} = -Wq\mathcal{D}\frac{dN_I}{dz} \approx -Wq\mathcal{D}\frac{N_I|_{drain} - N_I|_{source}}{L}, \tag{6.214}$$

which gives

$$\begin{aligned}
I_{Dsubthr} = {} & \frac{W\mu kT}{L}\left[1 - \exp\left(-q\frac{V_{DS}}{kT}\right)\right] \times \\
& \frac{\lambda_D N_C}{\sqrt{2}(q\psi_{s0}/kT)^{1/2}} \exp\left(\frac{qV_{T0}}{kT}\right) \exp\left(\frac{-qV_G}{kT}\right).
\end{aligned} \tag{6.215}$$

In the sub-threshold region, $V_G < V_{T0}$, and drain voltage is usually larger than thermal voltage. The above expression shows that the sub-threshold current would exponentially depend on the applied gate voltage, and the dependence on the drain voltage would be very weak. And indeed, just like MISFETs, HFETs show an exponential drain current–gate voltage dependence the in sub-threshold region with the sub-threshold gate swing $S = \ln 10/(d\ln I/dV_G)$, which is approximately 60 mV/decade at room temperature. An example of this calculation is shown in Figure 6.44.

6.8.2 Intrinsic Capacitances

We derive the capacitances using our approximate model by finding the total channel charge Q_T, and then finding its dependence on the gate-to-source potential and gate-to-drain potential, i.e., we will determine it from

$$C_{gs} = \frac{\partial Q_T}{\partial\,(V_G - V_S)}$$

$$\text{and}\quad C_{gd} = \frac{\partial Q_T}{\partial\,(V_G - V_D)}. \tag{6.216}$$

Recalling our remarks regarding charge partitioning and conservation, this calculation of capacitance is path-dependent and limited in scope. Following our definition of ξ,

$$\frac{d\,(V_G - V)}{dz} = 3\xi^2\frac{d\xi}{dz}. \tag{6.217}$$

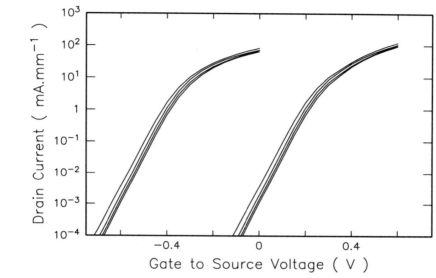

Figure 6.44: Theoretical behavior of sub-threshold current versus gate voltage for a depletion-mode (left set) and an enhancement-mode (right set)
$Ga_{1-x}Al_xAs/GaAs$ 1 μm gate-length HFETs. The drain bias is increased
in steps of 1 V from 0.5 V.

We have

$$
\begin{aligned}
Q_T &= W \int_0^L qN_I dz \\
&= W \int_{\xi_s}^{\xi_d} qN_I \frac{dz}{d\xi} d\xi \\
&= qW \int_{\xi_s}^{\xi_d} \frac{1}{q} \frac{C_{Al}\xi^3}{1 + \lambda\xi^{-1}} \frac{dz}{d\xi} d\xi.
\end{aligned}
\tag{6.218}
$$

The drain current, known already from current continuity in terms of ξ_d
and ξ_s, is also related to ξ and its derivative by

$$
I_D = -W\mu_0 C_{Al} \frac{\xi^3}{1 + \lambda\xi^{-1}} \frac{3\xi^2}{1 - (3\xi^2/\mathcal{E}_c)(d\xi/dz)} \frac{d\xi}{dz},
\tag{6.219}
$$

and hence the derivative is known as a function of ξ,

$$
\frac{d\xi}{dz} = \frac{1}{(3\xi^2/\mathcal{E}_c) - [3W\mu_0 C_{Al}\xi^6/I_D(\xi + \lambda)]},
\tag{6.220}
$$

and hence

$$
\begin{aligned}
Q_T &= WC_{Al} \int_{\xi_s}^{\xi_d} \left(\frac{\xi^4}{\xi + \lambda} \right) \left[\frac{3\xi^2}{\mathcal{E}_c} - \frac{3W\mu_0 C_{Al}\xi^6}{I_D(\xi + \lambda)} \right] d\xi \\
&= WC_{Al} \int_{\xi_s}^{\xi_d} \left[\frac{3\xi^6}{\mathcal{E}_c(\xi + \lambda)} - \frac{3W\mu_0 C_{Al}\xi^{10}}{I_D(\xi + \lambda)^2} \right] d\xi \\
&= WC_{Al} \left\{ \frac{3}{\mathcal{E}_c} \left[\sum_{i=1}^{6} \frac{(-1)^i}{i} \lambda^{6-i} \xi^i \Big|_{\xi_s}^{\xi_d} + \lambda^6 \ln(\xi + \lambda)|_{\xi_s}^{\xi_d} \right] + \right. \\
&\quad \frac{3W\mu_0 C_{Al}}{I_D} \left[\sum_{j=1}^{9} (-1)^j \lambda^{9-j} \left[\sum_{i=1}^{j} \frac{(-1)^i}{i} \lambda^{j-i} \xi^i \Big|_{\xi_s}^{\xi_d} + \right. \right. \\
&\quad \left. \left. \left. \lambda^j \ln(\xi + \lambda)|_{\xi_s}^{\xi_d} \right] + \frac{\lambda^{10}}{\xi + \lambda} \Big|_{\xi_s}^{\xi_d} \right] \right\}.
\end{aligned}
\tag{6.221}
$$

Substituting for I_D and using the limits,

$$
\begin{aligned}
Q_T &= WC_{Al} \left\{ \frac{3}{\mathcal{E}_c} \left[\sum_{i=1}^{6} \frac{(-1)^i}{i} \lambda^{6-i} \left(\xi_d{}^i - \xi_s{}^i \right) + \lambda^6 \ln \left(\frac{\xi_d + \lambda}{\xi_s + \lambda} \right) \right] + \right. \\
&\quad L \left[\sum_{i=1}^{6} \frac{(-1)^i}{i} \lambda^{6-i} \left(\xi_d{}^i - \xi_s{}^i \right) + \lambda^6 \ln \left(\frac{\xi_d + \lambda}{\xi_s + \lambda} \right) \right]^{-1} \times \\
&\quad \left\{ \sum_{j=1}^{9} (-1)^j \lambda^{9-j} \left[\sum_{i=1}^{j} \frac{(-1)^i}{i} \lambda^{j-i} \left(\xi_d{}^j - \xi_s{}^j \right) + \lambda^j \ln \left(\frac{\xi_d + \lambda}{\xi_s + \lambda} \right) \right] + \right. \\
&\quad \left. \left. \lambda^{10} \left(\frac{1}{\xi_d + \lambda} - \frac{1}{\xi_s + \lambda} \right) \right\} \right\}.
\end{aligned}
\tag{6.222}
$$

We may now determine the capacitances, since

$$
C_{gd} = \frac{1}{3\xi_d{}^2} \frac{\partial Q_T}{\partial \xi_d},
$$

$$
\text{and} \quad C_{gs} = \frac{1}{3\xi_s{}^2} \frac{\partial Q_T}{\partial \xi_s},
\tag{6.223}
$$

where

$$
\xi_s = (V_G - V_{T0})^{1/3}
$$

$$\text{and} \quad \xi_d = (V_G - V_{T0} - V_D)^{1/3}. \tag{6.224}$$

This will give us a rather complicated closed-form analytical solution. Here, noting that the magnitude of λ is sufficiently smaller than unity, and its higher powers even smaller, we only consider the contribution of the zero'th power in λ, i.e., we use the approximation for total charge Q_T,

$$Q_T = WC_{Al} \left\{ \frac{3}{\mathcal{E}_c} \left[\frac{1}{6} \left(\xi_d{}^6 \xi_s{}^6 \right) \right] + L \frac{2}{3} \frac{\xi_d{}^9 - \xi_s{}^9}{\xi_d{}^6 - \xi_s{}^6} \right\}, \tag{6.225}$$

giving

$$
\begin{aligned}
C_{gd} &= \frac{1}{3\xi_d{}^2} \frac{\partial Q_T}{\partial \xi_d} \bigg|_{\xi_s} \\
&= WC_{Al} \left[\frac{\xi_d{}^3}{\mathcal{E}_c} + \frac{2L}{3} \xi_d{}^3 \frac{\xi_d{}^3 + 2\xi_s{}^3}{\left(\xi_d{}^3 + \xi_s{}^3 \right)^2} \right]. \tag{6.226}
\end{aligned}
$$

As a function of voltages, ignoring parasitics so that $V_S = 0$, this yields

$$
\begin{aligned}
C_{gd} &= WC_{Al} \left[\frac{V_G - V_{T0} - V_D}{\mathcal{E}_c} + \right. \\
&\quad \left. \frac{L}{2} \frac{(V_G - V_{T0} - V_D) \left(V_G - V_{T0} - \frac{1}{3} V_D \right)}{\left(V_G - V_{T0} - \frac{1}{2} V_D \right)^2} \right]. \tag{6.227}
\end{aligned}
$$

Note that for large $\mathcal{E}_c$, e.g., in a constant mobility model, and for $V_D = 0$, the capacitance reduces to

$$C_{gd} = \frac{WC_{Al}L}{2}, \tag{6.228}$$

which is half of the total capacitance associated with the gate. Under conditions of no drain bias, half of the capacitance is associated with the source and half is associated with the gate, as expected. Elsewhere, some charge non-conservation effects occur due to the inadequacy of the analysis (see Problem 18).

An expression similar to the above, associated with the gate-to-source capacitance, is

$$
\begin{aligned}
C_{gs} &= \frac{1}{3\xi_s{}^2} \frac{\partial Q_T}{\partial \xi_s} \bigg|_{\xi_d} \\
&= WC_{Al} \left[\frac{\xi_s{}^3}{\mathcal{E}_c} + \frac{2L}{3} \xi_s{}^3 \frac{\xi_d{}^3 + 2\xi_s{}^3}{\left(\xi_d{}^3 + \xi_s{}^3 \right)^2} \right]
\end{aligned}
$$

$$= WC_{Al} \left[\frac{V_G - V_{T0} - V_D}{\mathcal{E}_c} + \frac{L}{2} \frac{(V_G - V_{T0})(V_G - V_{T0} - \frac{1}{3}V_D)}{\left(V_G - V_{T0} - \frac{1}{2}V_D\right)^2} \right],$$
$$(6.229)$$

which follows identical limits for $V_D = 0$ and $\mathcal{E}_c \to \infty$.

We now have the capacitive elements of the independent terms of our model. These characterize the channel charge storage relationship under quasi-static conditions. Let us now complete this elementary model.

6.8.3 Transconductance

We would like to characterize the device in terms of voltages V_{GS} and V_{DS}, by

$$I_D = I_D(V_{GS}, V_{DS}), \qquad (6.230)$$

and hence the small-signal transconductance is

$$g_m = \left. \frac{\partial I_D}{\partial V_{GS}} \right|_{V_{DS}}, \qquad (6.231)$$

and output conductance is

$$g_d = \left. \frac{\partial I_D}{\partial V_{DS}} \right|_{V_{GS}}. \qquad (6.232)$$

We will not determine these expressions, but they naturally follow from our derivation above in a similar manner as the capacitances. We have two capacitances to model the charge storage, one current source to model the output current, and one conductance to model the output conductance. All these are determinable in terms of V_{GS} and V_{DS}. The quasi-static equivalent circuit representing this is shown in Figure 7.45.

6.9 Quasi-Static Equivalent Circuit Refinements

Now consider refinements to the equivalent circuit model discussed. We have excluded the consequences of dipole capacitance in the channel in this analysis,[10] i.e., we have excluded the effects of the negative differential

---

[10]Should the major cause of this dipole be the negative differential mobility that was discussed in the treatment of MESFETs, its effect should be stronger, since the channel material is less doped in HFETs.

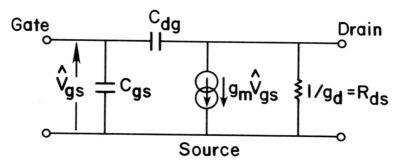

Figure 6.45: Elementary quasi-static small-signal equivalent circuit with inclusion of charge storage and output conductance effects; this model does not include input conductances, etc.

region of the velocity–field characteristics. We remarked earlier, in our discussion of the GaAs MESFET, that it leads to an accumulation and depletion region in the channel. Under quasi-static conditions, this region contributes a capacitive and a weak conductive element, which models the modulation of charge in the channel by the drain-to-channel bias occurring across the dipole region. If we model this channel contribution through a capacitance C_{dc}, it is charged from the source through the intrinsic channel resistance R_i. R_i also occurs in the charging path for the gate-to-source capacitance. This refinement leads to the quasi-static equivalent circuit shown in Figure 7.46.

Recall that the depletion and accumulation regions can move at most as rapidly as the movement of carriers. This occurs at approximately the saturated velocity ($\approx 1 \times 10^7$ cm.s^{-1}), so the current source can not respond instantaneously and has a phase delay of $\omega \tau_d$. The basis for this phase delay is similar to that of the MESFETs, it arises from the distributed transmission-line effects. Thus, the current source has a small-signal transconductance G_m given by

$$G_m = g_m \exp\left(-j\omega\tau_d\right). \qquad (6.233)$$

We should also include extrinsic resistances for the gate (R_g) due to gate metallurgy, etc., resistance from the source (R_s) due to ohmic contacts and parasitic source regions, etc., and resistance from the drain (R_d) due to ohmic contacts and parasitic drain regions, etc. There are also parasitic capacitances between the drain and the source electrode, which are dominated by the fringing capacitance through the semiconductor substrate. Our second-order extrinsic equivalent circuit model then appears as shown

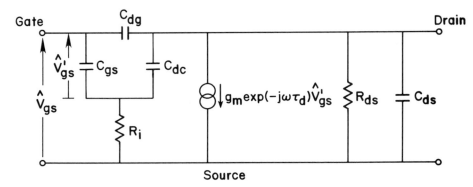

Figure 6.46: Quasi-static small-signal model with first-order refinements showing the inclusion of a capacitive element due to channel capacitance and the intrinsic resistance of the low-field region of the channel.

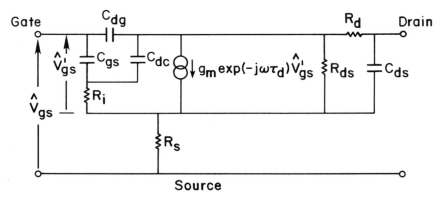

Figure 6.47: Second-order quasi-static small-signal model for HFETs incuding extrinsic resistances and the phase effects of current source.

in Figure 7.47.

In the related section for MESFETs, we had discussed the consequences of the existence of the dipole or the channel capacitance, should the parasitic capacitances and resistances be small. Most of those comments still hold for the HFET since the device equivalent circuit is quite similar. So, as gate lengths are shortened, if parasitics are extremely small, and provided that this method of quasi-static analysis and its equivalent circuit representation is valid at the frequencies where it is applied, a pole would occur at a resonant frequency, and following that the unilateral gain would have an increase in roll-off from 6 dB/octave to 12 dB/octave and the transistor would have a smaller f_{max} than predicted by low frequency measurements. Similar phenomena may occur in MESFETs, however their higher doping at short gate lengths makes the causes that lead to the dipole capacitance less likely.

There is an additional frequency-related delay effect that is of some concern in all fast FETs. This is due to gate transmission line effects and was discussed for MESFETs. Since the gate structure causes a distributed transmission of the signal along the gate width, it also introduces a phase delay in the signal leading to constructive and destructive build-up of signal along the drain line. One could analyze such a structure as a distributed connection of FETs with gradually increasing phase delay of the gate. Signals at the opposite end from the feed end have a larger delay. Usually, one strives to design devices so that the intrinsic device sets the frequency limit of the completed structure, and not the nature of the width of the device or the resistivity of the gate, etc., i.e., the propagation time constants, etc. In logic devices, where the widths are small, it is usually not a problem at μm sized dimensions. It can be a problem, however, at sub-0.5μm dimensions because of increased gate resistance per unit length.

6.10   Small-Signal Analysis

We have discussed the analysis and modelling of HFETs based on the quasi-static approximation, i.e., by invoking static solutions, and a static perturbation of the static solution to derive the response of the devices. This is likely to be inadequate, in spite of additions to allow the equivalent circuits to be valid at intermediate frequency ranges. In particular, this does call into question any attempt at predicting high-frequency effects within a decade in frequency of the limiting frequencies of the device, and the time scales corresponding to it. To adequately model this regime, we must resort to solutions of the small-signal equations. Unfortunately, even more so than the static equations, this involves approximations and idealizations. Such solutions are, however, quite instructive and quite adequate in some of the bias regions.

Our discussion of small-signal analysis of HFETs follows along similar lines as in the case of MESFETs.[11] First we consider a solution that can

---

[11]The small-signal analysis, which reformulates the current equation into a wave equation, has been applied extensively in non-quasi-static modelling of MOSFETs. Three references of particular interest are J. A. Geurst, "Calculation of High-Frequency Characteristics of Thin Film Transistors," *Solid-State Electronics*, **V8**, p. 88, 1965; J. A.

be found in a manner similar to the use of Weber's equation in MESFETs. The control equations for the transport, assuming unit gate width, are

$$I\left(z,t\right) = -q\mu C_{ins.}\left(V_g - V_T - V\right)\frac{\partial V}{\partial z}, \tag{6.234}$$

and

$$\frac{\partial I}{\partial z} = -\frac{\partial \rho}{\partial t}, \tag{6.235}$$

with

$$\rho\left(z,t\right) = -C_{ins}\left(V_g - V_T - V\left(z,t\right)\right). \tag{6.236}$$

As in the MESFET case, we make normalizations,

$$
\begin{aligned}
\xi &= \frac{z}{L}, \\
\kappa &= \frac{V_g - V_T - V}{V_{norm.}}, \\
\iota &= \frac{I}{I_0}, \\
\text{and} \quad \theta &= \frac{t}{t_0}, \tag{6.237}
\end{aligned}
$$

where

$$
\begin{aligned}
V_{norm.} &= \overline{V}_g - V_T, \\
I_0 &= \frac{\mu C_{ins} V_{norm.}^2}{2L^2}, \\
\text{and} \quad t_0 &= \frac{L^2}{\mu V_{norm.}}. \tag{6.238}
\end{aligned}
$$

The continuity equation, following these substitutions, is

$$\frac{1}{2}\frac{\partial^2 \kappa}{\partial \xi^2} = \frac{\partial \kappa}{\partial \theta}. \tag{6.239}$$

We resort to our standard approach to determining the solution by assuming the steady-state and small-signal variation to be given by

$$
\begin{aligned}
\kappa\left(\xi,\theta\right) &= \overline{\kappa}\left(\xi\right) + \tilde{\kappa}\left(\xi,\theta\right) = \overline{\kappa}\left(\xi\right) + \hat{\kappa}\exp\left(j\omega t_0\theta\right) \\
\text{and} \quad \iota\left(\xi,\theta\right) &= \overline{\iota}\left(\xi\right) + \tilde{\iota}\left(\xi,\theta\right) = \overline{\iota}\left(\xi\right) + \hat{\iota}\exp\left(j\omega t_0\theta\right). \tag{6.240}
\end{aligned}
$$

Van Nielen, "A Simple Accurate Approximation to the High-frequency Characteristics of IGFETs," *Solid-State Electronics*, **V12**, p. 826, 1969; and M. Bagheri, "An Improved MODFET Microwave Analysis," *IEEE Trans. on Electron Devices*, **ED-35**, p. 1147, July 1988.

The steady-state magnitudes of κ and ξ, in terms of parameters at the source end and the drain end, are

$$\bar{\iota} = \bar{\kappa}_d^2 - \bar{\kappa}_s^2$$

$$\text{and} \quad \xi = \frac{\bar{\kappa}^2 - \bar{\kappa}_s^2}{\bar{\kappa}_d^2 - \bar{\kappa}_s^2}. \tag{6.241}$$

Using these substitutions, the small-signal equations are

$$\frac{\partial}{\partial \bar{\kappa}} \left[\frac{1}{\bar{\kappa}} \frac{\partial}{\partial \bar{\kappa}} (\bar{\kappa} \tilde{\kappa}) \right] = \frac{4}{\bar{\iota}^2} \frac{\partial}{\partial \theta} (\bar{\kappa} \tilde{\kappa}), \tag{6.242}$$

and

$$\tilde{\iota} = \frac{\bar{\iota}}{\bar{\kappa}} \frac{\partial}{\partial \bar{\kappa}} (\bar{\kappa} \tilde{\kappa}). \tag{6.243}$$

We now substitute for the time dependence of the small-signal variation of

$$\tilde{\kappa}(\xi, \theta) = \hat{\kappa}(\xi, \omega_1) \exp(j\omega_1 \theta), \tag{6.244}$$

and

$$\tilde{\iota}(\xi, \theta) = \hat{\iota}(\xi, \omega_1) \exp(j\omega_1 \theta), \tag{6.245}$$

with $\omega_1 = \omega t_0$. Representing the common term in these equations by $\hat{v}$, i.e.,

$$\hat{v} = \bar{\kappa} \hat{\kappa}, \tag{6.246}$$

and using the substitution

$$z = j\bar{\kappa} \left(\frac{4\omega_1}{\bar{\iota}^2} \right)^{1/3}, \tag{6.247}$$

we can obtain the simplified equation

$$\frac{d}{d\hat{v}} \left(\frac{1}{z} \frac{d\hat{v}}{dz} \right) + \hat{v} = 0. \tag{6.248}$$

The solution of this equation has similarity with the solution of Stokes' equation,

$$\frac{d^2 \hat{v}}{dz^2} + z\hat{v} = 0. \tag{6.249}$$

The function $d\hat{v}/dz$ is the solution of our equation if $\hat{v}$ is the solution of Stokes' equation. Using the common-source configuration, the admittance parameters with the gate current phasor and drain current phasors are

$$\hat{I}_g = -\left. \hat{I} \right|_{z=0} + \left. \hat{I} \right|_{z=d} = y_{11}^s \hat{V}_g + y_{12}^s \hat{V}_d, \tag{6.250}$$

and

$$\hat{I}_d = y_{21}^s \hat{V}_g + y_{22}^s \hat{V}_d. \tag{6.251}$$

The admittance parameters follow from the above (see Problem 19) as

$$
y_{11}^s = \bar{g}_m \frac{|\varphi_s|^2}{2\Delta} \left(1 - \nu^2\right) \left[\begin{vmatrix} h_1(\varphi_s) & h_2(\varphi_s) \\ \dot{h}_1(\varphi_d) & \dot{h}_2(\varphi_d) \end{vmatrix} + \right.
$$
$$
\left. \nu \begin{vmatrix} h_1(\varphi_d) & h_2(\varphi_d) \\ \dot{h}_1(\varphi_s) & \dot{h}_2(\varphi_s) \end{vmatrix} - (1 + \nu) W(h_1, h_2) \right], \tag{6.252}
$$

$$
y_{12}^s = \bar{g}_m \frac{|\varphi_s|^2}{2\Delta} \left(1 - \nu^2\right) \nu \left[W(h_1, h_2) - \begin{vmatrix} h_1(\varphi_d) & h_2(\varphi_d) \\ \dot{h}_1(\varphi_s) & \dot{h}_2(\varphi_s) \end{vmatrix} \right], \tag{6.253}
$$

$$
y_{21}^s = \bar{g}_m \frac{|\varphi_s|^2}{2\Delta} \left(1 - \nu^2\right) \left[W(h_1, h_2) - \nu \begin{vmatrix} h_1(\varphi_d) & h_2(\varphi_d) \\ \dot{h}_1(\varphi_s) & \dot{h}_2(\varphi_s) \end{vmatrix} \right], \tag{6.254}
$$

and

$$
y_{22}^s = \bar{g}_m \frac{|\varphi_s|^2}{2\Delta} \left(1 - \nu^2\right) \nu \begin{vmatrix} h_1(\varphi_d) & h_2(\varphi_d) \\ \dot{h}_1(\varphi_s) & \dot{h}_2(\varphi_s) \end{vmatrix}. \tag{6.255}
$$

In these equations,

$$\bar{g}_m = \frac{\mu C_{ins}}{L} \left(\overline{V}_g - V_T\right), \tag{6.256}$$

$$\nu = \frac{\overline{V}_g - V_T - \overline{V}_d}{\overline{V}_g - V_T}, \tag{6.257}$$

$$\varphi_s = j \left[\frac{4\omega_1}{(1 - \nu^2)^2} \right]^{1/3}, \tag{6.258}$$

$$\varphi_d = j\nu \left[\frac{4\omega_1}{(1 - \nu^2)^2} \right]^{1/3}, \tag{6.259}$$

$$\Delta = \begin{vmatrix} \dot{h}_1(\varphi_s) & \dot{h}_2(\varphi_s) \\ \dot{h}_1(\varphi_d) & \dot{h}_2(\varphi_d) \end{vmatrix}, \tag{6.260}$$

and the Wronskian is

$$W(h_1, h_2) = \begin{vmatrix} h_1(\varphi_s) & h_2(\varphi_s) \\ \dot{h}_1(\varphi_d) & \dot{h}_2(\varphi_d) \end{vmatrix}. \tag{6.261}$$

In these equations h_1 and h_2 are Hankel functions,[12] which are the linearly independent solution of the homogeneous equation for this problem.

The control equation has been written in the constant mobility approximation. The control equation is suspect, therefore, in the region of saturation of the steady-state current. Following channel pinch-off, $\nu = 0$, and hence according to this formulation, both y_{12}, the feedback term, and y_{22}, the output term reduce to zero. This is a consequence of the steady-state analysis where the current following channel pinch-off is a constant. In actual practice, it would be different, and indeed we could find more accurate solution following a complicated analysis (see Problem 20). Our results are closed-form solutions. We could also attempt analysis in a series expansion form. This solution, written in terms of the known functions, is valid over a large frequency range, provided the other approximations made also for the steady-state equations are also valid.

Our general small-signal analysis, using series expansion in frequency and employed in Chapter 5 on MESFETs can be applied for the HFETs. The accuracy of the solution is limited now by the number of terms considered in the series expansion. We will consider it in the triode region of operation, and its similarity with the treatment of MESFETs will become clear, emphasizing the general framework of the small-signal treatment. The conductance at any position z in the channel is

$$G = \frac{\mu C_{ins.}}{L^2} \left(\overline{V}_G - V_T - \overline{V} \right), \tag{6.262}$$

where $\overline{V}$ is the channel voltage. In the triode region, the static drain current is

$$\overline{I}_d = \frac{\mu C_{ins}}{L} \left[\left(\overline{V}_g - V_T \right) \overline{V}_d - \frac{1}{2} \overline{V}_d^2 \right]. \tag{6.263}$$

Again, we employ normalizations for voltages,

$$\xi_s = \frac{\overline{V}_g - \overline{V}_s - V_T}{V_T}$$

$$\text{and} \quad \xi_d = \frac{\overline{V}_g - \overline{V}_d - V_T}{V_T}. \tag{6.264}$$

The characteristic frequencies in this analysis follow (see Problem 21) as

$$\frac{1}{\omega_0} = \frac{4}{15} \frac{L^2}{\mu V_T} \frac{\xi_s^5 - \xi_d^5 + 5\xi_s^2\xi_d^3 - 5\xi_s^3\xi_d^2}{\left(\xi_s^2 - \xi_d^2 \right)^3},$$

---

[12]See, e.g., M. Abramowitz and I. A. Stegun, *Handbook of Mathematical Functions with Formulas, Graphs, and Mathematical Tables*, U.S. Government Printing Office, Washington, D.C., p. 446, 1964. Hankel functions are Bessel functions of the third kind. Weber functions, used in the small-signal solution for MESFETs, are Bessel functions of the second kind.

$$\frac{1}{\omega_{11}} = \frac{4}{3}\frac{L^2}{\mu V_T}\frac{\xi_s^3 - \frac{3}{2}\xi_s^2\xi_d + \frac{1}{2}\xi_d^3}{\left(\xi_s^2 - \xi_d^2\right)^2},$$

$$\text{and} \quad \frac{1}{\omega_{21}} = \frac{2}{3}\frac{L^2}{\mu V_T}\frac{\xi_s^3 - 3\xi_s\xi_d^2 + 2\xi_d^3}{\left(\xi_s^2 - \xi_d^2\right)^2}. \tag{6.265}$$

At a bias point where channel pinch-off and current saturation occur, these give the characteristic frequencies as

$$\frac{1}{\omega_0} = \frac{4}{15}\frac{L^2}{\mu}\frac{1}{\overline{V}_g - \overline{V}_s - V_T},$$

$$\frac{1}{\omega_{11}} = \frac{4}{3}\frac{L^2}{\mu}\frac{1}{\overline{V}_g - \overline{V}_s - V_T},$$

$$\text{and} \quad \frac{1}{\omega_{12}} = \frac{2}{3}\frac{L^2}{\mu}\frac{1}{\overline{V}_g - \overline{V}_s - V_T}. \tag{6.266}$$

The first-order elements of the equivalent circuit now follow. For example, the gate capacitance is given (see Problem 22, which also considers the drain-to-gate capacitance) as

$$C_{gs} = \frac{2}{3}C_{ins}LW\frac{\xi_s\left(\xi_s^3 - 3\xi_s\xi_d^2 + 2\xi_d^3\right)}{\left(\xi_s^2 - \xi_d^2\right)^2}. \tag{6.267}$$

At current saturation, this capacitance is

$$C_{gs} = \frac{2}{3}C_{ins}LW, \tag{6.268}$$

consistent with the quasi-static analysis.

We will take a third alternate approach that uses the particular form of the current control equation to obtain a more accurate solution valid to higher frequencies. The transmission-line equations for this problem, with C_{ins} as the capacitance per unit area of the gate dielectric, are of the form

$$I = \mu W C_{ins}\left(V_g - V_T - V\right)\frac{\partial V}{\partial z}$$

$$\text{and} \quad \frac{\partial I}{\partial z} = W C_{ins}\frac{\partial V}{\partial t}. \tag{6.269}$$

Separating the steady-state and time-dependent terms for current and voltage,

$$I(z,t) = \overline{I}(z) + \hat{I}(z,\omega)\exp(j\omega t)$$

$$\text{and} \quad V(z,t) = \overline{V}(z) + \hat{V}(z,\omega)\exp(j\omega t). \tag{6.270}$$

Substituting in the transmission line equations, we obtain the sets of equations for steady-state and small-signal conditions. The steady-state equations are

$$\overline{I}(z) \;=\; \mu W C_{ins} \left(\overline{V}_g - V_T - \overline{V}\right) \frac{\partial \left(\overline{V}_g - V_T - \overline{V}\right)}{\partial z},$$

$$\text{and} \quad \frac{\partial \overline{I}}{\partial z} \;=\; 0, \tag{6.271}$$

and the small-signal equations are

$$\hat{I}(z,\omega) \;=\; \mu W C_{ins} \left[\left(\overline{V}_g - V_T - \overline{V}\right) \frac{\partial \left(\hat{V}_g - \hat{V}\right)}{\partial t} + \right.$$

$$\left. \left(\hat{V}_g - \hat{V}\right) \frac{\partial \left(\overline{V}_g - V_T - \overline{V}\right)}{\partial z} \right]$$

$$=\; \mu W C_{ins} \frac{\partial \left(\overline{V}_g - V_T - \overline{V}\right)\left(\hat{V}_g - \hat{V}\right)}{\partial z},$$

$$\text{and} \quad \frac{\partial \hat{I}}{\partial z} \;=\; j\omega W C_{ins.} \left(\hat{V}_g - \hat{V}\right). \tag{6.272}$$

For the steady-state solution, the boundary conditions for the problem are

$$\overline{V}(z = 0) \;=\; \overline{V}_s,$$
$$\overline{I}(z = 0) \;=\; \overline{I}_s,$$
$$\overline{V}(z = L) \;=\; \overline{V}_d,$$
$$\text{and} \quad \overline{I}(z = L) \;=\; -\overline{I}_d. \tag{6.273}$$

Note that, as in the case of MESFETs, the drain current is treated as positive when entering the port. The small-signal boundary conditions are similar, and given by

$$\hat{V}(z = 0) \;=\; \hat{V}_s,$$
$$\hat{I}(z = 0) \;=\; \hat{I}_s,$$
$$\hat{V}(z = L) \;=\; \hat{V}_d,$$
$$\text{and} \quad \hat{I}(z = L) \;=\; -\hat{I}_d. \tag{6.274}$$

The steady-state solution, due to continuity of steady-state current, is straightforward following integration and the use of the boundary conditions. Integration of the steady-state current equation results in

$$\frac{1}{2}\left(\overline{V}_g - V_T - \overline{V}\right)^2 = \frac{\overline{I}}{\mu W C_{ins}} z + \mathcal{A},\tag{6.275}$$

where the constant of integration $\mathcal{A}$ follows from the source boundary condition as

$$\mathcal{A} = \frac{1}{2}\left(\overline{V}_g - V_T - \overline{V}_s\right)^2.\tag{6.276}$$

The current at $z = L$, $-\overline{I}_d$, follows from

$$\begin{aligned}\overline{I}_d &= -\frac{\mu W C_{ins}}{L}\left[\frac{1}{2}\left(\overline{V}_g - V_T - \overline{V}_d\right)^2 - \frac{1}{2}\left(\overline{V}_g - V_T - \overline{V}_s\right)^2\right]\\ &= -\frac{\mu W C_{ins}}{L}\left[\frac{1}{2}\overline{V}_d^{\,2} - \frac{1}{2}\overline{V}_s^{\,2} - \left(\overline{V}_g - V_T\right)\left(\overline{V}_d - \overline{V}_s\right)\right],\end{aligned}\tag{6.277}$$

the triode equation where the potential reference is arbitrary.

In order to ease the appearances of the grouped terms, we make the following substitutions:

$$\begin{aligned}\overline{\Upsilon}(z) &= \overline{V}_g - V_T - \overline{V}(z)\\ \text{and}\quad \hat{\Upsilon}(z,\omega) &= \hat{V}_g - \hat{V}(z,\omega),\end{aligned}\tag{6.278}$$

and consequently the control equations are

$$\begin{aligned}\hat{I} &= \mu W C_{ins}\frac{\partial}{\partial z}\left[\overline{\Upsilon}\hat{\Upsilon}\right],\\ \text{and}\quad \frac{\partial \hat{I}}{\partial z} &= j\omega W C_{ins}\hat{\Upsilon}.\end{aligned}\tag{6.279}$$

The integral forms of these equations can be written as

$$\hat{\Upsilon}(z,\omega) = \frac{1}{\overline{\Upsilon}(z,\omega)}\left[\overline{\Upsilon}(L)\hat{\Upsilon}(L,\omega) - \frac{1}{\mu W C_{ins}}\int_z^L \hat{I}dz'\right],\tag{6.280}$$

and

$$\hat{I}(z,\omega) = \hat{I}(L,\omega) - j\omega W C_{ins}\int_z^L \hat{\Upsilon}(z',\omega)dz'.\tag{6.281}$$

We can obtain an expanded series solution to the problem using an iterative solution of the above. The steady-state solution allows us to write the potential term $\overline{\Upsilon}$ along the channel in terms of the current and the position following the treatment as before. The steady-state current in the channel is given by

$$\overline{I} = \frac{\mu W C_{ins}}{2L} \left[\overline{\Upsilon}^2(0) - \overline{\Upsilon}^2(L) \right] \tag{6.282}$$

in terms of the potentials, and the potential as a function of position is given by

$$\overline{\Upsilon}(z) = \left[\overline{\Upsilon}^2(0) - \frac{2\overline{I}z}{\mu W C_{ins}} \right]^{1/2}, \tag{6.283}$$

which may be more simply written as

$$\overline{\Upsilon}(z) = \overline{\Upsilon}(0) \left\{ 1 - \left[1 - \frac{\overline{\Upsilon}^2(L)}{\overline{\Upsilon}^2(0)} \right] \frac{z}{L} \right\}^{1/2}. \tag{6.284}$$

Note that when the current saturates and the channel pinches off, $\overline{\Upsilon}(L) = 0$; this is the largest bias for which this analysis is valid.

An iterative procedure allows us to determine both $\hat{I}(z,\omega)$ and $\hat{\Upsilon}(z,\omega)$ to increasing accuracy. Our equations allow us to determine the sinusoidal current in the channel as a function of position. So, either the source or the drain current is known in terms of the other, the necessary requirement for calculating the y-parameters. Also, since the gate, a highly conducting region, has an equi-potential across it of $\overline{V}_g + \hat{V}_g \exp(j\omega t)$, we may determine the gate current by integrating the current through the capacitor as a function of position,

$$\hat{I}_g = j\omega W C_{ins} \int_0^L \left(\hat{V}_g - \hat{V}(z) \right) dz, \tag{6.285}$$

i.e.,

$$\hat{I}_g = j\omega W C_{ins} \int_0^L \hat{\Upsilon}(z,\omega) dz. \tag{6.286}$$

We now use the iterative procedure to show how it may be applied to obtain an accurate solution of the problem. First consider the zero'th-order approximation of the current phasor as being independent of position. This is true in the limit of very low frequency; current continuity implies this for the steady-state term. So we first substitute $\hat{I} = \hat{I}(L,\omega)$ in the equation for potential (Equation 7.280). This leads to

$$\hat{\Upsilon}(z,\omega) = \frac{1}{\overline{\Upsilon}(z,\omega)} \left[\overline{\Upsilon}(L)\hat{\Upsilon}(L,\omega) - \frac{1}{\mu W C_{ins}} \hat{I}(L,\omega)(L-z) \right] \tag{6.287}$$

as the zero'th-order equation for the phasor for potential.

Knowing this, we may substitute back into the current equation (Equation 7.281) to obtain the first-order current phasor solution. The general form is

$$\hat{I}(z,\omega) = \hat{I}(L,\omega) - j\omega W C_{ins} \times$$
$$\left[\overline{\Upsilon}(L)\hat{\Upsilon}(L,\omega) \int_z^L \frac{dz'}{\overline{\Upsilon}(z')} - \frac{\hat{I}(L,\omega)}{\mu W C_{ins}} \int_z^L \frac{L-z}{\overline{\Upsilon}(z')} dz' \right].$$

$$(6.288)$$

This lengthy integral can be evaluated and the iterative process continued to allow us to determine the position dependence of the phasor of the channel current and the potential. Knowing the currents and potentials to the requisite accuracy, the y-parameters will follow, since the drain current is $\hat{I}_d = -\hat{I}(L,\omega)$, the source current is $\hat{I}_s = \hat{I}(0,\omega)$, and the gate current is given by the integral equation above. The calculation is lengthy, although a more simple form exists when the channel is pinched off at position $z = L$. Here $\overline{\Upsilon}(z = L) = 0$, and

$$\overline{\Upsilon}(z) = \overline{\Upsilon}(0) \frac{(L-z)^{1/2}}{L^{1/2}}, \qquad (6.289)$$

and hence a simplified result is obtained for the phasor of the potential as

$$\hat{\Upsilon}(z,\omega) = -\frac{L^{1/2}\hat{I}(L,\omega)}{\mu W C_{ins}\overline{\Upsilon}(0)}(L-z)^{1/2}. \qquad (6.290)$$

Iterating with Equation 7.281, we now obtain the first-order approximation of the phasor of the channel current as

$$\hat{I}(z,\omega) = \hat{I}(L,\omega) - \frac{j\omega L^{1/2}\hat{I}(L,0)}{\mu\overline{\Upsilon}(0)} \int_z^L \left(L-z'\right)^{1/2} dz'$$
$$= \hat{I}(L,\omega) \left[1 - \left(j\frac{\omega}{\omega_0}\right) \frac{2}{3} \frac{(L-z)^{3/2}}{L^{3/2}} \right], \qquad (6.291)$$

where

$$\frac{1}{\omega_0} = \frac{L^2}{\mu\overline{\Upsilon}(0)}. \qquad (6.292)$$

From this, one may again derive the potential phasor as

$$\hat{\Upsilon}(z,\omega) = -\frac{\hat{I}(L,\omega)}{\mu C_{ins}W\overline{\Upsilon}(0)} \left[L^{1/2}(L-z)^{1/2} + j\frac{4}{15}\frac{\omega}{\omega_0}\frac{(L-z)^2}{L} \right], \qquad (6.293)$$

and hence, the second-order expression for current as

$$\hat{I}(z,\omega) = \hat{I}(L,\omega)\left[1 + \left(j\frac{\omega}{\omega_0}\right)\frac{2}{3}\frac{(L-z)^{3/2}}{L^{3/2}} + \left(j\frac{\omega}{\omega_0}\right)^2\frac{4}{45}\frac{(L-z)^3}{L^3}\right].$$

(6.294)

When this expression is evaluated at $z = 0$, we obtain the source current phasor as a function of drain current phasor $-\hat{I}(L,\omega)$. This is only a function of gate-to-source potential. So,

$$y_{11} = y_{21}\left[1 + \left(j\frac{\omega}{\omega_0}\right)\frac{2}{3} + \left(j\frac{\omega}{\omega_0}\right)^2\frac{4}{45} + \cdots\right]$$

(6.295)

What we have shown is a general method that lets us determine the y-parameters to the desired accuracy. This procedure is actually functionally quite similar to that described generally in the MESFET section—indeed, one can see the fractional forms of the terms to be quite similar.

We end by summarizing, without proof, the general forms of the y-parameters for below and at pinch-off conditions. In the common-source configuration, the drain current and the gate current can be expressed as

$$\hat{I}_d = y_{dg}\hat{V}_{gs} + y_{dd}\hat{V}_{ds}$$

and

$$\hat{I}_g = y_{gg}\hat{V}_{gs} + y_{gd}\hat{V}_{ds}.$$

(6.296)

The expressions are

$$y_{gg} = \frac{j\omega}{\Delta_\omega}\frac{2}{3}WLC_{ins}\frac{6\alpha + (1-\alpha)^2}{(1+\alpha)^2}\left[1 + j\frac{\omega}{\omega_0}\frac{15\alpha + 2(1-\alpha)^2}{15(1+\alpha)\left(6\alpha + (1-\alpha)^2\right)}\right],$$

(6.297)

$$y_{gd} = -\frac{j\omega}{\Delta_\omega}\frac{2}{3}WLC_{ins}\frac{\alpha(2+\alpha)}{(1+\alpha)^2}\left[1 + j\frac{\omega}{\omega_0}\frac{3 + 12\alpha + 2(1-\alpha)^2}{15(1+\alpha)^2(2+\alpha)}\right],$$

(6.298)

$$y_{dg} = \frac{G_m}{\Delta_\omega} + y_{gd},$$

(6.299)

and

$$y_{dd} = \frac{G_d}{\Delta_\omega} - y_{gd},$$

(6.300)

where

$$G_m = \frac{\mu W C_{ins}\overline{V}_{ds}}{L},$$

(6.301)

$$G_d = \frac{\mu W C_{ins}\left(\overline{V}_{gs} - V_T - \overline{V}_{ds}\right)}{L},$$

(6.302)

and

$$\Delta_\omega = 1 + j\frac{\omega}{\omega_0}\frac{4}{15}\frac{5\alpha + (1-\alpha)^2}{(1+\alpha)^3} + \left(j\frac{\omega}{\omega_0}\right)^2\frac{1}{45}\frac{6\alpha + (1-\alpha)^2}{(1+\alpha)^4}.$$

(6.303)

In these expressions,

$$\alpha = \frac{\overline{V}_{gs} - \overline{V}_T - \overline{V}_{ds}}{\overline{V}_{gs} - \overline{V}_T}.$$

(6.304)

It varies from 1 to 0, with the latter value at channel pinch-off, i.e., $\overline{V}_{ds} = \overline{V}_{gs} - V_T$.

Note the low frequency limits of these expressions, as the case was with our treatment of MESFETs. They reduce to the quasi-static expressions for the triode case in the low frequency limit. Based on this relatively more accurate representation, we may derive an equivalent circuit that is relatively more accurate, as we had for intermediate frequencies in the treatment of MESFETs. Note the frequency dependence of the current source, which has a phase-lag term included in it due to the transmission-line nature of the charging of the capacitance in the channel.

6.11 Transient Analysis

In this section, we will make some qualitative observations on the transient behavior of the device. In order to obtain transient response, one usually has to resort to numerical techniques since the network response is complicated and can not be placed in the simple forms that are possible for frequency domain. If a response could be obtained in frequency domain for the input and output matching conditions of interest, an inverse Fourier transform gives us the time-dependent response. The behavior is quite analogous to the small-signal behavior. A gate turn-on pulse, e.g., causes carriers from the gate to be injected into the channel, and initially, these carriers supply the displacement current through the depleted control region of the gate. As the carrier build-up occurs in the channel, a build-up that has a time delay similar to the phase delay of the current-source in the small-signal analysis, the drain current begins to build up.

Figure 7.48 shows such a transient response with bias conditions similar to those applied in our discussion of transients in MESFETs. The initial delay corresponds to the time period required for the carriers to reach the

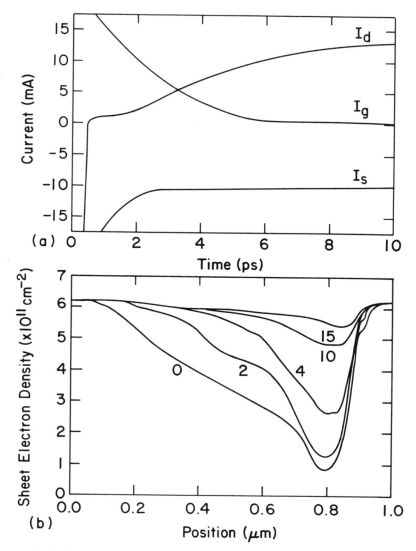

Figure 6.48: Transient of the drain current, the gate current, and the source current is shown in (a) in a 0.5 μm gate length GaAs MISFET when the gate is rapidly forward biased by 0.25 V from near the threshold conditions. (b) shows the sheet electron density in the channel at various instances of time in pico-seconds. This figure should be compared with the response of GaAs MESFET discussed in Chapter 5.

channel pinch-off region. This time period is the time period related to the charging of the channel region; the channel current is largely displacement current in the gate control region. Once carriers reach the pinch-off region, the carrier concentration there begins to change, and the drain current begins to increase. This leads to further delay, which is mostly related to the time constants associated with the gate-to-drain and the drain-to-source capacitances.

6.12   Hot Carrier Injection Effects

We now discuss some important issues related to the use of HFETs.[13] One particularly important limitation to the use of the structures comes about from the channel hot electron problem. As carriers pick up energy as they move along the channel under the influence of the drain bias, some acquire sufficient energy to emit over the heterostructure barrier into the large bandgap material,[14] as shown in Figure 7.49. So, the same carriers whose larger velocity results in the faster response of the device also cause an increase of injection into other regions of the device. Also, because this barrier to a doped-barrier region is very thin, usually limited by the spacer thickness used, some of these carriers can actually tunnel through. Hot carriers occur in the high-field region in this structure; they are injected cold from the source, hence, the injection becomes most pronounced near the drain end. This injection occurs both in the normal and the inverted structures, the implications in the inverted structure being stronger because a convenient path for collection of the injected carriers may not exist due to the conduction band discontinuity. The injected hot carriers in an inverted doped-barrier HFET thermalize in the well, where they then transport towards the drain in the larger gap material. In the normal structures, the hot carriers can be collected at the ohmic contacts. Note that this occurs because, following injection into the larger bandgap region, some scattering occurs, and so long as the electron energy is lower than the

---

[13]Our discussion here is quite general; the injection problem is of importance to both undoped-barrier and doped-barrier HFETs. Their implications, particularly those resulting from capture of carriers at the anomalous DX center, are strong for the doped-barrier devices.

[14]For an extended discussion of this, see D. J. Frank, P. M. Solomon, D. C. La Tulipe, Jr., H. Baratte, C. M. Knoedler, and S. L. Wright, "Excess Gate Current Due to Hot Electrons in GaAs-Gate FETs," *High-Speed Electronics: Basic Physical Phenomena and Device Principles*, Proc. of the International Conference, Stockholm, Sweden, Springer Series in Electronics and Photonics, **22**, Springer-Verlag, Berlin, p. 709 (1986), and M. S. Shur, D. K. Arch, R. R. Daniels, and J. K. Abrokwah, "New Negative Resistance Regime of Heterostructure Insulated Gate Transistor (HIGFET)," *IEEE Electron Device Letters*, **EDL-7**, p. 78, 1986.

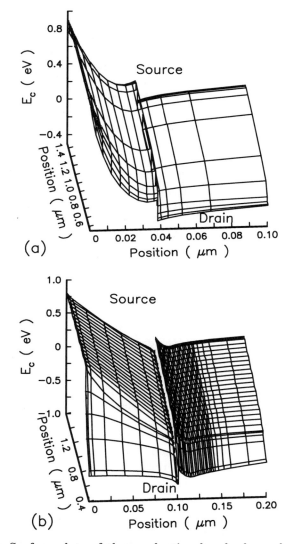

Figure 6.49: Surface plots of the conduction band edges of a normal (a) and inverted (b) doped-barrier HFET under 0.8 V bias on the drain and similar sheet electron density in the channel. In the pinch-off region, the conduction band edge changes rapidly, leaving a higher hot carrier tail with an energy larger than the barrier band edge. These carriers can emit over the barrier into the large bandgap semiconductor.

metal–semiconductor barrier height, the carriers are much less likely to end up in the gate. However, if sufficiently large drain potential and gate potential are applied, electrons suffering very few collisions in the large gap material will have sufficient energy to transport into the gate region. At this point the gate current also begins to rise rapidly, beyond the usual generation–recombination and diffusion currents.

In the case of a doped-barrier HFET, therefore, there is a problem of hot carrier injection into the larger gap material, and these injected carriers transport into the drain. Later, we discuss the trapping of these carriers in DX centers because of the anomalous nature of the behavior of donors in $Ga_{1-x}Al_xAs$.

If the larger gap material is undoped, as in the SISFET or MISFET structures, the problem of hot carrier injection is much more serious, because then there is an accelerating field for these carriers that aids their transit to the gate (see Figure 7.50), there being no metal barrier preventing the flow of these carriers. We discuss this effect for both a metal gate and a semiconductor gate structure employing the undoped large bandgap barrier in the $Ga_{1-x}Al_xAs/GaAs$ system. Examples of the output characteristics of both these devices are shown in Figure 7.51. There is a negative resistance behavior in the drain current–voltage characteristics that occurs at high gate voltages and moderate drain voltages. A high gate voltage implies that the gate is being pulled down substantially in energy with respect to the channel, and with the drain voltage being moderate, the gate region has a lower energy than the drain region. Towards the source end, most carriers are relatively cold, and the barrier prevents any significant injection into the barrier material. Let $q\varphi$ be the difference between the conduction band edge at the barrier in the large gap material and the Fermi energy. Along the channel, as carriers pick up energy, they occupy higher and higher energies in the Γ-related subbands until they can pick up enough energy to scatter into the L-related subbands or, since at this energy the quantization effects are minimal, the classical L band. This requires, at the minimum, ($\approx 0.36 - \xi_f/q$) of channel potential for GaAs if the initial carrier densities are non-degenerate. At a drain bias $V = 0.36$ V, the L band is being populated, and at $V = \varphi$ some carriers also may have enough energy to surmount the hetero-barrier. At a channel potential of $V = V_G - V_T$ larger than 0.36 V, the L band is significantly more populated, and many of these carriers can cross over into the gate. If the current contribution is strong enough, i.e., if the potential distribution and the discontinuity are appropriate (this is a function of gate length and material parameters) then I_G may rise rapidly and I_D may decrease, leading to the negative resistance. Figure 7.52 shows the behavior of gate current togther with that of drain current as a function of both gate voltage and drain voltage for a SISFET,

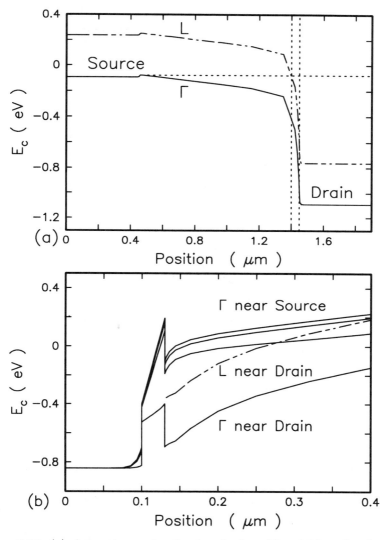

Figure 6.50: (a) shows the conduction band edges (Γ and L) at the channel interface of a GaAs/Ga$_{1-x}$Al$_x$As/GaAs SISFET employing an AlAs mole-fraction of 0.5 and biased at gate voltage of 0.8 V and drain voltage of 1.0 V. (b) shows the band edges through vertical cross-sections in the device. Cross-sections at the source end, in the channel, and at the drain end are shown.

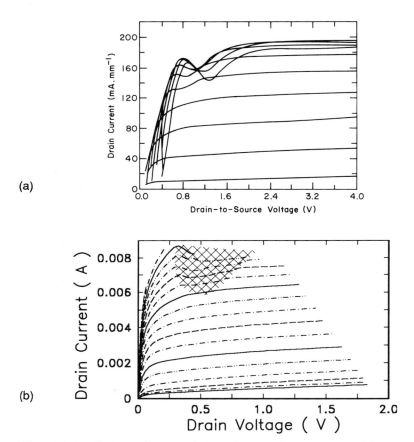

Figure 6.51: Output characteristics of a metal-gate undoped-barrier HFET (a) and a GaAs gate SISFET (b) showing regions of negative resistance in the output. After M. S. Shur, D. K. Arch, R. R. Daniels, and J. K. Abrokwah, "New Negative Resistance Regime of Heterostructure Insulated Gate Transistor (HIGFET)," *IEEE Electron Device Letters*, **EDL-7**, p. 78, 1986., and D. J. Frank, P. M. Solomon, D. C. La Tulipe, Jr., H. Baratte, C. M. Knoedler, and S. L. Wright, "Excess Gate Current Due to Hot Electrons in GaAs-Gate FETs," *High-Speed Electronics: Basic Physical Phenomena and Device Principles*, Proc. of the International Conference, Stockholm, Sweden, Springer Series in Electronics and Photonics, **22**, Springer-Verlag, Berlin (1986).

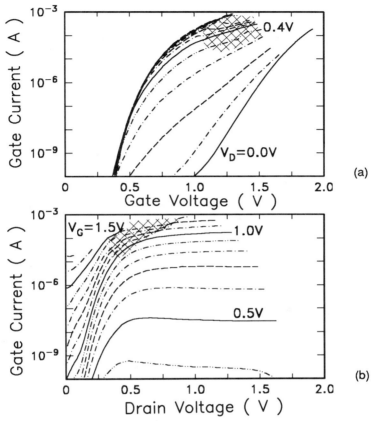

Figure 6.52: Gate and drain currents as a function of gate and drain voltages in the GaAs SISFET emphasizing their behavior in the region of negative output resistance. After D. J. Frank, P. M. Solomon, D. C. La Tulipe, Jr., H. Baratte, C. M. Knoedler, and S. L. Wright, "Excess Gate Current Due to Hot Electrons in GaAs-Gate FETs," *High-Speed Electronics: Basic Physical Phenomena and Device Principles*, Proc. of the International Conference, Stockholm, Sweden, Springer Series in Electronics and Photonics, **22**, Springer-Verlag, Berlin (1986).

confirming the origin of the negative resistance.

At large gate and large drain biases, however, the drain field may become large enough to divert the electron flux from the gate to the drain. Only in a smaller region of the device does the injected hot carrier flux go largely to the gate. So, the negative resistance may disappear at large drain biases, as is seen in the characteristics shown in Figure 7.51. The phenomenon is more serious in short channel devices where the injection portion (only a fraction of the gate length) is more significant because of the relative increase in the drain field–dominated portion. Parasitic resistances also play an important role in this phenomenon because they are important to the field distribution in the channel and in the barrier layer. Large resistances suppress the effect because they lead to less channel potential drop.

6.13   Effects Due to DX Centers

Even though hot electrons may not give rise to negative resistance because of the retarding field and higher gate barrier in doped-barrier HFETs, their injection into materials like $Ga_{1-x}Al_xAs$, materials which exhibit large trapping effects in conjunction with the n-type doping, leads to significant changes in device characteristics that are both bias- and time-dependent. These trapping centers, deep donor centers associated with the n-type doping, have been called DX centers for historical reasons. The origin of these centers is still subject to debate. They exhibit differing thermal emission and capture energies, large optical ionization energy, and multiple time constants of emission in pulsed capacitance measurements. We will discuss possible origins of the centers later in this section. From a device perspective, their effect is strong and well characterized, since it results directly from the capture and emission processes, which can be evaluated experimentally as a function of the AlAs mole-fraction.

The reason for the time- and bias-dependent changes of the device characteristics is that the carriers injected into $Ga_{1-x}Al_xAs$ (of moderate or high AlAs mole-fraction) get trapped into DX centers, the deep levels. This change of ionization state of the donors leads to a depletion of carriers at the drain end, the high-field region, and, hence, a collapse of the current–voltage characteristics due to a larger resistance. The effect is time-dependent since the carriers can de-trap on time scales that can range from μs to greatly exceeding seconds depending on the temperature. The effect is also dependent on other excitation processes present. For example, Figure 7.53 shows the effect in the dark and compares the same effect with illumination from a light source near the bandgap energy. Shining light recovers characteristics to a normal FET behavior. The trapping still takes place during the operation of the device, but the DX centers emit electrons very rapidly in the presence of light whose energy exceeds the optical ionization threshold energy. In the dark, because donors have captured electrons available due to electron injection, there are fewer carriers and hence smaller currents and

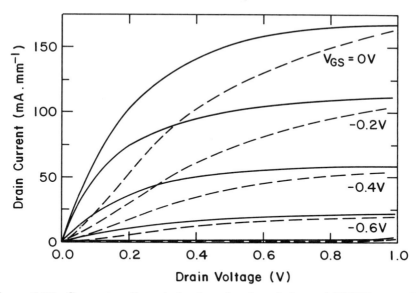

Figure 6.53: Current–voltage behavior of a doped channel HFET at 77 K in the dark (dashed lines) and with light (solid lines) from a tungsten lamp.

larger resistances. This is the origin of the collapse in current–voltage characteristics that is shown in the dark in Figure 7.53. Note that such emission and capture cause large-scale shifts in the threshold voltage. Since injection of electrons into $Ga_{1-x}Al_xAs$ is a function of the $Ga_{1-x}Al_xAs$ band bending and bias, etc., i.e., the complete parameter space of thickness, doping, metal barrier height, gate length, voltage, etc., the magnitude of the effect depends on both the material and bias parameters of the device and the geometry of the device. It is temperature-dependent because the characteristics of the DX center, in particular the process of emission of electrons after they are captured, are strongly temperature-dependent. Additionally, these characteristics are a function of the AlAs mole-fraction and, secondarily, the dopant species that gives rise to the DX center.

The effects on device behavior due to capture and emission in the DX center region are also a function of the way the biases are applied and hence the carriers injected because these determine the locale of the capture of carriers in the large bandgap material. Consider a metal–n-$Ga_{1-x}Al_xAs$–GaAs capacitor structure. A large thickness of the doped $Ga_{1-x}Al_xAs$ region, under conditions of large sheet charge density in the channel, has a stronger effect than a thinner doped $Ga_{1-x}Al_xAs$ region because of the thicker region over which electron capture can occur. Thus, thicker $Ga_{1-x}Al_xAs$ or other conditions leading to large sheet densities in the channel, such as a large forward bias, lead to a stronger effect from the electron capture process in $Ga_{1-x}Al_xAs$. The emission of carriers from the DX center becomes very inefficient with the lowering of temperature, the emission being a barrier-limited process. If electrons exist in the $Ga_{1-x}Al_xAs$ during the cooling process, they freeze into the DX center, leading to a lower threshold voltage of a device and a larger threshold voltage shift between room temperature and liquid nitrogen temperature. A smaller thickness, or equivalently a more enhancement mode–like device, has a smaller threshold shift because fewer carriers occur in the $Ga_{1-x}Al_xAs$. Thus, a smaller $Ga_{1-x}Al_xAs$ thickness sample shows a smaller threshold shift than a larger thickness structure, i.e., it shows a less positive shift of threshold voltage. So, both the gate and the drain biases lead to injection of electrons into the larger bandgap region and hence to changes in device characteristics. These changes may be viewed as a V_T shift of a device while it is going through a switching operation itself, and hence have an effect on the dynamic noise margin of the device.

We will look at this DX center simplistically—it continues to be a subject of investigation,[15] and many models exist that may fit many of the experimental observations made to date. The experimental observations of the properties of these centers include the following: Hall measurements indicate that the DX center population is proportional to the dopant concentration, which is mostly DX center–like at mole-fractions of AlAs in excess of ≈ 0.2. The DX center is associated with all substitutional donors used in $Ga_{1-x}Al_xAs$, e.g., Te, Sn, Si, and Se, and is independent of material growth techniques. It has a large thermal capture energy which is mole-fraction-dependent and varies with the dopant species, and an even larger thermal emission energy which is weakly mole-fraction-dependent. The photo-ionization energy is even larger than either of these. The thermal activation energy, measured using dependence of carrier concentration on temperature via steady-state Hall measurements, indicates the smallest activation energy of all. Pulsed capacitance measurements indicate three or four time constants associated with the emission of the carrier from the captured state.

One possible model explaining this behavior follows. The capture of electrons by a substitutional isolated donor (a hydrogenic state) leads to

---

[15]T. N. Theis and P. M. Mooney, "The DX Center: Evidence for Charge Capture via an Excited Intermediate State," *Mat. Res. Soc. Symp. Proc.*, **163**, p. 729, 1990 discusses the related arguments at length. Also, a discussion of the structural model for DX centers can be found in D. J. Chadi and S. B. Zhang, "Atomic Structure of DX Centers: Theory," *J. of Electronic Materials*, **20**, No. 1, p. 55, 1991.

local deformation of the lattice and lowering of the total energy so that
the changed configuration has a level in the energy gap for mole-fractions
greater than ≈ 0.2. This change occurs via an intermediate state. Most
likely, the intermediate state involves capture of one electron, and the DX
state state involves trapping of two electrons. This intermediate state is
also a localized substitutional state with an energy that follows the DX level
energies. Figure 7.54 shows a configuration coordinate energy diagram[16]
consistent with the observations of these properties and the above model
description. The corresponding capture process is shown schematically in
this figure. This process has an activation energy equal to the difference
between the thermal energy of the mobile electron and the band edge energy
of the intermediate state. The electron becomes localized following the
capture and has a lower state of energy. Capturing itself, however, requires
a minimum energy for the electron before the process can occur. The
capture and emission in a thermal process proceed via this intermediate
state, and a repulsive barrier exists for both of the processes.

The origin of this interesting but deleterious behavior apparently lies in
the distortion and associated change in lattice energy, the requirement that
the electron capture or emission process occur via an intermediate state,
and because the DX center contains two electrons. The lattice distortion
behavior depends on the specific donor, and the intermediate state corresponding to the distortion following capture of one electron has an energy
relationship that depends on both the mole-fraction of AlAs as well as the
specific dopant. This is a process where the energy associated with lattice
relaxation is of similar magnitude as the binding energy of the purely electronic part of the defect Hamiltonian. The electron at the defect cannot be
looked at as a screened hydrogen-like atom, as we do in the effective mass
theory, but has to include the energy associated with lattice relaxation.
An appropriate way to depict the energy of the electron, de-localized in
the conduction band, or localized at the defect, is to show the electronic
and the defect distortion energy as a function of the defect configuration
coordinate, which was considered in Figure 7.54.

So, an electron in the intermediate state has to have a minimum energy
before it can be localized by the defect. The intermediate state may also be
a localized state and not an extended state of conduction band or hydrogenic level of donors. When localized, the energy drops into the bandgap,

---

[16]Configuration coordinate energy diagram is a plot of the electron energy together
with the lattice energy. Normally, for hydrogenic dopants there is no change in lattice
energy due to capture or emission of electrons. However, if it is significant, and substitutional donors in $Ga_{1-x}Al_xAs$ are believed to have an associated distortion of the lattice
and a change in lattice energy, then the total energy does change. This is represented in
the configuration coordinate energy diagram.

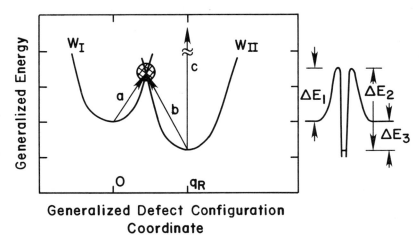

Generalized Defect Configuration Coordinate

Figure 6.54: An energy model of the DX center based on the observed activation energies for various processes. An electron capture process occurs through an intermediate state as shown schematically in transition a. The intermediate state is shown as a shaded region. It is not hydrogenic and not at the intersection of W_I and W_{II}. When an electron is captured at the center, lattice distortion results in the band W_{II} for total energy. A thermal emission requires the process b, and photo-ionization occurs via the process c. An associated energy barrier schematic between the two states is also shown. The energies for thermal capture ΔE_1, thermal emission ΔE_2, and dopant activation (ΔE_3, from Hall measurements) are also shown in the latter.

and a larger emission energy is required before it again becomes a conducting electron. Both these processes can take place via phonon emission and capture. The photo-ionization process, which does not have any associated change of momentum, requires significantly higher energy for transfer to the intermediate electron band. The weak alloy mole-fraction dependence follows from the behavior of the intermediate state and the configuration energy in the captured state. This observed dependence is shown in Figure 7.55. This figure qualitatively shows that low AlAs mole-fractions, since the energy in the defect-localized state (the DX state) is higher than in the de-localized state, should not show appreciable DX center–related collapse phenomena or photo-sensitivity. The apparent crossover to the lowering of the energy occurs near ≈ 0.2–0.23, at which point the DX center becomes of significant concern in the device.

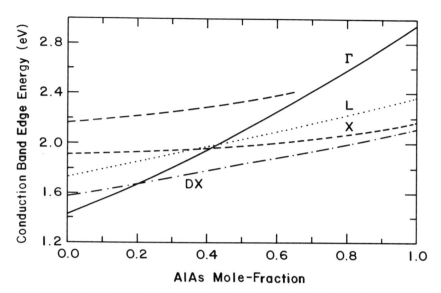

Figure 6.55: A qualitative picture of the mole-fraction dependence of the energy of the DX center (the most populated of the possibly four different types of distorted states associated with the donor) together with the changes in Γ, L, and X valley minima. The figure also plots the likely intermediate state energy schematically. After P. M. Mooney, "Deep Donor Levels (DX Centers) in III-V Semiconductors," *J. of Appl. Phys.*, **67**, No. 3, p. R1, 1990.

A behavioral picture of the DX center state in the energy band diagram as a function of mole-fraction, consistent with the prior discussion, is shown in Figure 7.56, together with a schematic of the capture processes. Since the relative positions of the band energies change as a function of the AlAs mole-fraction, there is a corresponding change in the DX center effect. At a low AlAs mole-fraction, the localized DX state has a higher energy, and the process is less likely. At a high AlAs mole-fraction, on the other hand, it becomes quite likely.

Time transients in capacitance due to pulsing of biases is a common technique to determine the characteristics of emission processes. For the DX center, the time transients appear to have at least three time constants. This may come about from lattice distortion caused by the capture of electrons at the donor site. In this distorted state, the donor atom can have distinctly different local environments depending on the different possible

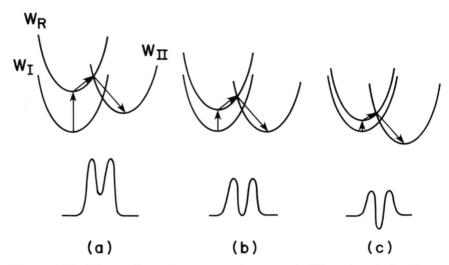

Figure 6.56: The configuration energy diagram at different mole-fractions, showing the process of capture via the intermediate state into the localized state at the DX center. (a) is at low AlAs mole-fractions, (b) is at an AlAs mole-fraction of ≈ 0.22, and (c) is at higher AlAs mole-fractions. After P. M. Mooney, "Deep Donor Levels (DX Centers) in III-V Semiconductors," *J. of Appl. Phys.*, **67**, No. 3, p. R1, 1990.

combinations of nearest group III atoms. Being a ternary compound, this can be various combinations of Ga and Al atoms. A different local environment should lead to different associated rates and, hence, would be consistent with the observation of different time constants. So, the possible DX levels are more than one, and quite likely four.

One can also note why, based on this description, the illuminating of a doped-barrier HFET allows one to recover the current–voltage characteristics of the device. When the device is cooled from room temperature, with the device turned off, in a sample with appropriate design so that the Fermi level is still well below the DX center state, most of the DX centers are ionized at room temperature, and they will remain so at low temperatures until the device is biased to pull the conduction band closer to the Fermi level. This occurs with forward biasing of gate or when hot electrons are injected in the current–voltage measurements. When this occurs, DX centers capture electrons, and the capacitance–voltage moves to an enhancement mode-like behavior because the DX center remains in this state unless provided energy through optical or thermal means. Shining light releases the captured electron, leading to the recovery of the current–voltage characteristics. The capture and emission processes can occur again when the device is biased appropriately. Note that hot electrons of sufficient energy are captured more readily because they have the requisite excess energy to surmount the capture barrier. The emission and capture time constants associated with the DX centers are relatively long compared to the device switching time constants. So, even though the device will switch rapidly, when left in a certain bias state, such as in logic, it will slowly reach a new state associated with the time constant effects of DX center. The large time constants of capture and emission, following the shutting off of the optical illumination, also result in persistent photoconductivity. Following photoemission of the captured electrons, there exists a larger electron population, both in the two-dimensional electron gas and in the $Ga_{1-x}Al_xAs$, which will be recaptured with rather large time constants. This results in persistent photoconductivity in the doped barrier channel. Note that this particular problem does not exist in the gate portion of the undoped-barrier HFET or in p-channel devices.

6.14 Off-Equilibrium Effects

In a fashion similar to our discussion of the off-equilibrium effects in the case of MESFETs, we discuss off-equilibrium phenomena in HFETs. Our discussion would be somewhat restricted since many comments relevant to MESFETs are equally applicable to HFETs. The regimes in which these off-equilibrium effects are so important that they just can not be ignored are similar in HFETs as in MESFETs because the scattering characteristics at higher energies are similar. In MESFETs a large doping is used to achieve large carrier concentrations; in HFETs a large carrier concentration occurs via transfer from remote donors or by inducing of a carrier gas. While impurity scattering is important for MESFETs, it is less so for HFETs. However, because of larger carrier densities due to confinement of the channel, carrier–carrier scattering and the effects of confinement on scattering behavior become important. These differences have little effect on the high energy behavior of scattering in the high field region of the device, Coulombic processes being inefficient at higher energies.

We will look at the overshoot effects in HFET structures by considering a simpler approximation of the insulator–semiconductor FET, where the insulator is an idealization of the large bandgap material. We assume a large barrier height, similar permittivity as the actual large bandgap material (e.g., $Ga_{1-x}Al_xAs$ for GaAs structures), and no confinement effects. As remarked before, in the operation of the device, confinement effects can be important towards the source end of the channel at 77 K. Although the device operation depends much more on the behavior close to the drain end, where the confinement is minimal, the behavior of carriers at the drain end is tied in with the behavior of carriers at the source end. In short devices, the presence of overshoot means limited scattering events, and hence the nature of the initial state remains important in overshoot, even though it could have been ignored in the long channel devices. We restrict our discussion, therefore, to 300 K operation.

Figure 7.57 shows overshoot and kinetic energy behavior in HFET structures operating at 300 K under similar bias conditions as the MESFET. Both the MESFETs and the HFETs being discussed are designed to have nominally similar enhancement mode thresholds near 0 V. Our discussion here will be comparative to elucidate the differences in their off-equilibrium behavior. The amount of velocity overshoot is somewhat larger than that of MESFETs. There is a noteworthy difference between the behavior of the boundary at which the carriers enter the channel region in this device and in the MESFET, and this simulation serves to point out the importance of such effects of boundary because the calculations are non-local. In the MESFET case, a built-in barrier existed. As in the case of our MISFET quasi-static theory, this barrier is quite small and therefore itself does not impede the flow of the necessary current. But because of the barrier, the average energy of carriers entering the channel is close to thermal and not close to the Fermi energy of the degenerately doped source. The average velocity of the carrier in the MESFET case is small, and the carrier velocity overshoot occurs from this low velocity condition. In the case of the HFET here, the choice of doping and compensation is such that carriers enter the channel from a degenerate electron gas; carriers entering the channel, i.e., those with the right orientation of momentum, also have a larger energy, and this is reflected in the initial velocity in the source and at the source end of the channel. Thus, overshoot, which is a function of scattering, the entry velocity, and the extent of the field region, is different, and almost correspondingly larger. The peak kinetic energy is similar in the case of the HFET to that of the MESFET, although the spatial distribution of the energy is broader. This broadness is reflected in the distribution function as a function of position; the carriers do heat more in this structure, but given the approximations of the simulations, including those regarding large barrier height at the insulator interface, the differences are insignificant.

Trends similar to those of MESFETs occur when comparing different materials with different scattering characteristics. Figure 7.57 also compares the overshoot behavior between the GaAs HFET, an InP HFET, and a $Ga_{1-x}In_xAs$ HFET. The overshoot in InP lags a little behind that of GaAs because of a larger effective mass and larger low energy scattering, but the

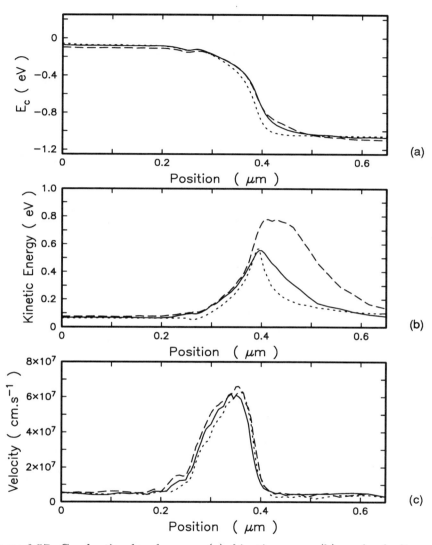

Figure 6.57: Conduction band energy (a), kinetic energy (b), and velocity along the channel (c) for 0.15 μm HFET structure. Solid lines are for GaAs channel, short dashed lines for InP channel, and long dashed lines are for a $Ga_{.47}In_{.53}As$ channel. An idealized insulator of 200 Å thickness with permittivity of $Ga_{.5}Al_{.5}As$ is assumed. The bias conditions are $V_{GS} = 0.5$ V and $V_{DS} = 1.0$ V.

peak overshoot velocities are nearly identical, and the peak kinetic energy is slightly larger than that of GaAs. However, all these peak kinetic energies are still lower than the applied drain-to-source bias. Only if the motion was ballistic would the peak kinetic energy have reached the drain-to-source bias. The $Ga_{1-x}In_xAs$ HFETs exhibit similar behavior as the GaAs HFET in overshoot closer to the source end, and the InAs HFETs exhibit a large overshoot nearly reaching the maximum group velocity, consistent with its very low scattering rate.

The scattering into secondary valleys is highlighted in Figure 7.58, which shows the valley distributions in GaAs and $Ga_{1-x}In_xAs$ for our HFET examples. Transfer to the secondary valleys is much more dominant in GaAs than in $Ga_{1-x}In_xAs$, corresponding to the inter-valley separations.

In all these examples, the source contact region was maintained at nearly identical injection conditions in order to allow an acceptable relative comparison. Changes in this would make differences in the observed overshoot behavior, much as it did in comparing the GaAs MESFET and GaAs HFET example. Another similarity in all these is that when the carriers reach the drain contact region, where there is a large carrier density $(2 \times 10^{18} \ cm^{-3})$, significant scattering occurs, the energy is rapidly lost, and if there exists a satellite valley (the exception here is InAs which has a satellite valley in excess of 1 eV), carriers scatter into the satellite valley and eventually decay back to the Γ valley.

In this analysis, we have ignored two-dimensional electron gas effects. In compound semiconductors, where the channel is usually formed in a region of low background doping, for simulations at 77 K, there is very clearly a large effect in low-field transport as reflected in the mobility. Thus, scattering effects become sufficiently different in regions of operation where electron–electron screening becomes important and where ionized impurity scattering is also important. In a field effect transistor, at the drain end of the channel the carriers are spread out further from the interface and carrier confinement is not substantial. Thus, the operation at the source end is the one that is affected by the carrier confinement. At 300 K, this effect should be minimal. A look at the differences in the velocity–field curves for highly pure GaAs (where carrier-carrier screening is small), and a two-dimensional electron gas confirms this. An actual simulation, without the carrier confinement effects reflected in the quantization effects, would still include these and therefore should be an adequate representation. On the other hand, at 77 K, where energy level separation becomes substantially larger than the thermal energy, this may be cause for concern for the representation at the source end. A combining of the Monte Carlo procedure with the quantum confinement effects, and inclusion of it in the scattering effects, is quite a daunting task.

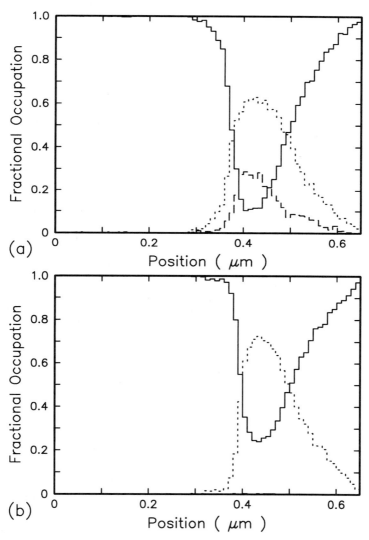

Figure 6.58: The fractional Γ (solid lines), L (short dashed lines), and X valley (long dashed lines) populations along the channel for the 0.15 μm gate-length HFET examples. Results for GaAs are shown in (a) and $Ga_{1-x}In_xAs$ are shown in (b).

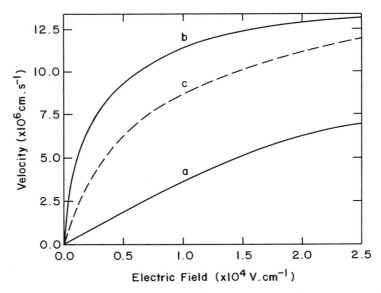

Figure 6.59: Hole velocity at 300 K (curve marked a) and 77 K (curve marked b) in GaAs at low ionized impurity density. Curve c shows observed two-dimensional hole gas velocity–field characteristics at 77 K.

6.15   p-channel Field Effect Transistors

We have discussed the n-channel device and its behavior in detail. Because of the unusual velocity–field characteristics, because of the lower electron effective mass and its effect on quantization, and because of the problems associated with DX centers and other hot electron effects, it is a structure which is considerably complex to analyze. p-channel devices, similar to n-channel devices, are also of interest. In the absence of scattering due to ionized impurities, low temperature mobilities can be high for holes in compound semiconductors.

Holes, with effective mass being large, and there being no dominant secondary valley effect, do not show a negative differential velocity effect, nor strong quantization effects. In many respects, this behavior is very similar to that of electrons and holes in silicon. The velocity–field characteristics for holes in low-doped GaAs and in heterostructures is shown in Figure 7.59. For comparison, the electron and hole velocity–field characteristics for silicon are also shown. Because they are similar, analyses similar to those of silicon devices are readily applicable to the p-channel structures.

For example, analyses based on the hyperbolic velocity–field relationship are applicable. Thus, our early analysis based on the hyperbolic relationship is applicable to p-channel devices (indeed, more accurately than it was for the n-channel devices). The practical limitations to these structures in compound semiconductors, unlike those of silicon, arise from low barrier heights to p-type material. This limits the largest gate voltages that can be applied before large forward gate conduction begins to occur.

A subject of particular interest related to these devices is methods of improvement in the hole transport. The light hole and heavy hole bands are degenerate at zero crystal momentum. Biaxial strain resulting from, e.g., growth of pseudomorphic structures ($Ga_{1-x}In_xAs$ with $Ga_{1-x}Al_xAs$ at low InAs mole-fraction is one example), distorts the conduction and valence band structure as discussed earlier in this chapter. This distortion in the valence band structure is believed to cause an improvement in the in-plane transport of holes.

6.16 Summary

This chapter developed the theory of operation of FETs based on insulators and heterostructures. We also studied parasitic phenomena that occur in these devices. At first, we considered the formation of heterostructures, the importance of heterostructure discontinuities, and the methods of analysis of band bending and statistics for the subbands that form at the interfaces involving an abrupt discontinuity. We developed criteria that would allow us to evaluate the approximations involved due to the use of Maxwell–Boltzmann statistics, the use of Fermi–Dirac statistics, the detailed inclusion of the density of states of the subbands, etc. We related, later in the chapter, the importance of these approximations to the analysis of the device in various bias and spatial regions of the device.

Initially, we considered an ideal insulator-based FET, a MISFET, with Maxwell–Boltzmann statistics. The theory developed, particularly suited for p-channel and silicon devices at 300 K, served as a basis to develop the concepts and details of the behavior of an idealized FET. We then considered the particular case of HFETs in compound semiconductors with large bandgap materials of limited resistivity and discontinuity. In this, we included the effects of subband formation but ignored the effects related to the flow of diffusive current. Like the case of the MESFET, we developed the small-signal theory for operation of the device, studied the operation of the device during a transient and due to off-equilibrium effects. One of the more common problems with undoped-barrier devices is related to injection of carriers across the small band discontinuities. We looked at the hot electron effects that cause this injection. Another problem, that occurs with n-type $Ga_{1-x}Al_xAs$ barrier devices is due to DX centers, centers with long emission time constants that are related to the incorporation of donors in the lattice. We considered the underlying principles of the behavior of these centers and their effect on device characteristics.

General References

1. T. Ando, A. B. Fowler, and F. Stern, "Electronic Properties of Two-dimensional Systems," *Reviews of Modern Physics*, **54**, No. 2, p. 437, April 1982.

2. D. L. Smith and C. Mailhiot, "Theory of Semiconductor Superlattice Electronic Structure," *Reviews of Modern Physics*, **62**, No. 1,p. 173, Jan. 1990.

3. R. S. C. Cobbold, *Theory and Applications of Field-Effect Transistors*, Wiley-Interscience, N.Y. (1970).

4. S. M. Sze, *Physics of Semiconductor Devices*, John Wiley, N.Y. (1981).

5. Y. P. Tsividis, *Operation and Modeling of the MOS Transistor*, Mc-Graw-Hill, N.Y. (1987).

6. M. S. Shur, *GaAs Devices and Circuits*, Plenum, N.Y. (1987).

7. J. R. Brews, "Physics of the MOS Transistor," in D. Kahng, Ed., *Applied Solid State Science*, Suppl. 2A, Academic Press, San Diego, CA (1981).

8. P. M. Mooney, "Deep Donor Levels (DX Centers) in III-V Semiconductors," *J. of Appl. Phys.*, **67**, No. 3, p. R1, 1 Feb. 1990.

Problems

1. Consider the quantization of hole energy in an inversion layer in silicon—the situation of compound structures is analogous because the valence band structure in most semiconductors is quite similar. Holes have multiple bands that should all be considered—the light hole band, the heavy hole band, and the split-off band. We will ignore the anisotropy of the bands, which is also substantial, so our results should be of interest for carriers whose energy is low in any of these subbands. Draw, taking account of the masses of the holes

at zone center (see Table 2.2), the positions of the minima in energy of a quantized hole inversion layer on a (100) surface of silicon. Also show schematically the density of state distribution.

2. As discussed, in silicon, the conduction band edge minimum occurs near the the X point in the Brillouin zone, and the conduction band exhibits a high degree of anisotropy, with the longitudinal mass considerably larger than the transverse mass (see Table 2.2). We considered inversion on a (100) surface in the text. What will the positioning of energy levels and the distribution of states be for inversion on a (110) surface and a (111) surface?

3. The Airy functions occur in solutions of differential equations involving a linear variation of the independent variable in the zero'th order term, i.e., for equations whose simplest homogeneous form can be expressed as

$$\frac{d^2\varphi}{dz^2} + z\varphi = 0. \tag{6.305}$$

Show, using the series method of finding a solution to this equation, that the solution of this occurs in the form described in the text.

4. The density of states for three-dimensional distributions were considered in Chapter 2. Using arguments based on orientation dependence of inversion and considering m_l and m_t as the longitudinal and transverse masses for the L valleys in germanium, find the density of state masses appropriate to determining the density of states distribution of electron inversion layers on a (100) surface.

5. We wish to evaluate the importance and necessity of considering quantization effects in calculations related to HFET structures. Consider inversion on the surface of p-type GaAs. Assume that one can manage to suitably adjust the position of the conduction band edge on the surface. At 300 K, consider a separation in energy of $3kT$, 0, and $-3kT$ between the conduction band edge and the electron quasi-Fermi level. Employing a consistent calculation that accounts for the effect of the acceptor charge in the depletion region, find the position of the zero'th energy level at acceptor dopings of 1×10^{14} cm^{-3}, 1×10^{15} cm^{-3}, and 1×10^{16} cm^{-3}, as well as the sheet density of carriers in this band. In which of these cases should one consider the effects of quantization of bands important for problems requiring accurate determination of carrier densities?

6. We have estimated that the accuracy of ignoring quantization effects is acceptable for conditions under which $E_0 < kT$. Find the sheet

carrier density in GaAs, for conditions when $E_0 = kT$, at 300 K and 77 K, for structures with acceptor doping of 1×10^{14} cm^{-3}, 1×10^{15} cm^{-3}, and 1×10^{16} cm^{-3}.

7. In Chapter 3, we evaluated the internal energy of a three-dimensional Fermi gas at absolute zero to be $W = 3n\xi_f/5$, where n is the carrier density. Show that for a two-dimensional system, the internal energy is given by $W = n\xi_f/2$.

8. The insulator–semiconductor FET considered was an n-channel device. The operation of a p-channel device is also instructive in emphasizing the behavior and concepts related to these devices. Consider a hypothetical p-channel GaAs MISFET fabricated using SiO$_2$ as an insulator and employing aluminum as the material for the gate with the device operating in the normal active mode of operation. Aluminum has a work-function of 4.36 eV.

   (a) What is the polarity of the bias of the gate, drain, and substrate with respect to the source?

   (b) For a cross-section running perpendicular to the surface, draw E_c, E_v, ξ_n, and ξ_p schematically, indicating the energies corresponding to the bias voltages at the source end and the drain end.

   (c) Draw a schematic three-dimensional perspective plot of how E_v and ξ_p vary, analogous to the figure of the n-channel device. However, include in this plot the band bending in the oxide and the gate region.

   (d) The n-type substrate has a background doping of 3×10^{14} cm^{-3}. What is the flat-band voltage, assuming no charge?

   (e) Draw schematically the capacitance–voltage characteristics of this MIS structure as a function of bias for low-frequency and high-frequency applied signals and for the structure going into deep-depletion. Identify which parts of the curves occur with the formation of an inversion region.

   (f) Now consider this schematic behavior in the presence of interface states, as shown in Figure 7.60. Draw, on an additional schematic plot, how the low-frequency and high-frequency characteristics get distorted from the ideal characteristic.

   (g) Assume a thickness of 400 Å for the insulator. Draw the accurate low-frequency capacitance–voltage characteristics between the gate and the substrate.

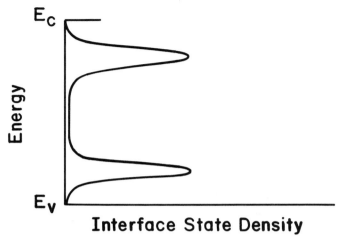

Interface State Density

Figure 6.60: Interface state distribution for an MIS structure.

(h) Show the nature of output characteristics for a gate length of 10 μm and 1 μm without including short channel effects. How do the latter characteristics change with the inclusion of short channel effects?

9. We wish to show that the capacitance associated with a semiconductor under flat-band conditions is given by

$$C_{FB} = \frac{\epsilon_s}{\lambda_D}. \qquad (6.306)$$

This is a problem important to high injection conditions in p–n junctions also, since near-flat-band conditions occur under those conditions. By considering the charge, field, and electrostatic potential in the semiconductor at a metal–semiconductor junction, we can determine the semiconductor capacitance contribution since all the charge on the metal electrode is terminated in the semiconductor. We ignore the effects of current flow, so this is strictly an MIS problem with an infinitely thin insulating layer. Considering mobile and immobile charge due to both electrons and holes, show that the capacitance of the semiconductor under flat-band conditions, i.e., for no change in electrostatic potential, reduces to the above. What is the general expression for charge and capacitance in terms of the arbitrary electrostatic potential ψ_S at the surface?

10. The flat-band voltage of an MIS structure is influenced by charges that can occur in a number regions of the device structures. The insulator can contain charge (often charge can be trapped in the insulator due to injection from the semiconductor during device operation), immobile charge also accumulates at the interface, and interface states can also exist as charged states. Consider an insulator with a sheet charge density of Q_{inss}, whose centroid is t_{inss} away from the interface, and a sheet charge density at the interface of Q_{ints}. Derive the modification to the threshold voltage and the flat-band voltage resulting from these charges. How does this influence the capacitance–voltage characteristics of the structure?

11. Show that the charge due to the carriers in the inversion layer in Equation 7.86 is negligible under conditions of band bending defined by Equation 7.92.

12. Show that an exact solution for the channel current of a MISFET, in constant mobility approximation, is given by Equation 7.103. Consider the MISFET structure at the minimum drain voltage necessary for channel pinch-off. Show that this accurate expression gives a similar result as the later derivation based on approximation of weak inversion charge.

13. The current–voltage characteristics of a p-channel MISFET can be best modelled using the procedures discussed in the text, since both the constant mobility approximation and the hyperbolic velocity–field approximation are relatively more accurate in the long channel and short channel limit than for n-channel devices. Consider $t_{ins} = 400$ Å with a dielectric constant of 11, and a device with a gate length of 2.5 μm. Derive the current–voltage characteristics assuming a substrate of 3×10^{14} cm^{-3} donor doping, assuming no velocity saturation and with velocity saturation. Assume a zero-field mobility of 350 cm^2.V^{-1}.s^{-1}, and a saturation velocity of 8×10^6 cm.s^{-1} for the hyperbolic velocity–field fit.

14. We have derived the expression for sheet mobile charge in an electron inversion layer for isotropic spherical conduction bands. The simplified form of this result occurred from identical density of states in each subband. Hole inversion layers exhibit complications of both differing masses and differing density of states. Find an expression similar to Equation 7.14 and 7.15 for the case of light and heavy holes in GaAs. The locations and density of states in hole inversion layer on the surface of silicon was considered in Problem 4.

15. We have compared the various approximations with rigorous calculations for the variation of electron sheet charge at the interface of $Ga_{1-x}Al_xAs$ and GaAs where the GaAs is doped 3×10^{14} cm^{-3}. We wish to see how this differs in 10^{15} cm^{-3} acceptor doping range for the 2/3'rd power law variation when compared with Fermi–Dirac and triangular well approximation. Consider the variation in the background acceptor density, the zero'th energy level following Equation 7.18, and sheet electron density in the $5 - 10 \times 10^{11}$ cm^{-2} range. Find fits for ξ_{f0} with $\gamma = 1.79 \times 10^{-9}$ eV.cm$^{4/3}$.

16. Consider Equation 7.207, which employs a parameterized hyperbolic velocity–field relationship for calculation of output characteristics. Saturation of current in high mobility transistor structures, of micronsized dimensions, arises from velocity saturation. A possible way of including this in calculations of output characteristics is to parameterize the electric field at the drain end at which this saturation occurs. For low-doped GaAs-based HFET structures, a longitudinal field near the peak velocity field should be the expected form. Consider characteristics based on the hyperbolic velocity model of a GaAs channel HFET structure, as well as that of a constant mobility model at 1 μm gate length. Find the electric field which would adequately characterize the occurrence of velocity saturation in the channel.

17. Consider an HFET structure that utilizes a metal gate and a doped barrier semiconductor of $Ga_{.7}Al_{.3}As$ on GaAs. The GaAs is doped p-type 3×10^{14} cm^{-3}, the $Ga_{.7}Al_{.3}As$ is doped 1.5×10^{18} cm^{-3} n-type and is 375 Å thick, and the metal gate has a barrier height of 0.8 eV and a gate length of 1 μm. Plot, as a function of the gate voltage, the sheet carrier density in GaAs and $Ga_{.7}Al_{.3}As$. Assume the Boltzmann approximation, we are interested in obtaining the approximate magnitude of the resulting threshold voltage of the HFET and the parasitic MESFET in the $Ga_{.7}Al_{.3}As$. At what voltage is there a substantial screening of the HFET channel by the electron charge in $Ga_{.7}Al_{.3}As$? What would be the easiest way of implementing this parasitic MESFET in a design model for the HFET useful at high forward gate biases?

18. We noted that the gate-to-source and gate-to-drain capacitances in Equations 7.229 and 7.227, as derived from the quasi-static analysis, are identical at $V_S = V_D = 0$. Consider constant gate bias for a 300 Å thick $Ga_{.7}Al_{.3}As$ barrier transistor structure which has a $V_{T0} = -5$ V and a gate length of 1 μm. Plot for $V_G = 0$ V and $V_G = 0.25$ V the gate-to-source and gate to drain capacitances as a function of drain

bias in the constant mobility approximation. Is the sum a constant as a function of drain bias? What is the estimate of deviation from charge conservation?

19. In the small-signal analysis we had outlined the procedure for deriving the common-source admittance parameters from the gate current phasor and drain current phasor using the Wronskian from Stokes' equation. Derive these results.

20. In the small-signal analysis of MESFETs, we showed how the effect of velocity saturation might be included in the extension of the admittance parameter solution. Describe how a similar methodology might be employed to extend the small-signal solution of the HFET admittance parameters.

21. Derive the characteristic frequencies of Equation 7.265 using the conductance and capacitance per unit length for an HFET, and the general integrals for frequency series expansion derived in Chapter 5.

22. Using the characteristic frequencies, derive the expressions for gate-to-source capacitance (Equation 7.267) and the gate-to-drain capacitance using the input and the output admittances.

23. Consider a p-channel metal–p-$Ga_{1-x}Al_xAs$–i-GaAs abrupt HFET. In the normal active mode of operation, draw schematically E_c, E_v, ξ_n, and ξ_p for the structure at the source end and the drain end for a cross-section running perpendicular to the surface.

24. Is a p-channel MIS capacitor structure based on $Ga_{1-x}Al_xAs$ as an insulator always in thermal equilibrium?

25. We derived the expression for current–voltage characteristics by lumping together both the drift and diffusion currents and using the quasi-Fermi level as the basis for analysis. We can determine the the drift and diffusion currents at any cross-section of the device since the requisite parameters are known. For the example chosen for illustration in the text, determine these components at the source end and the drain end, following the drain bias dependence at $V_{GS} = 1.0$ V.

26. Consider a doped-barrier HFET structure, with a metal of barrier height 0.9 eV, a 1×10^{18} cm^{-3} n-doped $Ga_{.7}Al_{.3}As$ region, and a 40 Å undoped $Ga_{.7}Al_{.3}As$ on a 10^{15} cm^{-3} p-doped GaAs substrate.

    (a) What is the thickness of this doped region in order to achieve a threshold voltage of 0.2 V if we assume the donors to be shallow?

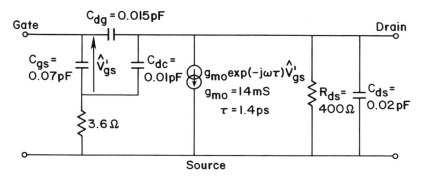

Figure 6.61: Equivalent circuit of a hypothetical HFET.

We assume a long channel device and the threshold voltage as that gate-to-source voltage necessary to obtain a carrier concentration of 5×10^{10} cm^{-2} in the conducting channel.

(b) Plot E_c, ξ_n, and the Fermi level in the metal at the condition of threshold. Draw these to be quantitatively accurate.

(c) If the donors are deep, assumed to be 60 meV, how do these parameters change? Plot again.

(d) What does the threshold voltage change to if the doping is increased to 1.1×10^{18} cm^{-3} or the thickness is increased by 10 Å of the doped Ga$_{.7}$Al$_{.3}$As region for both the shallow donors and deep donors?

27. Assuming appropriate V_{T0}, which accounts for ξ_{f0} (the parameter defined for non-linear charge control), $\gamma = 1.787 \times 10^{-9}$ eV.cm$^{4/3}$ for the non-linear variation of the Fermi level with the channel charge, and a hyperbolic velocity–field relationship with the fit for the parameters of $\mu_0 = 20000$ cm^2.V^{-1}.s^{-1}, and $\mathcal{E}_c = 1 \times 10^3$ V.cm^{-1}, plot output characteristics below the current saturation voltage for $V_G - V_T = 0.2$ V, 0.4 V, and 0.6 V. The gate length of the device is 1.0 μm, and ignore parasitic conduction in Ga$_{.7}$Al$_{.3}$As.

28. Determine the current gain and unilateral gain as a function of frequency for the sub-micron HFET whose equivalent circuit is shown in Figure 7.61. Calculate, using the approximate expressions of the text, the unity current gain frequency and the maximum frequency of oscillation. Compare and comment on discrepancies.

29. Estimate the drift and diffusion currents at the current saturation point in the channel of a 1 μm gate-length GaAs HFET at 300 K. Is it a good approximation to ignore the diffusive current? In what bias regions of the device operation is diffusive current important?

30. What current transport mechanism—drift or diffusion—dominates in the region that gives rise to current collapse in the characteristics of doped $Ga_{1-x}Al_xAs$ barrier HFETs?

31. Estimate the contribution of threshold voltage shift in doped-barrier $Ga_{.7}Al_{.3}As/GaAs$ HFETs due to change in density of states. Should the change lead to an increase in threshold voltage of an n-channel device with a lowering of the temperature of operation?

32. Given that HFETs generally have a larger capacitance than MES-FETs, should the transmission-line effects due to propagation of the signal along the width of the device be larger or smaller?

33. Estimate the differences in frequency limits in these transmission-line effects along the width of an HFET using a metal gate versus a semiconductor gate material.

34. How will the piezoelectric effect, discussed in the context of MES-FETs, affect the operation of HFETs?

35. This question pertains to the operational basis of FET structures—HFETs and MESFETs. Consider similar design of ideal structures, i.e., similar depleted control regions such as the larger barrier region of an HFET and the gate depletion region of a MESFET, as well as similar contact regions and substrate structures. For identical gate lengths,

    (a) which device should be more linear in its transfer characteristics, (i.e., I_{DS} versus V_{DS})?

    (b) which device should have a higher output conductance?

    (c) which device could be designed to conduct larger currents?

Chapter 7

Heterostructure Bipolar Transistors

7.1 Introduction

The bipolar transistor, so named to emphasize the importance of both electrons and holes in the base to its operation, is the first of the semiconductor transistors to be reduced to practice. It continues to be dominant in applications requiring the highest speeds but integration levels much lower than those of field effect transistors. The bipolar transistor has evolved significantly from its early days, and its most recent evolution has been towards the use of heterostructures, particularly in the emitter.

As the silicon bipolar transistor has shrunk in vertical and lateral dimensions to improve its speed and frequency of operation, low access resistances, higher current densities, and lower capacitances with reasonable current gains have been achieved with increases in doping in the emitter, base, and collector. Increased doping results in bandgap shrinkage, and since emitter injection efficiency is largely responsible for the current gain, the emitter doping has to be maintained considerably higher than that of the base for operation at 300 K. A smaller bandgap in the emitter results in a smaller effective hole barrier, resulting in larger hole currents and a smaller effective increase in injection efficiency as a result of larger emitter doping. Use of poly-silicon as an emitter material results in relatively efficient blocking of the hole transport without a significant effect on the electron transport, thus improving on the deleterious consequences of bandgap shrinkage in the emitter.

Another method of reducing the hole injection is the use of a larger

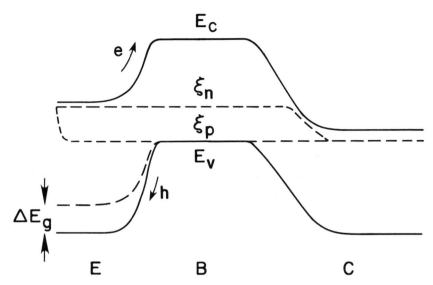

Figure 7.1: Energy band diagram of an n–p–n heterostructure bipolar transistor under forward bias at the base–emitter junction. ΔE_g is the increase in bandgap of the emitter material compared to the rest of the structure; the alloy composition at the junction has been assumed to be graded to avoid any discontinuities in the band edges.

bandgap emitter or a smaller bandgap base, e.g., silicon carbide for the former or a pseudomorphic silicon-germanium alloy for the latter in a silicon-based bipolar transistor. These embodiments exemplify the use of heterostructures in the bipolar transistor, and are now referred to as heterostructure bipolar transistors. Compound semiconductors, with the larger degree of freedom in choosing compositional variations that allow changes in bandgap while maintaining lattice integrity, offer additional degrees of freedom in addition to the advantages related to semiconductor transport.[1]

Consider Figure 7.1, where a larger bandgap material is employed for the emitter than the rest of the bipolar transistor structure. We had considered the p–n heterojunction diode problem, similar to the case here at the base–emitter junction, in Chapter 4, as an example of the application

---

[1]The article H. Kroemer, "Heterostructure Bipolar Transistors and Integrated Circuits," *Proc. of IEEE*, **70**, p. 13, 1982, comments extensively on the various aspects of the utility of heterostructures.

of the modified Gummel–Poon model. The increase in bandgap ΔE_g in the structure, under the modest injection conditions employed, appears in the valence band. This structure also assumes that alloy grading was accomplished over a long enough length scale to prevent appearance of any discontinuities in the band edges at the junction or anywhere else. In the absence of discontinuities, the conduction in this structure is by similar means as in the homojunction bipolar transistor, i.e., via drift and diffusion in the junction region, and limited by the transport in the quasi-neutral regions over much of the bias range of operation. Low-level injection implies that the quasi-Fermi levels are constant, the junction does not limit the transport, and hence the carrier concentration at the edges of the space regions are related exponentially through the barriers of the junctions for the corresponding carrier. To the first order, ignoring the density of state variations, etc., the barrier for electrons is still the same as in a homojunction bipolar transistor, while the barrier for holes has increased by ΔE_g. The hole concentration is correspondingly less by $\approx \exp\left(\Delta E_g/kT\right)$ (equivalently, hole concentration decreases proportionately to the change in n_i^2), and hence the diffusive minority carrier hole current in the emitter is correspondingly less, and hence the injection efficiency is correspondingly higher.

The larger latitude in controlling these features, the introduction of bandgap changes in the quasi-neutral base region to obtain modest quasi-fields for aiding carrier motion, and the advantages of fast carrier transport are some of the attributes appealing in their compound semiconductor implementations. There is one distinguishing difference between the silicon bipolar transistors and the compound semiconductor bipolar transistors, and this is related to the significantly poorer lifetime both in the bulk and at the surface due to their direct bandgap (the usual case of compound semiconductors of interest) and due to HSR recombination centers. Thus, whereas injection efficiency is the major hindrance in the scaling of silicon bipolar transistors and hence a reason to incorporate heterostructures, in compound semiconductors, they are a necessity since homojunction compound semiconductor bipolar transistors are limited both in gain and performance. There are several additional advantages accrued with the use of heterojunctions—the heterojunction barrier, e.g., allows lowering of the emitter doping while still maintaining injection efficiency, thus allowing lower input capacitances.

Our discussion of the compound semiconductor heterostructure bipolar transistor (HBT) is organized in a similar way as the discussion of MESFETs and HFETs. We discuss the quasi-static behavior of the devices first by expanding on the modelling of the p–n junction discussed earlier. Recall that the Gummel–Poon model discussion of this was made using the intrinsic carrier concentration as a parameter; to a first order this includes in it the bandgap variation since it is the one primarily responsible for variations in intrinsic carrier concentration. Subsequently, this quasi-static analysis will be expanded to discuss effects that are important and specific to compound semiconductors HBTs related to injection, collection, and recombination effects, and then we will follow it with discussions of small-signal behavior and off-equilibrium effects in these devices. The small-signal behavior, and the development of the network parameters and equivalent circuits in it, will point out some of the limitations of the quasi-static modelling. The discussion of off-equilibrium effects will point out the effects of the short dimensions and the limitations of using drift-diffusion in analyzing small structures.

7.2   Quasi-Static Analysis

Our discussion of the quasi-static modelling of HBTs will extend the development of p–n junction theory of Chapter 4. In our extended Gummel–Poon analysis of the p–n junction, we had included effects related to the variation of the intrinsic carrier concentration. Thus, the extended expressions, which explicitly involve intrinsic carrier concentration in the analysis, are applicable directly, provided drift-diffusion theory is applicable. The latter requires the absence of discontinuities, since transport in the presence of discontinuities requires consideration of transport limitations due to thermionic emission and tunneling. To a first order, the intrinsic carrier concentration varies as $\sqrt{N_C N_V} \exp\left(E_g/2kT\right)$; the pre-factor of this varies but is a smaller effect than the exponential half power dependence on the bandgap. Thus, the effective Gummel numbers incorporate the variation in bandgap, and the drift-diffusion approach is applicable so long as the changes are gradual and follow the constraints established in Chapter 4. Time-dependent modelling of the Gummel–Poon approach will lead us to equivalent circuit representation, applicable under quasi-static conditions, for the HBT.

The extended Gummel–Poon modelling approach is the preferred technique for quasi-static analysis of arbitrary structures. General approaches, by necessity, are also considerably more demanding in the use of numerical techniques. Computer-aided design models that use these analysis techniques, therefore, also adopt some simplifications. One such form is the Ebers–Moll[2] embodiment, which we will derive as a simplification of the extended Gummel–Poon modelling approach. This will lead us to the simpler equivalent circuits based on diodes, capacitors, and resistors, and will

---

[2]J. J. Ebers and J. L. Moll, "Large-Signal Behavior of Junction Transistors," *Proc. of IRE*, **42**, p. 1761, Dec. 1954.

be particularly useful in time domain analysis in the quasi-static limit.

As an introduction, we will consider here the behavior of an HBT (n–p–n $Ga_{1-x}Al_xAs/GaAs/GaAs$, with a 0.10 μm base doped to 5×10^{18} cm^{-3}). This device will serve as a standard example for our discussion of quasi-static and small-signal analysis. The band edges and quasi-Fermi levels, the electron and hole densities, and the electron and hole velocities, in this structure are shown in Figure 7.2, for a variety of bias conditions ranging from low injection to high injection. These are based on a numerical analysis of the problem using drift-diffusion and modelling ohmic contacts with infinite recombination velocities.

At low bias conditions, where low-level injection conditions prevail, the quasi-Fermi levels at the base–emitter junction are flat, Shockley boundary conditions prevail, and the electron density in the base and the hole density in the emitter change exponentially with bias. The hole density in the emitter is much smaller than the electron density in the base, even though the electron density in the emitter is smaller than the hole density in the base. This is a reflection of the two different barriers to the two different carriers, and is considered in Chapter 4. The holes see a barrier that is larger by ΔE_g, leading to an exponentially smaller carrier density. So, at low-level injection, at the base–emitter junction, Shockley boundary conditions prevail. Now consider the base–collector junction. At low-level injection any incident electron gets swept off by the base–collector electric field. At low-level injection there exists a very small electron concentration in the base–collector junction region. For normal operation, the electron diffusion current is from the collector to the base, and since the collector doping is low and the junction either biased to a low forward voltage, or zero, or reverse voltage, it is very small. A high forward bias voltage at the base–collector junction, as at the base–emitter junction, leads to the saturation region of the operation of the transistor with large injection of electrons into the base from the collector. This will be considered separately in our quasi-static discussion. Electron diffusion current being small, and the electric fields in the junction region being large, any incident electron in the depletion region gets swept away into the collector. The electron density in the collector depletion region is small, significantly smaller than the background donor or acceptor density, and the base–collector region is not perturbed to any significant extent by the presence of these carriers. The electron density at the edge of the quasi-neutral base is very small, significantly smaller than at the base–emitter junction edge. Since the carrier velocity in the base–collector depletion region is close to the saturation velocity, the highest velocities carriers can have in the drift-diffusion approximation, the carrier density needed to sustain the current in the low-level injection region is very small. For example, to sustain a current

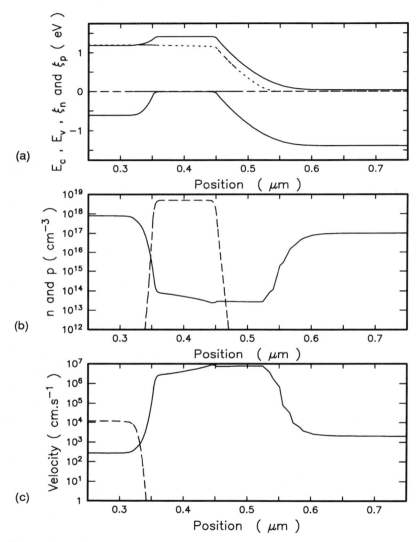

Figure 7.2: Band edges and quasi-Fermi levels (a), the electron and hole densities (b), and the electron and hole velocities (c) in an n–p–n HBT structure of $Ga_{1-x}Al_xAs/GaAs/GaAs$, with a 0.10 μm base doped to 5×10^{18} cm^{-3}. In (b) and (c) the electron characteristics are identified by a solid line and the hole characteristics by a dashed line.

density of 1×10^3 A.cm^{-2}, with a saturated velocity of 1×10^7 cm.s^{-1}, a carrier density of 6.25×10^{14} cm^{-3} is necessary. For this 1000 Å base device, assuming a mobility of 1000 cm^2.V^{-1}.s^{-1}, Einstein relationship, and a diffusive transport in the base region (the base is homogeneously doped with no compositional variation), the carrier density at the base–emitter junction edge should be 3.6×10^{15} cm^{-3}, sufficiently large that the carrier density at the base–collector junction edge can be ignored. For practical purposes, for modelling transport in the base, we may approximate the base–collector junction region with negligible carrier density in the forward mode of operation for carrier transport from the emitter to the collector. One could also look upon this as a boundary with a nearly infinite velocity. The concept of velocity was adopted with respect to the ohmic contact in the form of a recombination velocity. The base–collector junction differs from the ohmic contact in this one substantial aspect—there exists negligible recombination at the base–collector junction, even though the electron quasi-Fermi level still changes in the junction region just like the ohmic contact. So, unlike the ohmic contact, there is negligible majority carrier current at the base–collector junction under normal operating conditions. The quasi-Fermi levels are still separated by a substantial margin at the base–collector junction, and can only merge at the collector ohmic contact. The electron quasi-Fermi level in the junction depletion region decreases to maintain the appropriate electron density in the junction region.

As current density increases beyond the low-level injection limit, the hole density in the base also changes, and the carrier density at the base–collector junction begins to increase more rapidly to sustain this current density even though the carrier velocity in the base–collector depletion region may still be large. The hole density change in the quasi-neutral region of the base is noticeable since the electrons in the base are a fair fraction of the background hole concentration. Shockley boundary conditions are no longer representative of the conditions prevailing at the base–emitter junction. A change in the hole concentration leads to a finite drift field in the base with composition and doping homogeneity. This behavior in the base is analogous to the high injection condition of the p–n junction considered in Chapter 4.

As the injection is increased even beyond this condition, the electric field in the base–collector depletion region is reduced and the highest fields move towards the sub-collector junction. The electron density is now very large at the base–collector junction, comparable to or larger than the collector doping, leading to the phenomenon of push-out of the base to the sub-collector region, where the collector doping is sufficiently large to allow for the depletion region to form. This phenomenon, in homostructures, is known as the Kirk effect and is largely responsible for the drop in performance of devices at high current densities. Under these conditions, the quasi-Fermi levels do change as a function of position in the structure, in device regions critical to our analysis, largely as a result of pushing the current density in the structure to values close to the maximum limits available from the structure in the diffusion or drift limit.

The behavior of the structure in the medium and high current density limit is sufficiently complex that simpler formulations can only be employed for explaining and understanding specific phenomena alone. An example is the condition for onset of the Kirk effect. They cannot, however, be employed in developing a general theory of the device, which is best done using numerical approaches as in Figure 7.2.

The behavior of the HBT structure in low-level injection, however, does lend itself to general analysis. We may treat the base–emitter boundary condition as one where the quasi-Fermi level remains flat, and hence the Shockley boundary condition is still applicable. Since the control equations are linear, we may use superposition to describe the behavior of the transistor structure in the various bias regimes.

7.2.1   Extended Gummel–Poon Model

The application of the extended Gummel–Poon modelling approach for analyzing carrier transport was first considered in Chapter 4. It is an approach based on integrated charge densities and the parameters of the semiconductor and device structure and it was applied there in analysis of p–n junctions based on both homostructures and heterostructures. It is continued here as a natural extension to HBTs. In addition to its generality, extending to doping and compositional changes, one particular attribute of the extended approach, of special significance in compound semiconductor structures, is the consistent inclusion of recombination.

The p–n junction analysis allowed us to write, explicitly, the current equations for electron and hole transport in the two regions, provided we could mathematically model the boundary conditions at the contacts. We employed boundary conditions based on a finite recombination velocity at the contact (Problem 29 of Chapter 4), as well as infinite velocity boundary conditions (i.e., vanishing excess minority carrier concentration) as examples of application of the technique. Use of this technique in bipolar transistors follows in a similar manner, so long as we can formulate the boundary conditions. These follow in the light of our discussion of the operation of bipolar transistor above.

The problem of arbitrary biasing of the base–emitter and base–collector junctions can be broken into a superposition of the forward mode of operation (base–emitter forward biased, base–collector zero-biased), and the re-

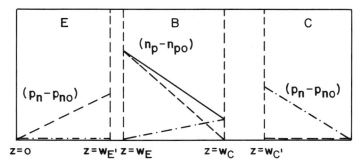

Figure 7.3: The breaking up of an arbitrary biasing condition (solid lines) of the bipolar transistor as a superposition of a forward mode (dashed lines) and reverse mode (dot-dashed lines) of operation. Excess carrier densities are shown schematically in this figure.

verse mode of operation[3] (base–collector forward biased, base–emitter zero biased), as shown in Figure 7.3. The reverse mode of operation, i.e., biasing of the base–collector junction, which has its most significant impact when it is forward biased and injects minority carriers into the quasi-neutral base region, can be looked upon as simply exchanging the roles of emitter and collector. Device design makes the reverse mode of operation, the collector acting as an emitter, highly inefficient. The superposition of the forward and reverse mode is a direct consequence of the linearity of the operating equations in quasi-static approximation.

Our derivation will also assume that the bandgap changes due to compositional variations all take place within the junction depletion regions. These are regions whose transport characteristics do not influence the behavior of the transistor directly, since the current transport in low-level injection is determined by the characteristics of the quasi-neutral region. We can write even more general equations (see Problem 1) where we may consider the compositional changes in the quasi-neutral regions such as that considered in the final example of Chapter 4. These lead to complicated equations; most problems of interest, however, occur with the compositional

---

[3]Forward active mode, by definition, refers to operation with forward bias at the base–emitter junction and reverse bias at the base–collector junction. Reverse active mode, by definition, refers to operation with forward bias at the base–collector junction and reverse bias at the base–emitter junction. Saturation mode refers to operation with both junctions forward biased.

changes in the high field regions of the junction depletion region. The equations describing this behavior are considerably more simplified and will be considered here.

Using the superposition, and under the approximations discussed above, our derivation is an extension of the p–n junction analysis, where we may consider the emitter–base part of the device as being akin to the p–n diode for minority carrier transport. For reverse mode, we may treat the collector as the emitter. Thus, we may write the equations for electron and hole currents at the edges of the junctions of the base–emitter space charge region. In our analysis, we assume, for simplicity, that the base and the collector are made of the same material, i.e., they have identical intrinsic carrier concentration n_{iB}, while the emitter has an intrinsic carrier concentration n_{iE}. Problem 2 considers the effect of different materials for collector in the analysis. The current densities, for forward mode, are

$$J_n(w_E) = \frac{q\overline{D}_{nB} n_{iB}^2}{\mathcal{G}\mathcal{N}_B'} \left[\exp\left(\frac{qV_{BE}}{kT}\right) - 1 \right]$$

$$\text{and} \quad J_p(w_{E'}) = \frac{q\overline{D}_{pE} n_{iE}^2}{\mathcal{G}\mathcal{N}_E'} \left[\exp\left(\frac{qV_{BE}}{kT}\right) - 1 \right]. \tag{7.1}$$

Here, our definitions of the effective Gummel numbers associated with the base and the emitter are

$$\mathcal{G}\mathcal{N}_B' = \frac{\mathcal{G}\mathcal{N}_B}{1 + \int_{w_E}^{w_C} \left(p J_{gr}^B(z)/q\mathcal{D}_n n_{iB}^2\right) dz}, \tag{7.2}$$

and

$$\mathcal{G}\mathcal{N}_E' = \frac{\mathcal{G}\mathcal{N}_E}{1 + \int_0^{w_{E'}} \left(n J_{gr}^E(z)/q\mathcal{D}_p n_{iE}^2\right) dz}. \tag{7.3}$$

The recombination current integrals and the base and emitter Gummel numbers are defined as

$$J_{gr}^B(z) = \int_{w_E}^z q\left[\mathcal{R}(z) - \mathcal{G}(z)\right] dz, \tag{7.4}$$

$$J_{gr}^E(z) = \int_0^z q\left[\mathcal{R}(z) - \mathcal{G}(z)\right] dz, \tag{7.5}$$

$$\mathcal{G}\mathcal{N}_B = \int_{w_E}^{w_C} p\,dz, \tag{7.6}$$

and

$$\mathcal{G}\mathcal{N}_E = \int_0^{w_{E'}} n\,dz. \tag{7.7}$$

Similarly, the effective diffusivities, following our earlier p–n junction analysis, are

$$\overline{\mathcal{D}}_{nB} = \frac{\int_{w_E}^{w_C} p\,dz}{\int_{w_E}^{w_C} (p/\mathcal{D}_n)\,dz},\tag{7.8}$$

and

$$\overline{\mathcal{D}}_{pE} = \frac{\int_0^{w_{E'}} n\,dz}{\int_0^{w_{E'}} (n/\mathcal{D}_p)\,dz}.\tag{7.9}$$

We may now write a similar set of equations for the reverse mode of operation, i.e., we now consider the base–collector junction as being biased by the potential V_{BC} which was ignored earlier, and assume that the base–emitter junction is short circuited. The equations are alike (see Problem 3) with the emitter and the collector reversed. The consequence of this is that we may now write the net electron and hole currents at the edge of the base–emitter depletion region with the quasi-neutral base and with the emitter. The current densities follow by superposition as

$$J_n(w_E) = \frac{q\overline{\mathcal{D}}_{nB}n_{iB}^2}{\mathcal{G}\mathcal{N}_B'}\left[\exp\left(\frac{qV_{BE}}{kT}\right) - \exp\left(\frac{qV_{BC}}{kT}\right)\right]$$

and
$$J_p(w_{E'}) = \frac{q\overline{\mathcal{D}}_{pE}n_{iE}^2}{\mathcal{G}\mathcal{N}_E'}\left[\exp\left(\frac{qV_{BE}}{kT}\right) - 1\right].\tag{7.10}$$

We have considered the processes of injection and collection occurring in the transistor structure under operating conditions. The net flow can be broken up into flow associated with forward and reverse conditions. The net current in the collector quasi-neutral region, which is also the current that is collected by the ohmic electrode, can be expressed in terms of products of ratios of the inflow and outflow of currents in the base–emitter space charge region, the quasi-neutral base region, and the base–collector space charge region as

$$J(w_{C'}) = \frac{J(w_{C'})}{J(w_C)} \times \frac{J(w_C)}{J(w_E)} \times \frac{J(w_E)}{J(w_{E'})} \times J(w_{E'}).\tag{7.11}$$

The current in the collector quasi-neutral region as well as in the base–collector space charge region is largely carried by electrons, the junction being at most modestly forward biased except when operating in saturation. As a result, the net ratio of outflow in the collector and the inflow in the emitter, the common-base current gain of the transistor (α), is written as

$$\alpha = \frac{J_n(w_{C'})}{J_n(w_C)} \times \frac{J_n(w_C)}{J_n(w_E)} \times \frac{J_n(w_E)}{J_n(w_E) + J_p(w_{E'})}.\tag{7.12}$$

The first ratio in this equation represents the effect of transport in the collector space charge region, to be called the collector transport factor and to be represented by ζ. Under quasi-static conditions, it may differ from unity predominantly due to carrier multiplication caused by avalanche multiplication or band-to-band tunneling at high fields. The second term in the equation represents the changes in electron current density due to transport in the quasi-neutral base region; it is commonly called the base transport factor and represented by α_T. The last term in the equation is the ratio of the electron current density injected into the quasi-neutral base to the total current density flowing in the base–emitter space charge region, it is called the emitter injection efficiency and denoted by γ. The common-base current gain α, is therefore

$$\alpha = \zeta \alpha_T \gamma. \tag{7.13}$$

Since we have expressed the current densities at the edges of the space charge regions at the base–emitter and the base–collector junctions, we can now write the expressions for these factors. Under quasi-static conditions, the collector transport factor deviates from unity mostly because of carrier multiplication, i.e., only under particular operating conditions such as biases which cause carrier avalanching and band-to-band tunneling in small bandgap semiconductors. These conditions are usually avoided since they can be irreproducible and lead to excess noise. Under high frequency conditions or at small time scales it deviates because of the large displacement currents $(d\mathcal{D}/dt)$ in the space charge regions. Thus, under the quasi-static conditions, we are principally interested in the injection efficiency and the base transport factor given by

$$\gamma = \frac{J_n(w_E)}{J_n(w_E) + J_p(w_{E'})} = \frac{1}{1 + \overline{\mathcal{D}}_{pE}\mathcal{G}\mathcal{N}'_B / \overline{\mathcal{D}}_{nB}\mathcal{G}\mathcal{N}'_E}, \tag{7.14}$$

and

$$\alpha_T = \frac{J_n(w_C)}{J_n(w_E)} = 1 - \frac{J^B_{gr}(w_C)}{J_n(w_E)}. \tag{7.15}$$

We now express the currents in the forward operating mode of the bipolar transistor in terms of the parameters described. The emitter current is the sum of the forward injection current of electrons and holes at the base–emitter junction if we ignore recombination effects within the junction. The collector current is the electron current reaching the collector junction, i.e., the base transport factor multiplied by the electron current at the base–emitter junction; the base current is the difference between the emitter current and the collector current, thus satisfying Kirchoff's current law. These currents, as functions of our parameters of the problem, are

$$
\begin{aligned}
J_E &= J_p(w_{E'}) + J_n(w_E) \\
&= \frac{q\overline{D}_{pE}n_{iE}^2}{\mathcal{G}\mathcal{N}_E'} \left[\exp\left(\frac{qV_{BE}}{kT} \right) - 1 \right] + \\
&\quad \frac{q\overline{D}_{nB}n_{iB}^2}{\mathcal{G}\mathcal{N}_B'} \left[\exp\left(\frac{qV_{BE}}{kT} \right) - \exp\left(\frac{qV_{BC}}{kT} \right) \right], \\
J_C &= \alpha_T J_n(w_E) \\
&= \alpha_T \frac{q\overline{D}_{nB}n_{iB}^2}{\mathcal{G}\mathcal{N}_B'} \left[\exp\left(\frac{qV_{BE}}{kT} \right) - \exp\left(\frac{qV_{BC}}{kT} \right) \right], \\
\text{and} \quad J_B &= J_E - J_C \\
&= \frac{q\overline{D}_{pE}n_{iE}^2}{\mathcal{G}\mathcal{N}_E'} \left[\exp\left(\frac{qV_{BE}}{kT} \right) - 1 \right] + \\
&\quad (1 - \alpha_T)\frac{q\overline{D}_{nB}n_{iB}^2}{\mathcal{G}\mathcal{N}_B'} \left[\exp\left(\frac{qV_{BE}}{kT} \right) - \exp\left(\frac{qV_{BC}}{kT} \right) \right].
\end{aligned}
$$

$$(7.16)$$

This set of equations is quite general, useful for conditions of arbitrary bandgap or intrinsic carrier concentrations in any of the device regions. The limitations on this set are those related to low-level injection, and also that the compositional changes take place over a sufficiently long length scale and within the junction space charge regions. More complex equations, where we relax the last of the conditions regarding the position of the compositional changes in the space charge regions, are considered in Problem 1. However, all these analyses still assume the validity of drift-diffusion transport, i.e., no discontinuities should appear in the structures and the transport must be limited by drift-diffusion processes in the quasi-neutral regions of the structures.

We have now derived the external currents, under quasi-static conditions, using the modified Gummel–Poon approach. Similar to our approach of perturbational modelling in MESFETs and HFETs, we may extend this approach to include time-dependent effects and thus be able to model changes and effects taking place over time scales that can be treated quasi-statically. This requires us to analyze the capacitive effects associated with storage of charge—the diffusion capacitance effects—and changes in space charge regions—the transition capacitance effects. Our goal is to generate an objective model of the bipolar transistor based on the charge storage concept that is valid at low and moderate frequencies.

The approach is to utilize the changes in the stored charges in various regions as the basis for the analysis. Its origin is in the fact that we can view the flow of carriers in minority carrier devices to be associated with minority carrier charge storage. The electron current at the base–collector junction, which occurs due to the injection at the base–emitter junction, requires storage of electron charge in the p-type base. In homogeneously doped base structures, the current flows by diffusion; thus, there exists a gradient of electron concentration, with the electron concentration decreasing from the emitter edge to the collector edge of the quasi-neutral base. The existence of electrons in the base, and its gradient, occur together with the flow of electrons into the collector. Thus, stored charge may be considered as the basis for the quasi-static time-dependent modelling. We consider one example to demonstrate the usefulness of this approach based on stored charge and to show its equivalence to the approach based on transport. The base time constant τ_B is, by definition, the time required by a carrier to transit the quasi-neutral base region. Consider the coordinate system with its origin in the quasi-neutral base region at the edge of the base–emitter depletion region.[4] Thus,

$$\tau_B = \int_0^{w_B} \frac{1}{v(z)} dz, \tag{7.17}$$

where v is the velocity of the carrier and w_B is the base width. We consider a homogenously doped base and assume recombination to be negligible. The electron current can be written as

$$J_n = q\mathcal{D}_n \frac{d(n_p(z) - n_{p0})}{dz} = q\mathcal{D}_n \frac{n_p(0) - n_{p0}}{w_B}, \tag{7.18}$$

because it is by diffusion, and because in one-dimensional analysis with recombination ignored, the gradient is a constant. This electron current can also be written as a product of excess carrier concentration and the velocity of the carrier,

$$J_n = q(n_p(z) - n_{p0})v(z) = q(n_p(z) - n_{p0})\frac{dz}{dt}. \tag{7.19}$$

---

[4]To simplify our mathematics, we will use two coordinate systems. The first, a more general coordinate system, used until this point, considers $z = 0$ as the emitter contact. This places $z = w_{E'}$ and $z = w_E$ as the emitter and the base edges of the emitter–base depletion region, and $z = w_C$ and $z = w_{C'}$ as the base and the collector edges of the base–collector depletion region. The second coordinate system, used when we are specifically focussing on transport in the quasi-neutral base region, moves the origin to the base edge of the emitter–base depletion region. The base edge of the base–collector depletion region, then, occurs at $z = w_B$. Note, $w_B = w_C - w_E$.

The two expressions allow us to write the velocity of the carriers, since current density is constant and minority carrier density varies linearly in the base in the absence of recombination. Thus, for $0 < z < w_B$,

$$v(z) = \frac{\mathcal{D}_n}{w_B} z, \tag{7.20}$$

and hence

$$\tau_B = \int_0^{w_B} \frac{z}{\mathcal{D}_n} dz = \frac{w_B{}^2}{2\mathcal{D}_n}. \tag{7.21}$$

While we estimated this based on homogeneous doping, inhomogeneously doped structures lead to a modified expression (see Problem 4) and an analysis similar to the treatment of the p–n junction (in Chapter 4) leads to the base time constant as

$$\tau_B = \frac{w_B{}^2}{\nu \overline{\mathcal{D}_n}} \tag{7.22}$$

for low-level injection and forward active mode of operation, with the factor ν determined from the transistor parameters by

$$\nu = \left\{ \frac{1}{w_B{}^2} \int_{w_E}^{w_C} \frac{1}{p(z)} \left[\int_z^{w_C} p(\eta) d\eta \right] dz \right\}^{-1}. \tag{7.23}$$

The problem of alloy grading also leads to similar results (see Problem 5) and will also be considered in our treatment of small-signal analysis.

Let us now explore this time constant, the time of transit in the base, using the stored charge approach, and show that this same result can be derived by charge control analysis. Again, we consider the forward active mode of operation. Most of the collector current is electron current in a conventional n–p–n bipolar transistor. The excess minority carrier stored charge in the base $\mathcal{Q}_{nB}$, assuming unit cross-section area is given by

$$\mathcal{Q}_{nB} = \int_{w_E}^{w_C} q n'_p(z) dz = \int_{w_E}^{w_C} q \left[n_p(z) - n_{p0} \right] dz, \tag{7.24}$$

where $n'_p(z)$ is the excess minority carrier population of electrons in the p-type base at any position z. Let this charge support the collector current in this structure. The time constant that relates the minority carrier charge $\mathcal{Q}_{nB}$ in the base with the collector current is the base time constant τ_B. Consider the uniformly doped base. We have

$$\mathcal{Q}_{nB} = \frac{1}{2} q w_B n'_p(w_E). \tag{7.25}$$

The collector current density is given by

$$J_n(w_C) = q\mathcal{D}_n \frac{n_p'(w_E)}{w_B}, \tag{7.26}$$

hence the base time constant is

$$\tau_B = \frac{\mathcal{Q}_{nB}}{J_n(w_C)} = \frac{w_B{}^2}{2\mathcal{D}_n}, \tag{7.27}$$

an identical result.

Now consider the same problem for a non-uniformly doped base. The collector current density, for the forward mode, is given as

$$
\begin{aligned}
J_n(w_C) &= \frac{q\overline{\mathcal{D}}_{nB} n_{iB}^2}{\mathcal{G}\mathcal{N}_B'} \left[\exp\left(\frac{qV_{BE}}{kT} \right) - \exp\left(\frac{qV_{BC}}{kT} \right) \right] \\
&\approx \frac{q\overline{\mathcal{D}}_{nB} n_{iB}^2}{\int_{w_E}^{w_C} p\,dz} \left[\exp\left(\frac{qV_{BE}}{kT} \right) - \exp\left(\frac{qV_{BC}}{kT} \right) \right], \tag{7.28}
\end{aligned}
$$

where the effect of recombination in the base has been ignored. The excess minority stored charge in the base is

$$\mathcal{Q}_{nB} = q \int_{w_E}^{w_C} n_p'(z)\,dz, \tag{7.29}$$

where the excess carrier concentration is a function of the bias. This leads to the base time constant

$$\tau_B = \frac{\int_{w_E}^{w_C} p(z)\,dz \int_{w_E}^{w_C} n_p'(z)\,dz}{\overline{\mathcal{D}}_{nB} n_{iB}^2 \left[\exp\left(qV_{BE}/kT \right) - \exp\left(qV_{BC}/kT \right) \right]}. \tag{7.30}$$

This is identical to the result based on carrier velocity (see Problem 4). Alloy grading also leads to similar conclusions and an identical interpretation of the base time constant (see Problem 5).

We have now established that under the assumption of quasi-static conditions, the two approaches give identical results. The second approach is intuitively more appealing and is analytically more tractable for the general problem since it is based on the extended Gummel–Poon approach. So, we now consider it in detail, and include the effect of other charges.

Current flow through the device supports the minority charge[5] in the base quasi-neutral region $\mathcal{Q}_{nB}$, which we just considered, in the emitter

---

[5]See J. te Winkel, "Past and Present of The Charge-Control Concept in the Characterization of the Bipolar Transistor," in L. Marton, Ed., *Advances in Electronics and Electron Physics*, Academic Press, N.Y. (1975) for a quasi-static treatment of small-signal, large-signal, and transient phenomena. R. S. Muller and T. I. Kamins, *Device Electronics for Integrated Circuits*, John Wiley, N.Y. (1986), an introductory reference, also treats this subject in detail.

quasi-neutral region $\mathcal{Q}_{pE}$, and in the collector quasi-neutral region $\mathcal{Q}_{pC}$. We will assume the devices to have unit cross-section area; $\mathcal{Q}$s are therefore integrated charge densities in unit volume. The base–emitter and the base–collector space charge regions can also be viewed to have space charge $\mathcal{Q}_{sE}$ and $\mathcal{Q}_{sC}$ associated with them. These are the excess charges on either side of the metallurgical junction. Changes in bias cause movements of the depletion region edge, which lead to either uncovering or compensation of the immobile charge; the net charge on either side of the space charge region either increases or decreases with changes occurring at the edges of the depletion region. Thus, the transition capacitances, characterizing the change in the charge of the space charge region, are

$$C_{tE} = \frac{\partial \mathcal{Q}_{sE}}{\partial V_{BE}} \qquad (7.31)$$

for the base–emitter depletion region, and

$$C_{tC} = \frac{\partial \mathcal{Q}_{sC}}{\partial V_{BC}} \qquad (7.32)$$

for the base–collector depletion region.

We consider the problem as a superposition of forward and reverse transport and storage. For the forward problem we consider $V_{BC} = 0$ and for the reverse problem we consider $V_{BE} = 0$. Thus, in forward mode the base–collector junction is short circuited, and no hole injection occurs into the collector, i.e.,

$$\mathcal{Q}_{pC}^F = 0, \qquad (7.33)$$

and based on similar arguments, in reverse mode, no hole injection occurs into the emitter quasi-neutral region and

$$\mathcal{Q}_{pE}^R = 0. \qquad (7.34)$$

Forward mode, thus, sustains minority carrier charge of electrons in the base region, and of holes in the emitter region. Thus, total minority carrier charge in the forward mode is given by

$$\mathcal{Q}_F = \mathcal{Q}_{nB}^F + \mathcal{Q}_{pE}^F. \qquad (7.35)$$

Similarly, for the reverse mode, minority carrier charge is sustained in the quasi-neutral base region and the quasi-neutral collector region, and is given by

$$\mathcal{Q}_R = \mathcal{Q}_{nB}^R + \mathcal{Q}_{pC}^R. \qquad (7.36)$$

We now introduce time constants to relate charge storage with currents. Let the forward time constant τ_F relate the total forward forward charge storage with the forward current, i.e.,

$$\tau_F = \frac{Q_F}{I_C^F}. \tag{7.37}$$

Note that the base time constant is a subset of this time constant. Using F as the superscript to identify the forward mode, the base time constant is related as

$$\tau_B^F = \frac{Q_{nB}^F}{I_C^F}. \tag{7.38}$$

Hence, the forward time constant is related to the forward base time constant as

$$\tau_F = \tau_B^F \frac{Q_F}{Q_{nB}^F} = \tau_B^F \frac{Q_{nB}^F + Q_{pE}^F}{Q_{nB}^F}. \tag{7.39}$$

This time constant relates forward current with forward charge.

We can also define a time constant that relates the forward mode base current with the forward charge

$$\tau_{BF} = \frac{Q_F}{I_B^F}. \tag{7.40}$$

Since the forward collector and base currents are related through the forward current gain β_F,

$$I_C^F = \beta_F I_B^F, \tag{7.41}$$

the forward time constants for the base current and the collector current are related as

$$\tau_{BF} = \beta_F \tau_F = \beta_F \tau_B^F \frac{Q_F}{Q_{nB}^F}, \tag{7.42}$$

with the forward time constant for base current being substantially larger than that for collector current. The time constants τ_F and τ_{BF} are more basic to our quasi-static modelling since they directly relate the forward collector current and the forward base current with the forward charge storage. The emitter current for forward mode follows directly from this in steady-state conditions as

$$I_E^F = Q_F \left(\frac{1}{\tau_F} + \frac{1}{\tau_{BF}} \right). \tag{7.43}$$

Similar time constants τ_R and τ_{BR} can be defined to relate the reverse charge storage with the reverse currents. Note that now, the conventional emitter, the emitter of the forward mode of the structure, becomes the collector for the reverse relations, and the conventional collector, the collector of the forward mode of the structure, becomes the emitter of the reverse relations. The steady-state can, therefore, be written as superposition of forward and reverse currents,

$$I_C = \frac{\mathcal{Q}_F}{\tau_F} - \mathcal{Q}_R \left(\frac{1}{\tau_R} + \frac{1}{\tau_{BR}} \right), \tag{7.44}$$

$$I_B = \frac{\mathcal{Q}_F}{\tau_{BF}} + \frac{\mathcal{Q}_R}{\tau_{BR}}, \tag{7.45}$$

and $$I_E = \mathcal{Q}_F \left(\frac{1}{\tau_F} + \frac{1}{\tau_{BF}} \right) - \frac{\mathcal{Q}_R}{\tau_R}. \tag{7.46}$$

Now consider the effect of time-dependent changes. Both the forward and reverse charges $\mathcal{Q}_F$ and $\mathcal{Q}_R$ change, as do the space charge region charges $\mathcal{Q}_{sE}$ and $\mathcal{Q}_{sC}$. If the Kirk effect, i.e., the base push out effect due to large minority carrier densities at the base-collector metallurgical junction, does not occur, then the change in the charge $\mathcal{Q}_F$ is due to charges flowing in or out of the base and the emitter electrodes, there being no forward charge in the collector quasi-neutral region. Any current resulting from the time-dependent change in $\mathcal{Q}_F$, the current $\partial \mathcal{Q}_F / \partial t$, flows through the emitter and the base terminal. Thus, the time-dependent expressions for I_E and I_B contain the current $\partial \mathcal{Q}_F / \partial t$ due to changes in the forward charge. Similarly, the term $\partial \mathcal{Q}_R / \partial t$ appears in the expressions for I_C and I_B.

Ignoring high injection effects, the base–emitter space charge region responds directly to any changes in the bias at the base–emitter junction, and the base collector space charge region to the changes in base–collector bias. Thus, the current $\partial \mathcal{Q}_{sE} / \partial t$ appears in emitter and base currents, and the current $\partial \mathcal{Q}_{sC} / \partial t$ appears in collector and base currents. Consolidating all these terms, we obtain the following equations for the quasi-static currents of the bipolar transistors:

$$I_C = \frac{\mathcal{Q}_F}{\tau_F} - \mathcal{Q}_R \left(\frac{1}{\tau_R} + \frac{1}{\tau_{BR}} \right) - \frac{\partial \mathcal{Q}_R}{\partial t} - \frac{\partial \mathcal{Q}_{sC}}{\partial t}, \tag{7.47}$$

$$I_B = \frac{\mathcal{Q}_F}{\tau_{BF}} + \frac{\mathcal{Q}_R}{\tau_{BR}} + \frac{\partial \mathcal{Q}_F}{\partial t} + \frac{\partial \mathcal{Q}_R}{\partial t} + \frac{\partial \mathcal{Q}_{sE}}{\partial t} + \frac{\partial \mathcal{Q}_{sC}}{\partial t}, \tag{7.48}$$

and

$$I_E = \mathcal{Q}_F \left(\frac{1}{\tau_F} + \frac{1}{\tau_{BF}} \right) - \frac{\mathcal{Q}_R}{\tau_R} + \frac{\partial \mathcal{Q}_F}{\partial t} + \frac{\partial \mathcal{Q}_{sE}}{\partial t}. \tag{7.49}$$

These equations can be easily visualized in the schematic circuit shown in Figure 7.4.

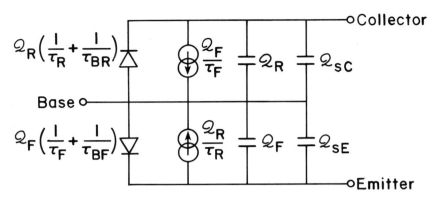

Figure 7.4: Schematic of a circuit representing the association of base, emitter, and collector currents with forward and reverse charges, and temporal change in the space charge region charge at the base–collector and the base–emitter junctions. The capacitances shown are associated with changes in charges $\mathcal{Q}_F$, $\mathcal{Q}_R$, $\mathcal{Q}_{sE}$, and $\mathcal{Q}_{sC}$.

The forward and reverse minority charges change with bias. Since the applied bias directly modulates the barrier, the carrier densities at the junction edges are directly related via the exponential in applied bias, hence for the simplest of the structures, homogeneous in composition and doping, the forward and reverse charges are related as

$$\mathcal{Q}_F = \mathcal{Q}_{F0} \left[\exp\left(\frac{qV_{BE}}{kT} \right) - 1 \right],$$ (7.50)

and

$$\mathcal{Q}_R = \mathcal{Q}_{R0} \left[\exp\left(\frac{qV_{BC}}{kT} \right) - 1 \right].$$ (7.51)

This is a quasi-static model, i.e., it assumes that the device is always in steady-state. Thus, it is valid only when applied signals do not vary too rapidly. One may quantify this time scale as exceeding the largest time constant of the system, i.e., the quasi-static model may be applied for time scales of $t \gg \tau_F$.

We can now derive the small-signal equivalent circuit valid for the quasi-static conditions. We use the perturbational analysis employed earlier for both MESFETs and HFETs. The perturbation method uses superposition of a perturbing signal on the static signal to derive circuit elements from the response to the perturbation. Thus, the transconductance g_m, output conductance g_d, etc., were the ratio of the small change in the drain current resulting from a small change either in the gate voltage or in the drain voltage. Our perturbational analysis of the time-dependent quasi-static equations will utilize the fact that circuit elements connecting two nodes can contribute to currents only in the corresponding nodes. Here, we consider the forward mode terms; the reverse mode terms follow using similar arguments. The applied voltages are

$$V_{BE} = \overline{V}_{BE} + \Delta V_{BE}, \tag{7.52}$$

and

$$V_{BC} = \overline{V}_{BC} + \Delta V_{BC}. \tag{7.53}$$

The forward mode equations, ignoring effects of $\mathcal{Q}_R$, which is small, are

$$I_C = \frac{\mathcal{Q}_F}{\tau_F} - \frac{\partial \mathcal{Q}_{sC}}{\partial t}, \tag{7.54}$$

$$I_B = \frac{\mathcal{Q}_F}{\tau_{BF}} + \frac{\partial \mathcal{Q}_F}{\partial t} + \frac{\partial \mathcal{Q}_{sE}}{\partial t} + \frac{\partial \mathcal{Q}_{sC}}{\partial t}, \tag{7.55}$$

$$\text{and} \quad I_E = \mathcal{Q}_F \left(\frac{1}{\tau_F} + \frac{1}{\tau_{BF}} \right) + \frac{\partial \mathcal{Q}_F}{\partial t} + \frac{\partial \mathcal{Q}_{sE}}{\partial t}, \tag{7.56}$$

where I_E is written to be positive if the current flow out of the emitter terminal.

First consider terms that are affected by the base–emitter bias. The second term of the collector equation is not a function of the base–emitter junction, but the first term is. Its perturbed component, i.e., that related to the perturbation in the base–emitter voltage, can be found by finding the derivative w.r.t. V_{BE} using the same techniques employed in the FET quasi-static analysis. We have

$$\begin{aligned} \frac{\partial I_C}{\partial V_{BE}} &= \frac{\partial}{\partial V_{BE}} \left(\frac{\mathcal{Q}_F}{\tau_F} \right) \\ &= \frac{1}{\tau_F} \frac{\partial}{\partial V_{BE}} \mathcal{Q}_{F0} \left[\exp \left(\frac{qV_{BE}}{kT} \right) - 1 \right] \\ &= \frac{\mathcal{Q}_{F0}}{\tau_F} \frac{q}{kT} \exp \left(\frac{qV_{BE}}{kT} \right) \approx \frac{qI_C}{kT} = g_m. \end{aligned} \tag{7.57}$$

Thus, the $\mathcal{Q}_F/\tau_F$ term, which occurs in both the emitter and the collector current, results in a conductance of g_m, i.e., the perturbation current in the collector ΔI_C varies as $g_m \Delta V_{BE}$. Now consider the base current. The first three terms depend on the base–emitter potential and hence result in circuit elements corresponding to those nodes, while the last results in a circuit element corresponding to the base–collector potential and nodes.

We look at these individually. The derivative of the first term with V_{BE} gives

$$\frac{\partial}{\partial V_{BE}}\left(\frac{Q_F}{\tau_{BF}}\right) \approx \frac{qI_C}{kT\beta_F} = \frac{g_m}{\beta_F} = \frac{1}{r_\pi}. \tag{7.58}$$

This is a conductance element between the base and the emitter terminal. The time-derivative term of Q_F can be expanded in terms of the derivative w.r.t. the impressed potentials V_{BE} and V_{BC} as

$$\frac{\partial Q_F}{\partial t} = \frac{\partial Q_F}{\partial V_{BE}}\frac{\partial V_{BE}}{\partial t} + \frac{\partial Q_F}{\partial V_{BC}}\frac{\partial V_{BC}}{\partial t}. \tag{7.59}$$

The first term has units of capacitance as its pre-factor. This capacitance is called the diffusion capacitance C_D, and it models, together with the forward current I_C^F, the time delay τ_B^F due to storage in the base.

$$C_D = \frac{\partial Q_F}{\partial V_{BE}} = \frac{\partial}{\partial V_{BE}}(I_C\tau_F) = \frac{\partial I_C}{\partial V_{BE}}\tau_F = g_m\tau_F. \tag{7.60}$$

The space charge term Q_{sE} responds only to the base–emitter voltage and hence corresponds to a capacitive term C_{tE} through

$$\frac{\partial Q_{sE}}{\partial t} = \frac{\partial Q_{sE}}{\partial V_{BE}}\frac{\partial V_{BE}}{\partial t} = C_{tE}\frac{\partial V_{BE}}{\partial t}. \tag{7.61}$$

So, associated with the perturbation in the base–emitter bias V_{BE}, the changes in the forward currents in the emitter, base, and collector can be represented by a current source $g_m\Delta V_{BE}$ between the collector and the emitter, a resistance $r_\pi = \beta_F/g_m$, and capacitances C_D and C_{tE} between the base and the emitter. The latter two are sometimes lumped together as C_π.

Now consider the terms affected by changes in the base–collector potential. The term Q_F/τ_F has a weak dependence on the base–collector potential due to the Early effect, i.e., due to changes in the quasi-neutral base region width.

$$\frac{\partial}{\partial V_{BC}}\left(\frac{Q_F}{\tau_F}\right) = \frac{\partial I_C}{\partial V_{BC}} = \frac{I_C}{V_A} = \frac{g_m kT}{qV_A} = -g_m\eta, \tag{7.62}$$

where V_A is the Early voltage and η is the magnitude of the Early factor. The Early effect arises from modulation of the quasi-neutral base width, which leads to a change in the gradient of the minority carrier charge density and hence in the collector current. It is related by

$$V_A = \frac{\int_{w_E}^{w_C} p\,dz}{p(w_C)\partial w_C/\partial V_{CB}}. \tag{7.63}$$

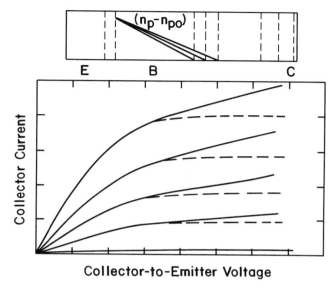

Collector-to-Emitter Voltage

Figure 7.5: A schematic of the effect of the Early voltage, the increase in gradient of the minority charge in the base due to reduction in the quasi-neutral base width by a reverse biasing of the base–collector junction. This results in an increase in collector current from the characteristics shown as dashed lines to those shown as solid lines.

Hence, an increase in the reverse bias voltage at the base–collector junction leads to an increase in the collector current, as shown schematically in Figure 7.5. This implies that both the collector and emitter perturbation currents ΔI_C and ΔI_E have a current component $-g_m\eta\Delta V_{BC}$ or $g_m\eta\Delta V_{CB}$ in them. Since $\Delta V_{CB} = \Delta V_{CE} - \Delta V_{BE}$, the perturbation effect of the base–collector potential results in a resistance $r_o = 1/g_m\eta$ between the collector and the emitter, and a current source of $g_m(1-\eta)\Delta V_{BE}$.

The dependence of the term $\mathcal{Q}_F/\tau_{BF}$ on the base–collector potential is small because it models the response of the base current due to the modulation of the base width. This is significantly smaller than that of the collector current, which is influenced directly by the change in the gradient of minority carrier distribution in the base. The change in base current occurs due to changes in recombination in the base, a small quantity. We have

$$\frac{\partial}{\partial V_{BC}}\left(\frac{\mathcal{Q}_F}{\tau_{BF}}\right) = \frac{\partial}{\partial V_{BC}}\left(\frac{\mathcal{Q}_F}{\beta_F\tau_F}\right)$$

$$\approx \frac{I_C}{\beta_F V_A}\left[1 - \frac{V_A}{\beta_F}\frac{\partial\beta_F}{\partial V_{BC}}\right]. \tag{7.64}$$

We can derive the magnitude of the derivative of the current as

$$\begin{aligned}
\frac{\partial\beta_F}{\partial V_{BC}} &= \frac{\partial}{\partial V_{BC}}\left(\frac{\alpha_F}{1-\alpha_F}\right)\\
&= \frac{1}{(1-\alpha_F)^2}\frac{\partial\alpha_F}{\partial V_{BC}}\\
&= \frac{1}{(1-\alpha_F)^2}\frac{w_B\alpha_T}{\mathcal{L}_n V_A}, \tag{7.65}
\end{aligned}$$

where $\mathcal{L}_n$ is the diffusion length of electrons in the base of the device. This simplifies (see Problem 6) to

$$\frac{\partial}{\partial V_{BC}}\left(\frac{Q_F}{\tau_{BF}}\right)\approx\frac{I_C}{V_A}\left[1-\alpha_T\left(1-\frac{\alpha_T w_B}{\mathcal{L}_n}\right)\right], \tag{7.66}$$

which is a resistance between the base and the collector electrode of the approximate magnitude

$$r_\mu = \frac{1}{1-\alpha_T}\frac{V_A}{I_C}. \tag{7.67}$$

The time derivative of the forward charge storage has a base–collector potential dependence

$$\begin{aligned}
\frac{\partial Q_F}{\partial V_{BC}} &= \frac{\partial}{\partial V_{BC}}(\tau_F I_C)\\
&= \frac{\tau_F I_C}{V_A} = -g_m\tau_F\eta = -\eta C_D, \tag{7.68}
\end{aligned}$$

where η is kT/qV_A. This term leads to a storage or diffusion capacitance whose magnitude is substantially mitigated by the small Early factor. This effect of change in storage occurs through the modulation of the edge of the quasi-neutral base region at the base–collector junction where the minority carrier charge concentration is very small. Thus, this capacitive effect between the base and the collector electrode, which is very similar to the diffusion capacitance between the base and the emitter electrode, is a small effect whose magnitude is smaller by the factor η. The space charge term Q_{sC} contributes a capacitive term C_{tC} between the base and the collector electrode similar to the capacitance C_{tE}.

We have now modelled all the elements of the equivalent circuit representing response of all of the terms of the quasi-static current equations based on charge control analysis. The forward mode elements of this are

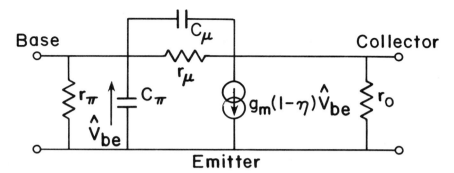

Figure 7.6: Hybrid-pi equivalent circuit for the forward mode of the bipolar transistor operation derived using the charge control analysis.

shown in Figure 7.6. This representation is usually called the hybrid-pi model of the transistor. Since $\eta \ll 1$, the current source in this figure is proportional to the small-signal voltage $\hat{V}_{be}$ with the transconductance g_m as the constant of proportionality. This figure is for forward mode; elements corresponding to reverse mode can be superposed on this in a similar manner to increase its accuracy in the saturation mode.

7.2.2 Ebers–Moll Model

The Gummel–Poon modelling approach utilized the integrated charge in various regions as the basis for modelling the transport and storage effects under quasi-static conditions. The Ebers–Moll model utilizes currents as the basis for the modelling, and we can derive it, in its simplest form, from the Gummel–Poon approach in a fairly straightforward fashion. The form we derive here, by calculating the currents associated with the charges, invokes all the approximations that formed the basis of our Gummel–Poon models. We consider the HBT with a base–emitter junction bias of V_{BE} and base–collector junction bias of V_{BC}. Let us consider the forward current in the emitter of this structure. Since $V_{BC}^{F} = 0$, we have

$$
\begin{aligned}
I_E^F &= \mathcal{A}\left[J_p^F(w_{E'}) + J_n^F(w_E)\right] \\
&= \mathcal{A}\frac{q\overline{D}_{pE}n_i^2}{\mathcal{G}\mathcal{N}_E'}\left[\exp\left(\frac{qV_{BE}}{kT}\right) - 1\right] + \\
&\quad\, \mathcal{A}\frac{q\overline{D}_{nB}n_i^2}{\mathcal{G}\mathcal{N}_B'}\left[\exp\left(\frac{qV_{BE}}{kT}\right) - 1\right],
\end{aligned}
\tag{7.69}
$$

where the first term represents the hole injection term in the emitter and the second term represents the electron injection term. This can be simply written as

$$I_E^F = I_{ES}\left[\exp\left(\frac{qV_{BE}}{kT}\right) - 1\right], \tag{7.70}$$

where

$$I_{ES} = \mathcal{A}\left(\frac{q\overline{\mathcal{D}}_{pE}n_i^2}{\mathcal{G}\mathcal{N}_E'} + \frac{q\overline{\mathcal{D}}_{nB}n_i^2}{\mathcal{G}\mathcal{N}_B'}\right). \tag{7.71}$$

In this same spatial junction region of the device, there exists a current due to the reverse mode of the device. This is the fraction of the electron current injected due to the bias V_{BC}, and collected following a base transport process which is characterized by the reverse base transport factor α_T^R whose magnitude is usually considerably smaller than unity. This reverse current is related as

$$\begin{aligned}
I_E^R &= \alpha_T^R \mathcal{A} J_n^R(w_C) \\
&= -\alpha_T^R \mathcal{A}\frac{q\overline{\mathcal{D}}_{nB}n_i^2}{\mathcal{G}\mathcal{N}_B'}\left[\exp\left(\frac{qV_{BC}}{kT}\right) - 1\right] \\
&= -I_{OE}\left[\exp\left(\frac{qV_{BC}}{kT}\right) - 1\right],
\end{aligned} \tag{7.72}$$

where

$$I_{OE} = \alpha_T^R \mathcal{A}\frac{q\overline{\mathcal{D}}_{nB}n_i^2}{\mathcal{G}\mathcal{N}_B'}. \tag{7.73}$$

This current is opposite in magnitude to the forward mode current, and by definition is related to the reverse injected current (occurring physically in the collector region and hence identified as I_C^R) by

$$I_E^R \equiv \alpha_R I_C^R. \tag{7.74}$$

So, in the physical emitter region of the device structure, two current sources may be identified. One is associated with the forward injection process, and consists of both electrons and holes, a diode-like current. The other is associated with the reverse collection process, and consists only of electrons. It is proportional to the reverse injection current from the physical collector region.

We may write, corresponding to this, the forward and reverse currents in the physical collector region. The forward current is composed of electrons only; it is the fraction of the electron current injected in the forward mode at the emitter, which is successfully transported through the base. As before, in all these we consider the collector transport factor as unity. This current is proportional to the diode current of the base–emitter junction. The reverse current consists of the injection process at the base–emitter junction and consists of both electrons and holes. This current, like the forward current in the emitter, is diode-like. The forward collector current can be written as

$$
\begin{aligned}
I_C^F &= \alpha_T^F A J_n^F(w_E) \\
&= A\alpha_T^F \frac{q\overline{D}_{nB}n_i^2}{\mathcal{G}\mathcal{N}_B'} \left[\exp\left(\frac{qV_{BE}}{kT}\right) - 1 \right] \\
&= I_{OC} \left[\exp\left(\frac{qV_{BE}}{kT}\right) - 1 \right],
\end{aligned}
\tag{7.75}
$$

where

$$
I_{OC} = A\alpha_T^F \frac{q\overline{D}_{nB}n_i^2}{\mathcal{G}\mathcal{N}_B'}.
\tag{7.76}
$$

Like the reverse emitter current, this current can be written as

$$
I_C^F = \alpha_F I_E^F.
\tag{7.77}
$$

The reverse injection current in the collector is

$$
\begin{aligned}
I_C^R &= A\left[J_p^R(w_{C'}) + J_n^R(w_C) \right] \\
&= -A\frac{q\overline{D}_{pC}n_i^2}{\mathcal{G}\mathcal{N}_C'} \left[\exp\left(\frac{qV_{BC}}{kT}\right) - 1 \right] - \\
&\quad A\frac{q\overline{D}_{nB}n_i^2}{\mathcal{G}\mathcal{N}_B'} \left[\exp\left(\frac{qV_{BC}}{kT}\right) - 1 \right] \\
&= -I_{CS} \left[\exp\left(\frac{qV_{BC}}{kT}\right) - 1 \right],
\end{aligned}
\tag{7.78}
$$

where

$$
I_{CS} = A\left(\frac{q\overline{D}_{pC}n_i^2}{\mathcal{G}\mathcal{N}_C'} + \frac{q\overline{D}_{nB}n_i^2}{\mathcal{G}\mathcal{N}_B'} \right).
\tag{7.79}
$$

The base current is interrelated with these; it consists of the hole injection current that is available through the ohmic electrode, and the current corresponding to the holes that recombine during the electron transit in the base region. So, the forward base current can be written as

$$
I_B^F = A\frac{q\overline{D}_{pE}n_i^2}{\mathcal{G}\mathcal{N}_E'} \left[\exp\left(\frac{qV_{BE}}{kT}\right) - 1 \right] +
$$

$$\mathcal{A}(1 - \alpha_T^F)\frac{q\overline{D}_{nB}n_i^2}{\mathcal{GN}_B'}\left[\exp\left(\frac{qV_{BE}}{kT}\right) - 1\right]$$

$$= \mathcal{A}\left[\frac{q\overline{D}_{pE}n_i^2}{\mathcal{GN}_E'} + (1 - \alpha_T^F)\frac{q\overline{D}_{nB}n_i^2}{\mathcal{GN}_B'}\right]\left[\exp\left(\frac{qV_{BE}}{kT}\right) - 1\right],$$

$$(7.80)$$

and the reverse base current can be written as

$$I_B^R = \mathcal{A}\left[\frac{q\overline{D}_{pC}n_i^2}{\mathcal{GN}_C'} + (1 - \alpha_T^R)\frac{q\overline{D}_{nB}n_i^2}{\mathcal{GN}_B'}\right]\left[\exp\left(\frac{qV_{BC}}{kT}\right) - 1\right]. \quad (7.81)$$

The total currents flowing through the emitter, collector, and base electrodes consist of the forward and reverse currents. These equations, formally known as the Ebers–Moll equations, are

$$\begin{aligned}
I_E &= I_E^F + I_E^R \\
&= I_{ES}\left[\exp\left(\frac{qV_{BE}}{kT}\right) - 1\right] - I_{OE}\left[\exp\left(\frac{qV_{BC}}{kT}\right) - 1\right], \quad (7.82) \\
I_C &= I_C^F + I_C^R \\
&= I_{OC}\left[\exp\left(\frac{qV_{BE}}{kT}\right) - 1\right] - I_{CS}\left[\exp\left(\frac{qV_{BC}}{kT}\right) - 1\right],
\end{aligned}$$

$$(7.83)$$

and

$$I_B = I_B^F + I_B^R. \quad (7.84)$$

In these, both I_{ES} and I_{CS} consist of electron and hole components; the form of the terms corresponding to these assumes a diode-like form with I_{ES} and I_{CS} identifying the saturation currents of the diodes. The equivalent circuit corresponding to the above set of equations is shown in Figure 7.7. The terms corresponding to I_{OE} and I_{OC} are dependent terms related to the injection terms through the forward and reverse current transfer ratios; they represent the coupling between the two regions, which is the basis for transistor action. Assuming unity collector transport factor,

$$I_{OC} = \alpha_F I_{ES} = \gamma_E^F \alpha_T^F I_{ES}, \quad (7.85)$$

and

$$I_{OE} = \alpha_R I_{CS} = \gamma_C^F \alpha_T^R I_{CS}. \quad (7.86)$$

This is what has been called the injection version of the Ebers–Moll model.

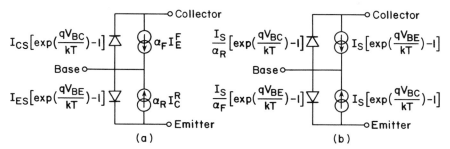

Figure 7.7: Equivalent circuits (a) and (b) based on the Ebers–Moll model. These are the simplest representations that model the current flow.

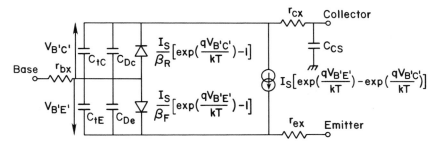

Figure 7.8: An extension of the equivalent circuit based on the Ebers–Moll model; this circuit is more adaptable to inclusion of higher-order effects such as the Early effect, recombination, and saturation effects due to heterostructures at the base–collector junction.

The injection version is not the only form in which the Ebers–Moll model is employed. Other forms exist; one example is shown in Figure 7.8. The appeal of these other adaptations is in their extendibility for inclusion of secondary or higher-order effects in modelling. They are also numerically more adaptable. The first of these extensions is the transport version; it follows from our derivation using a change of the reference currents from I_{ES} and I_{CS} terms to I_{OE} and I_{OC} terms, i.e., the currents that get transported across the base. These reference currents are characterized by one identical saturation current (see Problem 7) instead of two reference currents. In the highly parameterized environment common to computer-aided design models, this feature is particularly appealing. Another form (see Problem 8) can be viewed as a hybrid-pi version of the Ebers–Moll model. It utilizes the commonality of the basis in currents, and the properties of the transport version, to generate an equivalent circuit similar in appearance to the small-signal equivalent circuits of FETs. Indeed, small-signal and circuits based on perturbation analysis using linearization of this quasi-static model, reduce to nearly identical circuits. This last adaptation of the Ebers–Moll model has served most generally as the basis for extensions to include effects of resistances, high injection effects of transport and storage, etc.

7.3   Implications of Heterostructures and Alloy Grading

We will now consider some selected aspects of the theory of operation of HBTs. We will emphasize the distinguishing aspects of the operation vis-a-vis the quasi-static aspects of operation discussed earlier. Our earlier discussion is restricted to low-level injection conditions and is particularly suitable under the low-level injection conditions for the graded HBT and the homostructure bipolar transistor.

The particular difference between homojunction and heterojunction devices is the change in bandgap that can be achieved both on the atomic distance scale and on the device length scale. Thus, one may achieve an interface where the conduction band edge energy changes rapidly, as well as one where it changes very slowly. The former is the abrupt heterostructure device, the basis for successful HFETs, hot carrier devices, etc., while the latter is the graded heterostructure device, quite common in the HBTs. Figure 4.24 had shown various combinations of type I abrupt and graded heterojunctions between a large and a small bandgap material in thermal equilibrium.

In the abrupt case, the conduction band discontinuity ΔE_c is a given from experimental measurements. As we have discussed, although we do not yet quite completely understand its origin, its magnitude in abrupt heterostructures for many of the interesting cases has been precisely determined. It is not that which is predicted by electron affinity rule. At abrupt interfaces, the vacuum level has a discontinuous change and so does the electrostatic potential, which has been defined with respect to some reference position of the vacuum level at some location in the semiconductor. The abrupt structures require abruptness at a length scale closer to the lattice constant. The other extreme of this heterostructure is the graded heterostructure where no discontinuity occurs in the conduction or the valence band. This is in the limit where the alloy composition, a material parameter, is being changed on a much larger scale than the lattice constant scale, and the forces that give rise to the discontinuities vanish. Usually, a grading distance of the order of 100 to 300 Å, related to the Debye length, is sufficient to remove the discontinuity. The relationship to Debye length follows from arguments based on the alloy fields and the electrostatic fields, and was discussed in Chapter 4. The Debye length is related to the screening length and hence to the electrostatic field.

The electrostatic potential change across the Debye length is of the order of kT/q. The modelling of the two distinct extremes of heterostructure grading is quite straightforward. The abrupt case can be shown to be similar to a metal–semiconductor diode case, because the current that can be extracted by drift-diffusion in the quasi-neutral region is much larger than the current that can be supplied by the junction itself. The current transport is therefore limited by characteristics of the junction itself. In general, one would have to determine the thermionic and field emission components to determine this current. In this respect it is like the Schottky diode. Note that drift-diffusion in the junction does not determine the current if the mobility and diffusivity are high. In the abrupt heterojunction limit, therefore, the current is limited by thermionic emission over the barrier and field emission through the barrier for most compound semiconductors of interest. At the other extreme, in the graded case, the extracted current (by drift diffusion in the quasi-neutral base) is the limiting current. This is similar to our bipolar treatment to this point, and hence leads to a straightforward analysis. Finite element device programs actually look at the ΔE_c for successive mesh points and use the thermionic model if the gradient is large compared to the electrostatic field. In our treatment, we will emphasize the graded heterostructure case. The abrupt heterostructure case is a specialized case best treated by modelling programs that include the hot electron effects that occur in the abrupt heterostructure case. Considerations related to this were discussed in detail in Chapter 4.

While we considered grading in the base–emitter junction region, it can also be used in the base region to effectively increase the quasi-field for electrons of an n–p–n device. This leads to changes in storage and different signal frequency dependence that is of considerable interest.

First, we consider the quasi-static base transport behavior in the presence of a quasi-field. Our transport equations for the homojunction case have to be modified for the changes resulting from alloy grading. Here, we summarize our discussion of Chapter 3 on adaptation of the drift-diffusion equation approach to heterostructures. Recall that we can write the carrier concentration as a function of the quasi-Fermi level with the coordinate system chosen as in Figure 3.14.

$$n = N_C \exp\left[\frac{q(\psi + \phi_C - \phi_n)}{kT}\right], \qquad (7.87)$$

where ϕ_C is the conduction band edge potential with respect to the vacuum level reference. It follows, then, within the drift-diffusion approximation, that

$$
\begin{aligned}
J_n &= -qn\mu_n \nabla \phi_n \\
&= -qn\mu_n \nabla \left\{ \psi + \phi_C - \frac{kT}{q} \ln\left(\frac{n}{N_C}\right) \right\} \\
&= qn\mu_n \left\{ -\nabla\psi - \nabla\left[\phi_C + \frac{kT}{q}\ln(N_C)\right] \right\} + q\mathcal{D}_n \nabla n, \quad (7.88)
\end{aligned}
$$

and similarly,

$$J_p = qp\mu_p \left\{ -\nabla\psi - \nabla\left[\phi_V - \frac{kT}{q}\ln(N_V)\right] \right\} - q\mathcal{D}_p \nabla p. \qquad (7.89)$$

By defining

$$
\begin{aligned}
\phi_{Cn} &= \phi_C + \frac{kT}{q}\ln(N_C) \\
\text{and} \quad \phi_{Vp} &= \phi_V - \frac{kT}{q}\ln(N_V), \qquad (7.90)
\end{aligned}
$$

we account for variations in electron and hole affinity, effective masses, etc. We assumed Maxwell–Boltzmann statistics in deriving this. This is not necessary; the degeneracy effects and consequences of Fermi–Dirac statistics, which are related to use of high doping, can be directly included by a prefactor proportional to the Fermi integrals in the density of states terms. These additional terms are $(kT/q)\ln\left[F_{1/2}(\eta_{fn})/\exp(\eta_{fn})\right]$ for n-type and $(kT/q)\ln\left[F_{1/2}(\eta_{fp})/\exp(\eta_{fp})\right]$ for p-type. Thus, within the drift-diffusion approximation,

$$
\begin{aligned}
J_n &= qn\mu_n\left\{-\nabla\psi - \nabla\phi_{Cn}\right\} + q\mathcal{D}_n\nabla n \\
\text{and} \quad J_p &= qp\mu_n\left\{-\nabla\psi - \nabla\phi_{Vp}\right\} - q\mathcal{D}_p\nabla p \qquad (7.91)
\end{aligned}
$$

are the modified equations for transport in graded-alloy structures. These together with Poisson's equation and the current continuity equation allow us to solve many of the interesting problems in heterostructure bipolar transport. Quite often, we actually ignore the ∇N_C and ∇N_V terms of

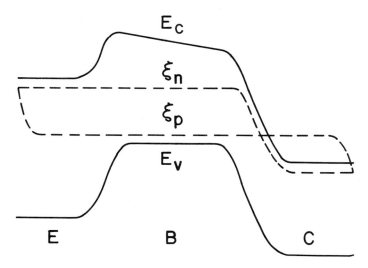

Figure 7.9: Band diagram of a graded alloy base HBT.

$\nabla \phi_{Cn}$ and $\nabla \phi_{Vp}$ because they tend to be small. One of these interesting problems is the static current through the base of a graded alloy base HBT shown in Figure 7.9. We will return to this problem in the section on small-signal analysis. Here we wish to analyze, for quasi-static conditions, the consequences of alloy grading on the base transport.

We are interested in the problem of determining J_n, given compositional grading in the base, and we are interested in determining the modifications to the relationship

$$J_n = \frac{q\mathcal{D}_n n_i^2 \exp\left(qV_{BE}/kT\right)}{\int_0^{w_B} N_A dz} \tag{7.92}$$

that exists for homojunction bipolar transistors. The denominator is one of the Gummel numbers. Recall that we had dealt with effective Gummel numbers, etc., that accounted for recombination, changes in bandgap, changes in doping, etc. The changes in bandgap can be ignored for the electron current in the base of a homojunction transistor[6] and hence the most basic form of Gummel number is utilized. The base time constant for this is given by the relation (see Equation 7.23 with the inclusion of a change in coordinate system; the base is assumed to extend from $z = 0$ to

---

[6]So long as bandgap narrowing is not significant.

$z = w_B$)

$$\tau_B = \frac{1}{\mathcal{D}_n} \int_0^{w_B} \frac{1}{N_A} \left\{ \int_z^{w_B} N_A d\zeta \right\} dz, \tag{7.93}$$

which determines the quasi-static base time constant when N_A is allowed to vary across the base. It reduces to

$$\tau_B = \frac{w_B^2}{2\mathcal{D}_n} \tag{7.94}$$

in the uniformly doped case, as expected.

We have

$$
\begin{aligned}
J_n &= -q\mu_n n \nabla \phi_n \\
\text{and} \quad J_p &= q\mu_p p \nabla \phi_p.
\end{aligned}
\tag{7.95}
$$

In the base of a bipolar transistor, p is large and J_p is small because a useful device exhibits current gain. Thus, $\nabla \phi_p = 0$, and

$$
\begin{aligned}
J_n &\approx -q\mu_n n \nabla (\phi_n - \phi_p) \\
&= -q\mu_n n \nabla (\phi_p - \phi_n).
\end{aligned}
\tag{7.96}
$$

By the definition of quasi-Fermi levels,

$$np = n_i^2 \exp\left[\frac{q(\phi_p - \phi_n)}{kT} \right], \tag{7.97}$$

where n_i^2 is now dependent on the position in the base. This expression can be recast for the purposes of determining current density as

$$\nabla(\phi_p - \phi_n) = \frac{kT}{q} \nabla \ln\left(\frac{np}{n_i^2}\right) = \frac{kT}{q} \frac{n_i^2}{np} \nabla\left(\frac{np}{n_i^2}\right). \tag{7.98}$$

Substituting for the gradient in the difference of quasi-Fermi levels,

$$\nabla \frac{np}{n_i^2} = -\frac{J_n}{q}\left(\frac{p}{\mathcal{D}_n n_i^2}\right), \tag{7.99}$$

and hence

$$\left. \frac{np}{n_i^2} \right|_{w_B} - \left. \frac{np}{n_i^2} \right|_z = -\frac{J_n}{q} \int_z^{w_B} \frac{p}{\mathcal{D}_n n_i^2} d\zeta. \tag{7.100}$$

At the base edge of the base–collector depletion region ($z = w_B$), $(np/n_i^2) \to 0$, at $z = 0$, the emitter edge of the quasi-neutral base region, the Shockley boundary condition yields

$$\left. \frac{np}{n_i^2} \right|_{z=0} = \exp\left(\frac{qV_{BE}}{kT}\right), \tag{7.101}$$

giving

$$J_n = \frac{q \exp\left(qV_{BE}/kT\right)}{\int_0^{w_B} (p/\mathcal{D}_n n_i^2)\, dz}. \tag{7.102}$$

This is the simpler form from our extended Gummel–Poon expression where we had allowed for variations in material parameters including effects such as bandgap narrowing.

In the more general form, the minority carrier density of electrons in the p-type base is

$$n(z) = \frac{J_n}{q} \frac{n_i^2}{p(z)} \int_z^{w_B} \frac{p}{\mathcal{D}_n n_i^2}\, d\zeta, \tag{7.103}$$

giving for the time constant of interest,

$$\tau_B = \frac{q \int_0^{w_B} n(z)\,dz}{J_n} = \int_0^{w_B} \left[\frac{n_i^2}{p} \int_z^{w_B} \frac{p}{\mathcal{D}_n n_i^2}\, d\zeta \right]. \tag{7.104}$$

As an example of changes in τ_B, compare an example of constant N_A, but varying bandgap

$$E_g = E_{g0} - q\mathcal{E}_e z, \tag{7.105}$$

where $\mathcal{E}_e$ is the magnitude of the quasi-electric field for electrons induced by alloy grading, e.g., a change in bandgap using a linear grading at small aluminum mole-fractions. E_{g0} is the bandgap at the emitter edge of the quasi-neutral base region, and within this base region the bandgap change varies linearly with position. Since the intrinsic carrier concentration varies exponentially with the bandgap, ignoring the second-order effective mass effects, etc., we may write the position-dependent intrinsic carrier concentration as

$$n_i^2(z) = n_{i0}^2 \exp\left(\frac{q\mathcal{E}_e z}{kT} \right), \tag{7.106}$$

where n_{i0} is the intrinsic carrier concentration associated with the bandgap E_{g0} at the emitter edge of the base quasi-neutral region. The corresponding quantities at the collector edge are n_{i1} and E_{g1}. The current density is given as

$$\begin{aligned}
|J_n| &= -\frac{q\mathcal{D}_n}{N_A} n_{i0}^2 \frac{q\mathcal{E}_e}{kT} \frac{\exp\left(qV_{BE}/kT\right)}{\left[\exp\left(-q\mathcal{E}_e w_B/kT\right) - 1\right]} \\
&= \frac{q^2 \mathcal{E}_e \mathcal{D}_n}{N_A kT} \frac{n_{i0}^2 n_{i1}^2}{n_{i1}^2 - n_{i0}^2} \exp\left(\frac{qV_{BE}}{kT} \right),
\end{aligned} \tag{7.107}$$

and

$$\tau_B = \frac{w_B kT}{q\mathcal{D}_n \mathcal{E}_e} \left\{ 1 - \frac{kT}{q\mathcal{E}_e w_B} \left(1 - \frac{n_{i0}^2}{n_{i1}^2} \right) \right\}. \tag{7.108}$$

As an example, consider a linear change in bandgap of ΔE_g across the base, then

$$n_{i1}^2 = n_{i0}^2 \exp\left(\frac{\Delta E_g}{kT}\right),$$ (7.109)

$$\Delta E_g = q\mathcal{E}_e w_B,$$ (7.110)

and

$$\tau_B = \frac{w_B^2}{\mathcal{D}_n} \frac{kT}{\Delta E_g} \left\{1 - \frac{kT}{\Delta E_g}\left[1 - \exp\left(-\frac{\Delta E_g}{kT}\right)\right]\right\}.$$ (7.111)

If the bandgap change is sufficiently large, then

$$\tau_B = \frac{w_B^2}{2\mathcal{D}_n}\left(\frac{2kT}{\Delta E_g}\right).$$ (7.112)

For a total change in the bandgap of 0.1 eV at 300 K, the base time constant reduces by a factor of ≈ 2 due to the bandgap grading.

This is a simplistic view of the transport across the quasi-neutral base region. The solution of this problem, for transport across a graded heterojunction where the grading may be insufficient, and hence may limit the transport, is considerably more complicated. As current density increases, or electrostatic potential decreases such as in high forward bias, the alloy potential gradient term related to $\nabla\phi_{Cn}$ may actually become important. At a large bandgap/small bandgap N–p junction, this alloy field would oppose the electrostatic field. This becomes obvious when one considers the case of p–P and p–N heterojunctions shown in Figure 7.10. Here we have adopted the convention of denoting the smaller bandgap material by lower case and the larger bandgap material by uppercase letters. Note that while there is no barrier for electron movement from the small bandgap to the large bandgap material for the p–N junction, one exists for the p–P case. In fact, this is used with success in double heterojunction lasers for carrier confinement. The p–P junction has a smaller change in the electrostatic potential as opposed to the p–N junction. In other cases, this low electrostatic field may come about because of charge transport in the junction. An example of this is the base–collector junction of a bipolar transistor at high current density. It can also come about directly due to applied bias, such as in the base–emitter junction of a bipolar transistor.

This discussion brings up another interesting question of what the appropriate grading is that results in a monotonic variation of both the conduction band and the valence band.[7]. Because under depletion conditions

---

[7]We discussed this question for isotype heterojunctions in Chapter 4. The reader

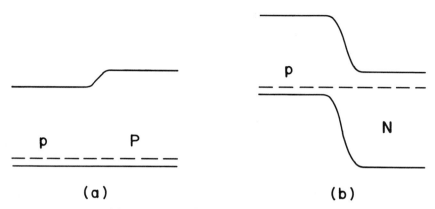

Figure 7.10: p–P (a) and p–N (b) graded heterojunctions at thermal equilibrium.

for a uniformly doped junction the electrostatic potential changes parabolically, a parabolic bandgap change will appear as an additional parabolic variation of the valence band energy at low biases. Effects related to doping in this variation are shown in Figure 7.11. However, this is only true under low bias and low current conditions. When the current density increases, the forward bias results in a decreased electrostatic potential. Then $\nabla\psi$ may become smaller than $\nabla E_g/q$, and some of the change in the alloy potential, which included both the electron affinity and the bandgap effect, may begin appearing in the conduction band. This occurs earlier in the lower doped junction because of the lower electrostatic field.

This behavior is particularly remarkable in the collector when a wide gap heterojunction collector is employed. When such a barrier appears in the collector, large carrier storage results in the base, i.e., a larger diffusion capacitance exists, and consequently there is an increase in recombination and a decrease in current gain of devices. Note that this occurs at high current densities because that is when reduction in the electrostatic field occurs in the base–collector junction due to injected charge.

Compound semiconductor HBTs exhibit several additional effects[8] that

is referred to D. T. Cheung, S. Y. Chiang, and G. L. Pearson, "A Simplified Model for Graded-Gap Heterojunctions," *Solid-State Electronics*, **18**, p. 263, 1975 and J. R. Hayes, F. Capasso, R. J. Malik, A. C. Gossard, and W. Wiegmann, "Optimum Emitter Grading for Heterojunction Bipolar Transistors," *Appl. Phys. Lett.*, **43**, No. 10, p. 949, 15 Nov. 1983.

[8]This discussion closely follows S. Tiwari and D. J. Frank, "Analysis of the Operation of GaAlAs/GaAs HBTs," *IEEE Trans. on Electron Devices*, **ED-36**, No. 10, p. 2105,

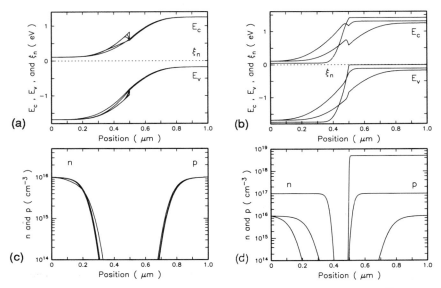

Figure 7.11: Effects of parabolic grading (grading length of 300 Å) in heterojunctions at thermal equilibrium for symmetric and one-sided junctions of $Ga_{1-x}Al_xAs/GaAs$ are shown in this figure. The parabolic grading occurs in the larger bandgap n-type semiconductor. (a) and (c) show the band edge profiles and carrier concentrations for the symmetric case. (b) and (d) show these for a variety of doping conditions. Monotonic change in the conduction band edge occurs for the largest doped case where the built-in field is the largest. In the junctions with 10^{17} cm^{-3} p-type doping and 10^{16} cm^{-3} n-type doping, there exists a small barrier that becomes more pronounced when the p-type doping is decreased to 10^{16} cm^{-3}.

we have not discussed. Use of heterojunctions leads to additional phenomena, particularly at high currents, which we have also not discussed. The primary operational differences are in the bias dependence of both charge transport and storage. These differences are caused by several processes that occur in the HBT: the injection of carriers at a varying bandgap heterostructure, the collection of carriers at a varying bandgap heterostructure, and hot carrier and quasi-drift field effects in the base. In addition to these primary effects, recombination at the surface is important as a parasitic effect, while it is virtually non-existent in silicon bipolar transistors. We will now look at these and other effects, within quasi-static and drift diffusion

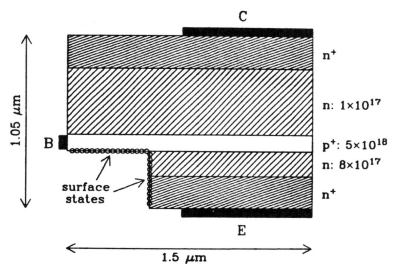

Figure 7.12: Schematic cross-section of an HBT structure used for the discussion of quasi-static characteristics. The 8×10^{17} cm^{-3} doped area is the wide gap emitter region and the 1×10^{17} cm^{-3} doped area is the wide gap collector region for those simulations in which the collector is a heterojunction. The extrinsic base length is 0.5 μm and base width is 1000 Å. From S. Tiwari and D. J. Frank, "Analysis of the Operation of GaAlAs/GaAs HBTs," *IEEE Trans. on Electron Devices*, **ED-36**, No. 10, p. 2105, ©Oct. 1989 IEEE.

approximation. Figure 7.12 shows the cross-section of the device that we will elaborate on. Our discussion will consider heterojunctions in the emitter as well as heterojunctions in the collector; the former is called the single heterostructure bipolar transistor and the latter the double heterostructure bipolar transistor. The latter usually employs a graded junction at the collector in order to suppress inefficiencies in the transport of injected carriers arising from any barriers.

7.3.1 Charge Transport and Storage in the Base–Emitter Junction

The current flow through a heterojunction is considerably more complicated than the flow through a homojunction. It depends on whether the heterojunction is graded or abrupt, since graded junctions can support drift-diffusion transport throughout the junction region, while abrupt junctions are limited by thermionic and thermionic field emission at the abrupt heterostructure. For both the graded and the abrupt junctions, the transport depends on details of the grading or abruptness. Graded junctions, e.g., have a changing alloy potential associated with the change in electron or hole affinity. This causes an alloy field, $\nabla \phi^A$, which adds to the usual electrostatic fields and modifies the current flow. Thus, parabolically and linearly graded junction regions exhibit differing total fields on the carrier as it transports through the graded region. These may be the limiting bottleneck of a transport. For the p–P graded heterojunction, the electron is confined to the narrow bandgap region by this alloy field, while the holes are free to move around. As mentioned, though, this phenomenon is a function of several other factors that influence the total field and the electrostatic potential change in the junction region. In the p–N graded heterojunction, the electron is confined in the n-type region largely by the same electrostatic considerations that apply to the homojunction. The holes, though, have both the electrostatic force and the force due to alloy potential change confining them to the p-type region. So, at low currents, with appropriate heterostructure grading, the built-in quasi-electric fields for electron transport remain essentially the same as in the homojunction transistor, except for minor second-order effects related to the density of states, donor ionization energy, etc. The built-in potentials are now different for electrons and holes. Heterostructure grading results in an increase of the hole potential barrier by the difference in bandgap and adjusted by the second-order effects mentioned above. The resulting suppression of hole injection allows the unique design advantage of allowing higher doping in the base while still maintaining good injection efficiency. A significant practical result of this is a reduction in recombination in the depletion region of the base–emitter junction even though compound semiconductors have low carrier lifetimes due to HSR recombination. The magnitude of this suppression is proportional to the hole suppression.

At large forward bias and current, both the electrostatic potential and the space charge region width are reduced, leading to anomalies in transport and storage. The anomalies occur because as the electrostatic field decreases, the relative importance of the alloy field in the conduction band increases. Furthermore, the hole suppression property of the alloy grading region becomes less effective as the space charge region width becomes shorter than the alloy grading width.

In this section, we discuss the charge transport effects at the base–emitter junction. Later on, we will expand on this to include transport effects in the quasi-neutral base and the base–collector regions, and the charge storage effects corresponding to them.

7.3.2 Alloy Grading, Doping Design and Transport at the Base–Emitter Junction

From the practical standpoint of using a bipolar transistor, the central requirements of the design of the alloy grading at a heterojunction are maximizing the current drive capability, maintaining the device gain, and minimizing the delays associated with the resistance, capacitance, and transit time. The transit time associated with the base–emitter junction is generally much smaller than the delays associated with the base–collector junction since it is a narrow region. This small base–emitter junction transit time will also form the basis for our assuming instantaneous equilibrium when deriving small-signal response of the bipolar transistor. Similarly, except at the largest forward bias, the time constant associated with the charging of the base–emitter depletion capacitance is smaller than the base–collector delays, because of the relatively large collector area. Thus, maximization of the injection current becomes a more important design criterion at the base–emitter junction. In a heterojunction, high currents occur at small differences in electrostatic potential between the two sides of the junction, i.e., at low electrostatic fields; this is the condition at which the effect of alloy potential and alloy field would begin to dominate.

In homojunction transistors such as in silicon, high-level injection effects at the base–emitter junction are primarily caused by a reduction in the forward bias voltage, due to lateral ohmic drop in the base, vertical ohmic drop in the emitter, and base push-out effect in the collector. In such transistors, secondary effects due to majority carrier–induced drift-field are kept low by choosing an adequate base doping level. This secondary effect is even more insignificant in the HBT because the doping levels are even higher than those of the homojunction transistor. Homojunction transistors are designed with emitter dopings in excess of $\approx 10^{20}$ cm^{-3} and base dopings exceeding $\approx 10^{18}$ cm^{-3} in order to maintain sufficient injection efficiency. Since HBTs have a much higher injection efficiency, they are usually designed with much smaller emitter dopings in order to reduce the emitter capacitance and the tunneling effects associated with heavier doping in both the emitter and the base. There is a minimum to this emitter doping because it limits the number of minority carriers that can be injected into the base as well as the resulting reduction in electrostatic field for any injected carrier density. Thus, low dopings and associated low electrostatic fields enhance the role of the opposing effect of the alloy potential and alloy fields.

This alloy potential effect is significant not only at the emitter heterojunction, but also at the collector heterojunction. At high electrostatic fields, the barrier due to the difference in the bandgaps appears in the valence band, but at low electrostatic fields (i.e., high biases) it also begins to appear in the conduction band. As a result, at significant forward bias, the alloy potential causes an "alloy barrier" to the flow of carriers.

This effect, which does not exist in the homojunction case (the smaller bandgap in the emitter causes an alloy field in the opposite direction from that in the heterojunction), has to be suppressed to obtain the desired current density of operation. For any current density in the base–emitter junction, suppression of this alloy barrier effect can be assured by having a monotonic increase in the total electron potential (electrostatic and alloy) from the emitter to the base. This implies that the alloy field should be maintained lower than the electrostatic field throughout the grading region. Using Poisson's equation and the knowledge that for a device with high injection efficiency the electron space charge is the dominant mobile charge, this condition can be written in the mathematical form

$$\int_{w_{E'}}^{z} \frac{q}{\epsilon_s} \left(N_D - \frac{J_{ne}}{qv} \right) dz \geq -\frac{d\phi_C^A(z)}{dz} \tag{7.113}$$

anywhere in the base–emitter depletion region. The limits of integration are from w_E, the emitter edge of the space charge region or the emitter edge of the graded region, whichever comes first, to the point z in the space charge region. Alloy barrier suppression requires that this inequality hold for all points z in the junction region.

By differentiating and solving for N_D, one obtains the following upper bound on the minimum N_D needed to guarantee the earlier inequality:

$$N_D(z) \geq -\frac{\epsilon_s}{q} \frac{d^2\phi_C^A(z)}{dz^2} + \frac{J_{ne}}{qv(z)}. \tag{7.114}$$

Lower values for N_D can in fact be sufficient if the space charge region is wider than the grading length, as will be discussed later in connection with the base–collector junction. On the right hand side of this equation, the first term is associated with the alloy barrier, and the second term is simply the concentration of mobile carriers in the depletion region. Although the first term is readily determined, the carrier concentration in a depletion region does not yield to analytic solution except in special cases such as the metal–semiconductor junction. In the situation of interest—high current density and substantial degeneracy—the carrier mobility and diffusivity are non-linearly dependent on the carrier density, the alloy composition, the quasi-electric field, and the doping concentration. To ascertain the practical implications of this constraint, we consider the case of a parabolically graded, uniformly doped junction, where the mole-fraction $u(z)$ for parabolic grading is given by

$$u(z) = u_f - u_f \left(\frac{z + z_0}{z_0} \right)^2 \qquad \text{for} \qquad -z_0 \leq z \leq 0. \qquad (7.115)$$

Here the origin is taken at the base–emitter junction, z_0 is the grading length, and u_f is the final mole fraction in the parabolic grading. Note that the largest alloy field for this parabolic grading occurs at the junction ($z = 0$). Using $\phi_C^A(z) = \gamma u(z)$ for the alloy potential of $Ga_{1-x}Al_xAs$ relative to GaAs, and this parabolic grading, gives

$$N_D \geq \frac{\epsilon_s}{q} \frac{2\gamma u_f}{z_0^2} + \frac{J_{ne}}{q v_{min}} \qquad (7.116)$$

as the minimum value for a uniform N_D in order to maintain proper heterojunction operation at high current density. Here v_{min} is the minimum carrier velocity in the graded region.

Since the J_{ne}/v_{min} term is strongly dependent on the base–emitter junction bias, numerical simulations must be used to clarify the importance of the individual terms. This is quite illuminating in itself—this velocity is the net particle velocity resulting from drift-diffusion effects; in high fields such as at the base–collector junction, it is close to the saturation velocity, but at low fields such as in a forward biased p–n junction, it can be a lot smaller since the carrier concentrations are very large. Representative results of conditions in forward biased operation for the conduction band edge and the velocity are shown in Figure 7.13. Note that the velocities are orders of magnitude smaller than the saturated velocity. This is, however, not necessarily at odds with our assumption of ignoring the related time constant. The space charge region is very small compared to all other dimensions, the quasi-neutral base region, and particularly the base–collector junction region, and hence we may ignore the associated time constant. Returning to the practical implications of constraints related to preventing an alloy potential–related barrier, the simulations show that for a $Ga_{1-x}Al_xAs/GaAs$ junction with parabolic alloy grading of 0.3 over 300 Å, the contributions of the two terms are approximately equal at a current density of 10^5 A.cm^{-2}. Because of the low velocity of injected electrons in the forward biased junction, emitter dopings exceeding 8×10^{17} cm^{-3} are needed in order to assure operation in the 10^5 A.cm^{-2} range.

Recall our discussion of Figure 4.25 which showed the behavior of a 2×10^{17} and a 8×10^{17} cm^{-3} doped n-type emitters as a function of forward bias. The lower doped emitter device showed an alloy barrier at 1.4 V

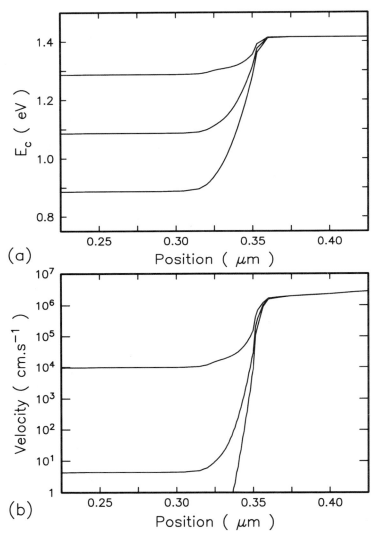

(a)

(b)

Figure 7.13: Conduction band edge (a) and the net velocity of carriers
(b) in forward biased $Ga_{1-x}Al_xAs/GaAs$ n–p junction. Forward biases of
0.9 V, 1.1 V, and 1.3 V are shown. The metallurgical junction is at 0.35 μm
with the emitter region extending from 0 to 0.35 μm.

forward bias, while the higher doped device did not. Clearly, an alloy barrier such as this can be expected to reduce the current which can be injected into the base. In this particular example, the current density through the device with 2×10^{17} cm^{-3} emitter doping is only $\approx 1.6 \times 10^4$ A.cm^{-2} at 1.4 V forward bias, while it is $\approx 6 \times 10^4$ A.cm^{-2} in the higher doped emitter device.

The effects of this alloy barrier and its emitter doping dependence are shown more completely in the device characteristics plotted in Figure 7.14, which shows the Gummel plots for the two emitter dopings and plots the excess voltage (i.e., the forward bias voltage needed in excess of the extrapolated low current exponential characteristics) and the current gain in the absence of surface recombination.

The low current fit of the Gummel plot shows an increase in current density varying at ≈ 60 mV/decade. This is a characteristic of the exponential dependence of the collector current on the base–emitter voltage. So long as the electrostatic barrier between the emitter and the base is directly modulated by the applied base–emitter bias, the carrier density injected in the base can be related by the exponential factor related to this bias. This is the origin of the $[\exp(qV_{BE}/kT) - 1]$ term in the Gummel–Poon model. A necessary condition for this direct modulation is the requirement that parasitic voltage drops such as ohmic drops be negligible, and for the graded junctions that the recombination effects be small enough that the quasi-Fermi levels are relatively flat. An interesting aside is the nature of low current characteristics in abrupt base–emitter heterojunctions. Here, since the injection mechanism is thermionic in most instances, it is naturally related by the exponential factor and hence the ≈ 60 mV/decade characteristic slope at 300 K. In the calculations of Figure 7.14, the surface recombination is set to zero. The rapid increase in excess voltage shows the onset of saturation in current. Note the larger deviation from the low current extrapolation in the lower doped device. This device has significantly smaller current handling capability. This is clearly illustrated in the lower part of Figure 7.14, where the rapid increase in excess voltage occurs at current densities which are ≈ 5 times lower in the lower doped device. The higher current gain of the lower doped device at lower current densities is a result of lower space charge region recombination.

A lower alloy field can be achieved at the junction by employing linear grading instead of parabolic grading. However, in this case the alloy field is larger at the emitter end of the grading region where the electrostatic fields are even lower. The barrier now appears at the emitter edge of the grading region. Figure 7.15 demonstrates this by plotting the conduction band edge and the quasi-Fermi level at $V_{BE} = 1.5$ V for a linearly graded and a parabolically graded junction for 8×10^{17} cm^{-3} emitter doping. The

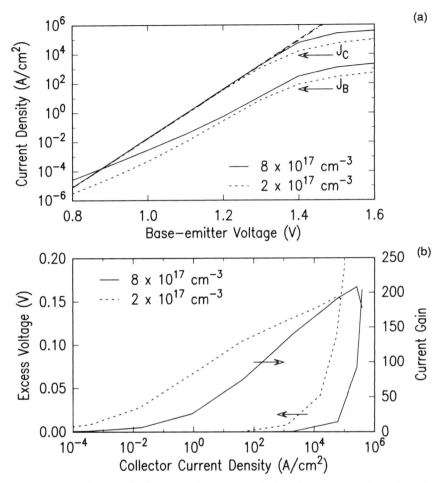

Figure 7.14: Gummel plots are shown in (a) for single heterojunction devices with emitter doping of 8×10^{17} cm^{-3} and 2×10^{17} cm^{-3} and parabolic grading. The collector current density is identified by J_C and the base current by J_B. The figure also shows a low current fit to the collector current density plot. This fit is used to derive the excess voltage shown in (b), the voltage beyond the low current extrapolation, of the base–emitter bias. From S. Tiwari and D. J. Frank, "Analysis of the Operation of GaAlAs/GaAs HBTs," *IEEE Trans. on Electron Devices*, **ED-36**, No. 10, p. 2105, ©Oct. 1989 IEEE.

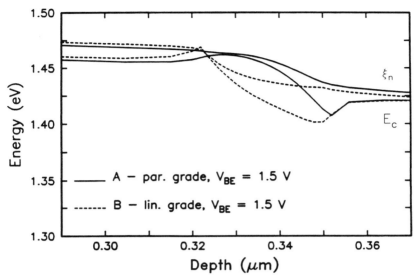

Figure 7.15: Conduction band edge (E_c) and quasi-Fermi level for electrons (ξ_n) at a base–emitter junction bias of 1.5 V for parabolic and linear grading in a single heterojunction transistor. The junction is located at 0.35 μm. The current density is $\approx 3 \times 10^5$ A.cm^{-2} through the parabolically graded device, and $\approx 2.3 \times 10^5$ A.cm^{-2} through the linearly graded device. From S. Tiwari and D. J. Frank, "Analysis of the Operation of GaAlAs/GaAs HBTs," *IEEE Trans. on Electron Devices*, **ED-36**, No. 10, p. 2105, ©Oct. 1989 IEEE.

linearly graded junction shows a barrier to injection at the emitter end of the grading region. In fact, this barrier to injection also exists at lower bias currents; Figure 7.16 shows an example at 1×10^4 A.cm^{-2} current density. One consequence of this barrier is that the current carried by the linearly graded device is less than that in the parabolically graded device for the same bias. A comparison between parabolically graded and linearly graded devices shows that the limiting current of the parabolically graded device is about a factor of three higher than that of the linearly graded device. But, it is important to recognize that the barrier due to alloy grading potential results in a dynamic resistance that restricts the current flow through the device.

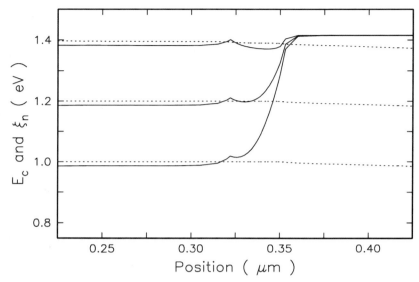

Figure 7.16: Conduction band edge and electron quasi-Fermi level for a linearly graded base–emitter junction biased at 0.9 V, 1.1 V, and 1.3 V of forward bias.

7.3.3   Base–Emitter Capacitance

The substantially faster transport in the base of the compound semiconductor HBT due to high diffusivity of electrons (i.e., a small base time constant τ_B) increases the relative importance of the emitter capacitance. Since the devices are operated at high currents and in large forward bias, this capacitance has to include both the immobile charge component (i.e., the depletion capacitance) and the mobile charge component (i.e., that due to the charges which are carrying the current—primarily electrons, but also holes at sufficiently large forward biases). The capacitances were introduced during our discussion of the Gummel–Poon and Ebers–Moll models. One limit of the capacitance is at the very lowest biases where the effect of mobile charge can be ignored, i.e., the limit of no diffusion capacitance with a depletion approximation for the junction space charge region. This capacitance is simply the parallel plate capacitance. Another limit is the capacitance associated with near flat-band conditions. This is the capacitance with the Debye length as the characteristic length scale encountered during our discussion of MIS structures in the absence of current flow. Transport of carriers and the presence of both mobile and immobile charge results in deviations from these two limits. Under active bias, and especially bias where the current density is large, we must again apply numerical techniques, using charge partitioning and bias perturbation. The capacitance can then be derived from the change of total charge in steady-state on the emitter side of the junction as a result of the change in bias. Such a perturbation technique can not separate the mobile and immobile charge contributions because they cannot be determined independently. The injected hole charge (which can be separately considered) is generally quite small and gives rise to a negligible diffusion capacitance. So, in such a technique, we can separate the depletion and diffusion capacitance (due to electron storage) in the base. The depletion region charge is the same as the net charge on the emitter side, and the storage charge is calculated by the integration of the increase in electron population in the base region. In the calculation of these capacitances in the bipolar transistor, the partitioning of the charge occurs quite naturally because of the p–n junctions, and hence is not subject to the inaccuracies that are natural to this approach. We now analyze the capacitance problem in the presence of heterostructure grading and high currents.

Figure 7.17 plots the total capacitance (depletion and diffusion) of the base–emitter junction for a single and a double heterojunction transistor. Since hole injection is significantly smaller than in the homojunction transistor, this capacitance is primarily determined by the depletion region thickness and the electron space charge in the depletion region. Both of these are in turn sensitive to the behavior of the potential in the presence of this injected charge. These junction characteristics for the single heterojunction device are illustrated in Figure 7.18, which shows the conduction band edge energy, the electron density, the hole density, and the total charge density versus depth for forward bias voltages of 0.8, 1.1, and 1.4 V. Note that the hole density is, indeed, very small because of the high injection efficiency. Even at a bias of 1.4 V, where the depletion region has become smaller than the grading region, the contribution of hole density continues to be small except in a region a few 10's of Å wide at the junction. Only at extremely low emitter dopings does the alloy barrier effect become large enough that hole injection becomes significantly important.

The behavior of the capacitance—first a rapid increase and then a drop-off at large forward bias—is very similar to that observed in homojunction transistors, with the maximum capacitance limited by Debye lengths at near flat-band conditions. As the applied bias becomes larger and the junction reaches or exceeds flat-band, the ability of the applied bias to modulate the current decreases, leading to a decrease in the charge modulation and, hence, a decreasing capacitance. As in the homojunction transistor, large mobile charge in the space charge region and voltage drops in the quasi-

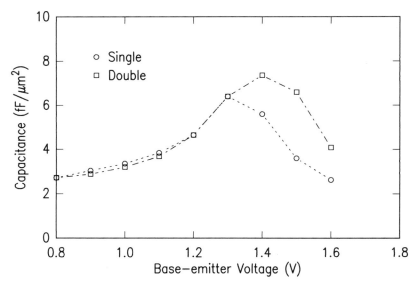

Figure 7.17: The total capacitance in fF.μm^{-2} at the base–emitter junction as a function of the applied base–emitter voltage. The base–collector voltage is 0 V. The alloy grading at the junction is parabolic, the doping in emitter is 8×10^{17} cm^{-3}, and that in the base is 5×10^{18} cm^{-3}. Capacitances for a single heterojunction device and a double heterojunction device are plotted. From S. Tiwari and D. J. Frank, "Analysis of the Operation of GaAlAs/GaAs HBTs," *IEEE Trans. on Electron Devices*, **ED-36**, No. 10, p. 2105, ©Oct. 1989 IEEE.

neutral regions both contribute to the decreasing efficacy of the applied bias. In the HBT, the alloy barrier also works to decrease the capacitance at high biases, by decreasing the modulation of the depletion region width. At high bias, the increasing voltage is accommodated by the relatively poor injection characteristics of the alloy barrier, and does not result in a large change in depletion region charge. Another way of viewing this is that the space charge region, under large forward bias conditions, has a strong conductivity modulation which limits the low of current. Indeed, if one were to continue increasing this bias even further, the charge effect would turn inductive because of this conductivity modulation.

HBTs employ lower emitter dopings than homojunction transistors because of their higher injection efficiency. The low doping gives rise to a lower emitter capacitance overall, and the effect of the alloy barrier at high bias

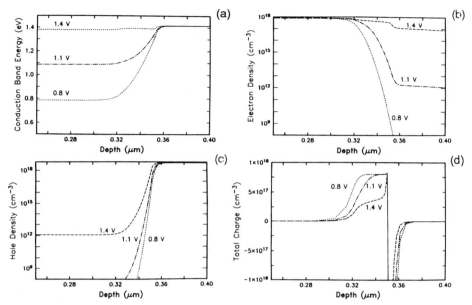

Figure 7.18: Plots of the conduction band edge energy (a), the electron density (b), the hole density (c), and the total charge density (d) at the base–emitter junction (located at 0.35 μm) for a single heterojunction device, with bias varying from low-level injection conditions (0.8 V) to moderately high-level injection conditions. From S. Tiwari and D. J. Frank, "Analysis of the Operation of GaAlAs/GaAs HBTs," *IEEE Trans. on Electron Devices*, **ED-36**, No. 10, p. 2105, ©Oct. 1989 IEEE.

is to reduce the peak capacitance even further. An increase in emitter doping leads to higher current drive capability, but causes higher capacitance throughout the bias range. In particular, the peak capacitance increases as a function of doping, and the peak occurs at a higher bias voltage because of a larger built-in voltage and a reduced effect of the alloy barrier.

The double heterojunction device capacitance–voltage curve in Figure 7.17 has a larger peak capacitance than the single heterojunction device. At these high bias levels—as we will show later, a large storage of electrons takes place in the quasi-neutral base. The increase in base charge influences the emitter injection behavior because a larger charge at the base–emitter junction also implies a larger space charge in the emitter depletion region. This increase in space charge causes most of the increase in the base–emitter capacitance at larger forward voltages, and also causes the barrier effect to be more prominent in the base–emitter depletion region.

The barrier modulation and the increase in majority carrier concentration due to the charge storage both lead to a larger hole diffusion capacitance for the double heterojunction device, which also contributes to the observed increase at sufficiently high bias.

7.3.4   Electron Quasi-Fields in Single Heterojunction Bipolar Transistors

We now consider the effect of alloy potential in the heterojunctions by discussing the electron quasi-fields, which include the effects of the electrostatic and the alloy components. Figure 7.19 shows the negative of the electron quasi-field as a function of applied forward bias at the base–emitter junction for the HBT being discussed. In this figure, the base region can be identified by the region of constant low electric field; the region of high field on the left is associated with the emitter, and the one on the right with the collector. We discuss the field variation in the base–collector junction in the next section, together with the differences between single and double heterojunction collectors. At low biases, the quasi-field varies nearly linearly with distance and is dominated by electrostatic considerations that are valid for depletion approximation. As the bias increases, the quasi-field decreases and the depletion region shrinks, until at a bias of about 1.3 V most of the region lies inside the base–emitter alloy grading region. At this bias, the alloy field $d\phi_C^A/dz$ acts to substantially reduce the electrostatic contribution to the net field. At higher biases the quasi-field, which is the sum of the electrostatic and alloy fields, decreases and even changes direction due to the influence of the alloy field (this is quite significant at the 1.5 V bias). The negative spikes in the quasi-field at the beginning and end of the grading region for 1.3 V bias and above are due to the abrupt changes in $d^2\phi_C^A/dz^2$ at the grading region edges, which cause short range breakdown of quasi-neutrality.

Although the quantitative magnitudes of the various emitter–base effects we have described can be changed by varying the modelling parameters, such as the diffusivity, mobility, etc., and their dependence on alloy composition, the qualitative variations remain the same. The largest difference in base–emitter junction behavior between HBTs and homojunction bipolar transistors is the influence of the alloy field in the junction region.

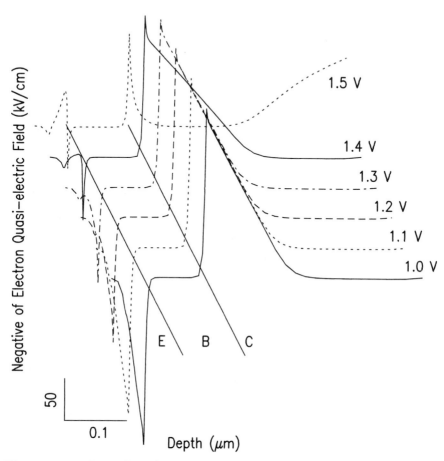

Figure 7.19: Quasi-field for an electron at the base–emitter junction, in the base, and at the base–collector junction as a function of base–emitter junction bias for the single heterojunction HBT. The base is 0.1 μm wide, and the different plots are shifted to allow a perspective view. The scales of distance and field are located in the lower left corner. The base can be identified as the region of near-zero electric field. From S. Tiwari and D. J. Frank, "Analysis of the Operation of GaAlAs/GaAs HBTs," *IEEE Trans. on Electron Devices*, **ED-36**, No. 10, p. 2105, ©Oct. 1989 IEEE.

7.4   High Current Considerations of the Base–Collector Junction

We have looked at, so far, the high current and high forward bias effects in the base–emitter junction; next we consider the base–collector junction. Our discussion of the base–emitter junction dwelt on designing a junction capable of injecting large currents. We also need to assure that the device is capable of collecting large currents with low storage and transit time to assure a desired speed or frequency characteristic. The following analyzes these properties for heterojunction collectors, and provides comparisons with homojunction collectors. In particular, we discuss the alloy barrier effects that occur in heterojunction collectors at high current densities, and the two-dimensional storage and transport characteristics which result consequently.

7.4.1   Barriers and their Influence in Heterojunction Collectors

Under low current operation, the homojunction theory of collector transport applies quite well to the heterojunction transistor. Since hole transport at the collector does not play a significant role, the junction being usually reverse biased, a heterojunction collector has little effect on the transport physics and the collector is an efficient transporter of electrons for the graded collector. In an abrupt collector, a barrier at the junction limits the flow of carriers since tunneling or thermionic emission occurs at the barrier. Both tunneling and thermionic emission are barriers to the flow of electrons, as opposed to the barrier-less situation of a graded collector.

In homojunction transistors, at medium and high current densities, various effects are known to become important, e.g., the Kirk effect (base push-out), and the Webster effect (base field and conductivity modulation). The Kirk effect is related to the compensation of the background immobile charge density by the mobile charge density; this results in a net decrease of the total charge density in the space charge region, and hence initially a broadening of the space charge region and ultimately a pushing out of the quasi-neutral base region in to the metallurgical junction region. At these high current densities, the newer collecting region occurs at the higher doped contact region of the collector, a region commonly referred to as the sub-collector region. The Webster effect, another high current effect, results from an increase in the minority carrier charge density in the quasi-neutral base region to a magnitude comparable to the majority carrier density. As discussed in Chapter 4 for p–n junctions under high injection, the consequences of this excess charge are excess potential drops. It also causes an increase in charge storage, which in turn causes an increased non-linearity from the ideal exponential modelled behavior in the Gummel–Poon analysis. For compound semiconductors with their low lifetimes, it would also result in a decrease in current gain of transistors. The Webster effect is, however, negligible in heterostructure bipolar transistors since the base doping is relatively large.

However, the Kirk effect remains equally important in HBTs. In addition, in an HBT with a heterojunction collector, the effect of the alloy barrier appears, and can be more important than the Kirk effect.

In discussing the existence of alloy barriers at high current in the emitter region of the HBT, we alluded to the importance of this effect in the heterojunction collector. The nature and origin of the effect in the collector are essentially the same as in the emitter: at low currents the alloy potential difference appears in the valence band, resulting in the usual hole-retarding barrier and field, while at higher currents it appears partially in the conduction band as a barrier to the flow of the minority carriers. The consequences of this barrier are excess storage of charge in the base and thus an increase in the associated capacitance, a decrease in the device current gain due to recombination, and a saturation in the collector current. The first of these has serious consequences because storage is exponentially related to the potential barrier. These effects worsen as the collector-base junction becomes forward biased, and thus limit the minimum usable value of V_{CE} in circuit applications.

Figure 7.20 shows two perspective plots of the conduction band edge energy in the base region of a double heterojunction transistor at $V_{BE} = 1.3$ V and 1.5 V, with $V_{BC} = 0$ V. The region plotted extends 500 Å into both the emitter and the collector. At the low bias condition, the potential profile is similar to that of a well-behaved homojunction bipolar transistor. However, at the higher injection voltage, a barrier due to both the alloy potential and the electron space charge appears at the collector junction. This barrier is the largest where the current density is the largest, i.e., opposite the emitter junction. The barrier is lowered towards and in the extrinsic part of the device because the current density is lower in those parts. There is a significant spreading of the electron flow towards the extrinsic base–collector junction area in this device because the alloy barrier varies proportionally with current density.

This consequence of the alloy barrier is a form of collector current spreading effect. The emitter crowding remains low in HBTs because the base doping is high. The spreading of the current to the extrinsic region results in an increased storage of carriers in the extrinsic part of the device. Before quantifying the magnitude of the storage as a result of this, we point

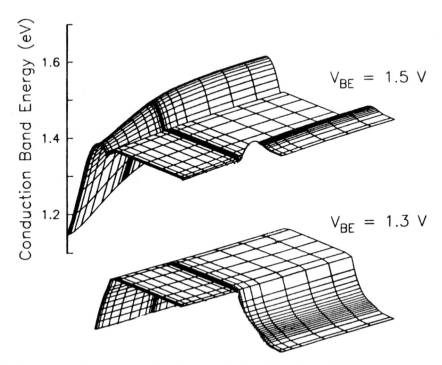

Figure 7.20: A surface plot of the conduction band edge in the base and at the junctions with the emitter and the collector for a forward bias voltage of 1.5 V and 1.3 V for a double heterojunction device. The emitter junction is located in the lower right, and the collector is the broader region in the background. The base is the constant energy region. The collector heterojunction is parabolically graded. From S. Tiwari and D. J. Frank, "Analysis of the Operation of GaAlAs/GaAs HBTs," *IEEE Trans. on Electron Devices*, **ED-36**, No. 10, p. 2105, ©Oct. 1989 IEEE.

out the serious consequence of this in a double heterojunction bipolar transistor where a wider gap junction is buried in the extrinsic base collector region. Such transistors are of some interest since they confine the flow of carriers in the graded heterojunction region and hence result in a symmetrical operation of the transistor at low current densities. Figure 7.21 shows the same two dimensional band edge plots as in Figure 7.20, except that now there is a buried extrinsic wider gap base–collector junction. The large barrier on the left is a result of the extrinsic p-type base region extending

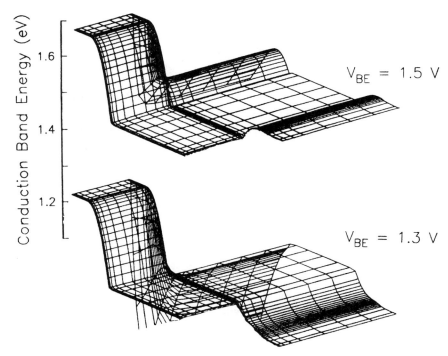

Figure 7.21: A surface plot of the conduction band edge in the base and at the junctions with the emitter and the collector for a forward bias voltage of 1.5 V and 1.3 V for a double heterojunction device that has a buried wide gap p-type region in the extrinsic region. Note that compared to a double heterojunction bipolar without a buried wide gap p-type region, there is a decrease in the effective collector area because electrons which diffuse out into the extrinsic base region cannot be collected. From S. Tiwari and D. J. Frank, "Analysis of the Operation of GaAlAs/GaAs HBTs," *IEEE Trans. on Electron Devices*, **ED-36**, No. 10, p. 2105, ©Oct. 1989 IEEE.

into the collector. This barrier, which is approximately the bandgap difference, is larger than the alloy barrier formed at high currents. The current is forced to transport to the collector in the intrinsic part, leading to a larger storage in the intrinsic base region, and also in the extrinsic base region due to diffusion of carriers. This transistor has even worse minority carrier storage and current saturation than the simpler double heterojunction bipolar transistor.

Figure 7.22 provides a comparison of the base storage effects in heterojunction and homojunction collector devices. Each figure plots the electron concentration along a lateral cross-section at the middle of the base for applied biases varying from 1.0 V to 1.6 V. The storage behavior is similar for the two devices at low biases (up to 1.4 V of bias), with an exponential decay length of ≈ 850 Å that is determined primarily by the base width of the device (1000 Å). At high biases, the homojunction collector device shows only the limitations of the injection process and maintains the exponential decay length. The heterojunction device, however, develops a much longer, non-exponential decay characteristic, which is limited by the designed spacing of 0.5 μm between the base ohmic contact and the intrinsic base region. Thus, the increased storage in the heterojunction collector bipolar transistor occurs both in the intrinsic and in the extrinsic part of the device.

The magnitude of this storage depends exponentially on the height of the barrier between the base and the collector. As in the emitter, the height of this barrier can be controlled by the use of doping, as discussed below. Alternately, since a substantial portion of the barrier is due to the alloy potential, one could reduce the base charge storage by reducing the AlAs mole-fraction u_f, and hence the alloy potential. The trade-off in so doing is that there will be more hole injection. This trade-off could be improved by finding and using a different semiconductor alloy system in which the alloy potential conduction band–to–valence band ratio ϕ_C^A/ϕ_V^A is smaller.

A comparison between the size of the alloy barrier effect and the size of the Kirk effect (electron space charge barrier) can be obtained by considering the collector junction version of Equation 7.114. If one assumes a base doping much higher than the collector doping, constant collector doping, parabolic grading, and a space charge region thickness W greater than the grading length z_0, then one can approximately solve the collector version of Equation 7.114 for N_D (rather than differentiating it as done for the base–emitter junction) to obtain

$$N_D > \frac{\epsilon_s}{q} \frac{2\gamma u_f}{z_0 W(N_D, V_{BC})} + \frac{J_C}{q\bar{v}}. \tag{7.117}$$

Here, $\bar{v} = \langle 1/v(z) \rangle^{-1}$ is the harmonic mean of the carrier velocity in the collector, and W depends on the doping and the bias voltage. This is a more relaxed requirement than that for the emitter, especially when the depletion region extends significantly farther than the grading region into the collector. If one further assumes that $J_C/\bar{v}$ is constant, one can evaluate W and solve for N_D explicitly,

$$N_D > \frac{\epsilon_s}{q} \frac{2\gamma^2 u_f^2}{z_0^2 \left(-V_{BC} + \psi_{j0} + \gamma u_f/q\right)} + \frac{J_C}{q\bar{v}}, \tag{7.118}$$

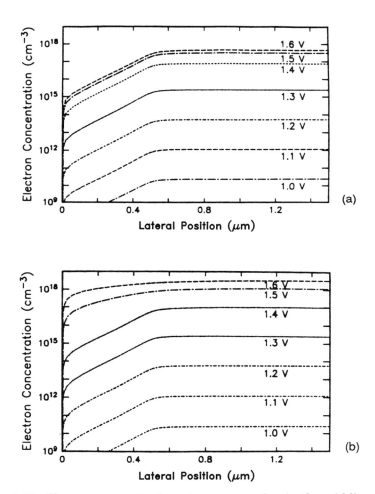

Figure 7.22: Electron concentration at a cross-section in the middle of the base, as a function of bias for a single heterojunction (a) and a double heterojunction (b) device. The figures differ at the highest biases, where the tail is much larger in the double heterojunction device ($\approx$ diffusion length) than in the single heterojunction device ($\approx$ base width). From S. Tiwari and D. J. Frank, "Analysis of the Operation of GaAlAs/GaAs HBTs," *IEEE Trans. on Electron Devices*, **ED-36**, No. 10, p. 2105, ©Oct. 1989 IEEE.

where ψ_{j0} is the built-in potential. To illustrate the characteristics of this constraint, we consider a specific example. For a current density of $J_C = 1 \times 10^5$ A.cm^{-2}, a mean velocity equal to the saturated velocity of 1×10^7 cm.s^{-1}, $\gamma u_f = 0.24$ V, $V_{BC} = 0$ V, $\psi_{j0} = 1.4$ V, and a parabolic grading length $z_0 = 300$ Å, at least some barrier appears at all doping levels below $\approx 1.2 \times 10^{17}$ cm^{-3}. The second term is 6.2×10^{16} cm^{-3}, and is due to the electrostatic effect from the mobile charge (the Kirk effect). The first term is due to the alloy grading and requires a compensation of at least 5.4×10^{16} cm^{-3} for this bias. Thus, the two effects, Kirk and alloy barrier, are approximately equal for this set of conditions. Since the doping requirement for a homojunction collector omits the first term, it would be about a factor of two lower for these conditions. On the other hand, for a heterojunction collector, the doping required by the first term increases as the maximum forward collector–base bias at which the barrier must be suppressed increases, until, $W < z_0$, when this analysis breaks down.

Choosing the grading in the collector requires slightly different considerations than those in the forward biased base–emitter junction. Most high speed circuits are designed so that the base–collector junction is not forward biased far enough that $W < z_0$ during operation. Hence, the electrostatic fields at and near the junction are larger than in the base–emitter junction, and the mobile charge moves significantly faster than in the emitter. As a result, we obtain the lower doping requirement in this case. Thus, designing junctions that do not show barrier effects is easier at the collector than at the emitter. Heterojunction collectors with linear grading that do not have alloy barriers are also possible, since for linear grading the grading field constraint occurs at the collector edge of the grading region. One may obtain the following relation for a linearly graded region:

$$N_D > \frac{\epsilon_s}{q} \frac{\gamma u_f}{z_0(W(N_D, V_{BC}) - z_0)} + \frac{J_C}{q\overline{v}}. \tag{7.119}$$

Solving this equation for the doping gives a much more complicated dependence on the alloy grading parameters and the bias than for parabolic grading. Typically there is only a limited range of N_D for which the inequality can be satisfied, and if the bias is increased in the forward direction, the size of the range decreases until it vanishes and the inequality can no longer be satisfied. Typically, also, the minimum N_D that satisfies the equation (when there is one) will be somewhat larger than for parabolic grading.

7.4.2 Collector Electron Quasi-Fields

In analyzing the electron quasi-electric fields in the collector, we first consider the homojunction collector HBT. The field in the base–collector region

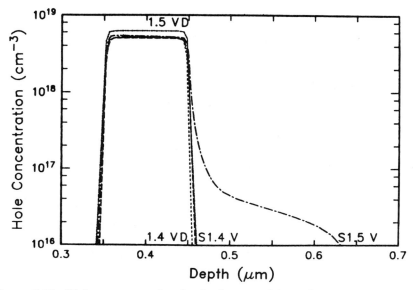

Figure 7.23: Hole concentration in the base and into the emitter and collector regions as a function of an applied bias of 1.4 V and 1.5 V for a single heterojunction (S) and double heterojunction (D) device. The base extends from 0.35 μm to 0.45 μm. The single heterojunction shows additional hole storage in the collector depletion region. From S. Tiwari and D. J. Frank, "Analysis of the Operation of GaAlAs/GaAs HBTs," *IEEE Trans. on Electron Devices*, **ED-36**, No. 10, p. 2105, ©Oct. 1989 IEEE.

decreases as the current density increases (due to the space charge of the electrons in the collector space charge region), until finally base push-out occurs (the Kirk effect). This push-out can actually be seen in the hole concentration, a reason why it is called base push-out since it extends the quasi-neutral region. This is shown in Figure 7.23 for both the single and the double heterojunction devices at 1.4 V and 1.5 V forward base–emitter bias. In the double heterojunction device, the holes are not significantly injected into the collector, even at high current density, but in the single heterojunction device, there is substantial hole injection at the higher current density (1.5 V) because of junction debiasing and the absence of a hole barrier. Thus, at the higher current density (1.5 V emitter–base bias), the single heterojunction device has an aiding field at the collector junction due to electrostatic considerations, followed by a decrease to low values because of the Kirk effect. Finally the field increases again near the sub-collector where the donor density is higher.

Figure 7.24 shows the variation in the electron quasi-electric field for a double heterojunction HBT in the base and the junction regions surrounding it. The collector junction behavior is very different from that of the single heterojunction device. Here, at high forward base–emitter bias, the quasi-electric field goes through a reversal in direction, and there is no high field region at the sub-collector. There are two major differences. First, there is a valence band alloy barrier which prevents injection of holes into the collector, thus allowing a rapid decrease in electrostatic field at the junction in the presence of large electron charge. Second, there is the conduction band alloy barrier. The influence of this electron alloy field, together with the decrease in the electrostatic field, results in the retarding field to electron motion that appears at the base–collector junction in Figure 7.24 for base–emitter biases of 1.4 V and 1.5 V. Unlike the situation in the homojunction collector, base push-out does not occur in this device at the highest current density bias point (1.5 V). This is so both because the hole injection is blocked and because the electron alloy barrier diminishes the numbers of electrons entering the collector depletion region by a factor between 1.5 and 2.

7.4.3   Diffusion Capacitances

We have described qualitatively the additional storage in the base that results from the alloy barrier in the heterojunction collector. Figure 7.22 showed the electron charge distribution in the middle of the base under various forward bias conditions. It is useful to show the stored minority carrier charge in the base and quantify it as a time constant in order to account for the two-dimensional nature of the effect and also for purposes of comparison. Figure 7.25 shows the integrated stored electron density (in cm^{-2}) and the base time constant (obtained by dividing total stored charge in the base by the collector current) for both the heterojunction collector and the homojunction collector. The stored charge in the base begins to differ substantially for the heterojunction collector case at bias levels exceeding 1.4 V (corresponding to a current density of 6×10^4 A.cm^{-2}). The charge storage can differ by as much as a factor of two at the highest bias conditions. The time constant associated with the base charge storage in the heterojunction collector is ≈ 3 ps at low currents, decreases with more forward bias because of minute aiding drift fields (the Webster effect), and then increases to greater than 12 ps under conditions of extreme forward bias where current densities reach $\approx 1 \times 10^5$ A.cm^{-2}. Under these conditions the base time constant certainly becomes a significant portion of the total time constant of the device. In the homojunction collector device, the

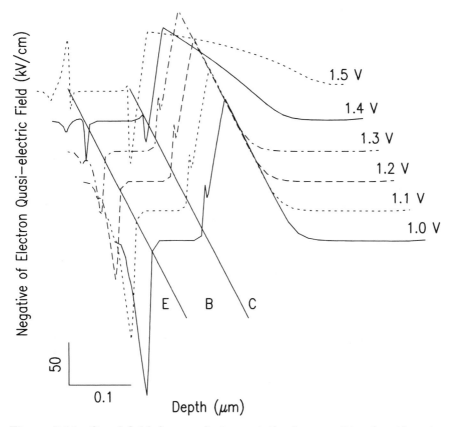

Figure 7.24: Quasi-field for an electron at the base–emitter junction, in the base, and at the base–collector junction as a function of base–emitter bias for the double heterojunction HBT. Field for a single heterojunction HBT was plotted in a previous figure for identical conditions. The base is 0.1 μm wide, and the different plots are shifted to allow a perspective view. The scales of distance and field are located in the lower left corner. From S. Tiwari and D. J. Frank, "Analysis of the Operation of GaAlAs/GaAs HBTs," *IEEE Trans. on Electron Devices*, **ED-36**, No. 10, p. 2105, ©Oct. 1989 IEEE.

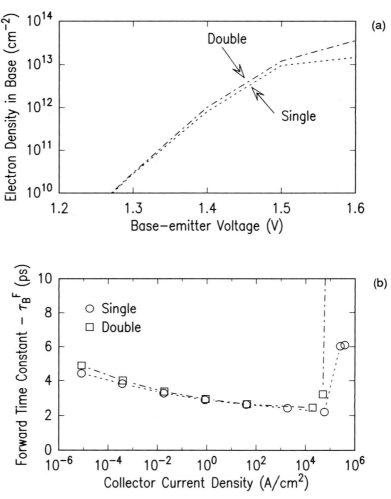

Figure 7.25: Part (a) shows total electron density in the base, and part (b) shows the base time constant as a function of applied forward bias at the base–emitter junction for a single heterojunction and a double heterojunction device. From S. Tiwari and D. J. Frank, "Analysis of the Operation of GaAlAs/GaAs HBTs," *IEEE Trans. on Electron Devices*, **ED-36**, No. 10, p. 2105, ⓒOct. 1989 IEEE.

low and medium current behavior is similar to that of the heterojunction collector device, and increases to ≈ 6 ps at high currents (current densities $\approx 4 \times 10^5$ A.cm^{-2}) due to the Kirk effect.

While the heterojunction collector leads to an increase in the base storage, it continues to suppress the hole injection. The homojunction collector does not have a hole alloy barrier, and hole injection into the collector does become important in these devices. Figure 7.23 showed a comparative hole distribution profile at the base–collector junction for the homojunction and the heterojunction collector case. It can be seen in this figure that $\approx 4 \times 10^{11}$ cm^{-2} of holes are stored in the collector in order to sustain a current density of 2.5×10^5 A.cm^{-2} (an additional time constant of $\tau \approx 0.3$ ps). Thus, while the heterojunction collector bipolar transistor had an increase in diffusion capacitance component associated with the electron storage in the base, the homojunction collector device has a hole storage component associated with the collector.

7.4.4 Current Gain Effects

The increase in base charge storage reduces the gain of HBTs because of increased base recombination current. Thus, in addition to both surface and bulk recombination effects and their voltage-dependent features, the HBT shows an anomalous decrease in gain at high current density when heterojunction collectors are used. This decrease is mostly due to quasi-neutral base recombination. However, the base charge storage also affects the transport of the carriers to the surface. Hence, surface recombination should also be increased slightly at high current densities by the presence of a heterojunction collector. In the absence of excess storage, the device gain continues to improve with current.

Figure 7.26 shows the current gain and the excess voltage at various collector dopings. The decrease in gain at ≈ 1.4 V for the 1×10^{17} cm^{-3} doped collector occurs because of increased base charge storage. Along with the increase in base current that this implies, there is a saturation of the collector current. The saturation of collector current is reflected in the rapid increase in the excess voltage as the current gain drops. The removal of this effect at 6×10^{17} cm^{-3} is correlated with the suppression of the alloy barrier effect. The saturation of the collector current density occurs at factors of six greater current density in the higher doped collector structure.

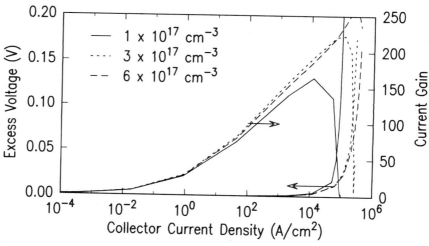

Figure 7.26: Excess voltage and current gain derived from Gummel plots for parabolic grading at an emitter doping of 8×10^{17} cm^{-3} and collector dopings of 1×10^{17} cm^{-3}, 3×10^{17} cm^{-3}, and 6×10^{17} cm^{-3} for a double heterojunction transistor. The rapid increase in excess voltage shows the onset of saturation in current. It occurs together with a decrease in gain and an increase in storage in the quasi-neutral base. From S. Tiwari and D. J. Frank, "Analysis of the Operation of GaAlAs/GaAs HBTs," *IEEE Trans. on Electron Devices,* **ED-36**, No. 10, p. 2105, ©Oct. 1989 IEEE.

7.5   Generation and Recombination Effects

Recombination is a more important parasitic effect in GaAs HBTs than in silicon bipolar transistors because of the significantly lower bulk recombination lifetime and higher surface recombination velocity of GaAs. At moderate and low doping levels, the bulk recombination life time is dominated by deep trap recombination centers (centers that can capture electrons and holes nearly equally efficiently), and lifetime can be as much as three orders of magnitude lower than in silicon at similar doping levels. At higher doping levels, the radiative lifetime may predominate , and since GaAs is a direct bandgap material this lifetime can be very low—even reaching 100's of picoseconds. In silicon, the lifetime is also low at high doping levels, because of Auger recombination. Surface recombination is higher for GaAs than for Si because the unpassivated GaAs surface contains a significant number of interface states (10^{13}cm^{-2} and higher), which lead to mid-gap pinning of the Fermi level at the surface, as well as recombination. As a result of

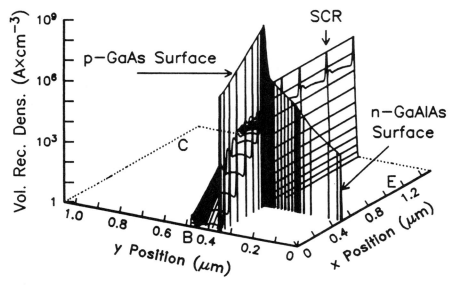

Figure 7.27: Perspective plot of volume recombination current density for a surface recombination velocity S of 2×10^5 cm.s^{-1}, for the HBT example of earlier figures. The base-to-emitter forward bias voltage is 1.3 V and base-to-collector voltage is 0 V. From S. Tiwari, D. J. Frank, and S. L. Wright, "Surface Recombination in GaAlAs/GaAs Heterostructure Bipolar Transistors," *J. of Appl. Phys.*, **64**, No. 10, p. 5009, 15 Nov. 1988.

the poor lifetime and surface pinning, the current–voltage behavior of p–n junctions and HBTs deviates substantially from the ideal due to recombination. We emphasize here the bulk and surface recombination behavior of graded junction devices at high surface recombination velocities.

The non-idealities in current–voltage behavior are introduced by generation and recombination of carriers at HSR centers in the space charge region of the p–n heterojunctions and at the interface states at the surface of the semiconductors. Figure 7.27 shows the relative recombination rate of these regions compared to the quasi-neutral recombination by plotting the volume recombination density of our HBT example. This figure shows that the electrons injected into the base recombine copiously at the surface of the GaAs base, and at the base–emitter junction, and there are holes injected from the base that recombine at the emitter surface. Of these, the base surface and the p–n junction recombination are the dominant components. The bulk recombination occurring in the quasi-neutral region is actually the smallest and is generally dominated by the surface recombination in this unpassivated device.

The bulk recombination taking place in the base–emitter space charge region contributes to degrading the injection efficiency. The non-idealities corresponding to these recombinations appear in the form of multiple exponential regions in the forward characteristics, and as non-saturating current in the reverse characteristics. The forward characteristics also have non-idealities due to the heterojunction injection phenomena, which we have considered, but these occur only at high currents. At low currents the generation–recombination effects dominate.

7.5.1   Bulk Effects

We first consider the bulk effect—the recombination taking place in the space charge region and in the quasi-neutral region. The former has been studied extensively for homojunctions. Simple theories that take into account the HSR recombination by integrating the net recombination rate in the junction show that it results in a "$2kT$" dependence, i.e., varying as $\exp(qV/2kT)$. We had treated this example in the discussion of Gummel Poon models as applied to p–n junctions in Chapter 4. The pre-exponential term of this exponential dependence is proportional to the volume of the recombination region (i.e., proportional to the depletion width), which has a weak voltage dependence. The "$2kT$" exponential dependence results from the fact that while current flow across the junction requires the carriers to cross the whole barrier, recombination requires each of the carriers to surmount only half of this barrier, on average. This results in an exponential dependence that varies as $V/2$, which is "$2kT$". In practice, even for homojunction devices, such recombination is rarely exactly "$2kT$", and has been shown to result from junction asymmetry, asymmetric capture statistics, position dependent trap distributions, and electric field dependence of capture statistics. This results in a recombination dependence that is "$2kT$"-like instead of being exactly "$2kT$". In heterojunctions, in addition to these effects, we have to consider the suppression of hole injection into the emitter. In fact, this is crucial to limiting space charge region recombination. Homojunctions in GaAs have orders of magnitude higher space charge region recombination than heterojunctions.

Figure 7.28 shows the volume recombination current density in the junction space charge region for the parabolically graded junction, and the ideality of this volume recombination current density (the factor n in $\exp(qV/nkT)$) as a function of position. The ideality is derived from the rate of exponential increase in the local volume recombination density, as determined by steady-state analysis of a small perturbation at the bias

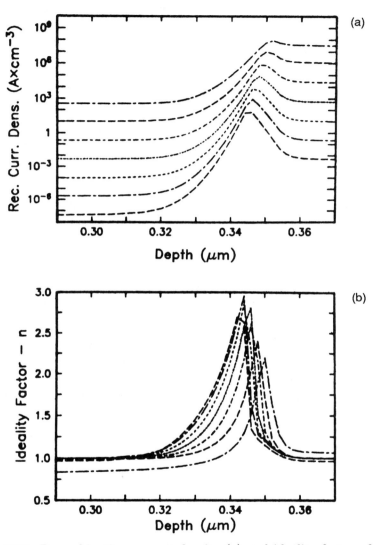

Figure 7.28: Recombination current density (a) and ideality factor of the recombination current density (b), versus depth, for biases varying from 0.8 V to 1.4 V. The base–emitter junction is located at 0.35 μm. From S. Tiwari and D. J. Frank, "Analysis of the Operation of GaAlAs/GaAs HBTs," *IEEE Trans. on Electron Devices*, **ED-36**, No. 10, p. 2105, ©Oct. 1989 IEEE.

point. The junction grading extends from 0.32 to 0.35 μm. At these biases, the peak in the volume recombination density occurs in the junction alloy grading region, within 100 Å—a region where the effective barrier is changing with bias because of varying bandgap. The ideality of the volume recombination current density peaks in this junction region but not at the same position as the peak of the volume recombination density. This is a result of the recombination process, which drives the system towards equilibrium (i.e, reducing the np product to n_i^2—a position-varying quantity). The np product does not peak at the junction itself in a heterojunction but close to it because of the efficient suppression of holes. The ideality of the recombination current depends on the carrier transport to the position of recombination. While further away from the junction space charge region edge, and at its edge, the ideality begins to approach the "kT"-like dependence; within the junction space charge region it rapidly increases. It can exceed "$2kT$" because of the heterojunction effect and junction asymmetry. Note in this figure that at the highest biases the quasi-neutral recombination (the "kT" recombination beyond the junction) nearly reaches the magnitude of the volume recombination density in the space charge region, and has a faster exponential dependence. Thus at sufficiently high biases the bulk recombination component will begin to dominate the space charge region component, resulting in a "kT"-like dependence.

We discussed earlier the Gummel plots and the current gain behavior of single and double heterojunction devices in the absence of surface recombination. That is, this past discussion included only bulk space charge region and quasi-neutral region contributions of recombination. For this condition, it is only in the high current region of a double heterojunction device that the bulk base transport factor becomes significant in determining the gain of the device. At low currents, the dominant factor is recombination in the space charge region and at the surface, and in the medium current range it is mostly surface recombination.

7.5.2 Surface Effects

Surface effects have been a dominant source of gain-degradation in GaAs HBTs. Experimental evidence shows that they give rise to a "$2kT$"-like dependence going towards a "kT"-like dependence at high bias conditions. The surface is a dominant recombination source in most common designs as a result of a high rate of recombination through surface states and the presence of Fermi level pinning at the surface. The first by itself would give rise to recombination dominated transport only in the low current regime, but the presence of Fermi level pinning causes an excess of electrons at the surface, leading to significantly higher recombination. The surface recombination can be minimized for the operating current densities of interest by various surface passivation techniques or by preventing the appearance of carriers with the use of a barrier from a wider bandgap semiconductor at the surface of the extrinsic base. This could be a p-type $Ga_{1-x}Al_xAs$ obtained by converting the polarity of the injecting emitter, or a depleted $Ga_{1-x}Al_xAs$ emitter region by thinning of the heterojunction emitter. Minimum lateral dimensions for such layers are discussed later. The use of a graded base layer can also reduce the surface recombination. In the following, we analyze surface recombination in the absence of surface recombination reduction techniques.

Surface recombination at medium and low current densities is dominated by a "$2kT$"-like recombination process, just as the bulk recombination process. The cause for this is the presence of Fermi level pinning. The argument consists of two parts. First, a high density of surface states (as required for surface pinning) causes the electron-to-hole ratio at the surface to remain constant and close to unity. Second, if there is essentially equilibrium between the surface and the bulk and only a small charge flow to the surface, such that the quasi-Fermi levels remain essentially flat, then the np product remains constant. Putting these two parts together yields a surface carrier concentration that depends on the square root of the bulk minority carrier density, $n_s \propto \sqrt{n_b \times p_b}$, and hence a "$2kT$" dependence for the surface recombination. At moderate recombination velocity, the rate-limiting process is the recombination velocity, and the assumptions used in deriving the "$2kT$" dependence are valid.

At high recombination velocity, however, significant deviations from the assumptions can occur. This was alluded to in our discussion of surface recombination in Chapter 3. In particular, numerical simulation of this problem in the limit of high surface recombination velocity shows that the electron quasi-Fermi level does not remain flat between the surface and the bulk. For example, at $S = 2 \times 10^6$ cm.s^{-1}, the quasi-Fermi level bends by more than 200 meV near the injecting junction for a bias of 1.2 V. Figure 7.29 shows the total surface recombination current density as a function of device current density for moderate-to-high surface recombination velocity using a single donor and single acceptor mid-gap trap at the surface. At low currents the surface recombination current density follows a "$1.8kT$" behavior. The higher current behavior is a function of surface recombination velocity. At high surface recombination velocity the ideality shows a deviation towards a "$1.2kT$" dependence. These high surface recombination velocity characteristics are in general agreement with experimental observations at high current densities. The lower surface recombination velocity characteristics are in general agreement with the simple analytic theory, as they ought to be. The high surface recombination velocity behavior occurs

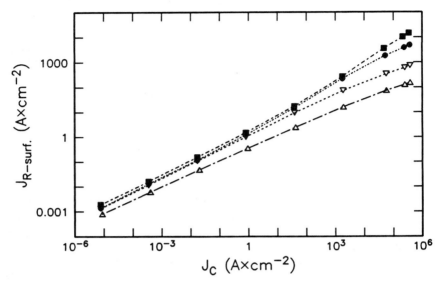

Figure 7.29: Surface recombination current density (normalized to the emitter area) as a function of collector density, with S as a parameter varying from 2×10^3 cm.s^{-1} for the lowest curve to 2×10^6 cm.s^{-1} for the highest curve in steps of a decade. The low current behavior in this plot has a "2kT"-like dependence and the high current behavior for the higher surface recombination velocities has a "kT"-like dependence. From S. Tiwari, D. J. Frank, and S. L. Wright, "Surface Recombination in GaAlAs/GaAs Heterostructure Bipolar Transistors," *J. of Appl. Phys.*, **64**, No. 10, p. 5009, 15 Nov. 1988.

because the recombination is dependent upon the rate at which carriers are provided to the surface, and this depends on device geometry, surface conditions, and the design of the injecting junction. The Fermi level of electrons is no longer flat between the surface and the bulk—and there is no simple relationship between n_s and n_b and p_b.

To better understand the high surface recombination velocity behavior, one needs to understand how the carriers reach the surface to recombine. Figure 7.30 plots the electron and hole current flow lines in the vicinity of the extrinsic base, showing the way in which the carriers get to the surface. Note that the electron current to the surface is almost entirely due to injection of carriers into a surface channel at the base–emitter junction. This surface electron channel is caused by the surface Fermi level pinning.

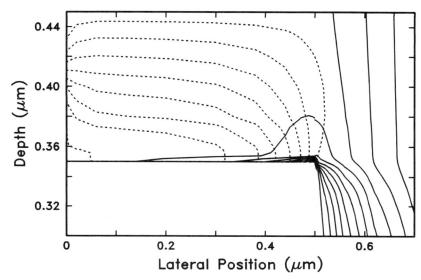

Figure 7.30: Current stream lines in our HBT example, showing the way in which the currents reach the surface of the extrinsic base, where recombination takes place. The solid lines are electron flow lines, and the dashed lines are hole flow lines. The base–collector junction is at the top of the figure, and the emitter–base junction is at a depth of 0.35 μm. The exposed surface of the extrinsic base, where surface recombination occurs, stretches between 0.0 and 0.5 μm laterally, and is at the same depth, 0.35 μm. From S. Tiwari and D. J. Frank, "Analysis of the Operation of GaAlAs/GaAs HBTs," *IEEE Trans. on Electron Devices*, **ED-36**, No. 10, p. 2105, ©Oct. 1989 IEEE.

Only a very small flux of electrons into this channel from the quasi-neutral base region is observed and it occurs at the intersection of the surface with the base–emitter junction, where two-dimensional effects are strong. Note also that the hole current for recombination is mostly perpendicular to the surface, and originates in the quasi-neutral base region.

Because of its close proximity to the surface, this electron injection phenomenon is a function of the boundary conditions that have been chosen. Entry into the channel is through a saddle point in the conduction band edge potential, as shown in Figure 7.31. This complex barrier shape is the result of the interface state distribution at the surface that caused the Fermi level pinning, the choice of band gap grading, and the choice of bias.

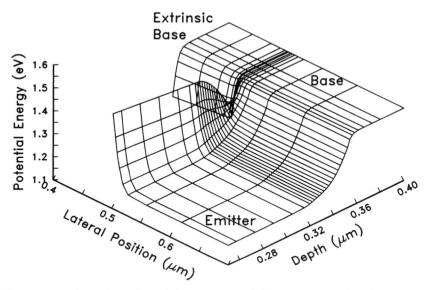

Figure 7.31: A surface plot of the negative of the conduction band energy at the intersection of the base–emitter junction with the base surface, showing the saddle point, which allows easier flow of the injected electrons into the surface depletion region. Note that this view is rotated about 90 degrees clockwise relative to the earlier figures. From S. Tiwari, S. L. Wright, and D. J. Frank, "Compound Semiconductor Heterostructure Bipolar Transistors," *IBM J. of Research and Development,* **34**, No. 4, p. 550, July 1990.

This barrier causes, at low bias conditions, a narrow constricted region through which it is energetically favorable for the carriers to stream into the surface channel. At high surface recombination velocity, this constriction is the rate-limiting step in the surface recombination, and the magnitude of current transported to the surface is determined by the barrier height and the cross-section of this saddle point, which are non-linearly dependent on the bias because of the changing junction depletion thickness.

Once the carriers reach the surface, the HSR statistics determine the spatial dependence of the recombination of these carriers. At high surface recombination velocities, these carriers recombine very rapidly in a very short region. At moderate surface recombination velocities, many of these carriers drift-diffuse down the channel before they recombine. Figure 7.32 shows the conduction band edge energy and the electron quasi-Fermi level as a function of position along the surface at two forward bias conditions, 1.0 V

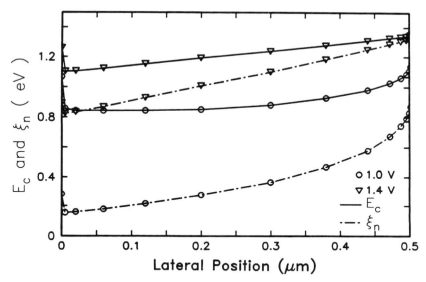

Lateral Position (μm)

Figure 7.32: Conduction band edge energy and electron quasi-Fermi level along the base surface, at 1.0 V and 1.4 V forward bias, with the base–emitter junction intersecting at 0.5 μm on the horizontal scale. Solid line is for E_c and the dot-dashed line is for ξ_n. From S. Tiwari and D. J. Frank, "Analysis of the Operation of GaAlAs/GaAs HBTs," *IEEE Trans. on Electron Devices*, **ED-36**, No. 10, p. 2105, ©Oct. 1989 IEEE.

and 1.4 V. At the low bias conditions further away from the intersection of the surface with the base–emitter junction (at 0.5 μm of the horizontal scale), the electron current is mostly diffusive, with drift only becoming important close to the junction. At the higher bias, however, drift and diffusion contribute approximately equally to the transport process. The rapid fall-off of the Fermi level shows that recombination leads to a very rapid depletion of electrons.

Just as the electron concentration rapidly diminishes with distance from the junction, so also the surface recombination density is highest at the injecting point and rapidly decreases thereafter. The length scale of this fall-off is ≈ 1250 Å at $S = 2 \times 10^5$ cm.s^{-1}, and ≈ 500 Å at $S = 2 \times 10^6$ cm.s^{-1}, giving between a square and cube root dependence on S. The effective ideality of the surface recombination current density (determined as for Figure 7.28) depends on recombination velocity, bias, and weakly on position. At $S = 2 \times 10^5$ cm.s^{-1}, the ideality varies from ≈ 1.7 at 0.95 V forward bias to ≈ 1.3 at 1.25 V bias. At $S = 2 \times 10^6$ cm.s^{-1}, the idealities behave similarly, but are somewhat lower. Figure 7.33 shows the surface recombination density and its effective ideality as a function of position.

These characteristics of the ideality and the magnitude of the surface recombination current depend not only on the value of surface recombination velocity, but also on the geometry, surface conditions, and design of the device, because of the strong two-dimensional nature of the phenomenon. Thus, surface recombination is quite variable. Thick bases allow a larger flux of carriers to the surface, quasi-fields that pull the carriers away from the junction cause a smaller flux of carriers, and abrupt heterojunctions cause a larger voltage barrier at the surface and hence cause a smaller flux of carriers. The total surface recombination current is proportional to the flux of electrons—hence, these devices show different surface recombination behavior.

7.5.3 Current Gain Behavior

Current gain behavior in the absence and presence of surface recombination is plotted in Figure 7.34 for single heterojunction, graded base (an electron quasi-field of 10 kV/cm created by grading the aluminum mole-fraction to GaAs at the collector), and double heterojunction bipolar transistors. In the absence of surface recombination, Figure 7.34 shows, at large current densities, the largest current gains in the graded base device followed by the single heterojunction device and then the double heterojunction device. The larger current gain in the graded base device is due to the aiding quasi-electric field for electrons, which reduces the storage of electrons in the base, and hence the neutral base recombination. The double heterojunction device shows a drop in gain at high currents due to the increased base storage, as discussed previously. Although the graded base device has a lower barrier to hole injection into the emitter, its gain actually continues to be comparable to that of the other devices at low current densities. The increased hole injection does not result in increased recombination in the base–emitter space charge region because it is compensated by a decrease in electron density due to the quasi-electric field in the neutral base. Figure 7.34 shows the behavior of these devices with surface recombination, again for a perimeter-to-area ratio of 1×10^4 cm^{-1}. The general trends of Figure 7.34 still apply, with the largest gains at the highest currents in the graded base device, and a decrease in gain of the double heterojunction device due to excess storage at high currents. The current gain of the graded base device is significantly higher than that of the single heterojunction device because the quasi-drift field results in fewer electrons to inject into the surface recombination region.

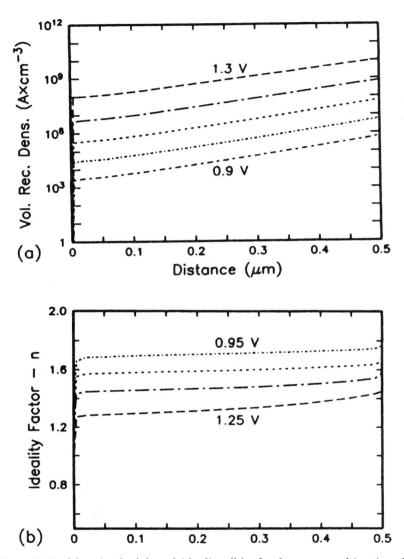

Figure 7.33: Magnitude (a) and ideality (b) of volume recombination density at the surface as a function of position along the surface of the base region. Different base–emitter forward bias conditions are shown.

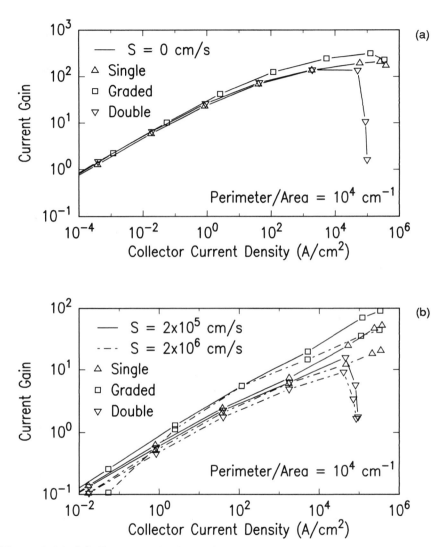

Figure 7.34: (a) Current gain dependence in the absence of surface recombination, for single heterojunction, graded base, and double heterojunction bipolar transistors. The perimeter-to-area ratio is 10^4 cm^{-1}. (b) Gain for the same devices at $S = 2 \times 10^5$ cm.s^{-1} and $S = 2 \times 10^6$ cm.s^{-1}. From S. Tiwari and D. J. Frank, "Analysis of the Operation of GaAlAs/GaAs HBTs," *IEEE Trans. on Electron Devices*, **ED-36**, No. 10, p. 2105, ©Oct. 1989 IEEE.

The dependence of current gain on perimeter-to-area ratio—the device size effect—is shown in Figure 7.35 at forward biases of 1.5 V and 1.4 V, respectively, for the graded base and single heterojunction devices. The current flowing through the graded base device is approximately a factor of ten smaller at 1.4 V, and a factor of two smaller at 1.5 V, due to the different injection characteristics of the two devices. At 1.4 V, where current saturation effects are not large, the graded device (which is operating at a lower current density) clearly shows less sensitivity to surface recombination velocity. The current gain behavior is in good agreement with the measured current gains of large area (small perimeter-to-area ratio) and small area (large perimeter-to-area ratio) devices that utilize no surface passivation. For small digital devices, with perimeter-to-area ratios of $\approx 2 \times 10^4$ cm^{-1}, the figure indicates current gains should be in the range of 10 to 40. Using a one-dimensional model, the highest current gain in these structures, if it were determined only by neutral base recombination, would be ≈ 380 ($\mathcal{D}_n \approx 38$ cm^2.s^{-1}, $\tau_n \approx 0.5$ ns, and hence the diffusion length $\mathcal{L}_n \approx 1.38$ μm). The current gains in Figure 7.34 approach this value, but are lower and are dependent on the device size because the base transport factor has to include recombination in the extrinsic region. The lateral extent of the extrinsic base that needs to be included is of order the base width in a single heterojunction device and the diffusion length in a double heterojunction device. Hence, the gain of HBTs shows perimeter dependence even in the absence of surface recombination.

7.6 Small-Signal Analysis

Bipolar transistors are high frequency devices with strong displacement current effects in the collector depletion region. Thus, during both fast transients and prediction and operation near the limit frequencies, non-quasi-static analysis is particularly important. We may actually draw some parallels, under these rapidly time-varying conditions, between the field effect transistors and the bipolar transistor. In the normal active mode of operations, the source end of the channel of an FET is a low field region ($< 10^4$ V.cm^{-1}) with a slow variation in time. The drain end of the channel is a high field region, with large displacement current changes during parts of a transient. Displacement current also flows in the gate control region during these time-dependent rapidly varying conditions. Thus, in the operation of the field effect transistor, the transport in the low field region near the source end of the channel and the transport in the high field region near the drain end and signal delay effects corresponding to them become important. The equivalent regions in the homogeneous base bipolar tran-

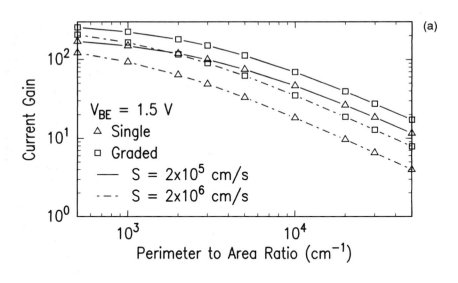

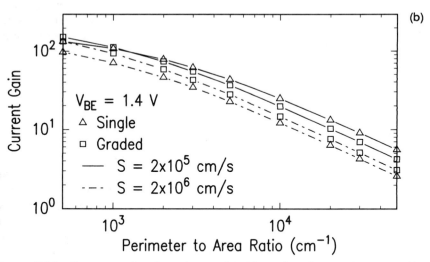

Figure 7.35: Current gain plotted as a function of perimeter-to-area ratio for $S = 2 \times 10^5$ cm.s^{-1} and $S = 2 \times 10^6$ cm.s^{-1}. (a) is for a base–emitter forward bias of 1.5 V and (b) is for 1.4 V. From S. Tiwari and D. J. Frank, "Analysis of the Operation of GaAlAs/GaAs HBTs," *IEEE Trans. on Electron Devices*, **ED-36**, No. 10, p. 2105, ©Oct. 1989 IEEE.

sistor are the quasi-neutral base region and the base–collector depletion region. Strong displacement current effects occur in the base–collector region during the transit of carriers. Another important consideration in high frequency modelling is the majority carrier transport in this minority carrier device. Note that the displacement current effects, in the base–collector depletion region, take place in a part of the device where the electrons are actually majority carriers.

Our analysis of small-signal operation of bipolar transistor follows early homojunction bipolar theory.[9] We will employ network parameters based on one-dimensional analysis. This is quite adequate since the high frequency bipolar transistor has a thin base; and in the emitter, base, and the collector, the intrinsic and critical behavior—transport of minority carriers—occurs largely one-dimensionally and orthogonally to the plane of the junctions. Transport of holes in the base, as well as of electrons in the quasi-neutral emitter and collector, can be introduced in this analysis as lumped elements without appreciably sacrificing accuracy.

7.6.1 Parameter Notation and Assumptions

We will use the admittance matrix (*y*-parameters) as the basis for our analysis, and like FETs where we considered common-gate configuration for the small-signal discussion, we will consider the common-base configuration. Other configurations, the common-emitter hybrid-pi model, e.g., as well as inclusion of parasitics, extrinsic elements, etc., will follow from this with suitable matrix manipulations. Like the FETs, the common-base *y*-parameters follow from the current equations, and lend themselves to physical insight following approximations for equivalent circuit modelling at high frequencies. These equivalent circuits, as well as the more accurate network parameters, can then be extended to the common-emitter configuration, which is of the most interest in applications. We wish to derive the four network parameters—input, output, and the two transfer elements of the *y*-matrix—and we base this on one-dimensional flow of carriers as shown in Figure 7.36. Adding extrinsic elements, parasitic elements, etc., extends this to two-dimensional or three-dimensional structures for frequencies where lumped parameter representation is still valid.

We considered the quasi-static analysis and the low frequency physics of the bipolar transistor in an earlier part of this chapter. The technique employed in the analysis of small-signal non-quasi-static response should become clear from the context; it is similar to that employed for FETs.

---

[9]See the general reference R. L. Pritchard, *Electrical Characteristics of Transistors*, McGraw-Hill, N.Y. (1967).

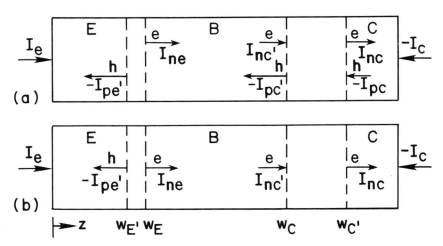

Figure 7.36: Current flow in an idealized transistor structure, showing the notations employed for small-signal analysis (a), together with its simplification where the smaller components are ignored (b).

The perturbational quasi-static solutions will be referred to in order to show similarities and differences between them and these small-signal solutions.

In evaluation of network parameters such as the y-parameters, the terminal currents are considered as going into a port. Our current density and current continuity equations, however, evaluate currents as flowing in the $+z$ direction. The emitter current at the emitter port $\tilde{I}_e$, ignoring recombination effects in the base–emitter depletion region, is composed of the electron and hole currents at the edges of the base–emitter depletion region ($\tilde{I}_{ne}$ and $\tilde{I}_{pe}$). Both the network parameter-based convention and the transport-based convention are identical for this. However, the network parameter convention for the collector current at the port is to treat it as flowing in, i.e., the $-z$ direction, while the electron and hole currents evaluated at the collector edge of the base–collector depletion region ($\tilde{I}_{nc}$ and $\tilde{I}_{pc}$) are treated as flowing in the $+z$ direction. Since this is also the current that flows in the quasi-neutral region of the collector during the small-signal conditions, displacement effects being absent in the quasi-neutral region, this is also the collector current ($\tilde{I}_c$). The current necessary for defining the y-parameters is $-\tilde{I}_c$. Hence our common-base y-parameters are based on the currents

$$\tilde{I}_e = \tilde{I}_{ne} + \tilde{I}_{pe}$$

$$\text{and} \qquad -\tilde{I}_c \;=\; -\tilde{I}_{nc} - \tilde{I}_{pc} \qquad\qquad (7.120)$$

determined from analysis of the device.

In analyzing the device, we will make some of the usual assumptions that were employed in the quasi-static analysis, as well as a few additional assumptions related to small-signal aspects. We consider one-dimensional carrier flow, normal active mode of operation, i.e., a forward biased base–emitter junction and a reverse biased base–collector junction. The injected carrier density in the base is considered small compared to the majority carrier density present in thermal equilibrium. This is the assumption of low-level injection and allows us to employ Shockley boundary conditions and treat the quasi-neutral base region as an equi-potential region. We also ignore transit time effects in the base–emitter region where the electric fields can be large. The rationale for this assumption is that the base–emitter depletion region is very short; and the delay effects, except in ultra-short structures, are substantially smaller than the delay effects in the base–collector depletion region, the quasi-neutral base region, or delays corresponding to the charging of transition capacitances. A base–emitter small-signal potential thus is assumed to instantaneously result in a new carrier distribution within the depletion region that is equal to the corresponding equilibrium distributions at the new electrostatic potential. This implies that, in the base–emitter space charge region, the holes and electrons are very nearly in equilibrium with the distributions governed by the potential barrier. For quasi-static conditions, because the electric field $\mathcal{E}$, the gradient in electron concentration ∇n, and the gradient in hole concentration ∇p are large, the changes involved cause a very small disturbance from equilibrium. The current extracted is much smaller than the individual drift and diffusion current components, and hence equilibrium is not disturbed significantly. Under small-signal conditions, this implies that the time and phase delay and attenuation are negligible, and the reason is that time constants other than those due to transit in the base–emitter depletion regions dominate. We also assume that the small-signal voltages applied are significantly less than the thermal voltage. Use of this assumption allows us to decouple the static and small-signal solutions. Breakdown of this assumption, i.e., large-signal analysis, is also important but will not be considered.

Initially we will ignore effects related to emitter injection efficiency. Both under quasi-static and small-signal conditions, the hole injection current into the emitter ($\tilde{I}_{pe}$) will be considered to be significantly smaller than the electron current injected into the base ($\tilde{I}_{ne}$). Subsequently, we will remove this assumption, and show that there can be important phase delay effects that result from the base–emitter injection process. We will also consider the multiplication process to be absent in the collector depletion region, i.e., for quasi-static conditions,[10]

$$\zeta = \frac{|\bar{I}_{nc}|}{|\bar{I}_{nc'}|} = \frac{|\bar{I}_c|}{|\bar{I}_{c'}|} = 1. \tag{7.121}$$

These simplify the analysis of our problem to that shown in the part (b) Figure 7.36, where a number of components shown in part (a) are now removed. We have

$$\tilde{I}_e = \tilde{I}_{ne}$$
$$\text{and} \qquad \tilde{I}_c = \tilde{I}_{nc}. \tag{7.122}$$

Under our small-signal conditions, $\tilde{I}_e$ and $\tilde{I}_c$ are related, using the base as a reference, to the emitter-base junction voltage V_{eb} and collector–base voltage V_{cb} by

$$\tilde{I}_e = y_{ee}^b \tilde{V}_{eb} + y_{ec}^b \tilde{V}_{cb}$$
$$\text{and} \qquad -\tilde{I}_c = y_{ce}^b \tilde{V}_{eb} + y_{cc}^b \tilde{V}_{cb}. \tag{7.123}$$

The four complex admittance parameters—the input admittance y_{ee}^b, the forward transfer admittance y_{ce}^b, the reverse transfer admittance y_{ec}^b, and the output admittance y_{cc}^b—are

$$y_{ee}^b \equiv \left. \frac{\tilde{I}_e}{\tilde{V}_{eb}} \right|_{\overline{V}_{cb}},$$

$$y_{ce}^b \equiv \left. \frac{-\tilde{I}_c}{\tilde{V}_{eb}} \right|_{\overline{V}_{cb}},$$

$$y_{ec}^b \equiv \left. \frac{\tilde{I}_e}{\tilde{V}_{cb}} \right|_{\overline{V}_{eb}},$$

$$\text{and} \qquad y_{cc}^b \equiv \left. \frac{-\tilde{I}_c}{\tilde{V}_{cb}} \right|_{\overline{V}_{eb}}. \tag{7.124}$$

Some of these parameters can be related to each other using factors of direct interest to us. For example, the forward transfer admittance,

$$y_{ce}^b = \left. -\frac{\tilde{I}_c}{\tilde{V}_{eb}} \right|_{\overline{V}_{cb}}$$

---

[10]Note that $I_{c'}$ is the current entering the base-collector depletion region from the quasi-neutral, i.e., at $z = w_C$, and I_c is the current exiting the base-collector depletion region into the quasi-neutral collector, i.e., at $z = w_{C'}$.

$$= -\frac{\tilde{I}_c}{\tilde{I}_e}\bigg|_{\overline{V}_{cb}} \times \frac{\tilde{I}_e}{\tilde{V}_{eb}}\bigg|_{\overline{V}_{cb}}$$

$$= -\tilde{\alpha}y_{ee}^b, \tag{7.125}$$

where $\tilde{\alpha}$ is the small-signal current transfer ratio, defined as

$$\tilde{\alpha} = \frac{\tilde{I}_c}{\tilde{I}_e}\bigg|_{\overline{V}_{cb}}, \tag{7.126}$$

and, as in quasi-static analysis, it can be written as

$$\tilde{\alpha} = \frac{\tilde{I}_{ne}}{\tilde{I}_e} \times \frac{\tilde{I}_{nc'}}{\tilde{I}_{ne}} \times \frac{\tilde{I}_{nc}}{\tilde{I}_{nc'}}$$

$$= \tilde{\gamma}\tilde{\alpha}_T\tilde{\zeta}. \tag{7.127}$$

All of these are the small-signal quantities, which can differ substantially from their quasi-static magnitude. Here, the first term $\tilde{\gamma}$ is the injection efficiency, the second term $\tilde{\alpha}_T$ is the base transport factor, and the third term $\tilde{\zeta}$ is the collector transport factor.

Initially, in our analysis, we are assuming negligible hole injection into the emitter, hence, the emitter admittance is associated only with electron injection into the base, i.e., with y_{ne}^b,

$$y_{ee}^b \approx y_{ne}^b$$

$$= \frac{\tilde{I}_{ne}}{\tilde{V}_{eb}}\bigg|_{\overline{V}_{cb}}. \tag{7.128}$$

For purposes of comparison, we consider the input conductance and base diffusion capacitance using perturbational analysis from the quasi-static solution first. These will serve as a reference for our discussions of differences in between the small-signal solution and its magnitude at the highest frequencies, and the small-signal solution and its magnitude at very low frequencies. We should expect the limit of low frequency to be the same as the quasi-static limit. Consider the emitter admittance, due to electron injection into the base. In the low frequency limit,

$$y_{ne} = \frac{dI_E}{d\overline{V}_{eb}}\bigg|_{\overline{V}_{cb}}$$

$$= \frac{d}{d\overline{V}_{eb}}\left\{I_{ES}\left[\exp\left(\frac{q\overline{V}_{eb}}{kT}\right) - 1\right] - \alpha_R I_{CS}\left[\exp\left(\frac{q\overline{V}_{cb}}{kT}\right) - 1\right]\right\}$$

$$= \frac{q}{kT}I_{ES}\exp\left(\frac{q\overline{V}_{eb}}{kT}\right) \approx \frac{q}{kT}\overline{I}_e \equiv g_e. \tag{7.129}$$

The inadequacy of the perturbational technique becomes evident here. This analysis indicates that assuming static conditions prevail, the input admittance should be a conductance, a conductance similar to that of a p–n junction diode. This is only true at zero frequency. Application of bias at the base–emitter junction also changes the amount of minority charge stored in the quasi-neutral region, a charge that is made available by the flow of current. Thus diffusion capacitances exist and are associated with the imaginary part of the input admittance. To extend the accuracy of this perturbation procedure to any finite frequency, we include the effects related to stored charges. This should result in a capacitance in parallel with the conductance derived. Note that implicit in deriving stored charge from quasi-static distributions is the assumption that the perturbation is such that steady-state has been achieved at all times, so that our steady-state equations are still valid. Small-signal analysis does not make this assumption. Under the quasi-static assumption, the response to a change in bias from V_{BE} to $V_{BE} + \Delta V_{BE}$, the electron concentration at the emitter–base junction changed from n_{BE} to $n_{BE} + \Delta n_{BE}$. The incremental charge $\Delta \mathcal{Q}_B$ is added by the external circuit, and the capacitance associated with this is the emitter diffusion capacitance C_{De},

$$
\begin{aligned}
C_{De} &= \mathcal{A} \times \frac{\Delta \mathcal{Q}_B}{\Delta V_{BE}} \\[2mm]
&= \mathcal{A} \times \frac{\frac{1}{2} q \Delta n_{BE} w_B}{\Delta V_{BE}} \\[2mm]
&= \frac{\mathcal{A} q w_B n_{p0} \left\{ \exp\left[q \left(V_{BE} + \Delta V_{BE} \right) / kT \right] - \exp \left(q V_{BE}/kT \right) \right\}}{2 \, \Delta V_{BE}} \\[2mm]
&= \mathcal{A} \frac{1}{2} q w_B \frac{1}{\Delta V_{BE}} n_{p0} \exp\left(\frac{q V_{BE}}{kT} \right) \frac{q}{kT} \Delta V_{BE} \\[2mm]
&= \frac{q}{kT} \frac{{w_B}^2}{2 \mathcal{D}_n} \frac{\mathcal{A} q \mathcal{D}_n n_{p0}}{w_B} \exp\left(\frac{q V_{BE}}{kT} \right) \\[2mm]
&\approx \frac{q}{kT} \frac{W_B^2}{2 \mathcal{D}_n} \overline{I}_{ne} \\[2mm]
&= g_e \frac{W_B^2}{2 \mathcal{D}_n}.
\end{aligned}
\tag{7.130}
$$

Hence, we have the model for the junction as a parallel assembly of a constant conductance and capacitance at low frequencies. As the frequency increases for the applied signals, we may not use our steady-state equations and solutions, and derivation of parameters and equivalent circuits at these higher frequencies is the objective of the following section.

7.6.2 Static and Small-Signal Solutions

The transport of electrons in the base is diffusive in low-level injection for a uniformly doped base. The current equation, under drift-diffusion approximation, is

$$J_n = q\mu_n n_p \mathcal{E} + q\mathcal{D}_n \frac{dn_p}{dz}$$

$$\approx q\mathcal{D}_n \frac{dn_p}{dz}. \qquad (7.131)$$

We also have, from the continuity equation,

$$\frac{\partial n_p}{\partial t} = \mathcal{G}_n - \mathcal{R}_n + \frac{1}{q}\frac{dJ_n}{dz}, \qquad (7.132)$$

which can be written as

$$\frac{\partial n_p}{\partial t} = \frac{n_p - n_{p0}}{\tau_n} + \frac{1}{q}\frac{dJ_n}{dz} \qquad (7.133)$$

for a single time constant approximation of carrier generation and recombination processes.

Since these are linear differential equations, under small-signal conditions, we may assume that the electron concentration $n_p(z,t)$ is separated into a steady-state component $\overline{n}_p(z)$ and a small-signal component $\tilde{n}_p(z,t)$, so

$$n_p(z,t) = \overline{n}_p(z) + \tilde{n}_p(z,t). \qquad (7.134)$$

Using phasor notation, with ω as the angular frequency of the small-signal,

$$n_p(z,t) = \overline{n}_p(z) + \hat{n}_p(z)\exp\left(j\omega t\right). \qquad (7.135)$$

We may now separate the current continuity equation into a time-independent and time-dependent part, each of which must be satisfied separately,

$$\frac{\partial n_p}{\partial t} = -\frac{n_p - n_{p0}}{\tau_n} + \frac{1}{q}\frac{dJ_n}{dz}$$

$$= -\frac{n_p - n_{p0}}{\tau_n} + \mathcal{D}_n\frac{d^2 n_p}{dz^2}$$

$$\Rightarrow j\omega\hat{n}_p\exp\left(j\omega t\right) = -\frac{\overline{n}_p - n_{p0}}{\tau_n} - \frac{\hat{n}_p(z)\exp\left(j\omega t\right)}{\tau_n} +$$

$$\mathcal{D}_n\frac{d^2\overline{n}_p}{dz^2} + \mathcal{D}_n\exp\left(j\omega t\right)\frac{d^2\hat{n}_p}{dz^2}. \qquad (7.136)$$

The steady-state part of this equation is

$$0 = -\frac{\overline{n}_p - n_{p0}}{\tau_n} + D_n \frac{d^2\overline{n}_p}{dz^2}, \tag{7.137}$$

and one form of the solution is

$$\overline{n}_p = n_{p0} + A\cosh(\frac{z}{\mathcal{L}_n}) + B\sinh(\frac{z}{\mathcal{L}_n}), \tag{7.138}$$

where $\mathcal{L}_n$ is the diffusion length ($= \sqrt{D_n\tau_n}$). The small-signal part of this equation is

$$j\omega\hat{n}_p(z) = -\frac{\hat{n}_p(z)}{\tau_n} + D_n\frac{d^2\hat{n}_p(z)}{dz^2}$$

$$\frac{d^2\hat{n}_p(z)}{dz^2} - \hat{n}_p(z)\left[\frac{j\omega}{D_n} + \frac{1}{D_n\tau_n}\right] = 0$$

$$\frac{d^2\hat{n}_p(z)}{dz^2} - \frac{\hat{n}_p(z)}{\mathcal{L}_n{}^2/(1+\omega\tau_n)} = 0. \tag{7.139}$$

Let us define the parameter

$$\varsigma = \frac{\sqrt{1+j\omega\tau_n}}{\mathcal{L}_n}, \tag{7.140}$$

then the solution of this second-order differential equation can be written in the form

$$\hat{n}_p(z) = C\cosh(\varsigma z) + D\sinh(\varsigma z). \tag{7.141}$$

So, the solution of the continuity equation (Equation 7.132) is of the form

$$n_p(z,t) = \overline{n}_p + \hat{n}_p \exp(j\omega t)$$

$$= n_{p0} + A\cosh\left(\frac{z}{\mathcal{L}_n}\right) + B\sinh\left(\frac{z}{\mathcal{L}_n}\right) +$$

$$[C\cosh(\varsigma z) + D\sinh(\varsigma z)]\exp(j\omega t). \tag{7.142}$$

We also need the boundary conditions to obtain the distribution of carriers in the quasi-neutral base region. At the base edge of the base–emitter junction (we define this for convenience as the point where $z = 0$), we assume instantaneous equilibrium of carriers and use the Shockley boundary conditions

$$n_p(0,t) = \overline{n}_p(0) + \hat{n}_p(0)\exp(j\omega t)$$

$$= n_{p0} \exp\left(\frac{qV_{be}}{kT}\right)$$

$$= n_{p0} \exp\left(\frac{q\overline{V}_{be}}{kT}\right) \exp\left(\frac{q\tilde{V}_{be}}{kT}\right)$$

$$= n_{p0} \exp\left(\frac{q\overline{V}_{be}}{kT}\right) \exp\left[\frac{q\hat{V}_{be}\exp\left(j\omega t\right)}{kT}\right]$$

$$\approx n_{p0} \exp\left(\frac{q\overline{V}_{be}}{kT}\right) \left[1 + \frac{q\hat{V}_{be}}{kT}\exp\left(j\omega t\right)\right]$$

$$= n_{p0} \exp\left(\frac{q\overline{V}_{be}}{kT}\right) +$$

$$n_{p0} \exp\left(\frac{q\overline{V}_{be}}{kT}\right) \frac{q\hat{V}_{be}}{kT} \exp\left(j\omega t\right). \tag{7.143}$$

The first term in the last equation is the boundary condition for steady-state analysis—it is the boundary condition we have employed in our quasi-static analysis—and the second term is the boundary condition for small-signal analysis.

At the collector edge, all the carriers are immediately swept out due to the high field. The carrier concentration is therefore significantly below that of the emitter edge during normal active mode of operation. Thus, the carrier concentration can be assumed to vanish at the collector edge of the quasi-neutral base region. Even though the carrier concentration vanishes, since output admittance is small, the movement of the base edge of the base–collector boundary can have a strong relative effect. Recall that this was the cause of Early effect in the bipolar transistor and gave rise to an output conductance. We need to introduce this boundary condition, which is a moving boundary condition, in our small-signal analysis. The variation of the base edge of the base–collector depletion region also follows the frequency of the small-signal. Let it vary as $w_B + \hat{w}\exp\left(j\omega t\right)$, where $\hat{w}$ involves the same reasoning that led to the Early voltage under quasi-static conditions. Under the applied collector bias our collector edge boundary conditions are

$$n_p(w_B, t) = n_p(w_B + \hat{w}\exp\left(j\omega t\right), t) = 0. \tag{7.144}$$

This equation does not imply that the static and small-signal component of the charge concentration are individually negligible. The sum vanishes and both the static and small-signal quantities are individually quite small. We can now determine the coefficients of our solution of $n_p(z, t)$ in terms of the device parameters. First, consider the base–emitter junction boundary conditions.

$$
\begin{aligned}
n_p(0,t) &= n_{p0}\exp\left(\frac{q\overline{V}_{be}}{kT}\right) + n_{p0}\exp\left(\frac{q\overline{V}_{be}}{kT}\right)\frac{q\hat{V}_{be}}{kT}\exp\left(j\omega t\right) \\
&= n_{p0} + A + C\exp\left(j\omega t\right). \tag{7.145}
\end{aligned}
$$

In order to satisfy this, both the steady-state and time varying parts of the carrier concentration have to be negligible, i.e.,

$$
\begin{aligned}
A &= n_{p0}\left[\exp\left(\frac{q\overline{V}_{be}}{kT}\right) - 1\right] \\
\text{and}\quad C &= n_{p0}\exp\left(\frac{q\overline{V}_{be}}{kT}\right)\frac{q\hat{V}_{be}}{kT}. \tag{7.146}
\end{aligned}
$$

At the collector edge,

$$
n_p\left(w_B + \hat{w}\exp(j\omega t), t\right) = 0, \tag{7.147}
$$

which implies

$$
n_{p0} + A\cosh\left[\frac{w_B + \hat{w}\exp(j\omega t)}{\mathcal{L}_n}\right] + B\sinh\left[\frac{w_B + \hat{w}\exp(j\omega t)}{\mathcal{L}_n}\right] +
$$
$$
\left[C\cosh\left(\varsigma w_B + \hat{w}\exp(j\omega t)\right) + D\sinh\left(\varsigma w_B + \hat{w}\exp(j\omega t)\right)\right]\exp\left(j\omega t\right) = 0 \tag{7.148}
$$

Expanding the hyperbolic terms,

$$
\begin{aligned}
0 = {} & n_{p0} + A\left[\cosh\left(\frac{w_B}{\mathcal{L}_n}\right)\cosh\left(\frac{\hat{w}\exp(j\omega t)}{\mathcal{L}_n}\right) + \right. \\
& \left. \sinh\left(\frac{w_B}{\mathcal{L}_n}\right)\sinh\left(\frac{\hat{w}\exp(j\omega t)}{\mathcal{L}_n}\right)\right] + \\
& B\left[\sinh\left(\frac{w_B}{\mathcal{L}_n}\right)\cosh\left(\frac{\hat{w}\exp(j\omega t)}{\mathcal{L}_n}\right) + \right. \\
& \left. \cosh\left(\frac{w_B}{\mathcal{L}_n}\right)\sinh\left(\frac{\hat{w}\exp(j\omega t)}{\mathcal{L}_n}\right)\right] + \\
& C\left[\cosh(\varsigma w_B)\cosh\left(\varsigma\hat{w}\exp(j\omega t)\right) + \right. \\
& \left. \sinh(\varsigma w_B)\sinh\left(\varsigma\hat{w}\exp(j\omega t)\right)\right]\exp\left(j\omega t\right) + \\
& D\left[\sinh(\varsigma w_B)\cosh\left(\varsigma\hat{w}\exp(j\omega t)\right) + \right. \\
& \left. \cosh(\varsigma w_B)\sinh\left(\varsigma\hat{w}\exp(j\omega t)\right)\right]\exp\left(j\omega t\right). \tag{7.149}
\end{aligned}
$$

The phasor of the base width modulation due to the small-signal $\hat{V}_{cb}$ is given by the following if we assume instantaneous response to the applied signal:

$$\hat{w} \approx \frac{dw_B}{dV_{CB}} \hat{V}_{cb}$$

$$\approx \frac{\hat{V}_{cb}}{V_A} w_B, \tag{7.150}$$

where V_A is the Early voltage. Since transistors have large Early voltages, and useful transistors also employ $w_B \ll \mathcal{L}_n$, the magnitude of the terms $\hat{w}\exp(j\omega t)/\mathcal{L}_n$ and $\varsigma\hat{w}\exp(j\omega t)$ are small, and we do a perturbation expansion of the hyperbolic terms, approximating,

$$\cosh\left(\frac{\hat{w}\exp(j\omega t)}{\mathcal{L}_n}\right) \approx 1,$$

$$\sinh\left(\frac{\hat{w}\exp(j\omega t)}{\mathcal{L}_n}\right) \approx \frac{\hat{w}\exp(j\omega t)}{\mathcal{L}_n},$$

$$\cosh\left(\varsigma\hat{w}\exp(j\omega t)\right) \approx 1,$$

$$\text{and} \quad \sinh\left(\varsigma\hat{w}\exp(j\omega t)\right) \approx \varsigma\hat{w}\exp(j\omega t). \tag{7.151}$$

Using these approximations, the boundary condition at the collector edge of the quasi-neutral base region assumes the form

$$\begin{aligned}
0 =\ & n_{p0} + A\left[\cosh\left(\frac{w_B}{\mathcal{L}_n}\right) + \sinh\left(\frac{w_B}{\mathcal{L}_n}\right)\frac{\hat{w}\exp(j\omega t)}{\mathcal{L}_n}\right] + \\
& B\left[\sinh\left(\frac{w_B}{\mathcal{L}_n}\right) + \cosh\left(\frac{w_B}{\mathcal{L}_n}\right)\frac{\hat{w}\exp(j\omega t)}{\mathcal{L}_n}\right] + \\
& C\left[\cosh(\varsigma w_B) + \sinh(\varsigma w_B)\varsigma\hat{w}\exp(j\omega t)\right]\exp(j\omega t) + \\
& D\left[\sinh(\varsigma w_B) + \cosh(\varsigma w_B)\varsigma\hat{w}\exp(j\omega t)\right]\exp(j\omega t). \tag{7.152}
\end{aligned}$$

Considering only the steady-state and the first harmonic terms (i.e., ignoring the second-order terms of $\exp(2j\omega t)$), we obtain

$$\begin{aligned}
0 =\ & n_{p0} + \left[A\cosh\left(\frac{w_B}{\mathcal{L}_n}\right) + B\sinh\left(\frac{w_B}{\mathcal{L}_n}\right)\right] + \\
& \left[A\sinh\left(\frac{w_B}{\mathcal{L}_n}\right)\frac{\hat{w}}{\mathcal{L}_n} + B\cosh\left(\frac{w_B}{\mathcal{L}_n}\right)\frac{\hat{w}}{\mathcal{L}_n}\right. \\
& \left. + C\cosh(\varsigma w_B) + D\sinh(\varsigma w_B)\right]\exp(j\omega t). \tag{7.153}
\end{aligned}$$

So, we obtain the coefficient B in terms of the coefficient A that has already been determined.

$$B = -\frac{n_{p0} + A\cosh(w_B/\mathcal{L}_n)}{\sinh(w_B/\mathcal{L}_n)}$$

$$
= -\frac{n_{p0} + n_{p0}\left[\exp\left(q\overline{V}_{be}/kT\right) - 1\right]\cosh\left(w_B/\mathcal{L}_n\right)}{\sinh\left(w_B/\mathcal{L}_n\right)}
$$

$$
\approx -n_{p0}\exp\left(\frac{q\overline{V}_{be}}{kT}\right)\coth\left(\frac{w_B}{\mathcal{L}_n}\right). \tag{7.154}
$$

Finally, we obtain D in terms of A, B, and C, which are now determined.

$$
D = -\left[A\sinh\left(\frac{w_B}{\mathcal{L}_n}\right)\frac{\hat{w}}{\mathcal{L}_n} + B\cosh\left(\frac{w_B}{\mathcal{L}_n}\right)\frac{\hat{w}}{\mathcal{L}_n} + C\cosh(\varsigma w_B)\right] \times
$$

$$
\frac{1}{\sinh(\varsigma w_B)}
$$

$$
= -\left\{ n_{p0}\left[\exp\left(\frac{q\overline{V}_{be}}{kT}\right) - 1\right]\frac{\hat{w}}{\mathcal{L}_n}\sinh\left(\frac{w_B}{\mathcal{L}_n}\right) - \right.
$$

$$
n_{p0}\exp\left(\frac{q\overline{V}_{be}}{kT}\right)\frac{\hat{w}}{\mathcal{L}_n}\coth\left(\frac{w_B}{\mathcal{L}_n}\right)\cosh\left(\frac{w_B}{\mathcal{L}_n}\right) +
$$

$$
\left. n_{p0}\exp\left(\frac{q\overline{V}_{be}}{kT}\right)\frac{q\hat{V}_{be}}{kT}\cosh(\varsigma w_B)\right\}\frac{1}{\sinh(\varsigma w_B)}
$$

$$
\approx -\left[n_{p0}\exp\left(\frac{q\overline{V}_{be}}{kT}\right)\frac{\hat{w}}{\mathcal{L}_n}\frac{\sinh^2(w_B/\mathcal{L}_n) - \cosh^2(w_B/\mathcal{L}_n)}{\sinh(w_B/\mathcal{L}_n)} + \right.
$$

$$
\left. n_{p0}\exp\left(\frac{q\overline{V}_{be}}{kT}\right)\frac{q\hat{V}_{be}}{kT}\cosh(\varsigma w_B)\right]\frac{1}{\sinh(\varsigma w_B)}
$$

$$
= -\left[-n_{p0}\exp\left(\frac{q\overline{V}_{be}}{kT}\right)\frac{\hat{w}}{\mathcal{L}_n}\frac{1}{\sinh(w_B/\mathcal{L}_n)} + \right.
$$

$$
\left. n_{p0}\exp\left(\frac{q\overline{V}_{be}}{kT}\right)\frac{q\hat{V}_{be}}{kT}\cosh(\varsigma w_B)\right]\frac{1}{\sinh(\varsigma w_B)}
$$

$$
= n_{p0}\exp\left(\frac{q\overline{V}_{be}}{kT}\right)\frac{\hat{w}}{\mathcal{L}_n}\operatorname{csch}\left(\frac{w_B}{\mathcal{L}_n}\right)\operatorname{csch}(\varsigma w_B) -
$$

$$
n_{p0}\exp\left(\frac{q\overline{V}_{be}}{kT}\right)\frac{q\hat{V}_{be}}{kT}\coth(\varsigma w_B), \tag{7.155}
$$

and the complete equation for carrier distribution is

$$
n_p(z,t) = n_{p0} + n_{p0}\left[\exp\left(\frac{q\overline{V}_{be}}{kT}\right) - 1\right]\cosh\left(\frac{z}{\mathcal{L}_n}\right) -
$$

$$
n_{p0}\exp\left(\frac{q\overline{V}_{be}}{kT}\right)\coth\left(\frac{w_B}{\mathcal{L}_n}\right)\sinh\left(\frac{z}{\mathcal{L}_n}\right) +
$$

$$n_{p0} \exp \left(\frac{q\overline{V}_{be}}{kT} \right) \frac{q\hat{V}_{be}}{kT} \cosh(\varsigma z) \exp (j\omega t) +$$

$$n_{p0} \exp \left(\frac{q\overline{V}_{be}}{kT} \right) \frac{\hat{w}}{\mathcal{L}_n} \mathrm{csch} \left(\frac{w_B}{\mathcal{L}_n} \right) \times$$

$$\mathrm{csch}(\varsigma w_B) \sinh(\varsigma z) \exp (j\omega t) -$$

$$n_{p0} \exp \left(\frac{q\overline{V}_{be}}{kT} \right) \frac{q\hat{V}_{be}}{kT} \coth(\varsigma w_B) \sinh(\varsigma z) \exp (j\omega t) . \tag{7.156}$$

The first two terms are the steady-state distribution of carriers while the last three terms constitute the small-signal distribution of carriers. In order to evaluate the matrix elements, we now evaluate the currents. From now on, for the sake of simplicity, the unity term occurring with $\exp \left(q\overline{V}_{be}/kT \right)$ will also be ignored because it is negligible in the active mode of operation for any appreciable current flow. The electron current in the base at the base–emitter junction,

$$
\begin{aligned}
I_{ne} &= \overline{I}_{ne} + \tilde{I}_{ne} \\
&= \overline{I}_{ne} + \hat{I}_{ne} \exp (j\omega t) \\
&= q\mathcal{D}_n \mathcal{A} \left. \frac{dn_p}{dz} \right|_{z=0} \\
&= q\mathcal{D}_n \mathcal{A} n_{p0} \exp \left(\frac{q\overline{V}_{be}}{kT} \right) \left[-\frac{1}{\mathcal{L}_n} \coth \left(\frac{w_B}{\mathcal{L}_n} \right) + \right. \\
&\quad \frac{\hat{w}}{\mathcal{L}_n} \varsigma \mathrm{csch} \left(\frac{w_B}{\mathcal{L}_n} \right) \mathrm{csch}(\varsigma w_B) \exp (j\omega t) - \\
&\quad \left. \frac{q\hat{V}_{be}}{kT} \varsigma \coth(\varsigma w_B) \exp (j\omega t) \right] .
\end{aligned}
\tag{7.157}
$$

The small-signal component of this equation, after substituting for $\hat{w} \approx (dw_B/dV_{CB}) \hat{V}_{cb}$, is

$$
\begin{aligned}
\tilde{I}_{ne} &= q\mathcal{D}_n \mathcal{A} n_{p0} \exp \left(\frac{q\overline{V}_{be}}{kT} \right) \frac{q}{kT} \varsigma \coth(\xi_b)\hat{V}_{eb} + \\
&\quad q\mathcal{D}_n \mathcal{A} n_{p0} \exp \left(\frac{q\overline{V}_{be}}{kT} \right) \frac{1}{\mathcal{L}_n} \frac{dw_B}{dV_{CB}} \mathrm{csch} \left(\frac{w_B}{\mathcal{L}_n} \right) \varsigma \mathrm{csch}(\xi_b)\hat{V}_{cb},
\end{aligned}
\tag{7.158}
$$

where we have defined a new variable ξ_b,

$$\xi_b = \varsigma w_B, \tag{7.159}$$

because it will appear as a parameter relating the frequency effect in the base in several of our network parameter relationships.

The electron current exiting the base at the base–collector junction is

$$
\begin{aligned}
I_{nc'} &= \overline{I}_{nc'} + \tilde{I}_{nc'} \\
&= \overline{I}_{nc'} + \hat{I}_{nc'} \exp\left(j\omega t\right) \\
&= q\mathcal{D}_n \mathcal{A} \frac{dn_p}{dz}\bigg|_{z=w_C} \\
&= q\mathcal{D}_n \mathcal{A} n_{p0} \exp\left(\frac{q\overline{V}_{be}}{kT}\right) \left[\frac{1}{\mathcal{L}_n} \sinh\left(\frac{w_B}{\mathcal{L}_n}\right) - \right. \\
&\quad \frac{1}{\mathcal{L}_n} \coth\left(\frac{w_B}{\mathcal{L}_n}\right) \cosh\left(\frac{w_B}{\mathcal{L}_n}\right) + \\
&\quad \frac{q\hat{V}_{be}}{kT}\varsigma \sinh(\varsigma w_B) \exp\left(j\omega t\right) + \\
&\quad \frac{\hat{w}}{\mathcal{L}_n} \operatorname{csch}\left(\frac{w_B}{\mathcal{L}_n}\right) \varsigma\operatorname{csch}(\varsigma w_B) \cosh(\varsigma z) \exp\left(j\omega t\right) - \\
&\quad \left. -\frac{q}{kT}\hat{V}_{be}\varsigma \coth(\varsigma w_B) \cosh(\varsigma z) \exp\left(j\omega t\right)\right]. \quad (7.160)
\end{aligned}
$$

The small-signal component of this equation, after substituting for $\hat{w} \approx (dw_B/dV_{CB})\hat{V}_{cb}$, is:

$$
\begin{aligned}
\hat{I}_{nc'} &= -q\mathcal{D}_n \mathcal{A} n_{p0} \exp\left(\frac{q\overline{V}_{be}}{kT}\right) \frac{q}{kT}\varsigma \left[\sinh(\varsigma w_B) - \right.\\
&\quad \left. \coth(\varsigma w_B) \cosh(\varsigma w_B)\right] \hat{V}_{eb} + \\
&\quad q\mathcal{D}_n \mathcal{A} n_{p0} \exp\left(\frac{q\overline{V}_{be}}{kT}\right) \frac{1}{\mathcal{L}_n} \frac{dw_B}{dV_{CB}} \operatorname{csch}\left(\frac{w_B}{\mathcal{L}_n}\right) \times \\
&\quad \varsigma\operatorname{csch}(\varsigma w_B) \cosh(\varsigma w_B)\hat{V}_{cb} \\
&= q\mathcal{D}_n \mathcal{A} n_{p0} \exp\left(\frac{q\overline{V}_{be}}{kT}\right) \frac{q}{kT}\frac{1}{w_B}\xi_b\operatorname{csch}(\xi_b)\hat{V}_{eb} + \\
&\quad q\mathcal{D}_n \mathcal{A} n_{p0} \exp\left(\frac{q\overline{V}_{be}}{kT}\right) \frac{1}{\mathcal{L}_n} \frac{dw_B}{dV_{CB}} \operatorname{csch}\left(\frac{w_B}{\mathcal{L}_n}\right) \times \\
&\quad \frac{1}{w_B}\xi_b \coth(\xi_b)\hat{V}_{cb}. \quad (7.161)
\end{aligned}
$$

We have now determined the currents at the emitter edge and the collector edge of the quasi-neutral base region in terms of the applied emitter–base and collector–base potentials. This was the necessary step in the determination of the admittance matrix for contributions from transport in the base. We can now utilize this essential result in determining the admittance matrix of the device taken in its entirety. As part of this discussion we also consider the approximations that lead to simplified modelling.

7.6.3 Network Parameters and their Approximations

The currents that we have determined, $\hat{I}_{ne}$ and $\hat{I}_{nc'}$, establish the effects of transport in the quasi-neutral base. Expressed as a function of the signals $\hat{V}_{eb}$ and $\hat{V}_{cb}$, these determine the admittance effects of electron transport. We need to add to these effects related to both the transport and the displacement effects in the base–collector depletion region and the emitter–base depletion region. For convenience, we will include these as additions to the admittance parameters associated with electron transport in the base, denoted by y_{nee}^b, y_{nec}^b, y_{nce}^b, and y_{ncc}, and representing the input, reverse transfer, forward transfer and the output admittances.

Consider y_{nee}^b first. It is the coefficient of $\hat{V}_{eb}$ in Equation 7.158. In the limit of $\omega \to 0$, the input admittance should reduce to the static conductance resulting from electron transport, i.e., g_{ne}. So, by definition

$$
\begin{aligned}
g_{ne} &\equiv q\mathcal{D}_n A n_{p0} \exp\left(\frac{q\overline{V}_{be}}{kT}\right) \frac{q}{kT} \varsigma \coth(\xi_b)\Bigg|_{\omega=0} \\
&= q\mathcal{D}_n A n_{p0} \exp\left(\frac{q\overline{V}_{be}}{kT}\right) \frac{q}{kT} \frac{1}{\mathcal{L}_n} \coth\left(\frac{w_B}{\mathcal{L}_n}\right) \\
&= \frac{q\overline{I}_{ne}}{kT}.
\end{aligned}
\tag{7.162}
$$

This is the same result we had derived from perturbation analysis, and it shows the contribution of electron transport. The conductance and admittance also have a contribution from hole transport in the emitter, which has not yet been included. The result that we have derived is, however, more general than this relationship in low frequency limit. The input admittance component due to base transport, for any arbitrary frequency, within the drift-diffusion approximation, can now be written as

$$
\begin{aligned}
y_{nee} &= g_{ne} \frac{1}{\coth(w_B/\mathcal{L}_n)} \frac{\mathcal{L}_n}{w_B} \xi_b \coth(\xi_b) \\
&= g_{ne} \xi_b \coth(\xi_b).
\end{aligned}
\tag{7.163}
$$

In deriving this, we have assumed that the bipolar transistor has reasonable gain at low frequencies, and hence $w_B \ll \mathcal{L}_n$. This implies that $(\mathcal{L}_n/w_B)/\coth(w_B/\mathcal{L}_n) \approx 1$. This is a very simple and yet general format of the complex equations that we have been dealing with.

Now consider the output admittance $y^b_{ncc'}$, which is the negative of the coefficient of $\hat{V}_{cb}$ in Equation 7.161 because network network parameters are defined with current flowing into a port. The limit of this at $\omega \to 0$ is a conductance which we denote by g_{nc}.

$$
\begin{aligned}
g_{nc} &= \left. -\frac{q\mathcal{D}_n A n_{p0}}{\mathcal{L}_n} \exp\left(\frac{q\overline{V}_{be}}{kT}\right) \frac{dw_B}{dV_{CB}} \operatorname{csch}\left(\frac{w_B}{\mathcal{L}_n}\right) \frac{1}{w_B} \xi_b \coth(\xi_b) \right|_{\omega=0} \\
&= -\frac{q\mathcal{D}_n A n_{p0}}{\mathcal{L}_n} \exp\left(\frac{q\overline{V}_{be}}{kT}\right) \frac{dw_B}{dV_{CB}} \operatorname{csch}\left(\frac{w_B}{\mathcal{L}_n}\right) \frac{1}{w_B} \frac{w_B}{\mathcal{L}_n} \coth\left(\frac{w_B}{\mathcal{L}_n}\right) \\
&= -\frac{q\mathcal{D}_n A n_{p0}}{w_B} \exp\left(\frac{q\overline{V}_{be}}{kT}\right) \frac{1}{w_B} \frac{dw_B}{dV_{CB}} \\
&= -\overline{I}_{ne} \frac{1}{w_B} \frac{dw_B}{dV_{CB}}.
\end{aligned}
\tag{7.164}
$$

Since the coefficient of $\overline{I}_{ne}$ in the above represents the Early factor, a small quantity, the output conductance is significantly smaller than the input conductance $q\overline{I}_{ne}/kT$ (see Problem 9). Also, it is positive because the base width decreases with an increase in collector-to-base voltage. The ratio of the two conductances,

$$
\frac{g_{nc}}{g_{ne}} = -\frac{kT}{qw_B} \frac{dw_B}{dV_{CB}} \ll 1.
\tag{7.165}
$$

This is to be expected since the collector junction is reverse biased and the base–emitter junction is forward biased. The same ratio holds for the admittances. The forward admittance is much larger than the reverse admittance. Using g_{nc} as the magnitude of the output admittance in the zero frequency limit, we have

$$
y_{nc'c'} = g_{nc} \xi_b \coth(\xi_b),
\tag{7.166}
$$

again a very simple relationship.

The coefficient of the second term of Equation 7.158 is the reverse admittance contribution of base transport, $y^b_{nec'}$. This can also be simplified, following the form above. The low frequency limit of this is a real term that is g_{nc} and given by

$$
\begin{aligned}
\left. y^b_{nec'} \right|_{\omega=0} &= q\mathcal{D}_n A n_{p0} \exp\left(\frac{q\overline{V}_{be}}{kT}\right) \frac{1}{\mathcal{L}_n} \frac{dw_B}{dV_{CB}} \frac{1}{w_B} \operatorname{csch}\left(\frac{w_B}{\mathcal{L}_n}\right) \times \\
&\qquad \left. \xi_b \operatorname{csch}(\xi_b) \right|_{\omega=0} \\
&\approx \frac{q\mathcal{D}_n A n_{p0}}{w_B} \exp\left(\frac{q\overline{V}_{be}}{kT}\right) \frac{1}{w_B} \frac{dw_B}{dV_{CB}} \\
&= -g_{nc}.
\end{aligned}
\tag{7.167}
$$

From this, the reverse admittance may be written as

$$y_{nec'}^b == -g_{nc}\xi_b\mathrm{csch}(\xi_b), \tag{7.168}$$

again a simple but accurate relationship.

The transfer admittance can be found from the coefficient of $\hat{V}_{eb}$ in Equation 7.161 together with the sign reversal to account for the definition of current for the network parameters.

$$\begin{aligned} y_{nc'e}^b &= -q\mathcal{D}_n\mathcal{A}n_{p0}\exp\left(\frac{q\overline{V}_{be}}{kT}\right)\frac{q}{kT}\frac{1}{w_B}\xi_b\mathrm{csch}(\xi_b) \\ &= -g_{ne}\xi_b\mathrm{csch}(\xi_b). \end{aligned} \tag{7.169}$$

We have now derived the simple relationships of the admittance parameter terms resulting from electron transport in the base. Summarizing them here,

$$\begin{aligned} y_{nee}^b &= g_{ne}\xi_b\coth(\xi_b), \\ y_{nec'}^b &= -g_{nc}\xi_b\mathrm{csch}(\xi_b), \\ y_{nc'e}^b &= -g_{ne}\xi_b\mathrm{csch}(\xi_b), \\ \text{and}\quad y_{nc'c'}^b &= g_{nc}\xi_b\coth(\xi_b). \end{aligned} \tag{7.170}$$

So far we have considered only the electron current part of the transport in the base. If the collector junction is not reverse biased into breakdown, the hole current is insignificant, and hence we may substitute $g_c = g_{nc}$, i.e., the output conductance at low frequency is entirely due to electron transport. We have not included the collector transition region capacitance, i.e., the modulation of the space charge region by changes in the bias. This is the capacitance C_{tC}, applicable both in the quasi-static and the high frequency conditions, because the dielectric relaxation time is significantly smaller than the times of even this model.

We also have to consider the effect of the collector transport factor, which relates the current collected at the collector ohmic contact as well as flowing through the quasi-neutral collector region with the electron current entering the base–collector depletion region. These are generally not identical, since a transiting carrier in a depletion region has displacement effects associated with it. The quasi-neutral regions abutting the depletion region can be viewed as the equi-potential surfaces of the base–collector capacitor. A moving charge between two parallel conducting electrodes, as it moves between the plates (see Figure 7.37), has a changing termination

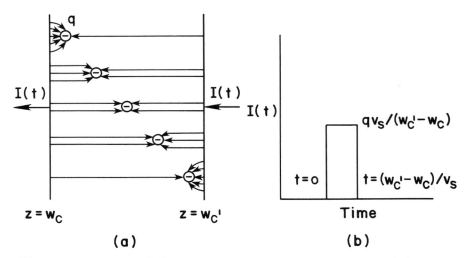

Figure 7.37: Transit of charge in between two parallel plates and the associated termination of fields during the transit (a), and a schematic of the time dependence of current in the external circuit (b).

in its associated electric field lines from one electrode to other. An electron very close to the quasi-neutral base region, e.g., terminates on a hole in the base at the edge of the base–collector depletion region. Similarly, an electron very close to the quasi-neutral collector region requires uncovering of an ionized donor. So while the electron is transiting, continuity of the total current is maintained at all cross-sections of the structure by the particle current in the quasi-neutral regions required to supply and collect charge flow, and by the particle current and the displacement current related to the changing of the electric field in the depletion region itself.

We consider the problem of current due to moving charges in the depleted space charge region, at the base–collector junction, using the current continuity equation under small-signal conditions. The electric field in the region, under the bias and small-signal conditions, is

$$
\begin{aligned}
\mathcal{E}(z) &= \overline{\mathcal{E}}(z) + \tilde{\mathcal{E}}(z) \\
&= \overline{\mathcal{E}}(z) + \hat{\mathcal{E}}(z) \exp\left(j\omega t\right),
\end{aligned} \tag{7.171}
$$

and the carrier density as a function of position is

$$
\begin{aligned}
n(z) &= \overline{n}(z) + \tilde{n}(z) \\
&= \overline{n}(z) + \hat{n}(z) \exp\left(j\omega t\right),
\end{aligned} \tag{7.172}
$$

in response to a collector bias of

$$V_{cb} = \overline{V}_{cb} + \hat{V}_{cb} \exp\left(j\omega t\right). \tag{7.173}$$

The total current density at any position z is

$$
\begin{aligned}
J(z) &= J_n + J_{nd} \\
&= J_n + \frac{\partial \mathcal{D}}{\partial t}.
\end{aligned}
\tag{7.174}
$$

The current J_n is the electron current density due to particle flow, i.e., the sum of the drift and diffusion components, and J_{nd} is the displacement current of electrons, $\mathcal{D}$ being the displacement vector. The small-signal component of this equation is

$$\hat{J} = \hat{J}_n + j\omega\epsilon\hat{\mathcal{E}}. \tag{7.175}$$

Integrating over the width of the depletion region,

$$\hat{J}(w_{C'} - w_C) = \int_{w_C}^{w_{C'}} \hat{J}_n dz + j\omega\epsilon \int_{w_C}^{w_{C'}} \hat{\mathcal{E}} dz. \tag{7.176}$$

The last integral is the small-signal voltage across the collector space charge region, $\hat{V}_{cb}$, and hence

$$\hat{J} = \frac{1}{(w_{C'} - w_C)} \int_{w_C}^{w_{C'}} \hat{J}_n dz + j\omega \frac{\epsilon}{(w_{C'} - w_C)} \hat{V}_{cb}. \tag{7.177}$$

Multiplying by the area $\mathcal{A}$, we recognize the fraction part of the last term as the transition capacitance of the base collector junction, C_{tC}. The current in the quasi-neutral collector region, corresponding to the collector electrode, follows as

$$\hat{I} = \frac{1}{(w_{C'} - w_C)} \int_{w_C}^{w_{C'}} \hat{I}_n dz + j\omega C_{tC}\hat{V}_{cb}. \tag{7.178}$$

The first term in this equation is usually referred to as the induced charge component and the last term is the transition charge component of currents. If we now assume that the electrons move with constant velocity v_s—the saturation velocity because the electric fields are high—then we may write the position-dependent carrier concentration as

$$\tilde{n}(z,t) = \hat{n}(0) \exp\left(-j\omega \frac{z}{v_s}\right), \tag{7.179}$$

and the position- and time-dependent carrier concentration as

$$\hat{n}(z,t) = \hat{n}(0)\exp\left[j\omega\left(t - \frac{z}{v_s}\right)\right]. \tag{7.180}$$

The particle current phasor is

$$\hat{I}_n = -q\mathcal{A}\hat{n}(0)\exp\left(-j\omega\frac{z}{v_s}\right)v_s, \tag{7.181}$$

and the particle current phasor at the base edge of the base–collector depletion region is

$$\hat{I}_{nc'} = -q\mathcal{A}\hat{n}(0)v_s. \tag{7.182}$$

The induced term can now be evaluated as

$$\frac{1}{(w_{C'} - w_C)}\int_{w_C}^{w_{C'}}\hat{I}_n dz$$

$$= -\frac{1}{(w_{C'} - w_C)}\int_{w_C}^{w_{C'}} q\mathcal{A}\hat{n}(0)\exp\left(-j\omega\frac{z}{v_s}\right)v_s dz$$

$$= -\frac{1}{(w_{C'} - w_C)}\frac{1}{(-j\omega/v_s)}q\mathcal{A}\hat{n}(0)\left[\exp\left(-j\omega\frac{w_{C'} - w_C}{v_s}\right) - 1\right]$$

$$= \hat{I}_{nc'}\frac{\sin(\omega\tau_c')}{\omega\tau_c'}\exp\left(-j\omega\tau_c'\right). \tag{7.183}$$

where

$$\tau_c' = \frac{(w_{C'} - w_C)}{2v_s}. \tag{7.184}$$

The ratio of particle current exiting the base–collector region at any instant to the particle current entering is the collector transport factor (ζ),

$$\zeta = \frac{\hat{I}_{nc}}{\hat{I}_{nc'}}$$

$$= \frac{\sin\omega\tau_c'}{\omega\tau_c'}\exp\left(-j\omega\tau_c'\right). \tag{7.185}$$

This time constant, τ_c', is the collector signal delay time, i.e., the delay in the exiting particle current signal with respect to that entering the base–collector depletion region. The actual time that any particle takes to traverse this collector depletion region is the collector transit time,

$$\tau_c = \frac{w_{C'} - w_C}{v_s}, \tag{7.186}$$

under our assumption of constant velocity transit. The signal delay is half that of the particle transit delay. Since,

$$\hat{I}_{nc} = \zeta \hat{I}_{nc'},\qquad(7.187)$$

we may now modify the admittance parameters related to the collector current by including the effect of base–collector region to our analysis of the transport in a quasi-neutral base. This results, following Equation 7.187, in both the output and forward admittance contributions of the base particle transport being modified to their product with ζ. Following the above analysis, the admittance of the collector transition capacitance can be included additively. These modified parameters are

$$y_{cc}^b = \zeta g_c \xi_b \coth(\xi_b) + j\omega C_{tC},\qquad(7.188)$$

and

$$y_{ce}^b = -\zeta g_c \xi_b \mathrm{csch}(\xi_b).\qquad(7.189)$$

For the emitter admittance, y_{pee}^b due to hole injection into the emitter, if substantial, has to be taken into account. In addition, we have to consider, similar to the case of base–collector depletion region transport, the effect of transit in the base–emitter depletion region and the admittance contribution of the emitter transition region capacitance. In our analysis, including the small-signal modelling, we have always assumed instantaneous equilibrium of carrier concentrations at the edges of the base–emitter depletion region. The basis of this assumption has been that the depletion region is considerably thinner than other transit regions under normal active mode operating conditions, and that other charging and transit effects overwhelm any contribution from this transit. This argument is still valid under the small-signal conditions, and hence we ignore the effect of transit in the base–emitter depletion region. We have discussed the quasi-static effects of the use of heterostructures at the base–emitter junction. It leads to an appreciable reduction of the hole injection under quasi-static conditions, and, by extension, to similar relative effects under small-signal conditions. The effect of small-signal operation should be to decrease the magnitude of the device, etc. Let the contribution of the hole part be called y_{pee}^b. Including this, and the effect of emitter transition capacitance C_{tE}, we obtain

$$y_{ee}^b = y_{pee}^b + g_{nc} \xi_b \coth(\xi_b) + j\omega C_{tE}.\qquad(7.190)$$

We now consider the relative magnitude of these parameters. First, consider y_{pee}^b for a heterostructure bipolar transistor. Since the hole injection has been decreased, the quasi-static magnitude of the injection efficiency is very high, and $y_{pee}^b \ll y_{nee}^b$ under most bias conditions. Consider the graded barrier heterostructure bipolar transistor. The change in bandgap of the materials ΔE_g appears as an excess barrier in the valence band. So, the difference between the heterostructure and homostructure is that under low-level injection conditions with the quasi-Fermi levels flat, the equilibrium hole concentration in the wide gap material has been reduced by this excess barrier. The hole concentration at the depletion region edge is given by the Shockley boundary conditions. It is lowered by the decrease in the intrinsic carrier concentration and reduction of p_{n0} from the homostructure case. Recall that the ratio of hole concentration between the heterostructure and the homostructure, all the other parameters remaining constant, is

$$
\frac{p_n(w_{E'})|_1}{p_n(w_{E'})|_2} = \frac{n_i^2|_1}{n_i^2|_2}
$$

$$
= \exp\left(-\frac{\Delta E_g}{kT}\right) \ll 1, \qquad (7.191)
$$

where 1 identifies the heterostructure and 2 the homostructure bipolar transistor. The hole carrier concentration is negligible at the ohmic contact or a higher doped region used to reduce emitter resistances. So this position serves as a boundary condition similar to that of the base–collector junction for base transport. The admittance equations are therefore identical, with the electron parameters replaced by hole parameters. We may write y_{pee}^b as

$$
y_{pee}^b = g_{pe}\xi_e \coth(\xi_e). \qquad (7.192)
$$

Here,

$$
\xi_e = \frac{w_{E'}}{\mathcal{L}_p}(1 + j\omega\tau_p)^{1/2}, \qquad (7.193)
$$

analogous to electron transport in the base, and

$$
g_{pe} = g_{ne} \exp\left(-\frac{\Delta E_g}{kT}\right), \qquad (7.194)
$$

which is the quasi-static conductance limit of the admittance. In these expressions, τ_p and $\mathcal{L}_p$ are the lifetime and the diffusion length in the n-type emitter. In view of the exponential factor, and limitations already placed on the applicability of our model, we will ignore the hole injection term in the following. For specialized cases, it can be included in the form described. We may therefore summarize our input admittance as

$$
y_{ee}^b = g_c\xi_b \coth(\xi_b) + j\omega C_{tE}. \qquad (7.195)
$$

Let us also consider the effect on injection efficiency. Since, by definition,

$$y_{nee}^b = \left. \frac{\hat{I}_{ne}}{\hat{V}_{eb}} \right|_{\hat{V}_{cb}=0}$$

and $\quad y_{pee}^b = \left. \frac{\hat{I}_{pe}}{\hat{V}_{eb}} \right|_{\hat{V}_{cb}=0}, \qquad (7.196)$

we obtain the small-signal injection efficiency $\tilde{\gamma}$ as

$$\tilde{\gamma} = \frac{\hat{I}_{ne}}{\hat{I}_{ne} + \hat{I}_{pe}}$$

$$= \frac{y_{nee}^b}{y_{nee}^b + y_{pee}^b}. \qquad (7.197)$$

Figure 7.38 shows a representative example of variation of the small-signal injection efficiency with frequency for an HBT structure; a structure that we will use as an example for rest of small-signal modelling. The parameters for this calculation are provided in the figure.

We have now considered transport at the emitter–base junction, in the quasi-neutral base, and at the base–collector junction. Using these, we can write the final form of our admittance parameters ignoring the effect of y_{pee},

$$\left[\hat{I}\right]_b = \begin{bmatrix} g_e\xi_b \coth(\xi_b) + j\omega C_{tE} & -g_c\xi_b \mathrm{csch}(\xi_b) \\ -\zeta g_e\xi_b \mathrm{csch}(\xi_b) & \zeta g_c\xi_b \coth(\xi_b) + j\omega C_{tC} \end{bmatrix} \left[\hat{V}\right]_b. \qquad (7.198)$$

These are straightforward in form and particularly suited for numerical approaches since they use complex mathematics. To gain insight, and to relate to our quasi-static discussion, we now consider some approximate ways of looking at this problem. These lend themselves to analysis in the form of poles, i.e., as resistive and capacitive elements and hence equivalent circuits of more general validity then the quasi-static derivations. Let us look at the input admittance term first.

$$y_{ee}^b = g_e\xi_b \coth(\xi_b), \qquad (7.199)$$

where

$$\xi_b = \frac{w_B}{\mathcal{L}_n}(1 + j\omega\tau_n)^{1/2}$$

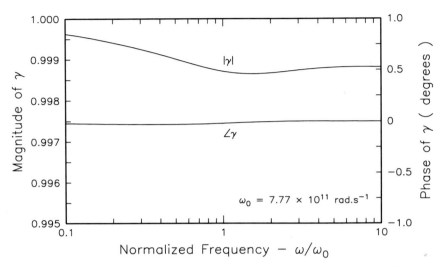

Figure 7.38: The magnitude and phase of small-signal injection efficiency of an idealized one-dimensional $Ga_.7Al_.3As/GaAs/GaAs$ n–p–n HBT. The device structure has a $Ga_.7Al_.3As$ emitter doped to 5×10^{17} cm^{-3} and a base of GaAs doped to 5×10^{18} cm^{-3}. The base thickness is 1000 Å. This device structure will also be used in discussing the behavior of other small-signal parameters. The collector of the structure is doped to 1×10^{17} cm^{-3}. ω_0 has the magnitude of 7.77×10^{11} rad.s^{-1}.

$$= \left[\left(\frac{w_B}{\mathcal{L}_n} \right)^2 + j\omega \frac{w_B{}^2}{\mathcal{D}_n} \right]^{1/2}$$

$$= \left[\left(\frac{w_B}{\mathcal{L}_n} \right)^2 + j2\omega\tau_B \right]^{1/2}. \tag{7.200}$$

The time parameter $\tau_B = w_B{}^2/2\mathcal{D}_n$ was encountered in quasi-static analysis as a time constant of charge storage in the quasi-neutral base, as well as the average transit time of carriers in the quasi-neutral base. We will use it as a parameter in our simplifications. The inverse of this time constant, which is the base transit frequency,

$$\omega_0 = \frac{1}{\tau_B}, \tag{7.201}$$

is also sometimes used as a parameter in our approximations.

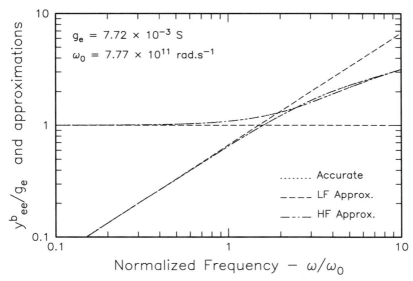

Figure 7.39: Normalized conductance and susceptance part of the input admittance in the common-base configuration, together with the low frequency and high frequency approximations. Note that the conductance is nearly a constant, and the susceptance varies linearly with frequency in the low frequency limit. ω_0 has the magnitude of 7.77×10^{11} rad.s^{-1}.

The low frequency and high frequency approximation of the general result can be obtained by expansion of the hyperbolic and the square root terms,

$$
\begin{array}{lll}
y_{ee}^b & = & g_e \left(1 + \frac{2}{3}\omega\tau_B\right) \qquad\qquad\qquad\;\; \omega\tau_B << \pi \\[4pt]
y_{ee}^b & = & g_e (2j\omega\tau_B)^{1/2} \coth\left((2j\omega\tau_B)^{1/2}\right) \quad \omega\tau_B >> \pi
\end{array}
\qquad (7.202)
$$

Figure 7.39 shows the conductance and the susceptance of the input admittance as a function of frequency, for the accurate expression and for the approximations. The low frequency limit exhibits a susceptance that is proportional to the frequency, and hence can be modelled by a capacitance. Thus, the input admittance, in the low frequency limit, is clearly amenable to an equivalent circuit model comprising of a resistor and a capacitor in parallel. The susceptive part corresponds to a capacitance (a diffusion capacitance that we denote as C'_{De}) whose magnitude corresponds to

$$
C'_{De} = g_e \frac{2}{3}\tau_B = \frac{g_e w_B{}^2}{3\mathcal{D}_n}. \qquad (7.203)
$$

This capacitance is two-thirds of the diffusion capacitance derived in our quasi-static modelling. This is, however, not in contradiction. This the capacitance corresponding to the input admittance in the common-base configuration. Our discussion of the quasi-static model was with respect to the common-emitter configuration. The diffusion capacitance derived here appears between the emitter and the base contacts when the collector-to-base voltage is kept constant. The earlier quasi-static calculation corresponds to a capacitance appearing between the base and the emitter with the collector-to-emitter voltage kept constant. The difference between these two should, therefore, appear between the emitter and the collector. Let us look at the forward admittance and its approximation to determine if this is so. The exact and the approximations, based on expansions for the base transport contribution to the forward transfer admittance in the low frequency limit, i.e., for $\omega\tau_B \ll \pi$, are

$$
\begin{aligned}
y_{nce}^b &= -g_e\xi_b\mathrm{csch}(\xi_b) \\
&\approx -g_e\left[\xi_b\left(\frac{1}{\xi_b} - \frac{1}{6}\xi_b\right)\right] \\
&\approx -g_e\left(1 - \frac{1}{6}\xi_b{}^2\right) \\
&\approx -g_e\left(1 - \frac{1}{3}\omega\tau_B.\right)
\end{aligned}
\qquad (7.204)
$$

So, in the low frequency limit, the susceptance can be modelled by a capacitance between the base and the collector $(C_{De}^{''})$, where

$$
C_{De}^{''} = g_e\frac{1}{3}\tau_B = \frac{g_e w_B{}^2}{6\mathcal{D}_n}. \qquad (7.205)
$$

In the low frequency limit of the small-signal admittances, the sum of this base–collector component of diffusion capacitance in the common-base configuration with the base–emitter component of diffusion capacitance is $g_e W_B^2/2\mathcal{D}_n$, identical to the quasi-static diffusion capacitance in the common-emitter configuration. The approximation of the base contribution to the forward transfer admittance, in high frequency limit $\omega\tau_B \gg \pi$, is

$$
\begin{aligned}
y_{nce}^b &= -g_e\xi_b\mathrm{csch}(\xi_b) \\
&\approx -g_e(2j\omega\tau_B)^{1/2}\mathrm{csch}\left((2j\omega\tau_B)^{1/2}\right).
\end{aligned}
\qquad (7.206)
$$

An estimate of the magnitude and the accuracy of these approximations for the forward transfer admittance can be obtained from Figure 7.40.

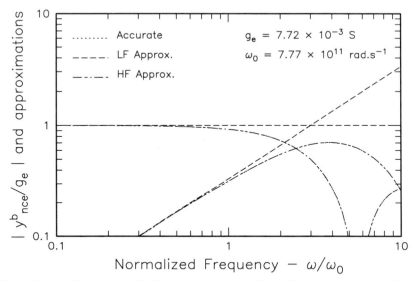

Figure 7.40: Base contribution of the normalized forward transfer admittance of the $Ga_{.7}Al_{.3}As/GaAs/GaAs$ HBT together with the low and high frequency approximations.

The approximations for the reverse transfer admittance and the output admittances follow in a similar manner. The low frequency approximation of the reverse transfer admittance, i.e., for $\omega\tau_B \ll \pi$, is

$$
\begin{aligned}
y_{ec}^b &= -g_e\xi_b\mathrm{csch}(\xi_b) \\
&\approx -g_e\left(1 - \frac{1}{3}\omega\tau_B\right).
\end{aligned}
\tag{7.207}
$$

It can be modelled as a resistance and capacitance also. The high frequency approximation of the reverse transfer admittance, i.e., for $\omega\tau_B \gg \pi$, is

$$
\begin{aligned}
y_{ec}^b &= -g_e\xi_b\mathrm{csch}(\xi_b) \\
&\approx -g_e(2j\omega\tau_B)^{1/2}\mathrm{csch}\left((2j\omega\tau_B)^{1/2}\right).
\end{aligned}
\tag{7.208}
$$

The low frequency approximation for the base transport contribution to the output admittance, i.e., for $\omega\tau_B \ll \pi$, is

$$
y_{cc'}^b = g_c\xi_b\coth(\xi_b)
$$

$$\approx \ g_c \left(1 + \frac{2}{3} \omega \tau_B \right), \tag{7.209}$$

and the high frequency approximation, i.e., for $\omega \tau_B \gg \pi$, is

$$
\begin{aligned}
y^b_{cc'} \ &= \ g_c \xi_b \coth(\xi_b) \\
&\approx \ g_c (2j\omega\tau_B)^{1/2} \coth \left((2j\omega\tau_B)^{1/2} \right).
\end{aligned}
\tag{7.210}
$$

These relationships, involving the collector current, i.e., the output and the forward admittance parameters, do not yet include the contribution from the transport in the base–collector depletion region. We will discuss this as part of our discussion of the transport factors. The forward transfer admittance contributes to the collector current as a product with the emitter–base small-signal voltage. Thus, it appears as a current source. The transport factors, therefore, appear as a necessary part of the modelling related to behavior of output port. The small-signal injection efficiency for an HBT is quite close to unity. It is best treated by considering the expression in its original form or approximated to unity, i.e.,

$$
\begin{aligned}
\tilde{\gamma} \ &= \ \frac{y^b_{nee}}{y^b_{nee} + y^b_{pee}} \\
&= \ \frac{g_e \xi_b \coth(\xi_b)}{g_e \xi_b \coth(\xi_b) + g_e \xi_e \coth(\xi_e)} \\
&\approx \ 1.
\end{aligned}
\tag{7.211}
$$

The base and the collector signal delays are still important. We have made a first-order analysis of the collector transport factor and the associated collector signal delay after assuming that the velocity of a carrier there is a constant. We will now look at the base transport factor, whose information is already contained in the parameter derivation of base transport. We are interested in the base transport factor α_T, its frequency dependence, and its approximations. By definition,

$$\tilde{\alpha}_T = \left. \frac{\hat{I}_{nc'}}{\hat{I}_{ne}} \right|_{\hat{V}_{cb}=0}. \tag{7.212}$$

These currents have been evaluated previously. Substituting,

$$
\begin{aligned}
\tilde{\alpha}_T \ &= \ \frac{q\mathcal{D}_n n_{p0} \exp\left(q\overline{V}_{be}/kT \right) \left(q\hat{V}_{be}/kT \right) \varsigma \mathrm{csch}(\xi_b)}{q\mathcal{D}_n n_{p0} \exp\left(q\overline{V}_{be}/kT \right) \left(q\hat{V}_{be}/kT \right) \varsigma \coth(\xi_b)} \\
&= \ \mathrm{sech}(\xi_b).
\end{aligned}
\tag{7.213}
$$

Recombination processes dominate the base transport effect, i.e., the most likely event occurring to a travelling electron is an annihilation process involving another hole, or an annihilation process involving an Auger process. Both of these can be approximated by the single time constant τ_n that we have employed. There are a few exceptions to this, e.g., punch-through operation during avalanching, but these are rarely of interest in practical structures. Thus $\tilde{\alpha}_T < 1$. In general, it is complex, with both magnitude and phase varying with frequency. The static value of this is

$$
\begin{aligned}
\alpha_T &= \operatorname{sech}\left(\frac{w_B}{\mathcal{L}_n}\right) \\
&\approx \frac{1}{1 + w_B{}^2/2\mathcal{L}_n{}^2} \\
&\approx 1 - \frac{w_B{}^2}{2\mathcal{L}_n{}^2}
\end{aligned}
\tag{7.214}
$$

for $w_B/\mathcal{L}_n \ll 1$.

Figure 7.41 shows this variation as a function of frequency for our bipolar transistor example. We now look at some approximations of this in various frequency ranges. For $\omega\tau_B \approx 1$,

$$
\tilde{\alpha}_T = \operatorname{sech}\left(\frac{w_B}{\mathcal{L}_n}\right) \operatorname{sech}\left((2j\omega\tau_B)^{1/2}\right).
\tag{7.215}
$$

Another approximation, which is often employed for intermediate frequencies, is to fit a pole that approximates the magnitude of $\tilde{\alpha}_T$ at the 3 dB point. This provides a good match to magnitude where $\tilde{\alpha}_T$ is down to half of its low frequency value. The frequency at which this occurs is commonly referred to as the alpha-cutoff frequency ω_{α_T}. Let us introduce the notation of $\tau_B^{\alpha_T}$, which denotes the inverse of the alpha-cutoff frequency. Using this,

$$
\tilde{\alpha}_T = \frac{\alpha_T}{1 + j\omega\tau_B^{\alpha_T}}.
\tag{7.216}
$$

The frequency at which α_T drops to half its static value is

$$
\omega_{\alpha_T} = \frac{2.43\mathcal{D}_n}{w_B{}^2} \approx 1.22\omega_0
\tag{7.217}
$$

for a homogeneous base transistor. On the other hand one may prefer to stay with our standard time constant τ_B, and obtain a simultaneous approximate fit to both the magnitude and the phase with

$$
\tilde{\alpha}_T \approx \frac{\alpha_T}{1 + j\omega\tau_B}.
\tag{7.218}
$$

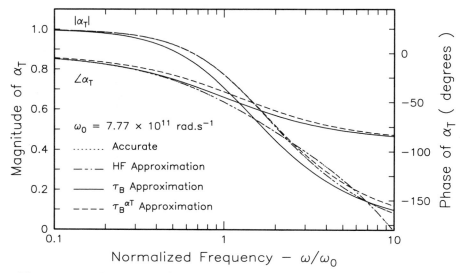

Figure 7.41: Variation of the magnitude of the small-signal base transport factor with frequency for the $Ga_{.7}Al_{.3}As/GaAs/GaAs$ HBT example together with the various approximations.

This expression actually slightly overestimates the phase delay. This is so because our accurate expression is actually a hyperbolic secant and does not lend itself to an easy approximation. A significantly more accurate expression that serves well for both low and high frequencies for the current transport factor is

$$\tilde{\alpha}_T \approx \alpha_T \frac{\exp\left(-\delta\omega\tau_B^{\alpha_T}\right)}{1 + j\omega\tau_B^{\alpha_T}}$$

$$\approx \alpha_T \frac{\exp\left(-\delta\omega\tau_B\right)}{1 + j\omega\tau_B}, \qquad (7.219)$$

where $\delta \approx 0.22$. Recall that from our previous treatment of these time constants,

$$\tau_B^{\alpha_T} \approx 0.82\tau_B. \qquad (7.220)$$

The phase delay at low frequencies looks like a pole response, but at frequencies in excess of $0.1\omega_{\alpha_T}$, an additional phase delay should be included. Figure 7.41 shows the variation of the exact expression together with the variation of these approximate expressions as a function of frequency for our bipolar transistor example.

Note that the complete current transport factor for HBTs is

$$\tilde{\alpha} = \tilde{\gamma}\tilde{\alpha}_T\tilde{\zeta} \approx \text{sech}(\xi_b)\frac{\sin(\omega\tau_c^{'})}{\omega\tau_c^{'}}\exp\left(-\omega\tau_c^{'}\right) \approx \frac{\exp\left(-\delta\omega\tau_B\right)\exp\left(-\omega\tau_c^{'}\right)}{1+j\omega\tau_B}.$$

(7.221)

Figure 7.42 shows the common-base current gain ($\tilde{\alpha}$) and its components for our transistor example. At frequencies near ω_0, the effect of transport in the base and the collector is quite similar in magnitude and is significantly larger than that of the injection efficiency.

These various approximations are summarized for the admittance parameters, including the collector transport effects, together with the accurate expressions, in Figure 7.43 for reverse and Figure 7.44 for output y-parameters in the common-base configuration.

Finally, we now use the approximations to obtain equivalent circuit models with limits of validity significantly beyond those of quasi-static models. Our complete y-parameters for this common-base condition result in a network model that can be represented as in Figure 7.45. We may now include the single pole approximation for the input and output admittances for this, resulting in Figure 7.46.

This theory is based on one-dimensional flow of carriers. Actual device structures are significantly more complicated, and require inclusion of bulk resistive effects such as due to base resistance. These parasitic and extrinsic effects are related to transport of majority carriers and charging of parasitic capacitances. Most of these, for the frequency ranges of interest, i.e., near and lower than the frequency limits, can be modelled as discrete lump elements in an equivalent circuit. For example, in the simplest implementations, the extrinsic base resistance may be included by adding it as a lumped resistance in series with our approximate model.

For applicability to medium frequency range, we could, finally, make most of the pole zero approximations, except in the current transport factor. This results in a more complicated current source representation but a simple equivalent circuit representation which has fairly broad validity.

Inclusion of other lumped elements or transformation to other network parameters can now be accomplished using procedures that have been discussed earlier; the transformations are included in the Appendix A. For all these we begin with our complete admittance parameter relationship (Equation 7.198) or its approximation. It can be manipulated to add whatever elements are desired, or to derive it in any other configuration. If one wished to add an emitter resistance to these common-base parameters, one could accomplish it by converting to an impedance matrix, and then the emitter resistance adds to the input impedance term. If one wanted to

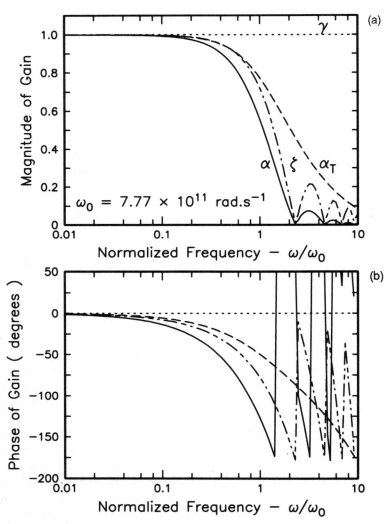

Figure 7.42: Variation of small-signal common-base current gain and its components with frequency for the $Ga_{.7}Al_{.3}As/GaAs/GaAs$ HBT. Part (a) shows the magnitude of the gain and part (b) shows the phase for the total and the components.

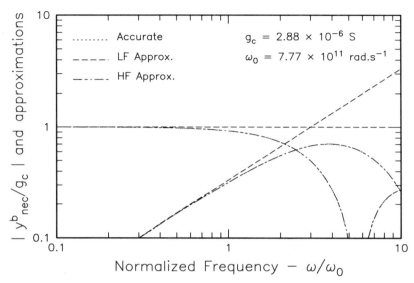

Figure 7.43: Variation of the normalized reverse admittance and its approximations with frequency for the $Ga_{.7}Al_{.3}As/GaAs/GaAs$ HBT example.

add the base resistance, however, one would have to do that by converting to common-emitter parameters. Conversion from one common node to another is another example of matrix manipulation, and is accomplished using indefinite matrices, as discussed in Appendix A, and small-signal analysis of FETs.

These transformations can affect one or more than one of the of parameters. Both an equivalent circuit and the matrix relationship indicate why many of the admittances are affected in some of these transformations. As an example, all admittances of the common-base mode are affected by inclusion of the base resistance because it has a negative feedback effect. But only the input term is affected for impedance due to the emitter resistance. On the other hand, all admittances of the common-emitter mode are affected by the emitter resistance because it has a negative feedback effect. But only the input term is affected for impedance in the common-base mode.

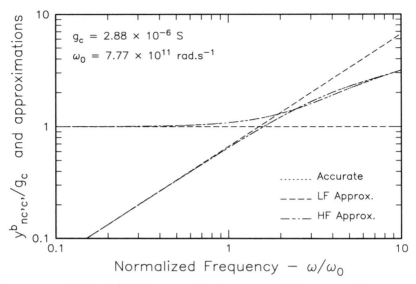

Figure 7.44: Variation of the normalized output admittance and its approximations with frequency for the $Ga_{.7}Al_{.3}As/GaAs/GaAs$ HBT example.

7.6.4   Frequency Figures of Merit

We now consider the evaluation of figures of merit of the HBT, basing these figures of merit on the analysis undertaken. However, we will attempt this only in the simplest of geometries; more complicated ones can best be done using numerically intensive methods, and we will give examples of results of these later. For determining the unity current gain frequency f_T of the transistor, we have to determine the hybrid parameter h_{21} in the common-emitter mode.

Hybrid parameters for the two-port transistor are defined from

$$\hat{V}_{be} = h_{11}\hat{I}_{be} + h_{12}\hat{V}_{ce}$$
$$\text{and} \qquad \hat{I}_{ce} = h_{21}\hat{I}_{be} + h_{22}\hat{V}_{ce}. \qquad (7.222)$$

f_T is the frequency at which the current gain is unity with a short-circuit at the output, i.e., it is the frequency when

$$\left.\frac{\hat{I}_{ce}}{\hat{I}_{be}}\right|_{\hat{V}_{ce}=0} = h_{21} = 1. \qquad (7.223)$$

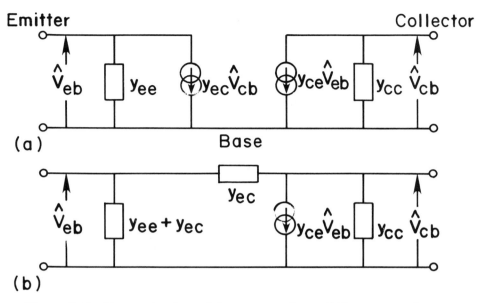

Figure 7.45: Two network model representations of the common-base y-parameters. Note that the model in (b) eliminates the current source in the input circuit.

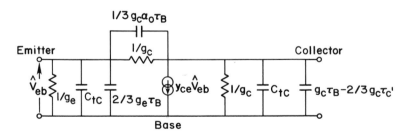

Figure 7.46: Common-base model with pole approximation to input and output admittances.

In terms of the common-base matrix of y-parameters,

$$[y]_b = \begin{bmatrix} y_{ee} & y_{ec} \\ y_{ce} & y_{cc} \end{bmatrix} \tag{7.224}$$

we obtain h_{21} in the common-emitter mode as

$$
\begin{aligned}
h_{21} &= \frac{-y_{ce} - y_{cc}}{y_{ee} + y_{ec} + y_{ce} + y_{cc}} \\
&= [\zeta g_e \xi_b \text{csch}(\xi_b) - \zeta g_c \xi_b \coth(\xi_b) - j\omega C_{tC}] \div \\
&\quad [g_e \xi_b \coth(\xi_b) + j\omega C_{tE} - g_c \xi_b \text{csch}(\xi_b) - \\
&\quad \zeta g_e \xi_b \text{csch}(\xi_b) + \zeta g_c \xi_b \coth(\xi_b) + j\omega C_{tC}].
\end{aligned} \tag{7.225}
$$

We will now make our pole zero approximations in the high frequency range with

$$
\begin{array}{lll}
(y_{nee}): & g_e \xi_b \coth(\xi_b) & \approx \quad g_e(1 + j\tfrac{2}{3}\omega\tau_B) \\
(y_{ncc}): & g_c \xi_b \coth(\xi_b) & \approx \quad g_c(1 + j\tfrac{2}{3}\omega\tau_B) \\
(y_{nec}): & -g_c \xi_b \text{csch}(\xi_b) & \approx \quad -g_c \alpha_0(1 - j\tfrac{1}{3}\omega\tau_B) \\
(y_{nce}): & -g_e \xi_b \text{csch}(\xi_b) & \approx \quad -g_e \alpha_0(1 - j\tfrac{1}{3}\omega\tau_B) \\
(\zeta): & \left(\sin\omega\tau_c'/\omega\tau_c'\right)\exp\left(-j\omega\tau_c'\right) & \approx \quad 1 - j\omega\tau_c' \\
(y_{ce}): & -\zeta g_e \xi_b \text{csch}\xi_b & \approx \quad -g_e\alpha_0(1 - j\tfrac{1}{3}\omega\tau_B - j\omega\tau_c') \\
(y_{cc}): & \zeta g_c \xi_b \coth\xi_b & \approx \quad g_e(1 + j\tfrac{2}{3}\omega\tau_B - j\omega\tau_c').
\end{array} \tag{7.226}
$$

Hence,

$$
\begin{aligned}
h_{21}|_e &\approx \left[\alpha_0 g_e\left(1 - j\tfrac{1}{3}\omega\tau_B - j\omega\tau_c'\right) - \right. \\
&\quad g_c\left(1 + j\tfrac{2}{3}\omega\tau_B - j\omega\tau_c'\right) - j\omega C_{tC}\right] \div \\
&\quad \left[g_e\left(1 + j\tfrac{2}{3}\omega\tau_B\right) + j\omega C_{tE} - \alpha_0 g_c\left(1 - j\tfrac{1}{3}\omega\tau_B\right) - \right. \\
&\quad \alpha_0 g_e\left(1 - j\tfrac{1}{3}\omega\tau_B - j\omega\tau_c'\right) + \\
&\quad \left. g_c\left(1 + j\tfrac{2}{3}\omega\tau_B - j\omega\tau_c'\right) + j\omega C_{tC}\right] \\
&\approx \left[\alpha_0\left(1 - j\omega\tau_c' - j\tfrac{1}{3}\omega\tau_B\right) - \frac{g_c}{g_e}\left(1 + j\tfrac{2}{3}\omega\tau_B - j\omega\tau_c'\right) - \right. \\
&\quad \left. j\frac{\omega C_{tC}}{g_e}\right] \div \left[\left(1 - \alpha_0 - \alpha_0\frac{g_c}{g_e} + \frac{g_c}{g_e}\right) + j\omega\left(\frac{2}{3}\tau_B + \frac{1}{3}\alpha_0\tau_B\frac{g_c}{g_e} + \right.\right.
\end{aligned}
$$

$$\alpha_0 \tau_c' + \frac{1}{3}\alpha_0 \tau_B - \tau_c' \frac{g_c}{g_e} + \tau_B \frac{g_c}{g_e} + \frac{C_{tC} + C_{tE}}{g_e} \Bigg) \Bigg].$$

(7.227)

Several of these factors are negligible compared to others for the frequencies of interest ($\omega \le 1/\tau_B$). We use the fact that $g_c/g_e \ll 1$ to eliminate the related terms, getting

$$
\begin{aligned}
h_{21}\big|_e &= \frac{\alpha_0 - j\omega\left(\alpha_0\tau_c' + \frac{1}{3}\alpha_0\tau_B + C_{tC}/g_e\right)}{1 - \alpha_0 + j\omega\left[\frac{2}{3}\tau_B + \alpha_0\tau_c' + \frac{1}{3}\alpha_0\tau_B + (C_{tE} + C_{tC})/g_e\right]} \\[2mm]
&\approx \frac{\alpha_0}{1-\alpha_0} \times \\[2mm]
&\qquad \frac{1 - j\omega\left[\frac{1}{3}\tau_B + \tau_c' + (1/\alpha_0)(C_{tC}/g_e)\right]}{1 + j(\omega/(1-\alpha_0))\left[\frac{2}{3}\tau_B + \alpha_0\tau_c' + \frac{1}{3}\alpha_0\tau_B + (C_{tE} + C_{tC})/g_e\right]} \\[2mm]
&\approx \frac{\alpha_0}{1-\alpha_0}\,\frac{1 - j\omega\left(\frac{1}{3}\tau_B + \tau_c' + C_{tC}/g_e\right)}{1 + j(\omega/(1-\alpha_0))\left[\tau_B + \tau_c' + (C_{tE} + C_{tC})/g_e\right]},
\end{aligned}
$$

(7.228)

where in the final expression we used the fact that $\alpha_0 \approx 1$. The imaginary term in the denominator is significantly larger than the imaginary term in the numerator for this reason. The current gain response is dominated by the imaginary term in the denominator at frequencies near f_T. A good approximation for h_{21}, near f_T, then, is

$$
\begin{aligned}
h_{21}\big|_e &\approx \frac{\alpha_0}{1-\alpha_0}\,\frac{1}{(j\omega/(1-\alpha_0))\left[\tau_B + \tau_c' + (C_{tE} + C_{tC})/g_e\right]} \\[2mm]
&\approx \frac{1}{j\omega\left[\tau_B + \tau_c' + (C_{tE} + C_{tC})/g_e\right]}.
\end{aligned}
$$

(7.229)

From this, by definition, the unity current gain frequencies are

$$\omega_T = \frac{1}{\left[\tau_B + \tau_c' + (C_{tE} + C_{tC})/g_e\right]},$$

(7.230)

and

$$f_T = \frac{1}{2\pi\left[\tau_B + \tau_c' + (C_{tE} + C_{tC})/g_e\right]}$$

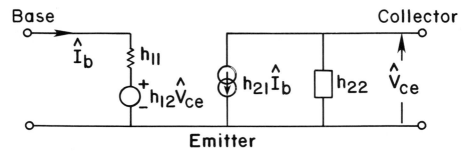

Figure 7.47: Equivalent circuit representation of the h-parameter equations.

$$= \frac{1}{2\pi \left\{ (W_B^2/2\mathcal{D}_n) + +[(w_{C'} - w_C)/2v_s] + (C_{tE} + C_{tC})/g_e \right\}}.$$
(7.231)

This result is independent of the base resistance because it is derived for a current drive at the input. If the assembly of the transistor structure was such that base and an extrinsic part of the collector capacitance had to be treated as a distributed network, or simple multiple stage equivalents, then it would have been a function of that too. In the h-parameter representation, our model appears as follows Figure 7.47. This is the hybrid-pi representation of an equivalent circuit and is similar in appearance of the hybrid model for FETs.

To calculate the maximum frequency of oscillation, one wants to find the unilateral gain of the transistor. Unilateral gain is the maximum gain of a device, obtained by using ideal passive elements to unilateralize the network, i.e., to compensate for the feedback terms. The frequency at which this unilateral gain goes to unity is the maximum frequency of oscillation, because this is the ultimate frequency at which the device would oscillate in an ideal circuit. A network is unilateralized by using loss-less reciprocal elements. In general, the unilateral gain for $y-$parameters is given by

$$U = \frac{\mid y_{21} - y_{12} \mid^2}{4 \left[\text{Re} \left[y_{11} \right] \text{Re} \left[y_{22} \right] - \text{Re} \left[y_{21} \right] \text{Re} \left[y_{12} \right] \right]},$$
(7.232)

which is quite a complicated equation. This would be quite difficult to evaluate if we wanted to include the base resistance. Let us first do this by an intuitive analysis. Since the bipolar is a highly unilateral device, we will ignore the feedback term in it (recall $g_c \ll g_e$). If we know the input and output impedance and the current gain, we can derive the frequency at which the power gain goes to unity. This is the maximum frequency of oscillation.

We have already found the current source, which at the highest frequency follows from Equation 7.228. Base resistance usually dominates the input impedance of the idealized model, the input admittance having resulted from a forward biased junction. So the input impedance is simply the input resistance r_b. The output resistance is given by h_{22},

$$
\begin{aligned}
h_{22} &= \frac{|y|}{\sum y} \\
&= \frac{y_{ee}y_{cc} - y_{ec}y_{ce}}{y_{ee} + y_{cc} + y_{ec} + y_{ce}} \\
&\approx \left\{ \left[g_e \left(1 + j\frac{2}{3}\omega\tau_B \right) + j\omega C_{tE} \right] \times \right. \\
&\quad \left[g_c \left(1 + j\frac{2}{3}\omega\tau_B - j\omega\tau_c' \right) + j\omega C_{tC} \right] - \\
&\quad \left. \left[\alpha_0 g_e \left(1 - j\frac{1}{3}\omega\tau_B - j\omega\tau_c' \right) \right] \left[g_c\alpha_0 \left(1 - j\frac{1}{3}\omega\tau_B \right) \right] \right\} \div \\
&\quad \left(j g_e \frac{\omega}{\omega_T} \right).
\end{aligned}
\tag{7.233}
$$

We have used the single pole and single zero approximations. The denominator uses the dominance of the imaginary part over the real part at the frequencies of interest. So,

$$
\begin{aligned}
h_{22} &\approx \left\{ g_e \left[1 + j\frac{2}{3}\omega\tau_B + j\omega\frac{C_{tE}}{g_e} \right] g_c \left[1 + j\frac{2}{3}\omega\tau_B - j\omega\tau_c' + j\omega\frac{C_{tC}}{g_c} \right] - \right. \\
&\quad \left. \alpha_0^2 g_e g_c \left[1 - j\frac{2}{3}\omega\tau_B - j\omega\tau_c' \right] \right\} \div \left(j g_e \frac{\omega}{\omega_T} \right).
\end{aligned}
\tag{7.234}
$$

Since $g_c \ll g_e$, it dominates the rest of the terms in the second bracket of the numerator, and hence

$$
\begin{aligned}
h_{22} &\approx \frac{j\omega C_{tC} \left[1 + j\frac{2}{3}\omega\tau_B + j\omega C_{tE}/g_e \right]}{j\omega/\omega_T} \\
&\approx \omega_T C_{tC} + j\omega \left[\frac{2}{3}\omega_T\tau_B C_{tC} + \omega_T \frac{C_{tE}C_{tC}}{g_e} \right].
\end{aligned}
\tag{7.235}
$$

So the output and input can be modeled at the frequencies close to f_T as in Figure 7.48. Here the conductances are $1/r_b$ in the input circuit and $\omega_T C_{tC}$ in the output circuit. The capacitive element in the output can be

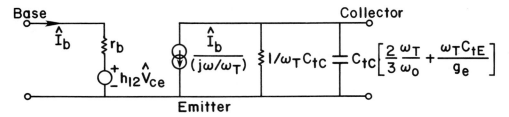

Figure 7.48: Equivalent circuit model approximation for determination of f_{max}.

tuned out by an inductance during unilateralization. We have ignored the feedback term inherent in h_{12}. In order to obtain the maximum available gain, we need to know the real parts of the input and output impedances only. Matching entails use of the complex conjugate of these impedances. So the output load will be a parallel network of a conductance $\omega_T C_{tC}$ and an inductance L,

$$L = \frac{1}{\omega^2 C_{tC} \left[\frac{2}{3}\omega_T \tau_B + \omega_T C_{tE}/g_e\right]}. \tag{7.236}$$

The power gain follows from this as

$$G = \frac{\left[(\hat{I}_b \omega_T)/(2\omega)\right]^2 (1/\omega_T C_{tC})}{\hat{I}_b^2 r_b}. \tag{7.237}$$

Only half of the current from the current source flows through the load resistance. Power gain is unity at ω_{max}, and hence

$$\frac{\omega_T^2}{4\omega_{max}^2 \omega_T C_{tC} r_b} = 1$$

$$\omega_{max} = \sqrt{\frac{\omega_T}{4r_b C_{tC}}}$$

$$\text{or} \quad f_{max} = \left(\frac{f_T}{8\pi r_b C_{tC}}\right)^{1/2}. \tag{7.238}$$

We have derived, including the base resistance, the equivalent circuit shown in Figure 7.49. In making this representation, both C_{tE} and C_{tC} were associated with the intrinsic device structure. This is clearly quite applicable to the emitter where the extrinsic fringing capacitances as well

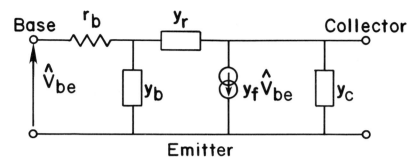

Figure 7.49: Equivalent circuit derived using the y-parameters and including the base resistance.

as capacitances associated with extrinsic base regions are smaller than the intrinsic capacitance. For the collector, however, this is an approximation. Conventional bipolar transistors incorporate the emitter in the region nearest to the surface, and the transistors are referred to as emitter-up structures. Collector-up structures are also possible, but are less common, their major appeal being in integration but usually with a lowering of device speed. The emitter-up bipolar transistor structure has a collector area that is larger than the emitter area. The injected electron current passes underneath the emitter into the collector. The attached extrinsic collector area does not have significant collector current flow through it. The division of the lumped elements, extrinsic and intrinsic base resistances, and collector capacitances can be complex, since both in the intrinsic and extrinsic part lumped elements are an approximation to a distributed network. A reasonably acceptable representation would be to divide the collector capacitance, connecting some of it at either of two ends of the extrinsic base resistance. One may also divide the extrinsic base resistance into two parts, at the least, combining the contact resistance part in one term and the extrinsic semiconductor resistance in the other.

The equivalent circuits derived can be cast in alternate forms. One example is shown in Figure 7.50 which is a common-base representation casted with a current source based on the the emitter current. These derived parameters may be simplified using the pole zero approximations and this results in a network that is a hybrid-pi equivalent circuit. It is similar to that of the FETs.

Recalling that the common-base representation can be converted to the common-emitter representation, we can obtain the common-emitter param-

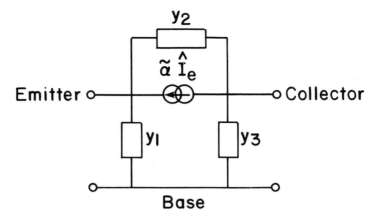

Figure 7.50: Common-base equivalent circuit based on y-parameters.

eters. The discussion of derivation of these parameters is continued in our treatment of transit time resonance effects later in this chapter together with a discussion of the inclusion of extrinsic resistances and parasitic capacitances. The current source term in these networks is related to the transit of carriers and delays associated with it. In our notations, it falls out naturally when transforming to hybrid parameters from the y-parameters as

$$\tilde{\alpha} = \tilde{\gamma}\tilde{\alpha}_T\tilde{\zeta}, \tag{7.239}$$

for which our excess phase approximation is

$$\tilde{\alpha} \approx \alpha_{T0}\frac{1}{1 + j\omega\tau_B}\exp\left(-j\delta\omega\tau_B\right)\exp\left(-j\omega\tau_c'\right). \tag{7.240}$$

7.7   Small-Signal Effects of Alloy Grading

The small-signal treatment considered thus far analyzed uniform doping in the base. The limitations placed in operational frequency or in transient delays for such structures can have a significant component of contribution from the transport in the base, after making the compromises between decreased base width and increased intrinsic resistance of the base. An additional improvement that can be made is to introduce an electric field in the base that accelerates the injected carriers from the base–emitter junction to the base–collector junction. In a non-heterostructure technology, this is accomplished by using a grading in the doping, with a higher doping at the base–emitter junction (see Problem 13). Such a gradient occurs naturally in implanted base transistors. Heterostructure growth techniques also allow this electric field to be introduced, in the form of a quasi-field due to the use of an alloy composition gradient. For minority carriers with low diffusivities, such as electrons in silicon or holes in most semiconductors, this can result in very significant improvements.

We will now consider modifications to our small-signal analysis by including this alloy grading field in the base through its effect on base transport. In view of the comments just made, the description of this problem is similar to that for bipolar transistors with a doping gradient in the base. With a suitable choice of parameterization, the underlying mathematical derivation is identical for both of the problems. We analyze the transport equation, within the drift-diffusion approximation, by considering the effect of the quasi-field $\mathcal{E}_e$, this field being the result of compositional variation in the base. This analysis, an extension of the uniformly-doped base analysis, with the same underlying assumptions, is again valid only at low-level injection conditions in the base.

In the base region, the control equations are

$$\frac{\partial n_p}{\partial t} = \mathcal{G} - \mathcal{R} + \frac{1}{q} J_n,$$

where

$$\mathcal{G} - \mathcal{R} = \frac{n_p - n_{p0}}{\tau_n},$$

and

$$J_n = qn\mu_n \left\{ -\nabla\psi - \nabla\phi_C - \frac{kT}{q}\nabla\ln(N_C) \right\} + qD_n\nabla n_p.$$

$$(7.241)$$

In the quasi-neutral base, which is uniformly doped with N_A acceptors,

$$-\nabla\phi_C = \mathcal{E}_e = -\frac{1}{q}\nabla E_g, \qquad (7.242)$$

and

$$\nabla\psi = 0. \qquad (7.243)$$

Since we assume low-level injection, the hole density is relatively unchanged, and hence the change in alloy composition with the resulting change in electron affinity results directly in the quasi-field $\mathcal{E}_e$. Let us consider the requirement for introducing a quasi-field of 1.0×10^3 V.cm^{-1}, a field adequate to cause an electron velocity close to 1×10^7 cm.s^{-1} in GaAs. For a 1000 Å base width, a change in bandgap of 10 meV is required. Since there is a direct dependence between the AlAs mole-fraction and the bandgap in this composition range, a linear change in the AlAs mole-fraction from $.010 \div 1.24 = .008$ at the base–emitter junction to GaAs at the base–collector junction would result in the requisite quasi-field. This is a quite minor change. With small base widths in the transistor structures, fairly large changes in mole-fractions can be introduced to obtain larger fields required for hole transport or for electrons in structures involving the use of SiGe alloys, without violating practical critical thickness requirements.

We will analyze conditions at which the alloy grading field is significantly larger than the gradient field resulting from density of states, i.e., for $(kT/q)\nabla N_C \ll \nabla E_g/q$. In order to obtain carrier transit in the base at the highest velocities possible, this field must be at least comparable to 1×10^3 V.cm^{-1}. Such a field adequately satisfies the constraint. Let $\mathcal{E}$ denote the electric field including the quasi-electric field. Then our current equation is

$$J_n = qn\mu_n\mathcal{E} + q\mathcal{D}_n\frac{\partial n}{\partial z}. \qquad (7.244)$$

We need not have made the above simplifying assumption. Its violation, following Chapter 3, with the inclusion of the density of state term in the first term of current equation, results only in a loss of the association with field. Our current continuity equation, following the above, is

$$\frac{\partial n_p}{\partial t} = -\frac{n_p - n_{p0}}{\tau_n} + \mu_n\frac{d(n_p\mathcal{E})}{dz} + \mathcal{D}_n\frac{d^2 n_p}{dz^2}. \qquad (7.245)$$

We consider uniform fields, hence,

$$\frac{d^2 n_p}{dz^2} + \frac{\mu_n\mathcal{E}}{\mathcal{D}_n}\frac{dn_p}{dz} - \frac{n_p - n_{p0}}{\mathcal{D}_n\tau_n} = \frac{1}{\mathcal{D}_n}\frac{\partial n_p}{\partial t}. \qquad (7.246)$$

Now consider the steady-state and sinusoidal expansion of $n_p(z)$.

$$n_p(z,t) = \overline{n}_p(z) + \hat{n}_p(z)\exp(j\omega t), \qquad (7.247)$$

giving rise to the steady-state equation

$$\frac{d^2\overline{n}_p}{dz^2} + \frac{q\mathcal{E}}{kT}\frac{d\overline{n}_p}{dz} - \frac{\overline{n}_p - n_{p0}}{\mathcal{D}_n\tau_n} = 0, \qquad (7.248)$$

and the small-signal equation

$$\frac{d^2\hat{n}_p}{dz^2} + \frac{q\mathcal{E}}{kT}\frac{d\hat{n}_p}{dz} - \frac{\hat{n}_p}{\mathcal{D}_n\tau_n} = \frac{j\omega}{\mathcal{D}_n}\hat{n}_p. \qquad (7.249)$$

We now define the unit-less normalization parameter

$$\kappa = \frac{q\mathcal{E}w_B}{kT}, \tag{7.250}$$

in order to simplify the mathematical description. This normalization parameter can also be used for the analysis of varying doping profile in the base, with the electric field as that resulting from the doping change. For the example of compositional variation, in order to obtain $\mathcal{E} = 10$ kV/cm in $w_B = 1000$ Å, κ is ≈ 3.9. We will choose a coordinate system, for mathematical convenience, where the base–collector junction is at $z = 0$ and the base–emitter junction at $z = w_B$ in our one-dimensional analysis.[11] The equation for steady-state is

$$\frac{d^2\overline{n}_p}{dz^2} + \frac{\kappa}{w_B}\frac{d\overline{n}_p}{dz} - \frac{\overline{n}_p - n_{p0}}{\mathcal{D}_n\tau_n} = 0, \tag{7.251}$$

with the boundary conditions

$$\begin{aligned}
\overline{n}_p(z = 0) &= 0 \\
\text{and} \quad \overline{n}_p(z = w_B) &= \overline{n}_{pE}
\end{aligned} \tag{7.252}$$

in our coordinate system.

Let $\overline{J}_{ne}$ be the steady-state injected electron current density at the base–emitter junction. Ignoring recombination effects in the base, but not the time-dependent phase-delay effects, the current continuity requires

$$\overline{J}_{ne} = q\mathcal{D}_n\frac{\kappa}{w_B}\overline{n}_p + q\mathcal{D}_n\frac{d\overline{n}_p}{dz}. \tag{7.253}$$

The minority carrier density has a solution of the form

$$\overline{n}_p(z) = A + B\exp\left(-\frac{\kappa z}{w_B}\right). \tag{7.254}$$

Using the boundary conditions, our solution is

$$\begin{aligned}
\overline{n}_p(z) &= \frac{\overline{J}_{ne}}{q\mathcal{D}_n}\frac{w_B}{\kappa}\left[1 - \exp\left(-\frac{\kappa z}{w_B}\right)\right], \\
\text{with} \quad \overline{n}_{pE} &= \frac{\overline{J}_{ne}}{q\mathcal{D}_n}\frac{w_B}{\kappa}\left[1 - \exp\left(-\kappa\right)\right]
\end{aligned} \tag{7.255}$$

---

[11]This coordinate system is the opposite of the coordinate system chosen in much of the treatment of base transport.

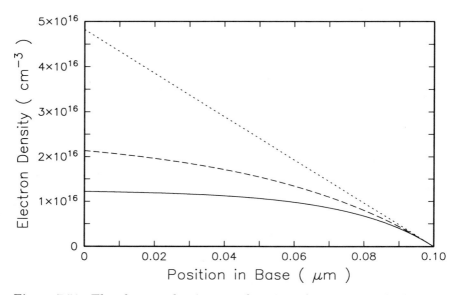

Figure 7.51: The electron density as a function of position in the base at 3×10^4 A.cm^{-2} current density. The dotted line is for no electric field, the dashed line is for a quasi-electric field of 5×10^3 V.cm^{-1}, and the solid line is for a quasi-electric field of 1×10^4 V.cm^{-1} in the base.

as the carrier density at edge of the emitter–base depletion region. The introduction of the field leads to a direct reduction in the carrier density closer to the emitter–base junction. This reduction is exponential in κ and the field. Note that the limitation on this derivation is that the velocity be proportional to the field, i.e., it is only valid at low fields. When the velocity saturates, further gains become marginal. Figure 7.51 shows this carrier density as a function of the current density for a graded alloy base Ga$_{1-x}$Al$_x$As transistor. This should be compared with the distribution without the presence of alloy grading also shown in this figure. These are for various alloy gradings such that the electrons experience a quasi-drift field up to 10 kV/cm. We will refer to this transistor in our subsequent examples of frequency effects also. With finite positive κ, the value of $\overline{n}_{pE}$ is reduced compared to the situation when no quasi-field is present. Also note that the carrier density expression and its position dependence reduce to the correct value of $\overline{n}_{pE}$ for the case when $\kappa \rightarrow 0$. In this situation, $\overline{n}_{pE} \rightarrow \overline{J}_{ne} w_B / q \mathcal{D}_n$, which is the correct value for diffusion-dominated transport.

The small-signal equation is

$$\frac{d^2\hat{n}_p}{dz^2} + \frac{\kappa}{w_B}\frac{d\hat{n}_p}{dz} - \hat{n}_p\left(\frac{j\omega}{\mathcal{D}_n} + \frac{1}{\mathcal{D}_n\tau_n}\right) = 0. \tag{7.256}$$

Again, the solution is of the form

$$\hat{n}_p(z) = A\exp\left(\frac{\lambda_1 z}{w_B}\right) + B\exp\left(\frac{\lambda_2 z}{w_B}\right), \tag{7.257}$$

where

$$\lambda_1 = -\frac{\kappa}{2} + \xi$$

and

$$\lambda_2 = -\frac{\kappa}{2} - \xi,$$

with

$$\xi = \left[\left(\frac{\kappa}{2}\right)^2 + \left(\frac{w_B}{L_n}\right)^2 + j\omega\frac{w_B^2}{\mathcal{D}_n}\right]^{1/2}. \tag{7.258}$$

The boundary conditions for the small-signal carrier density, similar to our earlier analysis, are as follows. At the collector ($z = 0$), the carrier density vanishes, i.e.,

$$\hat{n}_p = 0, \tag{7.259}$$

and at the emitter ($z = w_B$), the carrier density can be related by perturbative expansion assuming instantaneous equilibrium in the base–emitter depletion region, i.e.,

$$\hat{n}_p = \overline{n}_{pE}\frac{q\hat{V}_{be}}{kT}. \tag{7.260}$$

The boundary condition at the collector, $z = 0$, yields

$$A = -B, \tag{7.261}$$

and the boundary condition at the emitter, $z = w_B$, yields

$$A\exp(\lambda_1) + B\exp(\lambda_2) = \overline{n}_{pE}\frac{q\hat{V}_{be}}{kT}. \tag{7.262}$$

Consequently,

$$A = -B = \overline{n}_{pE}\frac{q\hat{V}_{be}}{kT}\frac{1}{\exp(\lambda_1) - \exp(\lambda_2)}, \tag{7.263}$$

the small-signal carrier density can be expressed as

$$\hat{n}_p(z) = \overline{n}_{pE}\frac{q\hat{V}_{be}}{kT}\frac{1}{\exp(\lambda_1) - \exp(\lambda_2)}\left[\exp\left(\frac{\lambda_1 z}{w_B}\right) - \exp\left(\frac{\lambda_2 z}{w_B}\right)\right], \tag{7.264}$$

and the small-signal current density can be expressed as

$$
\begin{aligned}
\hat{J}_n(z) &= q\mathcal{D}_n \frac{\kappa}{w_B} \hat{n}_p(z) + q\mathcal{D}_n \frac{d\hat{n}_p}{dz} \\
&= q\mathcal{D}_n \overline{n}_{pE} \frac{1}{w_B} \frac{q\hat{V}_{be}}{kT} \frac{1}{\exp(\lambda_1) - \exp(\lambda_2)} \left[(\lambda_1 + \kappa) \exp\left(\frac{\lambda_1 z}{w_B}\right) - (\lambda_2 + \kappa) \exp\left(\frac{\lambda_2 z}{w_B}\right) \right].
\end{aligned}
\tag{7.265}
$$

The current transport factor in the base $\tilde{\alpha}_T$, following this, is

$$
\tilde{\alpha}_T = \frac{\hat{J}_{nc}}{\hat{J}_{ne}} = \frac{\lambda_1 - \lambda_2}{(\lambda_1 + \kappa)\exp(\lambda_1) - (\lambda_2 + \kappa)\exp(\lambda_2)},
\tag{7.266}
$$

which can be rewritten as

$$
\tilde{\alpha}_T = \frac{\exp(\kappa/2)}{\cosh(\xi_b) + \kappa \sinh(\xi_b)/2\xi_b}.
\tag{7.267}
$$

Note that when $\kappa \to 0$, this expression reduces to the familiar zero field form. The effect of increasing κ is to reduce the delay in the base and increase the frequency range for large current gain. Figure 7.52 shows this for our alloy graded transistor. The collector signal delay becomes by far the dominant effect in this idealized structure. This should be compared with the same device without alloy grading discussed earlier. Similar to the earlier case, an approximation suitable for calculations is

$$
\tilde{\alpha}_T = \alpha_{T0} \frac{\exp(-j\delta\omega/\omega_{\alpha T})}{1 + j\omega/\omega_{\alpha T}},
\tag{7.268}
$$

where

$$
\omega_{\alpha T} = 2.43 \left[1 + \left(\frac{\kappa}{2}\right)^{4/3} \right]
$$

$$
\text{and} \quad \delta = 0.22 + 0.098\kappa.
\tag{7.269}
$$

Since we have found $\hat{J}_{ne}$, the phasor of small-signal injected current density at the base–emitter junction in terms of the phasor of the applied small-signal voltage $\hat{V}_{be}$, we can now derive the electron current contribution to the input admittance (see Problem 10) as

$$
y_{ne} = \frac{q}{kT} \frac{q\mathcal{D}_n \overline{n}_{pE}}{w_B} \left[\frac{\kappa}{2} + \xi_b \coth(\xi_b) \right],
\tag{7.270}
$$

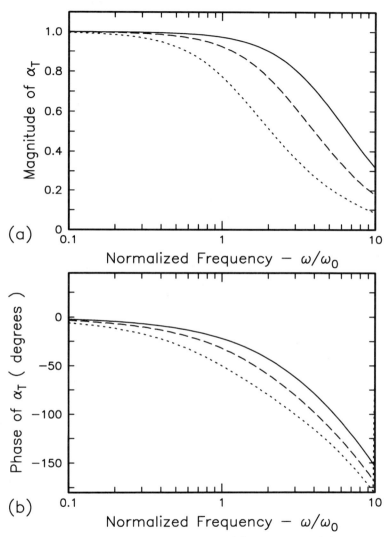

Figure 7.52: Variation of the magnitude (a) and phase (b) of $\tilde{\alpha}_T$ for no quasi-electric field (dotted line), a quasi-electric field of 5 kV/cm (dashed line), and a quasi-electric field of 10 kV/cm (solid line) for the HBT example.

which may be written in terms of the low frequency conductance (g_e, ignoring hole injection effects) as

$$y_{ne} = g_e \frac{(\kappa/2) + \xi_b \coth(\xi_b)}{(\kappa/2)\left[1 + \coth(\kappa/2)\right]}. \tag{7.271}$$

A suitable approximation for the diffusion capacitance (see Problem 11) is

$$C_{De} = g_e \frac{W_B^2}{2\mathcal{D}_n}\left\{\frac{2}{\kappa}\left[1 - \frac{1 - \exp\left(-\kappa\right)}{\kappa}\right]\right\}, \tag{7.272}$$

which shows the reduction in capacitance with increasing κ.

The forward transfer admittance (y_{ce}) can also be found following our earlier derivation technique as

$$\begin{aligned} y_{ce} &= \tilde{\alpha}_T y_{ne} \\ &= -g_e \frac{2}{\kappa} \frac{\exp\left(\kappa/2\right)}{1 + \coth\left(\kappa/2\right)} \xi_b \mathrm{csch}(\xi_b). \end{aligned} \tag{7.273}$$

This has the single pole approximation

$$y_{ce} = -\frac{\alpha_{T0} g_e}{1 + j(\omega/\omega_0)\left\{(2/\kappa)\coth\left(\kappa/2\right) - (2/\kappa)^2\right\}}, \tag{7.274}$$

which can also be approximated as

$$y_{ce} = -\alpha_{T0} g_e \left\{1 - j\frac{\omega}{\omega_0}\left[\frac{2}{\kappa}\coth\frac{\kappa}{2} - \left(\frac{2}{\kappa}\right)^2\right]\right\}. \tag{7.275}$$

The electronic contribution to the output admittance can also be shown using earlier techniques as

$$y_{nc} = g_c \left(\xi_b \coth(\xi_b) - \kappa\right), \tag{7.276}$$

where

$$g_c = -\frac{I_C}{w_B}\frac{dw_B}{dv_{CB}} \tag{7.277}$$

is the low frequency conductance. The transfer admittance, similarly,

$$y_{ec} = -g_c \frac{\kappa \exp\left(-\kappa\right)}{1 - \exp\left(-\kappa\right)}\frac{2}{\kappa}\frac{\exp\left(\kappa/2\right)}{1 + \coth\left(\kappa/2\right)} \xi_b \mathrm{csch}(\xi_b), \tag{7.278}$$

where we have employed

$$y_{ec} = -\eta \alpha_T y_{ne} = \eta y_{ce}, \tag{7.279}$$

with

$$\eta = \frac{kT}{q}\frac{1}{w_B}\frac{dw_B}{dv_{CB}}\frac{\kappa \exp\left(-\kappa\right)}{1 - \exp\left(-\kappa\right)}. \tag{7.280}$$

7.8 Transit Time Resonance Effects

We will look at our example of the $Ga_{1-x}Al_xAs/GaAs$ bipolar to understand the nature of the frequency dependence, current gain, and unilateral gain, since in these devices the transit time effect appears to be so important. We use the relations of y-parameters that we have derived. Although this implies that the overshoot effects that are present in the HBT are not included, these overshoot effects result in smaller transit times and hence even higher frequencies where transit time–related resonances could occur. Thus, drift-diffusion approximation results in relations that are suitable for judging the nature of roll-off. We already know the intrinsic common-base y-parameters (y_{11}^b, y_{12}^b, y_{21}^b, and y_{22}^b), which include the transition capacitances but exclude the intrinsic base resistance. As we have discussed, the hole injection effect is quite weak (even for a homojunction transistor, $\approx 10\%$ decrease in magnitude and $\approx \pi/5$ radians shift in phase at $2f_T$ for $\tilde{\gamma}$). It can practically be neglected for HBTs. The common-emitter y-parameters (y_{11}^e, y_{12}^e, y_{21}^e, and y_{22}^e) can be derived from this. The effect of the intrinsic base resistance (r_{bi}) can be included in this to obtain the intrinsic common-emitter y-parameters (see Problem 12),

$$y_{bbi} = \frac{y_{11}^e}{1 + r_{bi}y_{11}^e},$$

$$y_{bci} = \frac{y_{12}^e}{1 + r_{bi}y_{11}^e},$$

$$y_{cbi} = y_{21}^e \left(1 - \frac{r_{bi}y_{11e}}{1 + r_{bi}y_{11}^e}\right),$$

$$\text{and} \quad y_{cci} = y_{22}^e - \frac{r_{bi}y_{12}^e y_{21}^e}{1 + r_{bi}y_{11}^e}. \tag{7.281}$$

The simplest lumped representation for approximating the extrinsic elements of bipolar transistor is shown in Figure 7.53, where the usually large extrinsic contribution of the collector capacitance, and the parasitic resistances of the base, emitter, and collector are also included. The block within is characterized by the above intrinsic y-parameters. The extrinsic common-emitter y-parameters (y_{11x}^e, y_{12x}^e, y_{21x}^e, and y_{22x}^e) can now be derived.

The parasitic effects are estimated based on the material characteristics and our chosen device geometry discussed in quasi-static analysis. Figure 7.54 shows the modelled frequency dependence of the current gain and unilateral gain. The resonance in the unilateral gain of the intrinsic device occurs at 290 GHz due to the negative output resistance. This negative output resistance results from the phase delay associated with the tran-

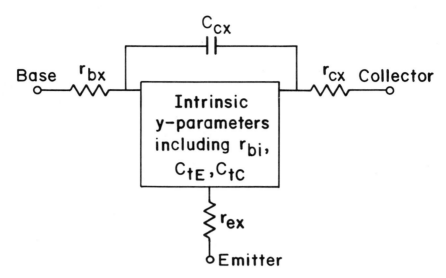

Figure 7.53: Schematic showing the inclusion of extrinsic parasitics to the intrinsic parameters derived from the transport theory of the bipolar transistor.

sit time in the collector and the transport and storage effects of the base (we have assumed negligible transit time in the emitter transition region). An approximate expression for predicting the frequency of this resonance is $\tan^{-1}(\omega W_B^2/2D_n) + 0.11 W_B^2/D_n + \tau_c' = \pi$. In this expression, the first two terms correspond to the phase delay in the base and the last term corresponds to the phase delay in the collector. Figure 7.54 shows that this resonance is, however, effectively suppressed when one includes the parasitics of the device.

The resonance is suppressed due to the combined effect of the base resistance and the collector resistance in conjunction with the extrinsic collector capacitance. The effect of the base resistance dominates over that of the collector resistance, as should be expected for a low input and high output impedance device. However, the effect of collector resistance is not negligible and can not be ignored. In this emphasis on the external parasitics, the resonance phenomenon is similar to that in our discussion of MESFETs and HFETs.

The accuracy of the f_T and f_{max} expressions derived earlier can also be assessed from this analysis. The f_T relation is accurate to within 10%, with the worst inaccuracy occurring when the parasitic collector resistance

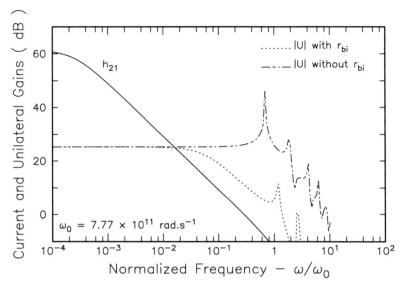

Figure 7.54: Intrinsic and extrinsic current gain ($|\, h_{21} \,|$) and unilateral gain ($|\, U \,|$) as a function of frequency for the HBT considered in our analysis.

is negligible. The f_{max} relation is more inaccurate under similar conditions. Part of this inaccuracy is due to the trans-admittance phase delay associated with collector transport. Part of the inaccuracy results from the lumping of the time constants of the network.

7.9 Transient Analysis

Our discussion of small-signal behavior of the bipolar transistor has considerable bearing on the transient analysis. In the section on off-equilibrium effects, we will discuss further the transient effects. Here, we will discuss some of the expected effects related to the model elements we noted in the small-signal analysis. A transient solution may be obtained by inverse Fourier transform, or using complex notation for inverse Laplace transform from the small-signal response of the transistor. Consider the response to an emitter current step if the output response was dominated by the base transport factor. For a current step in the emitter current of $\Delta I_e/s$, the collector current is

$$I_c(t) = \mathcal{L}^{-1} \left\{ \frac{\Delta I_e}{s} \operatorname{sech} \left[\frac{w_B}{\mathcal{L}_n} (1 + s\tau_n)^{1/2} \right] \right\}. \qquad (7.282)$$

If one may make the single pole approximation of the base transport factor, this lets us write the current response as

$$I_c(t) = \mathcal{L}^{-1} \left(\frac{\Delta I_e}{s} \frac{1}{1 + s\tau_B^{\alpha T}} \right) = \Delta I_e \left[1 - \exp \left(-\frac{t}{\tau_B^{\alpha T}} \right) \right]. \qquad (7.283)$$

The excess phase factor term of $\exp(-\delta \omega \tau_B^{\alpha T})$ in the base transport factor leads to a $\delta \tau_B^{\alpha T}$ delay term as a multiplier to the above. Its result is that, for a period of approximately $\delta \tau_B^{\alpha T}$, collector current does not change. Similarly, the collector transport factor with its approximately exponential dependence on the collector signal delay leads to an additional delay time equal to the collector signal delay in the transient response.

Figure 7.55 shows the transient response of the collector current, and the build-up of the electron density in the base and the collector space charge region, due to a forward biasing of the base–emitter junction of a GaAs bipolar transistor obtained using numerical simulation. The transient response of the collector, referenced to the forward biasing voltage, shows the delay effects with these transient effects.

7.10 Off-Equilibrium Effects

Off-equilibrium effects become important in bipolar transistors also, since the hot carriers can occur in the base–collector space charge region where the electric field changes suddenly over scales that are very short. It can also occur in the base, if carriers are injected hot using abrupt heterostructures as an emitter, e.g., at abrupt $Ga_{1-x}Al_xAs/GaAs$ junction, or are accelerated by quasi-electric fields produced by bandgap grading, or due to a very thin base and the natural filtering process of the injection phenomenon over a base–emitter junction barrier.

Figure 7.56 shows for a collector doping of 5×10^{16} cm^{-3} the electron energy distribution for a constant current density in various designs of an HBT. The base width in these structures is 1000 Å. The case (c) has a quasi-electric field of 1205 kV/cm and the case (d) an electric field of 25 kV/cm across the base.

A number of interesting features can be seen in these plots. In the $Ga_{.7}Al_{.3}As$ emitter, the electrons populate all the valleys—Γ, L, and X. While the Γ valley is the lowest, it has a low density of states; the higher L

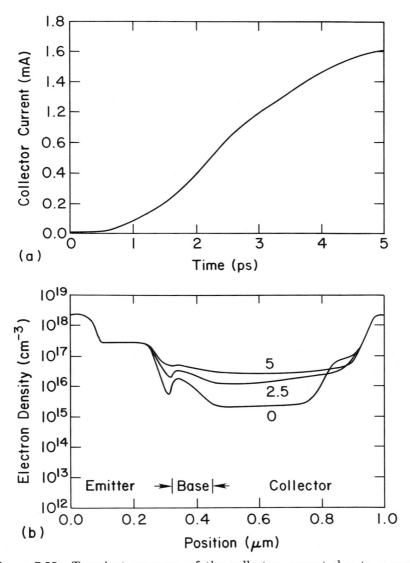

Figure 7.55: Transient response of the collector current due to a rapid change in the base current is shown in (a). The change in electron density as a function of position for various instances of time in ps is shown in (b).

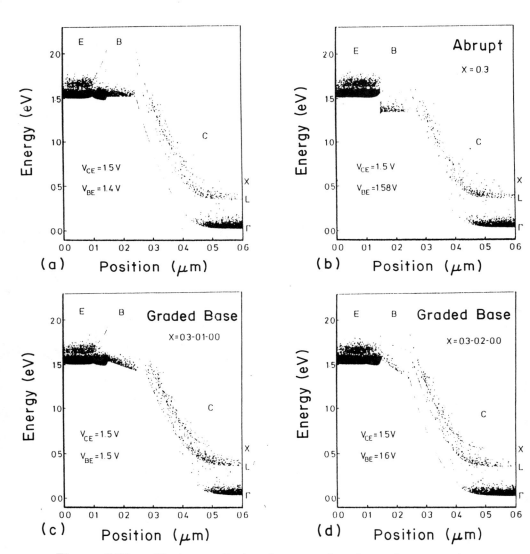

Figure 7.56:   Electrons displayed as a function of energy in a
$Ga_{1-x}Al_xAs$/GaAs HBT for a uniformly doped base (a), an abrupt injection barrier (b), a graded mole-fraction base (c), and a graded base with
higher electron quasi-electric field (d). From R. Katoh, M. Kurata, and
J. Yoshida, "Self-Consistent Particle Simulation of (Al,Ga)As/GaAs HBTs
with Improved Base–Collector Structures," *IEEE Trans. on Electron Devices,* **ED-36**, No. 5, p. 846, 1989.

and X valleys have a heavier effective mass and have six or eight equivalent minima resulting in significantly higher density of states. This results in large carrier populations in the L and X valleys also. In $Ga_{.7}Al_{.3}As$ the carriers are almost evenly divided between the valleys. In the base, the Γ, L, and X valleys are further apart, and the barrier to the X and L valleys is large. So, in device structures where drift-diffusion across a barrier (Cases (a), (c) and (d) of Figure 7.56) is important, the injected carriers in the base, are principally in the Γ valley. X and L valley electrons in the $Ga_{.7}Al_{.3}As$ emitter, by the time they pass through the depletion region into the base mostly end up in the Γ valley. Some of these carriers may actually be hot because they do not lose all their excess energy, and because surmounting the barrier filters away more of the low energy carriers. Figure 7.56 in case (a) shows electrons that are higher up in the Γ valley in the emitter at the injecting emitter–base junction. These carriers may enter the base hot if an abrupt heterojunction transition is used, as in case (b). Compared to the other cases, there are many more hot electrons at the emitter–base junction with the use of an abrupt barrier. For a Γ valley electron, this excess energy, if it is just thermionically emitted over the barrier using $Ga_{.7}Al_{.3}As$, is $\Delta E_c = 0.25$ eV, which corresponds to an approximate velocity of 3.7 $\times$ 10^7 cm.s^{-1}.

Carriers can pick up energy from the electron quasi-electric field in the base, as in cases (c) and (d), where as they approach the collector end of the base, there is a fair fraction of hot Γ electrons.

The collector depletion region has large electric fields in it resulting from the p–n junction, and at the base end it rises rapidly because of the heavy base doping. This results in rapid carrier heating in the depletion region in all the examples over a short distance of the order of 500 Å. Carriers heat up in the Γ valley, rapidly increasing in energy and velocity, and when they have sufficient energy they scatter into the X and L valleys, finally resulting in transport at the saturated velocity.

Figure 7.57 shows this velocity behavior for the structures of cases (a) through (d) of Figure 7.56. The velocity in the base varies from the drift-diffusion like velocity of 2-4 $\times$ 10^6 cm.s^{-1} in the uniformly doped base case, to higher velocities in the abrupt and graded base structures. The velocity in the collector rises very rapidly due to velocity overshoot and subsequently relaxes to the $\approx 1 \times 10^7$ cm.s^{-1} saturated velocity of GaAs. The peak velocities are near the maximum group velocity.

The interesting part, in addition to this off-equilibrium phenomenon, is the possible transfer of an L electron or an X electron in the $Ga_{1-x}Al_xAs$ emitter to a hot Γ electron in the GaAs base, even for a graded junction structure. Thus, although abrupt barriers are more efficient in injecting hot electrons into the GaAs base, hot electrons also occur due to the natural

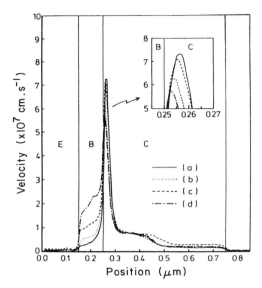

Figure 7.57: Electron drift velocity as a function of position for the various $Ga_{1-x}Al_xAs/GaAs$ HBTs of the previous figure. From R. Katoh, M. Kurata, and J. Yoshida, "Self-Consistent Particle Simulation of (Al,Ga)As/GaAs HBTs with Improved Base–Collector Structures," *IEEE Trans. on Electron Devices*, **ED-36**, No. 5, p. 846, 1989.

filtering process of injection across a potential barrier and because of the transfer in reciprocal space. These structures accomplished a reduction in base transit time by effectively increasing the velocity in the base. We also noted the large increases in velocity in the collector over a short distance. Since the transit delay times of both the emitter and the collector are important, we look into the off-equilibrium effects in the collector next by looking at various types of collector structures (see Figure 7.58). For cases (a) and (b), which employ a doping of 1×10^{17} cm^{-3} and 2×10^{17} cm^{-3} in the collector, the depletion region shrinks at the higher doping in the collector, with a corresponding increase in the electric field. This leads to transfer of electrons to the L and X valleys occurring over a shorter distance. Cases (c) and (d) are of more interest because they decrease the electric field in the collector depletion region as well as the rapid change in the field by employing a p-type extension of the base region. So, in addition to the 1×10^{19} cm^{-3} doping in the first 1000 Å of the base, a 5×10^{16} cm^{-3} doping over 2000 Å (case (c)) and 1×10^{17} cm^{-3} doping over 2000 Å (case

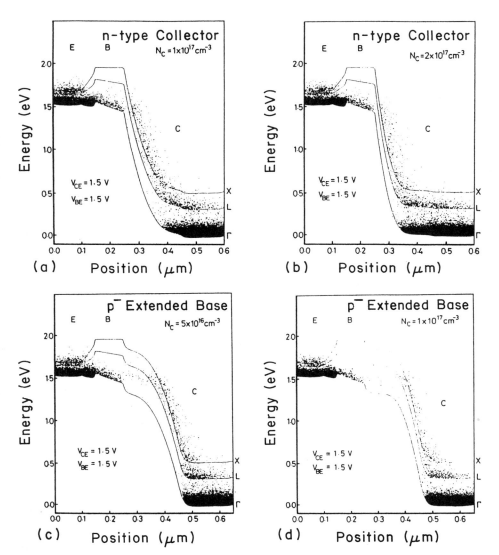

Figure 7.58: Electron energy as a function of position for a collector doping of 1×10^{17} cm^{-3} (a), 2×10^{17} cm^{-3} (b), a thicker 5×10^{16} cm^{-3} collector region with a p$^-$ extension of the base region (c), and a 1×10^{17} cm^{-3} collector with a p$^-$ extension of the base region. From R. Katoh, M. Kurata, and J. Yoshida, "Self-Consistent Particle Simulation of (Al,Ga)As/GaAs HBTs with Improved Base–Collector Structures," *IEEE Trans. on Electron Devices*, **ED-36**, No. 5, p. 846, 1989.

(d)) are also employed. The decrease in electric field, as a result, allows the carriers to overshoot over a longer distance, ≈ 1000 Å, because they do not pick up enough energy to transfer to the L and X valleys during the 1000 Å of travel. In these structures, during the transit through the collector depletion region, the average velocity can be much larger, and the corresponding transit delay significantly shorter.

We, however, need to design the p-type base region so that it is depleted under normal biasing conditions, including under the high bias condition when the Kirk effect is significant. The 1×10^{17} cm^{-3} p-type base, e.g., is undepleted over ≈ 1000 Å under the bias conditions of Figure 7.58(d). This gives rise to increased base charge storage. So, while the collector transit time is reduced, the base signal delay is increased. Thus, such structures need to be designed carefully, by including all the effects of the different biasing conditions. But they quite convincingly demonstrate the importance of the off-equilibrium effects in bipolar structures and also the importance of the secondary valleys and their energies w.r.t. the primary valleys in limiting the extent of it.

Clearly, materials like Ga$_{1-x}$In$_x$As, which have secondary valleys placed much higher than the primary valleys, may show stronger off-equilibrium effects than GaAs. We look at this further in the following discussion where we study both the steady-state and the switching characteristics of the off-equilibrium effects in the base–collector region of bipolar transistors. We will compare a few different materials for their scattering characteristics: GaAs, InP, Ga$_{.47}$In$_{.53}$As, and InAs. The small-signal as well as the transient response of the bipolar transistor depend on the transit of the carriers, the charging of both quasi-neutral and space charge regions, and the dispersive effects associated with these. The collector signal delay is a fraction of the collector transit time. It favors the behavior closer to the metallurgical junction where the field is the maximum. For a constant velocity transit through the collector space charge region, the collector signal delay is one-half of the collector transit time. If a constant velocity overshoot occurred throughout the collector space charge region during the change of bias in switching, then this signal delay relationship would still be useful. However, in reality, velocities are larger closer to the junction, and decrease towards the saturated velocity away from it. Moreover, these velocities are changing during the transient because of changes in the driving forces on the carrier. An extended lightly doped base region, e.g., was seen earlier to improve the relative length of the velocity overshoot region. It would be reasonable to conclude that one would obtain improved performance because of a decrease in collector transit time. However, such an improvement in overshoot length occurs with a decrease in the velocity at the metallurgical junction, and a poorer behavior at high current densities due to excess storage and increased Kirk effect.

Overshoot effects occur because the relaxation rate of momentum is larger than that of energy, and a major cause of these large rates is the large scattering rate resulting from secondary valley transfer. $Ga_{.47}In_{.53}As$, whose conduction bands were shown in Chapter 2, has been a material of interest because of this large Γ–L separation. In Chapter 2 we also showed the steady-state scattering rates in the low energy range. The maximum possible velocities are related to this and the effective mass of electrons. For large bandgap semiconductors, transfer to the secondary valley could quite well be prevented by use of a forward bias at the base–collector junction. This forward bias should be large enough to cause, at the base–collector junction, a band edge energy change which is less than the secondary valley separation and yet small enough to limit the saturation. It may be possible to accomplish this for small-signal applications, but is probably unlikely for digital applications such as emitter-coupled logic where sufficiently large collector swing is required. InP is of interest since it has a secondary valley separation similar to that of $Ga_{.47}In_{.53}As$, but a larger bandgap and hence a larger breakdown voltage. InAs is another contrasting choice because it has a very large secondary valley separation, an unusually low scattering rate and effective mass, but a small bandgap and hence a small breakdown voltage. The designs and choices are very extensive and rich.

We will consider a geometry quite similar to that considered in our quasi-static discussion; the emitter and the collector are cladded by highly doped layers, and the contacts to the regions are at the device edge. In this device, we will make perturbations such as in doping, etc., to remark on various nuances of the off-equilibrium phenomena. The results presented are averaged over the lateral extent of the emitter. In order to maintain comparable base resistance, the base width and doping have been kept constant at 750 Å and 5×10^{18} cm^{-3}, since the hole mobility in all these materials is relatively similar. In the case of the device with an extended lightly doped base region, the collector width is reduced to maintain comparable collector capacitance. The choice of collector doping is related to the breakdown voltages considered in Chapter 2. The breakdown voltages in InAs are the smallest but still useful in the low doping range. Even though it has a relatively large ionization coefficient at moderate fields, its electron ionization coefficient is only factors of five higher than that of $Ga_{.47}In_{.53}As$ at high fields. Another reason for the slightly higher than expected breakdown voltage in these indium-containing arsenides is that, at high fields, the hole ionization coefficient continues to be substantially lower, while in GaAs and InP it is actually either comparable or higher. InAs transistors are at the least theoretically feasible with smaller logic swings and turn-on voltages. In the discussion, they serve to contrast the limits of operation

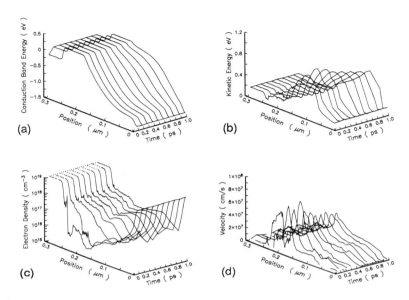

Figure 7.59: Transient evolution of the potential (a), the kinetic energy (b), the carrier density (c), and the velocity (d) for a switching in V_{BE} from 1.2 V to 1.4 V, with V_{BC} maintained at 0 V, for a GaAs bipolar transistor. From S. Tiwari, M. Fischetti, and S. E. Laux, "Transient and Steady-State Overshoot in GaAs, InP, Ga$_{.47}$In$_{.53}$As, and InAs Bipolar Transistors," *Tech. Dig. of International Electron Devices Meeting*, p. 435, Dec. 9–12, ©1990 IEEE.

achievable.

We first consider the case of switching in GaAs bipolar transistors. Figure 7.59 shows the transient evolution of the potential, the kinetic energy, the carrier density, and the velocity for a switching in V_{BE} from 1.2 V to 1.4 V, with V_{BC} maintained at 0 V. The collector doping is 1×10^{17} cm^{-3} with an epitaxial thickness of 0.175 μm, a compromise between excess storage delay due to the Kirk effect at the 1.4 V input bias, and collector signal delay. Indeed, at the high input bias, evidence of the Kirk effect is observable in the conduction band edge profiles of Figure 7.59. The electron transit, as a function of time, is observable in the collector space charge region. However, the time taken for this transit (the collector transit time) is larger than the delay time of the collector current in the quasi-neutral collector region (the collector signal delay). The collector transport factor

ζ is

$$\zeta = \frac{1}{w_{c'} - w_c} \int_{w_c}^{w_c'} \left\{ \exp\left[-j\omega \int_0^z \frac{dz'}{v(z')} \right] \right\} dz, \qquad (7.284)$$

where v is the velocity. This favors the velocity at the metallurgical junction ($z = 0$) as a consequence of the displacement current effect. The overshoot of the electron velocity as a function of position changes substantially over time, increasing at first and then decreasing. Thus, overshoot at both bias points is not an adequate representation of the nature of overshoot during the transient, and neither is the expression above of the actual signal delay. However, irrespective of these limitations, the consequence is that, in most transient and steady-state situations, it is desirable to use a compromise between a maximization of velocity at the junction and a broadening of the overshoot region.

Note that in Figure 7.59 a broader velocity overshoot region does develop due to the presence of a larger electron density in the collector space charge region at the higher bias. To generalize this, consider the InP steady-state characteristics shown in Figure 7.60. The peak velocity at the higher bias is smaller than at the lower bias. This large bias case is actually quite similar to that of an extended lower doped base, sometimes referred to as p$^-$ collector in the literature. Figure 7.61 shows the steady-state low bias and high bias results of the GaAs bipolar transistor together with similar results of an extended base device with a 1×10^{16} cm^{-3} 750 Å base. At a 1.2 V base–emitter bias, in the extended base device, there is a lowering of the peak electric field because of the broadening of the space charge region into the base. The overshoot region is broader. With the switching to a higher bias of 1.4 V at the base–emitter junction, the field in the base region decreases because of the higher electron density, and the high-field region shifts towards the sub-collector interface. The storage is larger, the collector signal delay increases, and the net delay during switching is larger. This occurs in spite of the collector transit time being smaller in this device. The collector signal delay is still larger. Therefore, both the analysis of the steady-state behavior alone and a strict emphasis on broadness of the overshoot can be deceiving.

Similar comments also hold regarding small-signal behavior, because a steady-state Monte Carlo calculation is a simulation of quasi-static behavior. An accurate analysis, similar to the transient analysis, requires the inclusion of dispersive effects associated with small-signal high-frequency operation throughout the device. This can only be elucidated by performing a small-signal Monte Carlo calculation of the problem where the signal oscillates in time, similar to the signal delay time of the device.

Steady-state behavior, however, does provide an adequate low-order de-

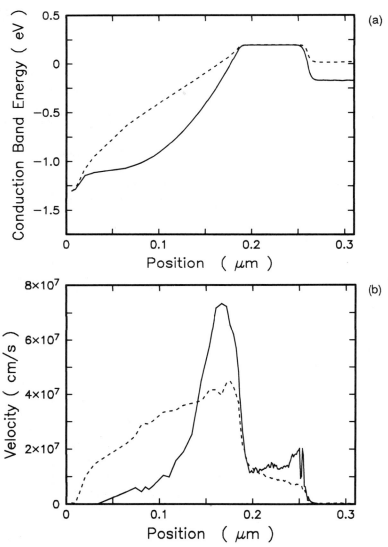

Figure 7.60: Conduction band edge energy (a) and the velocity in steady-state (b) for an InP bipolar transistor in low injection (solid line) and high injection (dotted line) conditions. From S. Tiwari, M. Fischetti, and S. E. Laux, "Transient and Steady-State Overshoot in GaAs, InP, $Ga_{.47}In_{.53}As$, and InAs Bipolar Transistors," *Tech. Dig. of International Electron Devices Meeting*, p. 435, Dec. 9–12, ©1990 IEEE.

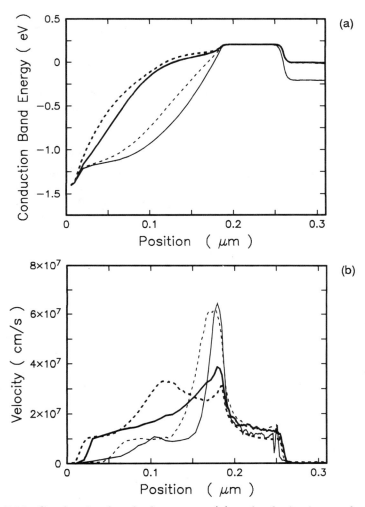

Figure 7.61: Conduction band edge energy (a) and velocity in steady-state
(b) for GaAs bipolar transistors in low injection and high injection conditions. The solid lines are for a uniform 1×10^{17} cm^{-3} doped collector
device and the dotted lines are for an extended p$^-$ base device where an
intervening lower doping is employed at the base–collector junction. From
S. Tiwari, M. Fischetti, and S. E. Laux, "Transient and Steady-State Overshoot in GaAs, InP, Ga$_{.47}$In$_{.53}$As, and InAs Bipolar Transistors," *Tech.
Dig. of International Electron Devices Meeting*, p. 435, Dec. 9–12, ©1990
IEEE.

scription of the device behavior, and is a good tool to compare the behavior of other materials and their behavior vis-a-vis GaAs. Figure 7.62 describes the steady-state behavior of the conduction band edge energy, the kinetic energy, and the velocity in GaAs, InP, $Ga_{.47}In_{.53}As$, and InAs under conditions of low Kirk effect. InP is of particular interest because of its larger secondary valley separation. This results in the broader but still peaked overshoot behavior. $Ga_{.47}In_{.53}As$ bipolar transistors have a secondary valley separation comparable to the bandgap, a smaller fraction of the carriers transfer at this bias condition, and a broader overshoot occurs. InAs is an extreme example in this comparison. It has a low scattering rate at low energies, a significantly larger secondary valley separation than the bandgap (> 1 eV separation compared to $\approx .4$ eV bandgap), and hence only a few carriers transfer to the secondary valleys. Consequently, InAs bipolar transistors show the broadest and highest overshoot features of all these devices.

The GaAs, InP, and $Ga_{.47}In_{.53}As$ transistors can all be forward biased at the collector in order to decrease this secondary valley transfer and to make the velocity overshoot broader. The compromises this entails are related to dopings, field profiles, capacitances, etc., that are suitable for logic or small-signal operation. In the larger bandgap materials, the doping can be increased beyond 1×10^{17} cm^{-3} in the collector to increase the field at the junction, and the device size shrunk to maintain low collector capacitance so that similar broad $Ga_{.47}In_{.53}As$-like overshoot features are observed. These compromises, while adequate in a paper exercise, are technologically challenging. The behavior of Figure 7.62 is thus a more representative picture if this technology can not be achieved. In this case, the overshoot is broad, but due to the decrease in field at the junction the peak overshoot is smaller.

The $Ga_{.47}In_{.53}As$ bipolar transistors have a secondary valley separation comparable to the bandgap and do not need this forward bias to show a large and broad overshoot. This large broadness of the overshoot is maintained over a variety of base–collector bias conditions, albeit with larger storage delays associated with the Kirk effect at its lower doping. For $Ga_{.47}In_{.53}As$ transistors, a consequence of the larger scattering rate of InP is likely to be that any compositional changes that include more phosphorous in the collector in order to increase the breakdown voltage are also likely to result in a decrease of the broadness in this overshoot. Both in $Ga_{.47}In_{.53}As$ and InAs bipolar transistors, the breakdown voltage places the largest constraint, although the large overshoots will lead to a decrease in ionization in the high-field overshoot region and thus to higher breakdown voltages than those of Chapter 2. For $Ga_{.47}In_{.53}As$, the doping has to be maintained below 8×10^{16} cm^{-3}, and for InAs bipolar transistor it

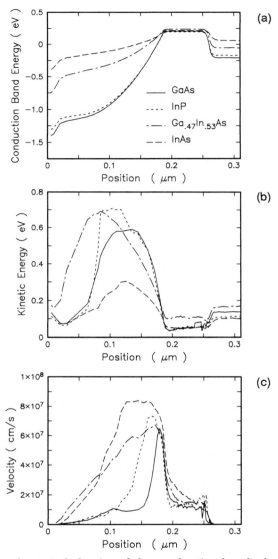

Figure 7.62: Steady-state behavior of the conduction band edge energy (a), the kinetic energy (b), and the velocity (c) in GaAs, InP, Ga$_{.47}$In$_{.53}$As, and InAs under conditions of low Kirk effect. From S. Tiwari, M. Fischetti, and S. E. Laux, "Transient and Steady-State Overshoot in GaAs, InP, Ga$_{.47}$In$_{.53}$As, and InAs Bipolar Transistors," *Tech. Dig. of International Electron Devices Meeting*, p. 435, Dec. 9–12, ©1990 IEEE.

has to be maintained below 3×10^{16} cm^{-3}, in order to obtain an adequate collector-to-emitter breadown voltages at adequate current gains.

To relate this steady-state behavior to the scattering of carriers, consider the scattering rates as a function of energy for these materials (see Chapter 2). This scattering rate is a function of many factors, primary among which is the density of states available for scattering. InAs and Ga$_{.47}$In$_{.53}$As bipolar transistors are superior in this respect. InP has a large secondary valley separation, however, it has a larger mass and a larger intra-valley LO phonon scattering. Both are not conducive to larger velocities. As a result, both GaAs and InP appear to be quite similar in their behavior.

Operation in a logic gate involves the change of both the base–emitter and the base–collector bias simultaneously. Changes in the base–collector bias, such as its forward biasing during the turn-on of the transistor, are also important. Usually, it occurs due to debiasing through the resistor by the increasing current flow. If this were fast, overshoot would be expected to occur at the sub-collector interface at the other end of the space charge region and in the opposite direction.

So, the velocity overshoot in the base–collector space charge region is a sensitive function of the transient conditions due to local and non-local effects. Since the component of the transistor delay due to transit of carriers, the collector signal delay, is more sensitive to carrier motion in the region around the metallurgical junction, a broad overshoot does not necessary result in shorter delay. Specific bias conditions and device designs may allow comparable small-signal operation in all these materials provided breakdown voltage constraints and device size constraints are satisfied. In logic applications, where breakdown is a significant constraint, broader velocity overshoot profiles and higher maximum overshoot velocities can be obtained over a wider bias range in Ga$_{.47}$In$_{.53}$As and InP bipolar transistors. InAs bipolar transistors show the largest overshoot and the smallest delay; however, because of their smaller bandgap, the operating bias range of these devices is restricted.

7.11   Summary

This chapter considered the quasi-static, small-signal, and transient operation of HBTs. We extended the Gummel–Poon model developed in Chapter 4 for p–n junctions to bipolar transistors, and showed its application in the presence of compositional grading. We then developed simpler forms of quasi-static models, the various levels of sophistication of Ebers–Moll models, that are useful in circuit oriented modelling. The quasi-static aspects of HBT operation were then analyzed by considering those aspects that are uniquely related to the use of heterostructures. As examples of this, we considered consequences of alloy grading, the behavior of quasi-fields in the junction regions, the effects that take place at high currents, i.e., under small changes of electrostatic potential across junction regions, and the combined effects of these in the behavior of capacitances and current gains of the device. Among the parasitic effects particularly important in compound semiconductors is the effect of poor lifetime in the bulk and at the surface. We considered the operating principles and consequences of this for the device.

We also developed a theory for the small-signal analysis of these devices, with an extension to include the effect of alloy grading. Bipolar transistors provide an interesting twist involving the flow of carriers in a depleted region—the base–collector depletion region—and hence involving interesting implications of displacement current for small-signal analysis. This leads to a difference between the time it takes for a carrier to transit this region (the transit delay) and the time it takes for the signal at the collector terminal to appear (the signal delay). We noted that the latter is shorter. We also looked at the transient behavior of the devices and how the carriers transit through the device. Finally, we considered consequences of off-equilibrium effects in the device.

General References

1. I. Getreu, *Modelling the Bipolar Transistor*, Tektronix, Beaverton, Oregon (1976).

2. D. J. Roulston, *Bipolar Semiconductor Devices*, McGraw-Hill, N.Y. (1990).

3. R. D. Middlebrook, *An Introduction to Junction Transistor Theory*, John Wiley, N.Y. (1957).

4. R. S. Muller and T. I. Kamins, *Device Electronics for Integrated Circuits*, John Wiley, N.Y. (1986).

5. R. M. Warner and B. L. Grung, *Transistors: Fundamentals for the Integrated-circuit Engineer*, John Wiley, N.Y. (1983).

6. R. L. Pritchard, *Electrical Characteristics of Transistors*, McGraw-Hill, N.Y. (1967).

7. S. M. Sze, *Physics of Semiconductor Devices*, John Wiley, N.Y. (1981).

Problems

1. In Chapter 4 we considered the extension of Gummel–Poon models to structures where the compositional changes may occur in quasi-neutral regions of the device structure. Write, for the general case of varying composition and including recombination effects, the electron and hole current for a p–n junction. Show that this reduces to the simplified case, covered in the text, where all compositional change occurs within the high field region at the junction.

2. Equation 7.1 is derived assuming an efficient collector where the carriers are swept away in a high field due to drift effect. A graded heterostructure collector also allows this under low-level injection conditions. An abrupt heterostructure collector, however, if it is low doped or has a large barrier, allows minimal tunneling current and hence excess storage results from the inefficient sweeping out of the carriers. Consider the collector junction as exhibiting small recombination velocity, and derive the equation of electron current at the emitter junction of an n–p–n HBT. Estimate what recombination velocity should be used for $Ga_{.7}Al_{.3}As/GaAs$ and $InP/Ga_{.47}In_{.53}As$ collector–base junctions with a 5×10^{18} cm^{-3} doped base and 5×10^{16} cm^{-3} doped collector.

3. The transport equations used in the analysis of the bipolar transistor are linear differential equations. Argue why a necessary consequence of this is the reversibility of emitter and collector transport. Under what conditions is the use of reversibility incorrect?

4. Show that the base time constant for an inhomogeneously doped base is of the form

$$\tau_B = \frac{w_B{}^2}{\nu \mathcal{D}_n}, \qquad (7.285)$$

where

$$\nu = \left\{ \frac{1}{w_B{}^2} \int_{w_E}^{w_C} \frac{1}{p(z)} \left[\int_z^{w_C} p(\eta)d\eta \right] dz \right\}^{-1}. \qquad (7.286)$$

Consider an exponentially decreasing doping with a grading length of 1000 Å for a 1000 Å thick GaAs base transistor structure. At the center of the base the doping is 5×10^{18} cm^{-3}. What is the difference in the base time constant between the exponentially doped and uniformly doped base structures?

5. Now consider the bipolar with a linearly decreasing bandgap from the emitter junction edge to the collector junction edge parameterized by a constant κ with units of eV.cm^{-1}. Derive the time constant using the Gummel–Poon expressions. Assuming a uniform doping of 5×10^{18} cm^{-3} p-type in the base, what is the base time constant? How does it compare with the time constants of Problem 4?

6. Consider the Gummel–Poon model applied to a uniformly doped base HBT structure. Show that

$$\frac{\partial}{\partial V_{BC}} \left(\frac{Q_F}{\tau_{BF}} \right) \approx \frac{I_C}{V_A} \left[1 - \alpha_T \left(1 - \frac{\alpha_T w_B}{\mathcal{L}_n} \right) \right]. \qquad (7.287)$$

7. Write the transport equations of the Ebers–Moll model using I_{OE} and I_{OC} as the reference currents. Show that these lead to one parameter in current, the saturation current, which is identical for both the base–emitter and the base–collector junction. This has sometimes been called the transport version of the Ebers–Moll model. What is the limitation on applying this to HBTs. Can it be used on an abrupt heterostructure emitter HBT?

8. Adapt the results of Problem 7 to derive a model that can be cast into a hybrid-pi equivalent circuit, similar to that encountered in our discussion of FETs. Show that this form can also be more generally adapted to include higher-order and parasitic effects.

9. In the quasi-static analysis we had introduced the parameter Early voltage, defined as

$$V_A = \frac{1}{w_B} \frac{\partial w_B}{\partial V_{BC}}. \qquad (7.288)$$

Its effect is most significant in the output conductance, since any change in quasi-neutral base width causes a change in the collector current. The ratio of low frequency output to input conductance is related as

$$\frac{g_{nc}}{g_{ne}} = \frac{kT}{qV_A}, \qquad (7.289)$$

following our analysis. Find the expression for Early voltage as a function of device parameters and find the ratio of the conductances for the GaAs transistor with a 1000 Å base doped to 5×10^{18} cm^{-3} p-type, and a 3500 Å wide collector doped to 1×10^{16} cm^{-3}.

10. For the small-signal analysis of the transistor with grading in the base, we derived an expression for carrier concentration as a function of position and electron current as a function of position. Derive, from this, an expression for the electron current density at the base–emitter junction $\tilde{J}_{ne}$ in terms of the applied small-signal voltage $\hat{V}_{be}$, and show Equation 7.270 from this.

11. Show that the susceptance component of the input admittance in Equation 7.270 corresponds to a diffusion capacitance of

$$C_{De} = g_e \frac{w_B^2}{2\mathcal{D}_n} \left\{ \frac{2}{\kappa} \left[1 - \frac{1 - \exp(-\kappa)}{\kappa} \right] \right\}. \tag{7.290}$$

12. Include the effect of base resistance at the input port to the intrinsic common-emitter y-parameters and derive the modified y-parameters claimed in the text.

13. Assuming low-level injection, show that the steady-state electron distribution in the base of a graded junction n–p–n HBT with doping variation in the base and in forward active bias is given by

$$n_p(z) \approx \frac{I}{q\mathcal{D}_n A N_A(z)} \int_0^{w_B} N_A(z)dz. \tag{7.291}$$

This equation is also valid for a homojunction device. Using this, find the improvement in the base time constant resulting from an exponentially graded doping profile in the base of the form $N_A(z) = N_{A0}\exp(-z/\ell)$. What is the electric field as a function of position for the exponentially decreasing base doping?

14. Derive the expression, similar to that of Problem 13, for the steady-state electron distribution in the base of a graded junction n–p–n HBT where the mole-fraction changes.

15. Consider a $Ga_{.7}Al_{.3}As/GaAs$ graded junction HBT with a GaAs collector. The emitter region is doped 2×10^{18} cm^{-3}, the grading is across a 300 Å and is assumed to be parabolic. The base is doped 1×10^{19} cm^{-3} and is 1000 Å in thickness, the collector is doped 5×10^{16} cm^{-3} and is 0.5 μm in thickness, and the sub-collector is doped 5×10^{18} cm^{-3} and is 0.2 μm in thickness. Assume one-dimensional geometry with hypothetical ideal contacts at the emitter, base, and sub-collector edges. The base contact can be assumed to not influence the minority carrier population, it being an idealization

of a contact that would be further away in an extrinsic region in a practical device.

(a) For this one-dimensional structure, draw the conduction band edge, the valence band edge, and the electron and hole quasi-Fermi levels in the four quadrant of operation: (a) $V_{BE} > 0$ and $V_{BC} < 0$, (b) $V_{BE} < 0$ and $V_{BC} < 0$, (c) $V_{BE} > 0$ and $V_{BC} > 0$, and (d) $V_{BE} < 0$ and $V_{BC} > 0$. Also draw these for the punch-through condition if it can occur in this structure.

(b) Can one define an Early voltage for this transistor? What is its approximate magnitude and range?

(c) At what current density will the Kirk effect become important? Assume a saturated velocity of 8×10^6 cm.s^{-1}.

(d) If the structure had Ga$_{.7}$Al$_{.3}$As collector with a linear grading over 500 Å, what would be the current density at which the effect due to the alloy barrier at the base–collector junction would dominate?

(e) Derive approximately the important time constants, and make a quasi-static estimation of the unity current gain frequency at 1×10^2, 1×10^3, 1×10^4, and 1×10^5 A.cm^{-2} at $V_{BC} = -1.5$ V.

16. Now consider this same device structure with the AlAs mole-fraction changing linearly from 0.1 at the base–emitter junction to 0 at the base–collector junction. Repeat the exercise and estimate the parameters of the device structure.

17. We have derived an expression for the collector transport factor given as

$$\zeta = \frac{\sin(\omega\tau_c')}{\omega\tau_c'} \exp\left(-j\omega\tau_c'\right). \tag{7.292}$$

(a) Is this the ratio of current carried by particles, the displacement current, or the total current?

(b) Doesn't current continuity in the the transistor structure imply $\zeta = 1$? Explain.

18. Should the f_T of an HBT increase or decrease with temperature? Explain.

19. Does surface recombination affect the injection efficiency or the base transport factor in an HBT?

Chapter 8

Hot Carrier and Tunneling Structures

8.1 Introduction

The charging of transition capacitances and the time constants associated with carrier transit in the base and signal delay in the collector depletion region constituted the most significant factors of the total delay in the operation of the bipolar transistor. The transition capacitances are nearly independent of the choice of the material and much more closely related to the design of the device structure. Scaling of the cross-section of the device structure leads to a significant reduction in these capacitances and the related time constants. However, both the base time constant τ_B and the collector signal delay τ_c' continue to be major factors of increasing importance in limiting the frequency response and the speed figures of merit. This base time constant is related to the average time a carrier spends in the base before being swept away into the collector depletion region. It is also the time constant that relates the minority charge stored in the base with the current carried by the device (similar storage of minority carriers also occurs in the emitter and the collector regions, but being small for the heterostructure devices, it is usually ignored). The other important transit-related time constant of collector signal delay comes from the displacement current effect due to transport of carriers through the collector depletion region. We discussed in Chapter 7, how hot electron effects can be utilized to decrease the time constants related to base and collector transport. The basis of these techniques is an increase in the velocity with which these carriers move. For collector transport, this occurs naturally due to the gradient

705

of electric field and the importance of the velocity in the short collector depletion region. For base transport, this is accomplished by hot electron injection and/or by an appropriate quasi-field that enhances the minority carrier motion. The increase in velocity decreases the amount of charge stored in the base to sustain the current, and equivalently the average time a transiting carrier spends in the quasi-neutral base (see Problem 1). All minority carrier devices require this minority carrier storage in some form in order to sustain the predominant current in the control region of the device. Additional storage effects also come about due to minority carrier injection into other parts of the device, and all these constitute a delay contribution that has its origin in the minority carrier–based operation of the device.

Since the early days of transistors, there has been interest in making such bipolar transistor–like devices, i.e., devices based on barrier-modulated current injection, which do not suffer from many of the minority carrier–related delay effects. Among the first of these transistors, an example of hot carrier transistors, is a hot electron transistor. Mead,[1] who recognized the advantages of such a majority carrier device, called his favored implementation a tunnel emission amplifier (see Figure 8.1).

We will use the nomenclature hot carrier transistor, and in particular hot electron transistor, as a general classification for devices that depend on hot carrier transport for usefulness of their operation. Bipolar transistors will operate independent of hot electron effects, but hot carrier transistors depend on the existence of the hot carriers for device amplification. The tunnel emission amplifier is one example of a hot carrier transistor. These devices, several of which were introduced when interest in tunnel diodes was high, were also meant to incorporate several of the transport-related major advantages recognized in tunneling. Tunneling forms a controlled source of carriers, and the tunneling junction can also serve to suppress processes involving injection of carriers in the reverse direction. Tunneling junctions also exhibit a frequency response limited by current density and capacitance of the junction, since tunneling transit times are short, commensurate with the length scale associated with de Broglie wavelength.

The tunneling emission device of Mead used a metal base (hence the name metal base transistor that is also applied to this structure) with the metal thickness kept shorter than the mean free path for carriers in it (see Figure 8.1). In its generic form, this metal is separated from the injecting emitter (another metal) and the collecting electrode (another metal) by insulators. In the aluminum-based device structures, Mead observed current amplification factors, in common-base configuration, of 0.1 to 0.3. The in-

---

[1] C. A. Mead, "The Tunnel-Emission Amplifier," *Proc. of IRE*, **48**, p. 359, 1960.

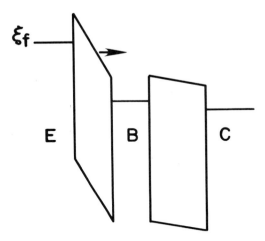

Figure 8.1: Tunnel-emission amplifier, an example of a hot carrier transistor as proposed by Mead. The structure in its original form uses an aluminum emitter, base, and collector, with anodized aluminum (a porous Al_2O_3 insulator) serving to isolate the three regions and allowing for application of bias.

jection of hot carriers need not be limited to tunneling processes. It can also be achieved using thermionic injection, another means for hot carrier injection. This basic concept of using tunneling and hot carrier emission in forming majority carrier barrier–controlled devices has been extended in several variations of these devices by others using different materials: metals and semiconductors for emitters, bases and collectors; and semiconductors, insulators, and semi-insulators for isolation. An example of a thermionic emission–based device[2] is shown in Figure 8.2; it uses a metal base of gold deposited on germanium as a collector and employs a silicon emitter. The emitter is a cantilevered silicon. For a gold film of ≈ 100 Å thickness, the common-base current amplification factor α_B increases to ≈ 0.46 from ≈ 0.3 of Mead's device. Common-base current amplification factors of less than one-half imply a common-emitter current amplification factor of less than unity, so even though the device affords some isolation between input and output, it does not allow for a current gain necessary to restore

---

[2]A thorough discussion of the early work of C. R. Crowell and S. M. Sze on this subject can be found in C. R. Crowell and S. M. Sze, "Hot Electron Transport and Electron Tunneling in Thin Film Structures," in G. Haas and R. E. Thum, Eds., *Physics of Thin Films*, **V4**, Academic Press, N.Y., p. 325 (1967).

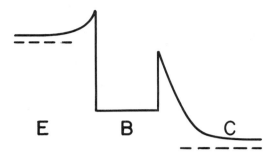

Figure 8.2: A hot carrier transistor example due to Sze and Crowell. It employs a thin gold film deposited on germanium collector as a base, and silicon as the emitter material.

Table 8.1: Mean free path as a function of energy.

Material	Energy $E - \xi_f$ (eV)	Energy Attenuation Length (Å)
Copper	0.6	280
	0.9	100
Gold	2.0	160
	4.0	70
Palladium	0.85	170
Silver	0.7	450

voltage levels in logic. A current gain of less than unity, i.e., negligible f_T also precludes utility at high frequencies. There are two major problems that cause this lack of gain. One is the length scale over the which the carrier rapidly loses energy (the energy attenuation length). This is shown as a function of hot carrier energy in Table 8.1. The mean free path for hot (0.85 eV) electrons in gold is 100 Å, so even though a gold metal film allows for high base conductivity, a substantial loss of energy occurs due to inelastic scattering during transit through the base region. The second problem is that even if the carrier mean free path were acceptable, a substantial quantum-mechanical reflection of the hot carrier occurs because of the large energy discontinuity at interfaces with metals. The metal base structures are particularly susceptible to this because it is more difficult to design a graded interface at a metal junction to reduce the quantum-mechanical reflection. This can now be achieved for some structures, e.g., $CoSi_2/Si$, by modern epitaxy techniques, but is considerably simpler to achieve at heterostructure–semiconductor interfaces. Heterostructures thus serve as a useful means for tunneling and hot carrier injection by other injection mechanisms. Semiconductors, with a larger mean free path than metals, also serve as an effective medium for the base of devices, albeit with the loss of advantages of conductivity.

Semiconductor heterostructures allow for both the coupling of barriers and quantization of momentum in the perpendicular direction when they approach the de Broglie wavelength in size. Such structures can show unusual features both in the presence and the absence of scattering-dominated transport. For very short structures, if the transmissive properties of the two barriers are identical for an electron that has an energy that allows it to tunnel to the quantized states in the well formed by the coupled barriers, a resonance occurs. So, even though individually the transmission coefficients are low, the transmission probability at this matched condition is large. A close analogue of this is the Fabry–Pérot interference in optics. An electron tunneling by such a coherent phenomenon may spend a considerable time in the well and exhibit a long decay time because of the constructive effect of interference, even though it has a large transmission probability. This raises the likelihood of scattering to other quantized states in the well (quasi-bound states because the electron eventually will tunnel out) and outside the well via an inelastic process. The electron loses coherence during this process, but exhibits a faster response determined now by scattering considerations. Both these phenomena, occur, and can be made relatively fast by a suitable choice of material parameters. Both these processes exhibit similar negative resistance current–voltage characteristics and an ability to supply electrons with a narrow energy spread. This can be made the basis for injection and intrinsic majority carrier device operation also, with short temporal scales.

Thus, there exist a number of heterostructure phenomena related to hot carrier injection and tunneling that are relevant to the majority carrier barrier mode devices. Our discussion in this chapter will concentrate on the physical principles underlying these. They are interesting for the unique semiconductor physics, and particularly for the physics of small dimensions, and this serves to help understand the limitations of the conventional field effect and bipolar transistors.

8.2 Quantum-Mechanical Reflections

We first consider this quantum-mechanical reflection problem for a free electron, incident at a barrier, as shown in Figure 8.3. The electron has

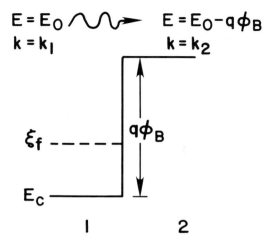

Figure 8.3: A free electron of energy E_0 incident at a barrier of height $q\phi_B$. Even if the energy of the carrier is larger than the height of the barrier, it may undergo reflection because of the effect of perturbation due to the barrier.

an incident energy E_0 and is incident at a barrier of barrier height $q\phi_B$, which exists between regions 1 and 2 of the system. We treat the problem as a single-body free electron problem, and ignore space charge effects, in order to estimate and understand the magnitude of the effect of quantum-mechanical reflection at a barrier. Let the wave functions that satisfy Schrödinger's equation be φ_I and φ_2 in the two regions. Our control equation is

$$-\frac{\hbar^2}{2m}\frac{d^2}{dz^2}(\varphi_1) = E_0\varphi_1 \tag{8.1}$$

in region 1, and

$$\left(-\frac{\hbar^2}{2m}\frac{d^2}{dz^2} + q\phi_B\right)\varphi_2 = E_0\varphi_2 \tag{8.2}$$

in region 2. Since we are considering free electrons, we may use the normalized plane wave functions as the basis for these wave functions. The wave functions, therefore, are of the form

$$\varphi_1 = a_1\exp(jk_1z) + b_1\exp(-jk_1z), \tag{8.3}$$

and

$$\varphi_2 = a_2\exp(jk_2z) + b_2\exp(-jk_2z), \tag{8.4}$$

where

$$k_1 = \sqrt{\frac{2mE_0}{\hbar^2}} \qquad (8.5)$$

and

$$k_2 = \sqrt{\frac{2m\left(E_0 - q\phi_B\right)}{\hbar^2}} \qquad (8.6)$$

are the wave vectors in the two regions. The first term in these wave functions is the incident wave going in the direction of positive z, and the second term is the reflected wave going in the direction of negative z. The coefficients and amplitudes of the incident and reflected waves are obtained by the matching of probability and momentum at the interface, i.e.,

$$\left.\varphi_1\right|_{z=z_0} = \left.\varphi_2\right|_{z=z_0} , \qquad (8.7)$$

and, because mass is assumed to be the same in both regions, that of the free electron,

$$\left.\frac{\partial\varphi_1}{\partial z}\right|_{z=z_0} = \left.\frac{\partial\varphi_2}{\partial z}\right|_{z=z_0} . \qquad (8.8)$$

These coefficients can be characterized by a matrix $\mathcal{R}$, a transmission-reflection matrix, given by

$$
\begin{aligned}
\mathcal{R} & \\
= & \begin{bmatrix} a_1 & b_1 \\ a_2 & b_2 \end{bmatrix} = \frac{1}{2k_1} \times \\
& \left\{ \begin{matrix} (k_1 + k_2)\exp\left[j\left(-k_1 + k_2\right)z_0\right] & (k_1 - k_2)\exp\left[j\left(-k_1 - k_2\right)z_0\right] \\ (k_1 - k_2)\exp\left[j\left(k_1 + k_2\right)z_0\right] & (k_1 + k_2)\exp\left[j\left(k_1 - k_2\right)z_0\right] \end{matrix} \right\} .
\end{aligned}
$$
$$(8.9)$$

We will use such matrices in more complicated problems, such as of multiple barriers, which will be encountered further in this chapter. This approach allows for a simple means to analyze problems where many such barriers may be cascaded by determining the new matrix as a multiplication of the matrix of individual barriers. In this respect, it is the equivalent of the *ABCD*-parameters used in the analysis of cascaded sections of networks. The reflection coefficient ϱ, for this particular problem of a single barrier, is

$$\varrho = \left|\frac{b_1}{a_1}\right|^2 = \left|\frac{k_1 - k_2}{k_1 + k_2}\right|^2 = \left[\frac{1 - (1 - q\phi_B/E_0)^{1/2}}{1 + (1 - q\phi_B/E_0)^{1/2}}\right]^2 . \qquad (8.10)$$

Let us now see the magnitude of this reflection coefficient in the metal base transistor with gold as the base and germanium as the collectors.[3]. For gold, $q\phi_B = 12$ eV, and for an electron capable of crossing the barrier, e.g., $E_0 - q\phi_B = 0.5$ eV, the reflection coefficient is still ≈ 0.4. This large reflection occurs because $q\phi_B$ is large for metals and the discontinuity results in a significant perturbation. Barriers with graded shape exhibit significantly less reflection. Consider, e.g., the barrier of shape[4] (also see Problem 2)

$$\phi_B(z) = \frac{\phi_1 \exp(z/\ell)}{1 + \exp(z/\ell)} + \frac{\phi_2 \exp(z/\ell)}{[1 + \exp(z/\ell)]^2}. \tag{8.16}$$

This is a graded barrier as shown in Figure 8.4. The reflection coefficient for this barrier can be solved exactly as

$$\varrho = \left| \frac{\Gamma\left(0.5 + j(\delta - k_1 - k_2)\ell\right) \Gamma\left(0.5 + j(-\delta - k_1 - k_2)\ell\right)}{\Gamma\left(0.5 + j(\delta + k_1 - k_2)\ell\right) \Gamma\left(0.5 + j(-\delta + k_1 - k_2)\ell\right)} \right|^2, \tag{8.17}$$

---

[3]The preceding model is correct for free electrons. We are now applying it to a semiconductor where free electron framework is inadequate. This is therefore approximate and only suggestive of what to expect.

[4]This is the Eckart barrier, which can be solved exactly, C. Eckart, "The Penetration of a Potential Barrier by Electrons," *Phys. Rev.*, **35**, p. 1303, 1930. Another example, mentioned by S. Luryi in F. Capasso and G. Margaritondo, Eds., *Heterojunction Band Discontinuities: Physics and Device Applications*, North-Holland, Amsterdam, (1987), is the barrier

$$\phi_B(z) = \phi_0 \left[1 + \exp\left(-\frac{z}{\ell}\right)\right]^{-1}. \tag{8.11}$$

This is an exponentially graded barrier, for which the reflection coefficient is

$$\varrho = \frac{\sinh^2\left[\pi\ell(k_1 - k_2)\right]}{\sinh^2\left[\pi\ell(k_1 + k_2)\right]}, \tag{8.12}$$

with

$$k_1 = \left(\frac{2mE_0}{\hbar^2}\right)^{1/2}, \tag{8.13}$$

and

$$k_2 = \left[\frac{2m(E_0 - q\phi_0)}{\hbar^2}\right]^{1/2}. \tag{8.14}$$

The subject of transmission has been treated rigorously in L. D. Landau and E. M. Lifshitz, *Quantum Mechanics: Non-Relativistic Theory*, Pergamon Press, Oxford (1977). This text treats the exponential barrier problem on p. 80. The reflection coefficient in this problem of exponentially graded barrier vanishes when

$$E_0 - q\phi_0 \gg \frac{\hbar^2}{2m\ell^2}. \tag{8.15}$$

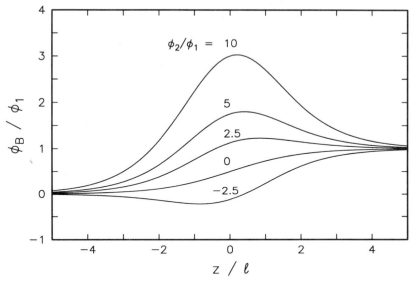

Figure 8.4: The Eckart barrier for various values of the parameter ratio ϕ_2/ϕ_1 as a function of z/ℓ.

where

$$k_1 = \left(\frac{2mE_0}{\hbar^2}\right)^{1/2},$$

$$k_2 = \left[\frac{2m\left(E_0 - q\phi_1\right)}{\hbar^2}\right]^{1/2},$$

$$\text{and} \quad \delta = \left[\frac{2m\left(q\phi_2 - \hbar^2/8m\ell^2\right)}{\hbar^2}\right]^{1/2}. \tag{8.18}$$

The reflection coefficient can be simplified to the form

$$\varrho = \frac{\cosh\left(2\pi(k_1 - k_2)\ell\right) + \cosh\left(2\pi\delta\ell\right)}{\cosh\left(2\pi(k_1 + k_2)\ell\right) + \cosh\left(2\pi\delta\ell\right)}. \tag{8.19}$$

This equation reduces to the previous analysis of an abrupt barrier when the grading length $\ell \to 0$. When the grading length is such that $k_1\ell, k_2\ell \gg 1$, i.e., the grading is over a distance scale that is larger than the atomic length scale, than the reflection coefficient tends to unity. Thus, the reflection coefficient is a function of grading length.

For the $CoSi_2/Si$ semiconductor–metal junction, the quantum reflection problem disappears if a grading distance of larger than about 5 Å is used. Formation of silicides naturally leads to occurrence of such grading distances.

This quantum-mechanical reflection problem is negligible or absent in the structures based on heterostructures formed using semiconductors. As an example, consider a $Ga_{.7}Al_{.3}As/GaAs$ structure, $q\phi_B$ is 0.25 eV, and for an electron with energy E_0 exceeding this by 0.2 eV, the reflection coefficient is less than 0.01, significantly smaller than the metal–semiconductor case. Reflection coefficients are thus intimately tied with Fermi energies. Metals have a very high Fermi energy (the conduction band is occupied); semiconductors have a low Fermi energy. The semiconductors are hence a very natural substitute in hot carrier structures. The disadvantage is that the large conductivity of metals is now lost, and if we try to achieve the large conductivity by doping heavily, we increase the scattering rate, and hence the likelihood of the hot carrier losing its energy and momentum in a scattering event during its transit in the base.

8.3   Hot Carrier Structures

We now consider hot carrier device structures that involve transport in a semiconductor base region.[5] Some examples of these structures are described in Figure 8.5. The first structure in Figure 8.5 was demonstrated initially, with useful current gains, using silicon for both the base and the collector of the device. It uses a clever but simple technique to form the barrier that isolates the collector for cold carriers, and still allows collection of hot carriers from the semiconductor base.

---

[5]For a historical and analytical perspective, see J. M. Shannon, "Hot Electron Camel Transistor," *IEE J. Solid State Electron Devices*, **3**, p. 142, 1979; J. M. Shannon and A. Gill, "High Current Gain in Monolithic Hot-Electron Transistors," *Electronics Letters*, **17**, No. 17, p. 620, 20 Aug. 1981; M. Heiblum, "Tunneling Hot Electron Transfer Amplifiers (THETA): Amplifiers Operating Up to the Infrared," *Solid-State Electronics*, **24**, p. 343, 1981; M. A. Hollis, S. Palmateer, L. F. Eastman, N. V. Dandekar, and P. M. Smith, "Importance of Electron Scattering with Coupled Plasmon–Optical Phonon Modes in GaAs Planar Doped Barrier Transistors," *IEEE Electron Device Letters*, **EDL-4**, p. 440, 1983; J. R. Hayes, A. F. J. Levi, and W. Wiegmann, "Hot Electron Spectroscopy," *Electronics Letters*, **20**, No. 21, p. 851, 11 Oct. 1984; S. Muto, K. Imamura, N. Yokoyama, S. Hiyamizu, and H. Nishi, "Sub-Picosecond Base Transit Time Observed in a Hot Electron Transistor (HET)," *Electronics Letters*, **21**, No. 13, p. 555, 20 June 1985; and M. Heiblum, M. I. Nathan, D. C. Thomas, and C. M. Knoedler, "Direct Observation of Ballistic Transport in GaAs," *Phys. Rev. Lett.*, **55**, No. 20, p. 2200, 11 Nov. 1985. The discussion in this section is based on concepts introduced and analyzed in these publications.

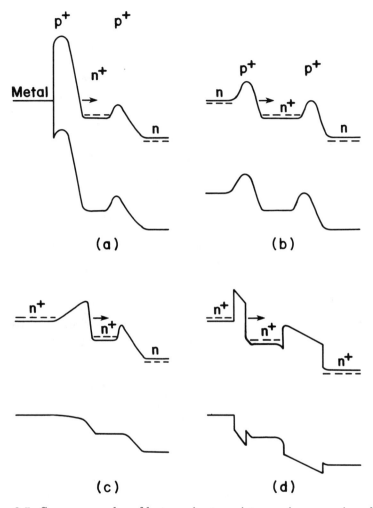

Figure 8.5: Some examples of hot carrier transistors using a semiconductor base. Examples of both thermionic and tunneling based injections are included. (a) employs a p$^+$-doped region for isolating the metal emitter and the semiconductor collector. (b) is a variation of this structure employing semiconductor as the emitter. (c) employs a graded heterostructure to obtain the isolation between the emitter and the collector with the base. (d) uses abrupt heterostructure for this isolation. The dominance of tunneling in these structures is dependent on temperature of operation and the characteristics of the barriers.

A limited sheet density of p$^+$ doping, which stays depleted at any bias at the junction, is employed in an n-type structure. For typical current densities of mid-10^4 A.cm^{-2}, this p$^+$ sheet doping is typically less than 0.5×10^{12} cm^{-2}. Since the region is always kept depleted, there is little hole injection, and hence negligible hole storage in the structure, even though both polarities of doping are utilized. This technique of forming a barrier of controllable small or large barrier energy (see Problem 24 of Chapter 4) has been found to be useful when interfaces with other materials are not desired during formation of a barrier. The technique has, therefore, found applicability in several other places to create barriers while still minimizing storage effects related to minority carriers. Examples of this are similar structures in compound semiconductors for achieving a similar hot carrier device, and in low barrier rectifying junctions for microwave detection. Due to the nature of the hump, it has been called a camel barrier, and due to the nature of doping in achieving it, it has also been called planar doped barrier. In the hot carrier transistor, the injection of carriers from the emitter occurs via both thermionic field emission and tunneling, with the latter being more important at lower temperatures. If the n$^+$ doped silicon region for the base is thin enough, i.e., less than or of the order of the mean free path, the hot carriers can be collected over the barrier. The relatively large current gains of this structure show evidence of limited scattering in this structure, and limited loss of energy and momentum during transit through the the base region.

An extension of this structure is the use of a semiconductor emitter formed in a similar way. Its advantage is that field emission and tunneling which inject carriers at lower energy into the base, are suppressed. The fraction of carriers that are hot is thus larger, and the device shows a correspondingly larger current gain. Similar structures have been implemented in compound semiconductors also, demonstrating both finite current gains and verifying the nature of transport with limited scattering in the base. All these structures, based on controlled doping of a single semiconductor, require a relatively reproducible technology of fabrication, since they involve precise control of doping charge. Base dimensions are critical between the barrier doping spikes; diffusion of these species has to be controlled, and difficult fabrication procedures have to be utilized in contacting the the injecting, controlling, and collecting regions.

The structures that have found increasing acceptance, therefore, are based on heterojunctions that can be processed with better control; examples of these are (c) and (d) of Figure 8.5. In these, the larger bandgap semi-insulating region is used to isolate, and to act as the medium for injecting hot carriers. In such structures, the potential drop at the base–emitter junction largely appears as the excess energy of the carrier being injected. By a suitable design of the heterostructure barrier, either of uniform or graded alloy composition, thermionic field emission or low energy tunneling can be made important (see Problem 3). By suitable choice of alloy composition in the collector region, an appropriate barrier can be formed. In all these cases, the barrier energy can be decoupled from the barrier width. Both barrier energy, which is dependent on alloy composition, and barrier width can be controlled accurately, giving greater freedom in device design. Such structures have shown relatively large current gains commensurate with the relatively large mean free paths at low dopings in the base region. They have all clearly verified the high velocity in the base, occurring due to a significant fraction of ballistic transport, and due to a significant maintenance of forward momentum even with scattering. In the other regions of the device (e.g., in the base–collector space charge region), which are relatively wide, collisions continue to be significant factors, and these structures exhibit significantly higher base resistances compared to both metal base structures and bipolar transistor structures because of the poorer sheet conductance at low dopings and small base widths.

Figure 8.6 shows representative output characteristics for the common-base mode of operation, together with the band edges at two bias points on these characteristics. A higher forward bias at the injecting junction results in a higher energy of the injected carrier stream. Even though the corresponding scattering rate is larger, recall that the scattering rate almost always increases with energy; the likelihood of a carrier crossing the base region with a forward momentum that takes it into the collector is large. During this transit a number of different scattering mechanisms are possible: alloy scattering in mixed crystals such as $Ga_{1-x}In_xAs$, ionized impurity scattering, carrier–carrier scattering, coupled carrier–phonon scattering (e.g., plasmon scattering), and phonon scattering are all quite likely, with phonon and plasmon scattering modes predominating. At low temperatures the relative intensity of phonon effects can be decreased, because optical phonon absorption can be suppressed due to freeze-out of optical phonons. Scattering rates can be maintained smaller by preventing inter-valley processes, so materials such as $Ga_{1-x}In_xAs$ allow for a larger bias range of operation with the high gain, albeit with increased alloy scattering as an additional scattering process. Both alloy and optical scattering have characteristics which mitigate their effects on the forward momentum of the carrier. Alloy scattering is less efficient at high energy than at low energy. Ion-core perturbation effects are akin to Rutherford scattering in this respect. Optical phonon scattering, dominated at low temperatures by emission processes, emphasizes the forward mode of scattering. So, even though there is a finite loss of energy, usually a fraction of the incident energy, the forward momentum is maintained to a significant extent.

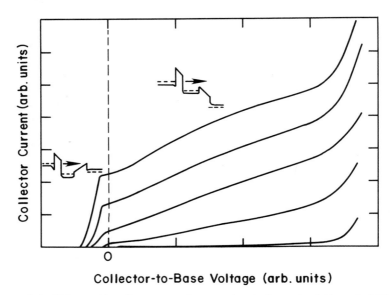

Figure 8.6: Schematic of common-base output characteristics at 4 K of a hot carrier transistor employing a uniform alloy mole-fraction emitter and collector barrier. The insets show the band edges in the saturation and the low output conductance regions of the device.

Consequently, even in the presence of scattering, at low temperatures, sufficiently high velocities can be maintained, and only plasmon scattering causes a major reduction in the current gain of device structures of less than 200 Å base widths.

We now consider some of the methods by which such velocities can be determined. Application of a magnetic field is an additional variable that can be introduced to extract the velocity information. One example of this is the use of Hall measurement, where a low magnetic field is applied perpendicular to the plane of transport. Use of higher magnetic fields, with the direction of the magnetic field perpendicular to that of the ensemble velocity of the carriers, would cause the carriers to also traverse in a direction orthogonal to their original direction, causing longer paths and hence higher probability of scattering. We will refer to this as the transverse direction of magnetic field. A magnetic field longitudinal with the velocity causes no change in the direction of the velocity, and hence leaves the current gain of device unchanged. So, a variable transverse magnetic field allows us to obtain the variation in path lengths through the variation in current gain,

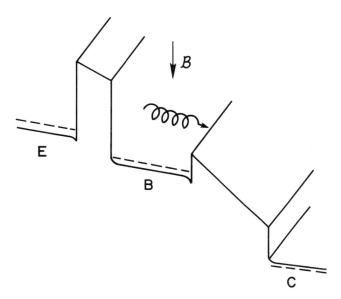

Figure 8.7: Carrier motion in the presence of an orthogonal magnetic field in the hot carrier transistor structure.

which can be used to determine the velocities.

Figure 8.7 shows the effect of the orthogonal magnetic field on the path of the electron. In the quasi-neutral base, we may treat the electric field as being negligible. For a magnetic field $\boldsymbol{\mathcal{B}}$, the wave vector $\boldsymbol{k}$ (or equivalently the momentum), the time rate of change of the wave vectors, follows from Lorentz equation as

$$\frac{d\boldsymbol{k}}{dt} = -q\boldsymbol{k} \times \frac{\boldsymbol{\mathcal{B}}}{m^*}. \tag{8.20}$$

The averaged wave vector $\overline{k}_z$ changes in the presence of the orthogonal field, and with it changes the base time constant τ_B, which may be defined in a way similar to that of the bipolar transistor. A large magnetic field can cause in the carrier a large enough change in the direction of its velocity that it may never even reach the collector in the absence of scattering. In general, the carrier follows the path of a cyclotron orbit. Clearly, at very high magnetic fields, the current gain of the device will be negligible. Thus, the common-base current gain becomes a parameterized function of the orthogonal magnetic field from which the average velocity of the carriers can be extracted. The basis for this extraction is that the rate of increase of the path length is linear in the magnetic field when scattering lengths

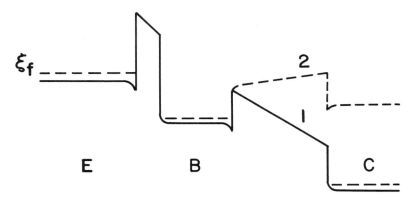

Figure 8.8: Modulation of the barrier at the base–collector junction in order to select the carrier momentum spread that will be collected in a hot carrier transistor. The bias condition, marked as 2, cuts off the low energy part of the injected distribution that would otherwise be collected in the bias condition marked as 1.

are larger compared to the base width. Hence, the common-base current gain varies inversely with the magnitude of the orthogonal magnetic field.

This method used a magnetic field to obtain a selection of carriers that reach the base–collector junction. An alternate method is to use the electrostatic effect at the base–collector junction to achieve the same objective. Thus, by modulating the position of the barrier at the base–collector junction, as shown in Figure 8.8, a spread in energy of the carriers at the collector can also be obtained. This method makes these devices excellent tools for analysis of carrier energy. The carrier density and the carrier velocity, at the collector, are reflected in the carrier current (the collector current I_C). Thus, the output conductance in common-base ($\partial I_C/\partial V_{BC}$) reflects the electron distribution function as a function of energy at the base–collector junction. Knowing the injecting bias and the mechanism, these can be related to the velocity of carriers in the base (see Problem 4). Tunneling-dominated transport at the injecting junction, e.g., causes injection of a narrow spread in the energy of electrons, which is shifted by the applied base–emitter bias. A peak occurring in $\partial I_C/\partial V_{BC}$ at this position indicates carriers that did not lose any energy and momentum, i.e., those that travelled without any scattering. Figure 8.9 shows a schematic plot of the output conductance for our earlier example. The narrowness of the peaks emphasizes the tunneling character of the injection, and the area un-

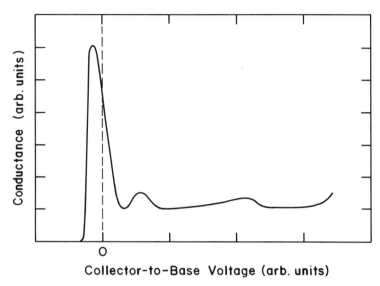

Figure 8.9: A schematic of output conductance, at 77 K for the hot carrier transistor example, shown as a function of collector–base bias.

der the peaks shows the relative ratio of carriers. The largest peak is due to carriers that did not suffer any scattering process; the gradual decays show the effect of low energy and low momentum loss scatterings from this peak.

The conductance peak corresponding to no scattering shifts towards the left with the higher current because injected energy is higher and a larger forward bias is required at the base–collector junction to screen the injected carriers. This occurs with relatively little change in the width of the peak, because of tunneling-dominated injection. The relative area under the peak does decrease with very high bias, because of increased inter-valley scattering effects.

The base current in these structures comes about from thermal and other leakage currents, and due to those carriers that lose sufficient energy in the base transit to be trapped in it and be collected by the base ohmic contact. To minimize this, base scattering effects have to be minimized. This requires maintaining low carrier density in the base to minimize plasmon scattering effects. The rationale behind the practical importance of these structures is the high carrier velocities and reduced storage effects compared to the bipolar transistor. However, the base resistance of these

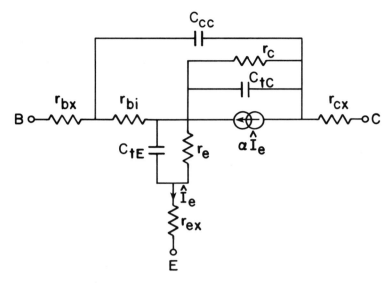

Figure 8.10: A small-signal equivalent circuit of the hot carrier transistor, cast in a form similar to that of the bipolar transistor.

structures has generally remained high because larger carrier densities result in larger plasmon scattering. In addition, larger current densities also require smaller barrier thicknesses, which, in turn, lead to an increase in transition capacitance. This is the dominant frequency-limiting mechanism for tunnel diodes. Corresponding to this, for hot carrier transistors, the effect of transition capacitances in the base–emitter and base–collector junctions begins to dominate. The emitter capacitance charges through the base resistance, and the collector capacitance through the collector current that results from the former. While conventional bipolar transistors have base Gummel numbers approaching 5×10^{13} cm^{-2} and HBTs have base Gummel numbers exceeding 5×10^{14} cm^{-2}, the hot electron transistors have generally been limited in Gummel number to $\approx 3 \times 10^{13}$ cm^{-2}. Thus, the base resistance of the hot carrier transistor can be nearly an order of magnitude larger than that of the bipolar transistor and the capacitances a few multiples of it.

The operation of the hot carrier transistor, as a natural evolution of the barrier-controlled operation of the bipolar transistor, also should be representable by similar equivalent circuits. An example of such an equivalent circuit is shown in Figure 8.10. The major difference between this and the bipolar transistor is the elimination of the effects related to the storage of minority carriers in the base. This is reflected in the elimination of the diffusion capacitance associated with the base–emitter and the base–collector junctions. We should also consider the effect of the differences in transport and storage on the output characteristics of the device. Hot carrier transistors generally exhibit a large output conductance, because the collected carriers are modulated by the collecting barrier. This shows up clearly in Figure 8.6. Voltage drops across large base resistances also lead to large shifts in the output characteristics in the saturation region with applied base–emitter bias, or base current, because of a negative feedback.

8.4 Resonant and Sequential Tunneling

We now extend our discussion of tunneling to multiple barriers, which may or may not couple with each other for carrier transit. The former, in which phase coherence of the tunneling carrier is maintained, is referred to as resonant tunneling, while the latter, in which phase coherence is lost through the multitude of scattering processes, is known as sequential tunneling. Resonant tunneling is the analogue of Fabry–Pérot interference encountered in optics; it requires certain barrier heights, barrier widths, well widths, and their matching, and lack of scattering. It requires building up of carrier densities associated with the resonance phenomenon. It therefore is also associated with larger time scales. This causes, whenever these time scales exceed a few pico-seconds, sequential tunneling to become the more important tunneling process, because scattering rates are usually of the order of 10^{13} s^{-1}, and because these inelastic tunneling processes in coupled barriers do not require specific matching conditions. We first develop a mathematical formalism to treat the general case of elastic tunneling in multiple barriers, i.e., the coherent case, and subsequently we will consider the effect of inelastic processes on it. An example of a coupled barrier structure is shown in Figure 8.11.

In our discussion of elastic tunneling in barriers, we developed a mechanism for studying the quantum-mechanical solution to the more general problem using the the transmission–reflection matrix $\mathcal{R}$.[6] A more complicated problem of multiple barriers can be characterized in terms of multiple transmission–reflection matrices. For example, in the problem of Figure 8.11, the terms of the matrix $\mathcal{R}$ can be used to write

$$a_1 = (\mathcal{R}_1 \mathcal{R}_2 \mathcal{R}_3 \mathcal{R}_4)_{11} a_5. \qquad (8.21)$$

---

[6]For a detailed and lucid discussion of this approach, see E. O. Kane, "Basic Concepts in Tunneling," in E. Burstein and S. Lundqvist, Ed., *Tunneling Phenomena in Solids*, Plenum, N.Y. (1969).

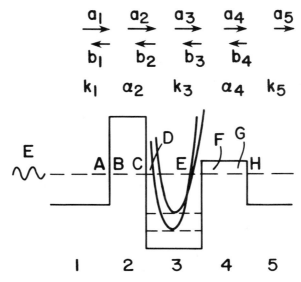

Figure 8.11: Two heterostructure barrier regions, in close proximity, with the spatial extents comparable to the de Broglie wavelength. The consequence of the close proximity is localization of states, i.e., the formation of subbands in the well formed by the adjoining barriers, and transmissions and reflections that emphasize the wave nature of carriers. The incident and reflected waves in this multiple-barrier structure are shown to be characterized by incident ($\cdots a_j \cdots$) and reflected waves ($\cdots b_j \cdots$).

In the barrier region, for particle energy smaller than the barrier energy, the wave is an evanescent wave, with a momentum $k = j\alpha$, where α is real, and where the k's are still related as

$$k = \left[\frac{2m^* \left(E - q\phi_B\right)}{\hbar^2}\right]^{1/2}, \tag{8.22}$$

E being the energy of the particle.

We consider the analysis in only one dimension, in the z direction. The probability current (flux), incident and transmitted, through this structure of multiple barriers can be written as

$$j_{inc.} = |a_1|^2 \frac{\hbar k_1}{m_1^*}, \tag{8.23}$$

and

$$j_{trans.} = |a_5|^2 \frac{\hbar k_5}{m_5^*}, \tag{8.24}$$

and similar expressions in other regions of the structure.

We will consider another simplification of equal mass in all the barrier regions, in order to simplify the mathematics. So, for $m_1^* = \cdots = m_5^* = m^*$, using the above relationships, we can write the ratio of the transmitted probability current and the incident probability current as

$$\frac{j_{trans.}}{j_{inc.}} = \frac{2^8 k_1 \alpha_2^2 k_3^2 \alpha_4^2 k_5}{\mathcal{M}\left(k_1^2 + \alpha_2^2\right)\left(\alpha_2^2 + k_3^2\right)\left(k_3^2 + \alpha_4^2\right)\left(\alpha_4^2 + k_5^2\right)}, \tag{8.25}$$

where

$$
\begin{aligned}
\mathcal{M} =\ & \exp\left(\alpha_2 w_2 + \alpha_4 w_4\right)\left\{\exp\left[j\left(-\phi_1 + \phi_2 + \phi_3 + \phi_4 + \phi_5\right)\right] - \right.\\
& \left.\exp\left[j\left(\phi_1 + \phi_2 - \phi_3 - \phi_4 + \phi_5\right)\right]\right\} + \\
& \exp\left(\alpha_2 w_2 - \alpha_4 w_4\right)\left\{-\exp\left[j\left(-\phi_1 + \phi_2 + \phi_3 - \phi_4 - \phi_5\right)\right] + \right.\\
& \left.\exp\left[j\left(\phi_1 + \phi_2 - \phi_3 + \phi_4 - \phi_5\right)\right]\right\} + \\
& \exp\left(-\alpha_2 w_2 + \alpha_4 w_4\right)\left\{-\exp\left[j\left(-\phi_1 - \phi_2 - \phi_3 + \phi_4 + \phi_5\right)\right] + \right.\\
& \left.\exp\left[j\left(\phi_1 - \phi_2 + \phi_3 - \phi_4 + \phi_5\right)\right]\right\} + \\
& \exp\left[-\alpha_2 w_2 - \alpha_4 w_4\right]\left\{\exp\left[j\left(\phi_1 - \phi_2 - \phi_3 - \phi_4 - \phi_5\right)\right] + \right.\\
& \left.\exp\left[j\left(-\phi_1 - \phi_2 + \phi_3 + \phi_4 - \phi_5\right)\right]\right\},
\end{aligned}
\tag{8.26}
$$

$$
\begin{aligned}
\phi_1 &= k_3 w_3, \\
\phi_2 &= \arctan\left(\frac{\alpha_2}{k_1}\right), \\
\phi_3 &= \arctan\left(\frac{\alpha_2}{k_3}\right), \\
\phi_4 &= \arctan\left(\frac{\alpha_4}{k_3}\right), \\
\text{and} \quad \phi_5 &= \arctan\left(\frac{\alpha_4}{k_5}\right).
\end{aligned}
\tag{8.27}
$$

We are interested in the condition when maximum current is transmitted. This is also the condition for resonance and occurs when the factor $\mathcal{M}$ is minimized. This requires that the first term in the above expression vanish; this is the only term with an exponential exponent in the evanescent regions that is the sum of two positive quantities. This occurs at

$$\exp\left[j\left(-\phi_1 + \phi_2 + \phi_3 + \phi_4 + \phi_5\right)\right] = \exp\left[j\left(\phi_1 + \phi_2 - \phi_3 - \phi_4 + \phi_5\right)\right],$$
(8.28)

which occurs when

$$-\phi_1 + \phi_2 + \phi_3 + \phi_4 + \phi_5 = \phi_1 + \phi_2 - \phi_3 - \phi_4 + \phi_5 - n2\pi, \qquad (8.29)$$

where n is an integer. This condition implies that a maximum in transmission occurs when

$$\phi_1 = \phi_3 + \phi_4 + n\pi, \qquad (8.30)$$

i.e., when

$$k_3 w_3 = \arctan\left(\frac{\alpha_2}{k_3}\right) + \arctan\left(\frac{\alpha_4}{k_3}\right) + n\pi. \qquad (8.31)$$

If, additionally, $\alpha_2 w_2 = \alpha_4 w_4$, i.e., the penetration coefficients of the two barriers are identical, then $\mathcal{M} \approx 1$ (slightly smaller for typical barriers), and transmission becomes independent of the widths w_2 and w_4. This condition, the resonant condition, allows near unity transmission coefficient, with attenuation occurring only due to reflections resulting from mismatches. This large transmission occurs even if the barrier widths are large—the only requirements of the above analysis are that elastic processes dominate, a one-body analysis be valid, and the above two equalities are satisfied.

Figure 8.12 shows the transmission coefficient for elastic tunneling for an example of symmetric barriers, using the above relationship. The peak, which corresponds to the above two conditions, results in a unity transmission coefficient. Note that this transmission occurs even though the energy of the electron is such that it sets up evanescent waves in the barrier regions. We will discuss the significance of this in greater detail later. Here, we point out our earlier analogy with the Fabry–Pérot interference of optics. The interference caused by the barriers in the waves is constructive, large barrier widths notwithstanding. There occurs a buildup of the probability of finding the particle in the well formed by the barrier via elastic tunneling processes. There is also a near unity probability, once this build-up has occurred (the situation we analyzed), that the carriers will transmit at the same rate as they appear in the well (a unity transmission), even though the time associated with the building up and the leaking out of this charge may be large. The large time may result from large barrier heights or large barrier widths, both of which decrease the evanescent carrier momentum.

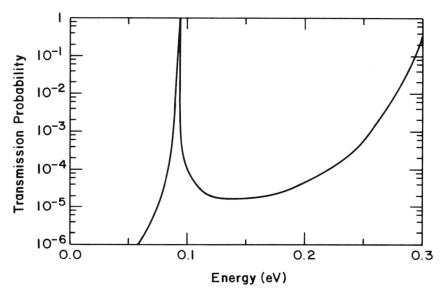

Figure 8.12: The transmission probability for the coupled barrier example as a function of electron energy; the barriers are assumed to be 25 Å in width, and 0.3 eV in height.

This efficient transmission occurring in the resonant condition is actually the basis of the Kronig–Penney model and the nearly free electron model[7] for periodic potentials such as in a semiconductor crystal. The series of identical barriers due to the lattice result in allowed energy bands where perfect transmission takes place. This is a resonant condition. In this example of double barriers, the condition of incident energy equal to quasi-bound state energy of the well characterizes the resonance condition where unity transmission occurs. Real problems are, of course, more complicated. We considered this as a one-body problem; real problems are many-body problems where the charge build-up in the well itself also causes a distortion of the well. In real problems, the observation of such resonance conditions, unlike the Fabry–Pérot case, requires biasing of the structure, which destroys the symmetry of the barrier structure. This symmetry is central to the resonance, and asymmetry causes reduction of the transmission coefficient. Real problems also involve inelastic scattering processes. We will include some of these effects later. For now, we consider simpler extensions of this analysis to understand resonant tunneling structures within the one-body

---

[7]Hence, strictly speaking, for the top and bottom of the bands.

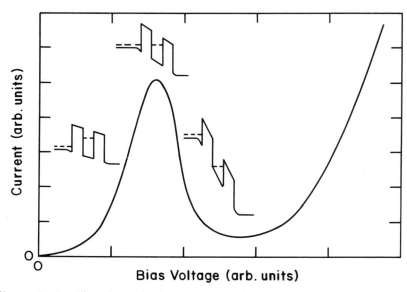

Figure 8.13: The three distinct regions of current–voltage behavior in a coupled barrier structure. The inset shows the band diagrams in these three regions; the first has the Fermi level below the bound state energy, the second has the Fermi level at the same energy as the bound state energy, and the third has the Fermi level above the bound state energy.

elastic approximation. We are interested in these for thin barriers with small heights, structures that can quite easily be implemented in semiconductors. An example is the use of thin barriers of undoped $Ga_{1-x}Al_xAs$ at a small aluminum arsenide mole-fraction with undoped GaAs well.

The current–voltage characteristics through such coupled barriers exhibit negative differential resistance. Figure 8.13 shows a schematic energy band diagram together with the current–voltage behavior of the structure. The peak currents through the structure occur when the electrons in the emitter can tunnel through its quasi-bound state in the well. When the bias is below this, the absence of states at comparable energy prevents tunneling through the structure. Above this bias, the current decreases because of the conservation of momentum parallel to the interface. The peak of the current depends on the barrier characteristics. An increase in the well width decreases the current, but may show tunneling through more than one quasi-bound state and hence multiple peaks. Examples of such are shown in Figure 8.14.

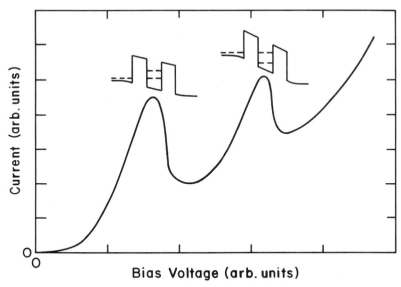

Figure 8.14: Current due to tunneling through a coupled structure with two bound states. The current now shows two corresponding peaks.

The complicated expression of Equation 8.25 was written with individual discontinuities as the basis. A barrier is formed with two such discontinuities and the transmission through this is the transmission coefficient of this barrier, Equation 8.25, e.g., being the transmission coefficient for the barriers of Figure 8.11. The expression of Equation 8.25 can then be recast in a more simple form. For the double barrier system, the transmission coefficient of the entire structure, T_{tot}, can be expressed in terms of that of the left barrier, T_l, and the right barrier, T_r, as (see Problem 5)

$$T_{tot} = \frac{C_0 T_l T_r}{C_1 (T_l T_r)^2 + C_2 T_l^2 + C_3 T_r^2 + C_4}. \tag{8.32}$$

Here, C_0, C_1, C_2, C_3, and C_4 are coefficients related to the phase factors of Equation 8.27. These vary slowly with energy. At resonance, $C_4 \to 0$, and either the C_2 or C_3 term dominates. The transmission at resonance T_{res} is then

$$T_{res} \approx \frac{C_0 T_r}{C_2 T_l}, \tag{8.33}$$

or

$$T_{res} \approx \frac{C_0 T_l}{C_3 T_r}. \tag{8.34}$$

The coefficients are usually close to unity, for transmissive barriers, and hence one may approximate, for barriers with sufficiently different transmissivity, that, at the resonance condition,

$$T_{res} \approx \frac{T_{min}}{T_{max}}, \tag{8.35}$$

where T_{min} and T_{max} are the minimum and maximum transmissivities. Off resonance, the C_4 term dominates and $T_l, T_r < 1$. The off-resonance transmission coefficient, $T_{off-res}$, then, can be approximated as

$$T_{off-res} \approx \frac{C_0}{C_4} T_l T_r \approx T_l T_r. \tag{8.36}$$

These relations show that in resonance and off resonance, the phase factors cause a small effect in the transmission coefficient of the double barrier. The behavior of the well, particularly important in off-resonance behavior, appears through the factor C_0/C_4, which is related to the phase delay.

We now consider Figure 8.15, where asymmetry has been introduced to the symmetric barrier structure, and we wish to understand the resonance behavior in these asymmetric conditions. Within the elastic approximation, the changes in the transmission coefficients T_l and T_r through these barriers with an electric field across them, and whose barrier potential varies as a function of position, are numerically complicated. We will summarize numerical solutions for the purposes of discussion.[8] Figure 8.15 establishes some of the relevant parameters of the problem: the electric fields, barrier heights, and barrier widths. We have the functional dependence of the barrier potentials as

$$\begin{aligned}
q\phi_1 &= q\phi_0 - \Delta E_1 = q\phi_0 - q\mathcal{E}_b d, \\
q\phi_2 &= q\phi_0 - \Delta E_1 - \Delta E_2 = q\phi_0 - q\left(\mathcal{E}_b d + \mathcal{E}_w w\right),
\end{aligned} \tag{8.37}$$

and

$$q\phi_3 = q\phi_0 - \Delta E_1 - \Delta E_2 - \Delta E_3 = q\phi_0 - q\left(2\mathcal{E}_b d + \mathcal{E}_w w\right), \tag{8.38}$$

where

$$\begin{aligned}
\mathcal{E}_b &= \frac{V}{2d + w\epsilon_b/\epsilon_w}, \\
\mathcal{E}_w &= \frac{\epsilon_b \mathcal{E}_b}{\epsilon_w}.
\end{aligned} \tag{8.39}$$

---

[8]WKB approximation should be used with care in such instances. It is not directly applicable in barrier regions with piece-wise discontinuity. Discontinuities give rise to a pre-factor to the exponential term of the WKB approximation.

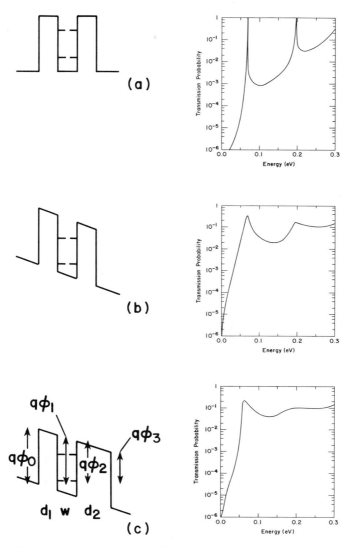

Figure 8.15: Transmission probability for a resonant tunneling under symmetric and asymmetric barrier conditions. Part (a) shows the structure and transmission probability for a 30 Å barrier of Ga$_{.7}$Al$_{.3}$As with a 35 Å well of GaAs. Part (b) shows the effect on transmission probability under bias. Part (c) shows the effect on transmission probability when the width of the second barrier in part (b) is increased to 50 Å. A schematic of the conduction band structure is shown on the left.

The simplified relationships of Equation 8.32 remain relevant under the approximations here, because the phase factors are slowly varying. Examples of exact calculations to determine the total transmission coefficient at resonance and off resonance are shown in part (b) of Figure 8.15. Part (a) of this figure shows the original transmission coefficient curve of the symmetric structure. Note the decrease in transmission probability with the application of an electric field and the smearing out of resonance features. The sharp peaks in the absence of an electric field are transformed into broad transmission maxima. The barrier widths also cause a stronger effect because the asymmetry can be accentuated by them. An example of this is shown in part (c) of Figure 8.15. This figure shows the change as a result of the increase in the width of the right barrier. This reduces the ideal transmission coefficient of the earlier example, shown in part (a) and causes additional off resonance effects beyond those of part (b).

We have considered asymmetry arising from the application of bias, within the one-body approximation. Flow of current, occurrence of resonance, and nuances of the design of the structure to compensate for technological effects in the making of such structures (e.g., dopant diffusion from the injecting and collecting regions), also give rise to asymmetries and limitations of this analysis. Current injection and resonance cause distortion in the shape of the well and the barriers, and it is practically quite difficult to maintain symmetry at the condition of resonance. This, therefore, limits the amount of transmission probability in such structures. Reduction in this transmission probability appears as a reduction in the peak current of the tunnel diode–like characteristics in the practical implementations of these structures.

In writing all these equations, we have continued to ignore scattering. The equations have been derived in one dimension by using conservation of energy and momentum parallel to the interface. The wave function and its derivative perpendicular to the interface were matched. The former follows from the continuity of finding a particle as a function of position in a finite energy system, and the latter matches the momentum perpendicular to the interface, at the interface, since the masses were assumed the same. In our discussion of Richardson's constant at a heterojunction discontinuity, we had not made this assumption, and as a result shown the proper form of it, independent of the masses. We can also implement differing masses here. Its consequence is more complex equations (see Problem 7) but the underlying principles remain the same. The momentum perpendicular to the interface varies—evanescent in the barrier region, and quantized in the well. This quantization is associated with the quasi-bound state referred to earlier. Conservation of energy and parallel momentum occurs if no scatter-

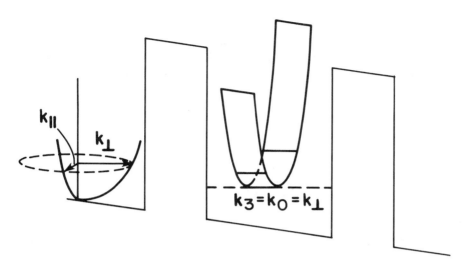

Figure 8.16: Tunneling in a coupled barrier system, from three-dimensional states in the injecting region to two-dimensional states in the well region. The figure is a schematic representation combining representation in real and reciprocal space.

ing processes are assumed. However, increasing times related to resonance bring in a higher probability of the occurrence of scattering in the well region. We have also referred to scattering occurring as part of the tunneling process itself, a process that allows indirect transitions, such as in indirect bandgap materials as well as in direct bandgap materials through indirect bands. This scattering leads to a change in momentum and, for inelastic scattering, a change in energy. This brings up the interesting problem of how scattering affects the behavior of transmission, resonance, the negative resistance characteristics, and the time constants of the system.[9].

Consider the tunneling in a coupled barrier system, together with occupation considerations of three-dimensional and two-dimensional states in the injector and the well, as shown in Figure 8.16. We still consider the tunneling to occur via elastic processes in the barrier regions, the barrier

---

[9]See A. D. Stone and P. A. Lee, "Effect of Inelastic Processes on Resonant Tunneling in One Dimension," *Phys. Rev. Lett.*, **54**, No. 11, p. 1196, 18 Mar. 1985; T. Weil and B. Vinter, "Equivalence between Resonant Tunneling and Sequential Tunneling in Double-Barrier Diodes," *Appl. Phys. Lett.*, **50**, No. 18, p. 1281, 4 May 1987; and M. Jonson and A. Grincwajg, "Effect of Inelastic Scattering Processes on Resonant and Sequential Tunneling in Double-Barrier Heterostructures," *Appl. Phys. Lett.*, **51**, No. 21, p. 1729, 23 Nov. 1987.

regions being quite thin. During the tunneling through the first barrier itself, the momentum parallel to the interface is conserved. In the well, the momentum perpendicular to the interface is quantized and restricted to a value k_0. Considering the occupation of carriers in the three-dimensional states of the injecting region, the momentum perpendicular to the interface is restricted to

$$k_z = k_0 = \left[\frac{2m\,(E_0 - q\phi_0)}{\hbar^2} \right]^{1/2}. \tag{8.40}$$

Since in the well, the perpendicular momentum is restricted to k_0, only those electrons with $k_z = k_0$ and

$$k_x^2 + k_y^2 + k_z^2 = k_F^2 \tag{8.41}$$

on the Fermi surface can tunnel through. This is to say that if momentum is conserved during tunneling, only those electrons lying on the above circle on the Fermi surface will tunnel. Consider now the occupation of states as a function of energy. E_0 is the minimum energy in the two-dimensional states in the well corresponding to the perpendicular momentum of k_0. For energy larger than E_0 in the injecting region, no carriers match this energy and the elastic tunneling current is reduced to zero. This leads to a decrease of current with bias and hence the tunnel diode–like characteristics. Before this occurs, the current increases due to a larger number of carriers on the circle available for the tunneling process. At any finite temperature, thermal effects will occur, causing a broadening of the transmission characteristics (see Problem 8). However, in this discussion, which showed negative differential current–voltage characteristics occurring, nowhere did we invoke the occurrence of resonance. The carrier can lose its coherence in the well; the decrease in transmission for energies higher than E_0 will still occur. The second barrier only served the purpose of restricting the perpendicular momentum. So, even in the presence of scattering, the structures exhibit similar current–voltage characteristics. In fact, they show identical current–voltage characteristics, as we will show presently.

The requirement for the decrease in current came about due to the restriction of perpendicular momentum. This can also occur through existence of a bandgap where no states exist and hence no momentum is allowed. An interesting example of this is the $Al_xGa_{1-x}Sb/InAs$ material system, as shown in Figure 8.17. Being dependent on the absence of states altogether, this is similar to the example of the tunnel diode, except that a similar principle is utilized with an isodoped structure.

The effect of thermal considerations on the behavior of these structures is through the occupation statistics, and scattering effects for tunneling

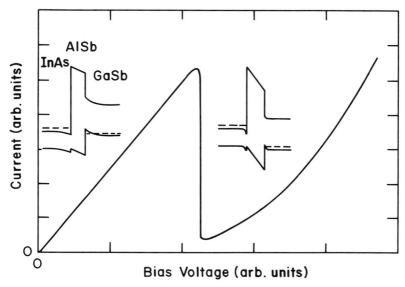

Figure 8.17: Schematic showing current–voltage characteristics and associated band bending under different bias conditions in a structure where the negative resistance arises due to existence of a bandgap between the spatially removed conduction and valence bands. The example utilizes a lattice matched InAs/AlSb/GaSb structure.

processes. Thermal effects also manifest themselves by increasing the likelihood of thermionic and thermionic field injection. An increasing voltage beyond the region of negative differential resistance shows a larger effect due to these injection processes because of the decreasing barrier to injection, and because the barriers are usually low in energy. Thus the resulting current–voltage characteristics show again an increase in current when a higher bias is applied beyond the region of the negative differential current–voltage characteristics.

We now consider other aspects of the resonant and the scattering-dominated behavior, their effects on the operation of the structure, in order to relate these to time constants of interest in high frequency structures. The current flow across a barrier can be written following equations that we applied in Chapter 4. The current flow in the direction of positive z, assuming a density of states of unity for mathematical simplicity, is given by:

$$J_{lr} = \frac{q}{4\pi^3} \int T(E_l)\frac{1}{\hbar}\frac{\partial E(k_l)}{\partial k_z} f_l(E)\left[1 - f_r(E + qV)\right] d^3k, \qquad (8.42)$$

where l and r identify the left and the right sides of the barrier, f is the Fermi–Dirac distribution function, and V is the applied bias of the structure shown in Figure 8.15. The particle velocity is explicitly included in the above expression. Similarly, the current flowing in the negative z direction is

$$J_{rl} = \frac{q}{4\pi^3} \int T(E_r)\frac{1}{\hbar}\frac{\partial E(k_r)}{\partial k_z} f_r(E)\left[1 - f_l(E + qV)\right] d^3k. \qquad (8.43)$$

The resulting net current in the structure is given by (see Problem 9)

$$J = \frac{qm^*}{2\pi^2\hbar^2} \int \left\{ \int T\left(E_\perp, E_\parallel\right)\left[f_l(E) - f_r(E + qV)\right] dE_\parallel \right\} dE_\perp. \qquad (8.44)$$

We called the state in the well a quasi-bound state since it is not stationary, there exists a finite time constant associated with it during which the carrier can leak out. The width in energy of the resonance, ΔE, is associated with this time constant through the uncertainty principle. When the temperature is high, $kT \gg \Delta E$, the transmission coefficient may be written as a delta function occurring at E_0, the resonance energy. The transmission probability under these conditions is given by

$$T(E_\perp) \approx T(E_0)\Delta E\delta\left(E_\perp - E_0\right), \qquad (8.45)$$

where $\delta\left(E_\perp - E_0\right)$ is the Dirac delta function located at energy E_0 and $T(E_0)$ is the parameter that can be determined from the characteristics of the barrier. At these temperatures the Fermi–Dirac distribution function is integrated in a straightforward form and the current is given as (see Problem 9)

$$J = \frac{qm^*}{2\pi^2\hbar^3}T(E_0)\Delta E kT \ln\left[1 + \exp\left(\frac{\xi_f - E}{kT}\right)\right]. \qquad (8.46)$$

At lower temperatures, with $kT \approx \Delta E$ or lower, where the Dirac delta approximation is inaccurate, we may consider a perturbative symmetric expansion in energy (broadening is symmetric in energy) of the transmission function, which still shows a sharp peak in energy at $E = E_0$, thus,

$$\frac{1}{T} = \frac{1}{T_{max}}\left[1 + \left(\frac{E - E_0}{\Delta E}\right)^2 + \cdots\right], \qquad (8.47)$$

and considering only the first term in energy, we obtain[10]

$$T(E) = \frac{T_{max}}{1 + [(E - E_0)/\Delta E]^2}.$$ (8.48)

This is the Lorentzian form $T = T_{max}\Lambda(E - E_0)$, where the Lorentzian function is

$$\Lambda(E - E_0) = \frac{1}{1 + [(E - E_0)/\Delta E]^2}.$$ (8.49)

Computation of the current now requires a more elaborate numerical integration, but is still straightforward. We remarked that this is assuming a unity density of states. The actual density of states can be included in a straightforward manner. For the case of $kT \gg \Delta E$, the transmission occurs in a small band of energy, where the density of states may be considered a constant and the appropriate value substituted. For the condition where use of the Lorentzian form is appropriate, we can substitute the density of state distribution function for two-dimensional distribution with its Lorentzian expansion (see Problem 10),

$$g(E) = \frac{1}{\pi\Delta E}\Lambda(E - E_0),$$ (8.50)

and proceed with the calculation numerically. We have not discussed the explicit form of the broadening ΔE. This follows, in resonant condition, by the explicit calculation of the barrier transmission–reflection matrix. Also, since this is the half-width of energy broadening, and the time constant associated with the decay of this state is τ_{res}, it follows from the uncertainty principle,

$$2\Delta E \tau_{res} = \hbar.$$ (8.51)

As an example, consider a quasi-bound state of half-width $\Delta E = 1$ meV, about 1/25th of the thermal energy at 300 K. The resonant lifetime associated with this is ≈ 0.3 ps, a fairly small number. However, for small broadening, which will occur with larger barriers in both energy and width, this can become significantly larger.

A direct relationship can also be established, consistent with the above, using wave propagation arguments assuming wave packets sufficiently broad in energy that the time scales can be quantified accurately. Such broad wave packets tunnel into the well region, and bounce back and forth between the

---

[10]This is a general result valid for $|E - E_0| \gg \Delta E$, not just a derivation from the formal expansion. See, e.g., D. Bohm, *Quantum Theory*, Dover, N.Y. (1989), which is a general reference for this chapter. It contains an incisive and general treatment of resonance phenomena.

discontinuities with a group velocity determined by the band structure as $v_g = (1/\hbar)\partial E/\partial k = (2E/m^*)^{1/2}$ for our nearly free electron model. For the structure, let t_l and t_r, and r_l and r_r, be transmission and reflection amplitudes for the left and the right barriers in the well. The total transmission amplitude is simply the product of the forward transmission and the negative feedback factor related to reflection, i.e,

$$t_{tot} = \frac{t_l t_r}{1 - r_r r_l} = t_l t_r \left[1 + (r_r r_l) + (r_r r_l)^2 + \cdots\right], \qquad (8.52)$$

where the nth term of the series expansion terms indicates the probability of transmission through the barrier following $n - 1$ round trips in the well. If resonance dominates, phase coherence is maintained, one round trip in the well changes the phase by 2θ, and resonance implies

$$\theta = \frac{\pi}{2}. \qquad (8.53)$$

The transmission and reflection probabilities of the barriers—T_l, T_r for transmission and R_l and R_r for reflection—are related to the above amplitudes as the squares, e.g.,

$$R_l = |r_l|^2, \qquad (8.54)$$

and

$$T_l = |t_l|^2, \qquad (8.55)$$

assuming the same group velocities in the non-barrier region. Then,

$$T_{tot} = \frac{T_l T_r}{\left[1 - (R_r R_l)^{1/2}\right]^2} \approx \frac{4 T_l T_r}{(T_l + T_r)^2}. \qquad (8.56)$$

Now consider the round trip of the broad wave packet in the well. It takes $2w/v_g$ of time delay, and each time it has a probability of being transmitted out of the well given by the corresponding term of the series expansion. The transmission probability after $n - 1$ such time intervals, at resonance, is given by the sum over n terms, i.e.,

$$T(n - 1) = T_{max} \left[1 - (R_r R_l)^n\right] \approx T_{max} \left[\frac{n}{2}(T_l + T_r)\right]. \qquad (8.57)$$

The build-up time for resonance corresponds to achieving maximum transmission probability, i.e., when the coefficient of T_{max} is unity. The amount of time corresponding to this is the number of round trips n given by the above equation multiplied by the time per round trip $2w/v_g$. This resonance build-up time is, in terms of parameters of the structure,[11]

$$\tau_{res} = \frac{n2w}{v_g} = \frac{4w}{v_g(T_l + T_r)}. \tag{8.58}$$

This resonant lifetime now allows us to also determine the quasi-bound state's energy broadening through the uncertainty principle. The lower the transmission probabilities of the barriers (larger width or larger height) the higher the resonance time. For example, a 50 Å barrier, of 400 meV energy, and a well width of 50 Å, has a τ_{res} of 70 ps, a time scale during which the scattering processes will surely destroy the coherence.

In the presence of inelastic scattering, the phase of the wave packet undergoes random changes during its stay in the well. Such a phase change introduces an additional phase term in the series expansion terms of the products of reflection amplitudes. The functional form of the expansion remains the same, and hence expanding around the resonant energy, one still obtains a Lorentzian expansion whose amplitude is smaller and width is larger (see Problem 12). Let T_{res}^{max} be associated with the maximum tunneling transmission probability of the coherent resonant process; then in the presence of inelastic processes,

$$T_{tot} = T_{res}^{max} \left(\frac{\Delta E_e}{\Delta E}\right)^2 \Lambda(E - E_0), \tag{8.59}$$

where the half width in energy is the sum of elastic and inelastic broadening, i.e.,

$$\Delta E = \Delta E_e + \Delta E_i. \tag{8.60}$$

The area under this Lorentzian curve is the same as before when inelastic processes were ignored. We may use this relationship to determine the current now. Since the area under the curve is a constant, the current is independent of whether elastic processes or inelastic processes dominate the tunneling in these coupled barriers. The differences between the two processes show up in the time responses.

The condition of resonance that we have derived mathematically is physically quite intuitive. It can be described as one in which the wave function has built up in the well. It is the condition at which the probability of finding the particle is highest in the well, together with the probability of the transmission through the coupled barrier structure (see Figure 8.18). The name resonance comes from the similarity between this and the Fabry–

---

[11]This result is identical to that derived in Equation 8.51, which was an example of the Breit–Wigner formula.

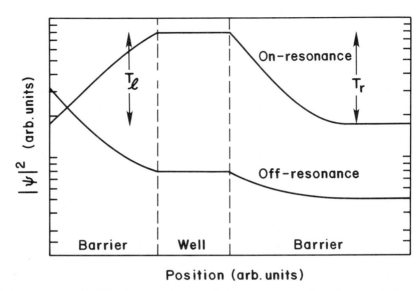

Figure 8.18: Build-up and decay of the wave function in coupled wells showing the resonant build-up from transmissions and reflections.

Pérot type behavior. In a particle picture, the electron could be considered as being trapped in the well. In the wave picture, the wave function has built up in the well sufficiently that the reflection of the incident wave is nearly cancelled out by the leaking out of the wave function in the well. Starting from an off-resonant state, the resonant state builds up in time as more and more incoming electrons are trapped. Thus, in the resonant state, a large population of carriers—or in the wave picture a large probability— exists in the well. This occurs together with a large current flowing through the structure. The building up of this charge can, under appropriate barrier and well conditions, take a significant amount of time. This time constant of resonance, which is approximately the lifetime or the decay time of the quasi-bound state, can be related to the resonance transmission width, as we have shown.

We have obtained the expressions appropriate to calculation of the resonant lifetime, based on arguments of continuous reinforcement (due to the phase matching) as reflections occur at the well barriers. We could have arrived at this same result directly, using the above intuitive picture of formation of resonance. The time it takes to leak out of a well can be estimated from the tunneling escape probability and the number of times the reflections occur per second. This argument leads to an estimate of

$$\tau_{res} = \frac{2w}{v_g T(E)}, \qquad (8.61)$$

where $T(E)$ is either T_l or T_r. This is identical to the previous estimate. Consider a 0.3 eV barrier of 30 Å barrier width, together with a 50 Å well. For an effective mass of GaAs, i.e., $m^* = 0.067m_0$, the resonant state is at an energy E_0 of 0.089 eV above the bottom of the conduction band. The time constant is 0.56 ps. However, if the barriers are 50 Å wide, $T(E)$ decreases significantly and τ is 5.6 ps. Similarly, if the barrier height is 400 meV, the τ increases to 70 ps, both of which are clearly too large for a useful high speed device. So barrier and well designs are very crucial to the performance of the coupled barrier structures.

The resonant time is also sometimes called the dwell time because it characterizes the time spent by each carrier in the well. Consider the inelastic scattering time constants for semiconductors, e.g., the inelastic scattering time is 0.4 ps at 300 K and 4 ps at 77 K in the highest purity GaAs. These time constants will dominate if resonant lifetimes exceed them.

The time constants, for the coherent or incoherent case, do not include the time for traversal of the barriers themselves. When the frequency of the signal is small enough, the traversal time for tunneling, in the barrier region, is given by

$$\tau_t = \int_{z_1}^{z_2} \sqrt{\frac{m^*}{2(q\phi - E)}} dz, \qquad (8.62)$$

which is an expression that follows from the classical interpretation of momentum. It is the time that a broadly spread packet in energy (or equivalently momentum) would take to traverse a barrier region. By considering motion of such wave packets, using the time-dependent Schrödinger equation from single and multiple barriers, the traversal times and the resonant times associated with charge storage can be distinguished. Traversal times depend on the energy of carriers; higher energy carriers may take longer to travel.

Inelastic scattering processes allow a change of momentum, hence barriers formed by band edges at different momenta in the Brillouin zone can all be involved in the tunneling process, just as they are in indirect bandgap tunnel diodes. This tunneling occurs via barriers in other valleys. For the GaAs system, with barriers formed by $Ga_{1-x}Al_xAs$, Γ, L, and X can all be important. Additionally, some of these other barriers can be lower than the barrier in energy at the same point in the Brillouin zone. Recall our discussion of the GaAs gate SISFET—the X valley was the lower energy barrier at the mole-fraction of $x \approx 0.4$ and hence the gate current occurred

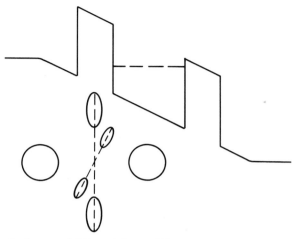

Sphere   Ellipsoids   Ring
Γ Valley   X Valleys   Γ Valley

Figure 8.19: Real and reciprocal space schematic showing the structure and possible valleys through which inelastic tunneling may occur under certain conditions such as high AlAs mole-fraction and wide barriers of $Ga_{1-x}Al_xAs$. The constant energy surfaces are identified in the three regions. The four-fold degenerate set of X valleys is most likely to take part in the inelastic tunneling process through $Ga_{1-x}Al_xAs$ at high AlAs molefractions.

largely due to it in structures with thick barriers. While the current was both thermionic and tunneling, the larger effective mass of the X valley was not sufficient to reduce it below the Γ band current. The same behavior is observed in coupled barrier tunneling devices also. For relatively thick barriers, the X barrier determines conduction with the Γ-to-X transfer requiring phonon emission. Like thermionic processes, this also occurs using the lower mass transverse X valleys of $Ga_{1-x}Al_xAs$ which are four-fold degenerate. The process therefore involves transfer of Γ electrons to X valleys and tunneling, a process that requires involvement of optical phonons and hence inelastic scattering. Figure 8.19 shows the X valleys that are most likely to contribute to this tunneling current.

Since the X point can be lower in $Ga_{1-x}Al_xAs$ than in GaAs for large aluminum mole-fractions, the X state in the well may then become the ground state in a well formed using $Ga_{1-x}Al_xAs$, even though the emitter is made of GaAs with Γ states being lower in energy. This may then give rise to tunneling current via X states instead of Γ states as considered in Chapter 4.

We have now discussed the origin of current–voltage behavior and the time constants associated with tunneling. Off-resonance, the time scales are very fast, because they are largely limited by traversal times in the barrier. These time constants are a couple of magnitudes lower than a pico-second. At resonance, the decay time can be large, and the filling time limited by the characteristics of the injecting barrier. If these times exceed about a ps, scattering begins to dominate, limiting the build-up and decay time-scales. However, once charge build-up corresponding to a resonance has occurred, the response can be fast, since any injected carrier pushes another one out, and the injection can be an efficient process. Thus, the speed of the structure, following the build-up, such as in small-signal applications, need not be slow. For thin and shallow barriers, the time constant for resonance—the resonance lifetime and the traversal time—can both be quite small. Hence, ignoring at least the transition capacitance component, the resonant tunneling phenomenon can be potentially fast. For thicker barriers, the time response is limited by scattering rates—processes that lead to loss of phase coherence. In either of these cases, the time scales involved are short.

8.5 Transistors with Coupled Barrier Tunneling

From the perspective of application in devices, the feature of coupled barrier tunneling that is considerably appealing is a fast transport mechanism. Coupled barrier structures also exhibit a negative resistance, a characteristic that may be of utility in generation of power in high frequency ranges utilizing relaxation oscillations. The negative resistance can possibly be a source for fabrication of device structures where current is low both in a low voltage state and a high voltage state as it traverses through a large current region during the transition. This could be of some appeal in fabricating low power circuits if the effects of instability due to negative resistance can be controlled. The current drive capability of structures is important for a circuit; the limited density of states of the quasi-bound states due to the limited dimensions of the devices restricts the current density that can be delivered through coupled barrier structures. However, this current is delivered at energies determined primarily by the quasi-bound states. Such characteristics can be of interest in three-terminal structures, e.g., an electron beam at a specified energy, with a narrow energy spread, could be of interest in hot carrier transistors. The most important limitations of these structures have continued to be instabilities of negative resistance for digital and microwave applications, the irreproducibility of device characteristics, which are sensitive to grown dimensions, grown compositions, and grown impurity distribution effects. However, features of coupled barriers are appealing, and of interest for small-signal and large-signal high frequency usage. Attempts have, therefore, continued to create structures based on conventional transistor structures, as well as on the intrinsic use of the coupling with a control of the characteristics of the well.

The simplest of the former include incorporation of the coupled barrier in the semiconductor contact regions of FETs, the base region of the bipolar transistor, etc. The incorporation in the quasi-neutral region of such structures results in a narrow range, the region in bias where the transmission is efficient, over which they are well-behaved FETs or bipolar transistors, enclosed between ranges where the current decreases. If the device performance does not suffer in these other regions, then these devices may utilize the lower power capability for digital applications. Usually, though, this is difficult to achieve, because of increased resistance effects due to inefficient transport in the coupled barrier structure.

In the latter category are devices where the coupled barrier structure actually plays a more central role. A doped well region whose electrostatic potential can be directly modulated is one example of this. Another example is the use of a coupled barrier structure in the base–emitter junction region such that it controls the injection of carriers and allows efficient injection only in a specific bias range. An example of such is given in Figure 8.20. The decrease of current in these structures, following the initial rise, is due to tunneling through the quasi-bound state in the base–emitter junction region. However, outside this region of bias, the resonant tunneling structure leads to inefficiency in transport and excess series resistance. This means of getting a selective energy beam of carriers is of particular interest in the tunneling emission transistor, where instead of the single tunneling $Ga_{1-x}Al_xAs$ barrier, a double barrier with a larger current capability has been applied. These devices have shown a larger current capability than the simpler transistor, and have shown larger current gains, too. An example of such a device in the various bias regions is shown in the inset of Figure 8.20. Arguably, all these structures have utilized the coupled barrier tunneling phenomenon as a series element, and only in part of the bias region as an intrinsic part of the device operation.

One good example of the intrinsic use of these coupled barrier structures over a wide bias range is the use of a quantum well as the base region of a bipolar transistor, as shown in the idealization of Figure 8.21. The

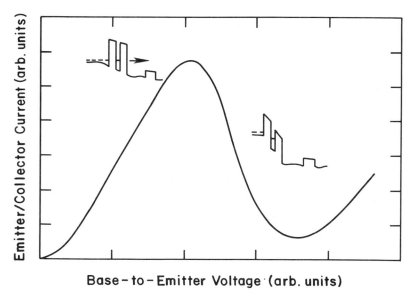

Base - to - Emitter Voltage (arb. units)

Figure 8.20: An example of resonant tunneling to control the injection characteristics at the base–emitter junction of a bipolar transistor, and the corresponding current–voltage characteristics.

base potential and the quasi-bound state energies are modulated with respect to the emitter; the doping in the base allows for ease in accomplishing this electrostatic modulation. By utilizing the p-doping in the abrupt and thin base, quasi-bound states are created in both valence band and the conduction band of the base. The injection of carriers from the emitter occurs through the quasi-bound states in the base to the collector. The base quasi-bound state potential can be directly modulated. This device therefore exemplifies the central use of resonance in device physics. Figure 8.21 is an idealization; the actual structure may be more complicated in order to properly design the position of the quasi-bound states and barrier energies.

This use of control of the electrostatic potential and quasi-bound state energy of the base and the quantum well can also be applied to a unipolar device. To accomplish this, one may use doping in the base of the same polarity as in the emitter and the collector regions. One may view this as a limit form of the hot carrier transistors. The problem with such a structure, of course, is technological, since dopant diffusion effects become important and one has to make contacts to the base regions while still maintaining

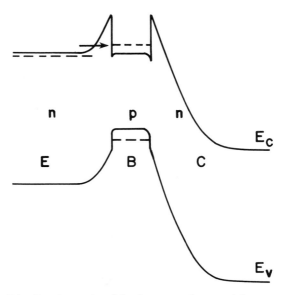

Figure 8.21: Idealization of a bipolar transistor with a very thin base, in which the transport in the base and the injection at the base–emitter junction depend on the quasi-bound states in the base well.

isolation in the parasitic regions of the device.

Use of smaller bandgap materials within the base and differing from the emitter and collector materials may allow a quasi-bound state that is occupied in thermal equilibrium. The base potential may then be modulated directly, and there will be less inelastic scattering in the structure due to the absence of dopants. However, these structures are more difficult to reverse isolate, e.g., at the base–collector junction, because of the symmetry of barrier. Another pseudomorphic heterostructure material with different barrier and transport characteristics could possibly be used to suppress leakage under the forward bias conditions of the base–collector junction. Such examples, by necessity, are technologically more difficult to achieve, they are also likely to reduce the transmission probabilities, and thus result in smaller current densities.

8.6 Summary

This chapter considered the behavior of structures based on hot carrier injection and quantum-mechanical tunneling, structures that utilize abrupt heterointerfaces to obtain unusual operating behavior. We developed, using Schrödinger's equation, the transmission and reflection behavior of carriers incident at abrupt discontinuities. This allowed us to understand how in certain structures one may reduce reflections by an appropriate shaping of the barrier region. This was particularly important in the operation of hot carrier structures at the base–collector junction. We showed how the behavior of these hot carrier devices depended on reduction of the total scattering events during transport in the base region, and how the high velocities could be inferred from measurements on the structure. We then considered the behavior of resonant and sequential tunneling in structures that are based on barriers that are closely coupled. Sequential tunneling-dominates in structures where phase coherence is destroyed by scattering in the well region. We analyzed the important time constants of these structures, and concluded that they could be both long and short; the nature of these time constants is very critically dependent on the structures themselves.

General References

1. E. Burstein and S. Lundqvist, Ed., *Tunneling Phenomena in Solids*, Plenum, N.Y. (1969).

2. E. H. Hauge and J. A. Støvneng, "Tunneling Times: A Critical Review," *Reviews of Modern Physics*, **61**, No. 4, p. 917, Oct. 1989.

3. B. Ricco and M. Ya. Azbel, "Physics of Resonant Tunneling: The One-Dimensional Double-barrier Case," *Phys. Rev. B*, **29**, p. 1970, 1984.

4. B. Ricco and M. Ya. Azbel, "Tunneling through a Multiwell One-Dimensional Structure," *Phys. Rev. B*, **29**, p. 4356, 1984.

5. P. J. Price, "Theory of Resonant Tunneling in Heterostructures," *Phys. Rev. B*, **38**, p. 1994, 1988.

6. D. Bohm, *Quantum Theory*, Dover, N.Y., (1989).

7. V. Rojansky, *Introductory Quantum Mechanics*, Prentice-Hall, N.Y. (1938).

8. C. B. Duke, *Tunneling in Solids*, in F. Seitz, D. Turnbull, and H. Ehrenreich, Eds., *Solid State Physics*, Suppl. **10**, Academic Press, N.Y. (1969).

Problems

1. We will relate the time that carriers spend on average in the quasi-neutral region of a bipolar transistor in order to compare this with the behavior of a hot carrier structure. Consider an HBT with a GaAs base, and let the acceptor doping in the base be either a constant 5×10^{18} cm^{-3} or given by $8 \times 10^{18} \exp(-z/3 \times 10^{-6})$, where z is referenced to the base–emitter junction and is in centimeters. Estimate the time spent, on average, by the minority carrier in a device with

   (a) a 300 Å wide base region, and

   (b) a 1000 Å wide base region.

2. Consider a barrier of the shape

$$\phi_B(z) = \phi_0 \left[1 + \exp\left(-\frac{z}{\ell}\right) \right]^{-1}, \tag{8.63}$$

   which is an exponentially graded barrier. Show that the reflection coefficient of this structure is

$$\varrho = \frac{\sinh^2\left[\pi\ell\left(k_1 - k_2\right)\right]}{\sinh^2\left[\pi\ell\left(k_1 + k_2\right)\right]}, \tag{8.64}$$

   with

$$k_1 = \left(\frac{2mE_0}{\hbar^2}\right)^{1/2}, \tag{8.65}$$

   and

$$k_2 = \left[\frac{2m\left(E_0 - q\phi_0\right)}{\hbar^2}\right]^{1/2}. \tag{8.66}$$

3. Consider the following n$^+$-GaAs/Ga$_{1-x}$Al$_x$As/n-GaAs structures in which the n$^+$ region is doped to 5×10^{18} cm^{-3}, the Ga$_{1-x}$Al$_x$As region is undoped, and the n-region is doped to 1×10^{17} cm^{-3}. Estimate, with suitable approximations and using the discussion of Chapter 4, the thermionic field emission currents and tunneling currents at 77 K for the following Ga$_{1-x}$Al$_x$As structures as a function of forward bias in the 0 to 0.6 V range:

   (a) a uniform composition region of 150 Å width and AlAs mole-fraction of 0.3, and

   (b) a region of Ga$_{1-x}$Al$_x$As with the AlAs mole-fraction increasing from 0 to 0.3 from the n$^+$-GaAs to the n-GaAs over 150 Å.

Also, estimate the currents for the same structure where the hetero-junction barrier region is replaced by a homojunction barrier region using a sheet doping of 2.5×10^{11} cm^{-2} of acceptor doping.

4. Consider the hot carrier transistor using a uniform aluminum mole-fraction for emitter and collector. In such structures, knowing the dopings, the Fermi level is known under thermal equilibrium conditions. Assuming an absence of collisions, and thus only observing the first peak in the dependence of $(\partial I_C / \partial V_{BC})$, the mean energy of the stream of carriers incident at the base–collector junction is known. In the first-order calculation, we may assume the absence of tunneling effects, etc., at the base–collector junction. Thus, a changing position of the energy of the barrier at the base–collector junction, obtained using a changing base–collector bias, can be employed to obtain the shape of the distribution with energy, as well as the energy. The energy can be related to the velocity. Determine the expression for this velocity.

5. For coupled square-well barriers, the transmission can be expressed in the form of Equation 8.27. Show that this can be simplified to the form of Equation 8.32. Show that this is actually a more general result for arbitrary barrier shapes.

6. Show, using a trapezoidal barrier, such as in the square barrier with electric field, that the transmission coefficient occurs with a pre-factor. If $q\phi_0$ and $q\phi_1$ are the two barriers with an electric field $\mathcal{E}$, show that the transmission coefficient may be written as

$$T = \mathcal{A} \exp \left[-\frac{4}{3} \frac{1}{\hbar} \sqrt{2m^*} \frac{(q\phi_0)^{3/2} - (q\phi_1)^{3/2}}{q\mathcal{E}} \right]. \qquad (8.67)$$

Find $\mathcal{A}$.

7. Consider tunneling in a single barrier structure with the tunneling current resulting from a barrier involving a different effective mass. Find an expression for the current in this trapezoidal barrier.

8. Show that the consequence of temperature on the transmission of carriers, from a three-dimensional distribution to a two-dimensional distribution in a quantized well, is a broadening of transmission characteristics. What is the characteristic energy of this broadening or decay of the transmission coefficient from its peak? What is the characteristic energy if the quantized well was a triangular well instead of a square well?

9. Consider the problem of transmission across a barrier; the current through such barriers was written quite generally as the following:

$$J = \frac{qm^*}{2\pi^2\hbar^2} \int \left\{ \int T\left(E_\perp, E_\parallel\right) [f_l(E) - f_r(E + qV)]\, dE_\parallel \right\} dE_\perp.$$
(8.68)

Here, the density of states is assumed to be unity for convenience. Show that for materials with isotropic spherical constant energy surfaces, the current can be expressed as

$$J = \frac{qm^*}{2\pi^2\hbar^3} T(E_0)\Delta E kT \ln\left[1 + \exp\left(\frac{\xi_f - E}{kT}\right)\right].$$
(8.69)

10. Consider tunneling through a barrier region into a square quantum well. Expanding the density of state distribution function, at an energy E, show that the density of state appropriate to calculation of current, such as a multiplication factor in Problem 9, is given by

$$g(E) = \frac{1}{\pi\Delta E}\Lambda(E - E_0),$$
(8.70)

11. Consider an electron of energy E incident on a potential barrier or potential well of height $q\phi$ and width w. Let the energy of the electron be greater than the barrier energy, i.e., $E > q\phi$. Show that transmission for the barrier case is given by

$$T = \frac{4E(E - q\phi)}{4E(E - q\phi) + q^2\phi^2 \sin^2(kw)},$$
(8.71)

where $k = \left[2m(E - q\phi)/\hbar^2\right]^{1/2}$. The maximum in this occurs at $kw = n\pi$, where n is an integer, and it results in unity transmission. The minimum occurs at $kw = n\pi/2$. What is the result for the well problem?

12. Consider transmission in the coupled-barrier problem in the presence of both elastic and inelastic scattering. Show that the transmission can be written to be related as

$$T_{tot} = T_{res}^{max}\left(\frac{\Delta E_e}{\Delta E}\right)^2 \Lambda(E - E_0),$$
(8.72)

where T_{res}^{max} is the maximum tunneling transmission probability of the coherent resonant process. Show that the consequence of inelastic scattering is to increase the width of characteristic energy compared to the problem with only elastic scattering (see Problem 8). Show, however, that the area under the curve is the same for elastic and mixed elastic-inelastic transmission.

13. Consider the problem of injection from a three-dimensional distribution to one-dimensional distribution instead of two-dimensional distribution considered in the text. Map the regions of allowed momentum states, considering conservation of energy and momentum.

14. Consider a 200 Å $Ga_{.7}Al_{.3}As$ barrier hot electron unipolar transistor with a 300 Å GaAs base. The emitter is doped to 4×10^{18} cm^{-3} and the base to 5×10^{17} cm^{-3}. The base–collector junction is sufficiently lower in barrier height to allow most carriers to transit to the collector, and the transistor is operated at 77 K. For the carriers elastically injected at the Fermi energy from the emitter, estimate the base transit time at a forward bias of 0.4 V assuming

   (a) no collisions, and

   (b) a mean free path of 500 Å with randomizing collisions.

   Compare this base transit time with that of a bipolar transistor of similar base doping of the opposite type.

15. In calculating the tunneling probability, results of the time-independent Schrödinger equation have been invoked. Since finite times are involved, shouldn't the time-dependent Schrödinger equation have been used?

16. Dielectric relaxation time is usually indicated as a limiting time constant for unipolar devices such as FETs. Is this a limiting time constant for the unipolar resonant tunneling and hot electron devices? If yes, why? If no, why not?

17. Show how one may obtain the conduction band edge profile of of Figure 8.22 in the $GaAs/Ga_{1-x}Al_xAs$ base–collector region of a hot carrier structure.

18. Consider the coupled barrier structures shown in Figure 8.23 and operation at 300 K.

   (a) Which of these cases can exhibit sequential tunneling?

   (b) Which of these cases can exhibit resonant tunneling?

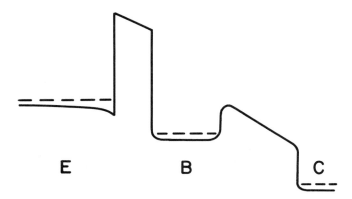

Figure 8.22: Conduction band edge profile for the hot carrier structure of Problem 17.

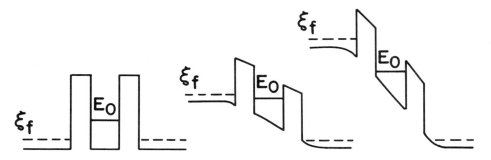

Figure 8.23: Conduction band edge profiles for the coupled barrier structures of Problem 18.

Assume that all other characteristics are conducive to the desired operation.

19. Consider the square quantum well shown in Figure 8.24. What does the Fermi surface for electrons look like in different regions? Draw.

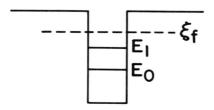

Figure 8.24: Square quantum well of Problem 19.

Chapter 9

Scaling and Operational Limitations

9.1 Introduction

Scaling is the conventional term used to describe the coordinated changing of device parameters with a reduction in their dimensions in order to obtain a commensurate improvement in performance. In this chapter we will take a general and subjective view of the operation and scaling of various devices. The treatment is designed to obtain a general understanding of the facets of operation of devices, and how they may be expected to change with reductions in dimensions, what particular problems such reductions would encounter, and what changes in operational characteristics may make them technically interesting. We will also look at general properties of materials and how they relate to and affect the scaling of device structures. Such a general framework, lacking rigor, does not necessarily allow us to make accurate predictions, because as device dimensions shrink, a more thorough application of physical principles must be incorporated. Our discussion, therefore, is directed to an understanding of the various technological limits of current implementations that need to be circumvented or solved by finding fundamentally new implementations and ways of operation.

Compound semiconductors have dominated high frequency applications by virtue of their superior transport for comparable-dimension devices implemented in silicon. This supérior transport usually results in higher average velocity of carriers in the device, and translates into a higher frequency of operation. Consider this transport characteristic within the limitations of saturated velocity (the highest velocity under conditions where drift-diffusion is a valid approximation). For an applied field of 5×10^3 V.cm^{-1}, a velocity of 1×10^7 cm.s^{-1} implies an energy transfer rate of 8×10^{-9} W. The time required to acquire a kT of energy at 300 K is 0.5 ps, a small time, but nonetheless a limitation because carriers must lose and gain energies even higher than this to be of practical import for both small-signal and digital usage. Such limitations narrow the gap between many of the compound semiconductor and silicon devices because of implications of high doping. Consider this argument further, in terms of length scales. Some of the operational length scales of interest include the mean free path, which is the average distance of travel between collisions, the de Broglie wavelength, the screening length, and the average distance between dopant atoms. All of these have implications on the operation of the device: the mean free path determines the suitability of applying drift-diffusion or other physical basis, the de Broglie wavelength stresses the importance of breakdown of conventional description and the importance of incorporating additional quantum-mechanical effects, the screening length—Debye in non-degenerate conditions and Thomas–Fermi in degenerate conditions— has strong consequences wherever quasi-neutrality is violated and charge packets exist, and the average distance between dopants affects band structure and transport-related material parameters. Larger mean free paths of compound semiconductors also maintained at larger dopings result in higher velocities beyond the saturation velocity considered in the argument above. The longer de Broglie wavelength of an electron, due to its smaller mass, implies a stronger role of quantum-mechanical effects. The effects of the rest are quite similar.

Mean free paths are of the order of a fraction of 1000 Å. Another characteristic length characterizing importance of similar scattering phenomena may be defined as the ratio of the maximum group velocity and the scattering rate. For the maximum group velocity of 8×10^7 cm.s^{-1}, and scattering rate of 1×10^{13} s^{-1}, this is 800 Å. Electron-based devices in sizes of the order of 2500 Å in compound semiconductors need to include this limited scattering, and hence our emphasis on off-equilibrium effects in this text. The de Broglie wavelength at the velocity of 1×10^7 cm.s^{-1} is of the order of 650 Å in GaAs, incorporation of tunneling and other quantum-transport effects are necessary to obtain accurate descriptions at this dimension. Screening lengths are also of a similar order of magnitude and become important in space charge regions. The average distance between atoms at a doping of 10^{19} cm^{-3} is of the order of 50 Å, causing a large wave function overlap effect that is reflected in heavy doping effects on transport, and in state distribution and occupation of bands. Clearly, some of these have a favorable effect in obtaining improvements in transport, and some are detrimental.

These arguments serve as an appropriate platform for discussing scaling and hence the technological limits that have to be challenged. The chapter first discusses some of the general features of the operation of devices, and follows it by a general discussion of scaling and limitations posed by the above due to the operational basis of the devices and due to technological and material restrictions.

9.2 Operational Generalities

We have been particularly interested in two basic types of devices, the bipolar transistor, where the diffusion of minority carrier plays a major part in current transport and operation, and the FET, where the drift of a majority carrier plays a major part in the transport and operation. In practice, there are variations from this. In bipolar transistors, drift is a significant factor in transport for graded doping or graded alloy composition structures; in field effect transistors diffusion can be significant in the pinched-off region of the device over a variety of bias conditions. Diffusion dominates in the sub-threshold region of operation of FETs.

The control in the bipolar transistor occurs through the injected base current, which allows us to modulate the injecting barrier at the base–emitter junction. The built-in voltages at the junctions play the role of isolating the controlled regions. In a bipolar transistor, the transport takes place through the barrier; the barrier, by virtue of its isolating property, allows resistive coupling to the emitter and base regions. The control in a FET occurs via the gate electric field, which controls the number of carriers affected. The region of this high electric field, the insulator–oxide or large-barrier semiconductor or the depleted metal–semiconductor junction region, allows no or little carrier transport across it, and hence allows a capacitive coupling of the channel transport region.

In a bipolar transistor, the control of the threshold is usually relatively straightforward because it depends on bandgap primarily, and doping, current density, etc., secondarily. The control of the threshold in a FET is more difficult because of charge density and thickness variations, in either the bulk, the interface, or the insulating regions of the device. In a bipolar transistor, the ease of maintaining charge neutrality, because of the high doping, allows a large carrier concentration to flow through the device, allowing larger current densities. Because of the built-in barrier of bandgap, the voltage levels of operation are also restricted to approximately a volt. The power dissipation of fast bipolar transistor structures is therefore generally larger than that of FETs where the currents are lower due to lower charge density, and voltage levels are similar. In digital circuits, the high current capability of a bipolar transistor, is suited to usages involving large loadings which tend to be dominated by capacitances and where the higher power dissipation can be acceptable. The lower power as well as simpler and easily integrable structure of field effect transistor is suited to usage involving higher integration levels than can be obtained with the complex bipolar technology.

While operationally, as seen in the above, the two device types can be distinguished in how the control is achieved, functionally, the effect of carrier transport from the injection electrode to the collection electrode takes a very similar form. Consider two short-dimension device structures, the bipolar transistor and field effect transistor shown in Figure 9.1. This figure shows, for a GaAs bipolar and an HFET structure with comparable short length scales, the behavior of the band edges and the velocities of transport. When a rapid change in the input signal occurs, at the base, or at the gate, the device response involves both particle current effects and displacement current effects. Both these devices have a low field region, in the quasi-neutral base of the bipolar transistor, and at the source edge of the channel underneath the gate. The displacement current in these regions is small; the particles moving with a smaller velocity support the injection into the collector and the charging of the gate displacement current. The pinched-off region of the HFET, and hence the drain current, can not change until carriers reach this region moving with the lower velocity and satisfying the displacement current requirements of the gate. This delay has an equivalent in the delay due to transport in the quasi-neutral base region of the bipolar transistor. Both carriers move with lower velocities than they do in the pinched-off region or the base–collector space charge region. The band edges in the pinched-off region and the base–collector space charge region are also quite similar, being representative of the bias drop of comparable voltage bias. The overshoot in these regions, given their comparable low and high energy transport characteristics, is also comparable. So, even though one is a minority carrier device and the other a majority carrier device, they have strong similarities in how the carriers that are injected get collected.

There are a number of devices that we have not referred to yet that may be said to fall in between. Some examples of these are the static induction transistor, the hot carrier transistor, the resonant tunneling transistor, etc. These devices have modes of operation where they either exhibit a bipolar-like or a FET-like mode in different biasing regimes or actually mix the two. Figure 9.2 shows examples of the devices and their broad classification. In this general classification, we refer to bipolar behavior as characterizing injection over a barrier and resistive coupling of control signal, and field effect behavior as characterizing injection that is coupled via ohmic means and

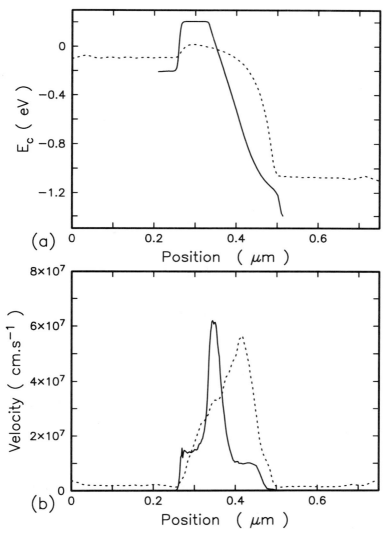

Figure 9.1: Operational similarity in carrier transport from injection to collection in a GaAs bipolar transistor (solid lines) and a field effect transistor (dashed lines). (a) shows the band structure under biased conditions and (b) shows the velocity along the path. The quasi-neutral base region of the bipolar transistor has similar transport effects as the source end of the channel. The base–collector region of the bipolar transistor has similar effects as the drain high field region of the channel in a field effect transistor.

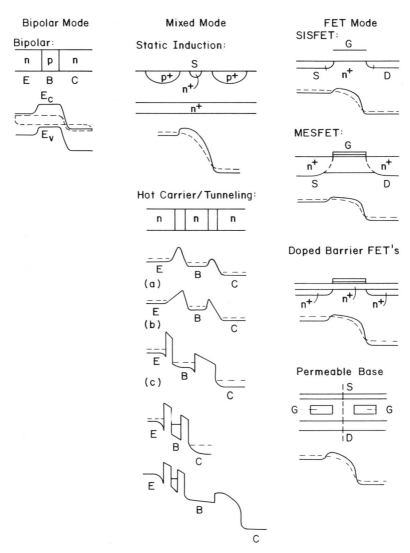

Figure 9.2: Examples of various classifications of devices and their mixes, with band diagrams in the control regions.

controlled via capacitive means. Examples of such mixed-mode behavior include certain designs of the permeable base transistor. In principle, all of these devices can be fabricated in all of the semiconductors of present-day interest—single or compound.

9.3 General Scaling Considerations

Scaling is the changing of device dimensions by appropriate proportions so that delay components reduce proportionally. This allows scaled devices to have a total delay that is proportionally smaller, with a proportionality factor given by the scaling factor. The advantage of resorting to the technique of scaling is that it allows us to shrink devices and circuits without having to perform a complete re-analysis of the device, and allows us to do that by following a standard set of rules. It remains valid so long as the governing rules chosen remain valid in prediction of the operation of the device. So, to scale a device for its static and dynamic behavior, we consider scaling of the dimensions, and consider the effect of this scaling on the governing equations. Consider dimensions larger than 0.25 μm for compound semiconductor devices. At these dimensions, the drift-diffusion approximation is applicable for most device conditions. The governing equations are Poisson's, the current, and the current conservation equations. Poisson's equation is a restatement of one of Maxwell's equations. The complete set of Maxwell's equations describes the behavior of signal transmission, both in circuit interconnection and within devices. We therefore consider Maxwell's equations first; they are of general import both for devices and circuit interconnections. Signal transmission along a transmission line is governed by the set of equations

$$
\begin{aligned}
\nabla \times \mathcal{E} &= -\frac{\partial \mathcal{B}}{\partial t}, \\
\nabla \cdot \mathcal{D} &= \rho, \\
\nabla \times \mathcal{H} &= J + \frac{\partial \mathcal{D}}{\partial t}, \\
\text{and} \quad \nabla \cdot \mathcal{B} &= 0.
\end{aligned}
\tag{9.1}
$$

Consider a scaling factor λ where $\lambda < 1$ represents shrinking of the corresponding device dimension or any other parameter. So, if a length parameter z is scaled to z' by a scaling factor λ, then the new scaled dimension $z' = \lambda z$. If we consider scaling without changing the combinations of the materials involved, so that the constants of the relationships are invariant, and if we consider scaling without changes in the signal swings, i.e., the electrostatic potentials remain constant too, then scaling of device dimensions by λ implies the scaling of all the fields ($\mathcal{E}$, $\mathcal{D}$, $\mathcal{H}$, and $\mathcal{B}$) in Maxwell's equations by the scaling factor $1/\lambda$. The field strengths increase as a result of the shrinking of device dimensions because of the increase in the gradient. All change by $1/\lambda$, i.e., they increase during shrinking of geometry because the derivative ∇ increases by this factor. We wish the time scale to also shrink by the same scaling factor, i.e., $t' = \lambda t$, so the scaled Maxwell equations, using prime to denote the scaled parameters, are

$$\nabla' \times \mathcal{E}' = -\frac{\partial \mathcal{B}'}{\partial t'},$$

$$\nabla' . \mathcal{D}' = \rho',$$

$$\nabla' \times \mathcal{H}' = J' + \frac{\partial \mathcal{D}'}{\partial t'},$$

$$\text{and} \quad \nabla' . \mathcal{B}' = 0. \tag{9.2}$$

The scaled set of equations (Equation 9.2) are invariant from the unscaled Equation 9.1, with a new charge density ρ' and a particle current density J' varying as $1/\lambda^2$. Note also that the displacement current term $\partial \mathcal{D}'/\partial t'$ also varies as $1/\lambda^2$. Thus, scaling of the interconnect lines results in a higher current density and charge along the lines. There are technological constraints to this, but so long as this scaling of all dimensions, including the insulator thicknesses, charge density, and current density parameters, is allowed to take place, then the interconnections will transmit the signals at a proportionally faster time scale. The reason for the more rapid increase of charge and current density is the requirement that the potential remain constant. If we allow the signal amplitude to scale, then the constraints on current and charge are more relaxed (see Problem 1).

Maxwell's second equation is our Poisson's equation, hence we have also scaled one of our important device equations. In terms of the fixed and immobile charge, we have Poisson's equation as

$$-\nabla^2 \psi = \frac{q}{\epsilon} \left(p - n + N_D - N_A \right). \tag{9.3}$$

For the scaling of the charge density as $1/\lambda^2$, the individual charge densities N_D, N_A, n, and p must all scale as $1/\lambda^2$. In a field effect transistor, this implies that the channel charge density (mobile and immobile) should vary as $1/\lambda^2$. This can be a considerably difficult proposition for HFETs since we wish to also avoid the parasitic effects of either excessive gate current in an undoped-barrier device, or of parasitic source-to-drain conduction in a doped-barrier device. In an insulator-based device, the breakdown fields and changes in transport properties of the carrier place constraints. In a

bipolar transistor, the largest of these needs to be increased in this proportion. The doping can be increased maintaining the scaling requirements
in the space charge regions, together with constraints placed by transport
effects such as injection efficiency (a reason for using HBTs) as well as parasitic conductions due to tunneling in various highly doped junction regions.
In the quasi-neutral region, the constraints from the space charge regions
can be extended maintaining the scaling since the rate of change of electrostatic potential is small. Traditionally, the current densities of bipolar
transistors are high, since the injected carrier concentration is maintained
high with small junction areas; an increase in the carrier density as $1/\lambda^2$
would place a very strong technological burden, and is not necessary since
this charge density becomes important to the scaling of Poisson's equation
mostly in the base–collector space charge region under high injection conditions. Thus, the increase in current density of $1/\lambda^2$ suffices, and the carrier
densities may be maintained nearly constant.

In the transport of carriers in these device structures at the dimensions
above 0.25 μm, the potentials of interest here are the quasi-Fermi potentials
(ϕ'_n and ϕ'_p). These do change slightly because

$$\phi'_n = \phi_n - \frac{kT}{q}\ln\left(\frac{n'}{n}\right)$$

$$\text{and} \quad \phi'_p = \phi_p + \frac{kT}{q}\ln\left(\frac{p'}{p}\right), \tag{9.4}$$

both of which have a perturbative term $2kT(1/q)\ln(1/\lambda)$. This dependence is logarithmic, smaller than the proportional changes, and stronger
for FETs than for bipolar transistors. Consider now the current density
equation for devices applicable at these dimensions,

$$J_n = -q\mu_n n\nabla\phi_n$$

$$\text{and} \quad J_p = q\mu_p p\nabla\phi_p. \tag{9.5}$$

Consider the case of FETs first; the scaling factor for current from this is
$(\mu'/\mu)1/\lambda^3$. For most useful carrier density and doping levels, the mobility
and diffusivity follow a weaker than the inverse concentration behavior, i.e.,
a weaker than λ^2 behavior. Hence, the particle current usually follows a
relationship closer to the weaker side of $1/\lambda^2$ dependence. Carrier concentration itself may not be scaled significantly, but the gradient does lead to
a stronger than μ'/μ dependence, a dependence that is stronger than $1/\lambda$.
For transport in high fields and thermionic injection across semiconductor
heterojunctions, the carriers move with a limited velocity v_l, which does not

scale; again, because of the concentration dependence, the particle current varies as $1/\lambda^2$.

Now consider another important controlling device equation, the current continuity equation,

$$\frac{\partial n}{\partial t} = \mathcal{G}_n - \mathcal{R}_n + \frac{1}{q}\boldsymbol{\nabla}.\boldsymbol{J}_n$$

$$\text{and} \quad \frac{\partial p}{\partial t} = \mathcal{G}_p - \mathcal{R}_p - \frac{1}{q}\boldsymbol{\nabla}.\boldsymbol{J}_p. \tag{9.6}$$

The rate of change of particle concentration in a given volume with time now varies as $1/\lambda^3$, as does the current transport contribution. The generation and recombination terms are of importance in bipolar transistors, although the devices are designed so that these are not the prominent terms of the current continuity equation. In the single time constant approximation, usually valid throughout most of the device regions of interest, it varies as $(n_p - n_{p0})/\tau_n$ in p-type material. The carrier concentration varies as $1/\lambda^2$ and the lifetime τ as λ to λ^2; the latter occurs at very high doping levels due to Auger recombination. As a result, the generation and recombination term also varies with close to $1/\lambda^3$ dependence. The current continuity equation, therefore, also scales with dimensions. So, we have seen that at least at dimensions exceeding 0.25 μm in compound semiconductors, the scaling of the control equations can be achieved consistently, albeit with technological constraints.

To observe that this scaling procedure does result in a scaling of the time scale in the operation of the devices, consider some of the time constants of device operation—that related to diffusion of carriers in the bipolar transistor, and that related to the charging of space charge capacitance in both bipolar and field effect transistors.

In the quasi-static approximation, the time constant due to diffusion of carriers in the base, the base time constant, is related as

$$\tau_B = \frac{Q}{I} \propto \frac{n\Delta X \Delta Y \Delta Z}{J\Delta X \Delta Y} \propto \lambda, \tag{9.7}$$

where, Q is the stored charge, and ΔX, ΔY, and ΔZ are the dimensions of the structure, i.e., the base time constant scales with the scaling factor. If the current does not scale exactly as λ^2 because of the doping dependence of the mobility and diffusivity (considered previously) then it varies as $(\mu/\mu')\lambda^2$, usually a slightly slower variation. So, storage effects do scale quite closely, as we should expect based on the scaling of the operating equations.

Now consider the time constants related to charging and discharging of space charge regions, and characterizing the displacement current effects, instead of the particle current effect in the above. This time constant τ is related with the capacitance (C), the voltage change across the junction (ΔV), and the current through the junction via

$$\tau = \frac{C\Delta V}{I} = \frac{\epsilon\Delta X\Delta Y}{\Delta Z}\frac{\Delta V}{J\Delta X\Delta Y} \propto \lambda, \qquad (9.8)$$

the expected time scaling. Like the diffusion behavior, if the current density does not have the exact proportionality to $1/\lambda^2$ due to mobility and diffusivity effects, then this time scale shows a weaker dependence of $(\mu/\mu')\lambda$.

9.4 Limits from Operational Considerations

We now consider, again within the general framework, limits on devices placed by their operational basis. The operational basis in turn leads to technological limitations as devices are scaled. We may consider these together to appreciate the difficulties and the advantages of scaling in the pursuit of higher performance.

The first of these is power dissipation in devices and how efficiently heat can be removed from the substrate to limit the rise in temperature to acceptable numbers. There is a fundamental limit to the minimum energy dissipated in processing of information, determined by thermodynamics and quantum mechanics, and of the order of the thermal energy. The amount of energy dissipated in devices of practical interest is higher than this since we wish to distinguish a state with very high probability. Circuits are designed to minimize the energy dissipated but with additional constraints requiring fast switching or high frequency of operation.

The need to drive loads—another device or a capacitive, inductive, or resistive element in acceptable time scales—requires the device to have a minimum current handling ability to make the switching between two distinctly identifiable states possible. The energy required in this process is significantly higher than the fundamental limit above. These states to be distinguished are usually levels in voltage, but they could as well be states in current, in which case a potential drive ability will be required. The potential changes required have to be sufficiently larger than the thermal voltage kT/q to prevent errors arising from thermal noise. Potential changes, in practice, are limited by bigger factors related to switching phenomena: one example is the effect of inductances when large changes in currents occur, normally called the LdI/dt effect, or changes in resistive drops in power and signal lines due to changes in current carried, and device variations such as in the turn-on and turn-off behavior.

A large scale circuit usually has inductances associated with lines that have low impedance terminations; power lines are example of this. A change in current of 1 mA in 25 ps with an inductance of 0.065 nH—an optimistic estimate for a power line—causes an inductive voltage change of ≈ 0.026 mV, i.e., of the order of thermal voltage. Such effects get particularly strong with fast switching, as well as for signal lines with low impedance loads or sources.

Resistive drop in power lines can be very strong since large densities are associated with large currents, even if the power dissipation per gate is small. Consider logic with 300 μW dissipation per gate, and employing a 1 V power supply voltage. For a 25,000 gate circuit, the total power dissipated is 7.5 W in an area likely to be of the size of a cm^2 and the current requirement is 7.5 A. The distribution of this power and current requires complex designs, so that resistive drops can be maintained low. There are additional line limitations; as the line dimensions decrease, fringing effects begin to dominate, line capacitances, e.g., are limited to nearly 5 fF.μm^{-1} due to the fringing effects of three-dimensional distribution of fields in a small line.

Heat dissipation in small geometries is limited by spreading effects (see Problem 2), and the average temperature rise at the surface for GaAs for this example is between 30 K and 70 K depending on the sparsity of the design. The upper limit on power dissipation is defined by electro-migration in metal lines and in semiconductors, which limits the current capability and reliability; the voltages that can be applied are limited by the operational basis of the device, parasitic effects such as injection effects in the gate, etc., and also the power that can be efficiently extracted from the chip so that temperature changes can be limited.

Threshold variations cause changes in voltage levels because of their effect on noise margins. Threshold voltage of the order of thermal voltage are common for FETs in compound semiconductors. With a decrease in device dimensions, the total number of impurity atoms within the current-limiting regions also generally increases. Since the impurity distributions are random and follow Poisson's probability distribution, fluctuations in charge contributions also increase with scaling of device dimensions. All these lead to a minimum threshold voltage variation because of fundamental constraints, and place additional operational constraints on the design of scaled devices and digital circuits.

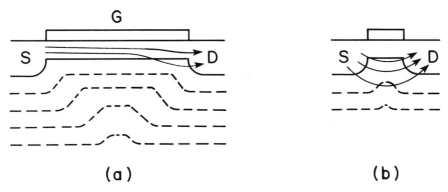

Figure 9.3: Current flow and potential contours showing the basis for short channel effects by comparing behavior in a long channel device (a) and a short channel device (b). The dashed lines show constant potential surfaces and the arrows indicate the flow of carriers.

9.5 Scaling and Operational Considerations of FETs

We now return to the question of limitations of devices and relate them to material properties and transport phenomena. In this discussion we will also include considerations of scaling below the dimensions defining the applicability of the drift-diffusion formalism.

We have considered, during the discussion of FETs, some of the effects of increases in output conductance, changes in threshold voltage, etc., that result from reduction of channel lengths. These effects can be broadly classified under the category of short channel effects. The short channel phenomenon, due to the two-dimensional distribution of fields, places a principal constraint on the scaling of FETs. The short channel phenomenon leads to weakly controlled current flow through longer paths from source to drain due to lowering of the source barrier (see Figure 9.3). This flow of carriers extending away from the surface or interface leads to a higher output conductance and poorer speed because of the longer path lengths, and a barrier mode of operation that is modulated by the drain voltage. These are all consequences of the breakdown of the nearly one-dimensional conditions that existed in the long channel device.

One of the limits to the shorter channel lengths is due to this punch-through phenomenon.[1] A rough analysis of the extent of depletion regions

---

[1]See, e.g., J. A. Cooper, Jr., "Limitations on the Performance of Field-Effect for Logic

at the source and the drain is

$$w_d = \left[\frac{2\epsilon\left(\psi_{j0} + V_D\right)}{qN_A}\right]^{1/2},$$

$$\text{and}\quad w_s = \left(\frac{2\epsilon\psi_{j0}}{qN_A}\right)^{1/2}, \tag{9.9}$$

assuming a one-dimensional analysis at the heavily doped regions themselves.

An overlap in these widths leads to an increase in the two-dimensional effects and barrier-limited injection of carriers into the substrate. The junction built-in voltage ψ_{j0} is nearly invariant, hence only w_d can be controlled. Hence, simultaneous reduction of w_d with channel length is achieved by reducing V_D. A consequence of this is poorer noise margin, and an increase of the acceptor doping to maintain good sub-threshold behavior and carrier confinement. Increasing N_A requires larger voltages or more doping in the large barrier region of a doped-barrier HFET to cause inversion. In a MESFET, an increase in the acceptor doping in the substrate has to be compensated by an additional increase in channel doping to maintain threshold voltages. Lowering V_D also requires lowering of the gate-to-source voltage (V_{GS}) in the logic gates, since one gate drives another, and hence barrier thicknesses have to be reduced to maintain channel charge. Barrier thicknesses for heterostructure field effect transistors are limited by gate leakage current, which is dependent on thermionic field emission and tunneling current. In insulator-based structures, the barrier thickness is limited by the insulator breakdown field. Like for HFETs, for MESFETs, the gate leakage characteristics are due to thermionic field emission at the heavily doped contact edges and in the central gate region. Limits related to these fields and currents are shown in Figure 9.4. This curve of breakdown-limited channel length is, however, an underestimate because it assumes a constant acceptor doping underneath the source and drain ohmic contacts. Lightly doped drain structures can be fabricated for HFETs, MISFETs, and MESFETs that prevent an increased doping in the gate–drain region, allowing this limitation to be lowered and thus allowing shorter channel lengths. An example of this is shown in Figure 9.5.

The typical short channel effects in FETs are summarized in Figure 9.6 and Figure 9.7 for scaled 1.0 μm, 0.5 μm, 0.25 μm, and 0.1 μm devices.

The short channel effects are virtually absent in the 1.0 μm devices, and gradually begin to appear in the shorter gate-length devices even with the use of scaling. The short channel effects occur in both HFETs and

---

Applications," *Proc. of IEEE*, **69**, No. 2, p. 226, Feb. 1981 for a discussion related to MOSFETs.

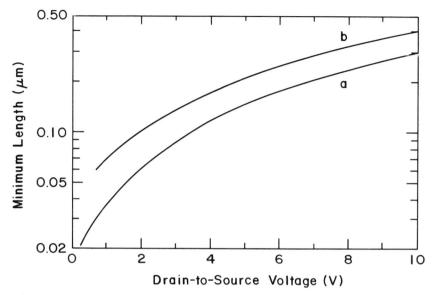

Figure 9.4: Channel length limits related to various breakdown phenomena for GaAs MESFETs. (a) shows limits due to gate leakage and breakdown with constant doping. (b) shows the increase in this by using lightly doped drain.

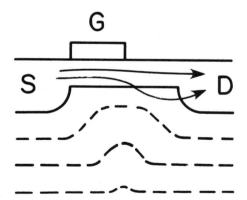

Figure 9.5: A GaAs MESFET structure with lightly doped drain region to obtain improvements in short channel phenomena. This should be compared with the earlier figure discussing short-channel effects.

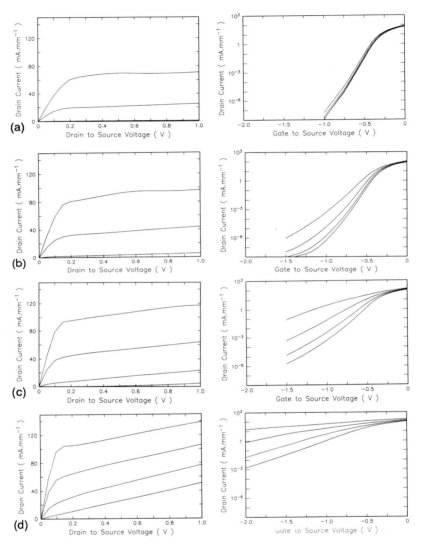

Figure 9.6: n-channel HFET output and sub-threshold characteristics for scaled 1.0, 0.5, 0.25, and 0.1 μm devices are shown in (a), (b), (c), and (d).

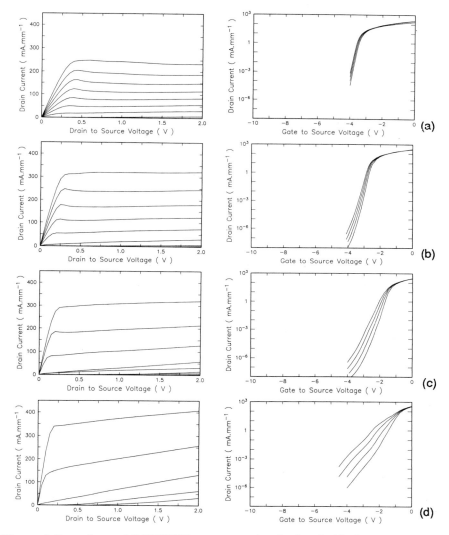

Figure 9.7: n-channel MESFET output and sub-threshold characteristics for scaled 1.0, 0.5, 0.25, and 0.1 μm devices are shown in (a), (b), (c), and (d).

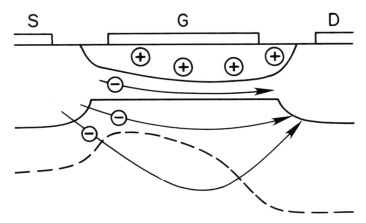

Figure 9.8: Mechanism of substrate conduction because of lowering of the barrier to the substrate in MESFETs. Electrons may travel through the substrate substantially away from the gate region because of the lowering of the back barrier due to proximity of the ohmic electrodes.

MESFETs, and are very pronounced in the one example of direct reduction of device channel length without any change in other device parameters.

The poor output conductance in these is increased by the lower barrier at the undoped or semi-insulating substrate. This mechanism is somewhat similar to the drain-induced barrier lowering mechanism due to poor control of channel by the gate (see Figure 9.8). The characteristics of the back interface, therefore, are quite important to limiting short channel effects during scaling. An abrupt barrier, e.g., utilizing heterostructures, reduces the short channel effect by limiting substrate injection. A consequence of this reduction of substrate injection, however, is a decrease in device current since the carrier densities are lowered by a stronger pinch-off behavior. Thus, additional charge has to be provided when limiting short channel behavior using a back barrier; this can possibly cause additional voltage variations.

Note that in an HFET, the inversion layer has a very small thickness and is separated by a semi-insulating or insulating barrier, while in a MESFET, the average separation of a gate from the conducting channel is bias-dependent. When the doped layer is thick (i.e., a is large), the gate has weaker control of the conducting channel because of increased two-dimensional effects, and the long channel models are increasingly invalid. If the conducting channel width is simultaneously modulated by both the gate and the drain, then the output conductance increases and other short channel effects also become important. Thus, just as in the HFET (i.e., similar to an increasing background doping N_A), MESFETs require an increasing doping in the channel (N_D) as gate lengths shrink. Usually, an additional constraint is placed on the channel thickness, $a < L/3$, to ensure that the electric field perpendicular to the interface is stronger than the electric field parallel to the interface, i.e., $\mathcal{E}_y > \mathcal{E}_z$. This allows for strong gate control, and hence good short channel behavior. Together with Figure 9.4, this limits the amount of channel charge that can be modulated (the maximum of this sheet charge is qN_Da) by the gate voltage. This limits the current. An example of poor output conductance due to large thickness is shown in Figure 9.9.

Conductance degradation due to similar thickness constraints occurs in HFETs also. While insulator thicknesses can be reduced to somewhere between 20–50 Å, the large gap semiconductor doped or undoped regions in HFETs (semi-insulating in character) can not be reduced due to tunneling currents resulting from lower barrier heights.

Operation at two logically distinct node voltages requires a consistent analysis that obtains the desired noise margin, speed, etc., while still minimizing power dissipation. Large switching speeds require that operating voltages differ amply enough to allow a significant current drive commensurate with fast switching. The lower voltage level, which is close to the threshold voltage (actually lower in the sub-threshold region), has to be sufficiently low to nearly cut off the current, and the threshold voltage should not be too large because proportionally larger power supplies will be required, together with larger power dissipation. Thus, for HFETs, MISFETs, and MESFETs, the sub-threshold cut-off should be rapid, i.e., as near to the ideal dictated by the device physics, and the threshold voltage should be the smallest value consistent with the above requirements. The power supply then is determined to allow necessary noise margin and circuit speed.

Some additional remarks regarding threshold voltage control are in order. The MOSFET in silicon with a poly-silicon gate of the opposite doping polarity as the substrate, or the SISFET with the same doping polarity gate as the inversion layer, have a naturally near-zero threshold voltage.

$$V_T = V_{FB} + 2\psi_B + \sqrt{2}\frac{kT}{q}\frac{C_{FB}}{C_{ins}}\left(2\frac{q\psi_B}{kT}\right)^{1/2} \qquad (9.10)$$

for a MISFET, and a similar variation of the equation occurs for SISFET as discussed in Chapter 6. Here, the flat-band voltage is the difference in the gate semiconductor work function, which is a negative quantity that

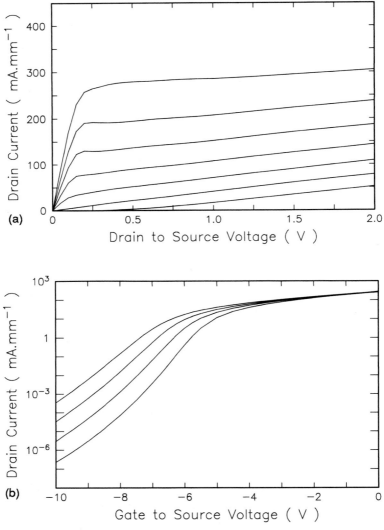

Figure 9.9: Output characteristics of a thick channel 0.5 μm gate-length device (a) and its sub-threshold characteristics (b) showing the poorer output conductance and sub-threshold control resulting from large channel thickness.

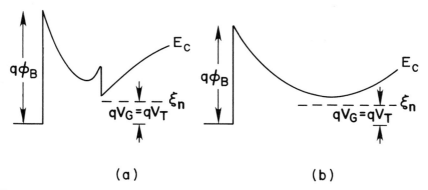

Figure 9.10: A comparison of band diagrams at threshold for HFETs and MESFETs.

compensates for twice the difference of the intrinsic and Fermi level in the bulk (the $2\psi_B$ term). Since C_{ins} is large, the last term is small, and hence the V_T is close to zero. Recall that in MOSFETs the usual technique of obtaining the appropriate threshold voltage is to do low dose implants to control ψ_B. In HFETs, the desire to obtain large mobilities usually leads to less emphasis on using substrate doping, and more emphasis on using doping in barriers, or use of differences in the composition of the gate material of a SISFET and the channel material to obtain the desired threshold voltage.

A MESFET, however, does not have compensating work functions, and the threshold voltage is a consequence of the delicate balancing of the built-in voltage, i.e., the Schottky barrier height, and potential drop across the doped channel region. A comparison of the band bending at threshold in the two situations is shown in Figure 9.10. For the MESFET, the threshold voltage can be approximately defined as

$$V_T \approx \phi_B + (\phi_C - \phi_F) - \frac{qN_D a^2}{2\epsilon}. \qquad (9.11)$$

MESFETs are relatively sensitive to channel parameter variations due to this delicate balancing, while MISFETs and SISFETs are relatively immune. Doped-barrier HFETs, however, are sensitive, due to similar reasons as the MESFETs. The sensitivities follow directly from the threshold equations above (see Problem 3). A change in charge in the MESFET channel directly influences the device threshold, while in the MISFET and SISFET it is reduced by the effect of drops across the insulating or semi-insulating barrier region, and hence is significantly smaller, although still tunable.

9.5.1 Limitations from Transport

We have discussed several material and device physics–related constraints so far, but one of the most significant properties of the material that we have not discussed is the velocity with which the carriers traverse. Specifically, we are interested in the velocities, or averages thereof, that are important to the device speed.

Consider the small-signal transconductance g_m of an FET in quasi-static approximation.

$$g_m = \left. \left| \frac{\Delta I_d}{\Delta V_{gs}} \right| \right|_{V_{DS}} = \frac{\Delta Q \times v}{\Delta V_{gs}} \tag{9.12}$$

in either the drift- or diffusion-dominated region, with negligible displacement current. This allows us to express velocity,

$$v = \frac{g_m}{C_g}, \tag{9.13}$$

where g_m is the small-signal quasi-static transconductance per unit width and C_g is the capacitance per unit area of the gate. This is an approximate relation, as our discussion of small-signal characteristics of field effect transistors showed, but it is of sufficient validity to make general arguments. Since current is constant, the source-end velocity may be expressed as

$$v_s = \frac{g_m}{C'_{gs}}, \tag{9.14}$$

where C'_{gs} is the capacitance per unit width at the source end.

This relation is based on control of the charge by the gate voltage, and hence drain bias dependence, i.e., two-dimensional effects, should remain small, otherwise, the resultant field will also influence the velocity. Now consider the steady-state velocity–field characteristics of various materials considered in Chapter 2. Examples of these various materials with their approximate source-end velocities are shown in Figure 9.11.

The source-end velocity regions on these curves correspond to g_m/C'_{gs} at the saturation current condition of the logical low state for an enhancement-depletion mode logic gate. At this condition, drift current dominates and the transistor is close to the resistive regime of operation, hence we may employ the average electric field (the ratio of the voltage and gate length) for comparison. The advantage of higher low-field mobility is clear at these gate lengths. It leads to a larger velocity of carriers as they enter at the

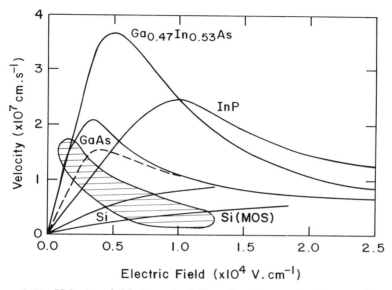

Figure 9.11: Velocity–field characteristics of various materials together with their source-end velocities in FETs marked as the shaded region.

source end of the channel. These higher velocities should translate to a natural speed advantage in materials with higher low-field mobility, provided the high-field velocities are either similar or better. Note that at high fields, most of the materials so far of interest have comparable steady-state velocities (approximately the saturated velocity). In the logical high state, much of the voltage drop occurs in a narrow region near the drain end for the micron sized devices. The fields are quite high. For 1.5 V supply voltages at 1 μm gate lengths, these will exceed the average field of 15 kV/cm in parts of the device. As gate lengths are shrunk further these fields would limit the velocity in parts of the device.

These general arguments are based on quasi-static and drift-diffusion considerations. Off-equilibrium and rapid changes can also be considered in an extension of these arguments. When rapid changes occur in time, in a FET, the current at the source end serves to charge or discharge the gate electrode during the initial part of the transient. The fields in this region are low; the low-field characteristics are important in determining the behavior of the response of the device at these time scales. Once the carrier density changes in the channel reach the drain end of the gate, the carrier densities there may change, and the drain current variations will occur. The velocities in this region can be large because of off-equilibrium effects, and this change can be rapid. Consequently, both low-field behavior and high-field behavior continue to be important in the switching transient as well as in the small-signal response of the transistor. Recall our initial arguments in this chapter regarding bipolar transistor and FET similarities. Such a dependence on the low-field and high-field properties of carrier transport continues to be important in a bipolar transistor also.

Our simple FET picture to this point has been derived based on a monotonically increasing velocity–field curve. It has to be modified in the context of the negative differential mobility that may exist in HFETs where channels are relatively undoped, or lightly doped ($< 5 \times 10^{17}$ cm^{-3}) GaAs or other compound semiconductor FETs. In these devices, the stationary space charge layer forms that we referred to as the dipole layer. It forms because any increase in fields beyond that for peak velocity decreases the velocity of the carriers. This local change in charge supports the large dipole field as well as the current continuity.

In the lower doped devices, one actually sees a negative resistance in output characteristics because of the decrease in velocity in the high field region with bias. At the usual dopings ($> 10^{17}$ cm^{-3}), this becomes smaller. However, one does observe a decrease in output and feedback capacitance because of increase in the width of the depletion region of the gate, and in the output conductance because of the effect of negative differential velocity is counter to the normal increase of output conductance due to short channel behavior. Both of these improve the behavior of compound semiconductor FETs. In particular, the effect in output and feedback capacitance is rather dramatic.

The implication from a device point of view, because C_{gs} represents the dominant input admittance element (there is very little conductance) and because C_{dg} represents the dominant feedback capacitance element (again there is very little feedback conductance), is that the ratio C_{gs}/C_{dg} might be considered as a figure of merit for high frequency devices. It characterizes the ratio of control to feedback in field effect devices. Comparative values of C_{dg} in between the silicon and compound semiconductor devices vary between two and eight. This discussion actually suggests that a very useful figure of merit for the purposes of gain may be $(g_m/C_{gs}) \times (C_{gs}/C_{dg}) = g_m/C_{dg}$.

9.6 Scaling and Operational Considerations of HBTs

Having studied the behavior and effects of grading, doping, current drive, recombination, forward base time constants, etc., we can now analyze and derive the design strategy for an HBT in the compound semiconductor system. Parabolic grading has been shown to give better current drive capability than linear grading, and hence is the preferred technique. Although space charge region recombination is lower for linear grading, it is dominated by quasi-neutral recombination at high current densities for all grading types, and hence is of secondary concern. Absence of doping in the grading regions gives lower space charge region recombination, but for the same reason as linear grading this is also not a preferable design technique. For collector alloy grading in heterojunction collector devices, linear grading is attractive because it reduces the alloy field near the junction compared to parabolic grading, and the larger space charge width allows for a design wherein the barrier at the end of the linear grading can be suppressed by a sufficiently large electrostatic field. The important design strategy is therefore to use dopings that are sufficiently high that both the conventional bipolar effects and the HBT alloy barrier are avoided.

We emphasize the digital aspects of the design—the high frequency aspects follow in a similar way. Digital operation requires the device to be capable both of high current drive with low capacitances and time constants and of sufficiently large current gains. The scaling of the device therefore differs in two important respects from that of the homojunction silicon transistor. First, the device has to be designed to compensate for the barrier effects, and second, the gain of the device has to be maintained by suppressing the surface recombination effect, and maintaining a sufficient base transport factor. The first constraint requires the emitter and the collector dopings to be increased as a function of current density. In silicon bipolar transistors, only the collector doping has to be increased; the emitter is already doped high to maintain good injection efficiency. The second constraint requires suppression of surface recombination and space charge region recombination. Surface recombination can be minimized by the use of the various techniques that place a barrier to minority carrier at the surface. Space charge region recombination is sufficiently small at current densities of interest, as we have seen, and remains so even with increasing emitter and base dopings.

In such a design, at high current densities the recombination current density in the quasi-neutral base becomes important, and this requires the inclusion of the base transport factor and thus the dependence of lifetime on doping. While injection efficiency is a constraint in silicon bipolar transistors, the base transport factor is the constraint in GaAs. The base doping can be significantly higher than in the homojunction bipolar. At base dopings of interest ($N_A > 10^{18}$ cm^{-3}), the lifetime varies inversely with the doping ($\tau \propto N_A^{-1}$). Ignoring two-dimensional effects, the gain varies as $2\mathcal{L}_n^2/w_B^2$. Thus, current gain $\beta \propto (N_A w_B^2)^{-1}$. The base time constant τ_B varies approximately as $w_B^2/2\mathcal{D}_n$. $\mathcal{D}_n$ varies inversely with the doping N_A because of ionized impurity scattering.[2]

A self-consistent method for obtaining scaling of the base time constant and maintaining gain requires the base width to be varied as the scaling factor λ, while varying the base doping between $1/\lambda^{1.33}$ and $1/\lambda$. This allows the base transport factor to vary between $\lambda^{1.33}$ and λ (slightly faster than the scaling factor), while maintaining the current gain in the device structure. This increase in base doping is not as rapid as in silicon bipolar transistors where the doping levels are lower. For constant voltage swing (usually chosen to be the minimal acceptable from noise and power constraints), and constant power supply voltages (determined by the bandgap of the base material and the logic swing), the currents remain constant and the current density increases as λ^{-2}. The collector and the emitter dopings also increase at this same rate. This results in the total delay due to intrinsic component scaling as λ. These scaling factors of the important epitaxial parameters are summarized in Table 9.1.

The minimum thicknesses of the emitter and collector regions depend on the doping, the junction voltage swing, and the minimization of the parasitic resistance. For the collector, the breakdown voltage must also be considered, and for the emitter, tunneling currents at low bias due to degenerate doping in the emitter and the base must be taken into account. While DX centers play a negligible role in the operation of HBTs, with regard to transient effects in current transport, their inefficient ionization leads to larger emitter resistance, which may become important at current densities of 10^5 A.cm^{-2}. Thus, although the $Ga_{1-x}Al_xAs/GaAs$ HBT has generally employed mole-fractions of AlAs close to 0.3, scaled submicron devices operating at high current densities may require lower AlAs mole-fractions.

9.7   Summary

This chapter discussed, in general, the limitations placed on miniaturization of devices by either the scaling framework followed or by operational con-

---

[2]The data on this are scarce, but in the regime dominated by ionized impurity scattering, this varies for majority carriers as $N_A^{-0.5}$ to $N_A^{-1.0}$.

Table 9.1: Scaling of compound semiconductor HBTs.

Vertical scaling for horizontal scaling of λ	
Current Density	$1/\lambda^2$
Emitter Doping	$J \approx 1/\lambda^2$
Emitter Thickness	λ
Base Thickness	λ
Base Doping	$1/\lambda$
Collector Doping	$J \approx 1/\lambda^2$
Collector Thickness	λ

straints. First, we considered the general scaling behavior of the underlying electromagnetic and transport equations. This allowed us to understand how the device fields and current densities may be expected to change in practice. We then considered operational constraints. An example of an operational constraint is breakdown voltage at short gate lengths in FETs due to an increase of substrate doping. Another operational constraint in FETs is the poor sub-threshold behavior at short dimensions. We then deduced possible scaling behavior for field effect and bipolar transistors assuming a certain set of ground rules for the scaling.

General References

1. S. I. Long and S. E. Butner, *Gallium Arsenide Digital Integrated Circuit Design*, McGraw-Hill, N.Y. (1990).

2. R. K. Watts, Ed., *Submicron Integrated Circuits*, John Wiley, N.Y. (1989).

3. H. S. Carslaw and J. C. Jaeger, *Conduction of Heat in Solids*, Oxford, N.Y. (1959).

4. R. W. Keyes, *The Physics of VLSI Systems*, Addison-Wesley, Wokingham, England (1987).

5. B. Hoenesin and C. A. Mead, "Fundamental Limitations in Microelectronics—I. MOS Technology," *Solid-State Electronics*, **15**, p. 819, Aug. 1972.

6. B. Hoenesin and C. A. Mead, "Fundamental Limitations in Micro-electronics—II. Bipolar Technology," *Solid-State Electronics*, **15**, p. 891, Sep. 1972.

7. P. M. Solomon, "A Comparison of Semiconductor Devices for High-speed Logic," *Proc. of IEEE*, **70**, No. 5, p. 489, May 1982.

Problems

1. Derive the scaled Maxwell equations by allowing the potential to also change. How do the current and charge scale under these conditions?

2. Consider the transport of heat on a chip. Consider circuits employing FETs of 1 μm gate length. The FETs are 10 μm wide and carry a current density of 0.1 A.mm^{-1}, with an average drain-to-source voltage of 0.75 V. These devices are made on GaAs substrates that are 250 μm thick. Consider an isolated device, with heat coming from a cylindrical source of an average 1 μm diameter (a typical gate, and gate-to-drain spacing). Find the temperature distribution assuming that the back side of the substrate is at 300 K. What would be the effect on the temperature if such heat sources were 10 μm apart? The general reference by Carslaw and Jaeger considers boundary conditions that are similar to those of this problem.

3. Let us compare the threshold voltage sensitivity to device parameters at long gate lengths. Consider the GaAs gate n-channel SISFET and the GaAs MESFET at threshold conditions. The GaAs gate n-channel SISFET utilizes a $Ga_{1-x}Al_xAs$ whose thickness is 300 Å and which exhibits an electron band edge discontinuity of 0.3 eV with GaAs. The substrate is doped to 5×10^{15} cm^{-3}, and the doping of gate material aligns the electron quasi-Fermi level with the conduction band edge away from the interface. The GaAs MESFET has a channel doping of 5×10^{17} cm^{-3}, a threshold voltage of 0.2 V, and a metal–semiconductor barrier height of 0.8 eV.

   (a) Find the percent change in the threshold voltage for a 10% change in the doping of the substrate and the channel for the SISFET and the MESFET.

   (b) Find the percent change in the threshold voltage for a 10% change in the thickness of the barrier semiconductor and the channel layer for the SISFET and the MESFET.

Appendix A

Network Parameters and Relationships

The network parameters can be represented in several forms associated with the input and output variables being represented. Admittance parameters (y-parameters), e.g., are associated with the currents being represented as a function of potentials; impedance parameters (z-parameters) are associated with the voltages being represented as a function of currents; $ABCD$-parameters are a combination of these, which are particularly suitable for prediction of the network parameters of a cascade of networks; hybrid parameters (h-parameters) are also a combination of current and voltage; and finally scattering parameters (S-parameters) relate power waves. For the n-port network represented in Figure A.1, consider the incident and reflected waves $[\hat{a}]$ and $[\hat{b}]$, and the currents $[\hat{I}]$ and voltages $[\hat{V}]$ as $n \times 1$ vectors.

In terms of the current vectors $[\hat{I}]$, the voltage vectors $[\hat{V}]$, and the characteristic impedence Z_0, the power waves are defined as

$$[\hat{a}] = \frac{1}{2\sqrt{Z_0}}[\hat{V}] + \frac{\sqrt{Z_0}}{2}[\hat{I}]$$

$$\text{and} \qquad [\hat{b}] = \frac{1}{2\sqrt{Z_0}}[\hat{V}] - \frac{\sqrt{Z_0}}{2}[\hat{I}] \qquad (A.1)$$

The network parameters, then, follow from the definitions:
y-parameters:

$$[\hat{I}] = [y][\hat{V}], \qquad (A.2)$$

z-parameters:

$$[\hat{V}] = [z][\hat{I}], \qquad (A.3)$$

783

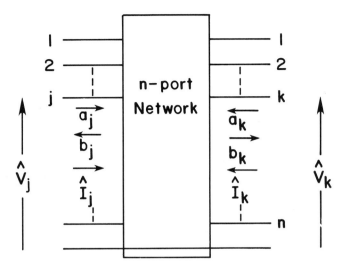

Figure A.1: An n-port network showing the incident and reflected power waves, currents, and voltages to relate the network parameters.

and S-parameters:

$$[\hat{b}] = [S][\hat{a}].\qquad (A.4)$$

h-parameters and $ABCD$-parameters are defined with combinations of currents and voltages as the independent and dependent variables. For a two-port network, these are
h-parameters:

$$\left[\begin{array}{c} \hat{V}_1 \\ \hat{I}_2 \end{array}\right] = \left[\begin{array}{cc} h_{11} & h_{12} \\ h_{21} & h_{22} \end{array}\right]\left[\begin{array}{c} \hat{I}_1 \\ \hat{V}_2 \end{array}\right],\qquad (A.5)$$

and $ABCD$-parameters:

$$\left[\begin{array}{c} \hat{V}_1 \\ \hat{I}_1 \end{array}\right] = \left[\begin{array}{cc} A & B \\ C & D \end{array}\right]\left[\begin{array}{c} \hat{V}_2 \\ -\hat{I}_2 \end{array}\right].\qquad (A.6)$$

The h-parameters get their name from the hybrid nature of the parameters. h_{11} is the input impedance under conditions of short circuit at the output port $\hat{V}_2 = 0$, h_{12} is the reverse voltage gain with open-circuit conditions at the input port, h_{21} is the forward current gain for short-circuit conditions at the output port, and h_{22} is the output admittance with the input port open-circuited.

These network parameters are definite parameters with a common reference terminal. This common reference terminal could have been used as an additional port, with corresponding network parameters not being independent. This extended set of parameters are called indefinite parameters. As an example, consider the two-port y-parameters. If we use the common terminal as an additional port 3,

$$\begin{bmatrix} \hat{I}_1 \\ \hat{I}_2 \\ \hat{I}_2 \end{bmatrix} = \begin{bmatrix} y_{11} & y_{12} & y_{13} \\ y_{21} & y_{22} & y_{23} \\ y_{31} & y_{32} & y_{33} \end{bmatrix} \begin{bmatrix} \hat{V}_1 \\ \hat{V}_2 \\ \hat{V}_3 \end{bmatrix}. \tag{A.7}$$

Kirchoff's law and the generality of the above for all applied $[V]$ results in the sum of all columns or rows being zero. For example, since the sum of the currents is zero, assuming a short circuit at port 2 and port 3 implies that the sum of the y-parameters in the first column is zero and likewise in the rest of the columns. The rows can be shown to follow this same behavior by assuming identical voltages at all the ports. Since this should result in zero current at all the ports, it implies that the sum of all parameters in a row is zero. Crossing out the row and column corresponding to one of the ports results in definite network parameters with that port common. Thus common-gate/base, common-source/emitter, and common-drain/collector parameters can be easily related to their other references. Transforming the common port is most conveniently achieved by determining the indefinite matrix within a particular type of parameter set.

Relationships between the parameters are determined by transformations using matrix manipulation and Kirchoff's laws. We summarize these here.

For y-parameters,

$$\begin{bmatrix} y_{11} & y_{12} \\ y_{21} & y_{12} \end{bmatrix} = \begin{bmatrix} 1/h_{11} & -h_{12}/h_{11} \\ h_{21}/h_{11} & \Delta_h/h_{11} \end{bmatrix} = \begin{bmatrix} z_{22}/\Delta_z & -z_{12}/\Delta_z \\ -z_{21}/\Delta_z & z_{11}/\Delta_z \end{bmatrix}, \tag{A.8}$$

for h-parameters,

$$\begin{bmatrix} 1/y_{11} & -y_{12}/y_{11} \\ y_{21}/y_{11} & \Delta_y/y_{11} \end{bmatrix} = \begin{bmatrix} h_{11} & h_{12} \\ h_{21} & h_{22} \end{bmatrix} = \begin{bmatrix} \Delta_z/z_{22} & z_{12}/z_{22} \\ -z_{21}/z_{22} & 1/z_{22} \end{bmatrix}, \tag{A.9}$$

and for z-parameters,

$$\begin{bmatrix} y_{22}/\Delta_y & -y_{12}/\Delta_y \\ -y_{21}/\Delta y & y_{11}/\Delta_y \end{bmatrix} = \begin{bmatrix} \Delta h/h_{22} & h_{12}/h_{22} \\ -h_{21}/h_{22} & 1/h_{22} \end{bmatrix} = \begin{bmatrix} z_{11} & z_{12} \\ z_{21} & z_{22} \end{bmatrix}, \tag{A.10}$$

where

$$\begin{aligned}
\Delta_y &= y_{11}y_{22} - y_{12}y_{21}, \\
\Delta_h &= h_{11}h_{22} - h_{12}h_{21}, \\
\text{and} \quad \Delta_z &= z_{11}z_{22} - z_{12}z_{21}.
\end{aligned} \qquad \text{(A.11)}$$

Transformation of reference ports and parameters is most easily accomplished by the use of an indefinite matrix. As an example, by obtaining the indefinite matrix in y-parameters, the y-parameter with any common reference is obtained. Any other parameter with this reference can be obtained by the matrix manipulation above. A summary of some of the more commonly encountered conversions from common-gate/base to common-source/emitter configurations is described below.

$$\begin{aligned}
y_{11}^{s/e} &= y_{11}^{g/b} + y_{12}^{g/b} + y_{21}^{g/b} + y_{22}^{g/b}, \\
&= \frac{\left(1 + h_{21}^{g/b}\right)(1 + \epsilon)}{h_{11}^{g/b}}, \\
y_{12}^{s/e} &= -\left(y_{22}^{g/b} + y_{12}^{g/b}\right), \\
&= -\frac{\epsilon\left(1 + h_{21}^{g/b}\right)}{h_{11}^{g/b}}, \\
y_{21}^{s/e} &= -\left(y_{22}^{g/b} + y_{21}^{g/b}\right), \\
&= -\frac{h_{21}^{g/b}(1 + \epsilon) + h_{12}^{g/b} + \epsilon}{h_{11}^{g/b}}, \\
\text{and} \quad y_{22}^{s/e} &= y_{22}^{g/b}, \\
&= \frac{h_{12}^{g/b} + \epsilon\left(1 + h_{21}^{g/b}\right)}{h_{11}^{g/b}}, \qquad \text{(A.12)}
\end{aligned}$$

and

$$\begin{aligned}
h_{11}^{s/e} &= \frac{h_{11}^{g/b}}{\left(1 + h_{21}^{g/b}\right)(1 + \epsilon)}, \\
h_{12}^{s/e} &= \frac{\epsilon}{1 + \epsilon}, \\
h_{21}^{s/e} &= -\frac{h_{21}^{g/b}(1 + \epsilon) + h_{12}^{g/b} + \epsilon}{(1 + \epsilon)},
\end{aligned}$$

$$\text{and} \quad h_{22}^{s/e} = \frac{h_{22}^{g/b}}{\left(1 + h_{21}^{g/b}\right)(1 + \epsilon)}, \tag{A.13}$$

where

$$\epsilon = \frac{h_{11}^{g/b} h_{22}^{g/b}}{1 + h_{21}^{g/b}} - h_{12}^{g/b}. \tag{A.14}$$

Experimental measurements usually provide S-parameter data, while device analysis is best carried out using parameters based on currents and voltages. The relationships between the S-parameters and other parameters are as follows:

$$s_{11} = \frac{(1 - y_{11})(1 + y_{22}) + y_{12}y_{21}}{(1 + y_{11})(1 + y_{22}) - y_{12}y_{21}} \qquad s_{12} = \frac{-2y_{12}}{(1 + y_{11})(1 + y_{22}) - y_{12}y_{21}}$$

$$s_{21} = \frac{-2y_{21}}{(1 + y_{11})(1 + y_{22}) - y_{12}y_{21}} \qquad s_{22} = \frac{(1 + y_{11})(1 - y_{22}) + y_{12}y_{21}}{(1 + y_{11})(1 + y_{22}) - y_{12}y_{21}}$$

$$s_{11} = \frac{(h_{11} - 1)(h_{22} + 1) - h_{12}h_{21}}{(h_{11} + 1)(h_{22} + 1) - h_{12}h_{21}} \qquad s_{12} = \frac{2h_{12}}{(h_{11} + 1)(h_{22} + 1) - h_{12}h_{21}}$$

$$s_{21} = \frac{-2h_{21}}{(h_{11} + 1)(h_{22} + 1) - h_{12}h_{21}} \qquad s_{22} = \frac{(1 + h_{11})(1 - h_{22}) + h_{12}h_{21}}{(h_{11} + 1)(h_{22} + 1) - h_{12}h_{21}}$$

$$s_{11} = \frac{(z_{11} - 1)(z_{22} + 1) - z_{12}z_{21}}{(z_{11} + 1)(z_{22} + 1) - z_{12}z_{21}} \qquad s_{12} = \frac{2z_{12}}{(z_{11} + 1)(z_{22} + 1) - z_{12}z_{21}}$$

$$s_{21} = \frac{2z_{21}}{(z_{11} + 1)(z_{22} + 1) - z_{12}z_{21}} \qquad s_{22} = \frac{(z_{11} + 1)(z_{22} - 1) - z_{12}z_{21}}{(z_{11} + 1)(z_{22} + 1) - z_{12}z_{21}}$$

$$y_{11} = \frac{(1 - s_{11})(1 + s_{22}) + s_{12}s_{21}}{(1 + s_{11})(1 + s_{22}) - s_{12}s_{21}} \qquad y_{12} = \frac{-2s_{12}}{(1 + s_{11})(1 + s_{22}) - s_{12}s_{21}}$$

$$y_{21} = \frac{-2s_{21}}{(1 + s_{11})(1 + s_{22}) - s_{12}s_{21}} \qquad y_{22} = \frac{(1 + s_{11})(1 - s_{22}) + s_{12}s_{21}}{(1 + s_{11})(1 + s_{22}) - s_{12}s_{21}}$$

$$h_{11} = \frac{(1 + s_{11})(1 + s_{22}) - s_{12}s_{21}}{(1 - s_{11})(1 + s_{22}) + s_{12}s_{21}} \qquad h_{12} = \frac{2s_{12}}{(1 - s_{11})(1 + s_{22}) + s_{12}s_{21}}$$

$$h_{21} = \frac{-2s_{21}}{(1 - s_{11})(1 + s_{22}) + s_{12}s_{21}} \qquad h_{22} = \frac{(1 - s_{11})(1 - s_{22}) - s_{12}s_{21}}{(1 - s_{11})(1 + s_{22}) + s_{12}s_{21}}$$

$$z_{11} = \frac{(1 + s_{11})(1 - s_{22}) + s_{12}s_{21}}{(1 - s_{11})(1 - s_{22}) - s_{12}s_{21}} \qquad z_{12} = \frac{2s_{12}}{(1 - s_{11})(1 - s_{22}) - s_{12}s_{21}}$$

$$z_{21} = \frac{2s_{21}}{(1 - s_{11})(1 - s_{22}) - s_{12}s_{21}} \qquad z_{22} = \frac{(1 - s_{11})(1 - s_{22}) + s_{12}s_{21}}{(1 + s_{11})(1 + s_{22}) - s_{12}s_{21}}. $$

$$\tag{A.15}$$

The relationships of the gains and stability factors are as follows: the maximum power gain when a two-port device has been made unilateral using only loss-less reciprocal elements—the unilateral power gain—is given by

$$U = \frac{|y_{21} - y_{12}|^2}{4\{\text{Re}[y_{11}]\,\text{Re}[y_{22}] - \text{Re}[y_{21}]\,\text{Re}[y_{12}]\}}$$

$$= \frac{|h_{21} + h_{12}|^2}{4\{\text{Re}[h_{11}]\,\text{Re}[h_{22}] + \text{Re}[h_{21}]\,\text{Re}[h_{12}]\}}$$

$$= \frac{|s_{11}s_{12}s_{21}s_{22}|}{\left|\left(1 - |s_{11}|^2\right)\left(1 - |s_{22}|^2\right)\right|}. \tag{A.16}$$

The Linvill stability factor is given by

$$C = -\frac{|y_{12}y_{21}|}{\text{Re}[y_{12}y_{21}] - 2\text{Re}[y_{11}]\,\text{Re}[y_{22}]}. \tag{A.17}$$

A network is unconditionally stable if $0 < C < 1$. The network is potentially unstable, i.e., oscillations may occur for passive terminations, if C is outside this range.

The transducer power gain, for arbitrary load and source, is

$$G_T = \frac{4|y_{21}|^2\text{Re}[y_l]\,\text{Re}[y_s]}{|(y_{11} + y_s)(y_{22}y_l) - y_{12}y_{21}|^2}, \tag{A.18}$$

and the maximum transducer power gain is

$$G_{T,max} = \frac{1}{2\text{Re}[y_{22}]}\left(\{2\text{Re}[y_{11}]\,\text{Re}[y_{22}] - \text{Re}[y_{12}y_{21}]\}^2 - |y_{12}y_{21}|\right)^{1/2}. \tag{A.19}$$

This is also referred to as the maximum available power gain and maximum operating power gain; it is meaningful only when the network is unconditionally stable, i.e., $0 < C < 1$. For conditionally unstable networks, quite often, one uses the power gain parameter called maximum stable gain,

$$G_{s,max} = \frac{|y_{21}|}{|y_{12}|}. \tag{A.20}$$

Finally, the definitions of frequency figures of merit are as follows. The maximum frequency of oscillation (f_{max}) is the frequency at which the unilateral gain is unity, and the transition frequency (f_T, also called short-circuit unity current gain frequency) is the frequency at which the current gain is unity under short-circuit conditions at the output. The physical meaning of these parameters is intuitive. A network would oscillate only until the frequency f_{max}, if a reciprocal passive matching network was employed, in order to obtain the oscillations. Likewise, the current gain can be obtained only up to f_T under short-circuit output conditions for the network.

Appendix B

Properties of Compound Semiconductors

The data are based on information from the following references:

1. Landolt-Börnstein; O. Madelung, M. Schulz, and H. Weiss, Eds., *Semiconductors*, **V 17**, Springer-Verlag, Berlin (1982).

2. Landolt-Börnstein; O. Madelung, M. Schulz, and H. Weiss, Eds., *Semiconductors*, **V 22**, Springer-Verlag, Berlin (1987).

3. J. S. Blakemore, "Semiconducting and Other Major Properties of Gallium Arsenide," *J. of Appl. Phys.*, **53**, No. 10, p. R123, Oct. 1982.

4. S. Adachi, "GaAs, AlAs, and $Al_xGa_{1-x}As$: Material Parameters for Use in Research and Device Applications," *J. of Appl. Phys.*, **58**, p. R1, 1985.

Table B.1: Material Properties of Compound Semiconductors

Material	Lattice	Space Group	Lattice Constant(Å)		Density (g.cm^{-3})
			300 K	77 K	300 K
BN	Hex	$P\overline{6}$	a-axis: 6.661	6.660	2.18
			c-axis: 2.504	2.503	—
BN	ZB	$F\overline{4}3m$	3.6155	—	3.487
BP	ZB	$F\overline{4}3m$	4.5383	4.5374	2.0
BAs	ZB	$F\overline{4}3m$	4.777	—	5.22
AlN	WZ	$P6_3mc$	a-axis: 3.112	—	3.23
			c-axis: 4.980	—	—
AlP	ZB	$F\overline{4}3m$	5.467	5.466	2.40
AlAs	ZB	$F\overline{4}3m$	5.661	5.660	—
AlSb	ZB	$F\overline{4}3m$	6.136	—	4.26
GaN	WZ	$P6_3mc$	3.160	—	6.095
GaP	ZB	$F\overline{4}3m$	5.4506	—	4.138
GaAs	ZB	$F\overline{4}3m$	5.6533	5.6419	5.3176
GaSb	ZB	$\overline{4}3m$	6.096	6.095	5.6137
InN	WZ	$P6_3mc$	a-axis: 3.544	3.543	—
			c-axis: 5.7034	5.7027	—
InP	ZB	$F\overline{4}3m$	5.8687	5.8680	4.81
InAs	ZB	$F\overline{4}3m$	6.0583	—	5.70
InSb	ZB	$F\overline{4}3m$	4.4794	—	5.7747

Table B.2: Material Properties of Compound Semiconductors

Material	Melting /Decomposition Temperature (K)	Debye Temperature (K)	Linear Expansion Coefficient at 300 K
BN_{hex}	2600	598	$\alpha_{\|\|} = 2.7 \times 10^{-6}$ $\alpha_{\perp} = 3.7 \times 10^{-6}$
BN_{ZB}	≈ 3000	1700	—
BP	1400	985	—
BAs	—	800	—
AlN	3273	—	$\alpha_{\perp} = 5.27 \times 10^{-6}$ $\alpha_{\|\|} = 4.15 \times 10^{-6}$
AlP	2823	588	—
AlAs	2013	690	—
AlSb	1338	292	—
GaN	> 2000	600	$\alpha_{\|\|} = 5.59 \times 10^{-6}$ $\alpha_{\perp} = 3.17 \times 10^{-6}$
GaP	1740	445	—
GaAs	1513	344	6.86×10^{-6}
GaSb	985	266	7.75×10^{-6}
InP	1335	321	4.75×10^{-6}
InAs	1215	262	4.52×10^{-6}
InSb	800	208	5.37×10^{-6}

Table B.3: Material Properties of Compound Semiconductors

Material	Thermal Conductivity $(W.K^{-1}.cm^{-1})$		Specific Heat $(J.g^{-1}.K^{-1})$	Poisson Ratio	Radiative Constant, B $(cm^3.s^{-1})$
	300 K	77 K	300 K		
BN_{ZB}	≈ 1	—	0.601	—	—
BP	≈ 8	—	0.64	—	—
BAs	—	—	0.367	—	—
AlN	≈ 3	—	0.490	—	—
AlP	—	—	0.816	0.32	—
AlAs	—	—	0.49	0.30	—
AlSb	≈ 0.7	≈ 1.8	0.437	0.33	—
GaN	1.3	≈ 0.8	0.440	—	—
GaP	.77	0.8	0.519	0.307	3.0×10^{-15}
GaAs	.455	$\approx 1\text{--}3$	0.35	0.31	—
GaSb	—	—	0.404	0.31	1.3×10^{-11}
InP	.68	5.3	0.41	0.36	—
InAs	.48	3.2	0.394	0.35	2.1×10^{-11}
InSb	.27	1.1	0.37	0.35	4.0×10^{-11}

Table B.4: Electronic Properties of Some Compound Semiconductors

Material	Direct/ Indirect	Energy Gap (eV)		Zone Center	
		300 K	77 K	LO Phonon Energy (meV)	TO Phonon Energy (meV)
BN_{hex}	Indirect	4.5	—	199.6	169.9
BN_{ZB}	Indirect	≈ 6.0	—	161.7	130.9
BP	Indirect	2.0	102.8	101.7	—
BAs	Indirect	—	—	—	—
AlN	Direct	6.2	—	112.8	82.7
AlP	Indirect	2.45	—	62.1	54.5
AlAs	Indirect	2.14	2.223	50.1	44.8
AlSb	Indirect	1.63	1.666	42.1	39.5
GaN	Direct	3.44	—	92.4	69.4
GaP	Indirect	2.268	2.338	50.0	45.5
GaAs	Direct	1.423	1.512	35.4	33.2
GaSb	Direct	0.70	—	—	—
InN	Direct	2.09	2.21	86.1	59.3
InP	Direct	1.3511	1.4135	42.8	37.7
InAs	Direct	.356	.414	29.6	26.9
InSb	Direct	.18	.228	23.7	22.3

Table B.5: Electronic Properties of Some Compound Semiconductors

Material	Electron Conduction Effective Mass in Units of m_0		Hole Conduction Effective Mass in Units of m_0		Average Binding Energy (meV)	
	300 K	77 K	300 K	77 K	Donor	Acceptor
BN_{hex}	—	—	—	—	—	—
BN_{ZB}	—	—	—	—	—	—
BP	—	—	—	—	—	—
BAs	—	—	—	—	—	—
AlN	—	—	—	—	—	—
AlP	—	—	—	—	—	—
AlAs	—	—	—	—	≈ 60	≈ 55
AlSb	—	—	0.4	—	70-160	≈ 40
GaN	—	—	—	—	—	—
GaP	—	—	—	—	—	—
GaAs	.067	.067	.5	.45	5.7–5.9	2.5–3.0
			.068	.068	—	—
GaSb	—	—	—	—	20–80	10–40
InN	—	—	—	—	—	—
InP	.073	.077	—	—	—	≈ 40–50
InAs	.027	.027	—	—	20	10–15
InSb	.013	.0145	.044	.045	50–100	≈ 10

Table B.6: Electronic Properties of Some Compound Semiconductors

Material	Electron Density of States Effective Mass in Units of m_0		Hole Density of States Effective Mass in Units of m_0	
	300 K	77 K	300 K	77 K
BN_{hex}	—	—	—	—
BN_{ZB}	—	—	—	—
BP	—	—	—	—
BAs	—	—	—	—
AlN	—	—	—	—
AlP	—	—	—	—
AlAs	—	—	—	—
AlSb	—	0.7	0.98	0.98
GaN	0.19	0.19	0.60	0.60
GaP	—	—	—	—
GaAs	.067	.067	.5	.45
			.068	.068
GaSb	0.7	0.729	0.4	0.4
InN	—	—	—	—
InP	0.077	0.077	0.64	0.64
InAs	0.023	0.023	0.4	0.4
InSb	0.015	0.015	0.4	0.4

Table B.7: Electronic Properties of Some Compound Semiconductors

| Material | Approximate Electron Mobility | | | | | |
| | at Low Doping of $< 10^{15}$ cm^{-3} ($\mathrm{cm^2.V^{-1}.s^{-1}}$) | | at Medium Doping of $\approx 10^{17}$ cm^{-3} ($\mathrm{cm^2.V^{-1}.s^{-1}}$) | | at Heavy Doping of $\approx 10^{19}$ cm^{-3} ($\mathrm{cm^2.V^{-1}.s^{-1}}$) | |
	300 K	77 K	300 K	77 K	300 K	77 K
BN_{hex}	—	—	—	—	—	—
BN_{ZB}	—	—	70	—	30	
BP	—	—	—	—	—	—
BAs	—	—	—	—	—	—
AlN	—	—	—	—	—	—
AlP	—	—	—	—	—	—
AlAs	—	—	400	—	40	—
AlSb	200	—	—	—	—	—
GaN	—	—	600	—	180	—
GaP	200	2800	160	1200	95	450
GaAs	8000	150000	4800	5200	2200	3000
GaSb	5000	—	—	—	—	—
InN	—	—	—	—	—	—
InP	5000	60000	3200	—	1800	—
InAs	30000	300000	20000	35000	13000	20000
InSb		400000	80000	150000	10000	80000

Table B.8: Electronic Properties of Some Compound Semiconductors

Material	Approximate Hole Mobility					
	at Low Doping of $< 10^{15}$ cm^{-3} ($\text{cm}^2.\text{V}^{-1}.\text{s}^{-1}$)		at Medium Doping of $\approx 10^{17}$ cm^{-3} ($\text{cm}^2.\text{V}^{-1}.\text{s}^{-1}$)		at Heavy Doping of $\approx 10^{19}$ cm^{-3} ($\text{cm}^2.\text{V}^{-1}.\text{s}^{-1}$)	
	300 K	77 K	300 K	77 K	300 K	77 K
BN_{hex}	—	—	—	—	—	—
BN_{ZB}	—	—	—	—	—	—
BP	—	—	—	—	30	20
BAs	—	—	—	—	—	—
AlN	—	—	—	—	—	—
AlP	—	—	—	—	—	—
AlAs	—	—	—	—	—	—
AlSb	450	—	375	—	175	—
GaN	—	—	—	—	—	—
GaP	150	—	120	—	30	—
GaAs	400	9000	320	—	50	—
GaSb	1500	—	880	2400	300	1000
InN	—	—	—	—	—	—
InP	180	1000	150	600	28	18
InAs	480	—	—	—	—	—
InSb	1500	9000	—	3000	—	700

Table B.9: Electronic Properties of Some Compound Semiconductors

Material	Dielectric Constant		Refractive Index in	Hall Factor at $\approx 10^{17}$ cm^{-3}	
	$\epsilon_r(0)$	$\epsilon_r(\infty)$	Visible Range	300 K	77 K
BN_{hex}	c-axis: 5.066	c-axis: 4.10			
	$\perp c$-axis: 6.85	$\perp c$-axis: 4.95			
BN_{ZB}	7.1	4.5	2.117	—	—
BP	11	—	3.10	—	
BAs	—	—	—	—	—
AlN	9.1	4.8	—	—	—
AlP	9.8	7.54	≈ 2.85	—	—
AlAs	10.06	8.16	—	—	—
AlSb	12.04	10.24	3.45	—	—
GaN	c-axis: 10.4	c-axis: 5.8	2.33	—	—
	$\perp c$-axis: 9.5	$\perp c$-axis: 5.35	—	—	—
GaP	11.1	9.075	3.17	—	—
GaAs	12.91	10.1	3.347	1.1	1.27
GaSb	15.69	14.44	3.82	1.03	—
InN	—	9.3	2.56	—	—
InP	12.61	9.61	3.327	≈ 1	≈ 1.1
InAs	15.15	12.25	3.516	≈ 1.3	≈ 1.5
InSb	16.8	15.68	5.13	—	—

Table B.10: Electronic Properties of Some Compound Semiconductors

Material	Valence Band Deformation Potential (eV)	Spin Orbit Splitting (eV)	Intrinsic Carrier Concentration (cm^{-3})	Ionic Charge $\div q$ —
BN_{hex}	—	—	—	—
BN_{ZB}	—	—	—	—
BP	—	—	—	—
BAs	—	—	—	—
AlN	—	0.012	—	—
AlP	—	0.060	—	—
AlAs	—	0.29	—	—
AlSb	−4.4	0.68	5.0×10^6	.19
GaN	—	0.011	—	—
GaP	−13	0.08	3.0×10^6	.24
GaAs	−12	0.34	1.8×10^6	.20
GaSb	−8.3	0.74	4.3×10^{12}	.13
InN	—	0.08		—
InP	−6.6	0.11	1.2×10^8	.26
InAs	−5.8	0.38	1.25×10^{15}	.22
InSb	—	0.82	2.0×10^{16}	.16

Appendix C

Physical Constants, Units, and Acronyms

Table C.1: Physical Constants

Physical Constant	Symbol	Value
Avogadro's number	N_A	6.022×10^{23} mole^{-1}
Gas constant	R	1.98719 cal.mole^{-1}.K^{-1}
		8.31441 J.mol^{-1}.K^{-1}
Boltzmann's constant	$k = R/N_A$	1.381×10^{-23} J.K^{-1}
Absolute electron charge	q	1.602×10^{-19} C
Electron mass at rest	m_0	9.11×10^{-31} kg
Permeability of free space	μ_0	1.257×10^{-8} H.cm^{-1}
Permittivity of free space	ϵ_0	8.854×10^{-14} F.cm^{-1}
Planck's constant	h	6.626×10^{-34} J.s
Reduced Planck's constant	$\hbar$	1.055×10^{-34} J.s
Speed of light in free space	c	2.998×10^{10} cm.s^{-1}
Thermal voltage	kT/q	0.02586 V at 300 K
		6.635×10^{-3} V at 77 K
Bohr radius	a_0	0.529×10^{-8} cm
Gravitational constant	G	6.6720×10^{-11} N.m^2.kg^{-2}
Magnetic flux quantum	$\phi_0 = h/q$	4.1357×10^{-15} J.s.C^{-1}
Stefan–Boltzmann constant	$\sigma = \pi^2 k^4 / 60\hbar^3 c^2$	5.67032×10^{-8} W.m^{-2}.K^{-4}

Table C.2: Conversion of Units

Symbol	Unit
Å	10^{-8} cm
eV	1.602×10^{-19} J
cal	4.184 J
inch	2.54 cm
Torr	133 Pa
	1000 μm Hg
bar	750 mm Hg
T	1 Wb.m^{-2}

Table C.3: Abbreviations of Units

Symbol	Unit
C	couloumb
F	farad
H	henry
Hz	hertz
J	joule
K	kelvin
Kg	kilogram
g	gram
m	meter
cm	centimeter
mol	mole
N	newton
Pa	pascal
s	second
T	tesla
W	watt
Wb	weber

Table C.4: Acronyms Used in the Text

Acronym	Full form
BTE	Boltzmann Transport Equation
FET	Field Effect Transistor
HBT	Heterostructure Bipolar Transistor
HFET	Heterostructure Field Effect Transistor
HSR	Hall–Shockley–Read
JFET	Junction Field Effect Transistor
MESFET	Metal–semiconductor Field Effect Transistor
MIS	Metal–insulator–semiconductor
MISFET	Metal–insulator–semiconductor Field Effect Transistor
MOSFET	Metal–oxide–semiconductor Field Effect Transistor
PHS	Pucel–Haus–Statz
SISFET	Semiconductor–insulator–semiconductor Field Effect Transistor
WKB	Wentzel–Kramers—Brillouin

Glossary

A symbol generated by using a tilde sign on a symbol, e.g., $\tilde{a}$ from a, is used to signify, explicitly, the complex time-varying quantity. The real part of this has a sinusoidal time variation. The phasor, or the amplitude of this time-varying component, is denoted by using the hat sign on the symbol, e.g., $\hat{a}$ for a. This notation is used in the context of small-signal variation. An exception to this nomenclature is the use of the hat symbol to denote a unit normal vector, e.g., $\hat{n}$ to denote the unit normal vector perpendicular to a surface. A lowercase subscript to an uppercase letter denotes a quantity which may have both a static and a time-varying component. In quasi-static approximation, these quantities are denoted by an uppercase subscript or an overline. Any other exceptions have been pointed out in context. This list defines the most frequently used symbols. The units are in representative form. The appropriate system of units should be employed for equations in the text.

Symbol	Symbol Definition	Unit
α_0	Static common-base current gain	—
α_F	Forward common-base current gain	—
α_n	Ionization coefficient of electrons	$cm^{-1}.s^{-1}$
α_p	Ionization coefficient of holes	$cm^{-1}.s^{-1}$
α_R	Reverse common-base current gain	—
α_T	Base transport factor	—
α_{T0}	Static base transport factor	—
α_T^F	Forward common-base base transport factor	—
α_T^R	Reverse common-base base transport factor	—
β_F	Forward common-emitter current gain	—
Γ_n	Gummel number for electrons	$cm^{-4}.s^{-1}$
Γ_p	Gummel number for holes	$cm^{-4}.s^{-1}$
ϵ_0	Permittivity of free space	$F.cm^{-1}$
ϵ_{Al}	Permittivity of $Ga_{1-x}Al_xAs$	$F.cm^{-1}$

ϵ_{ins}	Permittivity of the insulator	$F.cm^{-1}$
ϵ_s	Permittivity of the semiconductor	$F.cm^{-1}$
ϵ_{xx}	xx component of strain	—
ϵ_{yy}	yy component of strain	—
ς	Inverse of small-signal complex length	cm^{-1}
η_f	Ratio of Fermi energy and thermal energy	—
λ_D	Extrinsic Debye length	cm
μ_0	Drift mobility at zero field	$cm^2.V^{-1}.s^{-1}$
μ_n	Drift mobility of electrons	$cm^2.V^{-1}.s^{-1}$
μ_p	Drift mobility of holes	$cm^2.V^{-1}.s^{-1}$
ξ_b	Ratio of base width and small-signal complex diffusion length	—
ξ_d	Normalized potential at the drain	—
ξ_e	Ratio of emitter width and small-signal complex diffusion length	—
ξ_f	Fermi energy at thermal equilibrium	eV
ξ_n	Electron quasi-Fermi energy	eV
ξ_p	Hole quasi-Fermi energy; also, normalized potential at pinch-off point	eV
ξ_s	Normalized potential at the source	—
σ_n	Capture cross-section for electrons	cm^2
σ_p	Capture cross-section for holes	cm^2
τ_B	Base time constant	s
τ_B^{α}	Time constant for single pole approximation of common-base current gain	s
$\tau_B^{\alpha_T}$	Time constant for single pole approximation of base transport factor	s
τ_B^F	Time constant associated with stored integrated minority carrier density in the base due to forward transport component	s
τ_B^R	Time constant associated with stored integrated minority carrier density in the base due to reverse transport component	s
τ_{BF}	Time constant associated with stored integrated minority carrier density in the base associated with forward base current	s
τ_{BR}	Time constant associated with stored integrated minority carrier density in the base associated with reverse base current	s
τ_d	Drain current phase delay time constant	s
τ_F	Time constant associated with stored	

	integrated minority charge density in the base due to forward current	s
τ_R	Time constant associated with stored integrated minority charge density in the base due to reverse current	s
τ_c	Transit time in the base–collector depletion region	s
τ_c'	Signal delay in the base–collector depletion region	s
τ_n	Lifetime of electrons	s
τ_p	Lifetime of holes	s
ϕ_C	Potential of the conduction band edge	C
ϕ_F	Fermi level potential	V
ϕ_M	Metal–semiconductor barrier height	V
ϕ_V	Potential of the valence band edge	C
χ_1	Electron affinity of material 1	V
χ_2	Electron affinity of material 2	V
ψ_B	Electrostatic potential in the bulk	V
ψ_i	Intrinsic level potential with arbitrary reference	V
ψ_{j0}	Built-in voltage of a junction	V
ψ_H	Electrostatic potential at the surface referenced to the bulk intrinsic level when the sheet charge in the inversion layer is equal to the sheet depletion charge	V
ψ_S	Electrostatic potential at the surface referenced to the bulk intrinsic level	V
ψ_{sat}	Electrostatic potential at the surface referenced to bulk intrinsic level at sheet charge density saturation	V
ω	Radial frequency	rad.s^{-1}
ω_0	Radial frequency constant for pole approximation of the small-signal base transport factor	rad.s^{-1}
ω_{α_T}	Radial frequency constant for alternate approximation of the small-signal base transport factor	rad.s^{-1}
ω_{max}	Radial frequency at unity unilateral gain	rad.s^{-1}
ω_T	Radial frequency at unity short-circuit current gain	rad.s^{-1}
ω_q	Phonon frequency	rad.s^{-1}

$\mathcal{A}$	Effective area for current transport	cm^{-3}
$\mathcal{B}$	Probability of radiative recombination;	$cm^3.s^{-1}$
	also, magnetic induction	$Wb.m^2$
C_π	Input capacitance in the hybrid-pi model	F
C_{Al}	Capacitance per unit area with $Ga_{1-x}Al_xAs$	
	as the insulator	$F.cm^{-2}$
C_{DC}	Drain-to-channel capacitance	F
C_D	Emitter–base diffusion capacitance in the	
	common-emitter configuration	F
C_{De}	Emitter–base diffusion capacitance in the	
	common-base configuration	F
C_{depsat}	Capacitance per unit area associated with	
	the depletion region at channel charge	
	saturation	$F.cm^{-2}$
C_{dip}	Dipole capacitance	F
C_{dg}	Drain-to-gate capacitance	F
C_{dom}	Capacitance associated with the domain	F
C_{FB}	Capacitance per unit area of the structure	
	at flat-bands	$F.cm^{-2}$
C_g	Gate capacitance	F
C_{gs}	Gate-to-source capacitance	F
C_{ins}	Capacitance per unit area of insulator	$F.cm^{-2}$
C_{MIS}	Capacitance per unit area of the	
	metal–insulator–semiconductor structure	$F.cm^{-2}$
C_{sem}	Capacitance per unit area of semiconductor	$F.cm^{-2}$
C_{tC}	Capacitance associated with the	
	base–collector transition region	F
C_{tE}	Capacitance associated with the	
	base-emitter transition region	F
$\mathcal{D}$	Displacement field	$C.cm^{-2}$
$\mathcal{D}_{ins}$	Displacement vector in the insulator	$C.cm^{-2}$
$\mathcal{D}_n$	Diffusion coefficient of electrons	$cm^2.s^{-1}$
$\mathcal{D}_{nB}$	Diffusion coefficient of electrons in the base	$cm^2.s^{-1}$
$\overline{\mathcal{D}}_n$	Effective diffusion coefficient of electrons	$cm^2.s^{-1}$
$\overline{\mathcal{D}}_{nB}$	Effective diffusion coefficient of electrons	
	in the base	$cm^2.s^{-1}$
$\mathcal{D}_p$	Diffusion coefficient of holes	$cm^2.s^{-1}$
$\mathcal{D}_{pE}$	Diffusion coefficient of holes in the emitter	$cm^2.s^{-1}$
$\overline{\mathcal{D}}_p$	Effective diffusion coefficient of holes	$cm^2.s^{-1}$
$\overline{\mathcal{D}}_{pE}$	Effective diffusion coefficient of holes	
	in the emitter	$cm^2.s^{-1}$

$\mathcal{D}_{sem}$	Displacement vector in the semiconductor	$C.cm^{-2}$
$\mathcal{E}$	Electric field	$V.cm^{-1}$
E_0	Band edge energy of the zero'th subband	eV
E_1	Band edge energy of the first subband	eV
E_{10}	Energy difference between the first and zero'th quantized levels	eV
$\mathcal{E}_c$	Critical electric field	$V.cm^{-1}$
E_c	Conduction band edge energy	eV
$\mathcal{E}_e$	Quasi-electric field for electrons	$V.cm^{-1}$
ξ_{f0}	Fermi energy parameter for sheet carrier density modelling	eV
E_g	Bandgap energy	eV
E_{g0}	Bandgap energy at low doping level	eV
E_i	Energy of the intrinsic level	eV
$\mathcal{E}_{ins}$	Electric field in the insulator	$V.cm^{-1}$
$\mathcal{E}_m$	Maximum electric field	$V.cm^{-1}$
E_n	Band edge energy of the nth subband	
E_T	Threshold energy for ionization process; also, trap energy level	eV
E_v	Valence band edge energy	eV
$\mathcal{F}$	Force	N
f_{max}	Unity unilateral gain frequency	Hz
f_T	Unity short-circuit current gain frequency	Hz
$\mathcal{G}$	Generation rate	$cm^{-3}.s^{-1}$
g_d	Drain conductance	S
g_{ds}	Drain-to-source conductance	S
g_c	Low frequency limit of conductance of small-signal common-base output admittance	S
g_e	Low frequency limit of conductance of small-signal common-base input admittance	S
g_i	Intrinsic conductance	S
g_m	Transconductance	S
g_{ne}	Low frequency conductance limit of small-signal input admittance due to electron current in common-base operation	S
g_{nc}	Low frequency conductance limit of small-signal output admittance due to electron current in common-base operation	S
$\mathcal{G}\mathcal{N}_B$	Integrated density Gummel number for transport in the base	cm^{-2}
$\mathcal{G}\mathcal{N}_B'$	Effective integrated density Gummel number	

	for transport in the base	cm^{-2}
$\mathcal{GN}_C$	Integrated density Gummel number for transport in the collector	cm^{-2}
$\mathcal{GN}'_C$	Effective integrated density Gummel number for transport in the collector	cm^{-2}
$\mathcal{GN}_E$	Integrated density Gummel number for transport in the emitter	cm^{-2}
$\mathcal{GN}'_E$	Effective integrated density Gummel number for transport in the emitter	cm^{-2}
$\mathcal{GN}_n$	Integrated density Gummel number for electrons	cm^{-2}
$\mathcal{GN}'_n$	Effective integrated density Gummel number for electrons	cm^{-2}
$\mathcal{GN}_p$	Integrated density Gummel number for holes	cm^{-2}
$\mathcal{GN}'_p$	Effective integrated density Gummel number for holes	cm^{-2}
g_o	Output conductance	S
g_{sd}	Source to drain conductance	S
g_{pe}	Low frequency conductance limit of small-signal input admittance due to hole current in common-base operation	S
g_{pc}	Low frequency conductance limit of small-signal output admittance due to hole current in common-base operation	S
h	Planck's constant	J.s
$\hbar$	Reduced Planck's constant $(h/2\pi)$	J.s
$\hbar\omega_q$		—
I_B	Base current	A
$\hat{I}_b$	Phasor of sinusoidal current at base contact	A
I_B^F	Forward component of the base current	A
I_B^R	Reverse component of the base current	A
$\hat{I}_c$	Phasor of sinusoidal current at collector contact	A
I_C	Collector current	A
$\hat{I}_{cc}$	Phasor of sinusoidal current at the collector edge of the base–collector depletion region	A
I_C^F	Forward component of the collector current	A
I_C^R	Reverse component of the collector current	A
I_{CS}	Saturation current of the forward coupling base–collector diode in Ebers–Moll model	A
I_D	Drain current	A
I_{norm}	Normalization current	A

I_{DS}	Drain-to-source current	A
I_{DSS}	Drain-to-source current at saturation of current	A
$I_{Dsubthr}$	Drain current in sub-threshold region	A
$\hat{I}_e$	Phasor of sinusoidal current at emitter contact	A
I_E	Emitter current	A
$\hat{I}_{nc'}$	Phasor of sinusoidal electron current at base edge of base–collector depletion region	A
I_E^F	Forward component of the emitter current	A
I_E^R	Reverse component of the emitter current	A
I_{ES}	Saturation current of the forward coupling base-emitter diode in Ebers–Moll model	A
I_G	Gate current	A
$\hat{I}_n$	Phasor of sinusoidal electron current	A
$\hat{I}_{nc}$	Phasor of sinusoidal electron current at the collector edge of base–collector depletion region	A
I_{OE}	Saturation current of the reverse coupling base-emitter diode in Ebers–Moll model	A
I_{OC}	Saturation current of the reverse coupling base–collector diode in Ebers–Moll model	A
$\hat{I}_p$	Phasor of sinusoidal hole current	A
I_{substr}	Substrate current	A
$\hat{J}$	Phasor of sinusoidal current density	$A.cm^{-2}$
J_B	Base current density	$A.cm^{-2}$
J_C	Collector current density	$A.cm^{-2}$
J_E	Emitter current density	$A.cm^{-2}$
J_n	Electron current density	$A.cm^{-2}$
$\hat{J}_n$	Phasor of sinusoidal electron current density	$A.cm^{-2}$
J_{nd}	Displacement current density associated with electron flow in a depletion region	$A.cm^{-2}$
J_p	Hole current density	$A.cm^{-2}$
$\hat{J}_p$	Phasor of sinusoidal hole current density	$A.cm^{-2}$
J_r	Recombination current density	$A.cm^{-2}$
J_{scr}	Generation-recombination current density in a space charge region	$A.cm^{-2}$
k	Boltzmann's constant	$J.K^{-1}$
k, k	Wave vector	cm^{-1}
k_F	Fermi wave vector	cm^{-1}
k_x	Wave vector in the x-direction	cm^{-1}
k_y	Wave vector in the y-direction	cm^{-1}

k_z	Wave vector in the z-direction	cm^{-1}
L_m	Metallurgical gate length	cm
$\mathcal{L}_n$	Diffusion length of electrons	cm
$\mathcal{L}_p$	Diffusion length of holes	cm
m_0	Rest mass of a free electron	kg
m_e^*	Effective mass of electrons	kg
m_h^*	Effective mass of holes	kg
m_i^*	Effective mass of carriers in ith band	kg
m^*	Effective mass in the semiconductor	kg
m_{lh}^*	Effective mass of holes in the light hole band	kg
m_{hh}^*	Effective mass of holes in the heavy hole band	kg
m_l^*	Longitudinal conduction effective mass	kg
m_t^*	Transverse conduction effective mass	kg
m_d^*	Density of states effective mass	kg
n	Electron carrier concentration	cm^{-3}
$\widehat{n}$	Unit vector normal to the surface	—
n'	Excess electron density above its magnitude at thermal equilibrium	cm^{-3}
n_i	Intrinsic carrier concentration	cm^{-3}
N_A	Shallow acceptor density	cm^{-3}
N_A^-	Ionized acceptor density	cm^{-3}
N_C	Effective density of states in the conduction band	cm^{-3}
N_D	Shallow donor density	cm^{-3}
N_D^+	Ionized donor density	cm^{-3}
N_I	Sheet carrier density in the inversion layer	cm^{-2}
N_{I0}	Sheet carrier density in the inversion layer at $z = 0$	cm^{-2}
N_{IL}	Sheet carrier density in the inversion layer at $z = L$	cm^{-2}
$\overline{n}_n$	Static electron density in n-region	cm^{-3}
n_p	Electron density in p-type material	cm^{-3}
$\overline{n}_p$	Static electron density in p-region	cm^{-3}
n_{p0}	Electron density in p-type material at thermal equilibrium	cm^{-3}
N_s	Sheet carrier density in the semiconductor	cm^{-2}
N_V	Effective density of states in the valence band	cm^{-3}
n_q	Occupation probability of the state of energy	

p	Hole carrier concentration	cm^{-3}
p, $\mathbf{p}$	Momentum of carriers	kg.cm.s^{-1}
$\mathcal{P}$	Polarization field	C.cm^{-2}
p'	Excess hole density above its magnitude at thermal equilibrium	cm^{-3}
p_e	Electron momentum	kg.cm.s^{-1}
p_h	Hole momentum	kg.cm.s^{-1}
p_n	Hole density in n-type material	cm^{-3}
$\bar{p}_n$	Static hole density in n-region	cm^{-3}
p_{n0}	Hole density in n-type material at thermal equilibrium	cm^{-3}
$\bar{p}_p$	Static hole density in p-region	cm^{-3}
q	Magnitude of elementary charge	C
$\mathbf{q}$, q	Phonon wave vector	cm^{-1}
$\mathcal{Q}$	Integrated total charge density	C.cm^{-2}
$\mathcal{Q}_B$	Integrated base charge density	C.cm^{-2}
$\mathcal{Q}_{dom}$	Total charge in the depletion or accumulation region of a domain	C
$\mathcal{Q}_E$	Integrated emitter charge density	C.cm^{-2}
$\mathcal{Q}_C$	Integrated collector charge density	C.cm^{-2}
$\mathcal{Q}_{dep.}$	Sheet charge density in the depletion layer	C.cm^{-2}
$\mathcal{Q}_E$	Integrated emitter charge density	C.cm^{-2}
$\mathcal{Q}_F$	Integrated charge density associated with forward transport	C.cm^{-2}
$\mathcal{Q}_{F0}$	Saturation integrated charge density associated with forward transport	C.cm^{-2}
$\mathcal{Q}_g$	Sheet charge density on the gate electrode	C.cm^{-2}
$\mathcal{Q}_I$	Sheet charge density in the inversion layer	C.cm^{-2}
$\mathcal{Q}_n$	Integrated electron charge density	C.cm^{-2}
$\mathcal{Q}_{nB}$	Integrated electron charge density in base	C.cm^{-2}
$\mathcal{Q}_{nB}^F$	Integrated electron charge density in base associated with forward transport	C.cm^{-2}
$\mathcal{Q}_{nB}^R$	Integrated electron charge density in base associated with reverse transport	C.cm^{-2}
$\mathcal{Q}_p$	Integrated hole charge density	C.cm^{-2}
$\mathcal{Q}_{pC}$	Integrated hole charge density in the collector	C.cm^{-2}
$\mathcal{Q}_{pC}^F$	Integrated hole charge density in collector associated with forward transport	C.cm^{-2}
$\mathcal{Q}_{pC}^R$	Integrated hole charge density in collector	

	associated with reverse transport	$C.cm^{-2}$
Q_{pE}	Integrated hole charge density in the emitter	$C.cm^{-2}$
Q_{pE}^F	Integrated hole charge density in emitter associated with forward transport	$C.cm^{-2}$
Q_{pE}^R	Integrated hole charge density in emitter associated with reverse transport	$C.cm^{-2}$
Q_{pol}	Surface charge density associated with the polarization vector	$C.cm^{-2}$
Q_R	Integrated charge density associated with reverse transport	$C.cm^{-2}$
Q_{R0}	Saturation integrated charge density associated with reverse transport	$C.cm^{-2}$
Q_S	Sheet charge density in the semiconductor	$C.cm^{-2}$
Q_{sC}	Integrated charge density associated with base–collector depletion region	$C.cm^{-2}$
Q_{sE}	Integrated charge density associated with base-emitter depletion region	$C.cm^{-2}$
Q_{sem}	Sheet charge density in the semiconductor	$C.cm^{-2}$
$\mathcal{R}$	Recombination rate	$cm^{-3}.s^{-1}$
r_μ	Feedback resistance between the base and and collector in the hybrid-pi model	Ω
r_π	Input resistance in the hybrid-pi model	Ω
r_b	Base resistance	Ω
r_d	Drain resistance	Ω
r_g	Gate resistance	Ω
r_i	Intrinsic resistance	Ω
r_o	Output resistance	Ω
r_s	Source resistance	Ω
S	Surface recombination velocity	$cm.s^{-1}$
$\mathcal{S}$	Sub-threshold swing	$mV/decade$
t_{Al}	Thickness of $Ga_{1-x}Al_xAs$	cm
t_{ins}	Thickness of insulator	cm
t_{sp}	Thickness of undoped spacer layer	cm
$\mathcal{U}_{scr}$	Net recombination–generation rate in a space charge region	$cm^{-3}.s^{-1}$
$\mathcal{V}$	Normalized voltage in the channel	—
v_θ	Thermal velocity	$cm.s^{-1}$
$\mathcal{V}_{00}$	Normalizing voltage	V
V_A	Early voltage	V
V_{BC}	Applied voltage between base and collector	V

V_{bd}	Break down voltage	V
V_{BE}	Applied voltage between base and emitter	V
V_{BS}	Source-to-bulk potential	V
V_{CB}	Applied voltage between collector and base	V
V_{CE}	Applied voltage between collector and emitter	V
V_D	Drain potential	V
$\mathcal{V}_d$	Normalized voltage at the drain	—
V_{dom}	Voltage across the domain region	V
V_{dg}	Potential difference between the drain and the gate	V
V_{DS}	Potential difference between the drain and the source	V
V_{ds}	Potential difference between the drain and the source	V
V_{Dsat}	Drain voltage at drain current saturation	V
V_{DSS}	Voltage between drain and source at current saturation	V
V_{EB}	Applied voltage between emitter and base	V
V_{EC}	Applied voltage between emitter and collector	V
v_F	Fermi velocity	$cm.s^{-1}$
V_{FB}	Flat-band voltage	V
v_g	Group velocity	$cm.s^{-1}$
V_G	Gate potential	V
V_{GS}	Potential difference between the gate and the source	V
V_{gs}	Potential difference between the gate and the source	V
$\overline{v_i}$	Mean velocity of carriers in ith band	$cm.s^{-1}$
V_j	Applied voltage across a junction	V
v_l	Scattering-limited velocity	$cm.s^{-1}$
V_n	Partitioned voltage across n-type region of a junction;	V
	also, separation between conduction band edge and the electron quasi-Fermi level	V
V_p	Partitioned voltage across p-type region of a junction;	V
	also, separation between valence band edge and the hole quasi-Fermi level	V
$\mathcal{V}_p$	Normalized voltage at the point of channel	
v_s	Saturated velocity	$cm.s^{-1}$
V_S	Source potential	V
$\mathcal{V}_s$	Normalized voltage at the source	—

V_{sat}	Potential at the onset of saturation of carrier velocity	V
V_T	Threshold voltage, source reference	V
V_{T0}	Threshold voltage parameter	V
V_{TFL}	Trap-filled limit voltage	V
w_B	Width of quasi-neutral base region	cm
w_C	Position of base edge of base–collector depletion region	cm
$w_{C'}$	Position of collector edge of base–collector depletion region	cm
w_E	Position of base edge of base-emitter depletion region	cm
$w_{E'}$	Position of emitter edge of base-emitter depletion region	cm
$\overline{w_i}$	Mean energy of carriers in the ith band	J
$\overline{W_i}$	Mean total energy of carriers in ith band	J
w_n	Edge position of n depletion region	cm
w_p	Edge position of p depletion region	cm
y_{cc}	Output admittance parameter in common-base operation	S
y_{ce}	Forward admittance parameter in common-base operation	S
y_{ec}	Reverse admittance parameter in common-base operation	S
y_{ee}	Input admittance parameter in common-base operation	S
y_{nc}	Collector admittance parameter associated with electron current	S
y_{nce}	Forward admittance parameter due to electron current in common-base operation	S
y_{ne}	Emitter admittance parameter associated with electron current	S
y_{nec}	Reverse admittance parameter due to electron current in common-base operation	S
y_{pce}	Forward admittance parameter due to hole current in common-base operation	S
y_{pec}	Reverse admittance parameter due to hole current in common-base operation	S
z_n	Spatial position of contact to n-region	cm
z_p	Spatial position of contact to p-region	cm

Index

A

ABCD-parameters, 711, 783, 784
Abrupt heterojunction, 226, 280–284, 410, 413
Abrupt junction
    breakdown voltage, 87–91
    current transport, 164, 252–292
    generation–recombination current, 271–279, 616–629
Absorption of phonons, 52, 58, 129
Acceptor degeneracy, 32
Accumulation
    capacitance, 436, 448
    layer, 436
Acoustic
    phonon, 24
    phonon scattering, 52–54
Activation energy
    thermal, 530
    GaAs, 530
    $Ga_{1-x}Al_xAs$, 530–533
Admittance parameters
    forward transfer, 367, 371, 377, 378, 512, 634, 635, 645–658, 783–789
    input, 367, 371, 377, 378, 512, 634, 635, 647, 652–655, 783–789
    output, 367, 371, 377, 378, 512, 634, 635, 651, 664, 783–789
    reverse transfer, 367, 371, 377, 378, 512, 634, 635, 646, 647, 783–789
Airy functions, 235, 236, 417
Alpha cut-off frequency, 659
Aluminum arsenide
    band structure, 41
    properties, 792–801
Ambipolar diffusion coefficient, 298
Anderson's rule, 410
Anisotropy, 62, 77, 414
Anisotype heterojunction, 283
Aspect ratio, 321
Asymmetric junction, 252
Auger lifetime, 182
Auger recombination, 180–182
Avalanche
    breakdown, 78–89
    multiplication factor, 85–88
    process, 82, 85–89

B

Backgating, 300, 340
Ballistic transport, 139, 717
Bandgap, 16, 98
Bandgap narrowing, 95, 551
Band discontinuities
    conduction band, 223, 405, 410, 411
    valence band, 223, 405, 410, 411
Band structure
    AlAs, 41
    GaAs, 40

Ga.$_{47}$In.$_{53}$As, 42
Ge, 39
InAs, 41
InP, 40
Si, 29
Barrier height, 202, 449, 723, 730
Barrier lowering, 206
Base current, 562, 577, 721
Base resistance, 663, 668, 722
Base transit time, 565
Base transport factor, 635, 658–
        660
Bessel functions, 371, 513
Bohr radius, 93
Boltzmann factor, 28
Boltzmann transport equation, 106,
        107, 116, 122–124, 146–
        149
Breakdown condition, 87
Breakdown voltage, 88
Brillouin zone, 12, 36, 37
Built-in field, 146, 147,
Built-in potential, 165, 250, 304

C
Capacitance–voltage relationship
        HFET, 502–506
        HBT, 567, 572, 574, 598–602,
                612–615, 636
        MESFET, 327–329, 338, 339,
                384
        MIS structure, 447, 448,
Capture cross-section, 177
Carrier
        energy, 56–58
        intrinsic density, 35
        lifetime, 175, 179, 182, 780
        recombination, 173–190, 275–
                288, 616–629
        storage time, 564–566, 568–
                570, 586, 595, 654–656,
                705, 706

        transfer, 63, 70, 425–432, 528,
                529
Carrier–carrier scattering, 51, 130
Channel charge, 502–505
Channel conductance, 372
Channel confinement, 70, 71, 413,
        430
Channel depth, 310, 404
Channel length, 317–321, 482
Channel opening, 321
Charge storage, 258–260, 565–575,
        605–608, 705, 706
Charge transfer, 70, 426–432, 528,
        529
Collapse of current, 529
Collector
        capacitance, 567, 651
        current, 561–563, 569–571, 576–
                578, 608, 610
        depletion layer transit time,
                650, 693
        transport factor, 635, 647, 650,
                692, 693
        signal delay, 650, 692
Common-base current gain, 635,
        658–661
Common-emitter current gain, 568,
        615–615, 626–629, 666,
        667
Compound semiconductors, 1–4,
        38–44, 356, 755
Conductance
        HFET, 506, 767
        MESFET, 307, 326, 372, 767
        MISFET, 479, 480
Conduction band
        discontinuity, 223, 405, 410,
                411
        effective mass, 14, 19, 20, 62–
                66, 416
        minimum, 20, 21, 36–43
Conduction process, 54–78

Conservation equations, 122–128
Conservation of energy, 200, 230–232
Contact
    ohmic, 241–246
    tunneling, 244
Continuity equation, 160
Coulomb scattering, 70, 403
Coupled phonon-plasmon scattering, 71
Critical thickness , 407
Crystal momentum, 19, 62
Crystal plane , 12
Crystal structure, 12, 13
Current collapse, 529
Current continuity equation, 160
Current-density equation, 150
Current gain, 568, 615–615, 626–631, 635, 661, 666, 667
Current-voltage characteristics
    HBT, 560–571, 575–580
    heterojunction, 231
    HFET, 496–501
    MESFET, 305–310, 316, 317
    metal–semiconductor junction, 208–222
    MISFET, 459, 465–468, 476, 480–482
    p–n junction, 258–284

D

de Broglie wavelength, 99, 706, 756
Debye length, 251, 310, 756
Decay time, 528
Defect scattering, 49, 50
Degeneracy
    acceptor, 32
    carrier, 34, 95, 96
    donor, 32
Degenerate semiconductor, 94, 95
Delay time, 650, , 693

Density matrix approach, 107
Density of states
    conduction band, 34
    valence band, 35
Depletion approximation, 248–252
Depletion layer, 208, 248–252, 304, 318, 592, 648, 767, 768
Dielectric relaxation time, 110, 382
Diffusion coefficients, 76, 77
Diffusion current, 256, 269, 331
Diffusion length, 254, 272
Dipole formation, 332–339
Dipole layer, 332–339
Dipole voltage, 337–339
Dirichlet boundary conditions, 170–172
Dispersion, 43, 44
Distribution function, 106, 114–121
Donor, 29–32, 429
Donor ionization energy, 429
Dopants
    acceptors, 32
    donors, 29
Double barrier structure, 723–746
Double heterostructure bipolar transistor, 601
Drain conductance, 307, 326, 479, 506
Drain current, 305–310, 316, 317, 459, 465–468, 478, 480–482, 495–500
Drain-to-gate capacitance, 327, 339, 505
Drain-to-source capacitance, 329, 339
Drain voltage limitations, 767
Drift velocity, 50, 56, 59–61, 492, 540, 776, 777
Drift-diffusion current, 149–152
DX centers, 404, 528–535

E

Early effect, 572
Early voltage, 572, 641
Ebers–Moll model, 575–580
Effective density of states, 34, 35, 92–95
Effective mass, 14, 20, 62–66, 416
Effective Richardson constant, 213, 226, 284
Einstein relationship, 76
Elastic scattering, 49
Electron affinity, 151, 202, 410
Electron–hole scattering, 73–75, 130
Electron lifetime, 175, 179, 182, 616, 780
Electron mobility, 59–61, 66–68, 70–74, 109
Electron temperature, 144
Electron velocity, 59–61, 332, 492, 776
Electrostatic potential, 153, 201, 202, 577
Emission of phonons, 53, 58, 130
Emitter injection efficiency, 562, 653
Emitter–base capacitance, 567, 572, 598–602
Energy band
    coupled-barrier structures, 723, 728
    heterojunctions, 223–226, 247
    HBTs, 552, 555–557
    HFETs, 431, 484, 485
    MESFETs, 302, 303, 308, 309, 352–353
    metal–semiconductor junctions, 202, 203, 207
    MIS diodes, 433, 434, 450
    MISFETs, 451, 452, 462, 463
    p–n junctions, 249, 255, 281, 290
Energy bandgap, 16, 17

Energy conservation , 200, 230–232
Energy–momentum relationship, 19, 62
Energy relaxation time, 57, 125–127
Equivalent circuit
    coupled-barrier structures, 723
    HBTs, 570, 575, 579, 665, 668, 672
    HFETs, 506–509
    MESFETs, 338, 339, 378, 379, 386–388
Equi-energy surfaces
    ellipsoidal, 63–65
    spherical, 62, 63
    warped, 62, 63, 420

F

f_{max}, 388, 509, 670, 788, 789
f_T, 387, 667, 668, 788, 789
Fabry–Pérot, 723
Feedback capacitance, 334
Fermi–Dirac distribution, 28
Fermi integral, 34
Fermi level pinning, 2, 186–188, 242, 620–626
Fermi potential, 150
Fermi wave vector, 96
Flat-band condition, 449–451
Flat-band voltage, 449–451
Fletcher boundary conditions, 166–168
Fowler–Nordheim tunneling, 239, 240, 244
Frequency response, 365–389, 506–522, 629–683, 706
Fringing capacitance, 769, 766
Full channel current, 310, 317

G

Gallium aluminum arsenide, 49, 50, 404, 405, 429, 524–535, 684–690

Gallium antimonide, 242, 735

Gallium arsenide
band structure, 40
constant energy surface, 414, 742
diffusion coefficient, 77
drift velocity, 50, 56, 68, 69492, 540, 541, 776, 777
ionization energy, 429
ionization rate, 83
mobility, 59–61, 540, 541
phonon dispersion, 46
velocity–field behavior, 50, 56, 540, 777

Gallium indium arsenide, 42, 49, 55, 83, 87, 89, 394, 536–538, 689–698

Gamma function, 34

Gate current, 403, 522–528

Gate-to-source capacitance, 327, 505

Gate-to-drain capacitance, 327, 339, 505

Gauss's law, 172, 282, 417, 441, 495

Generation rate, 124, 173–190

Generation–recombination process, 173–190

Graded-base bipolar transistor, 583–586, 672–680

Graded heterojunction, 280–292, 586

Grading length, 224, 289, 580, 581, 608, 610

Gradual channel approximation, 302, 313, 372

Gummel number, 265–269, 278–280, 560

Gummel plot, 595

Gummel–Poon model, 257–280, 558–575

H

Hall constant, 66, 68, 113

Hall effect, 66-68, 112

Hall factor, 67, 113

Hall mobility, 66-68, 113

Hall voltage, 66

Heavy hole, 38, 410

Heterojunction
anisotype, 283
isotype, 226, 245
type I, 223
type II, 223
type III, 223

Heterostructure, 222–246, 395–425

High current effects, 252–257, 602–615

High field properties, 68, 69, 78–91, 392–394, 522–528, 535–539, 684–698

High frequency operation, 365–389, 509–520, 629–683

High injection condition, 252–257

High mobility, 70–72

Hybrid parameters, 385, 666–668, 784–787

Hole
light, 38, 180, 244, 408
heavy, 38, 180, 244, 408

Hole velocity, 540, 541

Hot electron injection, 522–538, 706–709, 714–725

Hot electron transistor, 706–709, 714–723

Hot electrons
HBT, 684–698
HFET, 522–538
MESFET, 392–394

HSR recombination, 175–180

Hybrid-pi circuit, 575, 671

I

Ideal contact, 202, 223, 224
Ideality factor, 618
Ideal MIS structure, 433–451
Image force, 204–206
Impact ionization
    coefficient, 79–89
    orientation-dependence, 79
Impedence parameters, 783, 787
Impurity
    acceptor, 32
    donor, 32
    freeze-out, 94, 95
    ionization energy, 429
    scattering, 49, 50, 70
Indirect transition, 90, 91
Indirect tunneling, 90, 91, 231, 232,
    741, 742
Indium arsenide, 41, 46, 84, 87,
    89, 690–698, 735
Indium phosphide
    band structure, 40
    ionization rate, 84
Inelastic scattering, 90, 91, 741,
    742
Interface states, 50, 223, 437, 544
Inter-valley scattering, 55, 56, 129
Intra-valley scattering, 55, 56
Intrinsic carrier concentration, 35
Intrinsic energy level, 150
Intrinsic transconductance, 308, 324–
    326, 479, 506, 571
Inverse current gain, 576
Inversion, 433–449
Inversion condition, 437, 438
Inversion layer, 433–443
Ionization coefficient, 83–85
Ionization integral, 87
Ionization rate, 83–85
Ionization threshold energy, 81, 90

Isotype heterojunction, 226, 245

J

Junction breakdown, 87–91

K

Kinetic approach, 107–113
Kirk effect, 605, 691
Kronig–Penney model, 14–20

L

Lifetime
    Auger, 182
    HSR, 179
    radiative, 175
Light-hole band, 38, 408, 410
Liouville's equation, 107
Long channel, 302–308
Longitudinal effective mass, 62–
    66
Longitudinal modes, 44–46
Longitudinal optical phonon, 44–
    46
Lorentz force, 66, 719
Low-level injection, 166

M

Majority carrier transport, 55–70
Maximum available gain, 788
Maximum current, 310, 317
Maximum stable gain, 788
Maximum frequency of oscillation,
    388, 509, 670, 788, 789
Maxwell–Boltzmann distribution
    function, 28
Maxwell's equations, 761
Mean free path, 89, 90, 708
MESFET, 299–401, 768–775
Metal base transistor, 706–709
Metal–semiconductor junction, 200–
    222
Metal–semiconductor ohmic con-
    tact, 241–246

Miller feedback, 387
Minority carrier
    density, 166–170
    lifetime, 175–182
    storage, 565, 705
    transport, 72–78
MIS diode, 432–451
Misawa boundary conditions, 168–170
MISFET, 299–483
Mobility
    behavior, 59–75
    drift, 59–66
    electron, 59–75
    Hall, 66–68
    hole, 540, 541
    temperature effect, 69
    two-dimensional electron gas, 70–72
    two-dimensional hole gas, 540, 541
Moments of Boltzmann transport equation, 122–128
Monte Carlo approach, 135–144
Mott transition, 94
Multiplication factor, 85–88

N
n-channel
    HFET, 426–432, 483–541
    MISFET, 432–483
Negative differential mobility, 332, 486
Neumann boundary condition, 172
Non-degenerate semiconductor, 28, 95, 421
Non-equilibrium condition, 3
Non-uniform doping, 153
Non-radiative recombination, 174

O
Off-equilibrium effects

HBT, 684–698
HFET, 535–539
MESFET, 392–394
Ohmic contact, 241–246
One-sided abrupt junction, 252
Optical
    phonon, 24
    phonon scattering, 52–54, 131
Output characteristics, 322, 432, 479, 717, 718
Output conductance
    HFET, 506, 767
    MESFET, 307, 326, 767
    MISFET, 479, 480
Overshoot, 3, 134, 393, 535, 536, 690, 691, 696

P
p-channel HFET, 540, 541
Parabolic bands, 16, 79, 142
Parasitic conduction, 483
Parasitics, 388, 389, 507, 681
Pauli's exclusion principle, 26
Peak electron velocity, 97
Permittivity, 109, 416
Phase delay, 365, 380, 507
Phonon-assisted tunneling, 91, 231, 232, 742
Phonon scattering
    acoustic, 52–54
    optical, 52–54
PHS model, 312–329
Physical constants, 803
Piezoelectric charge, 356–360
Piezoelectric scattering, 53
p–i–n diode, 298
Pinch-off condition, 315, 480
Pinning of Fermi level, 2, 186–188, 242, 620–626
Planar doping, 155, 310, 716
Plasma frequency, 110
Plasmon scattering, 51, 131

p-n product, 35, 165, 170, 621
Polar optical phonon scattering,
    53
Poole–Frenkel effect, 296

Q

Quantum-mechanical reflection, 709–
    714
Quasi-electric field, 156, 291, 585,
    600, 610–613, 676
Quasi-Fermi levels
    n-type, 150
    p-type, 150
Quasi-static analysis, 153
Quasi-neutrality, 153–157

R

Radiative lifetime, 175
Radiative process, 174, 175
Radiative recombination, 174, 175
Random distribution, 136–139
Reciprocal lattice, 12, 13
Recombination
    current, 271–279, 615–629
    lifetime, 176–182
    process, 173–190
    rate, 124, 173–190
    velocity, 172, 185, 188
Recombination–generation current,
    271–279, 615–629
Recombination–generation process,
    173–190
Reflection coefficient, 711–713
Relaxation time, 116–122
Relaxation time approximation, 116-
    122
Resonance frequency, 389, 682
Resonant Fowler–Nordheim tun-
    neling, 239–241
Resonant tunneling, 723–743
Reverse current, 276

Richardson's constant, 213, 226,
    284

S

Saturated velocity, 68, 69
Saturated velocity model, 308–311
Saturation mode, 559
Scaling, 755–780
Scattering
    acoustic phonon, 52–54
    carrier–carrier, 51, 130
    defect, 49, 50
    f-, 131
    g-, 131
    inter-valley, 55, 56, 129
    intra-valley, 55, 56
    ionized-impurity, 47, 130
    lattice, 52, 53
    optical phonon, 52–54, 131
    piezoelectric, 53
    plasmon, 51, 131
Scattering parameters, 784, 787
Schrödinger's equation, 9, 234, 416,
    710
Self-scattering, 139
Semiconductor crystal structure,
    13
Semiconductor–insulator–semicon-
    ductor FET, 5, 524–527
Semi-insulating substrate, 340–343
Sensitivity of threshold voltage, 773–
    775
Sequential tunneling, 723–743
Sheet carrier concentration, 420–
    424, 442, 443, 446, 451–
    454
Sheet carrier density, 420–424, 442,
    443, 446, 451–454
Sheet charge model, 432–424
Shockley boundary conditions, 163–
    166

Short channel effects, 480–482, 767–772

Sidegating, 340–356

Silicon
    band structure, 39
    constant energy surface, 62
    drift velocity, 50
    ionization rate, 84
    phonon dispersion, 45

Silicon dioxide (SiO_2), 358, 360, 422

Silicon–germanium, 410, 552

Small-signal
    $\tilde{\alpha}$, 635, 661
    $\tilde{\alpha}_T$, 635, 658, 678
    $\tilde{\gamma}$, 653, 658
    $\tilde{\zeta}$, 634, 650

Small-signal analysis, 157–162, 365–384, 509–520, 629–683

Small-signal equivalent circuit, 378, 379, 506–509, 665, 668, 672

Source resistance, 387, 507

Space charge recombination, 271, 616–620

S-parameters, 784, 787

Specific contact resistance, 242, 243

Spin-orbit splitting, 38

Split-off band, 38

Statistics
    of bands, 26–35
    of discrete levels, 27–32

Stability factor, 788

Stokes' equation, 511

Storage time, 564–571, 585, 705

Strained layer, 405–410

Strong inversion, 438

Subband, 412–425

Substrate bias, 456

Subthreshold current, 476, 501, 502

Subthreshold region, 474–478, 502

Subthreshold swing, 476, 477, 502

Surface potential, 438

Surface conversion, 354–356

Surface recombination velocity, 172, 185, 188

Surface space charge region, 354

Surface state density, 2, 186, 187, 620

Surface states, 2, 186, 187, 620

Switching delay, 390, 520–522, 684

T

Temperature rise, 766

Thermal conductivity, 126

Thermal equilibrium condition, 3, 76, 144, 165, 173, 433

Thermal velocity, 95, 111

Thermionic emission, 211–214, 226–231, 283, 284

Thermionic emission-diffusion, 219–222

Thermionic field emission, 214–219

Threshold voltage, 457, 458, 497

Threshold voltage sensitivity, 773–775

Time constant
    base, 564–566
    forward, 568
    for base current, 568
    reverse, 569

Total depletion width, 251

Transconductance
    HBT, 571
    HFET, 506, 507
    MESFET, 307, 324–326

Transfer characteristics, 470

Transient behavior, 148, 390, 520–522, 683, 684, 692

Transit time
    HBT, 650, 681, 692
    MESFET, 390

Transit time delay, 650, 681, 692

Transmission

coefficient, 729, 730
probability, 226, 236–238, 715, 736, 737
Transmission-line equation, 158, 367, 372
Transport, 54–78
Transverse effective mass, 62–66
Transverse modes, 44–47
Transverse momentum, 230
Transverse optical phonons, 44–47
Trap density, 175, 346
Trap energy level, 175, 342, 346
Trapping, 175–180, 528–535
Triangular barrier, 234–238, 245
Tunnel emission, 706–709, 744
Tunnel junction, 322
Tunneling current, 237, 238
Tunneling process, 90, 91, 99, 215, 216, 231–241, 706–709, 723–746
Tunneling time, 741
Tunneling transistor, 714–723, 743–746
Two-dimensional electron gas, 70–72, 403–506
Two-dimensional hole gas, 540, 541

U
Umklapp process, 26
Unilateral power gain, 388, 389, 681, 788

V
Valence
  band structure, 20, 38, 42, 62
  effective mass, 20, 65, 66
  band discontinuity, 223, 405, 410
Velocity
  surface recombination, 172, 185, 188

thermal, 95, 111
Velocity–field characteristics, 51, 56, 492, 540,
Velocity overshoot, 3, 134, 393, 535, 536, 690, 691, 696
Velocity saturation, 68, 69

W
Wave equation, 109
Wave vector, 9–21
Weber's equation, 371
Webster effect, 604
Wide bandgap collector, 605, 606
Wide bandgap emitter, 551, 552
WKB approximation, 215, 216, 238, 730
Work function, 202, 449

Y
y-parameters, 367, 371, 377, 378, 512, 519, 634, 635, 656, 657, 664, 783–789

Z
z-parameters, 783
Zener breakdown, 90, 91
Zener tunneling, 90, 91
Zinc blende lattice, 13